Life on Earth

Life on Earth

Teresa Audesirk   Gerald Audesirk

University of Colorado at Denver

Prentice Hall, Upper Saddle River, New Jersey 07458

Library of Congress Cataloging-in-Publication Data

Audesirk, Teresa.
    Life on earth / Teresa Audesirk, Gerald Audesirk.
       p.   cm.
    Includes bibliographical references and index.
    ISBN 0-13-570375-1 (alk. paper)
    1. Biology. I. Audesirk, Gerald. II. Title.
QH308.2.A936   1996
574—dc20

96-42128
CIP

Executive Editor: Sheri L. Snavely
Development Editor: Ray Mullaney
Production Editor: Debra A. Wechsler
Editor in Chief: Paul F. Corey
Editorial Director: Tim Bozik
Director, Production & Manufacturing: David W. Riccardi
Executive Managing Editor, Production: Kathleen Schiaparelli
Marketing Manager: Jennifer Welchans
Creative Director: Paula Maylahn
Art Director: Heather Scott
Interior Designer: Laura Ierardi
Cover Designer: Tom Nery
Copy Editor: Margo Quinto
Page Layout: Richard Foster
Manufacturing Manager: Trudy Pisciotti
Photo Research: Tobi Zausner, Beaura K. Ringrose, Melinda Reo
Photo Editor: Lorinda Morris-Nantz
Art Manager: Patrice Van Acker
Illustrators: Rolando Corujo, Hudson River Studios, Howard Friedman, David Mascaro,
       Edmund Alexander, Page Two Associates
Art Retouching: Todd Ware, Marita Froimson, Rolando Corujo, Mumtaz Hussain, Patrice Van Acker
Cover Illustrator: Dale Verzaal

©1997 by Prentice-Hall, Inc.
Simon & Schuster/A Viacom Company
Upper Saddle River, New Jersey 07458

Printed in the United States of America
10 9 8 7 6 5 4 3 2 1

ISBN   0-13-570375-1

Prentice-Hall International (UK) Limited, *London*
Prentice-Hall of Australia Pty. Limited, *Sydney*
Prentice-Hall Canada Inc., *Toronto*
Prentice-Hall Hispanoamericana, S.A., *Mexico*
Prentice-Hall of India Private Limited, *New Delhi*
Prentice-Hall of Japan, Inc., *Tokyo*
Simon & Schuster Asia Pte, Ltd., *Singapore*
Editora Prentice-Hall do Brasil, Ltda., *Rio de Janeiro*

Printed on Recycled Paper

Preface

Because we have a young daughter, we often hear the question "Why?" It's an exciting question, because it provides us with an opportunity to share some of what we know, or to learn more by looking up the answer to our daughter's question. The question "why?" begins a person's exploration of his or her world. For those fortunate enough to continue asking such questions, the exploration and the delight of learning become life-long and continuously enriching experiences. As teachers and scientists, we are continuously learning about life on Earth. In writing this textbook, we have had the good fortune both to learn more and to share our interest in biology with an extended classroom of thousands of students.

A course in introductory biology may be a student's first real exposure to the fascinating complexity of life. As teachers, we recognize how easily a student may become mired in the overwhelming number of new facts and terms, while losing sight of the underlying principles of biology and their relevance to daily life. *Life on Earth* has been carefully written to integrate the necessary biological facts into a broader conceptual framework that stresses unifying themes and the ways in which an understanding of biology can enrich and enlighten day-to-day living. Why study biology? Maybe we're biased, but what can be more fascinating than learning about life on Earth?

Life on Earth

Focuses on Concepts

The subheadings within each major chapter section have been cast as sentences so that they introduce key concepts in each section of text. Our "At a Glance" section opens each chapter with an outline that brings these statements together, presenting an overview of the chapter and its most important ideas. Figure titles tell the student at a glance what idea the figure illustrates, before the details are described in the caption. Concepts are further stressed in highlighted in-text summaries and in end-of-chapter summaries.

Makes Biology Relevant

There is a danger that students will stop asking the question "why?" if the equally important question "so what?" isn't answered. In fact, there is no science that can tell us more about ourselves and our world than can biology. Knowledge of biology illuminates everyday life: How can two brown-eyed parents have a blue-eyed child? Why do smokers cough so much? What causes a fever? A knowledge of biology is also crucial to understanding modern concerns and controversies such as population growth, the spread of diseases such as AIDS, the destruction of tropical rain forests, and the promise and perils of genetic engineering. In this text, we make a conscious effort to relate concepts in biology to everyday experiences and to important issues facing society. These applications are interwoven into the text, introduced in end-of-chapter essay questions, and highlighted in boxed essays such as:

 Earth Watch — a series of environmental essays addressing issues such as biodiversity, ozone depletion, and population growth.

 Health Watch — a series of essays investigating topics such as how ulcers form, how kidney dialysis works, and how drugs and alcohol can affect a developing child.

Communicates the Scientific Process

One cannot understand biology without knowing how science works. Biology is not just a compendium of facts and ideas; it is the outgrowth of a dynamic process of inquiry and human endeavor. In many places throughout the text, accompanying descriptions of biological facts and concepts, are descriptions of how those findings were achieved. The scientific process is further highlighted in Scientific Inquiry essays:

 Scientific Inquiry — students will learn, for example, how fossils are dated, how the structure of DNA was discovered, and how the gene for Huntington's disease was discovered.

Stresses Unifying Themes

Two themes are interwoven throughout the text: evolution and adaptation to the environment. As Theodosius Dobzhansky so aptly put it, "Nothing in biology makes sense, except in the light of evolution." Life on Earth has been forged in the crucible of evolution, and any modern biology text must reflect that fact. Many of the chapter introductions have an evolutionary theme, and many chapters end with a feature new to this edition:

 Evolutionary Connections — lively discussions that link chapter concepts into the broader perspective of evolution.

Adaptation to the environment flows naturally from the concept of evolution; the environment exerts selective pressure on organisms that helps shape their evolution. Nevertheless, not all structures are optimal solutions to environmental challenges. Many apparently unlikely biological designs are due to genetic drift, chance catastrophes, and the fact that the raw material for evolution consists of random mutations affecting previously existing structures. In the words of Sydney Brenner, "Anything produced by evolution is likely to be a bit of a mess," and we have tried to show that, too.

Our own concern for the environment also threads its way throughout this text. Whenever appropriate, we have tried to present students with the biological rationale for making sound environmental decisions.

Stimulates Exploration

For every "why?" that is answered, many more questions are raised. The knowledge and understanding of science that students gain from this text will allow them to continue to seek answers throughout their lives. Thought-provoking questions end each chapter, along with interesting and informative references, where topics can be explored in more detail. A free supplement for students, *The New York Times Themes of the Times* brings together a collection of recent, topical, biological articles from the pages of *The New York Times*. The *ABC News/Prentice Hall Video Library* for instructors features brief clips from programs like *20/20*, *World News Tonight*, and *Nightline*. Finally, the text challenges the student to enter the information age with its own home page on the World Wide Web. *Life on the Internet: Biology - A Student's Guide* is available as a supplement to help students navigate through the web to access the regularly updated resources on the *Life on Earth* home page:

 Net Watch — an icon at the end of each chapter highlights the specific address for each chapter's resources.

Responds to the Challenges of a One-Semester Course in Biology

Biology is absolutely unique in the diversity and range of its subject matter. This embarrassment of riches provides a special challenge to the instructor attempting to introduce biology, particularly in a one-semester course. Many share our concern that, when everything is covered superficially, explanation and real understanding of concepts may be sacrificed, reducing biology to a massive compendium of facts to be memorized and then quickly forgotten. In response to this dilemma, many instructors choose to cover fewer topics, but cover them well. *Life on Earth* is tailored to this approach. But which topics should be covered? The areas explored in this text have been chosen in response to surveys of introductory biology instructors nationwide. *Life on Earth* is designed to provide students, particularly non-biology majors, with a good working knowledge and understanding of areas of biology that will touch their lives.

We first introduce the basic chemistry of the molecules of life, cell structure, and the processes by which energy is captured and released. We then move on to inheritance, human genetics, and genetic engineering and biotechnology, followed by a unit that thoroughly explains the mechanisms and results of evolution. Our physiology unit emphasizes the human body, providing an owner's manual that explores crucial topics such as reproduction and development, birth control, and the immune system and diseases, including AIDS and cancer. Our complete ecology unit is a sort of user's guide to the Earth. In addition to the principles that govern population growth and the interactions that produce and sustain diverse natural ecosystems, we incorporate coverage of human impacts into each area of ecology.

Acknowledgments

Life on Earth is truly a team effort, and we wish to acknowledge and thank our teammates.

The dauntingly complex task of putting together a book of this magnitude has been handled with unflagging dedication and zeal by the skilled development team at Prentice Hall. Ray Mullaney, Editor in Chief of Engineering, Science, and Mathematics development, made sure the text was clear, consistent, and student-friendly. Ray continued to oversee the production of the text, contributing substantially to its design, photographic selection, and accuracy. Debra Wechsler, Senior Production Editor, supervised production of the text and patiently dealt with the minutiae of assuring accuracy in the art program and the translation from typed draft to final printed page. Tobi Zausner and Kathy Ringrose skillfully and doggedly tracked down excellent photos. Margo Quinto tackled the job of copyediting with meticulous care. We also wish to thank Heather Scott, our talented Art Director, for creatively guiding the text and cover design; Patrice Van Acker, who coordinated the art program; and Richard Foster, for his artful formatting of these pages. Mary Hornby competently and conscientiously has overseen the production of our print supplements, test bank, and transparencies. Linda Schreiber did a fantastic job of coordinating the multimedia program, especially our exciting new laser disc and CD-ROM.

Colleagues at other institutions have helped us enormously. Many, listed in the following pages, have stimulated us to rethink our presentation with careful, thoughtful reviews. Essays were contributed by Joseph Chinnici, Mark Shotwell, and Gisele Muller-Parker. A wonderful supplements package was engineered by Joseph Chinnici, Kristin Uthus, Rhoda Perozzi, Gayle Sauer, Linda Butler, Andrew Stull, Steven Brunasso, Mike McKinley, and Blanche Haning. We thank Bill Stark, Richard Mortensen, Gerald Summers, Rhoda Perozzi, Florence Juillerat, Jerri Lindsey, Scott Freeman, and Dan Doak for contributing thought-provoking end-of-chapter questions.

Marketing is being handled with talent and enthusiasm by Kelly McDonald, our Executive Marketing Manager. We also would like to thank Paul F. Corey, Editor in Chief of Science, for his insight and support. Finally, but most importantly, our editor, Sheri Snavely, has nurtured this project through thick and thin, smoothing rough spots and boosting our morale with her good humor, optimism, and energy.

Terry and Gerry Audesirk
Golden, Colorado

Reviewers

W. Sylvester Allred, *Northern Arizona University*
William Anderson, *Abraham Baldwin Agriculture College*
Bill Barstow, *University of Georgia, Athens*
Colleen Belk, *University of Minnesota, Duluth*
Marilyn Brady, *Centennial College of Applied Arts & Technology*
Arthur L. Buikema Jr., *Virginia Polytechnic Institute*
Kathleen Burt-Utley, *University of New Orleans*
Linda Butler, *University of Texas at Austin*
W. Barkley Butler, *Indiana University of Pennsylvania*
Joseph R. Chinnici, *Virginia Commonwealth University*
Ethel Cornforth, *San Jacinto College, South*
Lee Couch, *Albuquerque Technical Vocational Institute*
Donald C. Cox, *Miami University of Ohio*
Patricia B. Cox, *University of Tennessee*
Peter Crowcroft, *University of Texas at Austin*
Carol Crowder, *North Harris Montgomery College*
Robert A. Cunningham, *Erie Community College, North Campus*
Jerry Davis, *University of Wisconsin, La Crosse*
Jean DeSaix, *University of North Carolina, Chapel Hill*
Ed DeWalt, *Louisiana State University*
Daniel F. Doak, *University of California, Santa Cruz*
Susan A. Dunford, *University of Cincinnati*
Mary Durant, *North Harris College*
Ronald Edwards, *University of Florida*
Joanne T. Ellzey, *University of Texas, El Paso*
Carl Estrella, *Merced College*
Rita G. Farrar, *Louisiana State University*
Marianne Feaver, *North Carolina State University*
Linnea Fletcher, *Austin Community College, Northridge Campus*
Scott Freeman, *University of Washington*
Donald P. French, *Oklahoma State University*
Teresa Lane Fulcher, *Pellissippi State Technical Community College*

Martin E. Hahn, *William Paterson College*
Blanche C. Haning, *North Carolina State University, Raleigh*
Alison G. Hoffman, *University of Tennessee, Chattanooga*
Florence Juillerat, *Indiana University-Purdue University at Indianapolis*
Thomas W. Jurik, *Iowa State University*
Jerri K.Lindsey, *Tarrant County Junior College, Northeast Campus*
D. McWhinnie, *De Paul University*
Karen E. Messley, *Rockvalley College*
Richard Mortensen, *Albion College*
Gisele Muller-Parker, *Western Washington University*
Harry Nickla, *Creighton University*
Jane Noble-Harvey, *University of Delaware*
Rhoda E. Perozzi, *Virginia Commonwealth University*
Bill Pfitsch, *Hamilton College*
James A. Raines, *North Harris College*
Mark Richter, *University of Kansas*
Paul Rosenbloom, *Southwest Texas State University*
Edna Seaman, *University of Massachusetts, Boston*
Linda Simpson, *University of North Carolina, Charlotte*
Shari Snitovsky, *Skyline College*
Benjamin Stark, *Illinois Institute of Technology*
William Stark, *Saint Louis University*
Gerald Summers, *University of Missouri, Columbia*
William Thwaites, *San Diego State University*
Sharon Tucker, *University of Delaware*
Lloyd W. Turtinen, *University of Wisconsin, Eau Claire*
Kristin Uthus, *Virginia Commonwealth University*
Jyoti R. Wagle, *Houston Community College,Central Campus*
Roberta Williams, *University of Nevada, Las Vegas*
Bill Wischusen, *Louisiana State University*
Colleen Wong, *Wilbur Wright College*
Wade Worthin, *Furman University*

Multimedia Reviewers

Consultant:
Michael McKinley, *Glendale Community College*

Photo Coordinator:
Blanche Haning, *North Carolina State University*

Advisory Panel:
Arthur L. Buikema Jr., *Virginia Polytechnic Institute*
Thomas W. Jurik, *Iowa State University*
Frank Schwartz, *Cuyahoga Community College*
Charlene Waggoner, *Bowling Green University*

Reviewers:
Sarah Barlow, *Middle Tennessee State University*
Guy Cameron, *University of Houston*
Susan A. Dunford, *University of Cincinnati*
Florence Juillerat, *Indiana University/Purdue University*
Arnold Karpoff, *University of Louisville*

Terry Keiser, *Ohio Northern University*
Donald Reinhardt, *Georgia State University*
Dan Skean Jr., *Albion College*
Lloyd W. Turtinen, *University of Wisconsin, Eau Claire*
Robin Tyser, *University of Wisconsin, LaCrosse*

 INTERNET GUIDE
AND WORLD WIDE WEB
HOME PAGE REVIEWERS

Susan Brawley, *University of Maine*
Ronald Edwards, *University of Florida*
David Huffman, *Southwest Texas State University*
Jason Rosenblum, *Austin, Texas*
Patrick A. Thorpe, *Grand Valley State University*
Robin Tyser, *University of Wisconsin, La Crosse*
Roberta Williams, *University of Nevada, Las Vegas*

Supplements and New Media Tools

For the Instructor

Instructor's Examination Package (0-13-285041-9)

Instructor's Resource Manual
By Gayle Sauer, The Citadel
Contains a variety of lecture outlines and teaching tips.
(0-13-377987-4)

Transparency Pack
185 four-color, large type transparencies from the text. The balance of art is available as black line transparency masters.
(0-13-271800-6)

Slides
Same images as transparency acetates, available in slide format.
(0-13-378084-8)

Test Item File
By Joseph Chinnici, Rhoda Perozzi, Kristin Uthus, all of Virginia Commonwealth University
Contains over 2300 test questions. Makes use of a variety of question types so that users can customize tests and quizzes to incorporate conceptual, recall-oriented, applied, and critical thinking formats. (0-13-271784-0)

Prentice Hall Custom Test-IBM (0-13-271834-0)
Prentice Hall Custom Test-Macintosh (0-13-271842-1)
Available for Windows, Macintosh, and DOS, Prentice Hall Custom Text allows the educator to create and tailor the exam to their own needs. With the Online Testing option, exams can also be administered online and data can then be automatically transferred for evaluation. A comprehensive desk reference guide is included, along with online assistance.

ABC News /Prentice Hall Video Library for Biology
Coordinated by Linda Butler, University of Texas at Austin
This unique video library contains brief (5-20 minute) segments from award-winning shows such as 20/20, World News Tonight, and The American Agenda. This innovative resource shows biological principles at work and teaches students to critically analyze media messages based on their scientific knowledge.
(0-13-514191-5)

The Prentice Hall Laser Disc for Biology
The Prentice Hall Laser Disc for Biology is a visual encyclopedia created specifically for use with Life on Earth. The disc features over 1000 still images from the text and other sources, 30 minutes of video, and 25 minutes of broadcast-quality three-dimensional animations. The animations focus on topics that are challenging to students and particularly difficult to visualize (e.g., cell respiration, enzyme function, photosynthesis). The video segments focus on applications of biology to students' everyday world (e.g., DNA testing, CT and MRI technology).
(0-13-569583-X)

Prentice Hall CD-ROM Image Bank for Biology
This unique image bank makes available still images, animations, and video in a digitized format for use in the classroom or for students. It also includes a navigational tool to allow instructors to customize lecture presentations. Additional features include keyword searches and the ability to incorporate lecture notes based on custom presentations. (Windows version: 0-13-513987-2; Macintosh version: 0-13-552456-3)

Director Academic
Authorware Academic
By Prentice Hall and Macromedia
Available exclusively through Prentice Hall, these educational adaptations of the leading multimedia authoring tools put the power of multimedia development into your hands at a fraction of the cost. Both products also contain templates designed specifically for academic uses. Available for both Macintosh and Windows platforms. For further information and/or sales, contact Prentice Hall Multimedia Group at 1-800-887-9998. Saleable product only ($150 for Authorware Academic/ $99 for Director Academic).

For the Student

Audesirk *Life on Earth* World Wide Web Home Page
Developed by Andrew Stull and Steven Brunasso, California State University at Fullerton
This unique tool is the first to harness the resources of the Internet for easy integration into the classroom. Designed to launch student exploration, resources are regularly updated and linked to text chapters (see Net Watch icons at the end of each chapter).

Life on the Internet: Biology—A Student Guide
By Andrew Stull, California State University at Fullerton
The perfect tool to help your students take advantage of our Life on Earth home page on the World Wide Web and open up the world of the Internet for exploration! Tied specifically to our home page, this unique resource gives clear steps to access our regularly updated biology resource area as well as an overview of general navigation strategies. (0-13-244088-1)

Study Guide
By Joseph Chinnici, Virginia Commonwealth University
This essential study tool for students is written by the leading author of the test bank to ensure consistency in style. It includes a variety of study questions as well as chapter outlines and interactive exercises. (0-13-271859-6)

New York Times Themes of the Times Supplement
Coordinated by Linda Butler, University of Texas at Austin
This unique newspaper-format supplement brings together recent articles on dynamic biology applications from the pages of the world-renowned New York Times. This free supplement, available in quantity through your local representative, encourages students to make connections between the classroom and the world around them.

For the Laboratory

Explorations in Basic Biology, Seventh Edition
By Stanley Gunstream, Pasadena City College
This best-selling laboratory manual can be used with Life on Earth or any introductory biology text. It includes 40 self-contained, easy-to-understand experiments that blend traditional experiments with new investigative exercises. (0-13-372939-7; Instructor's Manual: 0-13-393927-8)

Biological Explorations: A Human Approach, Third Edition
By Stanley Gunstream, Pasadena City College
Specifically designed for introductory biology courses where the human organism is emphasized. (0-13-575770-3; Instructor's Manual: 0-13-717588-4)

How to Use this Text: A Guide for Students

Life on Earth has been designed as one of the tools to launch your exploration of the exciting world of biology! Our goal is to help you master the basic concepts of biology and then begin applying them to the world around you. Take a few moments now to see how our text can help you succeed.

At a Glance

At a Glance overviews the chapter ahead and helps you organize your study.

8 Harvesting Energy from Food: Glycolysis and Cellular Respiration

✦ AT A GLANCE

Glucose Metabolism: An Overview

Glycolysis
Glucose Must Be Activated by ATP before Its Energy Can Be Harvested

Fermentation
Some Cells Ferment Pyruvic Acid to Lactic Acid
Other Cells Ferment Pyruvic Acid to Alcohol

Cellular Respiration
Pyruvic Acid Is Transported into the Mitochondrial Matrix
Pyruvic Acid Is Broken Down by Reactions in the Mitochondrial Matrix
Energetic Electrons Are Carried to Electron Transport Systems in the Inner Mitochondrial Membrane
Chemiosmosis Captures Energy Stored in a Hydrogen Ion Gradient and Produces ATP
ATP Is Transported from the Mitochondrial Matrix and Diffuses into the Cell Cytosol
Cellular Metabolism Influences the Way Entire Organisms Function

When a hummingbird sips nectar from a flower, it eats the products of photosynthesis. Photosynthesis in the plant's leaves had converted the energy of sunlight into the chemical energy of organic molecules such as the sugar in nectar. To supply its enormous energy needs—the highest of any animal—the hummingbird must eat its weight in nectar daily. Then its cells must efficiently extract energy from the glucose in the nectar. When the hummingbird's cells break down the high-energy glucose molecules in the presence of oxygen, they produce the low-energy molecules the plant started with—carbon dioxide and water—making energy available for muscles and growth (Fig. 7-1).

However, the hummingbird's cells cannot directly use the chemical energy derived from this process. Its wing muscles require energy stored in the molecule adenosine triphosphate (ATP). Its brain uses ATP in conducting nerve signals, and its ovaries use ATP in making eggs. The plant, for its part, uses ATP to make the petals, pigments, and fragrance that attract the bird and the leaves and chlorophyll molecules that trap the sun's energy. In this chapter, we explore the metabolic processes by which organisms convert the energy of organic molecules into the usable energy of ATP, processes called glycolysis and respiration.

Most cells can metabolize a variety of organic molecules to produce ATP. We focus on the metabolism of the single molecule glucose for three reasons. First, virtually all single-celled organisms, and every cell of multicellular organisms, metabolize glucose for energy at least part of the time. Some, such as nerve cells in your brain, rely predominantly on glucose as a source of energy. Second, glucose metabolism is less complex than the metabolism of most other organic molecules. Finally, when cells use other organic molecules as energy sources, they usually convert the molecules to glucose or other compounds that enter the pathways of glucose metabolism (see "Health Watch: Metabolic Transformations—Why You Can Get Fat by Eating Sugar").

Glucose Metabolism: An Overview

During photosynthesis, photosynthetic organisms harvest and store the energy of sunlight in glucose. During glucose breakdown, that energy is released and converted to

An Anna's hummingbird feeds at a flower in California.

Opening Stories/Vignettes

Opening Stories/Vignettes in most chapters relate the opening photograph to the concepts that will be developed in the chapter.

Focus on Key Concepts

As you read, identify and focus on the key concepts.
Can you rephrase the main ideas in your own words?

Conceptual Headings

Conceptual Headings provide a context for the discussion to follow and give you a clear review of key concepts.

In-text Summaries

In-text Summaries help you check your understanding of complex processes, especially in the early chapters where the fundamental concepts of biology are laid out. Stop and read these summaries as you move through a chapter as a help in digesting new material.

---

see that cells have evolved elaborate mechanisms that ensure that each daughter cell inherits all the materials it needs to continue the flow of life.

The Essentials of Cellular Reproduction

When a cell divides, it must transmit to its offspring cells two essential requirements for life (Fig. 9-1): hereditary information to direct life processes and materials in the cytoplasm that the offspring need to survive and to utilize their hereditary information.

Cell Division Transmits a Complete Set of Hereditary Information to Each Daughter Cell

The hereditary information of all living cells is **deoxyribonucleic acid (DNA)**. Like many large biological molecules, a molecule of DNA consists of a long chain of smaller subunits (see Chapter 3). DNA has four different types of subunits, called nucleotides. Segments of DNA a few hundred to many thousand nucleotides long are the units of inheritance—the genes—that carry the genetic information to produce specific traits. The sequence of nucleotides in a gene encodes information for the synthesis of the RNA and protein molecules that are needed to build a cell and carry out its metabolic activities. We will see in Chapters 10 through 13 how DNA encodes genetic instructions and how a cell regulates which genes it uses at any given time.

For any cell to survive, it must have a *complete set of genetic instructions*. Therefore, when a cell divides, it can-

not simply split its set of genes in half and give each daughter cell half a set. Rather, the cell must first *duplicate its DNA*, much like making a photocopy of an instruction manual. Each daughter cell then receives a complete "DNA manual" containing all the genes.

Cell Division Transmits Essential Cytoplasmic Materials to Each Daughter Cell

Like the blueprints for a house, the in[formation carried] by DNA are useless without materials [and machinery. A] newly formed cell must receive the m[aterials it needs to] read its genetic instructions and to [acquire] enough to acquire new materials fro[m its environment] and to process them into new cellula[r structures. Fur-] thermore, as we described in Chapter 5[, cells cannot syn-] thesize either mitochondria or chlorop[lasts. Instead,] these organelles arise only by the div[ision of previously] existing mitochondria and chloroplas[ts. Therefore, as a] cell divides, its cytoplasm is divided ab[out equally between] the two daughter cells. This simple m[echanism usually] provides both daughter cells with all [the essential nu-] trients, enzymes, and other molecules [they require.]

The Activities of a Cell from One [Division to] the Next Constitute the Cell Cycle

Newly formed cells usually acquire n[utrients from their] environment, synthesize more of their [own materials, and] grow larger. After a variable amount of [time, depending on] the organism, the type of cell, and the [nutrients available,] the cell divides. This general descript[ion applies to both] eukaryotic and prokaryotic cells. Pr[okaryotic cells do]

Figure 9-1 A generalized cell cycle.

- cytoplasm
- DNA

1. Reduced to its essentials, a cell consists of genetic material (DNA) and cytoplasm.

2. The cell grows, adding more cytoplasmic constituents.

3. It replicates its DNA.

4. The cell divides, with one copy of the DNA and about half the cytoplasm going to each daughter cell. Each daughter cell may then repeat the cycle.

- (matrix)
- FADH₂
- NADH
- 2e⁻
- electron carriers
- energy to drive ATP synthesis
- (intermembrane compartment)
- 1/2 O₂ +2H⁺
- H₂O
- (inner membrane)
- H⁺

Figure 8-6 The electron transport system of mitochondria

(1) The electron carrier molecules NADH and FADH₂ deposit their energetic electrons with the carriers of the transport system located in the inner membrane. (2) The electrons move from carrier to carrier within the transport system. Some of their energy is used to pump hydrogen ions across the inner membrane from the matrix into the intermembrane compartment. This movement out of the inner membrane creates a hydrogen ion gradient that can be used to drive ATP synthesis (for details, see "A Closer Look at Chemiosmosis in Mitochondria"). (3) At the end of the electron transport system, the energy-depleted electrons combine with oxygen and hydrogen ions in the matrix to form water.

inner membrane, into the intermembrane compartment (see the section on chemiosmosis, below).

Finally, at the end of the electron transport system, oxygen and hydrogen ions accept the energetically depleted electrons: Two electrons, one oxygen atom, and two hydrogen ions combine to form water. This step clears out the transport system, leaving it ready to run more electrons through. Without oxygen, the electrons would "pile up" in the transport system, with the result that the hydrogen ions would not be pumped across the inner membrane. The hydrogen ion gradient would soon dissipate and ATP synthesis would stop.

Chemiosmosis Captures Energy Stored in a Hydrogen Ion Gradient and Produces ATP

Hydrogen ion pumping across the inner membrane generates a large concentration gradient; that is, a high concentration of hydrogen ions in the intermembrane compartment and a low concentration in the matrix. The inner membrane is impermeable to hydrogen ions except at protein pores that are part of ATP-synthesizing enzymes. In the process known as **chemiosmosis**, hydrogen ions move down their concentration gradient from the intermembrane compartment to the matrix through these ATP-synthesizing enzymes. The flow

of hydrogen ions provides the energy to synthesize 32 to 34 molecules of ATP from ADP and inorganic phosphate in the matrix. The box "A Closer Look at Chemiosmosis in Mitochondria" examines chemiosmosis in more detail.

ATP Is Transported from the Mitochondrial Matrix and Diffuses into the Cell Cytosol

The ATP that was synthesized from ADP and inorganic phosphate in the matrix during chemiosmosis is transported across the inner membrane from the matrix to the intermembrane compartment. It then diffuses out of the mitochondrion to the cytosol through the outer membrane, which is very permeable to ATP. These ATP molecules provide most of the energy needed by the cell. ADP simultaneously diffuses from the cytosol across the outer membrane and is transported across the inner membrane to the matrix, replenishing the supply of ADP.

▓ SUMMARY

ELECTRON TRANSPORT AND CHEMIOSMOSIS

Electrons from the electron carriers NADH and FADH₂ enter the electron transport system of the inner mitochondrial membrane. Here their energy is used to generate a hydrogen ion gradient. Movement of hydrogen ions down their gradient through the pores of ATP-synthesizing enzymes drives the synthesis of 32 to 34 molecules of ATP. At the end of the electron transport system, two electrons combine with one oxygen atom and two hydrogen ions to form water.

Cellular Metabolism Influences the Way Entire Organisms Function

Many students feel that the details of cellular metabolism are hard to learn and don't really help them to understand the living world around them. Have you ever read a murder mystery and wondered how cyanide could kill a person almost instantly? Cyanide reacts with one of the proteins in the electron transport system, immediately blocking the movement of electrons through the system and bringing cellular respiration to a screeching halt. Under more normal conditions, metabolic processes within individual cells have enormous impacts on the functioning of entire organisms. To take just two familiar examples, let's consider the migration of ruby-throated hummingbirds and track events in the Olympics.

Hummingbird Migration Requires Efficient Energy Storage and Use

As the days shorten in August, many birds in North America prepare to migrate to Central and South

---

Key terms

Key terms are boldfaced throughout.

Visualize Key Concepts

If you can visualize something related to a concept, you'll remember it more easily. Our art and photographic program will help you grasp concepts, so study the pictures as well as the text.

Micro to Macro Figures

To understand biology you must be able to visualize structures at several different scales. **Micro to macro figures** help you bridge this gap.

Compound Art

Compound art brings together different views of a topic.

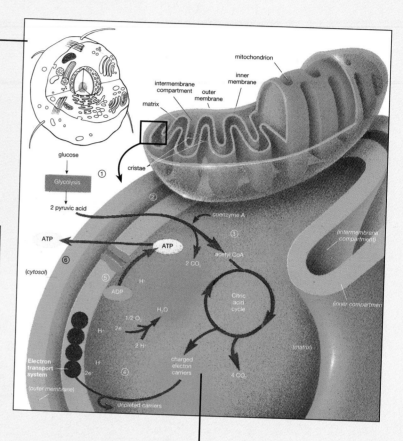

Process Art

Process art conveys key steps in processes, such as cellular respiration, as shown here.

Dynamic Photographs

Dynamic photographs provide vivid illustrations of concepts throughout.

Relate Concepts to the World Around You

Think about biology in the world around you. You have come into this course already knowing a lot about life. The concepts you are learning will give you a better understanding of your own life and of the world around you. Applications and stories are an important part of this text, and we hope that these features will spark your curiosity and make the study of biology as fascinating to you as it is to us.

Earth Watch

Earth Watch essays present environmental issues such as biodiversity, ozone depletion, and population growth.

EARTH WATCH

Has the Human Population Exceeded Earth's Carrying Capacity?

A glance at the age structure of less-developed countries, where most of the world's population resides, shows a tremendous momentum for continued growth. World population in the year 2010 is predicted to be over 7 billion and growing. A modest United Nations (UN) projection is that the human population may stabilize in the year 2150 at 11.5 billion (Fig. E28-3). Can Earth support over twice its current population?

Earlier we defined carrying capacity as the m[...] ulation that could be indefinitely sustained. [...] population requires that the ecosystem not b[...] ways that lower its ability to provide necessary [...] this definition we have already exceeded Earth' [...] pacity for people. The upper limit of the plan[...] capacity is determined by the ability of its plan[...] vest the energy from sunlight and produce high [...] ecules that organisms can use as food. Stanfo[...] biologist Peter Vitousek estimates that human a[...] already reduced the productivity of Earth's fore[...] lands by 12%. Each year, millions of acres of [...] tive land are being turned into desert throug[...] and deforestation, especially in less-develope[...] Earth's desert area is projected to increase by [...] year 2000 as a result of human activities. In a[...]

In the years between 1990 and 2010, the world population is projected to increase by 33%. During this same period, the available fish, cropland, pastureland, and forests will all decline by 10% to 30% per person. The UN Food and Agricultural Organization reports that Africa's food production for each person has dropped by 20% since 1960; another 30% drop in the next 25 years is predicted. Each year the United States is losing 1 million acres of

HEALTH WATCH

The Placenta Provides Only Partial Protection

When your grandparents were bearing children, doctors assumed that the placenta protected the developing fetus from most of the substances in maternal blood that could harm it. We know now that this is far from true. In fact, most medications and drugs and even some disease organisms readily penetrate the placental barrier and affect the fetus.

Infections May Cross the Placenta
The German measles virus can cross the placenta and attack the fetus, causing potentially severe retardation and other defects. The virus causing genital herpes (during active outbreaks) and the bacterium causing syphilis can cause mental or physical defects in the developing fetus. The virus causing AIDS may also cross the placenta, so infants may be born with this incurable disease.

Drugs Readily Cross the Placenta
A tragic example of drugs crossing the placenta is the tranquilizer thalidomide, commonly prescribed in Europe in the early 1960s (Fig. E27-3). Thalidomide's devastating effects on embryos were discovered only when many babies were born with missing or extremely abnormal limbs. More recently, in the late 1980s, the anti-acne drug Accutane was found to cause gross deformities in babies born to women using it. Although these are extreme examples, *any drug, including aspirin, has the*

Health Watch

Health Watch essays investigate issues such as prenatal diagnosis, sexually transmitted diseases, and osteoporosis.

The New York Times Themes of the Times

The New York Times **Themes of the Times** supplement brings together a collection of recent biology articles from this award-winning newspaper. Can you relate what you are learning in class to the issues discussed in recent articles?

The ABC News/Prentice Hall Video Library

The ABC News/Prentice Hall Video Library makes available brief clips from programs like *20/20*, *World News Tonight*, and *Nightline*. Will knowing more about the science behind the issues change the way you view the nightly news?

Understand the Process of Science

The processes by which scientists answer questions are as important as the questions themselves. *Life on Earth* emphasizes how we know what we know and gives you insight into the dynamic process of discovery.

Scientific Inquiry

Scientific Inquiry essays examine the people, the processes, and the methods of biology with topics such as DNA fingerprinting in forensics, Nancy Wexler on the trail of Huntington's disease, the use of isotopes in biology and medicine.

SCIENTIFIC INQUIRY

DNA Fingerprinting in Forensics—The Case of the Telltale Palo Verde

The polymerase chain reaction (PCR) procedure has caused a revolution in many fields of biology, but nowhere has the impact of this technique been felt more than in forensics, the collection of information to be used as evidence in court proceedings. The technique that is proving most useful has been nicknamed "DNA fingerprinting," because it may generate patterns of DNA fragments unique to each individual. This procedure involves the amplification of a set of DNA fragments of unknown sequence, a set that is likely to differ from one individual to the next. When these fragments are amplified and separated from one another on a gel, the pattern of bands is referred to as a DNA fingerprint. In theory, it represents a characteristic "snapshot" of the unique genome of an individual, allowing it to be distinguished from that of every other individual in the population. The quantity of DNA needed to analyze using PCR is exceedingly small; this amount of DNA can easily be obtained from a single spot of dried blood, from the cells clinging to the base of a hair shaft, from skin fragments under a victim's fingernails, or other small tissue samples collected at a crime scene.

Here is an example of the forensic use of DNA fingerprints that did not involve the suspect's DNA, but rather that of plants at the crime scene. On the night of May 2,

verdes show considerable genetic variation from one tree to the next. Otherwise, it would be impossible to unambiguously link a given pod to any particular tree. Fortunately, he found considerable differences in DNA between individual plants, allowing for discriminatory fingerprints.

Helentjaris then was given the two seed pods from the suspect's truck along with pods collected by investigators from 12 different palo verde trees near the factory. The investigators knew which of the 12 trees was near the body but did not tell Helentjaris. He extracted DNA from seeds taken from each pod and performed the PCR reaction, producing 10 to 15 distinct bands from each pod. The result of this experiment was unmistakable: The pattern from the pod found in the truck exactly matched the pattern from only one of the 12 trees, the one nearest the body. In an important additional test, Helentjaris found that this pattern was also different from that of 18 pods collected from trees at random sites around Phoenix. This result was admitted as evidence in the trial, the first time a DNA fingerprint of a plant had been used in a court case. This information was crucial in demolishing the suspect's alibi, and at the end of the 5-week trial the suspect was found guilty of first-degree murder. Riven the two seed pods from the suspect's truck along with pods col-

See the Connections

There are important patterns and themes that unify the discipline of biology. *Evolution* and *adaptation to the environment* are two related themes you'll revisit throughout the text. How do you see these themes at work?

Evolutionary Connections

Evolutionary Connections close many of the chapters, tying the subject matter of each chapter into the broader perspective of evolution. You'll also notice that many of the chapter introductions have an evolutionary theme.

EVOLUTIONARY CONNECTIONS

Kin Selection and the Evolution of Altruism

Altruism is any behavior that is potentially harmful to the survival and future reproduction of an individual organism but enhances the reproductive potential of other organisms. Altruism includes worker bees' rearing the offspring of queen bees. Note that altruism does not imply conscious, voluntary decisions to engage in selfless behavior. Rather, most altruistic behaviors have a strong instinctive component; that is, many animals have altruism programmed in their genes.

From an evolutionary viewpoint, how can this be? Surely, if a mutation arose that caused altruistic behavior, and the bearers of that mutation lost their lives or failed to reproduce because of their self-sacrificing behaviors, their "altruistic alleles" would disappear from the population. Maybe, or maybe not. To understand the evolution of altruism, we will need to introduce a new concept: **inclusive fitness**. As formulated by W. D. Hamilton, the inclusive fitness of an allele is the fitness conferred on all organisms that have the allele. Therefore, *if an altruist benefits related members of its own species that bear the same altruistic allele, then the altruistic allele may be favored by natural selection.*

To see how altruism might increase the inclusive fitness of an allele, let's consider the Florida Scrub Jay. Year-old jays usually do not mate and reproduce. Instead, these yearlings remain at their parents' nest and help out with next year's brood. Let's assume, for simplicity, that this altruistic behavior is controlled by a single "altruistic" allele and that in the distant past helper jays had the altruistic allele, and nonhelpers had another, "selfish," allele.

At least for one year, altruistic yearlings do not reproduce; some probably die from predation or accidents and never reproduce at all. How, then, can this behavior be adaptive? It all has to do with relatedness and the probability of successful reproduction. First, an animal's offspring inherit 50% of its genes (the other 50% of the offspring's genes come from the other parent). On the average, an animal also shares 50% of its genes with its siblings. Therefore, *a scrub jay is just as related to its siblings as it would be to its own offspring.* The second factor influencing scrub jay reproduction is that ideal habitat for jays is limited. Inexperienced yearling jays would probably have a hard time acquiring a good nest site and would be hard put to feed their offspring. Their best "reproductive bet," then, is to put their energy into helping their parents. Selfish yearling jays that try to nest on their own will probably contribute fewer genes to the next generation than the altruistic yearlings do. This phenomenon, whereby the actions of an individual increase the survival or reproductive success of its relatives, is called kin selection.

As this example suggests, *kin selection can favor the evolution of altruism if the altruistic behavior benefits relatives that bear the same altruistic allele.* In most cases, an animal will not know if another carries the altruism allele, but the animal must at least be able to distinguish relatives from strangers: Relatives stand a good chance of possessing the altruism allele, but you never can tell with strangers. A yearling jay that helped out at the nest of unrelated adult jays would probably waste its time and effort.

Identification of relatives isn't too hard to imagine in the case of jays and their parents. Many biologists objected to other proposed instances of altruistic behaviors, however, arguing that animals cannot evaluate degrees of relatedness. Two findings seem to address this objection. First, many social groups, including wolf packs and baboon troops, are actually family groups. Therefore, an animal would not have to identify relatives in order for its altruistic behaviors to benefit them the most. Second, many animals, including birds, monkeys, tadpoles, bees, and even tunicate larvae, can indeed identify relatives (Fig. 17-19). Given the choice between relatives and strangers, these animals preferentially associate with their relatives, *even if they were separated at birth and have never seen those relatives before.* If animals selectively form related groups, then altruistic behaviors will most likely benefit relatives. Although it is not the only mechanism, kin selection has been a powerful selective force in the evolution of altruism in many species, probably including humans.

Figure 17-19 Cannibalistic animals don't eat close relatives

Spadefoot toad tadpoles, found in transient water holes of the Arizona desert, are cannibalistic. Many of their prey, however, are released unharmed after being tasted briefly. Researchers have discovered that the tadpoles can indeed distinguish, and spit out, their own brothers and sisters, preferring to eat unrelated members of their own species.

Check Your Understanding.
Expand it. Explore!

SUMMARY OF KEY CONCEPTS

The Composition of Chromosomes
Eukaryotic chromosomes are composed of DNA and protein. DNA is the molecule that carries genetic information. It is composed of subunits called nucleotides, linked together into long strands. Each nucleotide consists of a phosphate group, the five-carbon sugar deoxyribose, and a nitrogen-containing base. Four different bases occur in DNA: adenine, guanine, thymine, and cytosine.

DNA: The Hereditary Molecule
Transformation in bacteria and infection of bacteria by bacteriophages provided the first evidence that DNA is the hereditary molecule. Griffith discovered that living bacteria can acquire genetic material from dead bacteria, transforming the live bacteria. Avery and his colleagues showed that the genes taken up by living bacteria during transformation are composed of DNA.

A bacteriophage consists solely of DNA and protein. When it infects a bacterium, a phage injects its genetic material into the bacterium, where it directs the synthesis of more phages. Hershey and Chase demonstrated that phages inject just DNA, but not protein, into bacteria. Therefore, DNA must be the genetic material of phages.

The Structure of DNA
The DNA of chromosomes is composed of two strands wound about one another in a double helix. The sugars and phosphates that link one nucleotide to the next form the backbone on each side of the double helix, while the bases from each strand pair up in the middle of the helix. Only specific pairs of bases, called complementary base pairs, can link together in the helix, held by hydrogen bonds: adenine with thymine and guanine with cytosine.

DNA Replication: The Key to Constancy
When a chromosome is replicated before mitosis or meiosis, the two DNA strands of each double helix unwind. DNA polymerase enzymes move along each strand, linking up free nucleotides into new DNA strands. The sequence of nucleotides in each newly formed strand is complementary to the sequence on a parental strand. As a result, two double helices are synthesized, each consisting of one parental DNA strand plus one newly synthesized, complementary strand that is an exact copy of the other parental strand. The two daughter DNA molecules are therefore duplicates of the parental DNA molecule.

Summary of Key Concepts
Summary of Key Concepts helps you begin your review.

Applying the Concepts
Applying the Concepts questions stretch your learning and ask you to apply your critical thinking skills to real-world scenarios.

KEY TERMS

bacteriophage   p. 171
base   p. 168
complementary base pair   p. 175
deoxyribonucleic acid (DNA)   p. 167
DNA polymerase   p. 178

double helix   p. 174
helix   p. 174
nucleotide   p. 168
purine   p. 168
pyrimidine   p. 168

replication for
semiconserva
transformatio

THINKING THROUGH THE CO

Multiple Choice
1. If you wanted to label amino acids but not DNA, which of the following radioactive isotopes would you use?
   a. ^{18}F b. ^{35}S c. ^{14}C
   d. ^{3}H e. ^{32}P
2. Whose pioneering work on the gene used the bacteriophage to show that DNA was the part of this virus that went into the bacterial cell for replication?
   a. Watson and Crick   b. Avery
   c. Hershey and Chase   d. Griffith
   e. Beadle and Tatum
3. Knowing what you now know about RNA, DNA, and DNA replication from this chapter and mitosis from the last chapter, sister chromatids
   a. each have one single strand of the two complementary DNA strands
   b. each have a fully formed double helix of DNA
   c. are the same as the homologous chromosomes of a diploid cell
   d. are present in bacteriophages, bacteria, and eukaryotes
   e. none of the above

4. How many different possible ba
   nucleotide chain 3 bases long?
   a. 1   b. 3
   d. 64   e. more than
5. Chargaff, in analyzing DNA, fo
   a. cytosine equaled that of guan
   b. cytosine equaled that of thym
   c. cytosine equaled that of aden
   d. each nucleotide was unrelate
   e. each nucleotide was equal to
6. Semiconservative replication re
   a. genetic information is contai
      molecule
   b. only the DNA of the bacteri
      cell
   c. certain bases pair with specifi
   d. one DNA strand remains wh
   e. point mutations sometimes i
      replication

Thinking Through the Concepts
Thinking Through the Concepts questions check your basic understanding. Both multiple-choice and short-answer questions are included.

Net Watch
Net Watch icons highlight the Internet's World Wide Web address of the *Life on Earth* Home Page where you can find resources and activities to aid your study and broaden your understanding of biology. If you have access to the World Wide Web (either though school or through a service like America Online) meet us on the Net!

Review Questions
1. Draw the structure of a nucleotide. Which parts are identical in all nucleotides, and which can vary?
2. Name the four types of nitrogen-containing bases found in DNA. Which bases are complementary to one another? How are they held together in the double helix of DNA?

3. What is transformation? In Griffith's experiments, why w the transformed bacteria able to synthesize capsules?
4. Diagram the life cycle of a bacteriophage.
5. Describe the structure of DNA in a chromosome. Wh are the bases, sugars, and phosphates in the structure?
6. Describe the process of DNA replication.

APPLYING THE CONCEPTS

1. Diagram and describe the Hershey-Chase experiment. Suppose they had found that *both radioactively labeled DNA and radioactively labeled protein* were injected into the bacteria. What would they have been able to conclude from this? (Note that if the bacteriophage chromosome was like the eukaryotic chromosome, that is, if it contained both DNA and protein, this is exactly what would have happened.)
2. A preview question for Chapter 11: Genetic information is encoded in the sequence of nucleotides in DNA. Let's suppose that the nucleotide sequence on one strand of a double helix encodes the information needed to synthesize a hemoglobin molecule. Do you think that the sequence of nucleotides on the other strand of the double helix would also encode useful information? Why or why not? An analogy to think about: Suppose that English were a "complementary language," with letters at opposite ends of the alphabet complementary to one another (for example, A is complementary to Z, B to Y, C to X, etc.). Would the complementary sentence to "All the world's a stage" make sense?
3. Suppose mammalian cells are grown and divide many times in a culture medium containing thymine made radioactive with a form of hydrogen (^{3}H). The cells are then removed

from the radioactive medium and allowed to replicate se al times in normal culture medium. Daughter chromoso are tested each generation to determine whether they con radioactive thymine. While in the radioactive medium, e strand of DNA of each chromosome has radioactive thym Assuming that each chromosome contains a single long ble-stranded molecule of DNA, predict the radioactive tus of daughter chromosomes after one, two, and three ro of cell division in normal medium. Explain how your dictions are consistent with the Watson-Crick explana of semiconservative DNA replication.
4. During the Inquisition, the church forced Galileo to avow some of his scientific statements. Today, scientif vances are being made at an astounding rate, and nowl is this more evident than in our understanding of the ogy of heredity. Using DNA as a starting point, do you th there are limits to the knowledge people should acquir
5. Historians debate the degree to which key people actu changed the course of history. From this chapter's histo perspective, discuss whether the pace or actual progre knowledge is influenced by the individual scientists invo

FOR MORE INFORMATION

Crick, F. H. C. "The Structure of the Hereditary Material." *Scientific American,* October 1954. It is entertaining to consider a paper on one of science's most fundamental results while the finding was still new and before the author won the Nobel Prize for Physiology and Medicine.
Holliday, R. "A Different Kind of Inheritance." *Scientific American,* June 1989. Methyl groups attached to DNA may alter its use in a cell. What's more, patterns of DNA methylation may be passed down from parent to offspring.
Judson, H. F. *The Eighth Day of Creation: Makers of the Revolution in Biology.* New York: Simon and Schuster, 1979. A remarkably readable history of the beginnings of molecular biology.
Kimura, M. "The Neutral Theory of Molecular Evolution." *Scientific American,* November 1979. Kimura persuasively argues that some evolutionary changes can occur without selection.

Mirsky, A. E. "The Discovery of DNA." *Scientific American,* J 1968. The early history of DNA research.
Radman, M., and Wagner, R. "The High Fidelity of DNA plication." *Scientific American,* August 1988. Faithful du cation of chromosomes requires both reasonably accurate tial replication of DNA sequences and final proofreadin
Rennie, J. "DNA's New Twists." *Scientific American,* Ma 1993. Relevant to this and several subsequent chapters, simple review covers some of the interesting modern c traditions that go beyond the standard textbook dogma
Watson, J. D. *The Double Helix.* New York: Atheneum Publish 1968. If you still believe the Hollywood images, that scien are either maniacs or cold-blooded, logical machines, be sur read this book. Although scarcely models for the behavior o ture scientists, Watson and Crick are certainly human enou

NET WATCH
On-line resources for this chapter are on the World Wide Web at:
http://www.prenhall.com/~audesirk (click on the table of contents link and then select Chapter 10).

For More Information
For More Information provides extra readings from books and the latest periodicals on chapter-related topics.

Brief Contents

Essays

EARTH WATCH

EVOLUTIONARY CONNECTIONS

HEALTH WATCH

SCIENTIFIC INQUIRY

Contents

Chapter 7

Capturing Solar Energy: Photosynthesis
122

Chapter 8

Harvesting Energy from Food: Glycolysis and Cellular Respiration   134

UNIT II   Inheritance   147

Chapter 9

Cellular Reproduction   148

Chapter 10

DNA: The Molecule of Heredity   166

Chapter 11

Gene Expression and Regulation   182

Chapter 12

Gene Exchange, Meiosis, and Eukaryotic Life Cycles   206

UNIT V   Ecology   553

Chapter 28

Population Growth and Regulation   554

1

An Introduction to Life on Earth

> *"Viewed from the distance of the moon, the astonishing thing about the earth, catching the breath, is that it is alive. The photographs show the dry, pounded surface of the moon in the foreground, dead as an old bone. Aloft, floating free beneath the moist, gleaming surface of bright blue sky, is the rising earth, the only exuberant thing in this part of the cosmos."*
>
> LEWIS THOMAS in *The Lives of a Cell* (1974)

Life on Earth is confined to a thin film encompassing Earth's surface: the biosphere. Earth, seen here from the moon, is an oasis of life in our solar system.

On your way across campus tomorrow morning, take a moment to look around you. Although you might not notice at first, an astonishing array of creatures thrives even in a place as domesticated as a college campus. Sparrows and squirrels scurry along branches, chirping and chattering. Trees, bushes, grass, and moss blanket the campus with green. Take a closer look. You may also see honeybees flitting from flower to flower, earthworms—or their winding tracks—at the edges of a small pond, a spider spinning its web, glistening mushrooms in the grass or beneath a bush. If you listen carefully, you may hear creatures you don't see; insects humming, birds calling in the distance, perhaps a frog croaking. But these are only the obvious organisms. Countless tiny ones, most too small to be seen with the naked eye, swim in the puddles left by last night's rain. A whole community of microscopic living things thrive in the soil. And living on, in, and around all these organisms, and the humans observing them, are billions of bacteria—simple, single-celled organisms that have survived nearly unchanged for billions of years.

How did such an astounding variety of living things evolve on Earth? In what ways do they interact? How are these bacteria, fungi, plants, and animals alike, and how do they differ? What processes must occur in each organism for it to survive in its environment and to reproduce? And perhaps the most basic question of all: How do the living and nonliving things differ? Common questions like these—asked by curious children and adults for centuries—form the basis of the science of biology. But they also relate to the very survival of the human species. For we too are organisms. We evolved in response to the same sorts of survival needs. Our bodies and those of all other living creatures consist of the same basic set of chemicals. The same processes allow us to survive and reproduce. As we journey through this book together, we hope you begin to share with us the excitement and awe that come from understanding the nature of Earth and its inhabitants and appreciating the beauty of it all.

The Characteristics of Living Things

To study the nature of life on Earth, we should begin at the beginning: *What is life?* If you look up *life* in a dictionary, you will find definitions such as "the quality that distin-

(a)

(b)

(c)

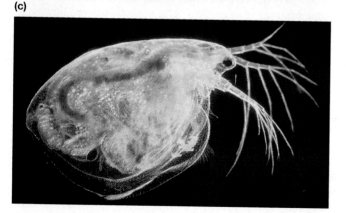

Figure 1-1   **Living organisms are complex and organized**

(a) Each crystal of table salt, sodium chloride, is a cube, showing great organization but minimal complexity. **(b)** The water and dissolved materials in the ocean represent complexity but very little organization. **(c)** Living things have both complexity and organization. The waterflea, *Daphnia pulex*, is only 1 millimeter (1/1000 meter) long, yet it has legs, a mouth, a digestive tract, reproductive organs, light-sensing eyes, and even a rather impressive brain considering its size.

guishes a vital and functioning being from a dead body," but you won't find out what that "quality" is.

We all have an intuitive understanding of what it means to be alive. Nevertheless, defining *life* is difficult, partly because living things are so diverse, and nonliving matter is sometimes so lifelike. A more fundamental difficulty in defining life is that living things cannot be described as the sum of their parts. The quality of life emerges as a result of the incredibly complex, ordered interactions among these parts. Because it is based on these **emergent properties**, life is a fundamentally intangible quality that defies any simple definition. We can, however, describe some of the characteristics of living things that, taken together, are not shared by nonliving objects. These characteristics are as follows:

1. Living things have a complex, organized structure based on organic (carbon-based) molecules.
2. Living things acquire materials and energy from their environment and convert them into different forms.
3. Living things actively maintain their complex structure and their internal environment, a process called homeostasis.
4. Living things grow.
5. Living things respond to stimuli from their environment.
6. Living things reproduce themselves, using a molecular blueprint called DNA.
7. Living things, taken as a whole, have the capacity to evolve.

Let's explore these characteristics in more detail.

Living Things Are Both Complex and Organized

Compared with nonliving matter of similar size, living organisms are highly complex and organized. A crystal of table salt (Fig. 1-1a), for example, consists of just two chemical elements, sodium and chlorine, arrayed in a precise cubical arrangement: The salt crystal is organized but simple. The oceans (Fig. 1-1b) contain some atoms of all the naturally occurring elements, but these atoms are randomly distributed: The oceans are complex but not organized. In contrast, even the tiny waterflea (Fig. 1-1c) contains dozens of different elements linked together in thousands of specific combinations that are further organized into ever larger and more complex assemblies to form structures such as eyes, legs, a digestive tract, and a brain.

Life on Earth is composed of a hierarchy of structures, with each level of the hierarchy based on the one below it and providing the foundation for the one above it (Fig. 1-2).

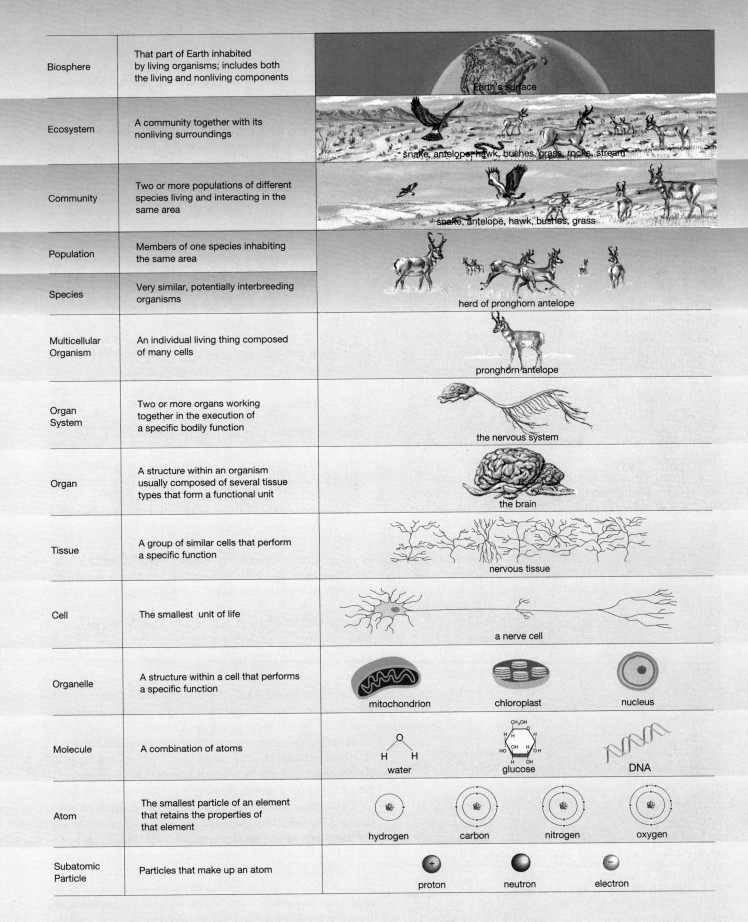

Biosphere	That part of Earth inhabited by living organisms; includes both the living and nonliving components	
Ecosystem	A community together with its nonliving surroundings	snake, antelope, hawk, bushes, grass, rocks, stream
Community	Two or more populations of different species living and interacting in the same area	snake, antelope, hawk, bushes, grass
Population	Members of one species inhabiting the same area	
Species	Very similar, potentially interbreeding organisms	herd of pronghorn antelope
Multicellular Organism	An individual living thing composed of many cells	pronghorn antelope
Organ System	Two or more organs working together in the execution of a specific bodily function	the nervous system
Organ	A structure within an organism usually composed of several tissue types that form a functional unit	the brain
Tissue	A group of similar cells that perform a specific function	nervous tissue
Cell	The smallest unit of life	a nerve cell
Organelle	A structure within a cell that performs a specific function	mitochondrion chloroplast nucleus
Molecule	A combination of atoms	water glucose DNA
Atom	The smallest particle of an element that retains the properties of that element	hydrogen carbon nitrogen oxygen
Subatomic Particle	Particles that make up an atom	proton neutron electron

Figure 1-2   **Levels of organization of matter on Earth**

All life has a chemical basis, but the quality of life itself emerges on the cell level. Interactions among the components of each level and the levels below it allow the development of the next-higher level of organization.

All of life is built on a chemical foundation, based on **elements**, each of which is a unique type of matter. An **atom** is the smallest particle of an element that retains the properties of that element. For example, a diamond is made of the element carbon. The smallest possible unit of the diamond is an individual carbon atom; any further division would produce isolated **subatomic particles** that would no longer be carbon. Atoms may combine in specific ways to form assemblies called **molecules**; for example, one carbon atom can combine with two oxygen atoms to form a molecule of carbon dioxide. Although many simple molecules form spontaneously, extremely large and complex molecules are manufactured only by living things. The bodies of living things are composed primarily of complex molecules. The molecules of life, which are based on carbon, are called **organic molecules**.

Although the chemical arrangement and interaction of atoms and molecules are the building blocks of life, the quality of life itself emerges on the level of the cell. Just as an atom is the smallest unit of an element, so too the **cell** is the smallest unit of life (Fig. 1-3). The difference between a living cell and a conglomeration of chemicals illustrates some of the emergent properties of life.

Fundamentally, all cells contain **genes** that provide the information needed to control the life of the cell; subcellular structures called **organelles**, miniature chemical factories that use the information in the genes and keep the

plasma membrane

nuclear envelope (double membrane)

nucleus

cell wall

mitochondrion (an organelle)

1 micrometer

Figure 1-3   **The smallest unit of life is the cell**

This electron micrograph clearly shows the plasma membrane that surrounds the cell, separating it from its environment; the nucleus that contains the cell's genes; and many other specialized structures, called organelles, that perform particular functions in the life of the cell. It also shows the cell wall present in many plant cells that surrounds and supports the cell while allowing free exchange of materials.

cell alive; and a thin **membrane** that encloses a watery medium (the cytoplasm) and separates the cell from the outside world. Some organisms, mostly microscopic, consist of just one cell, but larger organisms are composed of many cells whose functions are differentiated. In these multicellular organisms, cells of similar type form **tissues**, which perform a particular function. Examples of animal tissues are nervous tissue, composed of neurons (individual nerve cells), which produce electrical signals, and connective tissue, which often surrounds and supports other tissue types. Various tissue types combine to make up a structural unit, called an **organ** (for example, the brain). Several organs that collectively perform a single function are called an **organ system**; for example, together the brain, spinal cord, sense organs, and nerves form the nervous system. All the organ systems functioning cooperatively make up an individual living thing, the **organism**.

Beyond the individual organisms are broader levels of organization. A group of very similar, potentially interbreeding organisms constitutes a **species**. Members of the same species that live in a given area are considered a **population**. Populations of several species living and interacting in the same area form a **community**. A community plus its nonliving environment, including land, water, and atmosphere, constitute an **ecosystem**. Finally, the entire surface region of Earth inhabited by living things is called the **biosphere**.

The organization of this text will roughly follow the pattern of organization of life on Earth. We will begin with atoms and molecules, move to cells and principles of heredity, continue with the range of organisms and how they function, and conclude with the study of the interactions among organisms that are the focus of ecology.

Living Things Must Acquire and Use Materials and Energy

Organisms need materials and energy to maintain their high level of complexity and organization, to grow, and to reproduce (Fig. 1-4). The atoms and molecules of which all organisms are made may be acquired from the air, water, or soil or from other living things. Organisms extract these materials, called **nutrients**, from the environment and incorporate them into the molecules of their own bodies. The sum total of all the chemical reactions needed to sustain life is called **metabolism**.

Organisms obtain **energy**—the ability to do work, including carrying out chemical reactions, growing leaves in the spring, or contracting a muscle—in one of two basic ways. Plants and some single-celled organisms capture the energy of sunlight and store it in energy-rich sugar molecules, a process called **photosynthesis**. In contrast, neither fungi nor animals can photosynthesize, nor can most bacteria; these organisms must consume the energy-rich molecules contained in the bodies of other organisms. In either case, energy taken in is converted into a form that the organism can use or store for future use.

Ultimately the energy that sustains nearly all life comes from sunlight, captured by photosynthetic organisms and incorporated into energy-rich molecules. Organisms that cannot photosynthesize depend on photosynthetic life forms for food, either directly or indirectly. Thus, energy flows from the sun through nearly all forms of life and is eventually released again as heat.

Homeostasis Maintains Relatively Constant Internal Conditions

Complex, organized structures are not easy to maintain. Whether we consider the molecules of your body or the clothes in your closet, organization tends to disintegrate into chaos unless energy is used to sustain it (we will explore this tendency more fully in Chapter 4). To stay alive and function effectively, organisms must keep the conditions within their bodies fairly constant, a process called **homeostasis** (derived from Greek words meaning "to stay the same"). One of the many conditions regulated is body temperature. Among warm-blooded animals, for example, vital organs such as the brain and heart are kept at a warm, constant temperature despite wide fluctuations in environmental temperature.

Maintaining homeostasis is accomplished by a variety of automatic mechanisms; in the case of temperature regulation these include sweating during hot weather,

Figure 1-4 **Living organisms acquire energy and materials (nutrients) from their environment**

The plants in this Kenyan grassland capture energy from the sun and nutrients from the air, water, and soil. In contrast, the grazing Cape Buffalo extract both energy and nutrients from the plants. The lion will extract energy and nutrients from its animal prey, indirectly derived from plants. Without a continual supply of solar energy, nearly all life would cease.

metabolizing more food when it's cold, and behaviors such as basking in the sun or even (for modern humans) adjusting the thermostat in the room. In Chapter 21 in Unit IV, we will expand on the theme of homeostasis. Of course, not everything stays the same throughout an organism's life. Major changes, such as growth and reproduction, occur, but these are not failures of homeostasis. Rather, they are usually specific, genetically programmed parts of the organism's life cycle.

Growth Is a Property of All Living Organisms

At some time in its life cycle, every living thing becomes larger—that is, it grows. This feature is obvious for plants, birds, and mammals, all of which start out very small and undergo tremendous growth during their lives. Even single-celled bacteria, however, are small when they are first formed and grow to about double their original size before they divide. In all cases, growth involves the conversion of materials acquired from the environment into the specific molecules of the organism's own body.

Living Things Respond to Stimuli

Living organisms perceive and respond to stimuli in their internal and external environments. Animals have evolved elaborate sensory organs and muscular systems that allow them to detect and respond to light, sound, chemicals, and many other stimuli from their surroundings. Internal stimuli are perceived by receptors for stretch, pain, and various chemicals. For example, when you feel hungry, you are perceiving contractions of your empty stomach and low levels of sugars and fats in your blood. You then respond to external stimuli by choosing appropriate objects to eat, such as a piece of pie rather than the plate and fork. Animals, with their elaborate nervous systems and motile bodies, are not the only organisms to perceive and respond to stimuli. The plants on your windowsill grow toward the light, and even the bacteria in your intestine manufacture a different set of digestive enzymes depending on whether you drink milk, eat candy, or both.

Living Things Reproduce Themselves

The *continuity of life* occurs because organisms reproduce, giving rise to offspring of the same type (Fig. 1-5). The processes for producing offspring are varied, but the results—the perpetuation of the parents' genetic material—are the same. The *diversity of life* occurs in part because the offspring, though arising from the genetic material provided by the parents, are usually somewhat different from their parents, as explained briefly below, and in Chapters 16 and 17. The mechanism by which traits are passed from one generation to the next, through a "genetic blueprint" contained in molecules of DNA, produces these variable offspring.

Figure 1-5 **Living organisms reproduce**
As it grows, this baby orangutan will resemble, but not be identical to, its parents. The similarity and variability of offspring are crucial to the evolution of life.

DNA *Is the Molecule of Heredity*

All known forms of life use a molecule called **deoxyribonucleic acid**, or **DNA**, as the repository of hereditary information (Fig. 1-6). Much of Unit II will be devoted to exploring the structure and function of this remarkable molecule. For now, we will simply note that an organism's DNA is its "blueprint" or "molecular instruction manual," a guide to both the construction and, at least in part, the operation of its body. When an organism reproduces, it passes a copy of its DNA to its offspring. The accuracy of the DNA copying process is astonishingly high: Only about

one mistake occurs for every billion bits of information contained in the DNA molecule. But chance accidents to the genetic material also bring about changes in the DNA. The occasional errors and accidental changes, called **mutations**, are crucial. Without mutations, all life forms might be identical. Indeed, there is reason to believe that, without mutations, there would be no life at all (see Chapter 19). Mutations in DNA are the ultimate source of genetic variations. These variations, superimposed on a background of overall genetic fidelity, make possible our final property of life, the capacity to evolve.

Living Things Have the Capacity to Evolve

Although the genetic makeup of a single organism remains essentially the same over its lifetime, the genetic composition of a species as a whole changes over many lifetimes. Over time, mutations and variable offspring provide diversity in the genetic material of a species. In other words, the species evolves. The most important force in evolution is **natural selection**, the process by which organisms with adaptive traits (those that help organisms cope with the rigors of their given environmental conditions) survive and reproduce more successfully than do others that lack those traits. The adaptive traits arising from genetic mutation that enhanced survival are thereby passed on to the next generation.

The Diversity of Life

Although all living things share the general characteristics discussed earlier, evolution has brought forth an amazing variety of life forms. In the following brief description of the features used to classify living organisms, we introduce you to the diversity of life on Earth. We describe the classification of organisms in more detail in Chapter 20.

Living organisms are often grouped into five major categories, called **kingdoms**: Monera, Protista, Fungi, Plantae, and Animalia (Fig. 1-7). Although biologists believe that the groupings reflect evolutionary relationships among organisms, we must classify living things according to the characteristics we can see and measure today. There are exceptions to any simple set of criteria used to define the kingdoms, but three characteristics are particularly useful: cell type, the number of cells in each organism, and the mode of acquiring energy (Table 1-1).

Figure 1-6 **DNA**
A computer-generated model of DNA, the molecule of heredity.

Figure 1-7 **The five kingdoms**

(a)

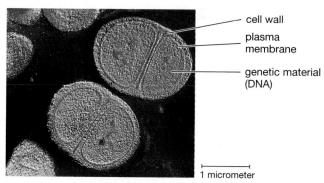

cell wall
plasma membrane
genetic material (DNA)

1 micrometer

(a) The Kingdom Monera. A color-enhanced electron micrograph of a dividing bacterium. Monerans are unicellular and prokaryotic, and most are surrounded by a thick cell wall. Although some can photosynthesize, most absorb food from their surroundings.

(b)

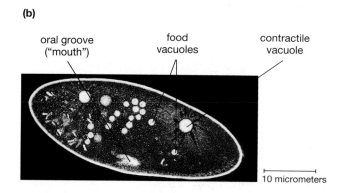

oral groove ("mouth")
food vacuoles
contractile vacuole

10 micrometers

(b) The Kingdom Protista. This light micrograph of a *Paramecium* illustrates the complexity of these large, usually single, eukaryotic cells. Some protists can photosynthesize, but others ingest or absorb their food. Many, including *Paramecium*, are mobile, moving with cilia or flagella.

(c)

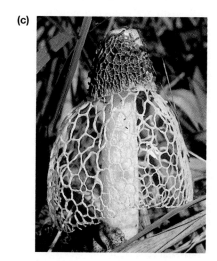

(c) The Kingdom Fungi. An exotic mushroom found in Peru. Most fungi are multicellular, often with distinct cell types specialized for feeding or reproduction. Fungi generally absorb their food, which is usually the dead bodies or wastes of plants and animals. The food is digested by enzymes secreted outside of the fungal body. Most fungi cannot move.

(d)

(d) The Kingdom Plantae. This butterfly weed represents the flowering plants, the dominant members of the kingdom Plantae. Flowering plants owe much of their success to mutually beneficial relationships with animals, such as these pearl crescent butterflies, in which the flower provides food and the insect carries pollen from flower to flower, fertilizing them. Plants are multicellular, nonmotile eukaryotes with considerable specialization of cell types. Plants acquire nutrients by photosynthesis.

(e) The Kingdom Animalia. This clown fish and the anemone with which it lives are representative animals. Animals are multicellular eukaryotes that usually ingest their food. Animal bodies contain a wide variety of tissues and organs, composed of specialized cell types. Most animals are motile and can respond rapidly to stimuli. The anemone represents the largest major group of animals: the invertebrates, which lack a backbone. The clown fish is a vertebrate, an animal that (like humans) has a backbone.

(e)

TABLE 1-1 ❖ *Some Characteristics of the Five Kingdoms*			
Kingdom	*Cell Type*	*Cell Number*	*Major Mode of Nutrition*
Monera	Prokaryotic	Unicellular	Absorb or photosynthesize
Protista	Eukaryotic	Unicellular	Absorb, ingest, or photosynthesize
Fungi	Eukaryotic	Most multicellular	Absorb
Plantae	Eukaryotic	Multicellular	Photosynthesize
Animalia	Eukaryotic	Multicellular	Ingest

Members of the Kingdom Monera Are Prokaryotic Cells; Members of All Other Kingdoms Are Composed of Eukaryotic Cells

There are two fundamentally different types of cells: **prokaryotic** and **eukaryotic**. *Karyotic* refers to the **nucleus** of the cell: a membrane-enclosed sac containing the genetic material (see Fig. 1-3). *Eu* means "true" in Greek, and eukaryotic cells are recognized by the presence of a "true," membrane-enclosed nucleus. Eukaryotic cells are larger than prokaryotic cells and contain a variety of other organelles, many surrounded by membranes. Prokaryotic cells do not have a nucleus; their genetic material resides in the cytoplasm of the cell. They are small, only one or two micrometers long, and lack membrane-bound organelles. *Pro* means "before" in Greek, because prokaryotic cells almost certainly evolved before eukaryotic cells (and eukaryotic cells probably evolved from prokaryotic cells, as we will see in Chapter 19). Members of the kingdom Monera consist of prokaryotic cells; the cells of the other four kingdoms are eukaryotic.

Members of the Kingdoms Monera and Protista Are Mostly Unicellular; Members of the Kingdoms Fungi, Plantae, and Animalia Are Primarily Multicellular

Organisms of the kingdoms Monera and Protista are usually single-celled, or **unicellular**, although a few live in strands or mats of cells with little communication, cooperation, or organization among cells. Most fungi, plants, and animals are **multicellular**; their lives depend on intimate cooperation among cells.

Members of the Different Kingdoms Have Different Ways of Acquiring Energy

All organisms need energy to live. Photosynthetic organisms capture energy from sunlight and store it in molecules such as sugars and fats. These organisms, including plants, some monerans, and some protists, are therefore called **autotrophs**, meaning "self-feeders." Organisms that cannot photosynthesize must acquire energy prepackaged in the molecules of the bodies of other organisms; hence these organisms are called **heterotrophs**, meaning "other-feeders." Many monerans and protists, and all fungi and animals, are heterotrophs. Heterotrophs differ in the size

of the food they eat. Some, such as bacteria and fungi, absorb individual food molecules; others, including most animals, eat whole chunks of food and break them down to molecules in their digestive tracts (ingestion).

Biology: The Science of Life

Biology is probably the most diverse of all the sciences. While some biologists are creating new forms of life by manipulating genetic material, others are probing the workings of the brain, tracing the complex interactions within ecosystems, or seeking out new forms of life from the tropical rain forest to the ocean floor. Through the study of biology you will become familiar with how your body works. You will learn how some diseases are spread and how they are fought by our natural defenses and by medications. You will explore the intricacies of development from single cell to whole human being. You will learn how plants capture the solar energy that ultimately sustains nearly all life. By learning about the relationships between people and other forms of life, you will be prepared to make informed choices about land use, waste disposal, family size, and many other issues that affect the environment that sustains us.

Biology is a science, and its principles and methods are the same as those of any other science. In fact, a basic tenet of modern biology is that living things obey the same laws of physics and chemistry that govern nonliving matter.

Scientific Principles Underlie All Scientific Inquiry

All scientific inquiry, including biology, is based on a small set of assumptions. Although we can never absolutely prove these assumptions, they have been so thoroughly tested and found valid that we might call them scientific principles. These are the principles of natural causality, uniformity in space and time, and common perception.

Natural Causality Is the Principle That All Events Can Be Traced to Natural Causes

The first principle of science is natural causality. Over the course of human history, two approaches have been taken to the study of life and other natural phenomena. The first assumes that some events happen through the intervention of supernatural forces beyond our understanding. The ancient Greeks believed that Zeus hurled thunderbolts from the sky and that Poseidon caused earthquakes and storms at sea. In contrast, science adheres to the principle of **natural causality**: All events can be traced to natural causes that are potentially within our ability to comprehend. For example, until relatively recently, epilepsy was commonly thought to be a visitation from the gods. Today we realize that epilepsy is a disease of the brain, in which groups of nerve cells fire uncontrollably.

The principle of natural causality has an important corollary: The evidence we gather about the causes of natural events has not been deliberately distorted to fool us. This corollary may seem obvious, yet not so very long ago some people argued that fossils are not evidence of evolution but were placed on Earth by God as a test of our faith. If we cannot trust the evidence provided by nature, then the entire enterprise of science is futile.

The Natural Laws That Govern Events Apply Everywhere and for All Time

A second fundamental principle of science is that natural laws, laws derived from nature, are uniform in space and time and do not change with distance or time. The laws of gravity, the behavior of light, and the interactions of atoms, for example, are the same today as they were a billion years ago and will hold just as well in Moscow as in New York or even on Mars. Uniformity in space and time is especially vital to biology, because many events of great importance to biology, such as the evolution of today's diversity of living things, happened before humans were around to observe them. Some people believe that all the different types of living organisms were individually created at one time in the past by the direct intervention of God, a philosophy called **creationism**. As scientists, we freely admit that we cannot disprove this idea. Creationism, however, is contrary to both natural causality and uniformity in time. The overwhelming success of science in explaining natural events through natural causes has led almost all scientists to reject creationism.

Scientific Inquiry Is Based on the Assumption That People Perceive Natural Events in Similar Ways

A third basic assumption of science is that, as a general rule, all human beings perceive natural events in fundamentally the same way and that these perceptions provide us with reliable information about the natural world. Common perception is, to some extent, a principle peculiar to science. Value systems, such as those involved in the appreciation of art, poetry, and music, do not assume common perception. We may perceive the colors in a painting in a similar way (the scientific aspect of art), but we do not perceive the aesthetic value of the painting identically (the humanistic aspect of art). Moral values also differ radically among peo-

ple, often owing to their culture or religious beliefs. Because value systems are subjective, not objective, science cannot solve certain types of philosophical or moral problems, such as the morality of abortion.

The Scientific Method Is the Basis for Scientific Inquiry

Given these assumptions, how do biologists study the workings of life? Scientific inquiry is a rigorous method for making observations of specific phenomena and searching for the order underlying those phenomena. Ideally, biology and the other sciences use the **scientific method**, which consists of four interrelated operations: observation, hypothesis, experiment, and conclusion. All scientific inquiry begins with an **observation** of a specific phenomenon. The observation, in turn, leads to questions, such as "how did this come about?" Then in a flash of insight, or more often after long, hard thought, a hypothesis is formulated. A **hypothesis** is an educated guess that a certain preceding cause produces the observed phenomenon. To be useful, a hypothesis must be testable by further observations, or **experiments**. These experiments produce results that either support or refute the hypothesis, and a **conclusion** is drawn about its validity. A single experiment is never an adequate basis for a conclusion; the results must be repeatable not only by the original researcher but by others as well.

Simple experiments test the assertion that a single factor, or **variable**, is the cause of a single observation. To be scientifically valid, the experiment must rule out a variety of other variables as the cause of the observation. So scientists design **controls** into their experiments, in which all the variables remain constant. Controls are then compared with the experimental situation, in which only the variable being tested is changed. In the 1600s, Francesco Redi used the scientific method to demonstrate that flies do not arise spontaneously from rotting meat (see "Scientific Inquiry: Does Life Arise Spontaneously?").

The scientific method can be used to solve everyday problems, as well as to generate new knowledge. Let's now consider an everyday situation in which you can apply the scientific method. Late to class, you rush to your car and make the *observation* that it won't start. Immediately, you form a *hypothesis*: The battery is dead. Quickly you design an *experiment*: You replace your battery with the battery from your roommate's new car and try to start your car again. The result seems to confirm your hypothesis; your car starts immediately. But wait! You haven't provided controls for several variables. Perhaps your battery was fine all along, and you just needed to try to start the car again. Or perhaps the battery cable was loose and simply needed to be tightened. Realizing the need for a good *control*, you replace your old battery, making sure the cables are secured tightly, and attempt to restart the car. If your car repeatedly refuses to

S C I E N T I F I C   I N Q U I R Y

Does Life Arise Spontaneously?

The experiments of the Italian physician Francesco Redi (1621–1697) beautifully demonstrate the scientific method and also help to illustrate the principle of causality, on which modern science is based. Redi investigated why maggots appear in spoiled meat. Before Redi, the appearance of maggots was considered to be evidence of **spontaneous generation**, the production of living organisms from nonliving matter.

Redi *observed* that flies swarm around fresh meat and that maggots appear in meat left out for a few days. He formed a testable *hypothesis:* The flies produce the maggots. In his *experiment*, Redi wanted to test just one variable: the access of flies to the meat. Therefore, he took two clean jars and filled them with similar pieces of meat. He left one jar open (the *control* jar) and covered the other with gauze to keep out flies (the *experimental* jar). He did his best to keep all the other variables the same (for example, type of jar, type of meat, and temperature). After a few days, he observed that maggots swarmed over the meat in the open jar, but no maggots appeared in the meat in the covered jar. Redi *concluded* that his hypothesis was correct and that maggots are produced by flies, not by the meat itself (Fig. E1-1). Only through controlled experiments could the age-old hypothesis of spontaneous generation be laid to rest.

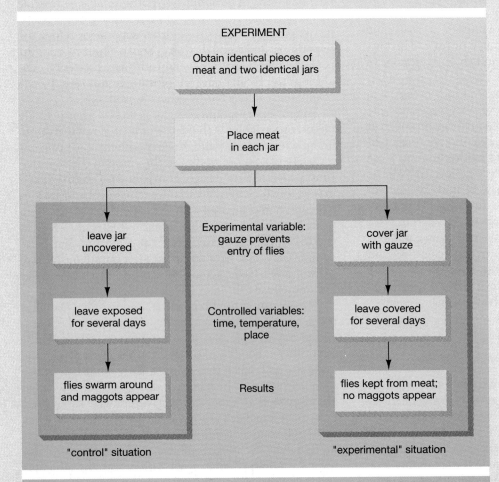

Figure E1-1 **The experiments of Francesco Redi**

start with the old battery, but then starts immediately when you put in your roommate's battery, you have isolated a single *variable*, the battery, and you can safely draw the *conclusion* that your old battery is dead.

It is important to recognize the limitations of the scientific method. In particular, scientists can seldom be sure that they have controlled *all* the variables other than the one they are trying to study. Therefore, scientific conclusions must always remain tentative and are subject to revision if new observations or experiments demand it.

Science Is a Human Endeavor

Scientists are real people. They are driven by the same ambitions, pride, and fears as other people, and they sometimes make mistakes. As you will read in Chapter 10, ambition played an important role in the discovery by James Watson and Francis Crick of the structure of DNA. Accidents, lucky guesses, controversies with competing scientists, and, of course, the intellectual powers of individual scientists contribute greatly to scientific advances. To illustrate what we might call "real science," let's consider an actual case.

A Good Scientist Is Prepared to Take Advantage of Chance Events

When microbiologists study bacteria, they must use pure cultures—that is, plates of bacteria that are free from contamination by other bacteria, molds, and so on. Only by studying a single type at a time can we learn about the properties of that particular bacterium. Consequently, at the first sign of contamination, a culture is usually thrown out, often with mutterings about sloppy technique. On one such occasion, however, in the late 1920s, the Scottish bacteriologist Alexander Fleming turned a ruined culture into one of the greatest medical advances in history.

One of Fleming's bacterial cultures became contaminated with a patch of a mold called *Penicillium*. Before throwing out the culture dish, Fleming noticed that *no bacteria were growing near the mold* (Fig. 1-8). Why not? Fleming hypothesized that perhaps *Penicillium* releases a substance that kills off bacteria growing nearby. To test this hypothesis, Fleming grew some pure *Penicillium* in a liquid medium, filtered out the *Penicillium* mold, and then applied the liquid in which the mold had grown to an uncontaminated bacterial culture. Sure enough, the bacteria were killed by the liquid. Further research into these mold extracts resulted in the production of the first antibiotic: penicillin, a bacteria-killing substance that has since saved millions of lives. Fleming's experiments are a classic in scientific methodology, proceeding from observation to hypothesis to experimental tests of the hypothesis, followed by a conclusion, further supported by experimental evidence. But scientific method

alone would have been useless without the lucky combination of accident and a brilliant scientific mind. Had Fleming been a "perfect" microbiologist, he wouldn't have had any contaminated cultures. Had he been less observant, the contamination would have been just another spoiled culture dish. Instead, it was the beginning of antibiotic therapy for bacterial diseases. As the French microbiologist Louis Pasteur said, "Chance favors the prepared mind."

Scientific Theories Have Been Thoroughly Tested

In scientific terminology, a **theory** is a hypothesis that has been supported by so many cases that few scientists seriously doubt its validity. Take a trivial example: Each time an apple falls *down* to Earth rather than *up* into the sky, the theory of gravity is supported. Thus, a scientific theory is similar to what we would call a "law" or "fact" in common English usage. Scientists, however, must be open-minded about even their most cherished theories, and they must be willing to discard them if new observations render their theories obsolete.

Probably the foremost theory in biology is evolution. Since its formulation by two English naturalists, Charles Darwin and Alfred Wallace, in the mid-1800s, the theory of evolution has been supported by fossil finds, geological studies, radioactive dating of rocks, genetics, molecular biology, biochemistry, and breeding experiments. People who refer to evolution as "just a theory" profoundly misunderstand what scientists mean by that term.

Figure 1-8   Penicillin kills bacteria

Penicillin diffuses outward from a penicillin-soaked disc of paper, creating a pronounced bald spot in the "lawn" of bacteria on this petri dish.

Evolution: The Unifying Concept of Biology

The most important concept in biology is evolution, a unifying theory that explains the origin of diverse forms of life as a result of changes in their genetic makeup. The theory of **evolution** states that modern organisms descended, with modification, from preexisting life forms. In the biologist Theodosius Dobzhansky's words, "Nothing in biology makes sense, except in the light of evolution." Why don't snakes have legs? Why are there dinosaur fossils but no living dinosaurs? Why are monkeys so like us, not only in appearance, but even in the structure of their genes and proteins? The answers to those questions, and thousands more, lie in the processes of evolution (which we will examine in detail in Chapters 16 through 18). Evolution is so vital to our understanding and appreciation of biology that we must briefly review its important principles before going further.

Three Natural Processes Underlie Evolution

In the mid-1800s, Darwin and Wallace formulated the theory of evolution that still forms the basis of our modern understanding. Evolution arises as a consequence of three natural processes: *genetic variation* among members of a population; *inheritance* of those variations by offspring of parents carrying the variation; and *natural selection*, the survival and enhanced reproduction of organisms with favorable variations.

Much of the Variability among Organisms Is Inherited

Look around at your classmates and notice how different they are. Although some of this variation is due to differences in environment and life-styles, it is mainly influenced by our genes: Most of us could pump iron for the rest of our lives and never develop a body like Arnold Schwarzenegger's. Where does genetic variation come from? The genetic instructions—the genes—of all living organisms are segments of molecules of deoxyribonucleic acid, or DNA. Occasionally, the DNA suffers an accident: Perhaps radiation strikes the DNA molecule just so, altering its information content. Mistakes in copying DNA during reproduction, although rare, also occur. These accidents and mistakes cause mutations, or changes in a gene, that may affect the organism's appearance or ability to function. Many mutations are harmless; some make the organism less able to function. But in rare cases mutations may actually improve an organism's ability to function. As a result of mutations, many of which occurred millions of years ago and have been passed from parent to offspring through countless generations, members of the same species are often slightly different from one another.

Natural Selection Tends to Preserve Genes That Help an Organism Survive and Reproduce

On the average, organisms that best meet the challenges of their environment will leave the most offspring. The offspring will inherit the genes that made their parents successful. Natural selection thus preserves genes that help organisms flourish in their environment. For example, a mutated gene providing instructions for larger teeth in beavers will be passed from parent to offspring. These offspring will be able to chew down trees more efficiently, build bigger dams and lodges, and eat more bark than "ordinary" beavers. Because these big-toothed beavers will obtain more food and better shelter than their smaller-toothed relatives, they will probably raise more offspring. The offspring will inherit their parents' genes for larger teeth. Over time, smaller-toothed beavers will become increasingly scarce, and after many generations all beavers will have large teeth.

Structures, physiological processes, or behaviors that aid survival and reproduction in a particular environment are called **adaptations**. Most of the features that we admire so much in our fellow life forms, such as the long clean limbs of deer, the wings of eagles, and the mighty columns of redwood trunks, are adaptations molded by millions of years of mutation and natural selection.

In the long run, however, what helps an organism to survive today may become a liability tomorrow. If environments change—for example, as ice ages come and go—then the genetic makeup that best adapts organisms to their environment will also change over time. When new mutations happen to appear that increase the fitness of an organism in the altered environment, these mutations in their turn will spread throughout the population.

Over millennia, the interplay of environment, variation, and selection results in evolution: the modification of the genetic makeup of species. In environments that are reasonably constant, such as the oceans, some well-adapted forms persist relatively unchanged and are sometimes called "living fossils." For example, sharks (Fig. 1-9) have retained essentially the same body form for tens of millions of years, as their sleek shape, powerful tail, acute sense of smell, and formidable teeth have made them superb predators. In changing environments, some species do not experience within an appropriate time frame the genetic changes that allow them to adapt. The rate of environmental change outstrips the rate of genetic changes, and those species become extinct. The petrified forests of Arizona (Fig. 1-10) are the remnants of giant trees that could not withstand the conversion of tropical lowland to high desert. Other species experience chance mutations that adapt them to meet new challenges. Fossil horses show gradual changes over time, caused by mutations that happened to produce adaptations to their changing environment (Fig. 1-11).

Figure 1-9 **Ancient adaptations**

This tiger shark, photographed in the clear waters of the Bahamas, possesses features that have characterized sharks for tens of millions of years: streamlined shape, a long, powerful tail, an acute sense of smell, and rows of sharp teeth.

Figure 1-10 **Logs in Petrified Forest National Park**

The petrified trunks scattered over the ground are remnants of vast forests of trees that lived here about 160 million years ago. As their habitat gradually changed from tropical lowland to high desert, the trees became extinct.

Figure 1-11 **The evolution of the horse**

The earliest horse, *Hyracotherium* (**a**; also called *Eohippus*) gave rise to several intermediate forms (**b, c**) culminating in the modern horse, *Equus* (**d**). The changing backgrounds through which the horses run depict the changing environments that molded horse evolution, from forests through open woodlands to hard, flat plains. The changing environments selected for larger, stronger hooves, stronger legs, and larger, harder teeth. See Chapter 16 for more on horse evolution.

The result of evolution is a tremendous variety of species. Within particular habitats, these species have evolved complex interrelationships with one another and with their nonliving surroundings. The diversity of species and the complex interrelationships that sustain them are encompassed by the term **biodiversity**. In recent decades, the rate of environmental change has been drastically accelerated by a single species, *Homo sapiens* (modern human). Few species are able to adapt to this change, and in habitats most affected by people, many species are being driven to extinction. This concept is explored further in the box "Earth Watch: Biodiversity Is Threatened by Human Activities."

Knowledge of Biology Illuminates Everyday Life

Some people regard science as a "dehumanizing" activity, feeling that too deep an understanding of the world robs us of vision and awe. Nothing could be further from the truth, as we repeatedly discover anew in our own lives. A few years ago, we watched a bee foraging at a spike of lupines. Lupines have a complicated structure, with two petals on the lower half of the flower enclosing the pollen-laden male reproductive structures (stamens) and sticky female pollen-capturing stigma (Fig. 1-12). In young lupine

EARTH WATCH

Biodiversity Is Threatened by Human Activities

Ever since the United Nations' 1992 "Earth Summit" in Rio de Janeiro, Brazil, the word *biodiversity* has jumped out at us from magazines and news articles. What is biodiversity, and why should we be concerned with preserving it? In his article "Defining Biodiversity," biologist and author Peter Raven describes biodiversity as the sum total of all the living organisms in a particular area and all of the interactions among them. As you learned in this chapter, this definition is similar to that of an ecological community. Why should we be concerned with preserving biodiversity and maintaining biological communities?

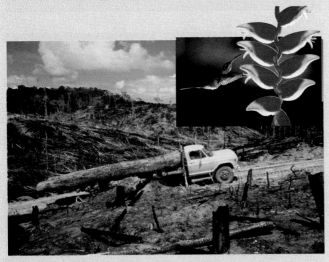

Figure E1-2 Biodiversity threatened
Destruction of tropical rain forests by indiscriminate logging threatens Earth's greatest storehouse of biological diversity. Interrelationships such as those that have evolved between this *Heliconia* and its hummingbird pollinator sustain these diverse communities and are threatened by human activities.

Over the 3.5-billion-year history of life on Earth, evolution has produced an estimated 8 to 10 million unique and irreplaceable species. Of these, scientists have named only about 1.4 million, and only a tiny fraction of this number has been studied. Evolution has not, however, merely been churning out millions of independent species. Over thousands of years, organisms in a given area have been molded by forces of natural selection exerted by other living species as well as by the nonliving environment in which they live. The outcome is the community, a highly complex web of interdependent life forms whose interactions sustain one another. By participating in the natural cycling of water, oxygen, and other nutrients, by producing rich soil and purifying wastes, these communities contribute to the sustenance of human life as well.

The tropics are home to the vast majority of all the species on Earth, perhaps 7 to 8 million of them, living in complex communities. The rapid destruction of habitats in the tropics as a result of human activities is producing high rates of extinction of many species (Fig. E1-2). Most of these species have never been named, and others never even discovered. Aside from ethical concerns over eradicating irreplaceable forms of life, as we drive unknown organisms to extinction we lose potential sources of medicine, food, and raw materials for industry. The concept of biodiversity has emerged as a result of our increasing concern over the loss of countless forms of life and the habitat that sustains them.

Conservationists have come to understand that the fate of an endangered species is intimately linked to the fate of its community, and that communities must remain intact to sustain the functioning web of life. As biologist and author E. O. Wilson observed: "The loss of species is the folly our descendants are least likely to forgive us."

flowers, the weight of a bee pushing on these petals compresses the stamens, pushing pollen onto the bee's abdomen. In older flowers, the stigma protrudes through the lower petals, and when a pollen-dusted bee visits, it usually leaves behind a few grains of pollen.

Did our new-found insights into the functioning of lupine flowers detract from our appreciation of them? Far from it. Rather, we now looked on lupines with new delight, understanding something of the interplay of form and function, bee and flower, that shaped the evolution of the lupine. A few months later we ventured atop Hurricane Ridge in Olympic National Park in Washington, where the alpine meadows fairly burst with color in August (Fig. 1-13). As we crouched be-

side a wild lupine, an elderly man stopped to ask what we were looking at so intently. He listened with interest as we explained the structure to him; he then went off to another patch of lupines to watch the bees foraging. He too felt the increased sense of wonder that comes with understanding.

We try to convey to you that dual sense of understanding and wonder throughout this text. We also emphasize that biology is not a completed work, but an exploration that we have really just begun. Lewis Thomas, physician and natural philosopher, described our situation in an elegant way: "The only solid piece of scientific truth about which I feel totally confident is that we are profoundly ignorant about nature. Indeed, I regard this as the major discovery of the past hundred years of biology…but we are making a beginning."

We cannot urge you strongly enough, even if you are not contemplating a career in biology, to join in the journey of biological discoveries throughout your life. Don't look upon biology as just another course to take, just another set of facts to memorize. Biology can be much more than that. It can be a pathway to a new understanding of yourself and the living Earth around you.

(a)

(b)

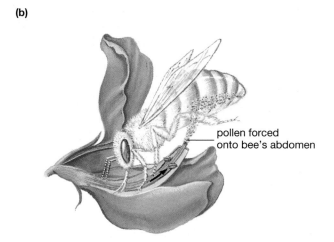

Figure 1-12   **Complex adaptations help assure pollination**
Lupines, like many other members of the pea family, have complex flowers. **(a)** The reproductive structures are enclosed within the lower petals. In young lupine flowers, the lower petals form a tube within which the stamens fit snugly. The stamens shed pollen within the tube. **(b)** When the weight of a foraging bee pushes on the lower petals, the stamens are thrust forward, and pollen is forced out the end onto her abdomen. Some pollen adheres to the abdomen and may come off on the sticky, pollen-receiving stigma of the next flower visited by the bee.

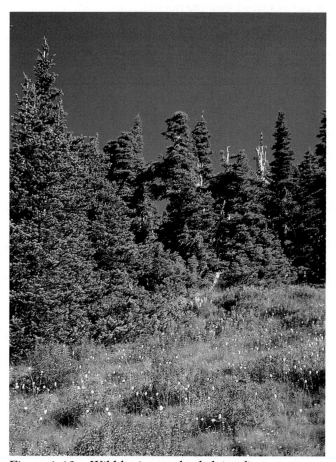

Figure 1-13   **Wild lupines and subalpine fir**
Thousands of people visit Hurricane Ridge in Olympic National Park each summer to gaze in awe at Mt. Olympus, but few bother to investigate the wonders at their feet.

✳ S U M M A R Y   O F   K E Y   C O N C E P T S

The Characteristics of Living Things
The quality of life emerges as the result of interactions of emergent properties. Living organisms usually possess the following characteristics: Their structure is complex and organized; they maintain homeostasis; they grow; they acquire energy and materials from the environment; they respond to stimuli; they reproduce; and they have the capacity to evolve.

The Diversity of Life
Living organisms can be grouped into five major categories, called kingdoms: Monera, Protista, Fungi, Plantae, and Animalia. Some features used to classify organisms into kingdoms are the type of cell the organism possesses, the number of cells in each organism, and the mode of acquiring energy.

Cell Type: Eukaryotic cells have their genetic material enclosed within a membrane-bound nucleus. Prokaryotic cells do not have a nucleus.

Cell Number: Organisms may consist of a single cell (unicellular) or of many cells bound together and working cooperatively (multicellular).

Energy Acquisition: Nearly all autotrophic organisms obtain energy by storing sunlight energy in energy-rich molecules through photosynthesis. Heterotrophic organisms obtain energy by eating energy-rich molecules

(food) synthesized in the bodies of other organisms. The food may be eaten in large chunks and broken down (ingestion) or may be absorbed molecule by molecule from the environment (absorption). The features of the five kingdoms are summarized in Table 1-1.

Biology: The Science of Life
Biology is based on the scientific principles of natural causality, uniformity in space and time, and common perception. These principles are assumptions that cannot be directly proved but that are validated by experience. Knowledge in biology is acquired through the application of the scientific method. First, an observation is made. Then a hypothesis is formulated that suggests a natural cause for the observation. The hypothesis is used to predict the outcome of further observations or experiments. A conclusion is then drawn about the validity of the hypothesis. A hypothesis becomes a scientific theory when repeated tests have confirmed it and none have refuted it.

Evolution: The Unifying Concept of Biology
Evolution is the theory that modern organisms descended, with modification, from preexisting life forms. Evolution occurs as a consequence of (1) genetic variation among members of a population, caused by mutation; (2) inheritance of those variations by offspring; and (3) natural selection of the variations that best adapt the organism to its environment.

✳ K E Y   T E R M S

adaptation   p. 12
atom   p. 4
autotroph   p. 8
biodiversity   p. 14
biosphere   p. 4
cell   p. 4
community   p. 4
conclusion   p. 9
control   p. 9
creationism   p. 9
deoxyribonucleic acid (DNA)   p. 6
ecosystem   p. 4
element   p. 4
emergent property   p. 2

energy   p. 4
eukaryotic   p. 8
evolution   p. 12
experiment   p. 9
gene   p. 4
heterotroph   p. 8
homeostasis   p. 5
hypothesis   p. 9
kingdom   p. 6
membrane   p. 4
metabolism   p. 4
molecule   p. 4
multicellular   p. 8
mutation   p. 6

natural causality   p. 9
natural selection   p. 6
nucleus   p. 8
nutrient   p. 4
observation   p. 9
organ   p. 4
organelle   p. 4
organic molecule   p. 4
organism   p. 4
organ system   p. 4
photosynthesis   p. 4
population   p. 4
prokaryotic   p. 8
scientific method   p. 9

species   p. 4
spontaneous generation   p. 10
subatomic particle   p. 4

theory   p. 11
tissue   p. 4
unicellular   p. 8

variable   p. 9

✖ THINKING THROUGH THE CONCEPTS

Multiple Choice

1. A requirement that is NOT involved in critical thinking is
   a. an intolerance for ambiguity
   b. questioning the facts
   c. examining the whole experimental system
   d. defining terms
   e. discovering biases
2. A scientist examines a group of cells under the microscope. She notices the presence of nuclei in these cells. Chemical tests reveal that each cell is surrounded by a cell surface covering composed of cellulose. She concludes that the cells come from an organism that is a member of the Kingdom
   a. Monera
   b. Protista
   c. Plantae
   d. Fungi
   e. Animalia
3. Which statement is correct?
   a. Genes are proteins that produce DNA.
   b. Each organism has its own unique DNA code.
   c. The DNA of humans is more similar to the DNA of birds than it is to that of apes.
   d. Differences among organisms reflect different nucleotide sequences in their DNA.
   e. Each DNA molecule is a single strand of nucleotides.
4. Choose the answer that best describes the scientific method.
   a. observation, hypothesis, experiment, absolute proof
   b. guess, hypothesis, experiment, conclusion
   c. observation, hypothesis, experiment, conclusion
   d. hypothesis, experiment, observation, conclusion
   e. experiment, observation, hypothesis, conclusion

5. Which of the following are characteristics of living things?
   a. They reproduce.
   b. They respond to stimuli.
   c. They are complex and organized.
   d. They acquire energy.
   e. all of the above
6. The three natural processes that form the basis for evolution are
   a. adaptation, natural selection, and inheritance
   b. predation, genetic variation, and natural selection
   c. mutation, genetic variation, and adaptation
   d. fossils, natural selection, and adaptation
   e. genetic variation, inheritance, and natural selection

Review Questions

1. What are the differences between a salt crystal and a tree? Which is living? How do you know? How would you test your "knowledge"? What controls would you use?
2. What is the difference between a theory and a hypothesis? Explain how each is used by scientists.
3. Define and explain the terms: natural selection, evolution, mutation, selective breeding, creationism, and population.
4. Starting with the cell, list the hierarchy of organization of life, briefly explaining each level.
5. Define homeostasis. Why must organisms continually acquire energy and materials from the external environment to maintain homeostasis?
6. Describe the scientific method. In what ways do you use the scientific method in everyday life?
7. What is evolution? Briefly describe how evolution occurs.

✖ APPLYING THE CONCEPTS

1. Natural occurrences in the heavily populated state of California, including earthquakes, heavy rains, and grass and forest wildfires have reduced the quality of life for many Californians. Look at it another way for a moment. What biomes are found in this state? What impacts have humans had on these natural ecosystems? Are changes in these areas the reason for fires and floods and mudslides? What needs to be done to alter the balance of "humans and nature"?
2. Design an experiment to test the effects of a new dog food, "Super Dog," on the thickness and water-shedding proper-

ties of the coats of golden retrievers. Include all the parts of a scientific experiment. Design objective methods to assess coat thickness and water-shedding ability.
3. Science is based on principles, including uniformity in space and time and common perception. Assume humans one day encounter intelligent beings from another galaxy who evolved under very different conditions. Discuss the two principles mentioned above and how they would affect (1) the nature of scientific observations on the different planets and (2) communications about these observations.

✜ FOR MORE INFORMATION

Attenborough, D. *Life on Earth.* Boston: Little, Brown, 1979. Gorgeously illustrated and beautifully written introduction to the diversity of life on Earth; the inspiration for our title.

Leopold, A. *A Sand County Almanac.* New York: Oxford University Press, 1949 (reprinted in 1989). A classic by a natural philosopher; provides an eloquent foundation for the conservation ethic.

Swain, R. B. *Earthly Pleasures.* New York: Charles Scribner's Sons, 1981. Insightful essays stress the interrelatedness and diversity of life.

Thomas, L. *The Medusa and the Snail.* New York: Bantam Books, 1980. The late Lewis Thomas, physician, researcher, and philosopher, shares his awe of the living world in a series of delightful essays.

NET WATCH

On-line resources for this chapter are on the World Wide Web at:

http://www.prenhall.com/~audesirk (click on the <u>table of contents</u> link and then select Chapter 1).

Unit 1
The Life of a Cell

Two types of white blood cells are shown inside a human lymph node in this false-color transmission electron micrograph. A large macrophage, colored in green and yellow, is surrounded by many smaller lymphocytes.

2 Atoms, Molecules, and Life

> *"If there is magic on this planet, it is contained in water."*
>
> LOREN EISELEY in *The Immense Journey*

Michelangelo's David, Accademia, Florence, Italy

Why do biology texts begin with a description of chemistry? Aren't we trying to learn about the biological equivalent of Michelangelo's famous sculpture *David* rather than rocks? About the cathedral of Notre Dame rather than stone and mortar? Yes we are, but to understand those works of art and architecture we must first understand basic principles. Why did Michelangelo sculpt *David* from fine-grained marble, not from slate? Why are the arches of the cathedral roof shaped the way they are, and why are the walls studded with support structures called flying buttresses? A sculptor needs an intimate knowledge of his materials, or else his statue may crack apart midway through its creation. An architect must know a good deal about engineering, or else her roof may collapse and her walls may buckle.

Similarly, to understand the structure and function of living organisms, we need a basic understanding of the atoms and molecules of which organisms are composed and how interactions among those atoms and molecules produce growth, movement, and all the other things that organisms do. How does photosynthesis convert the energy of sunlight into the energy of sugar molecules? How do nerve cells in your brain communicate with one another? What causes cancer? What is acid rain, and why does it endanger the life of many freshwater lakes?

The evolution of life, in fact, is restricted to certain paths by the structure of matter and the laws of physics that govern interactions among pieces of matter. When you learn about the forces that allow water to move to the top of a tree, you also learn one reason why trees don't grow a thousand meters tall (the columns of water extending from root to leaf would break apart under their own weight). When you learn about energy and chemical reactions, you also learn why there can be huge herds of herbivores (plant-eating animals), such as bison or wildebeest, but not similar-sized herds of wolves or lions (most of the energy that herbivores get from eating plants is used up in their own life processes and is not available to the predators that eat them).

This chapter will introduce some of the fundamental properties of energy and matter, the structure of atoms, and how atoms interact with one another to form molecules. Later chapters in this section of the text will describe the molecules that form the bodies of all living organisms; the structure and function of cells, which are remarkably similar in everything from bacteria to elephants; and the chemical reactions that allow the acquisition and use of energy by individual cells and entire organisms.

Matter and Energy

Technically speaking, matter and energy are interchangeable, as expressed by Albert Einstein's famous equation $E = mc^2$: energy equals matter times the speed of light $(c)^2$. For the chemical reactions that occur within living organisms, however, we can treat matter and energy as quite distinct from one another: **Matter** is the physical material of the universe; **energy** is the capacity to do work, usually manifested by moving pieces of matter from place to place.

Energy Exists in Several Forms That Can Be Interconverted

Energy can exist in several forms and may be converted from one form to another. The two major categories of energy are kinetic energy and potential energy. **Kinetic energy** is energy of movement. This includes not only movement of large objects such as a football hurtling through the air but also other forms of movement such as electrical energy (movement of electrons) and heat (movement of atoms and molecules). **Potential energy** is "stored" energy that can be released as kinetic energy under the right conditions. A coconut high in a palm tree has potential energy owing to its location. Its potential energy is converted into kinetic energy when it falls. The food you eat has chemical potential energy, some of which is converted into kinetic energy of movement and heat when you run (Fig. 2-1).

The Structure of Matter

An **element**, such as carbon or oxygen, is a substance with specific properties; it can neither be broken down nor converted to different substances by ordinary chemical reactions. Elements are the building blocks of all living and nonliving matter. Although there are 92 naturally occurring elements in our universe, most are quite rare, and only a relative handful are essential to life on Earth. Table 2-1 lists the most common elements in the universe, Earth, and the human body. Atoms, described below, are the smallest unit of an element that retain the element's properties.

A **compound** is a substance composed of precise proportions of two or more elements, joined together in a specific geometric pattern. Water, for example, consists of one part oxygen and two parts hydrogen, in a precise arrangement (see Fig. 2-6).

A **mixture** is composed of two or more compounds or elements, in variable proportions. For example, a can of soda

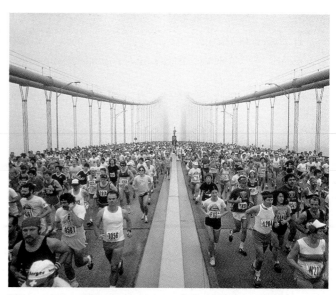

Figure 2-1 **The conversion of chemical potential energy into kinetic energy**

When 20,000 people run the New York marathon each year, they do more than just make the bridges quake. They also collectively convert over 50,000,000 kilocalories of food into movement, heat, and sore muscles—that's equivalent to burning off over 6400 kilograms (14,000 pounds) of fat or eating almost 14,000 kilograms (about 30,000 pounds) of chocolate cake.

is a mixture that contains, among other things, water, sugar, and variable amounts of carbon dioxide (which makes it fizz), depending on how long ago you popped open the top.

Atoms, the Smallest Units of Elements, Are Composed of Still Smaller Particles

If you took a diamond (a form of carbon) and cut it into pieces, each piece would still be carbon. If you could continue to make finer and finer divisions, you would eventually produce a pile of carbon **atoms**, the smallest possible particles of the element carbon (each atom, of course, would be too small to see even in a powerful microscope). Atoms themselves, however, are composed of two parts: a central **nucleus** and one or more outer **electron shells** (Fig. 2-2). Within the nucleus and electron shells reside subatomic particles. The nucleus contains heavy, positively charged **protons** and equally heavy but uncharged **neutrons**. The electron shells contain light, negatively charged **electrons**. The positive charge of a proton is designated by a plus (+) sign, and the negative charge of an electron by a minus (–) sign. An atom by itself has an equal number of electrons and protons and is therefore electrically neutral.

Nuclei and electron shells play complementary roles in the structure and function of atoms. The nuclei provide stability, and the electron shells allow interactions with other atoms. Nuclei resist disturbance by outside forces. Ordinary sources of energy, such as heat, electricity, and light, hardly

TABLE 2-1 :: Common Elements Important in Living Organisms

Element	Symbol	Atomic Number[a]	Percent in Universe[b]	Percent in Earth[b]	Percent in Human Body[b]
Hydrogen	H	1	91	0.14	9.5
Helium	He	2	9	Trace	Trace
Carbon	C	6	0.02	0.03	18.5
Nitrogen	N	7	0.04	Trace	3.3
Oxygen	O	8	0.06	47	65
Sodium	Na	11	Trace	2.8	0.2
Magnesium	Mg	12	Trace	2.1	0.1
Phosphorus	P	15	Trace	0.07	1
Sulfur	S	16	Trace	0.03	0.3
Chlorine	Cl	17	Trace	0.01	0.2
Potassium	K	19	Trace	2.6	0.4
Calcium	Ca	20	Trace	3.6	1.5
Iron	Fe	26	Trace	5	Trace

[a]Atomic number = number of protons in the nucleus.

[b]Approximate percentage of atoms of this element, by weight, in the universe, in Earth's crust, and in the human body.

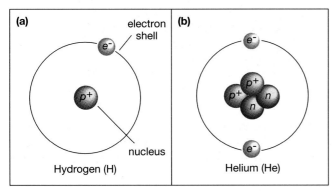

Figure 2-2   "Planetary" Atomic Models

Structural representations of the two smallest atoms, hydrogen (a) and helium (b). In these "planetary" models, the electrons are imagined as miniature planets, circling in precisely defined orbits around a nucleus that contains protons and neutrons. Although not strictly accurate, planetary models are easy to visualize and understand.

affect them at all. The stability of its nucleus is the reason why a carbon atom remains carbon whether it is part of a diamond, carbon dioxide, or sugar. Electron shells, however, are dynamic; as you will soon see, atoms interact with one another by gaining, losing, or sharing electrons.

The Atomic Nucleus Contains a Fixed Number of Protons; This Number Is a Unique Characteristic of Each Element

The number of protons in the nucleus, called the **atomic number**, is characteristic of each element. For example, every hydrogen atom has one proton in its nucleus, every carbon atom has six, and every oxygen atom has eight (see Table 2-1). Different atoms of a given element may, however, have different numbers of neutrons. Atoms with the same num-

ber of protons (that is, atoms of the same element) but with different numbers of neutrons are called **isotopes**. All the isotopes of an element are virtually identical in their chemical reactions but may vary in their physical properties. The nuclei of some isotopes spontaneously disintegrate, releasing radioactivity as they do. Radioactive isotopes are extremely useful as "labels" in studying biological processes (see "Scientific Inquiry: The Use of Isotopes in Biology and Medicine").

Electrons Orbit the Nucleus at Fixed Distances, Forming Electron Shells That Correspond to Different Energy Levels

At one time, it was believed that electrons circle the nucleus in well-defined orbits, like tiny planets orbiting the sun. But electrons are much more unpredictable than planets. They zip about outside the nucleus in regions of space called **orbitals**. At different times a given electron may be found anywhere within its orbital. Although it is not strictly accurate, here we will use the "planetary" model of electrons orbiting at fixed distances from their central nucleus within electron shells because it helps us to visualize how they interact with electrons of other atoms to form chemical bonds (Fig. 2-2).

Electrons repel one another, owing to their negative electrical charge. As you may know from experimenting with a magnet, like charges repel and opposite charges attract. Electrons, with their negative charge, are drawn to the positively charged protons of the nucleus. Because of their mutual repulsion, however, only limited numbers of electrons can occupy the space closest to the nucleus. Large atoms accommodate many electrons by having several orbits at increasing distances from their nucleus. Because the electrons orbit through a three-dimensional space, their orbits are called electron shells (Fig. 2-3).

The electron shell closest to the nucleus is the smallest and can hold only two electrons. The second shell can hold up to eight electrons. The electrons in an atom normally occupy the shells closest to the nucleus. Thus a carbon atom, with six electrons, has two electrons in the first shell, closest to the nucleus, and four electrons in its second shell (see Fig. 2-3).

Electrons in shells closest to the nucleus have the least energy and those farthest away have the most. For this reason, electron shells are often called **energy levels**.

Why do we speak of energy *levels* and *shells?* Because electron shells are found only at specific distances from the nucleus—distances that correspond to discrete energy levels. Most people aren't used to thinking about energy as occurring in steps, but perhaps an analogy with gravity can help. Picture yourself climbing a flight of stairs. You must spend a certain amount of kinetic energy to climb each step, and no intermediate amount will suffice. In other words, you can't lift your foot half a step and stand between steps. The energy used in climbing each step is stored as the potential energy; the higher the position relative to the bottom of the stairs, the greater the potential energy. This potential energy can be transformed back into kinetic energy by jumping down the steps: the higher up you start, the harder you land.

Energy levels in atoms are analogous to the staircase (Fig. 2-4). It takes energy to force an electron out of a low-energy shell into a higher-energy shell. An electron cannot occupy an intermediate position between shells, so it takes a certain minimum amount of energy to boost an electron to the next shell. But an atom is unstable when electrons have been boosted from lower- to higher-energy shells. Eventually, the electron returns to the low-energy shell, releasing the stored energy as it does (as you do when you jump down stairs you've just climbed). When plants photosynthesize, the energy of sunlight kicks electrons in chlorophyll molecules out of their low-energy shells to higher-energy shells. As the electrons

return to the low-energy shells, some of their "extra" energy is used to power the synthesis of sugar from carbon dioxide and water, as we will describe in Chapter 7.

Atoms Will React with Other Atoms Only When There Are Vacancies in Their Outermost Electron Shells

Just as an atom is the smallest possible particle of an element, so too a **molecule** is the smallest possible particle of a compound. A molecule consists of two or more atoms (of the same or different elements) held together by interactions among their outermost electron shells. The basic principles of atomic reactivity are these:

1. An atom is stable (that is, it will not react with other atoms) when its outermost electron shell is completely full or completely empty.
2. An atom is reactive when its outermost electron shell is only partially filled.

To illustrate these principles, consider the two smallest atoms, hydrogen and helium (see Fig. 2-2). Hydrogen has one proton in its nucleus, and one electron in its outermost electron shell, which can hold two. Helium has two protons in its nucleus and two electrons completely fill its outermost electron shell. Therefore, we predict that helium atoms, with a full shell, should be stable, and hydrogen atoms, with a half-empty shell, should be highly reactive. Our predictions are correct. Except for nuclear fusion reactions, helium is almost completely unreactive. Hydrogen, however, is highly flammable; that is, it reacts readily with oxygen in an explosive way. For example, the space shuttle and many other rockets use liquid hydrogen as fuel to power lift-off.

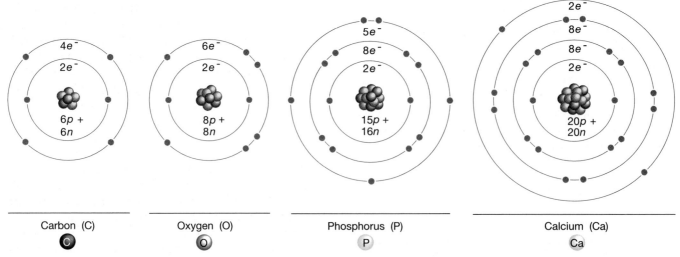

Carbon (C)   Oxygen (O)   Phosphorus (P)   Calcium (Ca)

Figure 2-3   **Electron shells in atoms**

Most biologically important atoms have at least two shells of electrons. The first shell, closest to the nucleus, can hold two electrons; the next three shells usually hold a maximum of eight electrons each.

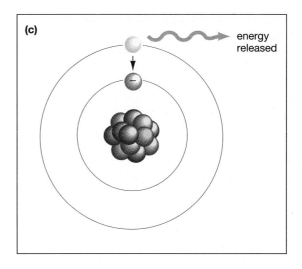

Figure 2-4   **Energy absorption and emission by an atom**

The electron shells of an atom are found at discrete energy levels; the lowest energy level is the one closest to the nucleus. **(a)** Energy (such as electrical energy or light) enters an atom and is absorbed by an electron. **(b)** Only the specific amounts of energy that boost the electron from a low-energy inner shell to a higher-energy outer shell can be absorbed. Intermediate amounts of energy are not absorbed. **(c)** The electron spontaneously returns to a low-energy shell. The "extra" energy is released. It may drive energy-requiring chemical reactions (for example, photosynthesis) or be emitted, often in the form of light.

Chemical Bonds: Joining Atoms to Make Molecules

An atom with an outermost electron shell that is partially full can gain stability by losing electrons (emptying the shell completely), gaining electrons (filling the shell), or sharing electrons with another atom (with both atoms behaving as though they had full shells). Losing, gaining, and sharing electrons result in forces called **chemical bonds** that hold atoms together in molecules. There are two major types of chemical bonds: ionic and covalent. Chemical bonds are not *things*, such as glue, but are *attractive forces* that hold atoms together. Each element has bonding properties that result from the configuration of electrons in its outer shell (see Table 2-2).

Atoms Can Become Stable by Gaining or Losing Electrons, Forming Charged Ions; the Attraction between Oppositely Charged Ions Produces Ionic Bonds

Some atoms have almost empty outermost electron shells, and some atoms have almost full outermost shells. Both types can become stable by losing electrons, completely emptying their outermost shells, or by gaining electrons, completely filling their outermost shells. The formation of table salt—sodium chloride—illustrates this principle. Sodium (Na) has an outermost electron shell containing only one electron, and chlorine (Cl) has an outer shell that lacks only one electron to be completely full (Fig. 2-5a). Sodium, therefore, can become stable by losing the electron from its outer shell, leaving that shell empty, and chlorine can fill its outer shell by gaining an electron. This is precisely what each of them does to form sodium chloride: Sodium loses an electron, becoming positively charged (Na^+), and chlorine picks up an electron, becoming a negatively charged chloride ion (Cl^-) (Fig. 2-5b). Atoms that have lost or gained electrons, altering the balance between protons and electrons, are charged. These charged atoms are called **ions**.

Opposite charges attract; therefore, sodium ions and chloride ions tend to stay near one another, forming crystals that contain repeating orderly arrangements of the two ions (Fig. 2-5c). The electrical attraction between oppositely charged ions that holds them together in crystals is called an **ionic bond**.

Atoms Can Become Stable by Sharing Electrons with Other Atoms, Forming Covalent Bonds

An atom with a partially full outermost electron shell can also become stable by sharing electrons with another atom in a **covalent bond**. To understand what this means, consider the hydrogen atom, which has one electron in a shell built for two. A hydrogen atom can become reasonably stable if it shares its single electron with another hydrogen atom, forming a molecule of hydrogen gas, H_2 (Fig. 2-6a). Because the two hydrogen atoms are identical, neither nucleus can exert more attraction and capture the other's electron. So both electrons spend part of the time in the shell of both hydrogen atoms, with the result that each hydrogen behaves almost as if it had two electrons in its shell.

The stability of a covalent bond depends on the atoms involved. Some covalent bonds, such as those in water (H_2O; Fig. 2-6c) and carbon dioxide (CO_2), are extremely stable—that is, it takes a lot of energy to break the bonds. Other bonds, such as those in oxygen gas (O_2; Fig. 2-6b) or gasoline, are less stable, coming apart more easily. When a chemical reaction occurs in which less stable bonds are broken and more stable bonds are formed (such as burning gasoline with oxygen to form carbon dioxide and water), energy is released (see Chapter 4).

Most Biological Molecules Utilize Covalent Bonding

The atoms in most biological molecules are joined by covalent bonds. Hydrogen, carbon, oxygen, nitrogen, phosphorus, and sulfur are the most common atoms found in biological molecules. Except for hydrogen, each of these atoms lacks two or more electrons to fill its outermost electron shell and can share electrons with two or more other atoms. Hydrogen can form a covalent bond with one other atom, oxygen and sulfur with two, nitrogen with three, and phosphorus and carbon with up to four other atoms. This diversity of bonding arrangements permits biological molecules to be constructed in almost infinite variety and complexity.

Sometimes, an atom will share two pairs of electrons with another atom, forming a **double covalent bond** (each atom contributes two electrons; see Fig. 2-6b). If atoms share three pairs of electrons, a **triple covalent bond** is formed. Double and triple bonds create further variety in the shapes and functions of biological molecules. Table 2-2 summarizes bonding patterns in biological molecules.

Polar Covalent Bonds Form When Atoms Share Electrons Unequally

Electron sharing in covalent bonds is not always equal. In hydrogen gas, the two nuclei are identical, and the shared electrons spend equal time near each nucleus. Therefore, not only is the molecule as a whole electrically neutral, but each end, or pole, of the molecule is also electrically neutral. Such an electrically symmetrical bond is called a **nonpolar covalent bond**. In many other molecules, one nucleus may attract the electrons more strongly than the other nucleus does because its nucleus has a larger positive charge. This situation produces a **polar covalent bond**. Although the molecule as a whole is electrically neutral, it has charged parts: The atom that attracts the electrons more strongly has a slightly negative charge (the negative pole of the molecule), and the other atom has a slightly positive charge (the positive pole). In water, for example, oxygen attracts electrons more strongly than hydrogen does, so the oxygen end of a water molecule is negative and each hydrogen is positive (Fig. 2-6c). Water with its charged ends can be called a polar molecule.

Hydrogen Bonds Are Weak Electrical Attractions between the Polar Parts of Molecules

Because of the polar nature of their covalent bonds, nearby water molecules attract one another, the partially negatively charged oxygens of some molecules attracting the

TABLE 2-2 ⚛ Bonding Patterns of Atoms Commonly Found in Biological Molecules				
Atom	Capacity of Outer Electron Shell	Electrons in Outer Shell	Number of Covalent Bonds Usually Formed	Common Bonding Patterns
Hydrogen	2	1	1	—H
Carbon	8	4	4	—C— —C— =C= —C≡
Nitrogen	8	5	3	—N— —N= N≡
Oxygen	8	6	2	—O— —O=
Phosphorus	8	5	5	—P=
Sulfur	8	6	2	—S—

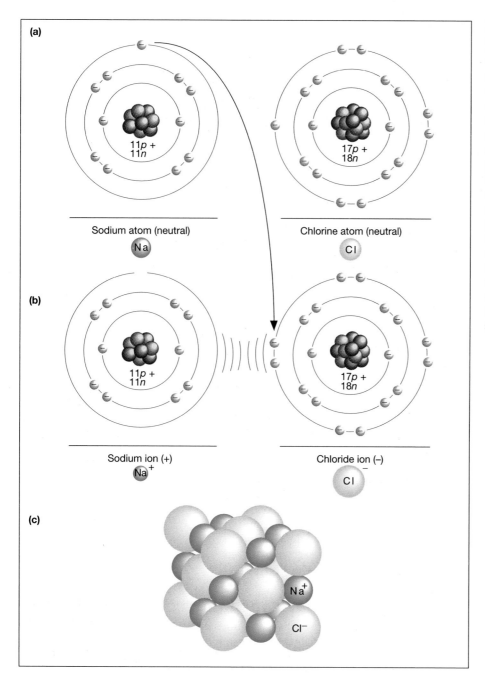

(a)

Sodium atom (neutral)
Na

Chlorine atom (neutral)
Cl

11p +
11n

17p +
18n

(b)

11p +
11n

17p +
18n

Sodium ion (+)
Na⁺

Chloride ion (–)
Cl⁻

(c)

Na⁺

Cl⁻

Figure 2-5 **The formation of ions and ionic bonds**

(a) Sodium has only one electron in its outer electron shell; chlorine has seven (lacking only one electron to fill the outer shell). **(b)** Sodium can become stable by losing an electron, and chlorine can become stable by gaining an electron. When this exchange happens, sodium has an empty and chlorine a full outer shell. Sodium becomes a positively charged ion, and chlorine a negatively charged ion. **(c)** Because oppositely charged particles attract one another, the resulting sodium and chloride ions nestle closely together in a crystal of salt, NaCl. Here we have used different-colored balls to represent entire ions: sodium (Na^+) and chloride (Cl^-). You will see balls representing atoms in several other figures as well.

partially positively charged hydrogens of other molecules. This electrical attraction is called a **hydrogen bond** (Fig. 2-7). As we will see shortly, hydrogen bonds between molecules give water several unusual properties that are essential to life on Earth.

Many biological molecules also form hydrogen bonds. Both nitrogen and oxygen atoms attract electrons more strongly than hydrogen atoms do. Therefore, the nitrogen or oxygen pole of a nitrogen-hydrogen or oxygen-hydrogen bond is slightly negative, and the hydrogen pole is slightly positive. The resulting polar parts of the molecules can form hydrogen bonds with water, with other biological molecules, or with distant, polar parts of the same molecule. While hydrogen bonds individually are quite weak, many of them working together are quite strong. As

we will see in Chapter 3, hydrogen bonds play crucial roles in shaping the three-dimensional structures of proteins and DNA.

Organic and Inorganic Molecules

Chemists classify molecules as organic or inorganic. The word *organic* originally signified that these molecules could be manufactured only within living organisms. Today, however, organic chemists can synthesize many of these molecules in the laboratory. Thus, the modern definition of an **organic molecule** is any molecule containing both carbon and hydrogen. Most organic molecules are large, with complex structures. **Inorganic molecules** include carbon dioxide and all molecules without carbon.

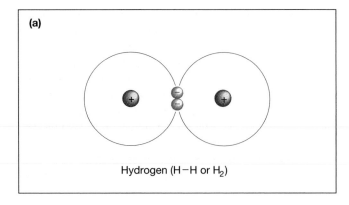

(a)

Hydrogen (H−H or H$_2$)

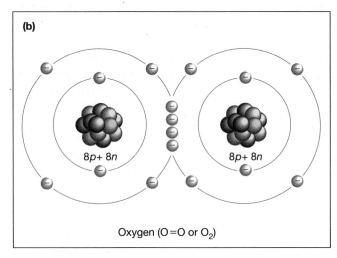

(b)

Oxygen (O═O or O$_2$)

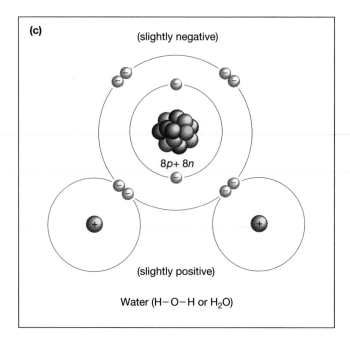

(c)

(slightly negative)

8p+ 8n

(slightly positive)

Water (H−O−H or H$_2$O)

Figure 2-6   **Covalent bonds**

Electrons are shared between atoms to form covalent bonds. **(a)** In hydrogen gas, one electron from each hydrogen atom is shared, forming a single covalent bond. The resulting bond is usually represented by a single dash connecting the atomic symbols, and the molecule of hydrogen gas may be represented as H–H or H$_2$. **(b)** In oxygen gas, two oxygen atoms share four electrons, making a double bond, usually represented by a double dash or equal sign (O═O or O$_2$). **(c)** Oxygen lacks two electrons to fill its outer shell, so oxygen can make two single bonds, one with each of two hydrogen atoms to form water (H–O–H or H$_2$O). Oxygen exerts a greater pull on the electrons than does hydrogen, and so the oxygen end of the molecule has a slight negative charge, and the hydrogen end has a slight positive charge. This is an example of polar covalent bonding, and the water molecule with its slightly charged ends is called a polar molecule.

Many inorganic molecules are important to living organisms, including, for instance, minerals in the soil and carbon dioxide in the air, which are the raw materials that plants use to construct their bodies. One inorganic molecule, however, is extraordinarily abundant on Earth, has unusual properties, and is so essential to life that it merits special consideration. That molecule is water.

Water and Life

Life almost certainly arose in the waters of the primeval Earth. The first life form was probably a small sac enclosing water with an array of dissolved enzymes and simple genet-

ic material. Living organisms are still composed of about 60% to 90% water, and all life on Earth depends intimately on the properties of water. Why is water so crucial to life?

Water Interacts with Many Other Molecules

Water is an extremely good **solvent**—that is, it is capable of dissolving a wide range of substances, especially salts. Recall that a crystal of table salt is held together by the electrical attraction between positively charged sodium ions and negatively charged chloride ions (Fig. 2-5). Because water is a polar molecule, it has positive and negative ends. If a salt crystal is dropped into water, the positively charged hydrogen ends of water molecules will be attracted to and sur-

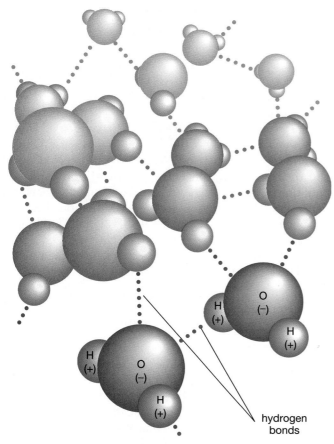

Figure 2-7   **Hydrogen bonds**

The partial charges on different parts of water molecules produce weak attractive forces called hydrogen bonds (dotted lines) between the hydrogens of one water molecule and the oxygens of other molecules.

round the negatively charged chloride ions, and the negatively charged oxygen ends of water molecules will surround the positively charged sodium ions. As water molecules enclose the sodium and chloride ions and shield them from interacting with each other, the ions separate from the crystal and drift away in the water—the salt dissolves (Fig. 2-8).

Water dissolves polar molecules in a similar fashion, as its positive and negative poles are attracted to oppositely charged regions of dissolving molecules. Charged and polar molecules are termed **hydrophilic** (Greek for "water-loving"), because of their electrical attraction for water molecules. Many biological molecules, such as sugars and amino acids, are hydrophilic and dissolve readily in water (Fig. 2-9). Water also dissolves gases such as oxygen and carbon dioxide.

Other liquids can dissolve some of these substances, but not all of them. Alcohol, for example, dissolves some sugars and proteins but not salts. By dissolving such a wide variety of molecules, the watery substance inside a cell provides a suitable environment for the countless chemical reactions essential to life on Earth.

Molecules that are uncharged and nonpolar, such as fats and oils, usually do not dissolve in water, and hence are

called **hydrophobic** ("water-fearing"). Nevertheless, water has an important effect on such molecules. Oils, for example, form globs when spilled into water. Why? In pure water, each water molecule can form hydrogen bonds with the maximum possible number of other water molecules. If two molecules of oil are added, at distant locations, each disrupts the hydrogen bonding of nearby water molecules. Any water molecule directly adjacent to the oil molecules cannot make hydrogen bonds in the direction of the oil.

Suppose now that the two oil molecules, randomly moving in the water, encounter one another. The instant they make contact, their nonpolar surfaces nestle closely together, surrounded by water molecules that form hydrogen bonds with one another but not with the oil. The two-molecule glob has a smaller surface area than the two separate molecules did, so there is less interference with hydrogen bonding among water molecules. To separate again, the oil molecules would have to shove apart the surrounding water molecules. Because it takes energy to break the hydrogen bonds interconnecting the water molecules, the water tends to prevent the oil molecules from moving apart. The tendency for hydrophobic molecules to clump together in watery solutions is often termed a **hydrophobic interaction**, although in fact it is the hydrogen bonding between water molecules that keeps the hydrophobic molecules together, rather than any special affinity among the hydrophobic molecules.

As we will see in Chapter 5, the membranes of living cells owe much of their structure to the organizing effect of water on hydrophobic molecules.

Figure 2-8   **Water as a solvent**

The polarity of water molecules allows water to dissolve polar and charged substances. When a salt crystal is dropped into water, the water molecules worm their way around the outsides of the sodium and chloride ions, surrounding them with oppositely charged ends of the water molecules. Thus insulated from the attractiveness of other molecules of salt, the ions float away, and the whole crystal gradually dissolves.

Figure 2-9   **Water dissolves many biological molecules**
Many biological molecules dissolve in water because they have polar parts, for example, OH⁻ (hydroxyl) groups, that can form hydrogen bonds with water molecules. This illustration depicts the hydrogen bonds that can form between the hydroxyl groups on a glucose molecule (a simple sugar) and surrounding water molecules.

Water Takes Part in Many Chemical Reactions

Water enters into many of the chemical reactions that occur in living cells. The oxygen that green plants release into the air is derived from water during photosynthesis. When your body manufactures a protein, fat, nucleic acid, or sugar, it produces water in the process, and, conversely, when you digest proteins, fats, and sugars in the foods you eat, water is used in the reactions.

The pH of a Solution Is a Measure of Its Concentration of Hydrogen Ions

Although water is generally regarded as a stable compound, individual water molecules constantly gain, lose, and swap hydrogen atoms. As a result, at any given time about two of every billion water molecules are ionized; that is, broken apart into hydrogen ions (H^+) and hydroxide ions (OH^-):

| water | hydroxide ion | hydrogen ion |
| (H_2O) | (OH^-) | (H^+) |

Pure water contains equal concentrations of hydrogen ions and hydroxide ions. In many solutions, though, the concentrations of H^+ and OH^- ions are not the same. If the concentration of H^+ ions exceeds the concentration of OH^- ions, the solution is **acidic**; if the concentration of OH^- ions is greater, the solution is **basic**. The degree of acidity is expressed on the **pH scale** (Fig. 2-10), in which neutrality (equal num-

bers of H^+ and OH^- ions) is assigned the number 7. Acids have a pH below 7; bases have a pH above 7. Pure water, with equal concentrations of H^+ and OH^-, has a pH of 7. Each unit on the pH scale represents a tenfold change in the concentration of H^+ ions. Thus, a cola drink (pH = 3) has 10,000 times the concentration of H^+ than does water (pH = 7).

An **acid** is a substance that gives off hydrogen ions. When hydrochloric acid (HCl), for example, is added to pure water, almost all of the HCl molecules separate into H^+ and Cl^- ions. Therefore, the concentration of H^+ ions greatly exceeds the concentration of OH^- ions, and the resulting solution is acidic. A **base** is a substance that combines with H^+ ions and thereby reduces their number. For instance, if sodium hydroxide (NaOH) is added to water, the NaOH molecules separate into Na^+ and OH^- ions. The OH^- ions combine with H^+ ions, reducing their number. The resulting solution is basic.

A Buffer Helps Maintain a Solution at a Relatively Constant pH

In most mammals, including humans, both the cell cytoplasm and the fluids that bathe the cells are nearly neutral (pH about 7.3 to 7.4). Small increases or decreases in pH may cause drastic changes in both structure and function, leading to the death of cells or entire organisms. Nevertheless, living cells seethe with chemical reactions that take up or give off H^+ ions. How, then, does the pH remain constant? The answer lies in the many buffers found in living organisms. A **buffer** is a compound that tends to maintain a solution at a constant pH by accepting or releasing H^+ ions in response to small changes in H^+ ion concentration. If the H^+ ion concentration rises, buffers combine with them; if the H^+ ion concentration falls, buffers release H^+. The result is that the concentration of H^+ ions is restored to its original level. Common buffers in living organisms include bicarbonate (HCO_3^-) and phosphate ($H_2PO_4^-$ and HPO_4^{2-}), both of which can accept or release H^+ ions, depending on the circumstances. If the blood becomes too acidic, for example, bicarbonate accepts H^+ ions to form carbonic acid:

$$HCO_3^- \quad + \quad H^+ \quad \rightarrow \quad H_2CO_3$$
(bicarbonate)   (hydrogen ion)   (carbonic acid)

If the blood becomes too basic, carbonic acid liberates H^+ ions that combine with the excess OH^- ions, forming water:

$$H_2CO_3 \quad + \quad OH^- \quad \rightarrow \quad HCO_3^- + H_2O$$
(carbonic acid) (hydroxide ion) (bicarbonate) (water)

In either case, the end result is that the blood pH remains near its normal value.

Water Moderates the Effects of Temperature Changes

Organisms can survive only within a limited temperature range: Both high and low temperatures can be fatal. High temperatures may damage the protein enzymes that guide the chemical reactions essential to life (see Chapter 4):

Most enzymes cease to function at temperatures well below boiling. Low temperatures are also dangerous, because enzyme action slows as the temperature drops. Subfreezing temperatures within the body are usually lethal, because cells are ruptured by spearlike ice crystals.

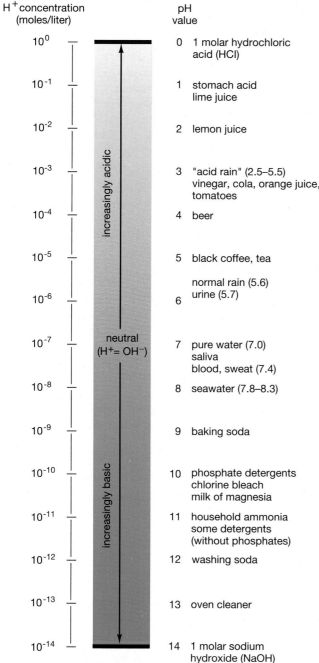

H⁺ concentration (moles/liter) and **pH value:**

- 10^{0} — 0 1 molar hydrochloric acid (HCl)
- 10^{-1} — 1 stomach acid / lime juice
- 10^{-2} — 2 lemon juice
- 10^{-3} — 3 "acid rain" (2.5–5.5) vinegar, cola, orange juice, tomatoes
- 10^{-4} — 4 beer
- 10^{-5} — 5 black coffee, tea
- 10^{-6} — 6 normal rain (5.6) / urine (5.7)
- 10^{-7} — 7 pure water (7.0) / saliva / blood, sweat (7.4) [neutral (H⁺ = OH⁻)]
- 10^{-8} — 8 seawater (7.8–8.3)
- 10^{-9} — 9 baking soda
- 10^{-10} — 10 phosphate detergents, chlorine bleach, milk of magnesia
- 10^{-11} — 11 household ammonia, some detergents (without phosphates)
- 10^{-12} — 12 washing soda
- 10^{-13} — 13 oven cleaner
- 10^{-14} — 14 1 molar sodium hydroxide (NaOH)

increasingly acidic / increasingly basic

Figure 2-10   The pH scale

The pH scale expresses the concentration of hydrogen ions in a solution on a scale of 0 (very acidic) to 14 (very basic). Each unit of change in pH on the pH scale represents a tenfold change in the concentration of hydrogen ions. Lemon juice, for example, is about 10 times more acid than orange juice, and the most severe acid rains in the northeastern United States are almost 1000 times more acidic than normal rainfall. Except for the insides of your stomach, nearly all the fluids in your body are finely adjusted to a pH of 7.4.

Water has three properties that moderate the effects of temperature changes. First, water has a high **specific heat**, the amount of energy needed to raise the temperature of 1 gram of a substance by 1° C. Temperature reflects the speed of molecules: the higher the temperature, the greater the average speed. Generally speaking, if heat energy enters a system, the molecules of that system move more rapidly, and the temperature of the system rises. Individual water molecules, however, are weakly linked to one another by hydrogen bonds (see Fig. 2-7). When heat enters a watery system such as a lake or a living cell, much of the heat energy goes into breaking hydrogen bonds rather than speeding up individual molecules. Thus, 1 **calorie** of energy will heat 1 gram of water 1° C, whereas it takes only 0.6 calorie per gram to heat alcohol 1° C, 0.2 calorie for table salt, and 0.02 calorie for common rocks such as granite or marble. As a result, because the human body is mostly water, a sunbather can absorb a lot of heat energy without sending his body temperature soaring.

Second, water moderates the effects of high temperatures through its great **heat of vaporization**, the amount of heat required to convert liquid water to water vapor. Water has one of the highest heats of vaporization known, 539 calories per gram. Again, this is due to the hydrogen bonds interconnecting individual water molecules. For a water molecule to evaporate, it must move quickly enough to break all the hydrogen bonds holding it to the other water molecules in the solution. Only the fastest-moving water molecules, carrying the most energy, can break their hydrogen bonds and escape into the air as water vapor. The remaining liquid is cooler for the loss of these high-energy molecules. As a sunbather's body temperature begins to rise, she perspires, covering her body with a film of water. Heat energy is transferred from her skin to the water and from the water to the vapor as the water evaporates. Evaporating just 1 gram of water cools 539 grams of her body 1° C, and so a great loss of heat can occur without much loss of water.

Third, water moderates the effects of low temperatures because it has a high **heat of fusion**, the energy that must be removed from molecules of liquid water before they form the precise crystal arrangement of ice. Therefore, water, both in a living organism or in a lake, freezes more slowly than other liquids at a given temperature.

Water Forms an Unusual Solid, Ice

Water, of course, will become a solid after prolonged exposure to temperatures below its freezing point. But even solid water is unusual. Most liquids become denser when they solidify, and the solid sinks. Ice is less dense than liquid water, so when a pond or lake starts to freeze in winter, the ice stays on top, forming an insulating layer that delays the freezing of the rest of the water. If ice sank, ponds and lakes in much of North America would freeze solid during the winter, eliminating fish and making liquid drinking water far less available to animals.

S C I E N T I F I C   I N Q U I R Y

The Use of Isotopes in Biology and Medicine

As you read this text, you will encounter many statements that may cause you to wonder, "How do they know *that*?" How do biologists know that DNA is the genetic material of cells (Chapter 10)? How do paleontologists measure the ages of fossils (Chapter 19)? These discoveries, and many more, have been possible only through the use of isotopes.

Although all atoms of a particular element have the same number of protons, the number of neutrons may vary. Neutrons don't affect the chemical reactivity of an atom very much, but they do make their presence felt in other ways. First, neutrons add to the mass of an atom, which can be detected by sophisticated instruments, such as mass spectrometers. Second, nuclei with "too many" neutrons break apart spontaneously, often emitting **radioactive** particles in the process. Those particles can also be detected, for example with Geiger counters.

A particularly fascinating and medically important application of radioactive isotopes is positron emission tomography, more commonly known as PET scans (Fig. E2-1). In a common application of PET scans, a subject is given glucose that has been labeled with a harmless radioactive isotope of fluorine, ^{18}F. When the nucleus of ^{18}F decays, it emits two bursts of energy that travel in opposite directions along the same line. Energy detectors arranged in a ring around the subject record the nearly si-multaneous arrival of the two energy bursts. A powerful computer then calculates the location within the subject at which the decay must have occurred and generates a map of the frequency of ^{18}F decays. Because the ^{18}F is attached to glucose molecules, this map reflects the glucose concentrations within the subject. The brain uses prodigious amounts of this sugar for energy; the more active a brain cell is, the more glucose it uses. How can this information be used in biological research?

Let's suppose that a neuroscientist is trying to locate the areas of the brain that are involved in memory. She might give ^{18}F-labeled glucose to a few volunteer subjects and then ask them to memorize a list of words that she reads aloud. Because brain regions active during this process would need more energy and would take up more ^{18}F-glucose molecules than would inactive regions, they would have more ^{18}F decays. PET scans taken during the memorization would pinpoint brain regions active in storing memories of words.

Physicians also use PET scans in the diagnosis of brain disorders. For example, brain regions that are the origin of epileptic seizures generally have excessively high glucose utilization and show up in PET scans as "hot spots." Many brain tumors also light up in PET scans (Fig. E2-2). Abnormal metabolism of certain brain regions may also appear in patients with some mental disorders, such as schizophrenia.

detector ring

decay of radioactive isotope

energy burst (gamma ray)

Figure E2-1   How positron emission tomography works

(a) A subject is given a substance containing a harmless radioactive isotope. The subject's head is inserted within a ring of detectors. Computerized analysis provides an image of the "slice" of the head that lies within the plane of the detector ring. (b) Inside the "head slice," radioactive decay releases energetic particles that activate the detectors. Computer analysis locates the site of the radioactive decay. (c) The number of decays at each location is color-coded and projected onto a television screen: Usually red and yellow mean many decays, green and blue mean few decays. The resulting image provides a visual representation of the activity and structure of the brain.

Figure E2-2   **A brain tumor**

A brain tumor consists of actively dividing, metabolically energetic cells. It usually has a rich blood supply and consumes both oxygen and glucose voraciously. Therefore, a brain tumor "lights up" in a PET scan. In this scan, a malignant brain tumor shows clearly as a red area.

Biology and medicine have profited immensely from close interactions with the other sciences, especially chemistry and physics. The development of PET scans required close cooperation with chemists (developing and synthesizing the radioactive probes), physicists (understanding positron-electron interactions and the geometry of the resulting energy emissions), and engineers (designing and building the electronic apparatus). Continued teamwork among scientists promises further advances in both fundamental understanding of biological processes and applications in medicine and agriculture.

Water Molecules Tend to Stick Together

Because of the hydrogen bonds that interconnect individual molecules, water has high **cohesion**—that is, water molecules tend to stick together. Cohesion among water molecules at the surface of a lake or pond produces **surface tension**, the tendency for the water surface to resist being broken. In general, objects that are denser than water sink. Because the water molecules at the surface of a pond stick to one another, however, the surface film acts almost as a solid, supporting relatively dense objects such as fallen leaves and water striders (Fig. 2-11a).

A more important role of cohesion occurs in the life of land plants (Fig. 2-11b). A plant absorbs water through its roots. How does the water reach the above-ground parts, especially if the plant is a hundred-meter-tall redwood? The water is pulled up by the leaves. Water fills tiny tubes that connect the leaves, stem, and roots. Water molecules that

(a)

(b)

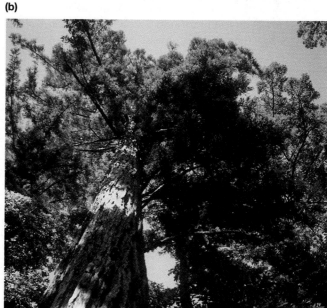

Figure 2-11   **Cohesion among water molecules**

(a) Cohesion among water molecules allows water striders to skate across the surface of still waters. **(b)** In giant redwoods, cohesion holds water molecules together in continuous strands from the roots to the topmost leaves even a hundred meters above the ground.

evaporate from the leaves pull water up the tubes, much like a rope being pulled up from the top. The system works because the hydrogen bonds interconnecting water molecules are stronger than the weight of the water in the tubes, even a hundred meters' worth, and so the water "rope" doesn't break. Without the cohesion of water, there would be no land plants as we know them, and terrestrial life would undoubtedly have evolved quite differently.

✚ SUMMARY OF KEY CONCEPTS

Matter and Energy

Matter is the physical material of the universe. Energy is the capacity to do work. Energy can exist in several forms that may be converted from one to another. The two major forms of energy are kinetic energy (the energy of motion) and potential energy (stored energy).

The Structure of Matter

An element is a substance that can neither be broken down nor converted to different substances by ordinary chemical means. The smallest possible particle of an element is the atom, which is itself composed of a central nucleus containing protons and neutrons and outer electron shells. All atoms of a given element have the same number of protons, which is different from the number found in the atoms of any other element.

Electrons are found in electron shells, at specific distances from the nucleus. Each shell can contain a fixed maximum number of electrons. The chemical reactivity of an atom depends on the number of electrons in its outermost electron shell; an atom is most stable, and therefore least reactive, when its outermost shell is either completely full or completely empty.

Chemical Bonds: Joining Atoms to Make Molecules

Atoms may combine to form molecules. The forces holding atoms together in molecules are called chemical bonds. There are two principal types of chemical bond, called ionic and covalent bonds. When one atom fills its outermost shell by acquiring electrons while another atom empties its shell by losing electrons, the results are negatively and positively charged particles called ions. Ionic bonds are electrical attractions between charged ions, holding them together in crystals.

Covalent bonds involve the sharing of electrons by two atoms, in which neither atom completely gains or loses an electron. In a nonpolar covalent bond, the two atoms share electrons equally. In a polar covalent bond, one atom may attract the electron more strongly than the other atom does; in this case, the strongly attracting atom bears a slightly negative charge, and the weakly attracting atom bears a slightly positive charge. Some polar covalent bonds give rise to hydrogen bonding, which is the attraction between charged regions of individual polar molecules or distant parts of a large polar molecule.

Organic and Inorganic Molecules

An organic molecule is one that contains both carbon and hydrogen; all other molecules are inorganic.

Water and Life

Properties of the water molecule that are important to the processes that occur within living organisms include its ability to dissolve many polar and charged substances, to force nonpolar substances to assume certain types of physical organization, to participate in chemical reactions, to maintain a fairly stable temperature in the face of wide temperature fluctuations in the environment, and to cohere to itself.

✚ KEY TERMS

✖ T H I N K I N G   T H R O U G H   T H E   C O N C E P T S

Multiple Choice

1. What is the purest form of matter that cannot be separated into different substances by chemical means?
   a. compounds   b. molecules
   c. hormones   d. elements
   e. electrons

2. What phrase best describes molecular bonds?
   a. physical bridges   b. energy links
   c. electronic glue   d. electronic orbitals
   e. all of these phrases are equally descriptive

3. When an atom ionizes, what happens?
   a. It shares one or more electrons with another atom.
   b. It emits energy as it loses extra neutrons.
   c. It completely gives up or takes up one or more electrons.
   d. It shares a hydrogen atom with another atom.
   e. none of the above

4. If electrons in water molecules were equally attracted to hydrogen nuclei and oxygen nuclei, water molecules would be
   a. more polar   b. less polar
   c. unchanged   d. triple bonded
   e. unable to form

5. Where are hydrogen bonds important to life?
   a. between water molecules
   b. between hydrogen atoms
   c. in ionic substances
   d. all of the above
   e. none of the above

6. What is the defining characteristic of an acid?
   a. It donates hydrogen ions.
   b. It accepts hydrogen ions.
   c. It will donate or accept hydrogen ions depending on the pH.
   d. It has an excess of hydroxide ions.
   e. It has a pH greater than 7.

Review Questions

1. What are the six most abundant elements in living organisms?
2. Distinguish among atoms and molecules; elements, compounds, and mixtures; protons, neutrons, and electrons.
3. Compare and contrast covalent bonds and ionic bonds.
4. Why can water absorb a great amount of heat with little increase in its temperature?
5. Describe how water dissolves a salt. How does this compare with the effect of water on a hydrophobic substance such as corn oil?
6. Define acid, base, and buffer. How do buffers reduce changes in pH when hydrogen ions or hydroxide ions are added to a solution? Why is this important in living organisms?

✖ A P P L Y I N G   T H E   C O N C E P T S

1. Many "over-the-counter" substances are sold to bring relief from "acid stomach" or "heartburn." What is the chemical basis for these compounds? Why do they work?
2. Fats and oils do not dissolve in water; polar and ionic molecules dissolve easily in water. Detergents and soaps help to clean dishes by dispersing fats and oils in water, so that they can be rinsed away. From your knowledge of the structure of water and the hydrophobic nature of fats, what general chemical structures (e.g., polar or nonpolar parts) must a soap or detergent have, and why?
3. What would the effects be for aquatic life if the density of ice were greater than that of liquid water? What would be the impact on all the rest of the living (terrestrial) organisms on Earth?
4. How does sweating help you to regulate your body temperature? Why do you feel hotter and more uncomfortable on a hot, humid day than on a hot, dry day?

✖ F O R   M O R E   I N F O R M A T I O N

Atkins, P. W. *Molecules*. New York: Scientific American Library, 1987. A layperson's introduction to atoms and molecules, with superb illustrations.

Morrison, P., and Morrison, P. *Powers of Ten*. New York: W. H. Freeman and Co., 1982. A fascinating journey from the universe to the nucleus of an atom.

Storey, K. B., and Storey, J. M. "Frozen and Alive." *Scientific American*, December 1990. By triggering ice formation here, suppressing it there, and loading up their cells with antifreeze molecules, some animals, including certain lizards and frogs, can survive with 60% of their body water frozen solid.

N E T   W A T C H

On-line resources for this chapter are on the World Wide Web at:
http://www.prenhall.com/~audesirk (click on the table of contents link and then select Chapter 2).

3

Biological Molecules

A computer-generated image of the shape of trypsin, a protein that serves as a digestive enzyme for protein

Living organisms are amazingly diverse: Bacteria, mushrooms, redwood trees, sea urchins, and people, at first glance, seem to have very little in common. However, this diversity is a veneer that covers an underlying unity. We have all evolved from common ancestors, a fact that can most readily be seen in the organic molecules of which our bodies are made. The organic molecules—the carbohydrates, lipids, proteins, and nucleic acids—are essentially the same in every organism. Over a couple of billion years of evolution, many differences have evolved between specific biological molecules in bacteria and bulldogs, but the similarities are striking. Why can people eat other organisms as diverse as chickens, corn, shrimp, and mushrooms? Because the building blocks that make up the biological molecules in these organisms are the same ones that we need to make our own molecules and because our bodies are able to break down many of their molecules to derive energy as well.

While the common structure and function of the types of organic molecules among organisms provides unity, the tremendous range of organic molecules accounts for the diversity of living organisms and the diversity of structures within single organisms and even within individual cells. This vast array of organic molecules, in turn, is possible because of the versatility of the carbon atom. A carbon atom has four electrons in its outermost shell, with room for eight. Therefore, carbon atoms are able to form many bonds. They become stable by sharing four electrons with other atoms, forming up to four single covalent bonds or smaller numbers of double or triple bonds. Molecules with many carbon atoms can assume complex shapes, including chains, branches, and rings.

Organic molecules are much more than just complicated skeletons of carbon atoms, however. Attached to the carbon backbone are groups of atoms, called **functional groups**, that determine the characteristics and chemical reactivity of the molecules. These functional groups are far less stable than the carbon backbone and are more likely to participate in chemical reactions. The common functional groups found in organic molecules are shown in Table 3-1.

The similarity among organic molecules found in all life on Earth is a consequence of two main features: the use

TABLE 3-1 ▓ *Important Functional Groups in Biological Molecules*

Group	Structure	Properties	Types of Molecules
Hydrogen (–H)		Polar or nonpolar, depending on what atom hydrogen is bonded to; involved in condensation and hydrolysis	Almost all organic molecules
Hydroxyl (–OH)		Polar; involved in condensation and hydrolysis	Carbohydrates, nucleic acids, alcohols, some acids, and steroids
Carboxyl (–COOH)		Acid; negatively charged when H^+ dissociates; involved in peptide bonds	Amino acids, fatty acids
Amino (–NH$_2$)		Basic; may bond an additional H^+, becoming positively charged; involved in peptide bonds	Amino acids, nucleic acids
Phosphate (–H$_2$PO$_4$)		Acid; up to two negative charges when H^+ dissociates; links nucleotides in nucleic acids; energy-carrier group in ATP	Nucleic acids, phospholipids
Methyl (–CH$_3$)		Nonpolar; tends to make molecules hydrophobic	Many organic molecules; especially common in lipids

of the same set of functional groups in virtually all organic molecules in all organisms and the use of the "modular approach" to synthesizing large organic molecules.

Synthesizing Organic Molecules: A Modular Approach

In principle, there are two ways to manufacture a large, complex molecule: One could synthesize the molecule atom by atom following an extremely detailed blueprint, or one could take preassembled smaller molecules and hook them together. Just as trains are made by coupling engines, boxcars, coal cars, and cabooses, so too life on Earth takes the modular approach. Small molecules (for example, sugars) are used as **subunits** with which to synthesize longer molecules (for example, starches), like cars in a train (Fig. 3-1a). The individual subunits are often called **monomers** (from Greek words meaning "one part"); long chains of monomers are called **polymers** ("many parts"). Most organic molecules in our bodies are in the form of either monomers or polymers.

Biological Molecules Are Joined Together or Broken Apart by Adding or Removing Water

Biological molecules almost always use the same type of chemical reaction, called a **dehydration synthesis** (literally "to form by removing water") to join subunits to one another. In a dehydration synthesis, a hydrogen (–H), removed from one subunit, and a hydroxyl (–OH), removed from a second subunit, combine to form a molecule of water (H_2O), as the remaining subunits are joined by a covalent bond (Fig. 3-1b). The reverse reaction, called **hydrolysis** (literally, "to break apart with water"), can split the molecule into individual subunits again (Fig. 3-1c). During hydrolysis, water is added back, a hydrogen to one subunit and a hydroxyl to the other.

The Principal Types of Biological Molecules

Although the bodies of living things often include thousands of different organic molecules, nearly all fall into one of four categories: carbohydrates, lipids, proteins, or nucleic acids (see Table 3-2).

Figure 3-1   **Synthesis and breakdown of organic molecules**
(a) A typical organic molecule is composed of similar or identical subunits linked together by covalent bonds. Here we see sugar subunits joined to form a starch. (b) In a dehydration synthesis, two subunits are joined by a covalent bond. Simultaneously, a hydroxyl group is removed from one subunit and combines with a hydrogen removed from a second subunit to form water. (c) Hydrolysis is the reverse of dehydration synthesis. Hydrogen and hydroxyl from water are added to the subunits as the large organic molecule is broken apart into its subunits.

TABLE 3-2 ▪ The Principal Biological Molecules

Class of Molecule	Principal Subtypes	Example	Function
Carbohydrate: Usually contains carbon, oxygen, and hydrogen, in the approximate formula $(CH_2O)_n$	Monosaccharide: Simple sugar	Glucose	Important energy source for cells; subunit of which most polysaccharides are made
	Disaccharide: Two monosaccharides bonded together	Sucrose	Principle sugar transported throughout bodies of land plants
	Polysaccharide: Many monosaccharides (usually glucose) bonded together	Starch	Energy storage in plants
		Glycogen	Energy storage in animals
		Cellulose	Structural material in plants
Lipid: Contains high proportion of carbon and hydrogen; usually nonpolar and insoluble in water	Triglyceride: Three fatty acids bonded to glycerol	Oil, fat	Energy storage in animals, some plants
	Wax: Variable numbers of fatty acids bonded to long-chain alcohol	Waxes in plant cuticle	Waterproof covering on leaves and stems of land plants
	Phospholipid: Polar phosphate group and two fatty acids bonded to glycerol	Phosphatidylcholine	Common component of membranes in cells
	Steroid: Four fused rings of carbon atoms with functional groups attached	Cholesterol	Common component of membranes of eukaryotic cells; precursor for other steroids such as testosterone, bile salts
Protein: Chains of amino acids; contain carbon, hydrogen, oxygen, nitrogen, and sulfur		Keratin	Helical protein, principal component of hair
		Silk	Pleated sheet protein produced by silk moths and spiders
		Hemoglobin	Globular protein composed of four subunit peptides; transport of oxygen in vertebrate blood
Nucleic acid: Made of nucleotide subunits; may consist of a single nucleotide or long chain of nucleotides	Long-chain nucleic acids	Deoxyribonucleic acid (DNA)	Genetic material of all living cells
		Ribonucleic acid (RNA)	Genetic material of some viruses; in living cells, essential in transfer of genetic information from DNA to protein
	Single nucleotides	Adenosine triphosphate (ATP)	Principal short-term energy carrier molecule in cells
		Cyclic adenosine monophosphate (cyclic AMP)	Intracellular messenger

Carbohydrates Are Sugars and Starches

All **carbohydrates** are either small, water-soluble **sugars** (glucose or fructose, for example) or chains, such as starch or cellulose, that are made by stringing sugar subunits together. If a carbohydrate consists of just one sugar molecule, it is called a **monosaccharide** (Greek for "single sugar"). When two or more monosaccharides are linked together, they form a **disaccharide** ("two sugars") or a **polysaccharide** ("many sugars").

Carbohydrates such as sugars and starches are important energy sources for most organisms. Other carbohydrates, such as cellulose and similar molecules, provide structural support for individual cells or even for the entire bodies of organisms as diverse as plants, fungi, bacteria, and insects.

Single sugars (monosaccharides) usually have a backbone of three to seven carbon atoms (Fig. 3-2a, left). Most of the carbon atoms have both a hydrogen (–H) and a hydroxyl group (–OH) attached to them. Therefore, the general formula for a monosaccharide is $(CH_2O)_n$, where n is the number of carbons in the backbone. This formula explains the origin of the name carbohydrate, which literally means "carbon plus water." When dissolved in water, such as in the cytoplasm of a cell, the carbon backbone of a sugar usually "circles up" into a ring (Fig. 3-2a, right). It is in this ring form that sugars link together to make disaccharides and polysaccharides (see Fig. 3-3).

Most small carbohydrates are soluble in water. As in water molecules, the O–H bond in a hydroxyl group is polar, because oxygen attracts electrons more strongly than hydrogen does. Hydrogen bonds between water molecules and the polar hydroxyl groups of the carbohydrate keep the carbohydrate in solution (see Fig. 2-9).

There Are a Variety of Monosaccharides with Slightly Different Structures

Glucose (Fig. 3-2a) is the most common monosaccharide in living organisms and is the subunit of which most polysaccharides are made. Glucose has six carbons, and hence has the

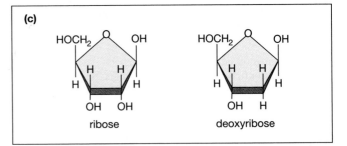

Figure 3-2   Monosaccharide structure

(a) The most common monosaccharides have "backbones" of either five or six carbon atoms (chain diagram of glucose, left). When dissolved in water, however, the chain bends upon itself to form a ring (center). The ring forms are usually drawn on paper as if you were looking at the edge of the ring (right). The thick edge projects out of the paper toward you, the thin edge recedes behind the paper, and the –OH, and –CH₂OH groups are perpendicular to the ring, in the plane of the paper. For convenience, carbon atoms at the corners of the polygons are omitted, but other atoms (such as oxygen) are usually shown. **(b)** Fructose and galactose have the same atomic composition as glucose but a different structure. **(c)** Ribose and deoxyribose are five-carbon monosaccharides that form parts of nucleic acids.

chemical formula $(CH_2O)_6$, or $C_6H_{12}O_6$. Many organisms also synthesize other monosaccharides that have the same chemical formula as glucose, but have slightly different structures (Fig. 3-2b). These include fructose (the "corn sugar" found in corn syrup) and galactose (part of lactose, or "milk sugar"). Some other common monosaccharides, such as ribose and deoxyribose, have five carbons (Fig. 3-2c). Ribose and deoxyribose are parts of the genetic molecules ribonucleic acid (RNA) and deoxyribonucleic acid (DNA), respectively.

Disaccharides Consist of Two Single Sugars Linked Together by Dehydration Synthesis

Monosaccharides, especially glucose and its relatives, have a short life span in a cell. Most are either broken down to free their chemical energy for use in driving needed cellular reactions or are linked together by dehydration synthesis to form disaccharides or polysaccharides (Fig. 3-3). Disaccharides are often used for short-term energy storage, especially in plants. Common disaccharides include **sucrose** (table sugar: glucose plus fructose), **lactose** (milk sugar: glucose plus galactose), and **maltose** (glucose plus glucose, formed during the digestion of starch). When energy is required, the disaccharides are broken apart again into their monosaccharide subunits by hydrolysis.

Polysaccharides Are Chains of Single Sugars

For long-term energy storage, monosaccharides, usually glucose, are joined together into polysaccharides, forming **starch** (in plants; Fig. 3-4) or **glycogen** (in animals). Starch, commonly found in roots and seeds, may occur as coiled, unbranched chains of up to 1000 glucose subunits, or, more frequently, as huge branched chains of up to half a million

glucose monomers. Glycogen, stored in the liver and muscles of animals like ourselves, is usually much smaller than starch and has branches every 10 to 12 glucose subunits. Having many small branches probably makes it easier to split off the glucose subunits for quick energy release.

Many organisms use polysaccharides as structural materials. One of the most important structural polysaccharides is **cellulose**, which makes up a large part of the cell walls of plants and about half the bulk of a tree trunk (Fig. 3-5). When you picture the vast fields and forests blanketing much of our planet, you will not be surprised to learn that there is probably more cellulose on Earth than all other organic molecules put together. Ecologists estimate that about a *trillion tons of cellulose* is synthesized each year!

Like starch, cellulose consists of glucose subunits bonded together; however, whereas most animals can easily digest starch, only a few microbes, such as those in the digestive tracts of cows and termites, can digest cellulose. Why is this the case, given that starch and cellulose both consist of glucose? In cellulose the orientation of the bonds between subunits is different, so that every other glucose is "upside down" (compare Fig. 3-5 with Fig. 3-4). This difference in orientation prevents the digestive enzymes of animals from attacking the bonds between glucose subunits. Wholly different enzymes synthesized by certain microbes, though, can break these bonds. As a result, cellulose is food for these microbes, but for most animals cellulose is *roughage* or *fiber*, material that passes undigested through the digestive tract.

Polysaccharides are also the starting point for the synthesis of many other important molecules. The hard outer coverings (exoskeletons) of insects, crabs, and spiders are made of **chitin**, a polysaccharide in which the glucose subunits have been chemically modified by the addition of a nitrogen-

Figure 3-3 Synthesis and breakdown of a disaccharide

(a) The disaccharide sucrose is synthesized by a dehydration synthesis reaction in which a hydrogen (–H) is removed from glucose and a hydroxyl group (–OH) is removed from fructose, forming a water molecule and leaving the two monosaccharide rings joined by single bonds to the remaining oxygen atom. **(b)** Hydrolysis of sucrose is just the reverse of its synthesis, as water is split and added back to the monosaccharides.

masses of
starch globules

100 micrometers

Figure 3-4   **Plants and animals store energy in polysaccharides made of glucose subunits**

(a) Most plants synthesize starch, which forms water-insoluble globules consisting of many starch molecules, such as these that make up most of the bulk of a potato. Here starch globules are seen inside individual potato cells. **(b)** Starch commonly occurs as branched chains of up to half a million glucose subunits. Here we see a small portion of a single starch molecule. **(c)** The precise structure of the blue highlighted portion of the starch molecule is shown here. Note the linkage between the individual glucose subunits for comparison with cellulose, shown in Figure 3-5.

containing group (Fig. 3-6). Interestingly, chitin also stiffens the cell walls of many fungi. Bacterial cell walls contain still other types of modified polysaccharides, as do the lubricating fluids in our joints and the transparent corneas of our eyes.

Many other molecules are partly carbohydrate. Perhaps the most important of these molecules are the nucleic acids (discussed later), the carriers of hereditary information in all organisms. Other examples include mucus, some hormones, and many molecules in the cell membrane, including "identification molecules," such as those on red blood cells that determine blood type.

Lipids Are Diverse Molecules That Are Insoluble in Water

Lipids are a diverse assortment of molecules, all of which share two important features. First, lipids contain large regions composed almost entirely of hydrogen and carbon, with nonpolar carbon-carbon or carbon-hydrogen bonds. Second, these nonpolar regions make lipids hydrophobic and insoluble in water. The various types of lipids serve a wide variety of functions. Some lipids are energy-storage molecules; some form waterproof coverings on both plant and animal bodies; some make up the bulk of all of the membranes of a cell; and still others are hormones.

Lipids are classified into three major groups: (1) oils, fats, and waxes, which are similar in structure and contain only carbon, hydrogen, and oxygen; (2) phospholipids, structurally similar to oils but also containing phosphorus and nitrogen; and (3) the fused-ring family of steroids.

Oils, Fats, and Waxes Are Lipids Containing Only Carbon, Hydrogen, and Oxygen

Oils, fats, and waxes are related in three ways: (1) They contain only carbon, hydrogen, and oxygen; (2) they contain one or more **fatty acid** subunits (long chains of carbon and hydrogen with a carboxyl group [–COOH] at one end; Fig. 3-7); and (3) they usually do not have ring structures. **Fats** and **oils** are formed by dehydration synthesis from one molecule of **glycerol** (a short, three-carbon molecule with one hydroxyl group [–OH] per carbon) and three fatty acid subunits. This structure of three fatty acids joined to one glycerol molecule gives fats and oils their chemical name, **triglyceride**. Fats and oils have a high concentration of chemical energy, about 9.3 Calories per gram, compared with 4.1 for sugars and proteins (a Calorie with a capital C equals 1000 calories; the Calorie is used in measuring the energy content of foods). Fats and oils are used for semipermanent energy storage—

1 micrometer   1 micrometer

hydrogen bonds
cross-linking
cellulose molecules

CH₂OH

individual
cellulose molecules

microfibril   cellulose fiber

Figure 3-5   Cellulose structure and function

Cellulose, like starch, is composed of glucose subunits, but the orientation of the bond between subunits in cellulose is different (compare with Fig. 3-4c) so that every other glucose molecule is "upside down." Unlike starch, cellulose has great structural strength, due partly to the difference in bonding and partly to the arrangement of parallel molecules of cellulose into long, cross-linked fibers. Plant cells often lay down cellulose fibers in layers that run at angles to each other, resulting in resistance to tearing in both directions. The final product can be incredibly tough, as this 3000-year-old bristlecone pine in California's White Mountains testifies.

Figure 3-6   Chitin: a unique polysaccharide

Chitin has the same "alternating upside down" bonding of glucose molecules as cellulose; the difference is that the glucose subunits are modified by replacement of one of the hydroxyl groups with a nitrogen-containing functional group (yellow). Tough, slightly flexible chitin supports the otherwise soft bodies of arthropods (insects, spiders, and their relatives) and fungi.

Figure 3-7   **Triglycerides**

Triglycerides (fats and oils) are synthesized by a dehydration reaction linking three fatty acids to a single glycerol molecule. A double bond in the topmost fatty acid creates a kink in the chain.

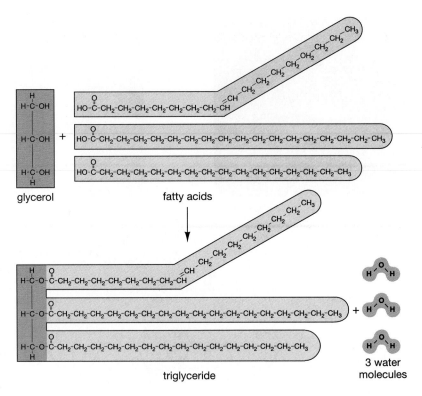

for example, bears that feast during summer and fall put on fat to tide them over during their winter hibernation.

The difference between a fat (which is a solid at room temperature) and an oil (liquid at room temperature) lies in their fatty acids. Fats have fatty acids with all single bonds in their carbon chains. Hydrogens occupy all the other bond positions on the carbons. The resulting fatty acid (for example, stearic acid; Fig. 3-8a) is called **saturated** because it is "saturated" with hydrogens—that is, it has as many hydrogens as possible. If there are double bonds between some of the carbons, and consequently fewer hydrogens, the fatty acid is called **unsaturated** (oleic acid; Fig. 3-8b). Oils have mostly unsaturated fatty acids. The saturated fatty acids of fats can nestle closely together, forming solid lumps at room temperature (beef fat; Fig. 3-8a). The double bonds in the unsaturated fatty acids of oils, on the other hand, form kinks in the fatty acid chains (linseed oil; Fig. 3-8b). The kinks keep oil molecules apart, with the result that an oil is liquid at room temperature. An oil can be converted to a fat by breaking the double bonds between carbons, replacing them with single bonds, and adding hydrogens to the remaining bond positions. This is the "hydrogenated oil" listed in the ingredients on a box of margarine.

Waxes are similar to fats and oils except that the fatty acids are linked to large, long-chain alcohols rather than to glycerol. Waxes form a waterproof coating over the leaves and stems of land plants. Animals also synthesize waxes, as waterproofing for mammalian fur and insect exoskeletons, and, in a few cases, to build elaborate structures such as beehives.

Phospholipids Have Water-Soluble "Heads" and Water-Insoluble "Tails"

The cell membrane that separates the inside of a cell from the outside world contains several types of **phospholipids**. A phospholipid is similar to an oil, except that one of the fatty acids is replaced by a phosphate group with a short, polar, often nitrogen-containing group attached to the end (Fig. 3-9). Unlike the fatty acid "tails," which are insoluble in water, the phosphate-nitrogen "head" is polar, or charged, and is water soluble. Thus a phospholipid has two dissimilar ends: a hydrophilic head attached to hydrophobic tails. As you will see in Chapter 6, this dual nature of phospholipids is crucial to the structure and function of the cell membrane.

Steroids Consist of Four Carbon Rings Fused Together

Steroids are structurally different from all the other lipids. All steroids are composed of four rings of carbon fused together, with various functional groups protruding from them (Fig. 3-10). Steroids include cholesterol, which is a vital component of the membranes of most eukaryotic cells and which is used by cells to synthesize other steroids. These other steroids include male and female sex hormones, hormones that regulate salt levels, hormones that assist in fat digestion, and insect hormones that stimulate the shedding of exoskeletons.

Proteins Play Many Roles in Living Organisms

Proteins are molecules composed of one or more chains of amino acids (see below). Proteins perform many functions. Protein **enzymes** guide almost all the chemical reactions that occur inside cells (see Chapter 4). Because each enzyme assists only one or a few specific reactions, cells usually contain hundreds of different enzymes. Other proteins are used for structural purposes, such as elastin, which gives skin its elasticity; keratin, the principal protein of hair, horns, and claws; and the silk of spider webs and silk moth cocoons (Fig. 3-11). Still other types of proteins are used for energy and material storage (albumin in eggs, casein in milk), transport (hemoglobin to carry oxygen in the blood), and cell movement (contractile proteins in muscle). Some hormones (insulin, growth hormone), antibodies that help to fight disease and infection, and many poisons (rattlesnake venom) are also proteins.

(a)   Beef fat

Figure 3-8   **Fats and oils differ in the saturation of their fatty acids**

(a) A saturated fatty acid, such as stearic acid, has all single bonds between carbons. In beef fat, which is composed of triglycerides with a high proportion of stearic acids, the fatty acid chains pack closely together, forming a solid at room temperature. **(b)** An unsaturated fatty acid, such as oleic acid, has one or more double bonds between carbons, which cause kinks in the chain. These kinks in the fatty acids that make up linseed oil prevent close packing; as a result, the oil is liquid at room temperature.

(b)

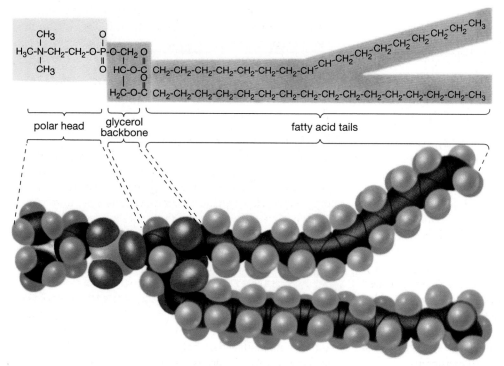

polar head   glycerol backbone   fatty acid tails

Figure 3-9   **Phospholipids**

Phospholipids are similar to fats or oils, except that only two fatty acid tails are attached to the glycerol backbone. The third position on the glycerol is occupied by a polar head composed of a phosphate group ($-PO_4^-$) to which is attached a second, often nitrogen-containing group. The phosphate group is negatively charged, and the nitrogen-containing group is positively charged.

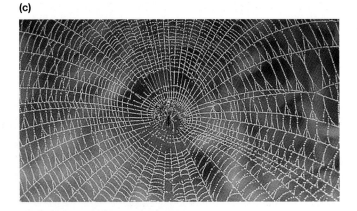

Figure 3-10   Steroids

Steroids are synthesized from cholesterol, and all have almost the same molecular structure (colored rings; note that the carbon atoms at the corners of the rings and the hydrogen atoms attached to the rings have been omitted from these drawings). Great differences in steroid function are a result of the different functional groups attached to the rings. The male sex hormone testosterone, the female sex hormone estradiol (a type of estrogen), and the insect hormone ecdysone, which causes exoskeleton shedding, share a common cholesterol base but differ in functional group attachments.

Proteins Are Formed from Chains of Amino Acids

Proteins are polymers of **amino acids** (Fig. 3-12). All amino acids have the same fundamental structure, consisting of a central carbon bonded to four different functional groups: a nitrogen-containing amino group ($-NH_2$); a carboxyl group ($-COOH$); a hydrogen; and a variable group (represented by the letter R):

Figure 3-11   Structural proteins

Common structural proteins include those of **(a)** hair, **(b)** horn, and **(c)** spider web silk.

Figure 3-12   Amino acid diversity

The diversity of amino acid structures is a consequence of differences in the variable R group (colored blue). Some amino acids are classified according to the variable R group as hydrophilic, others as hydrophobic. Cysteine stands in a class by itself. Two cysteines in distant parts of a protein molecule can form a covalent bond between their sulfur atoms, making a disulfide bridge that brings the cysteines very close together and bends the protein chain (see Fig. 3-15).

The R group differs among amino acids and gives each its distinctive properties (see Fig. 3-12).

The 20 amino acids commonly found in the proteins of living organisms have different properties based on the nature of the R group. Some amino acids are hydrophilic; their R groups are polar and soluble in water. Others are hydrophobic, with nonpolar R groups that are insoluble in water. Another type of amino acid, cysteine, has sulfur in its R group and can form bonds with other cysteines that can link protein chains together. These bonds between the R groups of cysteine are called **disulfide bridges**.

Twenty amino acids may seem like a small number of subunits from which to construct thousands of different proteins, but perhaps an analogy with language will help you understand protein structure and function more easily. The English language uses 26 letters to construct thousands of words, each with a distinct meaning based on the exact sequence of letters. But only a small fraction of all possible letter combinations are used as words; for example, the combination "protein" has meaning, but "nteiopr" does not. Similarly, living organisms construct thousands of different proteins—about 50,000 in your body—from an "alphabet" of 20 amino acids.

Amino acids differ in their chemical and physical properties—size, water solubility, electrical charge—because of their different R groups. Therefore, the exact sequence of amino acids dictates the function of each protein: whether it is water soluble or not, whether it is an enzyme or a hormone or a structural protein. Scrambled sequences of amino acids are useless. In some cases, just one wrong amino acid may cause a protein to function incorrectly.

Amino Acids Are Joined to Form Chains by Dehydration Synthesis

Like lipids and polysaccharides, proteins are formed by dehydration synthesis. The nitrogen of the amino group ($-NH_2$) of one amino acid is joined to the carbon of the carboxyl group ($-COOH$) of another amino acid by a single covalent bond (Fig. 3-13). This bond is called a **peptide bond**, and the resulting chain of two amino acids is called a **peptide**. More amino acids are added, one by one, until the protein is complete. Protein chains found in living cells vary in length from three to thousands of amino acids.

Figure 3-13   Protein synthesis

In protein synthesis, a dehydration synthesis joins the carbon of the carboxyl acid group of one amino acid to the nitrogen of the amino group of a second amino acid. The resulting covalent bond is called a peptide bond.

Often, the word *protein* is reserved for long peptides, say 50 or more amino acids in length. For simplicity, we will use both peptides and proteins to refer to amino acid chains.

Protein May Have Up to Four Levels of Structure

The phrase "amino acid chains" may evoke images of proteins as floppy, featureless structures, but this is not correct. Proteins are, instead, highly organized molecules that come in a variety of shapes. Biologists recognize four levels of organization in protein structure (Fig. 3-14). The **primary structure** is the sequence of amino acids that make up the protein (Fig. 3-15). This sequence is coded by the genes. Different types of proteins (for example, insulin versus growth hormone) have different sequences of amino acids. However, all molecules of the same type of protein, such as human insulin, have the same primary structure.

secondary structure
(helix)

tertiary structure
(folded individual peptide)

primary structure
(amino acid sequence)

quaternary structure
(aggregration of two or more peptides)

Figure 3-14   The four levels of protein structure

Levels of protein structure are represented here by hemoglobin, the oxygen-carrying protein in red blood cells. All levels of protein structure are determined by the amino acid sequence of the protein, interactions among the R groups of the amino acids (primarily hydrogen bonds and disulfide bridges between cysteines), and interactions between the R groups and their surroundings (usually water or lipids).

Figure 3-15   The primary structure of insulin

The insulin molecule actually consists of two small amino acid chains, whose sequences represent insulin's primary structure. Cysteines form disulfide bridges within and between the chains.

Hydrogen bonds cause many protein chains to form one of two simple, repeating **secondary structures**. Looking back at Figure 3-13, notice that every amino acid subunit in a peptide retains a –C=O from its carboxyl group and an –N–H from its amino group. Because oxygen attracts electrons more strongly than does carbon, the oxy-

gen is relatively negative. Similarly, nitrogen attracts electrons more strongly than does hydrogen, leaving the hydrogen relatively positive. Therefore, hydrogen bonds can form between the oxygen of the –C=O and the hydrogen of the –N–H groups of a protein chain. Many proteins, such as the hair protein keratin, have a coiled, springlike shape called a **helix** in which hydrogen bonds hold the turns of the coils together (Fig. 3-16a). Some proteins, such as silk, consist of many protein chains lying side by side, with hydrogen bonds holding adjacent chains in a **pleated sheet** arrangement (Fig. 3-16b).

It is rare for an entire protein to be a simple helix or pleated sheet. Most proteins assume complex, three-dimensional **tertiary structures** (see Fig. 3-14). Disulfide bridges formed between cysteine amino acids may bring otherwise distant parts of a protein close together. Internal stresses, owing to the size and properties of the functional groups on the particular amino acids present, also may contort a protein. For example, amino acids with very large R groups (such as phenylalanine; see Fig. 3-12) are too bulky to fit side by side in a simple helix. As a result, the helix bends.

Probably the most important influence on the tertiary structure of a protein is its cellular environment, specifically whether the protein is dissolved in the water of the cytoplasm, in the lipids of the membranes, or half in one and half in the other. Hydrophilic amino acids can form hydrogen bonds with nearby water molecules, whereas hydrophobic amino acids cannot. Therefore, a protein dissolved in water folds into an irregular glob, with its hydrophilic amino acids

Figure 3-16   Secondary structure of proteins: the helix and the pleated sheet

(a) In a helix, hydrogen bonds (dots) form between the oxygen of the carboxyl acid group of one amino acid and the hydrogen of the amino group of the third amino acid "up" the helix. These hydrogen bonds hold the protein chain in a spiral, in which about 3 ½ amino acids make up each turn of the helix. The variable R groups (green) project outward from the helix. **(b)** In a pleated sheet, several peptide chains lie side by side (running zigzag *across* the sheets left to right). Hydrogen bonds between peptides (lengthwise *along* the sheets top to bottom) hold each sheet together. The R groups (green) project alternately above and below the sheet. Despite its accordion-like appearance, each peptide chain is in its fully extended state and cannot easily be stretched further. For this reason, pleated sheet proteins such as silk are strong but not elastic.

facing the outside watery environment and its hydrophobic ones clustered in the center of the molecule.

Peptides may sometimes join into aggregations of two or more peptides to form the fourth level of protein organization, called **quaternary structure**. This fourth-level organization is maintained by the same types of noncovalent interactions, such as hydrogen bonds, that produce secondary and tertiary structures. A single molecule of hemoglobin, the oxygen-carrying pigment in red blood cells, exhibits all four structural levels. Hemoglobin consists of two pairs of very similar peptides, held together by hydrogen bonds (see Fig. 3-14). Each peptide holds an iron-containing organic molecule called a heme that can bind one molecule of oxygen. Although no one knows precisely why hemoglobin contains four peptides, it is known that a single peptide binds oxygen more tightly than the four-peptide molecule does. This characteristic would be great for picking up oxygen in the lungs, but, because the single peptides would not release the oxygen again under normal conditions, the rest of the body's cells would die from lack of oxygen. Interactions among the four peptides allow hemoglobin to bind oxygen tightly enough to acquire oxygen in the lungs but loosely enough to give it up again to the body tissues.

The Functions of Proteins Are Intimately Linked to Their Three-Dimensional Structure

The exact type, position, and number of different R groups of the amino acids of a protein determine both the structure of the protein and its biological function. In any given protein, however, some R groups are more important than others. In hemoglobin, for example, certain amino acids must be present in precisely the right places to hold the iron-containing heme group that binds oxygen. Many of the other amino acids are interchangeable to some extent, if they are functionally equivalent. For example, the amino acids on the outside of a hemoglobin molecule mostly serve to keep it dissolved in the cytoplasm of a red blood cell. Therefore, as long as they are hydrophilic, it doesn't matter too much exactly which amino acids are where. As we will see in Chapter 15, however, replacing a hydrophilic amino acid with a hydrophobic amino acid can have catastrophic effects on the solubility of the hemoglobin molecule; in fact such a substitution is the molecular cause of a painful and sometimes life-threatening disorder called sickle-cell anemia.

Nucleic Acids Form DNA and RNA, the Molecules of Heredity

As you will learn in later chapters, the amino acid sequence of every protein in your body is specified by the genetic instructions residing in the nuclei of your cells. In fact, that's precisely what most genes are: a set of instructions spelling out the amino acid sequences of your body's proteins. Genes are composed of the fourth major category of biological molecule, **nucleic acids**. Nucleic acids are long chains of similar but not identical subunits called **nucleotides**. All nucleotides have a three-part structure: a five-carbon sugar (ribose or deoxyribose), a phosphate group, and a nitrogen-containing base that differs among nucleotides:

Deoxyribose nucleotide

There are two types of nucleotides, the ribose nucleotides (containing ribose sugar) and the deoxyribose nucleotides (containing deoxyribose sugar). Four different nitrogen-containing bases—adenine, guanine, cytosine, and thymine—are found bonded to deoxyribose nucleotides. Similarly, there are four types of ribose nucleotides, with adenine, guanine, cytosine, or uracil as the base (see Chapters 10 and 11).

Nucleotides may be strung together in long chains, with the phosphate of one nucleotide covalently bonded to the sugar of another:

Nucleotide chain

Deoxyribose nucleotides form chains millions of units long called **deoxyribonucleic acid**, or **DNA**. DNA is found in the chromosomes of all living things, and its sequence of nucleotides, like the dots and dashes of a biological Morse

Figure 3-17   **A sampling of the diversity of nucleotides**

Individual nucleotides, constructed of a sugar, a phosphate group, and a nitrogen-containing base, are often modified by the addition of different functional groups and serve a variety of cellular functions.

code, spells out the genetic information needed to construct the proteins of each organism (see Chapter 10). Chains of ribose nucleotides, called **ribonucleic acid,** or **RNA,** are copied from the central repository of DNA in the nucleus of each cell. RNA carries DNA's genetic code into the cytoplasm and directs the synthesis of proteins (see Chapter 11).

Other Nucleotides Act as Intracellular Messengers, Energy Carriers, and Coenzymes

Not all nucleotides are part of DNA or RNA molecules. Some exist singly in the cell or occur as parts of other molecules. **Cyclic nucleotides,** such as cyclic adenosine monophosphate (cyclic AMP; Fig. 3-17a) are intracellular messengers that carry information from the cell membrane to other molecules in the cell. Cyclic AMP

is synthesized when certain hormones come into contact with the cell membrane. Cyclic AMP then stimulates essential reactions in the cytoplasm or nucleus .

Some nucleotides have extra phosphate groups. These diphosphate and triphosphate nucleotides, such as **adenosine triphosphate** (ATP; Fig. 3-17b), are unstable molecules that carry energy from place to place within a cell. They capture energy where it is produced (during photosynthesis, for example) and give it up to drive energy-demanding reactions elsewhere (say, to synthesize a protein). Finally, certain nucleotides assist enzymes in their role of promoting and guiding chemical reactions. These nucleotides are called **coenzymes** and usually consist of a nucleotide combined with a vitamin (Fig. 3-17c). You will learn more about energy-carrier nucleotides, enzymes, and coenzymes in Chapter 4, where we discuss energy production and use in the cell.

❈ S U M M A R Y   O F   K E Y   C O N C E P T S

Synthesizing Organic Molecules: A Modular Approach
Most large biological molecules are polymers synthesized by linking together many smaller subunit monomers. Chains of subunits are connected by covalent bonds through dehydration synthesis; the chains may be broken apart again by hydrolysis reactions.

The Principal Types of Biological Molecules
The most important organic molecules fall into one of four classes: carbohydrates, lipids, proteins, and nucleic

acids. Their major characteristics are summarized in Table 3-2.

Carbohydrates include sugars, starches, chitin, and cellulose. Sugars (monosaccharides and disaccharides) are used for temporary storage of energy and for the construction of other molecules. Starches and glycogen are polysaccharides that serve for longer-term energy storage in plants and animals, respectively. Cellulose and related polysaccharides form cell walls of bacteria, fungi, plants, and some microorganisms.

Lipids are nonpolar, water-insoluble molecules of diverse chemical structure that include oils, fats, waxes, phospholipids, and steroids. Lipids are used for energy storage (fats and oils), as waterproofing for the outside of many plants and animals (waxes), as the principal component of cellular membranes (phospholipids), and as hormones (steroids).

Proteins are chains of amino acids. The structure and function of a protein are determined by the sequence of amino acids in the chain. Proteins may be enzymes (which guide chemical reactions), structural molecules (hair, horn), hormones (insulin), or transport molecules (hemoglobin).

Nucleic acid molecules are chains of nucleotides. Each nucleotide is composed of a phosphate group, a sugar group, and a nitrogen-containing base. The two types of nucleic acids are deoxyribonucleic acid (DNA) and ribonucleic acid (RNA). Other nucleotides include intracellular messengers (cyclic AMP), energy-carrier molecules (ATP), and coenzymes.

✖ KEY TERMS

adenosine triphosphate   p. 51
amino acid   p. 46
carbohydrate   p. 40
cellulose   p. 41
chitin   p. 41
coenzyme   p. 51
cyclic nucleotide   p. 51
dehydration synthesis   p. 38
deoxyribonucleic acid (DNA)   p. 50
disaccharide   p. 40
disulfide bridge   p. 47
enzyme   p. 44
fat   p. 42
fatty acid   p. 42
functional group   p. 37
glucose   p. 40
glycerol   p. 42

glycogen   p. 41
helix   p. 49
hydrolysis   p. 38
lactose   p. 41
lipid   p. 42
maltose   p. 41
monomer   p. 38
monosaccharide   p. 40
nucleic acid   p. 50
nucleotide   p. 50
oil   p. 42
peptide   p. 47
peptide bond   p. 47
phospholipid   p. 44
pleated sheet   p. 49
polymer   p. 38
polysaccharide   p. 40

primary structure   p. 48
protein   p. 44
quaternary structure   p. 50
ribonucleic acid (RNA)   p. 51
saturated   p. 44
secondary structure   p. 49
starch   p. 41
steroid   p. 44
subunit   p. 38
sucrose   p. 41
sugar   p. 40
tertiary structure   p. 49
triglyceride   p. 42
unsaturated   p. 44
wax   p. 44

✖ THINKING THROUGH THE CONCEPTS

Multiple Choice

1. What is the main function of polysaccharides in living cells?
   a. energy storage
   b. storage of hereditary information
   c. catalytic activity
   d. formation of membranes
   e. only b and c

2. Characteristics of carbon that contribute to its ability to form an immense diversity of organic molecules include its
   a. tendency to form covalent bonds
   b. ability to bond with up to four other atoms
   c. capacity to form single and double bonds
   d. ability to bond together to form extensive, branched or unbranched "carbon skeletons"
   e. all of the above

3. Foods that are high in fiber are most likely to be derived from
   a. plants   b. dairy products
   c. meat   d. fish
   e. all of the above

4. Proteins differ from one another because
   a. the peptide bonds linking amino acids differ from protein to protein
   b. the sequence of amino acids in the polypeptide chain differs from protein to protein
   c. each protein molecule contains its own unique sequence of sugar molecules
   d. the number of nucleotides found in each protein varies from molecule to molecule
   e. the number of nitrogen atoms in each amino acid differs from the number in all others

5. Which, if any, of the following choices does not properly pair an organic compound with one of its building blocks (subunits)?
   a. polysaccharide–monosaccharide
   b. fat–fatty acid
   c. nucleic acid–glycerol
   d. protein–amino acid
   e. all are paired correctly

6. Which one of the following statements is NOT true with regard to glycogen, starch, and cellulose? All are
   a. complex carbohydrates
   b. composed solely of glucose monomers
   c. easily digested by animals
   d. polysaccharides
   e. found in plants, except glycogen, which is found in animals

Review Questions

1. Which elements are commonly found in biological molecules?

2. List the four principal types of biological molecules and give an example of each.

3. What roles do nucleotides play in living organisms?

4. One way to convert corn oil to margarine (solid at room temperature) is to add hydrogen atoms, decreasing the number of double bonds in the molecules of oil. What is this process called? Why does it work?

5. Describe and compare dehydration synthesis and hydrolysis. Give an example of a substance formed using each chemical reaction, and describe the specific reaction in each instance.

6. Distinguish among the following: monosaccharide, disaccharide, and polysaccharide. Give two examples of each, and their functions.

7. Describe the synthesis of a protein from amino acids, then describe primary, secondary, tertiary, and quaternary structures of a protein.

8. Most structurally supportive materials in plants and animals are polymers of special sorts. Where would one find cellulose? Chitin? In what way(s) are these two similar? Different?

9. What kinds of bonds or bridges between keratin molecules are altered when hair is (a) wet and allowed to dry on curlers; (b) given a permanent wave?

✖ APPLYING THE CONCEPTS

1. A preview question for Chapter 6: In Chapter 2 you learned that hydrophobic molecules tend to cluster when immersed in water. In this chapter, you discovered that a phospholipid has a hydrophilic head and hydrophobic tails. What do you think would be the configuration of phospholipids that are immersed in water?

2. Many birds must store large amounts of energy to power flight during migration. Which type of organic molecule would be the most advantageous for energy storage? Why?

3. Remember the nuclear incident at Chernobyl? A scientist suspects that the food in a nearby ecosystem may have been contaminated with radioactive nitrogen over a period of months. Which substances in plants and animals could be examined for radioactivity to test his hypothesis?

✖ FOR MORE INFORMATION

Atkins, P. W. *Molecules*. New York: W. H. Freeman and Co., 1987. A fascinating, nontechnical introduction to the beauties of chemistry.

Doolittle, R. F. "Proteins." *Scientific American*, October 1985. Proteins are the molecular tools that, directly or indirectly, carry out almost all the functions of a cell.

Dushesne, L. C., and Larson, D. W. "Cellulose and the Evolution of Plant Life." *BioScience*, April 1989. Chemical and natural history aspects of the most plentiful organic molecule on Earth.

Goodsell, D. S. *The Machinery of Life*. New York: Springer, 1993. Goodsell depicts the molecules of a cell in all of their three-dimensional, interactive glory. A great way to gain a feel for the beauty and intricacy of the organic molecules of life.

Hill, J. W. *Chemistry for Changing Times*. 7th ed. Englewood Cliffs, NJ: Prentice Hall, 1995. A chemistry textbook that is both clearly readable and thoroughly enjoyable, for nonscience majors.

Sharon, N., and Lis, H. "Carbohydrates in Cell Recognition." *Scientific American*, January 1993. Sugar-protein complexes on the surfaces of cells regulate cell identification and interaction between cells. See also Sharon's earlier article, "Carbohydrates," *Scientific American*, November 1980.

NET WATCH

On-line resources for this chapter are on the World Wide Web at:
http://www.prenhall.com/~audesirk (click on the table of contents link and then select Chapter 3).

4 Energy Flow in the Life of a Cell

"Disorder spreads through the universe, and life alone battles against it."

G. Evelyn Hutchinson

A chameleon consumes a grasshopper in Kenya. Both predator and prey convert the chemical energy of their food molecules into the kinetic energy of movement.

Among the fundamental characteristics of all living organisms is the ability to guide chemical reactions within their bodies along certain pathways. The chemical reactions serve many functions, depending on the nature of the organism: to synthesize the molecules that make up the organism's body, to reproduce, to move, even to think. Chemical reactions either require or release **energy**, which can be defined simply as *the capacity to do work*, including synthesizing molecules, moving things around, and generating heat and light. In this chapter we discuss the physical laws that govern energy flow in the universe, how energy flow in turn governs chemical reactions, and how the chemical reactions within living cells are controlled by the molecules of the cell itself. Chapters 7 and 8 focus on photosynthesis, the chief "port of entry" for energy into the biosphere, and glycolysis and cellular respiration, the most important sequences of chemical reactions that release energy.

Energy and the Ability to Do Work

As you learned in Chapter 2, there are two types of energy: **kinetic energy** and **potential energy**. Both types of energy may exist in many different forms. Kinetic energy, or *energy of movement*, includes light (movement of photons), heat (movement of molecules), electricity (movement of electrically charged particles), and movement of large objects. Potential energy, or *stored energy*, includes chemical energy stored in the bonds that hold atoms together in molecules, electrical energy stored in a battery, and positional energy stored in a diver poised to spring (Fig. 4-1). Under the right conditions, kinetic energy can be transformed into potential energy, and vice versa. For example, the diver converted kinetic energy of movement into potential energy of position when she climbed the ladder up to the platform; when she jumps off, the potential energy will be converted back into kinetic energy.

To understand how energy flow governs interactions among pieces of matter, we need to know two things: (1) the quantity of available energy and (2) the usefulness of

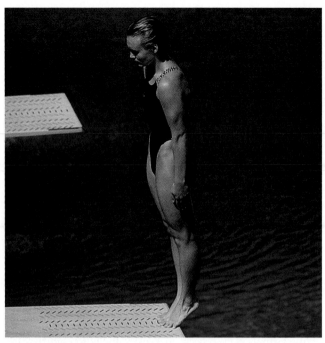

***Figure 4-1*   Potential energy**

The body of a diver perched atop the platform has potential energy, because of the different heights of the platform and the pool. The smacking sound and splashing water are manifestations of the conversion of this potential energy into kinetic energy when she dives into the water.

***Figure 4-2*   Thermodynamics of automobiles**

According to the first law of thermodynamics, the quantity of energy originally present in the gasoline equals the energy in the resulting products: the molecules in the exhaust, the moving car, and heat. The form of the energy changes from chemical energy into heat or energy of movement. According to the second law, some of the energy in the products is in a less concentrated, less usable form than the chemical energy of the gasoline. In cars, about 75% of the chemical energy of the gas is given off as waste heat to the environment, contributing to random movement of molecules in the air.

the energy. These are the subjects of the laws of thermodynamics, which we will now examine.

The Laws of Thermodynamics Describe the Basic Properties of Energy

All interactions among pieces of matter are governed by the two **laws of thermodynamics**, physical principles that define the basic properties and behavior of energy. The laws of thermodynamics deal with "isolated systems," which are any parts of the universe that cannot exchange either matter or energy with any other parts. Probably no part of the universe is completely isolated from all possible exchange with every other part, but the concept of an isolated system is useful in thinking about energy flow.

The First Law of Thermodynamics States That Energy Can Neither Be Created nor Destroyed

The **first law of thermodynamics** states that within any isolated system, energy can neither be created nor destroyed, although it can be changed in form (for example, from chemical energy to heat energy). In other words, within an isolated system *the total quantity of energy remains constant*. The first law is therefore often called the law of conservation of energy. To use a famil-

iar example, let's see how the first law applies to driving your car (Fig. 4-2). We can consider that your car (with a full tank of gas), the road, and the surrounding air roughly constitute an isolated system. When you drive your car, you convert the potential chemical energy of gasoline into kinetic energy of movement and heat energy. The total amount of energy that was in the gasoline before it was burned is the same as the total amount of this kinetic energy and heat.

An important rule of energy conversions is this: Energy always flows "downhill," from places with a high concentration of energy to places with a low concentration of energy. This is the principle behind engines. As we described in Chapter 2, temperature is a measure of how fast molecules move. The burning gasoline in your car's engine consists of molecules moving at extremely high speeds: a high concentration of energy. The cooler air outside the engine consists of molecules moving at much lower speeds: a low concentration of energy. The molecules in the engine hit the piston harder than the air molecules outside the engine do, so the piston moves upward, driving the gears that move the car. Work is done. When the engine is turned off, it cools down as heat is transferred from the warm engine to its cooler surroundings. The molecules on both sides of the piston move at the same speed, so the piston stays still. No work is done.

The Second Law of Thermodynamics States That the Amount of Useful Energy Always Decreases When Energy Is Converted from One Form to Another

The second law is related to the usefulness of energy. The **second law of thermodynamics** states that any change in an isolated system causes the quantity of concentrated, useful energy to decrease. Put another way, the second law states that all spontaneous changes result in a more uniform distribution of energy, reducing the energy differences that are essential for doing work; in other words, energy is always converted from more useful into less useful forms.

Let us continue our example of a car engine. The heat of the burning gasoline moves the piston, which moves the car. But the kinetic energy of the moving car is much less than the chemical energy that was originally contained in the gasoline (see Fig. 4-2). Where is the "missing" energy? Feel the engine and the exhaust: They are both hot. The burning gas not only moved the piston, it also heated up the engine, the exhaust system, and the air around the car. The friction of tires on pavement heated the road up a little, too. So, as the first law dictates, there really isn't any missing energy: The original quantity of energy in the gasoline still exists, only in different forms. However, some of the energy (the heat given off) merely increased the random movement of molecules in the engine block, the road, and the air. Such random movements aren't useful: You can't very well gather up the energy found in the warmed-up road and use it to drive your car a few more blocks.

The second law also tells us something about the organization of matter. Regions of concentrated energy are usually also regions of great orderliness. The eight carbon atoms in a single molecule of gasoline have a much more orderly arrangement than do the carbon atoms of the eight separate, randomly moving molecules of carbon dioxide that are formed when the gasoline burns.

Therefore, we can also phrase the second law in terms of the organization of matter: All processes in an isolated system result in an increase in randomness and disorder. This tendency toward loss of orderliness and high-level energy and an increase in randomness, disorder, and low-level energy is called **entropy**.

Eventually, several billion years from now, all the energy in the universe will be evenly dispersed as heat, and all the matter will be randomly distributed in small molecules. Without differences in energy, no further work will be possible. Life, therefore, will cease. (This rather gloomy prospect is explored in a marvelous science fiction short story by Isaac Asimov entitled "The Last Question": definitely recommended reading.)

Living Things Use the Energy of Sunlight to Create the Low-Entropy Conditions Characteristic of Life

If you stop to think about the second law of thermodynamics, you may wonder how life can exist at all. If chemical reactions, including those inside living cells, cause the amount of low-level, unusable energy to increase, and if matter tends toward increasing randomness and disorder, how can organisms accumulate the concentrated energy and precisely ordered molecules that characterize living things? The answer is this: The second law applies only to isolated systems. Living organisms, however, are not isolated systems. Nuclear reactions in the sun produce concentrated energy (sunlight) along with vast increases in entropy. Living things on Earth use that concentrated energy to synthesize complex molecules and maintain orderly structures. The small packets of low entropy that we call life do not violate the second law—they are achieved at the expense of an enormous loss of useful energy from sunlight. The entropy of the solar system as a whole constantly increases.

Energy Flow in Chemical Reactions

A chemical reaction begins with one set of substances, called the **reactants**, and converts them into another set, the **products**. All chemical reactions fall into one of two categories: exergonic and endergonic. A reaction is **exergonic** (Greek for "energy out") if the reactants have more energy than the products. Consequently, the reaction releases energy:

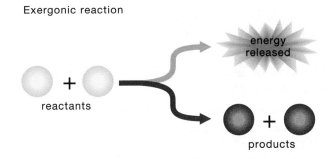

Conversely, a reaction is **endergonic** (Greek for "energy in") if the products have more energy than the reactants. Because, according to the first law of thermodynamics, the extra energy of the products must come from somewhere, energy input from outside the system is required to drive the reaction:

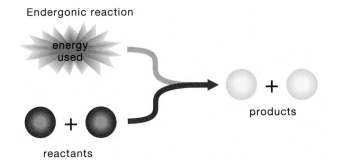

Let's look at two processes that illustrate these types of reactions: burning coal and photosynthesis.

Exergonic Reactions Release Energy

In an exergonic reaction, the reactants contain more energy than the products. For example, when coal (C) is burned it reacts with oxygen (O_2) to produce carbon dioxide (CO_2) and energy as described by this equation:

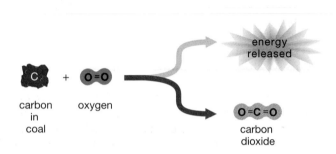

(To simplify things, we have assumed that coal is pure carbon.) This reaction illustrates two important concepts, diagrammed in Figure 4-3a. First, molecules of carbon and oxygen contain much more energy than molecules of carbon dioxide do, so the reaction releases energy. Energy release allows exergonic reactions to occur without a net input of energy. In our example, coal fires, once started, burn by themselves until the coal is consumed: You don't have to keep the fire going by adding outside energy from a flamethrower. (It may be helpful if you think of exergonic reactions as being "downhill," both in the graphical sense, as shown in Figure 4-3a, and in the everyday sense that cars and bicycles can coast downhill without engines or legs propelling them.)

Although burning coal releases energy, a lump of coal doesn't burst into flames by itself. This observation leads to the second point: All chemical reactions require an initial input of energy, called the **activation energy**, to get started. Chemical reactions require activation energy because atoms and molecules are surrounded by a cloud of negatively charged electrons. For two molecules to react with each other, their electron clouds must be forced together, despite their mutual electrical repulsion. Forcing them together requires energy. The usual source of activation energy is kinetic energy of movement. Only molecules moving with sufficient speed collide hard enough to force their electron clouds to mingle and react. Because molecules move faster at higher temperatures, most reactions occur more readily at high temperatures. The heat provided by setting coal on fire causes the carbon atoms to move faster; some atoms move so fast that they vaporize (become gaseous) and are no longer part of the solid lump of coal. These energetic carbon atoms collide with oxygen molecules and react to form carbon dioxide (CO_2), releasing heat. The heat causes more carbon atoms to vaporize and collide with oxygen molecules, producing more CO_2 and releasing more heat, which sustains the reaction until the coal is completely consumed.

Endergonic Reactions Require an Input of Energy

In contrast to what happens when coal is burned, many reactions in living systems result in products that contain more energy than the reactants. These reactions therefore require an input of energy into the low-energy reactants and a net input of energy into the reaction (Fig. 4-3b). As we shall see in Chapter 7, photosynthesis in green plants

Figure 4-3   **Energy relations in exergonic and endergonic reactions**

(a) An exergonic ("downhill") reaction proceeds from high-energy reactants, such as coal, to low-energy products, such as CO_2. The energy difference between the chemical bonds of the reactants and products is released as heat. To start the reaction, however, an initial input of energy—the activation energy—is required. **(b)** An endergonic ("uphill") reaction, such as photosynthesis, proceeds from low-energy reactants such as CO_2 and H_2O to high-energy products, such as glucose, and therefore requires a net input of energy.

takes low-energy water and carbon dioxide and produces high-energy sugar and oxygen:

Photosynthesis requires energy, which plants obtain from sunlight. We might call endergonic reactions "uphill," in the same sense that cars and bicycles need inputs of energy to go uphill.

Coupled Reactions Link Exergonic with Endergonic Reactions

Because endergonic reactions require energy from other sources, they do not violate the rule that energy always flows downhill, from high concentrations of energy to lower concentrations of energy. Photosynthesis is an example of a **coupled reaction**. In a coupled reaction, an exergonic reaction provides the energy needed to drive an endergonic reaction.

In photosynthesis, the exergonic reaction occurs in the sun, and the endergonic reaction occurs in the plant:

Sun: Nuclear fusion of hydrogen → helium + large amount of solar energy released

Plants:

$$6\,CO_2 + 6\,H_2O + \text{small} \rightarrow C_6H_{12}O_6 \text{ (glucose)} + 6\,O_2$$
low-energy   solar   higher-energy
reactants   energy   product
input

The reactions occurring in the sun liberate energy as light. The sunlight captured by a plant possesses much more energy than is needed to drive photosynthesis. Therefore, the overall process, if we include the sun, is exergonic.

The fact that sunlight contains more energy than is needed to drive photosynthesis is an example of a general rule of coupled reactions. According to the second law of thermodynamics, the amount of useful energy decreases during every chemical reaction. This means that not all the energy released by an exergonic reaction can be used to drive an endergonic reaction; some energy is lost to the environment as heat and random movement of molecules. Therefore, in coupled reactions, the exergonic reaction must always release more energy than is required to drive the endergonic reaction (Fig. 4-4a).

Living organisms are the ultimate chemists, constantly using the energy given off by exergonic reactions (such as the chemical breakdown of food) to drive essential en-

dergonic reactions (such as brain activity, movement, or the synthesis of complex molecules; Fig. 4-4b). The exergonic and endergonic halves of coupled reactions often occur in different places, so there must be some way to transfer the energy from the exergonic reaction to the endergonic reaction. In photosynthesis, sunlight carries energy from exergonic reactions in the sun to the endergonic reactions in plants. In coupled reactions occurring within cells, energy is usually transferred from place to place by *energy-carrier* molecules, of which the most common is *adenosine triphosphate*, or *ATP*. We will examine the synthesis and use of ATP later in this chapter.

Chemical Reactions Are Reversible

So far, we have treated chemical reactions as if they always proceed in one direction, "forward" from reactants to products, and then are finished. This is usually not the case. Most chemical reactions are reversible: They may proceed spontaneously in either direction under the appropriate conditions of reactants, products, and energy. Furthermore, all reactions, if left to themselves, eventually reach a steady state called a **chemical equilibrium**, in which the reaction proceeds at an equal rate in both directions. At equilibrium, "forward" and "backward" reactions occur at exactly equal rates, with the result that, although *individual molecules* are constantly undergoing chemical reactions, the *net concentrations of molecules* do not change.

To understand the reversibility of chemical reactions and the nature of chemical equilibria, let's look at the reversible combination of oxygen with hemoglobin in the blood (see Chapter 3 for more about hemoglobin). As blood circulates through your lungs, hemoglobin picks up oxygen and becomes oxyhemoglobin. When the blood passes through the other organs of the body, the oxygen is released. Why?

Blood that enters the lungs has been depleted of oxygen during its journey through the body, so few of its hemoglobin molecules have oxygen bound to them. Therefore, in the high-oxygen environment of the lungs, oxygen binds to hemoglobin, forming oxyhemoglobin:

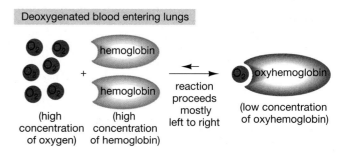

As more and more hemoglobin molecules become oxygenated, a few give up their oxygens. Finally, an equilibrium point is established: Just as many oxygen molecules

(a)

(b)

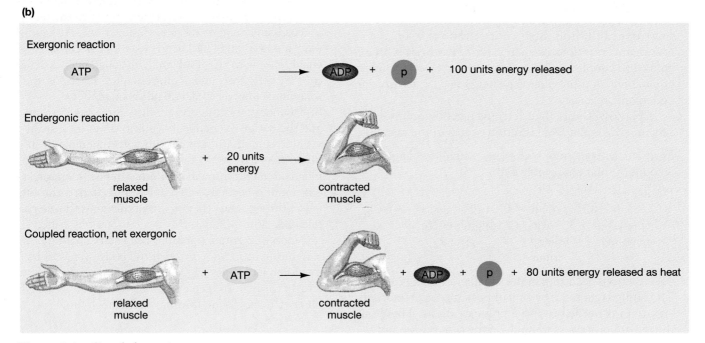

Figure 4-4   **Coupled reactions**

(a) Burning gasoline is an exergonic reaction that can be coupled to the endergonic reaction of moving a car. According to the second law of thermodynamics, the quantity of concentrated, useful energy decreases in the process. Therefore, the exergonic reaction must actually release more energy than the endergonic reaction requires. The excess energy is given off to the environment as heat. **(b)** Muscle movement is an endergonic reaction coupled to the exergonic reaction of ATP breakdown. The energy released by ATP as it is broken down into ADP plus phosphate (P) exceeds the energy put into muscle contraction, and the overall reaction is exergonic.

leave the air and bind to hemoglobin as leave the hemoglobin and reenter the air:

At equilibrium, the reaction proceeds in both directions at the same rate.

Chemical equilibria are affected by the concentrations of the molecules involved: The high concentration of oxygen in the lungs produces a high concentration of oxyhemoglobin. In contrast, body tissues such as exercising muscles use up oxygen, so the tissue concentration of oxygen is very low. When oxygenated blood enters a muscle, the overall reaction reverses direction, with oxyhemoglobin in the blood losing oxygen to the muscle cells:

Because the muscle uses up oxygen about as fast as the hemoglobin releases it, almost all the oxygen leaves the hemoglobin and enters the muscle cells. A new equilibrium is established, this time with most of the hemoglobin in the deoxygenated state:

Eventually the hemoglobin returns to the lungs, and the cycle repeats.

The hemoglobin-oxygen reaction illustrates an important principle of reactions occurring in living organisms. To remain alive, cells are constantly synthesizing some molecules and destroying others. By adding and removing the reactants and products, cells can control and automatically adjust the rates and directions of chemical reactions to meet their needs.

Controlling the Metabolism of Living Cells

Cells are miniature, incredibly complex chemical factories. The **metabolism** of a cell is the sum of all its chemical reactions. Many of these reactions are linked in sequences called **metabolic pathways** (Fig. 4-5). Photosynthesis is such a pathway. Glycolysis, the series of reactions that begin the digestion of glucose (see Chapter 8), is another. Different metabolic pathways may exchange molecules, with the re-

Figure 4-5   **A simplified view of metabolic pathways**

The original reactant molecule undergoes a series of reactions, each catalyzed by a specific enzyme. The product of each reaction serves as the reactant for the next reaction in the pathway. Metabolic pathways are often interconnected so that the product of a step in one pathway may serve as a reactant either for the next reaction in the same pathway or for a reaction in another pathway.

sult that all the reactions and all the molecules of a cell are actually interconnected in a single, enormously complicated metabolic pathway.

The chemical reactions within a cell are governed by the same laws of thermodynamics that control any other reactions. How, then, do orderly metabolic pathways arise? The biochemistry of cells is finely tuned in three ways:

1. Cells regulate chemical reactions through the use of protein catalysts called enzymes.
2. Cells couple reactions together, driving energy-requiring endergonic reactions with the energy released by exergonic reactions.
3. Cells synthesize energy-carrier molecules that capture energy from exergonic reactions and transport it to endergonic reactions.

At Body Temperatures, High Activation Energy Makes Spontaneous Reactions Proceed Too Slowly to Sustain Life

The laws of thermodynamics tell us that energy-releasing reactions can occur spontaneously, but they don't tell us how fast they will occur. In general, the rate of a reaction is determined by its activation energy, which is a measure of the initial amount of energy required to speed up reactant molecules enough to force their electron clouds together (Fig. 4-6). Reactions with low activation energies, such as the binding of oxygen to hemoglobin, can proceed swiftly at fairly low temperatures. Reactions with high activation energies, such as the reaction of coal with oxygen, where lots of energy is required to force the clouds together, will be very slow at low temperatures. Therefore, the rate of most reactions is limited by the speed of reactant molecules—the faster they move, the faster the reaction goes. Most reactions can be accelerated by raising the temperature, thereby increasing the speed of molecules.

Without high temperatures, many "spontaneous," exergonic reactions proceed extremely slowly. The reaction of sugar with oxygen to yield carbon dioxide and water is exergonic, but it has an enormous activation energy. Even quite high temperatures do not provide enough energy to overcome the activation energy barrier. As any cook knows, you can boil a sugar solution for hours and hardly any of the sugar breaks down. At the temperatures found in living organisms, sugar and many other energetic molecules would almost never break down spontaneously and give up their energy. Because of the catalytic enzymes produced by living cells, however, sugar is an important energy source for life on Earth. Let's see how enzymes and other catalysts influence chemical reactions.

Catalysts Reduce Activation Energy

Catalysts are molecules that speed up the rate of a reaction without themselves being used up or permanently altered. Catalysts speed up reactions by reducing the acti-

Figure 4-6 **Activation energy controls the rate of chemical reactions**

High activation energy (black line) means that reactant molecules must collide very forcefully in order to react. Only very fast-moving molecules will collide hard enough to react, so reactions with high activation energies proceed slowly at low temperatures, where most molecules move relatively slowly. Catalysts lower the activation energy of a reaction (red line), so a much higher proportion of molecules move fast enough to react when they collide. Therefore, the reaction proceeds much more rapidly.

vation energy (Fig. 4-6). As an example of catalytic action, let's consider the catalytic converters in car engines. When gasoline is burned completely, the final products are carbon dioxide and water:

$$2\,C_8H_{18}\ (\text{octane}) + 25\,O_2 \rightarrow 16\,CO_2 + 18\,H_2O + \text{energy}$$

However, flaws in the combustion process generate other substances, including poisonous carbon monoxide (CO). Carbon monoxide reacts spontaneously but slowly with oxygen in the air to form carbon dioxide:

$$2\,CO + O_2 \rightarrow 2\,CO_2 + \text{energy}$$

In large cities, so many cars emit so much CO that the spontaneous reaction of CO with O_2 can't keep up, and unhealthy levels of carbon monoxide accumulate. Enter the catalytic converter. The platinum catalysts in the converter hasten the conversion of CO to CO_2, thereby reducing air pollution.

There are four important principles to note about all catalysts:

1. Catalysts speed up reactions.
2. Catalysts cannot cause energetically unfavorable reactions to occur; that is, they can speed up only those reactions that would occur spontaneously anyway, although at a much slower rate.
3. Catalysts do not change the equilibrium point of a reaction. Catalytic converters, for example, cannot reduce the concentration of CO below the naturally oc-

curring minimum level. If fresh mountain air is put through a converter, the concentration of CO in the air coming out is the same as the concentration going in.

4. Catalysts are not consumed in the reactions they promote. No matter how many reactions they accelerate, the catalysts themselves are not permanently changed. Converters in cars seldom have to be replaced unless they become contaminated by other materials.

Enzymes Are Biological Catalysts

Enzymes are catalysts synthesized by living organisms. Almost all enzymes are proteins. Just a few years ago, biology texts stated unequivocally that *all enzymes are proteins*. During the 1980s, however, Thomas Cech and Sidney Altman discovered that certain molecules of ribonucleic acid also function as enzymes. These molecules, dubbed **ribozymes**, catalyze reactions involved in processing genetic information for use by a cell and may have been crucial during the early evolution of life (see Chapter 19). Ribozymes notwithstanding, in the rest of this chapter we will discuss only the actions of protein enzymes.

Enzymes possess the four characteristics of catalysts just described. But enzymes have two additional attributes that set them apart from inorganic catalysts:

1. An enzyme is usually very specific, promoting only a few types of chemical reactions. An inorganic catalyst, in contrast, can usually speed up many different reactions. In most cases, an enzyme catalyzes a single reaction involving one or two specific molecules, while leaving even quite similar molecules untouched. You may recall from Chapter 3, for example, that animals have enzymes that break apart starch molecules but leave cellulose intact, despite the fact that starch and cellulose are both composed of glucose subunits.

2. Enzyme activity is regulated—that is, enhanced or suppressed—often by the very molecules whose reactions they promote.

The Structure of Enzymes Allows Them to Bind Specific Molecules and Catalyze Specific Reactions

Why are enzymes specific, and how are they regulated? Enzyme function is intimately related to enzyme structure. Enzymes are proteins with complex three-dimensional shapes (Fig. 4-7). Each enzyme has a dimple or groove, called the **active site**, into which reactant molecules, called **substrates**, can enter.

The active site of each enzyme has a distinctive shape and distribution of electrical charge that is complementary to its substrate, analogous to the fit between a lock and

(a)

(b)

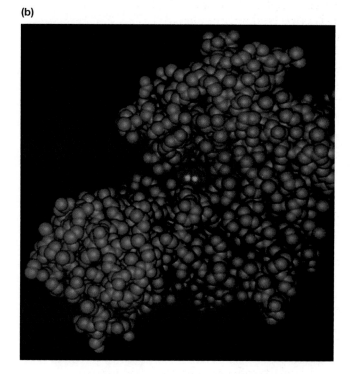

Figure 4-7 **How enzymes interact with substrates**

Computer-generated models of the enzyme hexokinase and its substrate glucose (hexokinase is the enzyme that catalyzes the first step in glucose metabolism). **(a)** The active site of hexokinase is a groove or dimple into which a glucose molecule may enter. **(b)** Entry of glucose into the active site causes both the glucose and hexokinase molecules to change shape (changes in the glucose molecule cannot be seen because it is too small and is buried) and is essential for the enzyme action to take place.

its key. The enzyme allows only a few extremely similar molecules to enter, much as the size and shape of a lock allow only one or a few keys to fit in. However, enzymes and substrates are actually much more flexible than locks and keys. For at least some enzymes, changes in the shape of both the enzyme and substrate are essential for enzyme action to occur (see Fig. 4-8).

The size, shape, and charge of the active site confer great specificity on an enzyme, allowing only certain molecules to enter and react, while rejecting even fairly similar molecules. For example, there are several protein-digesting enzymes in your intestines that break the peptide bonds that join amino acids. But each enzyme is quite specific. No single enzyme can digest every type of protein, because only a protein with the right amino acid sequence can fit into the active site of any individual enzyme. Other proteins, with amino acids that are too large or too small or have the wrong charge, cannot get in and consequently cannot be digested. Complete digestion of all the proteins in the human diet, therefore, requires several enzymes.

Conversely, some molecules may be able to enter the active site of an enzyme but do not have chemical bonds upon which the enzyme can act, so no reaction occurs. Many poisons, including some insecticides, enter the active site of enzymes essential to brain function and never leave again. The enzymes remain plugged up, and the reactions they normally promote cannot occur. Parts of the brain turn off, or become wildly hyperactive, sometimes with fatal consequences.

How does an enzyme promote a reaction? First, the shape and charge of the active site force substrates to enter the enzyme in specific orientations (Fig. 4-8, step 1). Second, when substrates enter the active site, both substrate

and active site change shape (step 2). Certain amino acids that form the active site may make temporary chemical bonds to atoms of the substrates, or electrical interactions between active site and substrates may distort the chemical bonds within the substrates. The combination of substrate selectivity, substrate orientation, temporary chemical bonds, and bond strain promotes the specific chemical reaction catalyzed by a particular enzyme. When the final reaction between the substrates is finished, the product(s) no longer fit properly into the active site and are expelled (step 3). The temporary changes in shape, charge, and bonding patterns within the enzyme revert to their original configuration, and the enzyme is ready to accept another set of substrates (back to step 1).

Why does this sequence of events speed up the rate of chemical reactions? The interactions between substrate and enzyme, which are like minireactions with very low activation energies, allow the overall reaction to bypass its otherwise high activation energy barrier.

The Activity of Enzymes Is Influenced by Their Environment

Protein enzymes have very complex, three-dimensional structures that are sensitive to environmental conditions. Each enzyme has evolved to function optimally at a particular pH, temperature, and salt concentration. Some also require the presence of other molecules called coenzymes, often derived from water-soluble vitamins, in order to function.

Most enzymes function optimally at a pH between 6 and 8, the level found in most body fluids and maintained within cells. An exception is the protein-digesting enzyme pepsin. Pepsin is converted from an inactive to an active form by the highly acid (pH 2) conditions within the stomach. At this pH, the excess of hydrogen ions causes hydrogen to attach to certain locations on the protein, altering its configuration and exposing the active site. In proteins that function best at neutral pH (7), such interactions would distort the structure and destroy normal function.

Before refrigeration, foods such as meat were often preserved using concentrated salt solutions, which kill most bacteria, partly by interfering with enzyme function. Salts dissociate into ions, which form bonds with enzymes that interfere with the enzymes' normal three-dimensional structure. This process destroys enzyme activity, which requires a precise protein configuration. Dill pickles are very well-preserved in a vinegar-salt solution, which combines both salty and acidic conditions. Organisms that live in highly salty environments, as you might predict, have enzymes whose configuration depends on the presence of salt ions.

Temperature also affects the rate of enzyme-catalyzed reactions. Because molecules move more rapidly at higher temperatures, their random movements are more likely to bring them into contact with the active site of an appropriate enzyme. Thus these reactions are accelerated by higher temperatures and slowed by lower temperatures.

Figure 4-8   The cycle of enzyme-substrate interactions

A child who fell through the ice was rescued and survived unharmed after 20 minutes under water; the lowering of his body temperature slowed all his metabolic reactions and drastically reduced his need for oxygen. When temperatures rise too high, the bonds that regulate enzyme shape may be broken apart by the excessive molecular motion.

Some enzymes require helper molecules called **coenzymes** to function. These organic molecules bind to the enzyme and interact with the substrate molecule. Coenzymes help weaken the bonds of the substrate, allowing it to react with another molecule. Many water-soluble vitamins (such as the B vitamins) are essential because they are used by the body to synthesize coenzymes.

Cells Regulate the Amount and Activity of Their Enzymes, Thus Precisely Regulating Their Metabolic Reactions

Speeding up reactions is not always desirable. For example, you would not want your body to break down every glucose molecule you eat immediately. For one thing, you might starve to death overnight if you couldn't store any energy between supper and breakfast. What's more, glucose is an important ingredient in essential body chemicals, so some must be retained for synthesizing molecules such as hormones and components of cell membranes.

Cells have evolved the following three ways of regulating enzyme activity:

1. A cell may regulate the amounts of enzymes that it contains. Obviously, reactions will occur only if the necessary enzymes are available. As we will see in Chapter 11, cells often regulate the synthesis of enzymes to meet their changing needs.
2. A cell may synthesize an enzyme in an inactive form and activate it only when needed. Certain cells in your digestive system, for example, produce enzymes that digest food molecules such as proteins and lipids.

These enzymes could just as easily digest the proteins and lipids of the cells themselves. They don't because the enzymes are synthesized in an inactive form, with the active site blocked off. Within your stomach and small intestine, the interfering parts are cut off, thus activating the enzymes (see Chapter 23).

3. A cell can temporarily activate or inhibit enzymes, depending on the conditions at any given time. For example, the enzyme threonine deaminase begins the metabolic pathway that converts the amino acid threonine to the amino acid isoleucine. A cell needs suitable concentrations of both amino acids to manufacture proteins. The concentrations are regulated automatically by **feedback inhibition**, in which the activity of an enzyme is inhibited by its product or a later product of the metabolic pathway (Fig. 4-9). If enough isoleucine is present, it inhibits the activity of threonine deaminase, preventing further conversion of threonine.

On the molecular level, there are a variety of mechanisms of enzyme regulation. Here we will discuss two: allosteric regulation and competitive inhibition (Fig. 4-10). In **allosteric regulation**, enzyme action is inhibited or enhanced when molecules bind to a binding site on the enzyme that is distinct from the active site. This separate binding region is called the allosteric regulatory site (Fig. 4-10a). When the allosteric regulatory site is occupied, the enzyme changes shape and its activity may be enhanced or inhibited as a result (*allosteric* literally means "other shape"). Feedback inhibition frequently works through allosteric inhibition (Fig. 4-10c). The reaction product binds to the allosteric regulatory site, blocking enzyme activity and inhibiting further production of the product.

The second inhibitory mechanism is called **competitive inhibition**, a situation in which two or more molecules compete for entry into the enzyme's active site (Fig. 4-10b). Obviously, if one molecule occupies the active site, another cannot. Some poisons exert their effect be-

Figure 4-9   **Enzyme regulation by feedback inhibition**

In this example, the first enzyme in the metabolic pathway that converts threonine (substrate) to isoleucine (end product) is inhibited by high concentrations of isoleucine. If a cell lacks isoleucine, the first enzyme is not inhibited, and the pathway proceeds rapidly. As isoleucine concentrations build up, the isoleucine binds to the first enzyme to gradually shut down the pathway. When concentrations of isoleucine drop again and fewer molecules are around to inhibit the enzyme, the pathway resumes production.

(a) Enzyme structure

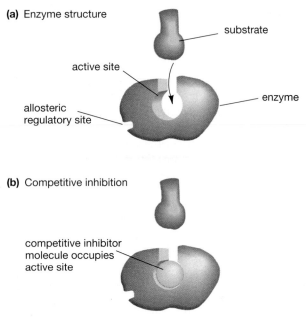

(b) Competitive inhibition

(c) Allosteric inhibition

Figure 4-10   **Enzyme regulation by competitive and allosteric inhibition**

Although they operate by very different mechanisms, both forms of inhibition prevent the substrate from entering the active site of the enzyme. **(a)** Many enzymes have an active site into which the substrate enters and an allosteric regulatory site on another part of the enzyme molecule. The enzyme is functional when the substrate can enter the active site. **(b)** In competitive inhibition, an inhibitor molecule competes with the substrate for entry into the active site: Whenever the inhibitor occupies the active site, the substrate cannot get in. **(c)** In allosteric inhibition, an inhibitor molecule binds to the allosteric regulatory site. This causes the shape or charge of the active site to change, thereby preventing the substrate from entering.

cause they resemble the normal substrate for an enzyme and bind to the active site, preventing normal metabolism. Two types of alcohol, methanol and ethanol, compete for the active site on the enzyme alcohol dehydrogenase. Methanol breakdown by this enzyme produces formaldehyde, which can cause blindness. Doctors administer ethanol intravenously to patients who have ingested methanol, because the ethanol competes with the methanol for the enzyme active site and blocks the production of formaldehyde.

Allosteric regulation and competitive inhibition can regulate concentrations of molecules in a cell because reactions are reversible. This includes the binding of inhibitor molecules (either allosteric or competitive) to enzymes. If inhibitors are scarce in the cell, the enzyme may be active most of the time. If the concentration of inhibitor molecules is high, then as soon as one leaves the enzyme, another is likely to take its place, and the enzyme will be inactive most of the time. In feedback inhibition, this means that when there is very little product in the cell there is very little enzyme inhibition, so the enzymes rapidly synthesize the product. As the product accumulates, it gradually shuts down the enzymes.

The ability of an enzyme to catalyze reactions is controlled by many factors, including the concentration of active enzyme, the concentration of inhibitor molecules, and the concentration of substrates. The interactions among these molecules usually maintain suitable concentrations of both substrates and products within a cell.

Coupled Reactions and Energy-Carrier Molecules

As we pointed out earlier, cells control energy flow by coupling reactions so that the energy released by exergonic reactions is used to drive endergonic reactions. Energy transfer is the role of **energy-carrier molecules**. Energy carriers work something like rechargeable batteries, picking up an energy charge at an exergonic reaction, moving to another location in the cell, and releasing the energy to drive an endergonic reaction.

ATP Is the Principal Energy Carrier in a Cell

The most common energy carrier molecule in living cells is **adenosine triphosphate**, or **ATP**. As you learned in Chapter 3, ATP is a nucleotide composed of a nitrogen-containing base, adenine; a sugar, ribose; and three phosphate groups (Fig. 4-11). Energy released in the cell through glucose metabolism is used to drive the synthesis of ATP from **adenosine diphosphate (ADP)** and inorganic phosphate:

ATP synthesis: Energy is stored in ATP

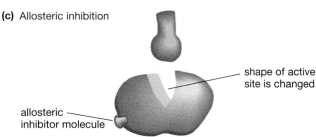

ATP: Principal energy carrier of the cell

Figure 4-11   **ADP and ATP structure**

A phosphate group is added to ADP (adenosine diphosphate) to make ATP (adenosine triphosphate). In most cases, only the last phosphate group and its high-energy bond are used to carry energy and transfer it to endergonic reactions within a cell. In the remainder of the text, we will usually employ the shorthand representations of ATP and ADP.

Most of the chemical bonds in ADP and ATP are "ordinary" covalent bonds, but the bonds joining the last two phosphate groups of ATP to the rest of the molecule are usually called high-energy bonds. Under most circumstances, only the last high-energy bond of ATP (the one that joins phosphate to ADP to make ATP) carries energy from exergonic to endergonic reactions. ATP is an unstable molecule, and its phosphate bond is easily broken, releasing energy that can be used to drive endergonic reactions.

The term *high-energy bond* is somewhat misleading, because, in fact, the bonds do not require an enormous amount of energy to form, nor do they release an enormous amount of energy when they are broken. Rather, a high-energy bond's relationship to a chemical reaction is a bit like Goldilocks's opinion of baby bear's porridge: It's just right. When your cells digest glucose, many of the reactions release just enough energy to add a phosphate to ADP to form ATP. Similarly, when ATP is broken down to ADP and inorganic phosphate, it releases an amount of energy that is just right to drive many essential cellular reactions. Nevertheless, the phrase "high-energy bond" is so common that we will continue to use it here.

ATP carries this energy to various sites in the cell that perform energy-requiring reactions, such as muscle contraction (see Fig. 4-4b). The ATP is then broken down to form ADP and inorganic phosphate:

ATP hydrolysis: Energy of ATP is released

The energy released by ATP hydrolysis may drive the energy-requiring reaction. The ADP and inorganic phosphate are usually recycled back to energy-generating reactions that resynthesize ATP.

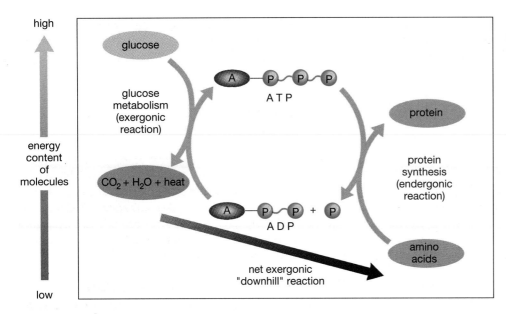

Figure 4-12   Coupled reactions within living cells

Exergonic reactions (for example, glucose metabolism) drive the endergonic reaction of ATP synthesis from ADP. The ATP molecule moves to another part of the cell, where ATP hydrolysis liberates some of this energy to drive an essential endergonic reaction (such as protein synthesis). The ADP and inorganic phosphate are recycled back to the exergonic reactions, where they will be converted to ATP once again. The overall reaction is "downhill": More energy is produced by the exergonic reaction than is needed to drive the endergonic reaction. The extra energy is released as heat.

Most uses of energy in living cells involve pairs of coupled reactions linked by ATP (Fig. 4-12). In the first coupled reaction, energy released by an exergonic reaction drives ATP synthesis; in the second, ATP hydrolysis drives an energy-requiring endergonic reaction. Each coupled reaction is guided by a specific enzyme that positions the molecules and channels the ATP energy properly. Coupled reactions catalyzed by enzymes provide the energy and the specificity necessary to construct the different types of molecules needed by the cell.

So much ATP is used by living organisms that the life span of any given ATP molecule is very short. For example, a human runner can use up as much as half a kilogram (a pound) of ATP each minute (obviously, the ADP

thus produced must be quickly converted back to ATP, or it will be a very brief run). As a result, ATP is *not* a long-term energy-storage molecule. More stable molecules, such as sucrose, glycogen, starch, or fat, are used to store energy for hours, days, or months.

Electron Carriers Also Transport Energy within Cells

Energy may also be transported around a cell by other carrier molecules. In some exergonic reactions, including both glucose metabolism and the light-capture stage of photosynthesis, some energy is transferred to electrons. These energetic electrons (sometimes along with hydrogen atoms)

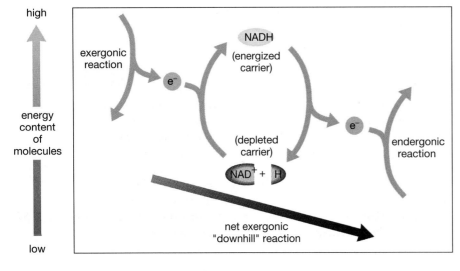

Figure 4-13   Electron carriers

Electron-carrier molecules such as nicotinamide adenine dinucleotide (NAD^+) pick up electrons generated by exergonic reactions and hold them in high-energy outer electron shells. Hydrogen atoms are often picked up simultaneously. The electron is then deposited, energy and all, with another molecule to drive an endergonic reaction, often the synthesis of ATP.

may be captured by **electron carriers** (Fig. 4-13), such as nicotinamide adenine dinucleotide (NAD⁺) and its relative flavin adenine dinucleotide (FAD). Loaded electron carriers then donate the electrons, along with their energy, to other molecules. You will see more about electron carriers and their role in cellular metabolism in Chapters 7 and 8.

SUMMARY OF KEY CONCEPTS

Energy and the Ability to Do Work

Kinetic energy is the energy of movement (light, heat, electricity, movement of large particles). Potential energy is stored energy (chemical energy, positional energy). The flow of energy among atoms and molecules obeys the laws of thermodynamics. The first law of thermodynamics states that within an isolated system the total amount of energy remains constant, although it may change in form. The second law of thermodynamics states that any change within an isolated system causes a decrease in the quantity of concentrated, useful energy and an increase in the randomness and disorder of matter. Entropy is a measure of disorder within a system.

Energy Flow in Chemical Reactions

Chemical reactions fall into two categories. In exergonic reactions, the product molecules have less energy than the reactant molecules, so the reaction releases energy. In endergonic reactions, the products have more energy than the reactants, so the reaction requires an input of energy. Exergonic reactions can occur spontaneously, but all reactions, including exergonic ones, require an initial input of energy, called the activation energy, to overcome electrical repulsions between reactant molecules. Exergonic and endergonic reactions may be coupled together so that the energy liberated by an exergonic reaction drives the endergonic reaction. Living organisms couple exergonic reactions, such as light-energy capture or sugar metabolism, with endergonic reactions, such as the synthesis of organic molecules.

Both exergonic and endergonic reactions are reversible and can proceed in either direction, given suitable inputs of products, reactants, and energy. All reactions, if left to themselves, eventually reach a steady state called a chemical equilibrium. Living organisms provide the necessary inputs in controlled ways that drive reversible reactions in the directions necessary to maintain life.

Controlling the Metabolism of Living Cells

High activation energies slow many reactions, even exergonic ones, to an imperceptible rate under normal environmental conditions. Catalysts lower the activation energy and thereby speed up chemical reactions. Catalysts are not permanently altered during the reaction. Living organisms synthesize protein catalysts called enzymes that promote one or a few specific reactions. The reactants temporarily bind to the active site of the enzyme; the binding strains their original chemical bonds and makes it easier to form the new chemical bonds of the products. Enzyme action is regulated on a cellular level in three ways: (1) by altering the rate of enzyme synthesis, (2) by activating previously inactive enzymes, and (3) by inhibiting enzyme activity through allosteric and feedback inhibition.

Coupled Reactions and Energy-Carrier Molecules

Energy released by chemical reactions within a cell is captured and transported about the cell by energy-carrier molecules such as ATP and electron-carrier molecules. These molecules are the major means whereby cells couple exergonic and endergonic reactions occurring at different places within the cell.

KEY TERMS

activation energy   p. 58
active site   p. 63
adenosine diphosphate (ADP)   p. 66
adenosine triphosphate (ATP)   p. 66
allosteric regulation   p. 65
catalyst   p. 62
chemical equilibrium   p. 59
coenzyme   p. 65
competitive inhibition   p. 65
coupled reaction   p. 59

electron carrier   p. 69
endergonic   p. 57
energy   p. 55
energy-carrier molecules   p. 66
entropy   p. 57
enzyme   p. 63
exergonic   p. 57
feedback inhibition   p. 65
first law of thermodynamics   p. 56
kinetic energy   p. 55

laws of thermodynamics   p. 56
metabolic pathway   p. 61
metabolism   p. 61
potential energy   p. 55
product   p. 57
reactant   p. 57
ribozyme   p. 63
second law of thermodynamics   p. 57
substrate   p. 63

✖ T H I N K I N G   T H R O U G H   T H E   C O N C E P T S

Multiple Choice

1. According to the first law of thermodynamics, the total amount of energy in the universe
   a. is always increasing
   b. is always decreasing
   c. varies up and down
   d. is constant
   e. cannot be determined

2. What is predicted by the second law of thermodynamics?
   a. Energy is always decreasing.
   b. Disorder cannot be created or destroyed.
   c. Systems always tend toward greater states of disorder.
   d. All potential energy exists as chemical energy.
   e. all of the above

3. Which statement about exergonic reactions is true?
   a. The products have more energy than the reactants.
   b. The reactants have more energy than the products.
   c. They will not proceed spontaneously.
   d. Energy input reverses entropy.
   e. none of the above

4. ATP is important in cells because
   a. it transfers energy from exergonic reactions to endergonic reactions
   b. it is assembled into long chains that make up cell membranes
   c. it acts as an enzyme
   d. it accelerates diffusion
   e. all of the above

5. How does an enzyme increase the speed of a reaction?
   a. by changing an endergonic to an exergonic reaction
   b. by adding activation energy
   c. by lowering activation energy requirements
   d. by decreasing the concentration of reactants
   e. by increasing the concentration of products

6. Which of the following statements about enzymes is (are) true?
   a. They interact with specific reactants (substrates).
   b. Their three-dimensional shapes are closely related to their activities.
   c. They change the shape of the reactants.
   d. They have active sites.
   e. all of the above

Review Questions

1. Explain why living organisms do not violate the second law of thermodynamics. What is the ultimate energy source for most forms of life on Earth?

2. Define metabolism and explain how reactions can be coupled to one another.

3. Are chemical equilibria ever reached in living systems? Why or why not?

4. What is activation energy? How do catalysts affect activation energy? How does this change the rate of reactions?

5. Describe some exergonic and endergonic reactions that occur in plants and animals very regularly.

6. Describe the structure and function of enzymes, and compare them with inorganic catalysts. How is enzyme activity regulated?

✖ A P P L Y I N G   T H E   C O N C E P T S

1. A preview question for ecology: When a brown bear eats a salmon, does it acquire all the energy contained in the body of the fish? Why or why not? What implications do you think this would have for the relative abundance (by weight) of predators and their prey? Does the second law of thermodynamics help to explain the title of the book *Why Big Fierce Animals Are Rare?*

2. Many people in sub-Saharan Africa have experienced the effects of malnutrition and starvation, but the very young are most severely affected. Some suffer permanent disability even if food is provided. How could a lack of food intake interfere with functions of individual cells and tissues? Which tissues are likely to suffer the most irreversible damage?

3. As you learned in Chapter 3, the subunits of virtually all organic molecules are joined by condensation reactions and can be broken apart by hydrolysis reactions. Why, then, does your digestive tract have separate enzymes to digest proteins, fats, and carbohydrates, and in fact several of each?

4. Suppose someone tried to disprove evolution with the following argument: "According to evolutionary theory, organisms have increased in complexity through time. However, such evolution of increased biological complexity contradicts the second law of thermodynamics. Therefore, evolution is impossible." If you were to support the theory of evolution, what would be your response to this argument?

✖ F O R   M O R E   I N F O R M A T I O N

Baker, J. J. W., and Allen, G. E. *Matter, Energy, and Life*. 4th ed. Reading, MA: Addison-Wesley, 1981. An excellent introduction to chemical-energy principles for those interested in biology.

Dickerson, R. E. "Cytochrome c and the Evolution of Energy Metabolism." *Scientific American*, March 1980. Cytochrome c is an electron carrier in mitochondria, and one of the best-studied proteins. Dickerson points out the evolutionary heritage of this molecule in diverse organisms and also discusses its role in energy acquisition by cells. See also Dickerson's earlier article on cytochrome c, "The Structure and History of an Ancient Protein." *Scientific American*, April 1972.

Fenn, J. *Engines, Energy, and Entropy*. New York: W. H. Freeman and Co., 1982. Elegantly simple introduction to the laws of thermodynamics and their relationship to everyday life.

Koshland, D. E., Jr. "Protein Shape and Biological Control." *Scientific American*, October 1973. Enzyme function is intimately related to structure. Koshland discusses enzyme specificity and regulation in terms of its protein structure.

Sackheim, G. *Introduction to Chemistry for Biology Students*. Redwood City, CA: Benjamin/Cummings Publishing Co., 1991. This book is designed specifically for biology students who don't have much chemistry background.

NET WATCH

On-line resources for this chapter are on the World Wide Web at:
http://www.prenhall.com/~audesirk (click on the <u>table of contents</u> link and then select Chapter 4).

5

Cell Structure and Function

"It is all too easy now to underestimate cells. We have known about them for such large fractions of our lives that, for the most part, we cease being aware of how remarkable they really are."
    BRUCE ALBERTS, DENNIS BRAY, JULIAN LEWIS, MARTIN RAFF, KEITH ROBERTS, AND
    JAMES WATSON from the Prologue to *Molecular Biology of the Cell*

This electron micrograph reveals the complexity of a single cell. Within the blue cytoplasm is a red and green nucleus; orange structures include round lipid granules, elongated mitochondria, and threadlike cut membranes of the Golgi complex and endoplasmic reticulum.

A vast period of evolutionary time separated the appearance of the first biological molecules from the first structures that could truly be considered alive. Evolution remained severely constrained until complex aggregates of these molecules became separated from their environment. Proteinlike molecules sometimes spontaneously form hollow balls, called microspheres, in water. Perhaps, several billion years ago, some of these microspheres trapped other biological molecules inside, making a sort of pre-cell. The first collections of molecules that could be considered alive surely possessed a thin covering that separated life from nonlife, a simple cell membrane.

The Development of Cell Theory

Human understanding of the cellular nature of life came slowly. In 1665, English scientist and inventor Robert Hooke reported observations with a primitive microscope. He aimed his instrument at an "exceeding thin...piece of Cork" and saw "a great many little Boxes" (Fig. 5-1). Hooke called the boxes "cells," because he thought they resembled the tiny rooms, or cells, occupied by monks. Cork comes from the dry outer bark of the Mediterranean oak. Hooke wrote that in the living oak and other plants, "These cells [are] fill'd with juices." What these cells were, and how they were related to the lives of entire plants, Hooke did not say.

In 1673, the Dutch inventor Anton van Leeuwenhoek reported to Britain's Royal Society on his observations of red blood cells, sperm, and myriad microscopic "animalcules" in pond water. But more than a century passed before biologists began to understand the role of cells in life on Earth. Microscopists first noted that many plants consist entirely of cells. The thick wall surrounding all plant cells made their observations easier. Animal cells, however, escaped notice until the 1830s, when German zoologist Theodor Schwann saw that cartilage contains cells that "exactly resemble—[the cells of] plants." In 1839, after studying cells for years, Schwann was confident enough to publish his theory, calling cells the elementary

Figure 5-1 **Early observations of cells**

Robert Hooke's drawings of the cells of cork, as he viewed them with an early light microscope similar to the one shown here. The cells of cork are not living, and only the cell walls remain to outline the cell.

particles of both plants and animals. By the mid-1800s, German botanist Mattias Schleiden had further refined science's view of cells when he wrote: "It is...easy to perceive that the vital process of the individual cells must form the first, absolutely indispensable fundamental basis" of life.

Within a few years, several microscopists had noted that living cells could grow larger and divide into smaller but still living cells. In the late 1850s, the Austrian pathologist Rudolf Virchow wrote, "Every animal appears as a sum of vital units, each of which bears in itself the complete char-

acteristics of life." Furthermore, Virchow predicted, "All cells come from cells." The three principles of modern cell theory evolved directly from Virchow's statements.

1. Every living organism is made up of one or more cells.
2. The smallest living organisms are single cells, and cells are the functional units of multicellular organisms.
3. All cells arise from preexisting cells.

An Overview of Cell Structure and Function

In Chapter 1, we defined cells as the smallest units of life. Each cell contains genes that control the synthesis of molecules within the cell and allow it to reproduce and organelles that act as microscopic chemical factories that guide the chemical reactions within the cell, all enclosed within a thin plasma membrane that separates the cell from its environment. Most cells are small, ranging from about 1 to 100 micrometers (millionths of a meter) in diameter (Fig. 5-2). Living organisms may consist of just one cell (bacteria, protists) or aggregates of interacting, cooperating cells (fungi, plants, and animals). Because of the small size of nearly all cells, biologists use several types of microscopes to assist their studies of cell structure and function. Some of the diversity of images produced by modern microscopes is shown in "Scientific Inquiry: Viewing the Cell—A Gallery of Microscopic Images."

All cells have at least three components: a plasma membrane, genetic material, and cytoplasm (Fig. 5-3, p. 78).

The Plasma Membrane Isolates the Cell but Allows Interactions with the Environment

The **plasma membrane** (also called the *cell membrane*) is a double layer of phospholipids (see Chapter 3) in which are embedded a wide variety of proteins. The plasma membrane performs functions: (1) It isolates the cytoplasm from the external environment; (2) it regulates the flow of materials between the cytoplasm and its environment, for example, acquiring nutrients and expelling wastes; and (3) it allows interaction with other cells.

The Genetic Material Provides a Cellular "Blueprint"

Each cell has its own hereditary blueprint, in which is stored the instructions for making all the other parts of the cell and for producing new daughter cells. In all living cells, the genetic material is **deoxyribonucleic acid (DNA)**. In eukaryotic cells (plants, animals, fungi, and protists), the DNA is contained within a separate, membrane-bound structure called the **nucleus**. In prokaryotic

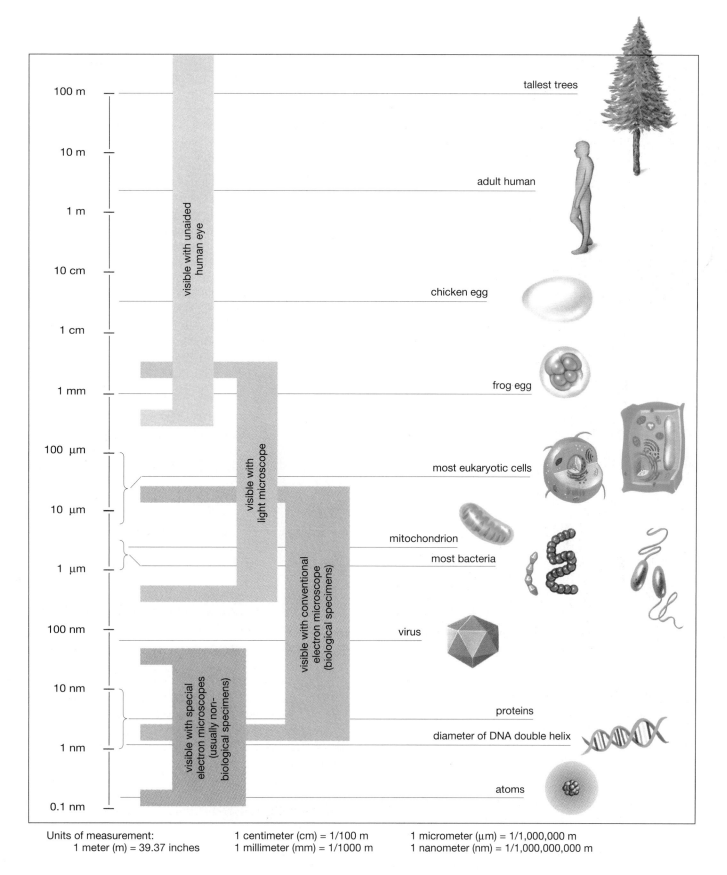

Figure 5-2 **Relative sizes**

Dimensions commonly encountered in biology range from about 100 meters (the height of the tallest redwoods) through a few micrometers (the diameter of most cells) to a few nanometers (the diameter of many large molecules). Note that, in the metric system (used almost exclusively in science), separate names are given to dimensions that differ by factors of 10, 100, and 1000.

SCIENTIFIC INQUIRY

Viewing the Cell—A Gallery of Microscopic Images

Biologists, physicists, and engineers have collaborated in the development of a variety of microscopes to examine structures too small to see with the naked eye. In some cases, biologists use microscopes to look at living cells (Fig. E5-1a, b). In most instances, however, the specimen must be carefully prepared, usually sliced thin, and stained (Figs. E5-1c, d, E5-2, and E5-3).

Light microscopes use lenses, usually made of glass, to focus and magnify light rays that either pass through or bounce off a specimen. Light microscopes provide a wide range of images, depending on how the specimen is illuminated (for example, from the top [dark field] or the bottom [bright field]) and whether it has been stained (Fig. E5-1a–c). The resolving power of light microscopes—that is, the smallest structure that can be seen—is about 1 micrometer (a millionth of a meter).

Electron microscopes use beams of electrons instead of light. The electrons are focused by magnetic fields rather than by lenses. Some types of electron microscopes can resolve structures as small as a few nanometers (billionths of a meter). *Transmission electron microscopes* (TEMs) pass electrons through a thin specimen and can reveal minute subcellular structures (Fig. E5-1d), including organelles and plasma membranes (Fig. E5-2). *Scanning electron microscopes* (SEMs) bounce electrons off specimens that have been coated with metals and provide three-dimensional images. SEMs can be used to view structures ranging in size from entire insects down to cells and even organelles (Fig. E5-3).

(a)

(b)

(c)

(d)

0.5 micrometer

Figure E5-1   **Images of *Paramecium* seen through different types of microscopes**

Living *Paramecia* seen with **(a)** bright field and **(b)** dark field illumination. **(c)** This *Paramecium* has been stained to reveal some subcellular structures and photographed with a bright field microscope. **(d)** A transmission electron micrograph of the contractile vacuole of the *Paramecium*, the star-shaped organelle visible in (a) and (c). The roles of contractile vacuoles are discussed in Chapter 6. Note the increased detail compared with the specimens photographed through a light microscope. Because the magnification is greater, the entire *Paramecium* is not shown.

0.01 micrometer

Figure E5-2   **The transmission electron microscope**

Transmission electron microscopes can magnify much more than light microscopes. This photo shows a pair of plasma membranes that have been cut perpendicularly (like cutting a sheet of paper with scissors and looking at the paper edge on). In micrographs such as this one, membranes appear as a pair of dark lines (the phospholipid heads) separated by a thin layer of paler material (the phospholipid tails; see Fig. 5-11, p. 86). A plasma membrane is only about 7 billionths of a meter in thickness.

10 micrometers

5 micrometers

Figure E5-3 **Scanning electron microscopes provide images that look three-dimensional**

(a) A high-tech ant carries off a computer chip. **(b)** An SEM photograph of *Paramecium* shows the pattern of hairlike cilia that cover the surface of the cell. **(c)** This SEM photograph shows part of the inside of a cell. The spherical structures, many partially sliced open, are mitochondria (see Fig. 5-17, p. 91).

2 micrometers

Figure 5-3   **An animal cell**

An animal cell as seen with a transmission electron microscope. Note the plasma membrane (thin line surrounding cell), cytoplasm (material enclosed by the plasma membrane), and the nucleus, containing the genetic material, surrounded by the nuclear envelope.

cells (bacteria), the DNA, although localized to a particular place within the cytoplasm called the **nucleoid**, is not separated by membranes from the rest of the cytoplasm.

Cytoplasm Fills Each Cell and Surrounds the Nucleus

The **cytoplasm** consists of all the material inside the plasma membrane and, in eukaryotic cells, outside the nucleus, which is itself contained by a membrane. The cytoplasm includes water, salts, an assortment of organic molecules including many enzymes that catalyze reactions and other proteins that form an intracellular support system. The cytoplasm also contains a variety of discrete structures, called **organelles**, each of which performs a distinct cellular function.

Cell Function Limits Cell Size

Why are all living things composed of cells? Looking over the essential components of cells described in the previous section should enable you to make a pretty good guess at why microscopic organisms are cellular: The plasma membrane separates the complex life processes in the cytoplasm from the chaos of the outside world. The genetic material is the databank that directs those life processes.

What may be less obvious is why large organisms consist of many cells rather than one large cell. To answer that question, we must consider the physical constraints that limit cell size. All living organisms exchange nutrients and wastes with their external environment through the plasma membrane. These activities would cause two major problems if a large organism were composed simply of a single cell (Fig. 5-4): Exchange would be limited both by the distance from the center of the cell to its surface and by the surface area of the cell.

First, as a cell becomes larger, its innermost regions become farther away from the membrane. Most nutrients and wastes move into, through, and out of cells by **diffusion**, the net transport of molecules from places of high concentration to places of low concentration (we will discuss diffusion more thoroughly in Chapter 6). Diffusion works well for moving molecules over very short distances (a few micrometers), but it is exceedingly slow for long-distance transport. Relying on diffusion alone, oxygen molecules would take more than *200 days* to reach the center of a cell 20 centimeters in diameter (8.5 inches; about the thickness of your chest). Clearly not much life could go on inside a cell that large!

The second difficulty arises from geometry: As a cell enlarges, its volume increases more rapidly than its surface area does. A cell that doubles its radius becomes eight times greater in volume but only four times greater in surface area. Thus, compared with their volume, large cells have a smaller surface area than small cells do. In general, as a cell increases in volume, more chemical reactions occur. Thus more nutrients and oxygen are needed, and more waste products must be eliminated—all through the plasma membrane. In a very large cell the surface area of the membrane would be too small to keep up with the cell's metabolic needs. The ratio of the cell surface to the cell volume thus sets the upper limit on individual cell size. Large organisms, therefore, consist of billions of coordinated cells, with structural and functional arrangements (such as respiratory and circulatory systems) that allow internal cells to exchange materials with the external environment.

distance to center (r)	1.0	2.0	3.0	1.0
surface area ($4\pi r^2$)	12.6	50.3	113.1	339.4
volume ($4/3\pi r^3$)	4.2	33.5	113.1	113.1
area/volume	3.0	1.5	1.0	3.0

Figure 5-4 **Geometric considerations limit the size of cells**

As a cell enlarges, the distance from the center of the cell to the outside world increases. Further, the volume increases much more rapidly than the surface area. As these spherical cells illustrate, doubling the radius halves the ratio of surface area to volume. Thus each unit volume of cytoplasm within the cell has only half the membrane area available to exchange nutrients and wastes with the external environment. However, if the volume of the largest sphere is divided among cells the size of the smallest sphere, the overall ratio of surface area to volume of the resulting "multicellular organism" remains large.

Types of Cells: Prokaryotic and Eukaryotic

As we briefly discussed in Chapter 1, there are two fundamentally different types of cells. The first type, represented by the bacteria, is called **prokaryotic**, Greek for "before the nucleus." The second type of cell, which evolved from the prokaryotic cell and is today found in protists, plants, fungi, and animals, is called **eukaryotic** ("true nucleus"). As their names imply, perhaps the most striking difference between prokaryotic cells and eukaryotic cells is that eukaryotic cells have their genetic material contained within a membrane-limited structure, the nucleus, whereas the genetic material of prokaryotic cells is not enclosed within a membrane. A variety of other features also differ between the two types of cells, making a closer look worthwhile.

Prokaryotic Cells Are Relatively Simple

Prokaryotic cells are usually very small (less than 5 micrometers long), with a relatively simple internal structure (Fig. 5-5; compare with the eukaryotic cell shown in Fig. 5-3). Most prokaryotic cells are surrounded by a relatively stiff cell wall. The substances that make up the cell

wall are secreted by the cell itself. Many bacteria live in fluid environments, from ponds to the human bloodstream, from which water tends to enter the bacterial cell (see Chapter 6). Without a cell wall, and without specialized energy-intensive mechanisms to excrete water, water entering from the environment would cause the cell to swell and explode; in fact, penicillin and some other antibiotics fight bacterial infections by inhibiting cell wall synthesis, causing the bacteria to rupture.

In some bacteria, such as the one shown in Figure 5-5, the cell wall has a polysaccharide coating that prevents white blood cells (part of the human body's immune system) from engulfing the bacterium. Although it plays an important supportive role, the cell wall and its polysaccharide coat are quite porous. Movement of materials into and out of a prokaryotic cell is regulated by the plasma membrane that lies just beneath the cell wall.

The cytoplasm of most prokaryotic cells is relatively homogeneous in appearance (although some photosynthetic bacteria have elaborate internal membranes). The DNA is usually coiled, attached to the plasma membrane, and concentrated in a region of the cell called the nucleoid. It is not, however, physically separated from the rest of the cytoplasm by a membrane.

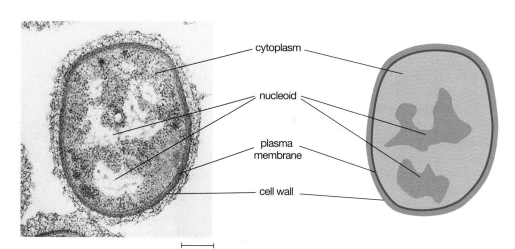

cytoplasm

nucleoid

plasma membrane

cell wall

0.2 micrometer

Figure 5-5 **A prokaryotic cell**

A transmission electron micrograph of a bacterium, a prokaryotic cell. The pale area occupying much of the cell, called the nucleoid, is the location of the cell's DNA; it is not separated from the rest of the cytoplasm by a membrane. In this bacterium, a cell wall surrounds the plasma membrane. This prokaryotic cell is far simpler in structure than the eukaryotic cell in Figure 5-3.

Eukaryotic Cells Are More Complex

Eukaryotic cells differ from prokaryotic cells in many ways. In addition to being larger than prokaryotic cells (often more than 10 micrometers in diameter), eukaryotic cells contain a variety of membranous organelles that lend structural and functional organization to the cell. Technically, the material within the plasma membrane of a eukaryotic cell is divided into the nucleus, an organelle consisting of a double layer of membrane that encloses the genetic material, and the cytoplasm, which includes everything else. The cytoplasm in turn is composed of several types of organelles, occupying as much as half the volume of the cell, and a fluid matrix, the **cytosol** (literally, "cell solution"), in which the organelles reside. The cytosol is a watery solution of salts, sugars, amino acids, proteins, fatty acids, nucleotides, and other materials. Giving shape and organization to the cytoplasm is a network of protein fibers, the **cytoskeleton**. Many of the organelles and even individual molecules of the cytoplasm are thought to be attached to the cytoskeleton.

With a few rare exceptions, all eukaryotic cells contain the following organelles, each with its own structural and functional specialization:

1. **Nucleus.** Contains the cell's genetic information
2. **Endoplasmic reticulum (ER).** A network of membranous channels, often studded with small protein/ribonucleic acid (RNA) particles called **ribosomes**; ribosomes synthesize many different types of proteins, whereas protein enzymes in the membrane of the endoplasmic reticulum synthesize lipids
3. **Golgi complex.** Stacks of membranous sacs that modify proteins and lipids, synthesize carbohydrates, and package molecules for transport
4. **Vesicles.** Membranous sacs that store and transport molecules around the cell
5. **Mitochondrion.** Captures some of the chemical energy of food molecules as high-energy bonds of adenosine triphosphate (ATP)

Eukaryotic cells, however, are not all alike. Figures 5-6 and 5-7 illustrate the structures that are found in animal and plant cells, although few individual cells contain all the features shown in the drawings. Each type of cell possesses a few unique organelles not found in the other. You may want to refer to these illustrations as we describe the structures of the cell in more detail. As a reference and study guide, Table 5-1 lists the major characteristics of prokaryotic and eukaryotic cells, including a brief description of the functions of the organelles.

In the following pages, we will introduce you to the major components of eukaryotic cells. We will emphasize the union of *structure* and *function* that is crucial to understanding biology. In studying the anatomy of the cell, remember that each structure originally evolved, and persists today, because it performs a specific function that is essential to the survival and reproduction of the cell.

The Nucleus: Control Center of the Cell

Deoxyribonucleic acid (DNA) is the genetic material of all living cells. A cell's DNA is the repository of information needed to construct the cell and direct the countless chemical reactions necessary for life and reproduction. Just as a blueprint is used selectively, one part at a time, to build a house, so too the hereditary information in DNA is used selectively by the cell, depending on its stage of development and its environmental conditions. In eukaryotic cells, the DNA is housed within the nucleus.

The nucleus consists of three readily distinguishable components (Fig. 5-8, p. 84). The **nuclear envelope** separates the nuclear material from the cytoplasm. Inside the nuclear envelope, the nucleus contains a granular-looking material called **chromatin** and a darker region called the **nucleolus**.

The Nuclear Envelope Allows Selective Exchange of Materials

The nucleus is isolated from the rest of the cell by a nuclear envelope that consists of two membranes riddled with pores (Fig. 5-8). Nuclear pores are complex structures consisting of at least eight protein subunits with a small channel through the middle. Water, ions, and small molecules such as ATP can pass freely through the central channel, but the passage of large molecules, particularly proteins and RNA, is regulated, presumably as the protein subunits move to enlarge and restrict the channel. Consequently, the pores help control the flow of information to and from the DNA.

Chromatin Consists of DNA and Its Associated Proteins

The bulk of the nucleus appears granular in an electron micrograph (see Figs. 5-3 and 5-8), with darker and lighter regions but no obvious structure. Since the nucleus is highly colored by several common stains used in light microscopy, early microscopists named the nuclear material chromatin, meaning simply "a colored substance." Biologists have since learned that chromatin consists of DNA associated with proteins. Although you can't see them in ordinary electron micrographs, eukaryotic DNA and its associated proteins form long strands called **chromosomes** ("colored bodies"). When cells divide, each chromosome coils upon itself, becoming thicker and shorter. The resulting "condensed" chromosomes are easily visible even with light microscopes (Fig. 5-9, p. 84).

Figure 5-6   **A "typical" animal cell**

In fact, the entire range of organelles depicted here seldom occurs in a single animal cell.

microtubules
(part of cytoskeleton)

mitochondrion

chromatin

nuclear envelope

nuclear pore

nucleolus

nucleus

centriole

vesicle

lysosome

Golgi complex

cytosol

flagellum

rough
endoplasmic
reticulum

ribosomes

smooth
endoplasmic
reticulum

plasma membrane

microtubules
(part of cytoskeleton)

plastid

mitochondrion

chloroplast

vesicle

Golgi complex

central vacuole

smooth
endoplasmic
reticulum

plasmodesma

rough
endoplasmic
reticulum

cell wall

plasma membrane

cytosol

nucleolus

nuclear pore

nucleus

chromatin

nuclear envelope

ribosomes

TABLE 5-1 :: Cell Structures, Their Functions, and Their Distribution in Living Cells

Structure	Function	Prokaryotes[a]	Plants	Animals
Cell surface				
Cell wall	Protects, supports cell	Present	Present	Absent
Plasma membrane	Isolates cell contents from environment; regulates movement of materials into and out of cell; communicates with other cells	Present	Present	Present
Organization of genetic material				
Genetic material	Encodes information needed to construct cell and control cellular activity	DNA	DNA	DNA
Chromosomes	Contain and control use of DNA	Single, circular, no proteins	Many, linear, with proteins	Many, linear, with proteins
Nucleus	Membrane-bound container for chromosomes	Absent	Present	Present
Nuclear envelope	Encloses nucleus; regulates movement of materials into and out of nucleus	Absent	Present	Present
Nucleolus	Synthesizes ribosomes	Absent	Present	Present
Cytoplasmic structures				
Mitochondria	Produce energy by aerobic metabolism	Absent	Present	Present
Chloroplasts	Perform photosynthesis	Absent	Present	Absent
Ribosomes	Provide site of protein synthesis	Present	Present	Present
Endoplasmic reticulum	Synthesizes membrane components and lipids	Absent	Present	Present
Golgi complex	Modifies and packages proteins and lipids; synthesizes carbohydrates	Absent	Present	Present
Lysosomes	Contain intracellular digestive enzymes	Absent	Present	Present
Plastids	Store food, pigments	Absent	Present	Absent
Central vacuole	Contains water and wastes; provides turgor pressure to support cell	Absent	Present	Absent
Other vesicles and vacuoles	Contain food obtained through phagocytosis; contain secretory products	Absent	Present (some)	Present
Cytoskeleton	Gives shape and support to cell; positions and moves cell parts	Absent	Present	Present
Centrioles	Synthesize microtubules of cilia and flagella; may produce spindle in animal cells	Absent	Absent (in most)	Present
Cilia and flagella	Move cell through fluid or move fluid past cell surface	Absent[b]	Absent (in most)	Present

[a]Many structures are listed as "absent" in prokaryotes; however, their functions are often essential to the life of any cell. In prokaryotes, these functions still occur, but not in discrete structures. For example, prokaryotes synthesize digestive enzymes on ribosomes attached to the cell membrane and immediately secrete the enzymes, where they digest food outside the cell. Therefore, although endoplasmic reticulum, secretory vesicles, and lysosomes are not found in prokaryotes, the functions of these organelles are still carried out.

[b]Many prokaryotes have structures called flagella, but these are not made of microtubules and move in a fundamentally different way than eukaryotic cilia or flagella do.

Cellular processes (the chemical reactions responsible for growth and repair, nutrient and energy acquisition and use, and reproduction) are governed by the information encoded in DNA. Because the DNA stays in the nucleus, while most of the chemical reactions that it controls occur in the cytoplasm, information molecules must be exchanged between the nucleus and the cytoplasm. Genetic information is copied from DNA into molecules of RNA, which move through the pores of the nuclear envelope into the cytoplasm. This information is then used to direct the synthesis of proteins (see below). These pro-

teins include enzymes that regulate chemical reactions, membrane proteins that govern interactions between the cell and its environment, and a variety of structural proteins. Some of these proteins pass from the cytoplasm into the nucleus and regulate the transfer of information from DNA to RNA, depending on what is happening in the cytoplasm and the extracellular environment. We take a closer look at these processes in Chapter 11.

The Nucleolus Is the Site of Ribosome Assembly

Most eukaryotic nuclei have one or more darkly staining regions called nucleoli ("little nuclei"; one nucleolus is shown in Fig. 5-8a), which are the sites of ribosome synthesis. A ribosome is a small particle composed of RNA and protein

Figure 5-7   A "typical" plant cell

Not all of these structures occur in every plant cell. Compare this drawing with that of an animal cell in Figure 5-6.

Figure 5-8   The structure of the nucleus and its envelope

(a) The eukaryotic nucleus consists of an outer double membrane, the nuclear envelope, enclosing the genetic material, DNA. The DNA is associated with proteins to form chromatin. One region of the chromatin, the nucleolus, stains much more darkly than the rest of the nucleus. (b) A transmission electron micrograph of the nuclear envelope clearly shows that it is composed of a double membrane that is fused around the lips of the nuclear pores. Micrographs such as this one can only hint at the structure of the pore, which is shown more clearly in the accompanying drawing.

Figure 5-9   Chromosomes

Chromosomes, seen here in a light micrograph of a dividing cell in an onion root tip, are the same material as the chromatin seen in nondividing cells (DNA and proteins) but in a more compact state.

that serves as a "workbench" for the synthesis of proteins (Chapter 13). Like many workbenches, ribosomes are non-specific—the same set of tools can be used to construct many different objects. Any ribosome can be used to synthesize any of the hundreds of proteins made by a cell. In electron micrographs, ribosomes appear as dark granules, either distributed in the cytosol (Fig. 5-10) or clustered along the membranes of the nuclear envelope and the endoplasmic reticulum (see Fig. 5-12).

The nucleolus consists of ribosomal RNA, proteins, ribosomes in various stages of synthesis, and DNA (bearing genes that specify the blueprint for ribosomal RNA). It stains darkly because of this accumulation of material. During cell division, ribosome synthesis slows. The chromosomes bearing the ribosomal RNA genes move away from one another. Consequently, the nucleolus disappears until after cell division is complete and the daughter cells begin synthesizing more ribosomes themselves.

The Membrane System of the Cell

All eukaryotic cells have an elaborate system of **membranes**. This membrane system is composed of the plasma membrane and several organelles within the cytoplasm, including the endoplasmic reticulum, nuclear envelope, Golgi complex, and a variety of membranous sacs such as lysosomes, which exchange membrane material with one another (see Figs. 5-6 and 5-7). All cellular membranes are composed of lipids and proteins that are synthesized in the endoplasmic reticulum. The membrane components are then sorted and modified in the Golgi complex and sent off to their appropriate destinations within the cell as small membranous sacs called vesicles. We will study this membrane system in three steps. First, we will provide a capsule summary of the structure of membranes, a topic covered in detail in Chapter 6. Second, we will examine the structure and function of individual components of the membrane system. Finally, we will trace the flow of membrane throughout the system.

Membranes Consist of a Double Layer of Phospholipids in Which Proteins Are Embedded

All the membranes of eukaryotic cells are composed of a double layer (bilayer) of phospholipids in which are embedded cholesterol and several kinds of proteins (Fig. 5-11). The exact types and amounts of phospholipids and proteins vary among membranes that enclose different types of organelles, according to the function of the organelle. In general, phospholipids and cholesterol form the structural framework of membranes, while proteins carry out specialized functions, such as transporting molecules across the membrane, catalyzing chemical reactions, or attaching one membrane to another.

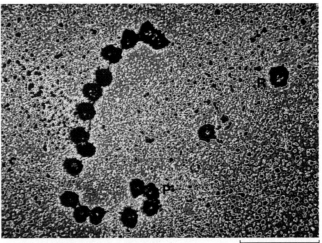

0.05 micrometer

Figure 5-10 **Ribosomes**

Ribosomes may be found free in the cytoplasm either singly (R) or in groups called polyribosomes (P). Free ribosomes synthesize proteins that will be used within the cell. Other ribosomes are attached to the endoplasmic reticulum (see Fig. 5-12).

The Cell's Membrane System Includes the Plasma Membrane, Endoplasmic Reticulum, the Golgi Complex, and Lysosomes

The Plasma Membrane Both Isolates the Cell and Allows Selective Interactions between the Cell and Its Environment

The plasma membrane forms the outer boundary of the living part of a cell. It is a marvelously complex structure that must perform the seemingly contradictory functions of separating the cytoplasm of the cell from the outside environment while simultaneously providing for transport of selected substances into or out of the cell. We will discuss plasma membrane structure and function in Chapter 6.

The Endoplasmic Reticulum Forms Membrane-Enclosed Channels within the Cytoplasm

The endoplasmic reticulum (ER) is a series of interconnected membranous tubes and channels in the cytoplasm (Fig. 5-12). Most eukaryotic cells have two forms of ER—rough and smooth. Numerous ribosomes stud the outside of **rough endoplasmic reticulum; smooth endoplasmic reticulum** lacks ribosomes.

The different structures of smooth and rough ER reflect different functions. Enzymes embedded in the membranes of the smooth ER are the major site of lipid synthesis, including the phospholipids of the ER and other membranes. In some cells the smooth ER synthesizes other types of lipids as well, such as the steroid hormones testosterone and estrogen produced in the reproductive organs of mammals. The ribosomes on the outside of rough ER synthesize proteins, including membrane proteins. There-

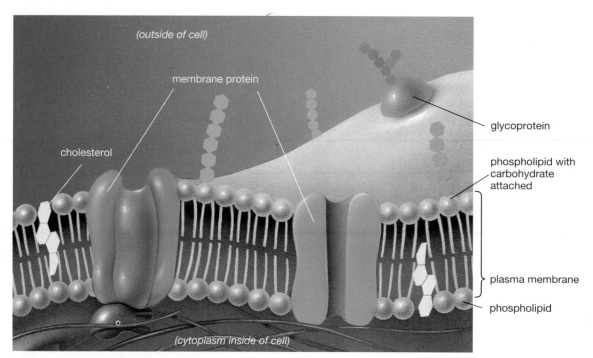

Figure 5-11   A biological membrane

The general structure of all biological membranes is similar: a phospholipid bilayer in which are embedded cholesterol and a variety of membrane proteins. Carbohydrates are sometimes attached to certain proteins to form glycoproteins and to phospholipids to form glycolipids.

Figure 5-12   Endoplasmic reticulum

There are two types of endoplasmic reticulum: rough ER, coated with ribosomes, and smooth ER without ribosomes. Although the ER looks like a series of tubes and sacs in electron micrographs, it is actually a maze of folded sheets and interlocking channels. In many cells the rough and smooth ER are thought to be continuous, as depicted in the drawing. Ribosomes (red) stud the cytoplasmic face (purple) of the rough ER membrane.

fore, the ER can synthesize itself, both lipid and protein components. Although most of the membrane synthesized in the ER forms new or replacement ER membrane, some of it moves inward to replace nuclear membrane or outward to form the Golgi complex, lysosomes, and the plasma membrane (see Fig. 5-15).

Ribosomes on rough ER also manufacture the proteins that some secretory cells export into their surroundings, including digestive enzymes and protein hormones (for example, insulin secreted by cells of the pancreas). As these proteins are synthesized by the ribosomes on the outside of the ER, they are simultaneously transported into the channels inside. The proteins then move through the ER and accumulate in pockets at the ends, especially the ends near Golgi complexes. These pockets then "bud off," forming membrane-bound vesicles that migrate to the Golgi complex.

The Golgi Complex Sorts, Chemically Alters, and Packages Important Molecules in Membranous Sacs

The Golgi complex is a specialized set of membranous sacs derived from the endoplasmic reticulum. In fact, the Golgi looks very much like a stack of smooth ER that has been stepped on, flattening the middle and making the ends bulge out (Fig. 5-13). Vesicles that bud off from the smooth ER fuse with the sacs on one side of the Golgi stack, adding their membrane to the Golgi and emptying their contents into the Golgi sacs. Other vesicles bud off

the Golgi on the opposite side of the stack, carrying away proteins, lipids, and other complex molecules. The Golgi complex performs the following three major functions:

1. It separates proteins and lipids received from the ER according to their destinations; for example, the Golgi separates digestive enzymes that are bound for lysosomes from hormones that will be secreted from the cell.
2. It modifies some molecules—for instance, adding sugars to proteins to make glycoproteins.
3. It packages these materials into vesicles that are then transported to other parts of the cell or to the plasma membrane for export.

Lysosomes Serve As the Cell's Digestive System

Some of the proteins manufactured in the ER and sent to the Golgi are intracellular digestive enzymes that can break down proteins, fats, and carbohydrates into their component subunits. In the Golgi, these enzymes are packaged in membranous vesicles called **lysosomes** (Fig. 5-14). The major function of lysosomes is to digest food particles, which range from individual proteins to complete microorganisms.

As we will see in Chapter 6, many cells "eat" by phagocytosis; that is, by engulfing extracellular particles with extensions of the plasma membrane. The food particles are then moved into the cytoplasm, enclosed within membranous sacs, now called **food vacuoles**. Lysosomes recog-

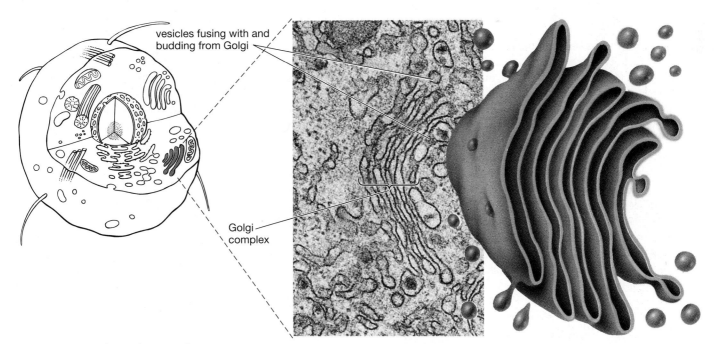

Figure 5-13   The Golgi complex

The Golgi complex is a stack of flat, membranous sacs derived from the endoplasmic reticulum. Vesicles constantly bud off from and fuse with the Golgi and ER, transporting material in several directions: from the ER to the Golgi and back again, and from the Golgi to plasma membrane, lysosomes, and vesicles.

lysosome containing
defective mitochondrion

Golgi complex
with budding
lysosome

0.5 micrometer

mitochondrion almost
completely digested

Figure 5-14   **Lysosomes bud from the Golgi complex**

Lysosomes are enzyme-filled vesicles that bud off from the Golgi complex. These enzymes
digest food and worn-out organelles, such as mitochondria.

nize these food vacuoles and fuse with them. The contents of the two vesicles mix, and the lysosomal enzymes digest the food into amino acids, monosaccharides, fatty acids, and other small molecules. These simple molecules then diffuse out of the lysosome into the cytosol to nourish the cell. Cell biologists continue to search for the key to how lysosomes recognize these food vacuoles.

Lysosomes also digest defective or malfunctioning organelles, such as mitochondria or chloroplasts. After identifying these organelles, the cell encloses them in vesicles made of membrane from the ER. These vesicles fuse with lysosomes, and digestive enzymes within the lysosome enable the cell to recycle valuable materials from the defunct organelles. How the cell identifies organelles that have outlived their usefulness is a topic of continuing research.

Membrane Synthesized in the Endoplasmic Reticulum Flows through the Membrane System of the Cell in an Orderly Way

The nuclear envelope, rough and smooth ER, Golgi complex, lysosomes, food vacuoles, and plasma membrane all form an integrated membrane system. Membrane is

synthesized in the ER and flows back and forth among these structures in an orderly way. As an example, let's look at the movement of materials destined for inclusion in the plasma membrane (Fig. 5-15). The ER synthesizes the phospholipids and proteins that make up the plasma membrane and simultaneously synthesizes proteins and lipids that belong in the ER itself.

How are these sorted out? The ER buds off a vesicle whose membrane is a mixture of ER and plasma membrane components. The vesicle fuses with the Golgi complex, adding its "mixed membrane" to the Golgi membrane. In the Golgi, ER and plasma membrane components are separated. ER material is removed to serve as "recycle" vesicles that pinch off and return to the ER. Plasma membrane material continues on through the Golgi, where it may be modified, for example by adding sugars to make glycoproteins or glycolipids (lipids to which a sugar is attached). Eventually the plasma membrane material becomes an "outward-bound" vesicle that buds off the far side of the Golgi and moves to the cell surface. The vesicle fuses with the plasma membrane, replenishing and enlarging the membrane.

The Golgi complex processes and packages all membrane-enclosed materials produced by the cell. Many of

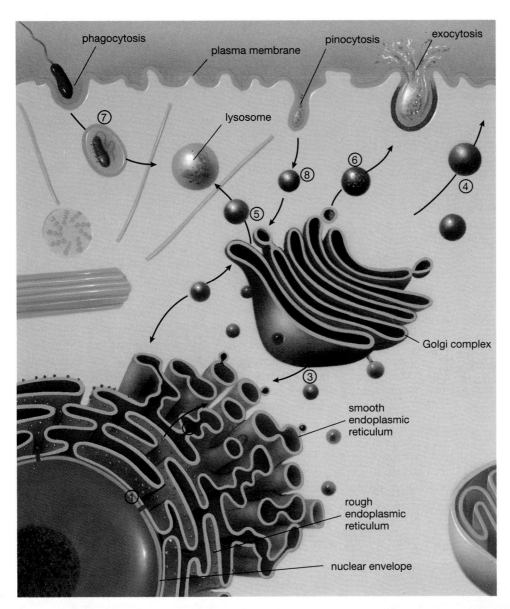

phagocytosis
plasma membrane
pinocytosis
exocytosis
lysosome
Golgi complex
smooth endoplasmic reticulum
rough endoplasmic reticulum
nuclear envelope

Figure 5-15 **The flow of membrane and protein within a cell**

Membrane, regardless of its destination, is synthesized by the endoplasmic reticulum. ① Some of this membrane moves inward to form new nuclear envelope, while most of it moves outward to form ② smooth ER and ③ Golgi membrane. From the Golgi, membrane moves to form ④ new plasma membrane and ⑤ the membranes that surround other cell organelles such as lysosomes. Many proteins are synthesized in the rough ER and move through the smooth ER to the Golgi. Here the proteins are sorted according to function; some remain as parts of the Golgi membrane, some are returned to the ER, some are packaged in vesicles bound for the plasma membrane where they will be ⑥ secreted from the cell, and some are packaged in lysosomes. ⑦ Lysosomes may fuse with food vacuoles for intracellular digestion of food particles. ⑧ Plasma membrane brought into the cell is recycled through the Golgi complex. The processes by which plasma membrane moves from the surface to the interior of the cell (called phagocytosis and pinocytosis) and by which the membrane of vesicles fuses with the plasma membrane (called exocytosis) are described in Chapter 6.

the vesicles pinched off from the Golgi contain secretory products (for example, hormones) that are released outside the cell. Lysosomes contain digestive enzymes and often fuse with food vacuoles to carry out intracellular digestion. How these diverse materials are recognized, purified, modified properly, separated out, and individually packaged remains a challenge to cell biologists.

Chloroplasts and Mitochondria: Energy Capture and Extraction

Every cell has prodigious energy needs: to manufacture materials, to pick things up from the environment and throw other things out, to move, and to reproduce. As Lewis Thomas put it, "There are structures squirming inside each of our cells that provide all the energy for living." These structures are the chloroplasts and mitochondria, "essentially foreign creatures" thought to have evolved from bacteria that took up residence long ago within a fortunate eukaryotic cell.

Chloroplasts and mitochondria are similar to each other in many ways (see Figs. 5-16 and 5-17). Both are usually oblong, about 1 to 5 micrometers long, and are surrounded by a double membrane. Both have enzyme assemblies that synthesize ATP, although the systems are used in a very different manner (see Chapters 7 and 8). Finally, both have many characteristics, including their own DNA, that seem to be remnants of their probable evolution from free-living organisms. However, they also have many differences, corresponding to their vastly different roles in cells: Chloroplasts capture the energy of sunlight during photosynthesis and store it in sugar, whereas mitochondria convert the energy of sugar into ATP for use by the cell.

Chloroplasts Are the Site of Photosynthesis

Chloroplasts (Fig. 5-16) are found only in plants and certain protists, notably the unicellular algae. Chloroplasts are surrounded by two membranes, although there is very little space between them. The inner membrane encloses a semifluid material called the **stroma**. Embedded within the stroma are interconnected stacks of hollow membranous sacs. The individual sacs are called **thylakoids**, and a stack of sacs is a **granum** (plural, **grana**). During chloroplast development, thylakoids probably bud off from the inner membrane into the stroma; in mature chloroplasts, thylakoids are not connected to the inner membrane.

The thylakoid membranes contain the green pigment **chlorophyll** (which gives plants their green color) as well as other pigment molecules. During photosynthesis, chlorophyll captures the energy of sunlight and transfers it to other molecules in the thylakoid membranes. These molecules in turn transfer the energy to ATP and other energy-carrier molecules. The energy carriers diffuse into the stroma, where their energy is used to drive the synthesis of sugar from carbon dioxide and water. Photosynthesis is described in more detail in Chapter 7.

Mitochondria Produce ATP Using Energy Stored in Food Molecules

Whereas chloroplasts convert solar energy into chemical energy (some of which is stored in food molecules such as sugar and starch), mitochondria extract energy from food molecules and store it in the high-energy bonds of ATP. Almost all eukaryotic cells have mitochondria, which are often called the "powerhouses of the cell." As you will see in Chapter 8, different amounts of energy can be released from a food molecule, depending on how it is metabolized. The breakdown of food molecules begins in the cytosol, but the cytosol lacks the enzymes needed to use oxygen to break down food. This **anaerobic** (without oxygen) metabolism does not convert very much food energy into ATP energy. Mitochondria are the only places in a cell where oxygen can be used in food breakdown. **Aerobic** metabolism reactions are much more effective in generating energy than are the anaerobic reactions; 18 or 19 times more ATP is generated by aerobic metabolism in the mitochondria than by anaerobic metabolism in the cytosol. Not surprisingly, mitochondria are found in large numbers in metabolically active cells, such as muscle, and are less abundant in more inert cells, such as those of bone and cartilage.

Mitochondria are round, oval, or tubular sacs made of a pair of membranes (Fig. 5-17). Although the outer mitochondrial membrane is smooth, the inner membrane loops back and forth to form deep folds called **cristae** (singular, **crista**, meaning "crest"). As a result, the mitochondrial membranes enclose two fluid-filled spaces, the **intermembrane compartment** between the inner and outer membranes and the **matrix**, or inner compartment, within the inner membrane. Some of the reactions of food metabolism occur in the fluid matrix contained within the inner membrane, while the rest are conducted by a series of enzymes attached to the membranes of the cristae.

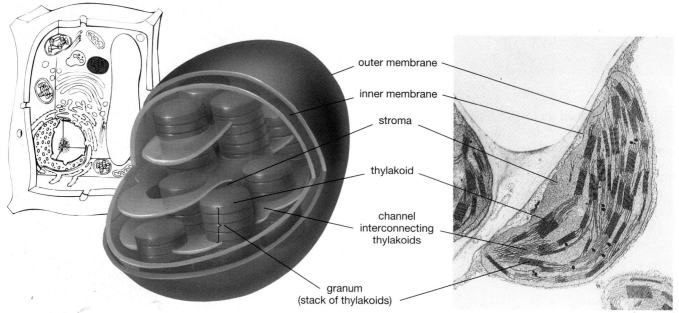

outer membrane
inner membrane
stroma
thylakoid
channel interconnecting thylakoids
granum (stack of thylakoids)

1 micrometer

Figure 5-16   **The chloroplast**

Chloroplasts are surrounded by a double membrane, although the inner membrane is not usually visible in electron micrographs. Enclosed by the inner membrane is the semifluid stroma, in which are embedded stacks of sacs collectively referred to as grana. The individual sacs of the grana are called thylakoids. Chlorophyll is embedded in the membranes of the thylakoids.

Plastids and Vacuoles: Storage and Elimination

If a cell finds itself in a favorable environment, some of the molecules that it synthesizes and some of the food that it acquires and digests will be more than it needs at the moment. Cells have evolved organelles in which to store such valuable, if temporarily surplus, molecules. Other materials, such as the waste products of digestion, must be eliminated, and still other organelles serve that function.

Many Types of Plastids Store Food in Plants

Besides mitochondria and nuclei, plant cells have other organelles with double outer membranes. These are the **plastids**; we have already met the most important plastid, the chloroplast. Other types of plastids are used as storage containers for various types of molecules, including pigments that give ripe fruits their yellow, orange, or red colors. Especially important, particularly for perennial plants (which continue growing year after year), are plastids that store photosynthetic products from the summer for use during the following winter and spring. Plants usually convert the sugars made during photosynthesis into starch and store it in plastids (Fig. 5-18). Potatoes, for example, are masses of cells stuffed with starch-filled plastids.

Vacuoles Serve Many Functions, Including Support and Storage and Elimination of Food and Wastes

Most cells contain sacs, called **vacuoles**, that are bound by a single membrane. Eukaryotic cells may contain several types of vacuoles, both temporary and permanent, that serve a variety of functions. The central vacuole of many plant cells, for example, provides support for the cell and also serves as a storage site for metabolic wastes. Other vacuoles store food or assist the cell in eliminating excess water that diffuses inside. Because many of the functions of vacuoles are intimately connected with the movement of water across membranes, we will describe vacuoles in Chapter 6, which discusses membrane structure and function in more detail.

The Cytoskeleton: Shape, Support, and Movement

Organelles do not drift about the cytoplasm haphazardly. Rather, most organelles are attached to a network of protein fibers called the cytoskeleton (Fig. 5-19). Even individual enzymes, which are often parts of complex metabolic pathways, may be fastened in sequence to the cytoskeleton, so that molecules can be passed from one enzyme to the next. Several types of protein fibers, including thin **microfilaments**, medium-sized **intermediate filaments**, and thick **microtubules** make up the cytoskeleton (Table 5-2).

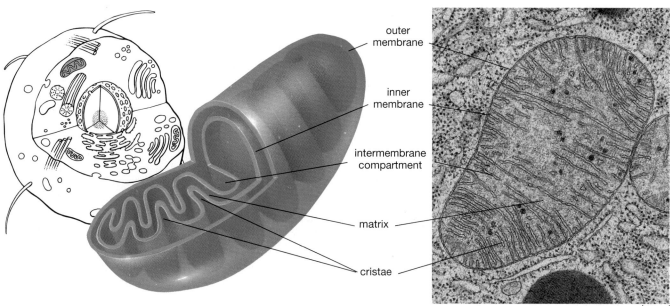

outer membrane
inner membrane
intermembrane compartment
matrix
cristae

0.2 micrometer

Figure 5-17 **The mitochondrion**

Mitochondria consist of a pair of membranes enclosing two fluid compartments, the intermembrane compartment between the outer and inner membranes and the matrix within the inner membrane. The outer membrane is smooth, but the inner membrane loops back and forth to form deep folds called cristae. Mitochondria are the site of aerobic metabolism.

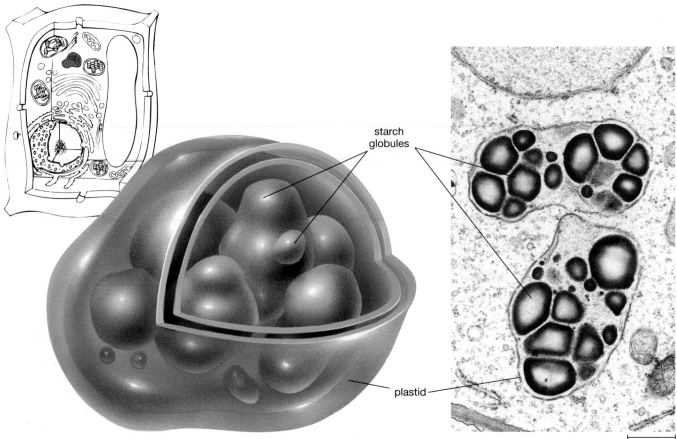

starch
globules

plastid

0.5 micrometer

Figure 5-18   The plastid

Plastids are organelles, found in the cells of plants and plantlike protists, that are surrounded
by a double outer membrane. Chloroplasts are the most familiar plastids; other types store
various materials, such as the starch filling these plastids in potato cells.

The cytoskeleton performs the following important
functions:

1. *Cell shape.* In cells without cell walls, the cytoskeleton, especially networks of intermediate filaments, determines the shape of the cell.
2. *Cell movement.* The assembly, disassembly, and sliding
   of microfilaments and microtubules cause cell movement. Cell movement includes both the familiar
   "crawling" of amoebae and white blood cells and the
   migration and shape changes that occur during the
   development of multicellular organisms.
3. *Organelle movement.* Microtubules and microfilaments
   move organelles from place to place within a cell. For example, microfilaments attach to vesicles formed during
   endocytosis (see Chapter 6), when large particles are engulfed by the plasma membrane and pull the vesicles
   into the cell. Vesicles budded off the ER and Golgi complex are probably guided by the cytoskeleton as well.
4. *Cell division.* Microtubules and microfilaments are essential to cell division in eukaryotic cells (see Chapter
   9). First, when eukaryotic nuclei divide, microtubules
   move the chromosomes into the daughter nuclei. Second, in animal cells, division of the cytoplasm of a sin-

gle parent cell into two new daughter cells results from
the contraction of a ring of microfilaments that pinch
the "waist" of the parent cell around the middle.

Microfilaments Allow Cells to Change Shape and Guide the Movement of Organelles

Microfilaments are strands that consist mostly of the protein actin, sometimes in association with a second protein,
myosin. Myosin has small extensions that can grab onto
actin and flex, sliding the actin filament past the myosin,
much like a sailor pulling up an anchor line hand over
hand. The most familiar function of actin and myosin is
muscle contraction, but actin and myosin in microfilaments also contribute to shape changes and organelle
movement in most, if not all, eukaryotic cells.

Intermediate Filaments Provide a Supportive Framework

There are at least five types of intermediate filaments, each
composed of a different protein. Each variety of intermediate filament is normally found in only one or a few cell
types. Intermediate filaments serve a variety of functions,

TABLE 5-2 ▪ Components of the Cytoskeleton

	Microfilaments	Intermediate Filaments	Microtubules
Structure	Solid strands about 7 nm in diameter; may be several cm long (muscle cells)	Solid strands 8 to 10 nm in diameter, 10 to 100 μm in length	Hollow tubes about 25 nm in diameter; may be more than 50 μm in length
Protein	Actin (most) and/or myosin	At least five different proteins	Tubulin
Function	Muscle contraction; changes in cell shape, including division of cytoplasm in dividing animal cells; cytoplasmic streaming; movement of pseudopodia	Maintenance of cell shape; attachments of microfilaments in muscle cells; support of nerve cell processes (axons)	Movement of chromosomes during cell division; movement of organelles within cytoplasm; movement of cilia and flagella

(a)   microfilament   intermediate filament   microtubule

50 micrometers   25 micrometers   10 micrometers

(b)   tubulin subunit

(c)

plasma membrane

endoplasmic reticulum

ribosomes

microfilaments and intermediate filaments

microtubule

mitochondrion

Figure 5-19   The cytoskeleton

Cells are given shape and organization by the cytoskeleton, which consists of three types of proteins: microfilaments, intermediate filaments, and microtubules. (a) Fluorescent substances that bind specifically to each of the cytoskeletal proteins are used to show the abundance, organization, and orientation of these proteins within living cells. (b) The structures of the cytoskeletal proteins. (c) An artist's interpretation of the cytoskeleton. Many organelles are probably attached to the cytoskeleton, which also reinforces the plasma membrane. Chemical reactions, movement of cell parts, and acquisition and release of materials are all coordinated by the cytoskeleton.

with one common feature: They are usually permanent frameworks that provide shape to cells and anchor various cell parts together. They do not assemble and disassemble. For example, the long axon of a nerve cell is supported by an internal skeleton of intermediate filaments in association with microtubules. In many cells, at least some of the proteins of the plasma membrane are attached to intermediate filaments. Finally, intermediate filaments anchor the microfilaments of actin in muscle cells, ensuring that the cells don't tear themselves apart during strong contractions.

Microtubules Help Position, Anchor, and Move Organelles and Alter the Shapes of Cells

Microtubules are hollow cylinders made from many dumbbell-shaped subunits of the protein tubulin (see Fig. 5-19). Some microtubules are relatively permanent features, such as those that make up the cilia and flagella by which many cells move (discussed shortly). Many microtubules, however, are transitory, being assembled and disassembled as required by the cell.

Both temporary and permanent microtubules are generated by **microtubule organizing centers** in various locations within the cell. All eukaryotic cells have a microtubule organizing center near the nucleus. During cell division, this center produces a football-shaped array of microtubules, called the **spindle apparatus**, that moves the chromosomes into the two daughter cells (see Chapter 9).

In animal cells, a prominent pair of **centrioles** is found at the microtubule organizing center near the nucleus (Fig. 5-20). Centrioles are short, barrel-shaped rings of microtubules. When a cell divides, the centrioles are duplicated, and one pair moves into each daughter cell. In ciliated cells, the centrioles multiply, and "offspring centrioles" move to the surface of the cell. Here they become anchored just beneath the plasma membrane and give rise to the microtubules of the cilia. Because of their location at the base of the cilia, these centrioles are often called **basal bodies**.

Microtubules are often associated with small proteins that travel along them, like engines on a railroad track. These proteins attach to various organelles, drag them along the microtubule to appropriate places in the cell, then anchor them in place using the microtubule as a scaffold. The endoplasmic reticulum and Golgi complex are positioned and held in place by microtubules. Short-term movements of organelles and movements of chromosomes during cell division are also produced by proteins moving along microtubules.

Microtubules can also slide past one another, like escalators moving in opposite directions. Microtubule sliding is responsible for the whiplike action of cilia and flagella (see below). Microtubules can change position in still a third way: by adding tubulin subunits at one end and removing them from the other. This activity effectively moves the microtubule along in one direction. As-

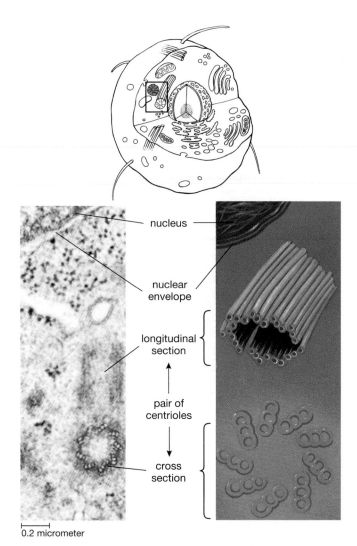

nucleus

nuclear envelope

longitudinal section

pair of centrioles

cross section

0.2 micrometer

Figure 5-20   **Centrioles**

In animal cells, a pair of centrioles is found at the microtubule organizing center near the nucleus. The centrioles, always lying at right angles to one another, are composed of nine triplets of short microtubules. A centriole is identical to the basal body of a cilium or flagellum (see Fig. 5-21).

sembly and disassembly of microtubules contribute to changes in the overall shapes of cells, for example when single nerve cells grow from your spinal cord out to the muscles of your legs.

Cilia and Flagella Move the Cell or Move Fluid Past the Cell

Both cilia and flagella are slender extensions of the plasma membrane. Each cilium and flagellum contains a ring of nine fused pairs of microtubules, with an unfused pair of microtubules in the center of the ring (forming what is called a "9 + 2" arrangement; Fig. 5-21). This pattern of

fused pair of
microtubules

protein "arms"

central unfused
pair of microtubules

0.1 micrometer

cilium

plasma membrane

basal body

0.1 micrometer

Figure 5-21   Cilia and flagella

Both cilia and flagella contain microtubules arranged in an outer ring of nine fused pairs of microtubules surrounding a central unfused pair (the 9 + 2 arrangement). The nine outer pairs have "arms" made of protein that interact with adjacent pairs to provide the force for bending (see Fig. 5-22). Cilia and flagella arise from basal bodies (centrioles) located just beneath the plasma membrane. Basal bodies have nine fused triplets, from which the fused pairs arise, and an indistinct central hub. It is not known how the basal body organizes the somewhat different structure of the cilium or flagellum.

microtubules is produced by a basal body (centriole) located just beneath the plasma membrane. A basal body consists of a ring of nine triplets of short microtubules. Two of the members of each triplet give rise to the pairs of microtubules in the cilium or flagellum.

Figure 5-22 diagrams how cilia and flagella bend. Tiny protein "arms" project out from each pair of microtubules in the outer ring. These arms attach to the neighboring pair of microtubules and flex, thereby moving the first pair along relative to the second. However, the basal body firmly anchors the "bottom" of all the microtubules in the entire cilium or flagellum. Therefore, adjacent microtubules can slide past one another only if the whole cilium or flagellum bends. (Think of a paint brush: When the bristles are straight, their tips are lined up evenly. When you flex the brush, the bristles slide a bit relative to one another, so that the tips of the bristles on the inside of the bend extend past the tips of the bristles on the outside of the bend.)

ATP energy powers the movement of the protein arms during microtubule sliding. Cilia and flagella often move almost continuously and consequently require enormous supplies of ATP, which are generated by mitochondria that are usually found in abundance near the basal bodies.

The main differences between cilia and flagella lie in their number, length, and the direction of the force they generate. In general, **cilia** (Latin for "eyelash") are short (about 10 to 25 micrometers long) and numerous. They provide force in a direction parallel to the plasma membrane, like the oars in a canoe. This is accomplished through a fairly stiff "rowing" motion during the power stroke, with most of the bending occurring at the base of the cilium, and a flexible return stroke (Fig. 5-23a).

Figure 5-22   **How cilia and flagella work**

Cilia and flagella bend when the arms of one pair of microtubules temporarily attach to the shaft of an adjacent pair and flex, "walking" along the shaft. If the microtubules were free, rather than anchored, this sequence of bending and attaching would result in one pair sliding past the other pair. But because the microtubules are anchored in place by the basal body, the action causes the cilium or flagellum to bend.

Figure 5-23   **Characteristic movement patterns of cilia and flagella**

(a) Cilia usually "row" along, providing a force of movement parallel to the plasma membrane, just as oars provide movement parallel to the sides of a rowboat. **(b)** Flagella often move in a wavelike motion, with a continuous bending that starts at the base and moves up to the tip. This motion provides a force of movement perpendicular to the plasma membrane. In this way a flagellum attached to a sperm can move the sperm straight ahead.

Flagella (Latin for "whip") are long (50 to 75 micrometers), usually few in number, and provide force perpendicular to the plasma membrane, like the motor on a motorboat. Flagella undulate with a continuous bending wavelike motion, without distinct power and return strokes (Fig. 5-23b).

Some unicellular organisms, such as *Paramecium* and *Euglena*, use cilia or flagella to move about. Most animal sperm also rely on flagella for movement. In multicellular animals, cilia are occasionally used for locomotion, moving the animal through a fluid. Many small aquatic invertebrates, for example, swim by the coordinated beating of rows of cilia, like the oars on a Roman galley. More often, however, the organism stays put, and its cilia move fluids and suspended particles past a surface. Ciliated cells line such diverse structures as the gills of oysters (moving food- and oxygen-rich water), the oviducts of female mammals (moving the eggs along from the ovary to the uterus), and the respiratory tracts of most land vertebrates (clearing debris and microorganisms from the windpipe and lungs).

Prokaryotic cells may also bear slender protrusions that undulate or spin, thereby enabling the cell to move about. However, the "flagella" of prokaryotic cells do not contain microtubules and have no evolutionary relationship to eukaryotic flagella or cilia.

✖ S U M M A R Y   O F   K E Y   C O N C E P T S

The Development of Cell Theory
The principles of cell theory are:

1. Every living organism is made up of one or more cells.
2. The smallest living organisms are single cells, and cells are the functional units of multicellular organisms.
3. All cells arise from preexisting cells.

Cell Function Limits Cell Size
Cells are limited in size by two constraints. First, if a cell were too large, the rate of diffusion of essential materials from the outer surface of the cell to the center of the cell would be too slow to sustain life. Second, as a cell enlarges its volume increases more rapidly than its surface area. Therefore, the surface area of a very large cell would be too small to service the metabolic needs of the cell cytoplasm.

Types of Cells: Prokaryotic and Eukaryotic
All cells are either prokaryotic or eukaryotic. Prokaryotic cells lack membrane-bound organelles; specifically, the DNA of prokaryotic cells is not enclosed within a membrane-limited nucleus. Eukaryotic cells have several types of membrane-bound organelles, including a nucleus containing DNA. The cytoplasm of eukaryotic cells consists of a fluid cytosol and organelles located outside the nucleus. With rare exceptions, all eukaryotic cells contain a nucleus, endoplasmic reticulum, Golgi complex, various types of vesicles, and mitochondria (see Table 5-1).

The Nucleus: Control Center of the Cell
Genetic material (DNA) is contained within the nucleus, which is bounded by the double membrane of the nuclear envelope. Pores in the nuclear envelope regulate the movement of molecules between nucleus and cytoplasm. The genetic material of eukaryotic cells is organized into linear strands called chromosomes, which consist of DNA and proteins. The nucleolus consists of the genes that code for ribosome synthesis, together with ribosomal RNA and ribosomal proteins. Ribosomes are particles of ribosomal RNA and protein that are the sites of protein synthesis.

The Membrane System of the Cell
The membrane system of a cell consists of the plasma membrane, endoplasmic reticulum (ER), Golgi complex, and vesicles derived from these membranes. Endoplasmic reticulum with ribosomes, called rough ER, manufactures many cellular proteins. Endoplasmic reticulum without ribosomes, called smooth ER, manufactures lipids. The ER is the site of all membrane synthesis within the cell. The Golgi complex is a series of membranous sacs derived from the ER. The Golgi processes and modifies materials synthesized in the rough and smooth ER. Some substances in the Golgi are packaged into vesicles for transport elsewhere in the cell. Lysosomes are vesicles budded off the Golgi that contain digestive enzymes, which digest food particles and defective organelles.

Chloroplasts and Mitochondria: Energy Capture and Extraction
All eukaryotic cells contain mitochondria, organelles that use oxygen to complete the metabolism of food molecules, capturing much of their energy as ATP. Cells of plants and some protists contain chloroplasts, which capture the energy of sunlight during photosynthesis, enabling the cells to manufacture organic molecules, particularly sugars, from simple inorganic molecules. Both mitochondria and

chloroplasts probably originated from bacteria that were captured by pre-eukaryotic cells over a billion years ago and have become incorporated into the normal functioning of the eukaryotic cell.

Plastids and Vacuoles: Storage and Elimination

Plastids are organelles with double membranes, including both chloroplasts and storage plastids that contain pigments or starch, that are found in plant cells. Many cells contain sacs bounded by a single membrane, called vacuoles, that may store food or wastes, excrete water, or support the cell. The size and function of many vacuoles change as a result of water movement across their membranes.

The Cytoskeleton: Shape, Support, and Movement

The cytoskeleton organizes and gives shape to eukaryotic cells and moves and anchors organelles. The cytoskeleton is composed of microfilaments, intermediate filaments, and microtubules (see Table 5-2). Cilia and flagella are whiplike extensions of the plasma membrane that contain microtubules in a characteristic 9 + 2 pattern. Cilia and flagella bend as a result of flexing movements of the microtubules.

Study Note

Figures 5-6 and 5-7 illustrate the overall structure of animal and plant cells, respectively. Table 5-1 lists the principal organelles, their functions, and their occurrence in prokaryotic cells, animal cells, and plant cells.

❈ K E Y   T E R M S

aerobic   p. 90
anaerobic   p. 90
basal body   p. 94
centriole   p. 94
chlorophyll   p. 90
chloroplast   p. 90
chromatin   p. 80
chromosome   p. 80
cilium   p. 96
crista   p. 90
cytoplasm   p. 78
cytoskeleton   p. 80
cytosol   p. 80
deoxyribonucleic acid (DNA)   p. 74
diffusion   p. 78
endoplasmic reticulum (ER)   p. 80

eukaryotic   p. 79
flagellum   p. 97
food vacuole   p. 87
Golgi complex   p. 80
granum   p. 90
intermediate filament   p. 91
intermembrane compartment   p. 90
lysosome   p. 87
matrix   p. 90
membrane   p. 85
microfilament   p. 91
microtubule   p. 91
microtubule organizing center   p. 94
mitochondrion   p. 80
nuclear envelope   p. 80
nucleoid   p. 78

nucleolus   p. 80
nucleus   p. 80
organelle   p. 78
plasma membrane   p. 74
plastid   p. 91
prokaryotic   p. 79
ribosome   p. 80
rough endoplasmic reticulum   p. 85
smooth endoplasmic reticulum   p. 85
spindle apparatus   p. 94
stroma   p. 90
thylakoid   p. 90
vacuole   p. 91
vesicle   p. 80

❈ T H I N K I N G   T H R O U G H   T H E   C O N C E P T S

Multiple Choice

1. The outermost boundary of an animal cell is the
   a. plasma membrane   b. nucleus
   c. cytoplasm   d. cytoskeleton
   e. cell wall

2. Which organelle contains the cell's genetic material?
   a. Golgi complex   b. ribosomes
   c. nucleus   d. mitochondria
   e. chloroplast

3. Most of the cell's ATP is synthesized in the
   a. Golgi complex   b. ribosomes
   c. nucleus   d. mitochondria
   e. chloroplast

4. Which organelle sorts, chemically modifies, and packages newly synthesized proteins?
   a. Golgi complex   b. ribosomes
   c. nucleus   d. mitochondria
   e. chloroplast

5. Membrane-enclosed digestive organelles that contain enzymes are called
   a. lysosomes
   b. smooth endoplasmic reticulum
   c. cilia
   d. Golgi complex
   e. mitochondria

6. A series of membrane-enclosed channels studded with ribosomes are called
   a. lysosomes
   b. Golgi complex
   c. rough endoplasmic reticulum
   d. mitochondria
   e. smooth endoplasmic reticulum
7. Which of the following is NOT a true statement about the cytoskeleton?
   a. It helps support cells.
   b. Once formed, the cytoskeleton is permanent and unchanging.
   c. It is composed of microfilaments, microtubules, and intermediate filaments.
   d. It plays an important role in cell movement.
   e. It is composed of proteins.

Review Questions

1. Diagram "typical" prokaryotic and eukaryotic cells, and describe their important similarities and differences.
2. Which organelles are common to both plant and animal cells, and which are unique to each?
3. Define cytosol, stroma, and matrix.
4. Describe the plasma membrane and explain which structures give the membrane its unique properties.
5. Describe the nucleus, including the nuclear envelope, chromatin, chromosomes, DNA, and the nucleolus.
6. What are the functions of mitochondria and chloroplasts? In each case, explain how the internal structure of these organelles supports its function.
7. What is the function of ribosomes? Where in the cell are they typically found?
8. Describe the structure and function of the endoplasmic reticulum and Golgi apparatus.
9. How are lysosomes formed? What is their function?
10. Diagram the structure of cilia and flagella.

✵ A P P L Y I N G   T H E   C O N C E P T S

1. If muscle biopsies (samples of tissue) were taken from the legs of a world-class marathon runner and a typical couch potato, which would you expect to have a higher density of mitochondria? Why? What about a muscle biopsy from the biceps of a weight lifter?
2. One of the functions of the cytoskeleton in animal cells is to give shape to the cell. Plant cells have a fairly rigid cell wall surrounding the plasma membrane. Does this mean that, for a plant cell, having a cytoskeleton is superfluous? Defend your answer in terms of other functions of the cytoskeleton.
3. Most cells are very small. What physical and metabolic constraints limit cell size? What problems would an enormous cell encounter? What adaptations might help a very large cell survive?

✵ F O R   M O R E   I N F O R M A T I O N

Glover, D. M., Gonzalez, C., and Raff, J. W. "The Centrosome." *Scientific American*, June 1993. The centrosome is the origin of much of a cell's cytoskeleton, and therefore regulates cell shape, movement, and cell division.

Goodsell, D. S. *The Machinery of Life*. New York: Springer, 1993. Wonderful, drawn-to-scale images of the organelles and molecules of the cell.

Kiester, E., Jr. "A Bug in the System." *Discover*, February 1991. Mitochondria are the descendants of bacteria that live within our cells. Although mitochondria are essential for human life, defects in mitochondrial genes can cause some devastating diseases.

Murray, M. "Life on the Move." *Discover*, March 1991. Many cells, including *Amoeba* and white blood cells, crawl about using tiny protein "motors" to manipulate their cytoskeleton.

Rothman, J. "The Compartmental Organization of the Golgi Apparatus." *Scientific American*, September 1985. Although they look homogeneous in electron micrographs, the sacs of the Golgi actually form three biochemically distinct processing units.

Symmons, M., Prescott, A., and Warn, R. "The Shifting Scaffolds of the Cell." *New Scientist*, February 18, 1989. Good overview showing the dynamic nature of the cytoskeleton.

Weber, K., and Osborn, M. "The Molecules of the Cell Matrix." *Scientific American*, October 1985. Biochemistry and microscopy are used to probe the structure and function of the cytoskeleton.

N E T   W A T C H

On-line resources for this chapter are on the World Wide Web at:
http://www.prenhall.com/~audesirk (click on the table of contents link and then select Chapter 5).

6 Cell Membrane Structure and Function

"To stay alive, you have to be able to hold out against equilibrium, maintain imbalance, bank against entropy, and you can only transact this business with membranes in our kind of world."
LEWIS THOMAS in *The Lives of a Cell* (1974)

Diatoms, seen here under the light microscope. Different species of diatoms are found in both fresh and salt water. A glassy outer shell and a selective cell membrane help cells maintain relatively constant internal conditions.

*C*onsider the challenge presented to a unicellular organism drifting about in a tide pool, stranded by the receding tide (Fig. 6-1). It inhabits an environment whose conditions change quickly and often. As the tide recedes, the sun warms the pool. As the day wears on, some of the water evaporates, boosting the concentration of dissolved salts. A sudden rain or the returning tide dilutes the water again. When night falls, the water temperature falls sharply. Just surviving requires a protective barrier between the inside and outside of the cell.

Merely withstanding environmental changes is not enough, however. The cell must absorb sunlight for photosynthesis or engulf other organisms to obtain the nutrients it needs to make new cellular components. To main-

Figure 6-1 A challenging environment
A tidepool contains countless life forms, including the photosynthetic single-celled diatom in the opening photograph. All cells face formidable challenges in coping with their external environment. The plasma membrane, which both isolates the cell from its environment and regulates interactions with that environment, is the focus of this chapter.

tain a consistent internal environment in which its components function best, the cell must shuttle salt, water, and other molecules in and out. It must excrete waste products. It must detect chemical clues to the location of food or to healthy and unhealthy environmental conditions. Finally, if it reproduces sexually, it must communicate with at least one other cell of its own species sometime during its life.

A lipid membrane only two molecules thick stands between the delicate inner workings of the cell and its hostile environment. The membrane also selectively permits the absorption and release of chemicals, sometimes using cellular energy. It allows chemical reactions to occur between one molecule outside the cell and another inside the cell. And it may deform to engulf food or enable the cell to move. All these processes rely on a variety of proteins embedded in the lipid membrane or attached to its surfaces.

Some single-celled organisms, such as the amoeba, survive with no barrier between life and the environment but a cell membrane. Cells of plants, bacteria, and many protists, such as diatoms, have walls outside the cell membrane that lend mechanical support. Cells of multicellular animals live surrounded by other cells and bathed in a bloodlike fluid. But all these cells possess very similar cell membranes and carry on many of the same processes across that membrane.

In this chapter, we explore the cell membrane structures that make defense, exchange of materials with the environment, and communication between cells possible.

Cell Walls

The outer surfaces of the cells of bacteria, plants, fungi, and some protists are covered with stiff, nonliving coatings called **cell walls**. Plant cell walls are composed of cellulose and other polysaccharides (see Fig. 3-5), while fungal cell walls are made of the modified polysaccharide **chitin** (see Fig. 3-6). Bacterial cell walls have a chitinlike framework to which short chains of amino acids and other molecules are attached.

Cell walls are produced by the cells they surround. In plants, membranous vesicles filled with sticky polysaccharides such as pectin (the ingredient that congeals grape juice into jelly) line up across the middle of a dividing cell. The vesicles fuse together, forming the new plasma membranes that separate the two daughter cells (see Chapter 9). The pectins formerly contained in the vesicles glue the daughter cells together, forming the **middle lamella** (Fig. 6-2). The two cells then secrete cellulose through their plasma membranes, underneath the middle lamella, forming the **primary cell wall**. Many plant cells later secrete more cellulose and other polysaccharides beneath the primary wall to form a

thick **secondary cell wall**. In some plant cells, the secondary wall may become thicker than the entire rest of the cell.

Cell walls support and protect otherwise fragile cells. For example, cell walls allow plants and mushrooms to resist the forces of gravity and wind and to stand erect on land. Tree trunks are the ultimate proof of the strength of cell walls, being composed almost entirely of cellulose and other materials laid down over the years and capable of supporting impressive loads (see Fig. 3-5 for more on the structure and properties of cellulose).

Although strong, cell walls are usually porous, permitting easy passage of small molecules such as minerals, water, oxygen, carbon dioxide, amino acids, and sugars (otherwise, the cell within would soon die). However, the structure that really governs the interactions between a cell and its external environment is the plasma membrane.

The Plasma Membrane

The **plasma membrane** of a cell can be thought of as a gatekeeper, allowing only specific substances in or out and passing messages from the external environment to the cell's interior. As gatekeeper, the plasma membrane must perform several specific functions:

1. Isolate the cell cytoplasm from the external environment.
2. Regulate the exchange of essential substances between the cytoplasm and the external environment.
3. Communicate with other cells.
4. Identify the cell as belonging to a particular species and a particular individual member of that species. In multicellular organisms, specific cell types often also bear unique molecular labels on their cell surfaces.

These are formidable tasks for a structure so thin that 10,000 stacked atop one another would scarcely equal the thickness of this page. The key to membrane function lies in membrane structure. Membranes are not simply homogeneous sheets: They are complex, heterogeneous structures, with different parts, indeed even individual molecules, performing very distinct functions.

As we pointed out in Chapter 5, all the membranes of a cell have a similar structure. Therefore, although this chapter focuses on the plasma membrane, the information presented here applies to other cellular membranes as well.

The Fluid Mosaic Model Describes the Structure of the Plasma Membrane

According to the **fluid mosaic model** of cellular membranes, developed by cell biologists S. J. Singer and

Figure 6-2   Plant cell walls

During cell division, the fusion of pectin-filled vesicles forms the middle lamella, which glues together the plasma membranes of the two daughter cells. Each cell then secretes cellulose and other carbohydrates to form its primary cell wall, between the middle lamella and the plasma membrane. Later, many cells may secrete additional cellulose and other carbohydrates beneath the primary wall, forming a secondary cell wall, pushing the primary cell wall and middle lamella farther away from the plasma membrane.

G. L. Nicolson in 1972, a membrane, when viewed from above, looks something like a lumpy, constantly shifting mosaic of tiles (Fig. 6-3). A bilayer of phospholipids forms a viscous, fluid "grout" for the mosaic, while an assortment of proteins are the "tiles," often sliding sluggishly about within the phospholipid bilayer. Thus, the overall image slowly changes over time, though the components remain relatively constant. As strange as this model may seem, it captures something of the dynamic quality of real membranes. With this model in mind, let's look more closely at the structure of membranes.

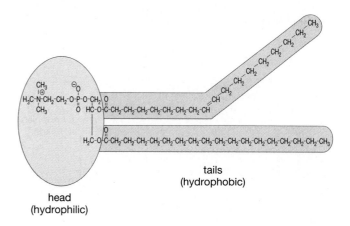

The Fluid Portion of Membranes Is Produced by a Double Layer of Phospholipids: The Phospholipid Bilayer

As you learned in Chapter 3, a phospholipid consists of two very different parts, a polar hydrophilic head and a pair of nonpolar hydrophobic tails:

All living cells are surrounded by water, whether it is the pond in which an *Amoeba* spends its life or the extracellular fluid that bathes the cells of animals. The cell cytoplasm is also mostly water. Plasma membranes therefore separate a watery cytoplasm from a watery external envi-

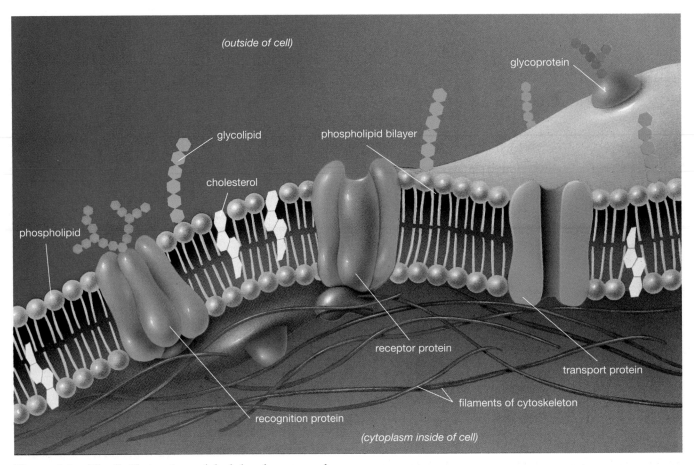

Figure 6-3   The fluid mosaic model of the plasma membrane

According to this model, the plasma membrane is a bilayer of phospholipids in which are embedded various proteins. Many proteins and lipids have carbohydrates attached to them, forming glycoproteins and glycolipids, respectively. The wide variety of membrane proteins fall mostly into three categories: transport proteins, receptor proteins, and recognition proteins.

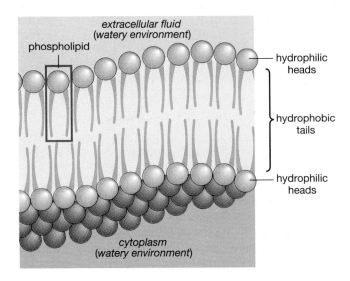

ronment. Under these conditions, phospholipids spontaneously arrange themselves in a double layer called a **phospholipid bilayer:**

Hydrogen bonds can form between water and the phospholipid heads, so the heads face the cytoplasm or the extracellular fluid. Hydrophobic interactions (see Chapter 2) cause the phospholipid tails to hide inside the bilayer. Because individual phospholipid molecules are not bonded to one another, this double layer is quite fluid, and individual molecules move about easily.

Most biological molecules, such as salts, amino acids, and sugars, are polar and water soluble. They therefore cannot pass easily through the nonpolar, hydrophobic fatty acid tails of the phospholipid bilayer. Because most substances that contact a cell are water soluble, the phospholipid bilayer acts as a barrier to the entrance of these molecules. It is largely responsible for the first of the four membrane functions that we listed earlier—isolating the cell cytoplasm from the external environment.

The phospholipid bilayer of membranes also contains cholesterol. The membranes of mitochondria have just a few cholesterol molecules here and there, but some plasma membranes have as many cholesterols as they do phospholipids. Cholesterol affects membrane structure and function in several ways. It makes the bilayer stronger, more flexible but less fluid, and less permeable to water-soluble substances such as ions or monosaccharides.

The flexible, somewhat fluid nature of the bilayer is very important for membrane function. As you breathe, move your eyes, and turn the pages of this book, cells in your body change shape. If their plasma membranes were stiff instead of flexible, cells would break open and die. Further, as we saw in the previous chapter, vesicles of membrane constantly bud off from the endoplasmic reticulum, merge with the Golgi complex, bud off again, and fuse with the plasma membrane. Membranes can bud off vesicles and seal up again, or fuse with vesicles and smooth out again, because of the fluid nature of the bilayer.

A Variety of Proteins in the Cell Membrane Form a Protein Mosaic

A variety of proteins are embedded within or attached to the surface of a membrane's phospholipid bilayer. You will recall from Chapter 2 that some of the amino acids that make up proteins are hydrophilic whereas others are hydrophobic. Many membrane proteins are held in the lipid bilayer by the interactions between their hydrophobic amino acids and the hydrophobic tails of the phospholipids. Hydrophilic parts of the proteins may protrude either into the cytoplasm or into the extracellular fluid. Many of the proteins in plasma membranes also have carbohydrate groups attached to them, especially to the parts that stick outside the cell. These proteins are called **glycoproteins**.

Many proteins can move about within the relatively fluid phospholipid bilayer. Other membrane proteins, however, are anchored in place by a network of protein filaments connected to the cytoskeleton. In animal cells, which lack cell walls, the attachments between plasma membrane proteins and the underlying cytoskeleton produce the characteristic shapes of the cells, from the dimpled discs of red blood cells to the elaborate branching nerve cells.

Membrane Proteins Serve a Variety of Functions That Allow the Cell to Interact with Its Environment

There are three major categories of membrane proteins, each of which serves a different function: transport proteins, receptor proteins, and recognition proteins (see Fig. 6-3).

Transport proteins regulate the movement of water-soluble molecules through the plasma membrane. Some, called **channel proteins**, form pores or channels that allow small water-soluble molecules to penetrate through the membrane (see Fig. 6-3). These proteins form a structure something like a sleeve with a lining: Hydrophobic amino acids (the outer material of the sleeve) anchor the protein in the phospholipid bilayer and hydrophilic amino acids form the inside of the channel (the lining of the sleeve). Every plasma membrane bears a large assortment of protein channels, each lined with specific amino acids that allow certain molecules, usually ions such as potassium (K^+), sodium (Na^+), and calcium (Ca^{2+}), to pass through. Other transport proteins, called **carrier proteins**, have binding sites, much like the active sites of enzymes, that can grab onto specific molecules on one side of the membrane. The transport protein then changes shape, in some cases through the use of cellular energy, and moves the molecule across the membrane. We will discuss the mechanisms whereby transport proteins move molecules across the membrane in the next section.

Receptor proteins are molecular triggers that set off cellular responses when specific molecules in the extracellular fluid, such as hormones or nutrients, bind to them. Most cells bear dozens of different types of receptors on their plasma membranes. When activated by the appropriate molecule, some receptors set off elaborate sequences of cellular changes, such as increased metabolic rate, cell division, movement toward a source of nutrients, or secretion of hormones. Other receptors act like gates on channel proteins; activating the receptor opens the gates, allowing ions to flow through the channels. For example, these receptor-activated channels allow nerve cells in your brain to communicate with one another (see Chapter 26).

Recognition proteins and glycoproteins serve as identification tags and cell-surface attachment sites. The cells of your immune system, for example, recognize a *Salmonella* bacterium as a foreign invader and target it for destruction. These same immune cells ignore the trillions of cells of your own body because of different identification glycoproteins on their surfaces. During development, the growth of nerve fibers from your spinal cord down to the muscles in your feet is guided by attachments between recognition proteins on the nerve cell and the other cells it traverses on its way to the muscle.

As you can see from these brief descriptions, membrane proteins are largely responsible for the second, third, and fourth membrane functions in our list: moving substances across the membrane, communicating with other cells, and identifying the cell.

Transport across Membranes

Nearly all living cells are bathed in liquid, which may be the extracellular fluid of a human body, the pond water in which the single-celled amoeba crawls, or the water-saturated cell walls of a young plant. The plasma membrane separates the fluid cytoplasm of the cell from its fluid environment. Let's begin our study of membrane transport, then, with a brief look at the characteristics of fluids, starting with a few definitions.

1. A **fluid** is a liquid or a gas; that is, any substance that can move or change shape in response to external forces without breaking apart.
2. The **concentration** of molecules in a fluid is the number of molecules in a given unit of volume.
3. A **gradient** is a physical difference between two regions of space, so that molecules tend to move from one region to the other. Cells frequently encounter gradients of concentration, pressure, and electrical charge.

Molecules in Fluids Move in Response to Gradients

The individual molecules in a fluid move continually, bouncing off one another in random directions. Superimposed atop these random collisions, however, may be a very nonrandom *net movement* of particular types of molecules in the fluid, or even of the entire fluid, in response to gradients. Molecules move from regions of high concentration to low concentration (for example, sugar dissolving in coffee) or from high pressure to low pressure (such as air flowing out of a bicycle pump when you push down the handle). Charged molecules (ions) move in response to electrical gradients, toward unlike charges or away from like charges. The gradient of concentration, pressure, or electrical charge provides the potential energy that can drive the net movement of the molecules. By analogy with gravity, we will refer to such movements as going "down" the gradient.

Movement across Membranes Occurs by Both Passive and Energy-Requiring Transport

Because the cytoplasm of a cell is very different from the extracellular fluid, gradients of concentration, electrical charge, and occasionally pressure span the plasma membrane. In its role as gatekeeper of the cell, the plasma membrane provides for two types of movement (Table 6-1).

During **passive transport**, substances move down gradients of concentration, electrical charge, or pressure. This movement by itself requires no expenditure of energy. The gradients provide the potential energy that drives the movement and controls the direction of movement, into or out of the cell. The plasma membrane acts as a filter; the lipids and protein pores regulate which molecules can cross and may influence the rate and timing of movement, but they do not influence the direction of movement.

In energy-requiring transport, substances move *against* gradients of concentration, electrical charge, or pressure. In this case, transport proteins in the plasma membrane do control the direction of movement, using chemical energy from cellular metabolism to drive movement against the gradients.

A helpful analogy for understanding the difference between these types of transport is to consider what happens when you ride a bike. If you don't pedal, you can go only downhill, as in passive transport. However, if you put enough energy into pedaling, you can go uphill as well.

Passive Transport: Movement down Concentration Gradients

Although gradients of pressure and electrical potential (see Chapter 26) are important in regulating the movement of certain substances in living organisms, most materials that move passively across plasma membranes do so in response to **concentration gradients**.

Molecules Diffuse from Areas of High Concentration to Areas of Low Concentration

Diffusion is the net movement of molecules in a fluid from regions of high concentration to regions of low concentration, driven by the concentration gradient. Diffusion can occur from one part of a fluid to another or across a membrane separating two fluid compartments. We will first examine the simplest case, the diffusion of molecules within a fluid.

To see how concentration gradients cause molecules to diffuse from one place to another, let's consider what happens when a drop of dye is placed in a glass of water (Fig. 6-4). With time, the drop will seem to become larger and paler, until eventually, even without stirring, the whole glass of water will be uniformly faintly colored. Why?

TABLE 6-1 *Transport across Membranes*	
Passive transport	Movement of substances across a membrane, going down a gradient of concentration, pressure, or electrical charge. Does not require the expenditure of energy by the cell.
Simple diffusion	Diffusion of water, dissolved gases, or lipid-soluble molecules through the phospholipid bilayer of a membrane.
Facilitated diffusion	Diffusion of (usually water-soluble) molecules through a membrane, assisted by membrane proteins.
Osmosis	Diffusion of water across a differentially permeable membrane—that is, a membrane that is more permeable to water than to dissolved solutes.
Energy-requiring transport	Movement of substances across a membrane, usually against a concentration gradient, using cellular energy.
Active transport	Movement of individual small molecules or ions through membrane-spanning proteins, using cellular energy, usually ATP.
Endocytosis	Movement of large particles, including large molecules or entire microorganisms, into a cell by a process in which the plasma membrane engulfs extracellular material, forming membrane-bound sacs that enter the cytoplasm.
Exocytosis	Movement of materials out of a cell by enclosing the material in a membranous sac that moves to the cell surface, fuses with the membrane, and opens to the outside, allowing its contents to diffuse away.

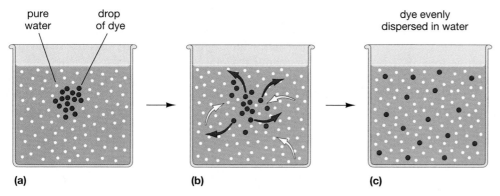

Figure 6-4   **Diffusion of a dye in water**

(a) A drop of pure dye (red dots) is placed in a glass of pure water (white dots in blue background). (b) Although individual molecules move at random, concentration gradients cause diffusion of dye molecules into the water and water molecules into the drop of dye. (c) Eventually, dye and water are both evenly dispersed. Individual molecules still move, but there is no longer any concentration gradient and therefore no further diffusion.

A drop of dye consists of many individual dye molecules moving in all directions. Those molecules that move but stay within the drop don't change its composition, which was all dye to begin with. But some molecules along the border of the drop, simply due to random motion, go out into the water. Thus, the net movement of dye molecules is from the region of high dye concentration (the drop) to the region of low dye concentration (the water). The same thing happens with water molecules. Random motion causes some to intersperse with the dye molecules, and the net movement of water is from the high water concentration outside the drop into the low water concentration inside the drop.

To the observer, dye molecules that move beyond the original border of the drop make the drop appear larger; water molecules that invade the drop dilute the dye, making the drop paler. At first, when the drop is pure dye and the water is pure water, there is a very steep concentration gradient, and the dye diffuses rapidly. As the concentration differences lessen, the dye diffuses more and more slowly. However, as long as the concentration of dye within the expanding drop is greater than the concentration of dye in the rest of the glass, the net movement of dye will be from drop to water, until the dye becomes uniformly dispersed in the water. Then, with no concentration gradient of either dye or water, diffusion stops. Individual molecules still move about, but there are no changes in concentration of either water or dye.

As you can appreciate from this imaginary experiment, diffusion cannot move molecules rapidly over long distances. Although the drop of dye immediately begins to diffuse into the water, uniform dispersion may take many minutes. As described in the previous chapter, the slow rate of diffusion over long distances is one of the reasons cells are small.

▓ S U M M A R Y

THE PRINCIPLES OF DIFFUSION

1. Diffusion is the net movement of molecules down a gradient from high to low concentration.
2. The greater the concentration gradient, the faster the rate of diffusion.
3. If no other processes intervene, diffusion will continue until the concentration gradient is eliminated.
4. Diffusion cannot move molecules rapidly over long distances.

Molecules Diffuse across Membranes down Their Concentration Gradients

Many molecules cross plasma membranes by diffusion, driven by differences between their concentration in the cytoplasm and in the external environment. Molecules cross the plasma membrane at different locations and at different rates, depending on the properties of the molecule in question. Therefore, plasma membranes are said to be **differentially permeable**: They allow some molecules to pass through, or **permeate**, more rapidly than others.

Some Molecules Move across Membranes by Simple Diffusion

Water, dissolved gases such as oxygen and carbon dioxide, and lipid-soluble molecules such as ethyl alcohol and vitamin A easily diffuse across the phospholipid bilayer. This process is called **simple diffusion** (Fig. 6-5a). Generally, the rate of simple diffusion is a function of the concentration gradient across the membrane, the size of the molecule, and how easily it dissolves in lipids (its lipid solubility): Large concentration gradients, small molecule size, and high lipid solubility all increase the rate of simple diffusion.

Other Molecules Cross the Membrane by Facilitated Diffusion, with the Help of Membrane Proteins

Most water-soluble molecules, such as ions (K^+, Na^+, Ca^{2+}), amino acids, and monosaccharides, cannot move through the phospholipid bilayer. These molecules can diffuse across only with the aid of two types of transport proteins, channel proteins and carrier proteins, which we discussed earlier in this chapter. This process is called **facilitated diffusion**.

Channel proteins form permanent pores, or channels, in the lipid bilayer through which certain ions can cross the membrane (Fig. 6-5b). Usually, each channel protein has a specific interior diameter and distribution of electrical charges that allow only particular ions to pass through. Nerve cells, for example, have separate channels

for sodium ions, potassium ions, and calcium ions. As we will see in Chapter 26, many of these channels have gates on one end, which open and close in response to electrical or chemical signals.

Carrier proteins bear groups of amino acids that bind specific molecules from the cytoplasm or extracellular fluid (Fig. 6-5c). Binding triggers a change in the shape of the carrier that allows the molecule to pass through the protein and thus across the plasma membrane. Carrier proteins that do not use cellular energy move molecules only down their concentration gradients (see the section on active transport, below).

Plasma membranes contain relatively few channels or carriers through which any given molecule can pass. The rate of facilitated diffusion is therefore a function of both the concentration gradient and the amount of transport

(a)

(b)

(c)

Figure 6-5   **Diffusion through the plasma membrane**

(a) Simple diffusion through the phospholipid bilayer: Water, gases such as oxygen and carbon dioxide, and lipid-soluble molecules can diffuse directly through the phospholipids. **(b)** Facilitated diffusion through a channel: Water-soluble molecules cannot pass through the phospholipid bilayer. Protein channels (pores) allow some water-soluble molecules, principally ions, to penetrate the cell. **(c)** Carrier-mediated facilitated diffusion: Carrier proteins may bind specific molecules, and, as a result, the carrier proteins change their shape, passing the molecule through the middle of the protein to the other side of the membrane.

proteins. Molecules that cross the membrane by facilitated diffusion usually do so more slowly than those that cross by simple diffusion through the lipid bilayer.

Osmosis Is the Diffusion of Water across Membranes

Water, like any other molecule, moves by diffusion from regions of high water concentration to regions of low water concentration. However, the diffusion of water across differentially permeable membranes has such dramatic consequences that it has been given a special name: **osmosis**. Let's investigate osmosis in another thought experiment.

A very simple kind of differentially permeable membrane consists of an impervious sheet perforated with tiny pores. The pores allow water molecules to pass through, but not larger molecules such as sugar. Suppose we make a bag out of such a membrane, fill it with a sugar solution, tie off the top, and place the bag in a glass of pure water. The bag will swell up, and, if it is weak enough, it will burst (Fig. 6-6). Why?

If you could see individual molecules, you would notice that there are two categories of water molecules in the sugar solution inside the bag (Fig. 6-6): "free" water molecules, well separated from the sugars; and "bound" water molecules, held to the sugars by hydrogen bonds (this is why sugars dissolve in water; see Fig. 2-9). In the pure water outside the bag, of course, there are only free water molecules. Now, free water molecules can diffuse through the pores in the membrane,

but the bound water molecules cannot, because they are attached, at least temporarily, to the bulky sugars.

Therefore, the concentration of free water molecules is lower inside the bag than in the pure water outside. This water concentration gradient favors the movement of free water molecules from the pure water outside the bag to the sugar solution inside the bag. The bag swells up as more water molecules enter the bag than leave it. The sugar can't escape at all, so the free water concentration inside the bag is always lower than in the pure water outside. Water continues to enter the bag until it bursts.

Now let's redo our experiment but with one change. Suppose we tie a piece of the same differentially permeable membrane across the end of a glass tube, suspend the tube in pure water, and put the sugar solution in the tube (Fig. 6-7). Water will diffuse across the membrane from the pure water into the solution in the tube. Because the upper end of the tube is open, as the water enters it will raise the level of the solution in the tube. Eventually, the solution in the tube will stop rising. Why?

As the solution in the tube rises above the level of water in the glass, its weight applies a pressure to the membrane across the bottom of the tube. Therefore, two opposing gradients are set up: a concentration gradient moving water *into* the tube and a pressure gradient pushing water *out* of the tube. When the two gradients are equal, no further net movement of water will occur across the membrane. The physical pressure that exactly balances the osmosis of water

(a)

differentially permeable membrane

sugar molecule

pure water

(b)

differentially permeable membrane

"free" water molecule: can fit through pore

"bound" water molecules clustered around sugar: cannot fit through pore

pore

bag bursts apart

Figure 6-6 **Osmosis**

The membrane is differentially permeable to free water molecules (white dots) but not to larger molecules such as sugar (yellow hexagons) or water molecules held to the sugars by hydrogen bonds. **(a)** If a bag made of such a membrane is filled with a sugar solution and suspended in pure water, free water molecules will diffuse down their concentration gradient from the high concentration in the pure water outside the bag to the lower concentration in the sugar solution inside the bag. The bag swells up as water enters. If the bag is weak enough, the increasing water pressure will cause it to burst. **(b)** Membrane pores allow "free" water molecules to pass through; sugar molecules are too large. "Bound" water molecules, attracted to the sugars by hydrogen bonds, are thus also kept from passing through the pore.

differentially permeable membrane

sugar molecule

pure water

(a) water enters sugar solution across membrane by osmosis

(b) solution rises in tube as water enters

(c) solution stops rising when pressure caused by column of water equals osmotic pressure

Figure 6-7   Osmotic pressure

(a) A membrane permeable to free water molecules (white dots) but not to sugar (yellow hexagons) is fastened across the bottom of a glass tube. A sugar solution is placed in the tube and lowered into a glass of pure water until the level of the sugar solution is even with the surface of the water. **(b)** Water moves by osmosis across the membrane into the tube. As a result, the solution rises within the tube. **(c)** When the pressure of the column of solution in the tube forces water out through the membrane as fast as osmosis causes water to enter, the column stops rising. The resulting hydrostatic pressure equals the osmotic pressure of the solution.

due to the concentration difference between a solution and pure water is defined as the **osmotic pressure** of the solution. A solution with a high osmotic pressure has a low concentration of free water; a solution with a low osmotic pressure has a high concentration of free water.

As we have emphasized, osmosis is driven by differences in *water concentration* between solutions. Therefore, osmosis does not depend on the *type of dissolved molecules* in a solution, but only on their *concentration*. We can produce the same osmotic pressure by adding sucrose, glucose, amino acids, sodium ions, or a mixture of all of these to a solution, as long as the concentration of all types of particles that cannot permeate a membrane stays the same.

SUMMARY

THE PRINCIPLES OF OSMOSIS

1. Osmosis is the diffusion of water across a differentially permeable membrane.
2. Water moves across a membrane from a high concentration of free water molecules to a low concentration of free water molecules, or from high pressure to low pressure.
3. Dissolved substances, regardless of what they are, lower the concentration of free water molecules in a solution.

Osmosis across the Cell Membrane Plays an Important Role in the Lives of Cells

Most plasma membranes are highly permeable to water. Because all cells contain dissolved salts, proteins, sugars, and so on, the flow of water across the plasma membrane depends on the concentration of water in the liquid that bathes the cells. The extracellular fluids of animals are usually **isotonic** ("having the same strength") to the insides of the body cells; that is, the concentration of water inside is the same as that outside the cells so there is no net tendency for water to either enter or leave the cells. Note that the *types of dissolved particles* are seldom the same inside and outside the cells, but the *total concentration of all dissolved particles* is equal, with the result that the water concentration inside is equal to that outside the cells.

If red blood cells, for example, are taken out of the body and immersed in salt solutions of varying concentrations, the effects of the differential permeability of the plasma membrane to water and dissolved particles become dramatically apparent (Fig. 6-8). If the solution has a higher salt concentration than the cytoplasm (that is, if the solution has a lower water concentration), water will leave the cells by osmosis. The cells will shrivel up until the concentrations of water inside and outside become equal. Solutions that cause water to leave by osmosis are called **hypertonic** ("having greater strength").

In the opposite situation, if the solution has little or no salt, water will enter the cells, causing them to swell. If the solution has little enough salt, the cells will burst. Solutions that cause water to enter by osmosis are called **hypotonic** ("having lesser strength").

Osmosis across plasma membranes is crucial to the functioning of many biological systems, including water uptake by plant roots, absorption of dietary water from your intestines, and reabsorption of water and minerals in your kidneys (Chapter 23).

Osmosis will occur across any membrane, including the membranes surrounding organelles within a cell, if the water concentration differs on the two sides of the membrane. Osmosis across internal membranes, such as those that surround vacuoles, can also be important in the lives of certain cells, as described later in this chapter.

Energy-Requiring Transport across Membranes

All cells need to move some materials "uphill" across their plasma membranes, against diffusion gradients. For example, every cell requires some nutrients that are less concentrated in the environment than in the cell cytoplasm. Therefore, diffusion would cause the cell to lose, not gain, these nutrients. Other substances, such as sodium and calcium ions in your brain cells, must be maintained at much lower concentrations inside the cells than in the extracellular fluid. When these ions diffuse into the cells, they must be pumped out again against their concentration gradients. Finally, many cells acquire or expel some substances, such as whole bacteria or large proteins, that are too large to diffuse across a membrane regardless of concentration gradients. Cells have evolved several processes that use cellular energy to move materials into or out of the cell.

Active Transport Uses Energy to Move Molecules across Membranes

In **active transport**, membrane proteins use cellular energy to move individual molecules across the plasma membrane, usually against their concentration gradient (Fig. 6-9). Active transport proteins span the membrane and have two active sites. One active site (which may be either on the face of the plasma membrane in contact with the cytoplasm or on the face in contact with the extracellular fluid, depending on the active transport protein) recognizes a particular molecule, say a calcium ion, and binds it. The second site (always on the inside of the membrane) binds an energy-carrier molecule, usually adenosine triphosphate (ATP). The ATP donates energy to the protein, causing it to change shape and move the calcium ion across the membrane. Active transport proteins are often called *pumps*, in analogy to water pumps, because they use energy to move molecules "uphill" against a concentration gradient. Plasma membrane pumps are vital in mineral uptake by plants, mineral absorption in your intestines (Chapter 23), and maintaining concentration gradients essential to nerve cell functioning (Chapter 26).

Cells Engulf Particles or Fluids by Endocytosis

Cells can acquire fluids or particles, especially large proteins or entire microorganisms such as bacteria, by a process called **endocytosis** (Greek for "into the cell"). Dur-

(a) Isotonic solution

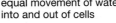

equal movement of water into and out of cells

(b) Hypertonic solution

net water movement out of cells

(c) Hypotonic solution

net water movement into cells

10 micrometers

Figure 6-8 **The effects of osmosis**

Red blood cells are normally suspended in the fluid environment of the blood and lack the ability to regulate water flow across their plasma membranes. **(a)** If red blood cells are immersed in an isotonic salt solution, which has the same concentration of dissolved substances as the blood cells, there is no net movement of water across the cell membrane, and the red blood cells keep their characteristic dimpled disc shape. **(b)** A hypertonic solution, with too much salt, causes water to leave the cells, shriveling them up. **(c)** A hypotonic solution, with less salt than the cells, causes water to enter, and the cells swell.

ing endocytosis, the plasma membrane engulfs the particle or fluid droplet and pinches off a membranous sac called a **vesicle**, with the particle or fluid inside, into the

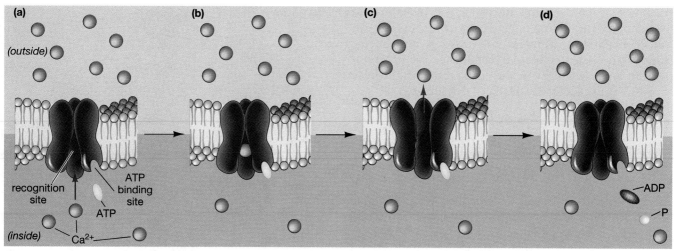

Figure 6-9   **Active transport**

Active transport uses cellular energy to move molecules across the plasma membrane, often against a concentration gradient. **(a)** A transport protein (blue) has a recognition site for the molecule to be transported, calcium ions (Ca^{2+}) in this case, and an ATP binding site. **(b)** The transport protein binds ATP and Ca^{2+}. **(c)** Energy from ATP changes the shape of the transport protein and moves the Ca^{2+} ion across the membrane. **(d)** The carrier releases the Ca^{2+} ion and the remnants of the ATP and resumes its original configuration.

cytoplasm (Fig. 6-10). Three types of endocytosis can be distinguished on the basis of the size of the particle acquired and the method of acquisition.

Pinocytosis Moves Liquids into the Cell

In **pinocytosis** ("cell drinking"), or **fluid-phase endocytosis**, a very small patch of plasma membrane dimples inward as it surrounds extracellular fluid and buds off into the cytoplasm as a tiny vesicle (Fig. 6-10a). Pinocytosis moves a droplet of extracellular fluid, contained within the dimpling patch of membrane, into the cell. Therefore, the cell acquires materials in proportion to their concentration in the extracellular fluid.

Receptor-Mediated Endocytosis Moves Specific Molecules into the Cell

Cells can take up certain molecules (cholesterol, for example) more efficiently by a process known as **receptor-mediated endocytosis** (Fig. 6-10b). Most plasma membranes bear many receptor proteins on their outside surfaces, each bearing a binding site for a particular nutrient molecule. In most cases, these receptors move through the phospholipid bilayer and accumulate in depressions of the plasma membrane called coated pits (Fig. 6-11). If the right molecule contacts a receptor protein in one of these coated pits, it attaches to the binding site. The coated pit deepens into a U-shaped pocket that eventually pinches off into the cytoplasm as a coated vesicle. Both the receptor-nutrient complex and a bit of extracellular fluid move into the cell in the coated vesicle.

Phagocytosis Moves Large Particles into the Cell

Phagocytosis ("cell eating") is used to pick up large particles, including whole microorganisms (see Fig. 6-10c).

When an *Amoeba*, for example, senses a tasty *Paramecium*, it produces extensions of its surface membrane, called **pseudopodia** (Latin for "false foot"; singular, **pseudopod**). The pseudopodia surround the luckless *Paramecium*, their ends fuse, and the prey is carried into the interior of the *Amoeba* for digestion. The resulting vesicle, called a food vacuole, fuses with lysosomes whose enzymes digest the prey (see Chapter 5). White blood cells also use phagocytosis and intracellular digestion to engulf and destroy bacteria that have invaded your body (Chapter 24).

Exocytosis Moves Material Out of the Cell

The reverse of endocytosis, called **exocytosis** (Greek for "out of the cell"), is often used by cells to dispose of unwanted materials, such as the waste products of digestion, or to secrete materials, such as hormones, into the extracellular fluid (Fig. 6-12). During exocytosis, a vesicle created by the Golgi complex moves to the cell surface, where the membrane of the vesicle fuses with the plasma membrane. The vesicle opens to the extracellular fluid and its contents diffuse out.

Transport across Intracellular Membranes

As we mentioned earlier, all the membranes of a cell, including those that surround such diverse organelles as chloroplasts, mitochondria, the nucleus, and vacuoles, are similar in structure and function to the plasma membrane. The transport processes that we have described in this chapter, from simple diffusion to active transport,

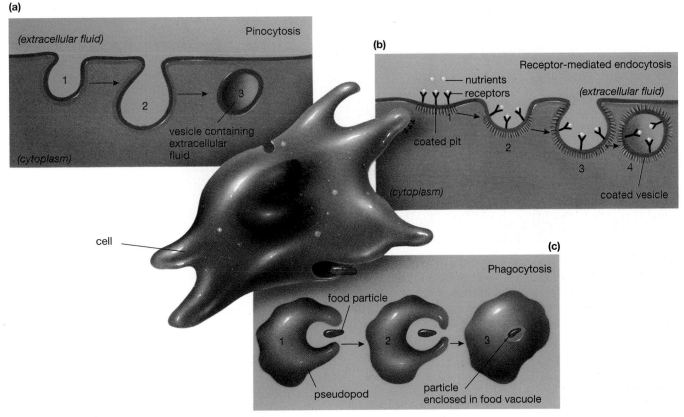

Figure 6-10   Three types of endocytosis

(a) Pinocytosis: A dimple in the plasma membrane deepens and eventually pinches off as a fluid-filled vesicle. The vesicle contains a random sampling of the extracellular fluid. (b) Receptor-mediated endocytosis: Receptor proteins selectively bind molecules (for example, nutrients) in the extracellular fluid. The receptors migrate along the fluid lipid bilayer of the plasma membrane to dimpling sites (coated pit). The membrane dimples inward, carrying the receptor–captured molecule complexes with it. The end of the coated pit buds off a coated vesicle into the cytoplasm of the cell. The vesicle contains both extracellular fluid and a high concentration of the molecules that bind to the receptors. (c) Phagocytosis: Extensions of the plasma membrane, called pseudopodia, encircle an extracellular particle (for example, food). The ends of the pseudopodia fuse, forming a large vesicle (a food vacuole) containing the engulfed particle.

Figure 6-11   Receptor-mediated endocytosis

These electron micrographs illustrate the sequence of events in receptor-mediated endocytosis through coated pits. (a) This type of endocytosis begins with a shallow depression in the plasma membrane, coated on the inside with a protein called clathrin (dark fuzzy substance in the micrographs) and bearing receptor proteins on the outside (not visible in the micrographs). The pit deepens (b), probably pulled in by elements of the cytoskeleton, and eventually pinches off as a coated vesicle (c, d). The clathrin protein is eventually recycled back to the plasma membrane, restarting the cycle.

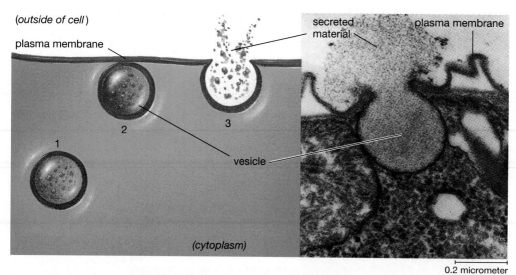

Figure 6-12   **Exocytosis**

Exocytosis is functionally the reverse of endocytosis. The material to be ejected from the cell is encapsulated into a membrane-bound vesicle that moves to the cell membrane and fuses with it. As the vesicle opens to the outside, the material within leaves by diffusion.

also occur across the internal membranes of a cell. For example, we will see in Chapters 7 and 8 that chloroplasts and mitochondria have specific protein channels that are permeable to hydrogen ions and that are crucial to ATP production in those organelles. Calcium ions are actively transported across certain internal membranes in muscle cells. And, if you look back at the section on membrane flow within cells in Chapter 5, you will see vesicles budding from and fusing with the membranes of the endoplasmic reticulum and Golgi complex, processes virtually identical to endocytosis and exocytosis. Transport across internal membranes is just as vital to the life of a cell as is transport across the plasma membrane, and similar mechanisms are used, as illustrated by transport in and out of cell vacuoles.

Vacuoles Help Regulate Cell Volume

Most cells contain one or more **vacuoles**, which are fluid-filled sacs surrounded by a single membrane (see Chapter 5). Some, such as the food vacuoles that form during phagocytosis, are temporary features of cells. Many cells, however, contain permanent vacuoles that have important roles in maintaining the integrity of the cell, especially by regulating the cell's water content.

Contractile Vacuoles Are Found in Freshwater Microorganisms

Freshwater protists such as *Paramecium* often contain complex **contractile vacuoles** composed of collecting ducts, a central reservoir, and a tube leading to a pore in the plasma membrane (Fig. 6-13). Because fresh water is hypotonic to

the cytosol of these organisms, water constantly enters the cell by osmosis. The increasing volume of incoming water might soon burst the fragile creature if it did not have a mechanism to excrete the water. Water is pumped into the collecting ducts and drains into the central reservoir (a process that requires cellular energy). When the reservoir is full, it contracts, squirting the water up the exit tube and out through the pore in the plasma membrane.

Central Vacuoles Are Found in Plant Cells

Three-quarters or more of the volume of many plant cells is occupied by a large **central vacuole** (see Fig. 5-7). The central vacuole has several functions. It provides a dump site for hazardous wastes, which plant cells often cannot excrete. Some plant cells store extremely poisonous substances, such as sulfuric acid, in their vacuoles; these substances deter animals from munching on the otherwise tasty leaves. Vacuoles may also store sugars and amino acids not immediately needed by the cell. Blue or purple pigments stored in central vacuoles are responsible for the colors of many flowers.

These dissolved substances make the vacuole contents hypertonic to the cell cytoplasm, which in turn is usually hypertonic to the extracellular fluid bathing the cells. Water therefore enters the vacuole by osmosis, which tends to make it swell. The pressure of the water within the vacuole, called **turgor pressure**, pushes the cytosol up against the cell wall with considerable force (Fig. 6-14). Primary cell walls are usually somewhat flexible, so the overall shape and rigidity of the cell depend on turgor pressure within the cell. Turgor pressure thus provides support for the nonwoody parts of plants. If you forget to water

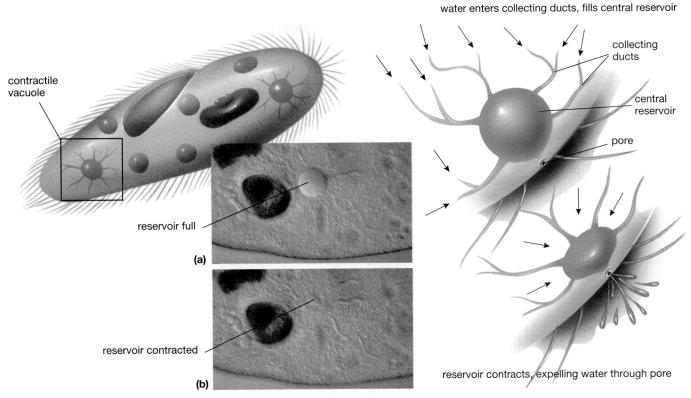

water enters collecting ducts, fills central reservoir

collecting ducts

central reservoir

pore

reservoir contracts, expelling water through pore

contractile vacuole

reservoir full

(a)

reservoir contracted

(b)

Figure 6-13   Contractile vacuoles

Many freshwater protists contain contractile vacuoles. **(a)** Osmosis causes water to constantly enter the cell, where it is taken up by collecting ducts and drains into the central reservoir of the vacuole. **(b)** When full, the reservoir contracts, expelling the water up the exit tube and outside the cell through a pore in the plasma membrane.

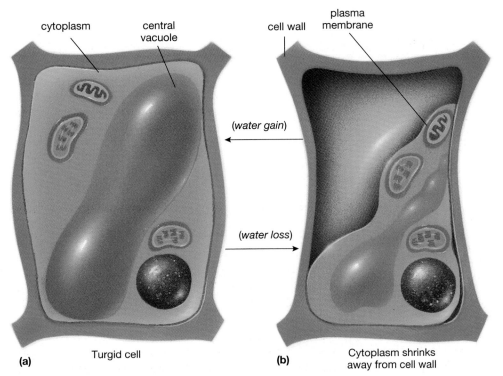

cytoplasm

central vacuole

cell wall

plasma membrane

(water gain)

(water loss)

(a)   Turgid cell

(b)   Cytoplasm shrinks away from cell wall

Figure 6-14   The role of turgor pressure in plant cells

(a) Water enters the central vacuole by osmosis, so pressure builds up within the vacuole. This pressure pushes the cytoplasm up against the cell wall, helping the cell to maintain its shape.
(b) When the central vacuole loses water and shrinks, a plant wilts. The cytoplasm pulls away from the cell wall, and the whole cell may become soft and shrunken.

your houseplants, the central vacuoles and cytosol lose water and the cell shrinks away from its cell walls. Just as a balloon goes limp when its air leaks out, so too the plant droops as its cells lose turgor pressure.

Cell Connections and Communication

In multicellular organisms, plasma membranes also function in holding together clusters of cells and in providing avenues through which cells can communicate with their neighbors. Depending on the organism and the cell type, one of four types of connection may occur between cells: desmosomes, tight junctions, gap junctions, and plasmodesmata.

Desmosomes Attach Cells Together

Animals, as you know, tend to be flexible, mobile organisms. Many of an animal's tissues are stretched, compressed, and bent as the animal moves about. If the skin, intestines, stomach, urinary bladder, and other organs are not to tear apart under the stresses of movement, their cells must adhere firmly to one another. Such animal tissues have junctions called **desmosomes** that hold adjacent cells together (Fig. 6-15). In a desmosome, the membranes of adjacent cells are glued together by proteins and carbohydrates. Intermediate filaments attached to the insides of the desmosomes extend into the interior of each cell, further strengthening the attachment.

Tight Junctions Leakproof the Cell

The animal body contains many tubes or sacs that must hold their contents without leaking: A leaky urinary bladder would spell disaster for the rest of the body. The spaces between the cells lining such sacs are sealed with strands of protein to form **tight junctions** (Fig. 6-16). The membranes of adjacent cells nearly fuse along a series of ridges, effectively forming waterproof gaskets between cells. Continuous tight junctions sealing each cell to its neighbors prevent molecules from escaping between cells.

Gap Junctions and Plasmodesmata Allow Communication between Cells

Multicellular organisms must coordinate the actions of their component cells. In animals, many cells, including heart muscle cells, most gland cells, some brain cells, and every cell of very young embryos, communicate through protein channels directly connecting the insides of adjacent cells (Fig. 6-17). These cell-to-cell channels are clustered in specialized regions called **gap junctions**. Hormones, nutrients, ions, and even electrical signals can pass through the channels at gap junctions.

Virtually all of the living cells of plants are connected to one another by much larger channels called **plasmodesmata** (Fig. 6-18). Each plasmodesma is a tube, lined with plasma membrane, that penetrates through the cell wall from the cytoplasm of one cell to the cytoplasm of its neighbor. Many plant cells have thousands of plasmodesmata. As a result, water, nutrients, and hormones pass quite freely from one cell to another.

Figure 6-15   **Desmosomes**
Cells lining the small intestine are firmly attached to one another by desmosomes. Intermediate filaments bound to the inside surface of each desmosome extend into the cytoplasm and attach to other elements of the cytoskeleton, strengthening the connection between cells.

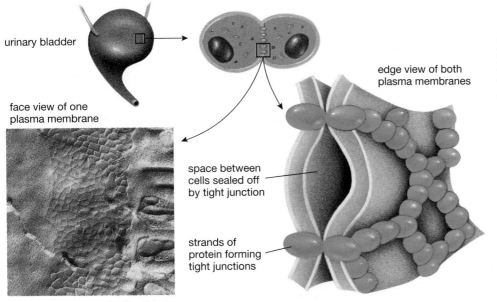

urinary bladder

face view of one
plasma membrane

edge view of both
plasma membranes

space between
cells sealed off
by tight junction

strands of
protein forming
tight junctions

Figure 6-16   Tight junctions

Leakage between cells of the
urinary bladder is prevented by
close-fitting tight junctions.

liver

face view of one plasma
membrane at gap junction

edge view of both
plasma membranes

pairs of
channels
connect
insides of
adjacent cells

masses of
gap junction
channels

plasma membranes

Figure 6-17   Gap junctions

Gap junctions contain cell-to-cell
channels that interconnect the
cytoplasm of adjacent cells.

root

upper cell

cell wall

plasmodesmata

lower cell

0.2 micrometer

plasma membranes

Figure 6-18   Plasmodesmata

Plant cells are widely interconnected by rather large cell-to-cell
pores called plasmodesmata.

EVOLUTIONARY CONNECTIONS

Natural Selection Has Produced Cell Membranes with Diverse Compositions

Although the membranes of all cells have a similar structure, membrane function varies tremendously from organism to organism and from cell to cell within a single organism. This diversity arises largely from the different proteins and phospholipids in the membrane, which have evolved under differing selection pressures.

Our discussion of membranes emphasized the unique functions of the membrane proteins. Consequently, you may think that the phospholipids are just a waterproof place for the proteins to reside. This isn't quite true, as we can see by examining the plasma membrane phospholipids in the legs of caribou, animals that live in very cold regions of North America (Fig. 6-19). During the long arctic winters of these regions, temperatures plummet far below freezing, and for caribou to keep their legs and feet really warm would waste precious energy. These conditions have favored the evolution of specialized arrangements of arteries and veins in caribou legs that allow the temperature of their lower legs to drop almost to freezing (0° C). The upper legs and main trunk of the body, in contrast, remain at about 37° C. Further, the phospholipids in the membranes of cells in the upper legs of caribou are very different from those near the hooves.

Remember, the membrane of a cell needs to be somewhat fluid to allow the proteins to move to sites where they are needed. The fluidity of a membrane is a function of the fatty acid tails of its phospholipids: Unsaturated fatty acids remain more fluid at lower temperatures than saturated fatty acids do (see Chapter 3). Caribou have a range of phospholipids in the plasma membranes of the cells in their legs. The membranes of cells near the chilly hoof have lots of unsaturated fatty acids, whereas the membranes of cells near the warmer trunk have more saturated ones. This arrangement gives the plasma membranes throughout the leg the proper fluidity despite great differences in temperature.

As important as the phospholipids are, the membrane proteins probably play the major roles in determining cell function and in governing the interactions between a cell and its neighbors. The extreme complexity of the protein molecule makes it more susceptible than other types of molecules to mutations that alter its amino acid composition, its shape, and hence its function. Over billions of years, an incredible diversity of proteins has evolved. Every nerve cell in your body, for instance, has membrane proteins essential for producing electrical signals and conducting them along the nerves to various parts of the body. Other membrane proteins receive chemical messages from neighboring nerve cells or from hormones and other chemicals in the blood. Each cell in the brain has a specific set of membrane proteins, allowing it to respond to some stimuli while ignoring others. In fact, your ability to read this page depends on the proteins residing in the membranes of your brain cells.

As we progress through this book, we shall return many times to the concepts of membrane structure presented in this chapter. Understanding the diversity of membrane lipids and proteins is the key to understanding not just the isolated cell, but entire organs, which function as they do largely because of the properties of the membranes of their component cells.

Figure 6-19   Caribou browse on the Alaskan tundra

The lipid composition of the membranes in the cells of a caribou's legs varies with distance from the trunk. Unsaturated phospholipids predominate in the lower leg, while more saturated phospholipids are found in the upper leg.

�ખ S U M M A R Y   O F   K E Y   C O N C E P T S

Cell Walls

The outer surface of some protist cells and of each bacterial, plant, and fungal cell is surrounded by a rigid cell wall outside the plasma membrane. The cell wall is produced by the cell that it surrounds. It protects and supports the cell.

The Plasma Membrane

The plasma membrane has four major functions: (1) isolate the cytoplasm from the external environment; (2) regulate the flow of materials into and out of the cell; (3) communicate with other cells; and (4) identify the cell. The membrane, according to the fluid mosaic model, consists of a bilayer of phospholipids in which are embedded a variety of proteins. There are three major categories of membrane proteins: (1) transport proteins, which regulate the movement of most water-soluble substances through the membrane; (2) receptor proteins, which bind molecules in the external environment, triggering changes in the metabolism of the cell; and (3) recognition proteins, which identify cells as to species and cell type.

Transport across Membranes

Particles in a fluid tend to move in response to gradients of concentration, pressure, or electrical charge. In cells, passive transport is the movement of substances across the plasma membrane down concentration gradients. Cellular energy is not used. Energy-requiring transport processes move substances across the membrane, usually against concentration gradients, through the use of cellular energy.

Passive Transport: Movement down Concentration Gradients

Diffusion is the movement of particles from regions of higher concentration to regions of lower concentration. In simple diffusion, water, dissolved gases, and lipid-soluble molecules diffuse through the phospholipid bilayer. In facilitated diffusion, water-soluble molecules cross the membrane through protein channels or with the assistance of protein carriers. In both cases, molecules move down their concentration gradients and cellular energy is not required.

Osmosis is the diffusion of water across a differentially permeable membrane, down its concentration gradient. Dissolved solutes decrease the concentration of free water molecules. Osmosis does not require cellular energy.

Energy-Requiring Transport across Membranes

In active transport, protein carriers in the membrane use cellular energy (ATP) to drive the movement of molecules across the plasma membrane, usually against concentration gradients. Large molecules (for example, proteins), particles of food, microorganisms, and extracellular fluid may be acquired by endocytosis, either by pinocytosis, receptor-mediated endocytosis, or phagocytosis. The secretion of substances such as hormones and the excretion of wastes from a cell are accomplished by exocytosis.

Transport across Intracellular Membranes

The intracellular membranes that surround organelles such as the endoplasmic reticulum, Golgi complex, mitochondria, chloroplasts, and vacuoles are fundamentally similar in structure and function to the plasma membrane. Substances are transported across these membranes by most of the same mechanisms that are involved in transport across the plasma membrane.

Cell Connections and Communication

Cells may be connected to one another by a variety of junctions. Desmosomes attach cells firmly to one another, preventing tearing of a tissue during movement or stress. Tight junctions seal off the spaces between adjacent cells, leakproofing organs such as the urinary bladder. Gap junctions in animals and plasmodesmata in plants are the locations at which the cytoplasm of two adjacent cells are interconnected by pores through adjoining plasma membranes.

✚ K E Y   T E R M S

active transport p. 111	desmosome p. 116	gap junction p. 116
carrier protein p. 105	differential permeability p. 107	glycoprotein p. 105
cell wall p. 102	diffusion p. 106	gradient p. 105
central vacuole p. 114	endocytosis p. 111	hypertonic p. 110
channel protein p. 105	exocytosis p. 112	hypotonic p. 110
chitin p. 102	facilitated diffusion p. 108	isotonic p. 110
concentration p. 105	fluid p. 105	middle lamella p. 102
concentration gradient p. 106	fluid mosaic model p. 102	osmosis p. 109
contractile vacuole p. 114	fluid-phase endocytosis p. 112	osmotic pressure p. 110

✖ T H I N K I N G   T H R O U G H   T H E   C O N C E P T S

Multiple Choice

1. Active transport through the plasma membrane occurs through the action of
   a. diffusion
   b. membrane proteins
   c. DNA
   d. water
   e. osmosis

2. The following is a characteristic of a cell membrane.
   a. It separates the cell contents from its environment.
   b. It is permeable to certain substances.
   c. It is a lipid bilayer with embedded proteins.
   d. It contains sodium-potassium pumps.
   e. all of the above

3. If an animal cell is placed into a solution whose concentration of solutes is higher than that inside the cell
   a. the cell will swell
   b. the cell will shrivel
   c. the cell will remain the same size
   d. the solution is described as hypertonic
   e. both b and d are correct

4. Small, nonpolar hydrophobic molecules such as fatty acids
   a. pass readily through a membrane's lipid bilayer
   b. diffuse very slowly through the lipid bilayer
   c. require special channels to enter a cell
   d. are actively transported across cell membranes
   e. must enter the cell via endocytosis

5. Which of the following would be least likely to diffuse through a lipid bilayer?
   a. water
   b. oxygen
   c. carbon dioxide
   d. sodium ions
   e. the small, nonpolar molecule butane

6. Which of the following pieces of evidence would show that a substance enters a cell by active transport rather than diffusion?
   a. A carrier protein moves the substance across the membrane.
   b. The substance enters the cell when its concentration is higher outside than inside.
   c. The breakdown of ATP is required for the substance to move into the cell.
   d. The substance moves through a channel protein.
   e. all of the above

Review Questions

1. Describe the process of cell wall formation in plant cells.

2. Describe and diagram the structure of a plasma membrane. What are the two principal types of molecules in plasma membranes? What are the four principal functions of plasma membranes?

3. What are the three types of proteins commonly found in plasma membranes, and what is the function of each?

4. Define diffusion and compare it with osmosis. How do these two forces help plant leaves to remain firm? What is turgor pressure?

5. Define hypotonic, hypertonic, and isotonic. What would be the fate of an animal cell immersed in each of the three types of solution?

6. Describe the following types of transport processes: simple diffusion, facilitated diffusion, active transport, pinocytosis, receptor-mediated endocytosis, phagocytosis, and exocytosis.

7. Name four types of cell-to-cell junctions and the function of each. Which allow communication between the interiors of adjacent cells?

✖ A P P L Y I N G   T H E   C O N C E P T S

1. Different cells have somewhat different plasma membranes. The plasma membrane of a *Paramecium*, for example, is only about 1% as water permeable as the plasma membrane of a human red blood cell. Referring to our discussion of the effects of osmosis on red blood cells and the role of contractile vacuoles in *Paramecium*, what do you think is the function of the low water permeability of *Paramecium*? What molecular differences do you think might account for this low water permeability?

2. A preview question for Chapter 24: The integrity of the plasma membrane is essential for cellular survival. Could this fact be utilized by the immune system to destroy foreign cells that have invaded the body? How might cells of the immune system disrupt membranes of foreign cells? (Two hints: Virtually all cells can secrete proteins, and some proteins form pores in membranes.)

3. Red blood cells will swell up and burst when placed in a hypotonic solution like pure water. Why don't we swell up and burst when we go for a swim in water that is hypotonic to our cells and body fluids?

4. Protists such as *Paramecia*, which locomote using cilia, and *Amoebae*, which locomote by extending pseudopods and ingest food particles by phagocytosis (see Figs. 6-13 and 6-10) have elaborate mechanisms to eliminate the water that flows into their bodies by osmosis. Plant cells, in contrast, are surrounded by rigid cell walls and use the water that flows in by osmosis to provide additional support. What problems would *Amoebae* and *Paramecia* have if their bodies were encased in cell walls? Why does this adaptation work well for plants?

✷ F O R M O R E I N F O R M A T I O N

Bretscher, M. S. "The Molecules of the Cell Membrane." *Scientific American*, October 1985. Beautifully illustrated, this article explores the structure and function of cell membranes, with special attention to cell junctions.

Dautry-Varsat, A., and Lodish, H. "How Receptors Bring Proteins and Particles into Cells." *Scientific American*, May 1984. Receptor-mediated endocytosis is an important pathway both for cell nutrition and for cell-to-cell signaling.

McNeil, P. L. "Cell Wounding and Healing." *American Scientist*, May-June 1991. The fluidity of plasma membrane phospholipids makes cells able to withstand minor damage.

Sharon, N., and Lis, H. "Carbohydrates in Cell Recognition." *Scientific American*, January 1993. Carbohydrates, usually attached to proteins as part of glycoproteins, identify cells, serve as parts of receptors in hormone binding, and regulate the attachment and movement of cells.

N E T W A T C H

On-line resources for this chapter are on the World Wide Web at:
http://www.prenhall.com/~audesirk (click on the table of contents link and then select Chapter 6).

7

Capturing Solar Energy: Photosynthesis

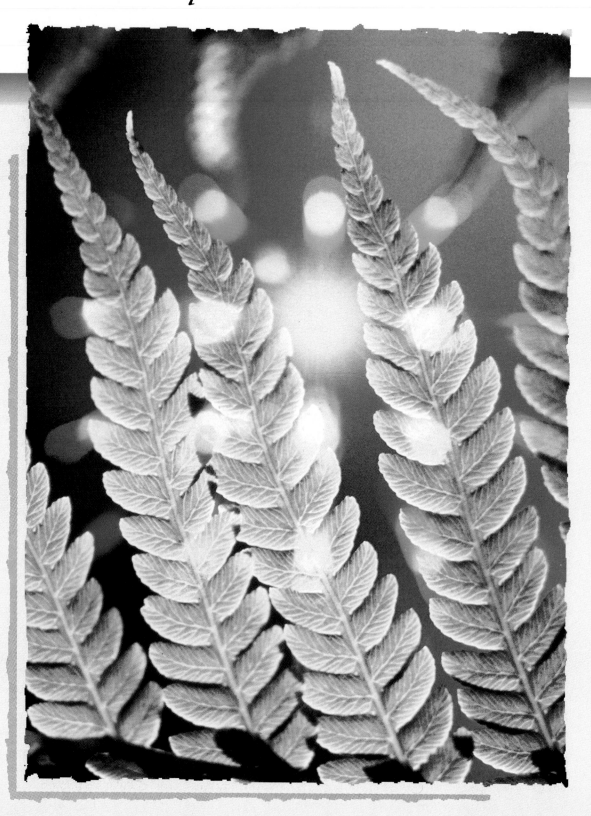

> "... the cells of life
> Bound themselves into clans, a multitude of cells
> To make one being—as the molecules before
> Had made of many one cell. Meanwhile they had invented
> Chlorophyll and ate sunlight, cradled in peace
> On the warm waves."
>
> ROBINSON JEFFERS, *The Beginning and the End*

A sunburst through the leaves of a fern depicts the harvesting of light energy by photosynthesis, the ultimate source of energy for nearly all life on Earth.

The flow of energy through life on Earth begins with the sun, but this was not always so. About 4.5 billion years ago, Earth formed as chunks of matter collided and fused, transforming their energy of movement into heat. Storms and volcanic eruptions released still more energy on the newly formed planet, but no organisms existed to harness the enormous fluctuations in energy, nor is it likely that any could have withstood the violence of that time. Energy-rich organic molecules were formed, however, their synthesis driven by heat and lightning (see Chapter 19). As Earth cooled and calmed, living cells arose, and for the first billion years or so, they fed on the soup of organic molecules provided by the earlier chemical cauldron. But the cells gradually consumed the organic molecules, and the soup thinned. Sources of organic energy became scarce.

All the while, another source of energy bathed the planet: light from the sun. Through chance mutations in their molecules, some cells acquired the ability to harness the energy in sunlight. They combined simple inorganic molecules—carbon dioxide and water—into more complex organic molecules such as glucose. In a process called photosynthesis, these cells captured a small fraction of the sunlight's energy, storing it as chemical energy in these complex organic molecules (Fig. 7-1, left). Exploiting this new source of energy without competition, photosynthetic cells filled the seas with their progeny. As evolution continued, several types of photosynthesis arose. The most common released oxygen as a by-product. This new element in the atmosphere was harmful to many organisms because oxygen combines readily with many molecules and disrupts stable chemical arrangements. But the endless variation produced by random mutation eventually included some cells that could survive in oxygen and, later, cells that could use oxygen to break down glucose in a new, more efficient process.

Before oxygen was pumped into the atmosphere by photosynthetic organisms, cells (including photosynthetic cells) broke down glucose in a process, called glycolysis, that does not utilize oxygen (Fig. 7-1, upper right). But glycolysis is inefficient; it reclaims only a small fraction of the chemical energy of glucose to drive essential cellular reactions. When cells evolved the enzymes to utilize oxygen to further break down the products of glycolysis in a process called cellular respiration, they were able to ex-

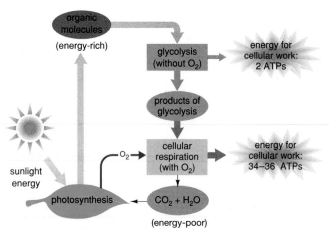

Figure 7-1   The relationship among photosynthesis, glycolysis, and cellular respiration

Chloroplasts in green plants use the energy of sunlight to synthesize high-energy carbon compounds such as glucose from low-energy molecules of water and carbon dioxide. Plants themselves, and other organisms that eat plants or one another, extract energy from these organic molecules by glycolysis and cellular respiration, yielding water and carbon dioxide once again. This energy in turn drives all the reactions of life. The orange arrows trace the flow of energy through a photosynthetic organism. Note that cellular respiration takes the products of glycolysis and extracts considerably more energy from them by combining them with oxygen.

tract about 18 times more energy from each glucose molecule (Fig. 7-1, lower right). Cells that carried on cellular respiration grew faster and reproduced more rapidly than cells relying on glycolysis alone.

The complementary reactions of photosynthesis and cellular respiration are perhaps 2 billion years old; together they drive the flow of energy through individual organisms and ecosystems. This chapter investigates photosynthesis, the process by which almost all useful energy enters the biosphere. Chapter 8 examines glycolysis and cellular respiration, by which organisms extract the energy captured by photosynthesis, conveying it to adenosine triphosphate (ATP), the principal energy-carrier molecule in living cells.

Photosynthesis: An Overview

Photosynthesis uses the energy of sunlight to synthesize energy-rich products—glucose and oxygen—from energy-poor reactants—carbon dioxide and water. Photosynthesis thus converts the electromagnetic energy of sunlight into chemical energy stored in the bonds of glucose and oxygen. The overall chemical reaction for photosynthesis is:

$$6\,CO_2 + 6\,H_2O + \text{light energy} \rightarrow C_6H_{12}O_6 + 6\,O_2$$

Photosynthesis occurs in plants, algae, and certain types of bacteria. In this chapter, we will restrict our discussion of pho-

tosynthesis to plants, with an emphasis on land plants. Photosynthesis in plants takes place within the chloroplasts, most of which are contained in leaf cells. Let's begin, then, with a brief look at the structures of leaves and chloroplasts.

Leaves and Chloroplasts Are Adaptations for Photosynthesis

The leaves of most land plants are only a few cells thick (Fig. 7-2a). The upper and lower surfaces of a leaf each consist of a layer of transparent cells, the epidermis. The outer surface of both epidermal layers is covered by a waxy, waterproof covering, the cuticle, that reduces the evaporation of water from the leaf. A leaf obtains CO_2 for photosynthesis from the air; adjustable pores in the epidermis, called **stomata**, open and close at appropriate times to admit CO_2. Inside the leaf are a few layers of cells collectively called **mesophyll** (which means simply "middle of the leaf"). The mesophyll cells contain the vast majority of a leaf's chloroplasts (Fig. 7-2b), and consequently photosynthesis occurs principally in these cells. Vascular bundles, or veins, supply water and minerals to the mesophyll cells and carry the sugars they produce to other parts of the plant.

As we discussed in Chapter 5, chloroplasts are organelles that consist of a double outer membrane enclosing a semifluid medium, the **stroma** (Fig. 7-2c, d). Embedded in the stroma are disc-shaped, interconnected membranous sacs called **thylakoids**. In most chloroplasts, the thylakoids are piled atop one another in stacks called **grana** (singular, **granum**).

The seemingly simple chemical reaction of photosynthesis actually involves dozens of enzymes catalyzing dozens of individual reactions. Conceptually, however, photosynthesis can be thought of as a pair of reactions coupled together by energy-carrier molecules (Fig. 7-2e). Each reaction occurs in a different site in the chloroplast.

1. In the **light-dependent reactions**, chlorophyll and other molecules in the membranes of the thylakoids capture sunlight energy and convert some of it into the chemical energy of energy-carrier molecules (ATP and NADPH).
2. In the **light-independent reactions**, enzymes in the stroma use the chemical energy of the carrier molecules to drive the synthesis of glucose or other organic molecules.

The Light-Dependent Reactions: Converting Light to Chemical Energy

The light-dependent reactions convert the energy of sunlight into the chemical energy of two different carrier molecules: the familiar energy carrier ATP and the electron carrier nicotinamide adenine dinucleotide phosphate (NADPH).

Figure 7-2   **An overview of photosynthesis**

(a) In land plants, photosynthesis usually occurs in the leaves. (b) Cells in the middle of the leaf, called mesophyll cells, contain chloroplasts, the organelles of photosynthesis (c, d). (e) Within the chloroplasts, sunlight energy is absorbed by pigments in the membranes of the thylakoids, and this energy is used to drive the synthesis of the energy carrier ATP and electron carrier NADPH (yellow). These are the light-dependent reactions. In the light-independent reactions, enzymes in the stroma of the chloroplast then use ATP and NADPH to synthesize carbohydrates from CO_2 and H_2O. In the process, ADP and $NADP^+$ (gray) are regenerated and pass back to the light-dependent reactions to be recharged.

During Photosynthesis, Light Is First Captured by Pigments in Chloroplasts

The sun emits energy in a broad spectrum of electromagnetic radiation, from short-wavelength gamma rays, through ultraviolet, visible, and infrared light, to very long-wavelength radio waves (Fig. 7-3a). As you may know, light and the other types of radiation are composed of individual packets of energy called **photons**. The energy of a photon corresponds to its wavelength: Short-wavelength photons are very energetic, whereas longer-wavelength photons have lower energies. Visible light consists of wavelengths with energies that are strong enough to alter the shape of certain pigment molecules but weak enough not to damage crucial molecules such as DNA.

Three processes may occur when light strikes an object such as a leaf: The light may be absorbed, reflected (bounced back again), or transmitted (passed through). Light that is absorbed can heat up the object or drive biological processes, such as photosynthesis. Light that is reflected or transmitted gives an object its color.

Chloroplasts contain several types of molecules that absorb different wavelengths of light. **Chlorophyll**, the key light-capturing molecule in thylakoid membranes, strongly absorbs violet, blue, and red light but reflects green, and therefore appears green (Fig. 7-3b).

Thylakoids also contain other molecules, called **accessory pigments**, that capture light energy and transfer it to chlorophyll. **Carotenoids** absorb blue and green light and appear yellow, orange, or red; **phycocyanins** absorb green and yellow and appear blue or purple. Because all wavelengths of light are absorbed to some degree by either chlorophyll, carotenoids, or phycocyanins, all wavelengths can drive photosynthesis to some extent (Fig. 7-3c).

The Light-Dependent Reactions Occur in Clusters of Molecules Called Photosystems

In the thylakoid membranes, chlorophyll, accessory pigment molecules, and electron-carrier molecules form highly organized assemblies called **photosystems**. Each thylakoid contains thousands of copies of two different kinds of photosystems, named photosystem I and photosystem II (Fig. 7-4a). Each consists of two major parts, a light-harvesting complex and an electron transport system.

The **light-harvesting complex** is composed of about 300 chlorophyll and accessory pigment molecules. These molecules absorb light and pass the energy to a specific chlorophyll molecule called the **reaction center**. By analogy with television reception, the light-absorbing pigments are called antenna molecules, because they gather energy and transfer it to the energy-processing reaction center. The reaction center chlorophyll is located next to the **electron transport system**, which is a series of electron carriers embedded in the thylakoid membrane.

When the reaction center chlorophyll receives energy from the antenna molecules, one of its electrons absorbs the energy, leaves the chlorophyll, and jumps over to the

Figure 7-3   Light, chloroplast pigments, and photosynthesis

(a) Visible light, a small part of the electromagnetic spectrum, consists of wavelengths that correspond to the colors of the rainbow. **(b)** The first step in photosynthesis is the absorption of light by pigment molecules in the thylakoid membranes of chloroplasts. Different types of pigments selectively absorb certain colors; the height of each line represents the ability of each pigment to absorb light of each color. Chlorophyll (green line) strongly absorbs violet, blue, and red light and therefore looks green. The other pigments in chloroplasts absorb other colors of light. **(c)** Photosynthesis is driven to some extent by all colors of light, owing to the absorption of light by several thylakoid pigments.

electron transport system. This energetic electron moves from one carrier to the next. At some of the transfers, the electron releases energy that drives reactions resulting in the synthesis of ATP from ADP or NADPH from NADP⁺ (NADP is the electron carrier NAD, described in Chapter 4, plus a phosphate group). This sequence of events is called **photophosphorylation**, because light energy (the "photo" in the name) is used to phosphorylate ("add a phosphate group to") ADP, thus forming ATP. With this overall scheme in mind, let's look a little more closely at the actual sequence of events in the light-dependent reactions (Fig. 7-4b).

(a)

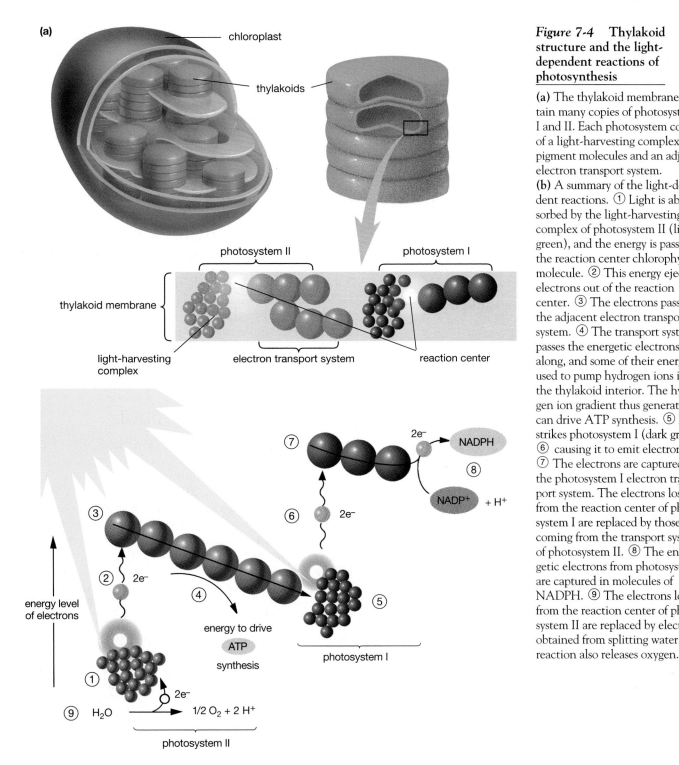

Figure 7-4 **Thylakoid structure and the light-dependent reactions of photosynthesis**

(a) The thylakoid membranes contain many copies of photosystems I and II. Each photosystem consists of a light-harvesting complex of pigment molecules and an adjacent electron transport system. **(b)** A summary of the light-dependent reactions. ① Light is absorbed by the light-harvesting complex of photosystem II (light green), and the energy is passed to the reaction center chlorophyll molecule. ② This energy ejects electrons out of the reaction center. ③ The electrons pass to the adjacent electron transport system. ④ The transport system passes the energetic electrons along, and some of their energy is used to pump hydrogen ions into the thylakoid interior. The hydrogen ion gradient thus generated can drive ATP synthesis. ⑤ Light strikes photosystem I (dark green), ⑥ causing it to emit electrons. ⑦ The electrons are captured by the photosystem I electron transport system. The electrons lost from the reaction center of photosystem I are replaced by those coming from the transport system of photosystem II. ⑧ The energetic electrons from photosystem I are captured in molecules of NADPH. ⑨ The electrons lost from the reaction center of photosystem II are replaced by electrons obtained from splitting water. This reaction also releases oxygen.

Photosystem II Generates ATP

For historical reasons, the photosystems are numbered "backward," and the usual process of light-energy capture is most easily understood by starting with photosystem II. The light-dependent reactions begin when a photon of light is absorbed by an antenna molecule in photosystem II (step 1 in Fig. 7-4b). The photon's energy passes from molecule to molecule until it reaches the reaction center, where it boosts an electron completely out of the chlorophyll molecule (step 2, which shows two electrons being boosted by two photons of light, for reasons explained below). The first electron carrier of the adjacent electron transport system instantly accepts these energized electrons (step 3). The electrons move from carrier to carrier, releasing energy as they go. Some of the energy is used to produce a hydrogen ion gradient within the thylakoid. This gradient drives the synthesis of ATP by a process known as **chemiosmosis** (step 4).

Photosystem I Generates NADPH

Meanwhile, light rays have also been striking the light-harvesting complex of photosystem I (step 5), each photon ejecting an electron from its reaction center chlorophyll (step 6, which shows two electrons being boosted by two photons of light). These electrons jump to photosystem I's electron transport system (step 7). Photosystem I's reaction center chlorophyll immediately obtains replacements for its lost electrons from the last electron carrier in photosystem II's electron transport system. Photosystem I's high-energy electrons move through its electron transport system to the electron carrier NADP⁺. Each NADP⁺ molecule picks up two energetic electrons and one hydrogen ion, forming NADPH (step 8). NADP⁺ and NADPH are both water-soluble molecules dissolved in the chloroplast stroma.

Splitting Water Maintains the Flow of Electrons through the Photosystems

Overall, electrons flow from the reaction center of photosystem II, through the photosystem II electron transport system, to the reaction center of photosystem I, through the photosystem I electron transport system, and on to NADPH. To sustain this one-way flow of electrons, photosystem II's reaction center must continuously be supplied with new electrons to replace the ones it gives up. These replacement electrons come from water (step 9). In a series of reactions that scientists are still working to uncover, photosystem II's reaction center chlorophyll attracts electrons from water molecules within the thylakoid compartment, causing the water molecules to split apart:

$$H_2O \rightarrow 1/2\ O_2 + 2\ H^+ + 2\ e^-$$

For every two photons captured by photosystem II, two electrons are boosted out of the reaction center chlorophyll and are replaced by the two electrons obtained by splitting one water molecule. As water molecules are split, their oxygen atoms combine to form molecules of oxygen gas, O_2. The oxygen may be used directly by the plant in its own respiration or it may be given off to the atmosphere.

✳ SUMMARY

LIGHT-DEPENDENT REACTIONS

(1) The light-dependent reactions begin with the absorption of light by the light-harvesting complex of photosystem II. (2) The light energy energizes electrons from the reaction center of the complex, causing them to be ejected. (3) The electrons are transferred to photosystem II's electron transport system. As the electrons pass through the transport system, they release energy. (4) Some of the energy is used to create a hydrogen ion gradient that drives ATP synthesis. (5) Meanwhile, light is absorbed by the light-harvesting complex of photosystem I. (6) The light energy ejects electrons from the reaction center that

(7) are picked up by photosystem I's electron transport system. Electrons lost from the reaction center are replaced by those coming from the transport system of photosystem II. (8) Some of this energy is captured as NADPH. (9) Finally, the "electron-deprived" chlorophyll of photosystem II attracts electrons from water molecules. A water molecule splits, donating electrons to the photosystem II chlorophyll and generating oxygen as a by-product.

The Light-Independent Reactions: Securing Chemical Energy in Glucose Molecules

The ATP and NADPH synthesized during the light-dependent reactions are dissolved in the stroma. Here, they provide the energy to power the synthesis of glucose from carbon dioxide and water. The reactions that eventually produce glucose are called the light-independent reactions, because they can occur independently of light as long as ATP and NADPH are available.

The C₃ Cycle Captures Carbon Dioxide

Carbon dioxide capture occurs in a set of reactions known as the **Calvin-Benson cycle** (after its discoverers), or the **C₃ (three-carbon) cycle**, because some of the important molecules in the cycle have three carbon atoms in them (Fig. 7-5). The C₃ cycle requires (1) CO_2 (normally from the air); (2) a CO_2-capturing sugar, ribulose bisphosphate (RuBP); (3) enzymes to catalyze all the reactions; and (4) energy in the form of ATP and NADPH (usually from the light-dependent reactions).

The C₃ cycle is most easily understood if we mentally divide it into three parts: carbon fixation, synthesis of phosphoglyceraldehyde (PGAL), and regeneration of ribulose bisphosphate. Keep track of the carbons as you follow along in Figure 7-5.

1. **Carbon fixation.** The C₃ cycle begins (and ends) with a five-carbon sugar, ribulose bisphosphate (RuBP). RuBP combines with CO_2 from the atmosphere to form an extremely unstable six-carbon compound. This compound spontaneously reacts with water to form two three-carbon molecules of phosphoglyceric acid (PGA), which gives the C₃ cycle its name. Capturing CO_2 is called carbon fixation because it "fixes" gaseous CO_2 into a relatively stable organic molecule (step 1 in Fig. 7-5).

2. **Synthesis of phosphoglyceraldehyde.** In a series of enzyme-catalyzed reactions, energy donated by ATP and NADPH (generated during the light-dependent reactions) is used to convert PGA to phosphoglyceraldehyde (PGAL; step 2).

3. **Regeneration of ribulose bisphosphate.** Through a complex series of reactions requiring ATP energy, 10

Figure 7-5   **The C$_3$ cycle of carbon fixation**

Carbon atoms are gray balls; depleted carrier molecules ADP and NADP$^+$ are gray ovals; and energized carriers ATP and NADPH are yellow ovals. (1) Six molecules of ribulose bisphosphate (RuBP) react with 6 molecules of CO$_2$ and 6 molecules of H$_2$O to form 12 molecules of phosphoglyceric acid (PGA). This reaction is carbon fixation, the capture of carbon from CO$_2$ into organic molecules. (2) The energy of 12 ATPs and the electrons and hydrogens of 12 NADPHs are used to convert the 12 PGA molecules to 12 phosphoglyceraldehydes (PGALs). (3) Energy from 6 ATPs is used to rearrange 10 PGALs into 6 RuBPs, completing one turn of the C$_3$ cycle. The remaining 2 PGAL molecules are further processed into glucose or other organic molecules such as glycerol, fatty acids, or the carbon skeleton of amino acids, depending on the needs of the plant.

molecules of PGAL (10 × 3 carbons) can regenerate the 6 molecules of RuBP (6 × 5 carbons; step 3) used at the start of carbon fixation.

Carbon Fixed during the C$_3$ Cycle Is Used to Synthesize Glucose

Obviously, if the C$_3$ cycle starts with RuBP, adds carbon from CO$_2$, and ends one "turn" with RuBP again, there is carbon left over from the captured CO$_2$. Using the simplest "carbon accounting" numbers shown in Figure 7-5, if you start and end one turn of the cycle with six molecules of RuBP, there are two molecules of phosphoglyceraldehyde left over. These two PGAL molecules (three carbons each) are combined to form one molecule of glucose (six carbons). Later, glucose may be broken down during cellular respiration or linked together in chains to form starch (a storage molecule) or cellulose (a major component of

cell walls), or modified further into amino acids, lipids, and other cellular constituents.

�֎ S U M M A R Y

LIGHT-INDEPENDENT REACTIONS

For the synthesis of one molecule of glucose through one turn of the C$_3$ cycle: Six molecules of CO$_2$ are captured by six molecules of RuBP. A series of reactions driven by energy from ATP and NADPH produces 12 molecules of PGAL. Two PGAL molecules join to become one molecule of glucose. Further ATP energy is used to regenerate the 6 RuBP molecules from 10 PGAL molecules.

The Relationship between the Light-Dependent and the Light-Independent Reactions

As Figure 7-2e illustrates, the light-dependent and light-independent reactions are closely coordinated. The light-dependent reactions in the thylakoids use light energy to "charge up" the energy-carrier molecules ADP and NADP$^+$ to form ATP and NADPH. These energized carriers move to the stroma, where their energy is used to drive glucose synthesis by means of the light-independent reactions. The depleted carriers, NADP$^+$ and ADP, then return to the light-dependent reactions for recharging to ATP and NADPH.

Water, CO$_2$, and the C$_4$ Pathway

Photosynthesis requires light and carbon dioxide, and lack of either one reduces the rate of photosynthesis. The light-dependent reactions of photosynthesis cannot occur in the dark, regardless of CO$_2$ levels; the light-independent reactions of carbon fixation cannot occur without a supply of CO$_2$, regardless of the light intensity. Therefore, you might think that an ideal leaf should have a large surface area to intercept lots of sunlight and be very porous to allow lots of CO$_2$ to enter the leaf from the air. For land plants, however, being porous to CO$_2$ also allows water vapor to be evaporated out of the leaf. Loss of water from the leaves is a prime cause of stress, and may be fatal, to land plants.

Many plants have evolved leaves that are a compromise between obtaining adequate CO$_2$ supplies and reducing water loss: large, waterproof leaves, with adjustable pores, the stomata, that admit carbon dioxide (see Fig. 7-2). When water supplies are adequate, the stomata open, letting in CO$_2$. If the plant is in danger of drying out, the stomata close. Closing the stomata reduces evaporation but at the cost of reducing CO$_2$ intake as well. Further, it also restricts the release of O$_2$, which is produced during photosynthesis.

When Stomata Are Closed to Conserve Water, Wasteful Photorespiration Occurs

What happens to carbon fixation when the stomata close? As it happens, the enzyme that catalyzes the reaction of RuBP with CO_2 is not very selective: It can cause *either* CO_2 *or* O_2 to combine with RuBP (Fig. 7-6). The reaction of O_2 with RuBP is the first step in a series of reactions, collectively called **photorespiration**, that generate CO_2, using carbon from RuBP. In photorespiration, oxygen instead of carbon dioxide becomes attached to RuBP. Thus photorespiration prevents carbon fixation by combining O_2 instead of CO_2 with RuBP. (Photorespiration, by the way, received its name because it uses up O_2 and produces CO_2. Unlike cellular respiration, it does not produce any useful cellular energy, such as ATP.)

Whether this matters very much depends on the relative amounts of carbon fixation versus photorespiration. The atmosphere contains much more O_2 (21%) than CO_2 (0.03%). In addition, the light reactions of photosynthesis release O_2 within the leaf. Therefore, some photorespiration occurs all the time, even under the best of conditions. During hot or dry weather, moreover, the stomata seldom open. The CO_2 that was initially present within the leaf is rapidly depleted by photosynthesis, new CO_2 from the air can't get in, and the O_2 generated by photosynthesis can't get out. With CO_2 levels low and O_2 levels very high, photorespiration dominates (Fig. 7-7a). Plants, especially seedlings, may die during hot, dry weather because respiration (both cellular respiration and photorespiration) exceeds photosynthesis, and the plants consume all their carbohydrate supplies by respiration.

C₄ Plants Reduce Photorespiration Using a Two-Stage Carbon Fixation Process

Some plants, described as C_4 plants, have evolved a way to reduce photorespiration and boost photosynthesis during dry weather. In the leaves of "regular" (C_3) plants, almost all the chloroplasts reside in the mesophyll cells, whereas the **bundle-sheath cells**, which surround the veins, lack chloroplasts. In plants adapted to hot, dry conditions, such as corn and crabgrass, both the mesophyll cells and the bundle-sheath cells contain chloroplasts (Fig. 7-7b). These plants use a two-stage carbon fixation pathway, the **C₄ pathway.**

The First Stage Forms a Four-Carbon Molecule That Gives C₄ Plants Their Name

In C_4 plants, the mesophyll cells contain a three-carbon molecule called phosphoenolpyruvate (PEP), instead of RuBP. CO_2 reacts with PEP to form a four-carbon molecule of oxaloacetic acid (hence the name, C_4 plants). This reaction is highly specific for CO_2 and is not hindered by high O_2 concentrations. This initial capture of carbon can continue for quite a while even when the stomata are closed and CO_2 concentrations within the leaf become

very low. Oxaloacetic acid is used as a shuttle to transport carbon from the mesophyll to the bundle-sheath cells.

The Second Stage Shuttles Carbon into the C3 Cycle

Oxaloacetic acid is transported into the bundle-sheath cells, where it breaks down, releasing CO_2 again. The high CO_2 concentration created in the bundle-sheath cells allows the RuBP + CO_2 reaction to fix carbon in the regular C_3 cycle. The remnant of the shuttle molecule returns to the mesophyll cells, where ATP energy is used to regenerate the PEP capture molecule.

C₃ and C₄ Plants Are Adapted to Specific Environmental Conditions

The C_4 pathway is advantageous when lots of light and little water are available. The C_4 pathway is not *always* better, though, because it uses more energy as it regenerates PEP. In some cases, the straightforward C_3 carbon fixation pathway is more efficient. For example, if water is plentiful the stomata of C_3 plants can stay open and let in lots of CO_2. Or, if light levels are low, photosynthesis proceeds slowly and C_4 plants may not be able to generate the extra energy they require.

A plant with the structures and enzymes of the C_4 pathway always uses the C_4 method of carbon fixation, regardless of environmental conditions. C_4 plants thrive, therefore, in deserts and in midsummer in temperate climates, when light energy is plentiful but water is scarce. C_3 plants have the advantage, however, in cool, wet, cloudy climates, where their exclusive use of the C_3 pathway is more energy efficient. This is why your lawn of lush Kentucky bluegrass (a C_3 plant) may be taken over by spiky crabgrass (a C_4 plant) during a long, hot summer.

Figure 7-6   Photorespiration

Carbon fixation through the C_3 cycle and photorespiration are alternative pathways for RuBP. Some photorespiration always occurs. When the stomata of a leaf close during hot, dry weather, CO_2 levels drop while O_2 levels remain high. Under these conditions, photorespiration predominates and the plant loses carbon.

(a) C₃ Plant

(b) C₄ Plant

Figure 7-7 **A comparison of C₃ and C₄ plants**

Under hot or dry conditions, plants have their stomata closed most of the time, restricting both CO_2 entry and O_2 emission. In these circumstances, C₄ plants fix more carbon, with less photorespiration, than C₃ do. **(a)** In C₃ plants, only the mesophyll cells carry out photosynthesis. All carbon fixation is by the C₃ pathway. With low CO_2 and high O_2 levels, photorespiration dominates in C₃ plants, because the enzyme that should catalyze the RuBP + CO_2 reaction catalyzes the RuBP + O_2 reaction instead. **(b)** In C₄ plants, both the mesophyll cells and bundle-sheath cells contain chloroplasts and participate in photosynthesis. The initial carbon fixation step in the mesophyll cells is a reaction between phosphoenolpyruvic acid (PEP) and CO_2, with which O_2 does not compete. A four-carbon molecule of oxaloacetic acid is produced, giving the C₄ pathway its name. The oxaloacetic acid then releases CO_2 in the bundle-sheath cells, thus maintaining a high CO_2 concentration in their chloroplasts. Higher CO_2 levels allow efficient carbon fixation in the C₃ pathway of the bundle-sheath cells with little photorespiration. Notice that the regeneration of PEP requires energy: Two phosphates are removed from ATP to produce AMP (adenosine monophosphate).

✖ SUMMARY OF KEY CONCEPTS

Photosynthesis: An Overview
Photosynthesis uses the energy of sunlight to convert low-energy inorganic molecules of carbon dioxide and water into high-energy organic molecules such as glucose. In plants, photosynthesis takes place in the chloroplasts, in two major steps: the light-dependent and the light-independent reactions. Photosynthesis is summarized in Figure 7-8.

The Light-Dependent Reactions: Converting Light to Chemical Energy
In the light-dependent reactions, light excites electrons in chlorophyll molecules and transfers the energetic electrons to electron transport systems. The energy of these electrons drives three processes:

1. Some of the energy from the electrons is used to pump hydrogen ions into the thylakoid sacs. The hydrogen

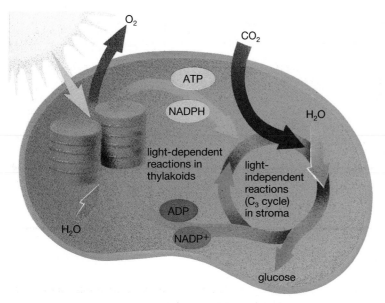

Figure 7-8   **A summary diagram for photosynthesis**
The light-dependent reactions in the thylakoids convert the energy of sunlight into the chemical energy of ATP and NADPH. Part of the sunlight energy is also used to split H_2O, forming O_2. In the stroma, the light-independent reactions (C_3 cycle) use the energy of ATP and NADPH to convert CO_2 and H_2O to glucose. The depleted carriers, ADP and $NADP^+$, return to the thylakoids to be recharged by the light-dependent reactions.

ion concentration is therefore higher inside the thylakoids than in the stroma outside. Hydrogen ions diffuse down this concentration gradient through ATP-synthesizing enzymes in the thylakoid membranes, providing the energy to drive ATP synthesis.

2. Some of the energy, in the form of energetic electrons, is added to electron-carrier molecules of $NADP^+$ to make the highly energetic carrier NADPH.

3. Some of the energy is used to split water, generating electrons, hydrogen ions, and oxygen.

The Light-Independent Reactions: Securing Chemical Energy in Glucose Molecules

In the stroma of the chloroplasts, ATP and NADPH provide the energy that drives the synthesis of glucose from CO_2 and H_2O. The light-independent reactions occur in a cycle of chemical reactions called the Calvin-Benson, or C_3, cycle. The C_3 cycle has three major parts:

1. In the carbon fixation step, CO_2 and water combine with ribulose bisphosphate (RuBP) to form phosphoglyceric acid (PGA).

2. PGA is converted to phosphoglyceraldehyde (PGAL), using energy from ATP and NADPH. PGAL may be used to synthesize organic molecules such as glucose.

3. 10 molecules of PGAL are used to regenerate 6 molecules of RuBP, again using ATP energy.

The Relationship between the Light-Dependent and the Light-Independent Reactions

The light-dependent reactions produce the energy carrier ATP and the electron carrier NADPH. The energy from these carriers is used in the synthesis of organic molecules during the light-independent reactions. The depleted carriers, ADP and $NADP^+$, return to the light-dependent reactions for recharging.

Water, CO_2, and the C_4 Pathway

The enzyme that catalyzes the reaction between RuBP and CO_2 is not very selective. It may also catalyze a reaction, called photorespiration, between RuBP and O_2. Photorespiration prevents carbon fixation and does not generate ATP. If CO_2 concentrations drop too low, or O_2 concentrations rise too high, photorespiration may exceed carbon fixation. Some plants have evolved an additional step for carbon fixation that minimizes photorespiration. In the mesophyll cells of these C_4 plants, CO_2 combines with phosphoenolpyruvic acid (PEP) to form the four-carbon molecule oxaloacetic acid. Oxaloacetic acid is transported into adjacent bundle-sheath cells, where it releases CO_2, thereby maintaining a high CO_2 concentration in the bundle-sheath cells. This CO_2 is then fixed in the C_3 cycle.

✷ K E Y   T E R M S

accessory pigments   p. 126
bundle-sheath cell   p. 130
Calvin-Benson cycle   p. 128
carbon fixation   p. 128
carotenoid   p. 126
C_3 cycle   p. 128
chemiosmosis   p. 127
chlorophyll   p. 126
C_4 pathway   p. 130

electron transport system   p. 126
granum   p. 124
light-dependent reactions   p. 124
light-harvesting complex   p. 126
light-independent reactions   p. 124
mesophyll   p. 124
photon   p. 126
photophosphorylation   p. 126
photorespiration   p. 130

photosynthesis   p. 124
photosystem   p. 126
phycocyanin   p. 126
reaction center   p. 126
stoma   p. 124
stroma   p. 124
thylakoid   p. 124

✖ T H I N K I N G   T H R O U G H   T H E   C O N C E P T S

Multiple Choice

1. Photosynthesis is measured in the leaf of a green plant exposed to different wavelengths of light. Photosynthesis is
   a. highest in green light   b. highest in red light
   c. highest in blue light   d. highest in red and blue light
   e. the same at all wavelengths

2. Where do the light-dependent reactions of photosynthesis occur?
   a. in the guard cells of the stomata
   b. in the chloroplast stroma
   c. within the thylakoid membranes of the chloroplast
   d. in the leaf cell cytoplasm
   e. in leaf cell mitochondria

3. The oxygen produced during photosynthesis comes from
   a. the breakdown of CO_2
   b. the breakdown of H_2O
   c. the breakdown of both CO_2 and H_2O
   d. the breakdown of oxaloacetic acid
   e. photorespiration

4. The role of accessory pigments is to
   a. provide an additional photosystem to generate more ATP
   b. allow photosynthesis to occur in the dark
   c. prevent photophosphorylation
   d. donate electrons to chlorophyll reaction centers
   e. capture additional light energy and transfer it to the chlorophyll reaction centers

5. The generation of ATP by electron transport in photosynthesis and cellular respiration depends upon
   a. a proton gradient across a membrane
   b. proton pumps driven by electron transport chains
   c. an ATP-synthesizing enzyme complex
   d. a, b, and c are all required for ATP generation
   e. none of the above are required for ATP generation

6. Where do the light-independent, carbon-fixing reactions occur?
   a. in the guard cell cytoplasm
   b. in the chloroplast stroma
   c. within the thylakoid membranes
   d. at night in the thylakoids
   e. in mitochondria

Review Questions

1. Write the overall equation for photosynthesis. Does it differ between C_3 and C_4 plants?

2. Draw a labeled diagram of a chloroplast. Explain specifically how chloroplast structure is related to its function.

3. Briefly describe the light-dependent and light-independent reactions. In what part of the chloroplast does each occur?

4. What is the difference between carbon fixation in C_3 and C_4 plants? Under what conditions does each mechanism of carbon fixation work most effectively?

✖ A P P L Y I N G   T H E   C O N C E P T S

1. Many lawns and golf courses are planted with bluegrass, a C_3 plant. In the spring, the bluegrass grows luxuriously. In the summer, a weed called crabgrass, a C_4 plant, often appears and spreads rapidly. Explain this sequence of events, given the normal weather conditions of spring and summer and the characteristics of C_3 versus C_4 plants.

2. You continuously monitor the photosynthetic oxygen production from the leaf of a higher plant illuminated by white light. Explain what will happen (and why) if you place (a) red, (b) blue, and (c) green filters between the light source and the leaf.

3. A plant is placed in a CO_2-free atmosphere in bright light. Will the light-dependent reactions continue to generate ATP and NADPH indefinitely? Explain how you reached your conclusion.

4. You are called before the House Ways and Means Committee to explain why the U.S. Department of Agriculture should continue to fund photosynthesis research. How would you justify an expensive project to genetically engineer the enzyme that catalyzes the reaction of RuBP with CO_2 to prevent it from reacting with oxygen as well as CO_2. What are the potential applied benefits of this research?

✖ F O R   M O R E   I N F O R M A T I O N

Bazzazz, F. A., and Fajer, E. D. "Plant Life in a CO_2-Rich World." *Scientific American*, January 1992. Burning fossil fuels is increasing CO_2 levels in the atmosphere (see Chapter 45). This increase could tip the balance between C_3 and C_4 plants.

Govindjee, and Coleman, W. J. "How Plants Make Oxygen." *Scientific American*, February 1990. The generation of oxygen during photosynthesis is just beginning to be understood.

Grodzinski, B. "Plant Nutrition and Growth Regulation by CO_2 Enrichment." *BioScience*, 1992. How higher CO_2 levels influence plant metabolism.

Mooney, H. A., Drake, B. G., Luxmoore, R. J., Oechel, W. C., and Pitelka, L. F. "Predicting Ecosystems' Responses to Elevated CO_2 Concentrations." *BioScience*, 1994. What effects will CO_2 enrichment of the atmosphere due to human activities have on ecosystems?

N E T   W A T C H

On-line resources for this chapter are on the World Wide Web at:
http://www.prenhall.com/~audesirk (click on the table of contents link and then select Chapter 7).

8

Harvesting Energy from Food: Glycolysis and Cellular Respiration

An Anna's hummingbird feeds at a flower in California.

Whenh a hummingbird sips nectar from a flower, it eats the products of photosynthesis. Photosynthesis in the plant's leaves had converted the energy of sunlight into the chemical energy of organic molecules such as the sugar in nectar. To supply its enormous energy needs—the highest of any animal—the hummingbird must eat its weight in nectar daily. Then its cells must efficiently extract energy from the glucose in the nectar. When the hummingbird's cells break down the high-energy glucose molecules in the presence of oxygen, they produce the low-energy molecules the plant started with—carbon dioxide and water—making energy available for muscles and growth (Fig. 7-1).

However, the hummingbird's cells cannot directly use the chemical energy derived from this process. Its wing muscles require energy stored in the molecule adenosine triphosphate (ATP). Its brain uses ATP in conducting nerve signals, and its ovaries use ATP in making eggs. The plant, for its part, uses ATP to make the petals, pigments, and fragrance that attract the bird and the leaves and chlorophyll molecules that trap the sun's energy. In this chapter, we explore the metabolic processes by which organisms convert the energy of organic molecules into the usable energy of ATP, processes called glycolysis and respiration.

Most cells can metabolize a variety of organic molecules to produce ATP. We focus on the metabolism of the single molecule glucose for three reasons. First, virtually all single-celled organisms, and every cell of multicellular organisms, metabolize glucose for energy at least part of the time. Some, such as nerve cells in your brain, rely predominantly on glucose as a source of energy. Second, glucose metabolism is less complex than the metabolism of most other organic molecules. Finally, when cells use other organic molecules as energy sources, they usually convert the molecules to glucose or other compounds that enter the pathways of glucose metabolism (see "Health Watch: Metabolic Transformations— Why You Can Get Fat by Eating Sugar").

Glucose Metabolism: An Overview

During photosynthesis, photosynthetic organisms harvest and store the energy of sunlight in glucose. During glucose breakdown, that energy is released and converted to

ATP. The chemical equations for glucose formation by photosynthesis and the complete metabolism of glucose back to CO_2 and H_2O (the original reactants in photosynthesis) are almost perfectly symmetrical:

Photosynthesis

$$6\,CO_2 + 6\,H_2O + \text{sunlight energy} \rightarrow C_6H_{12}O_6 + 6\,O_2$$

Complete Glucose Metabolism

$$C_6H_{12}O_6 + 6\,O_2 \rightarrow 6\,CO_2 + 6\,H_2O + \text{energy}$$

This symmetry might lead you to think that a cell can convert all of the chemical energy contained in a glucose molecule to high-energy bonds of ATP. Unfortunately, according to the second law of thermodynamics, "you can't break even"; in other words, the conversion of energy into different forms always results in the decrease of the amount of concentrated, useful energy. In fact, most of the "energy" listed on the right side of the second equation is heat, not ATP. Nevertheless, a cell can extract a great deal of ATP energy from glucose if the glucose molecule is completely broken down to CO_2 and H_2O.

Figure 8-1 summarizes the major steps of glucose metabolism in eukaryotic cells. The first stage, glycolysis, does not require oxygen and proceeds in exactly the same way under both aerobic (with oxygen) and anaerobic (without oxygen) conditions. Glycolysis splits apart a single glucose molecule (a six-carbon sugar) into two three-carbon molecules of pyruvic acid. This splitting releases a small fraction of the chemical energy stored in the glucose, some of which is used to generate two ATP molecules. The presence of oxygen becomes an issue only in the processes that follow glycolysis. Under anaerobic conditions, the pyruvic acid is usually converted by fermentation into lactic acid or ethanol. Fermentation does not produce any more ATP energy. Both glycolysis and fermentation occur in the cell cytosol.

The pyruvic acid produced by glycolysis may also enter the mitochondria. Here, if oxygen is available, cellular respiration uses oxygen to break pyruvic acid down completely to carbon dioxide and water, generating an additional 34 or 36 ATP molecules (the amount differs among cells). The extra ATP produced by cellular respiration is so important to most organisms that anything that inter-

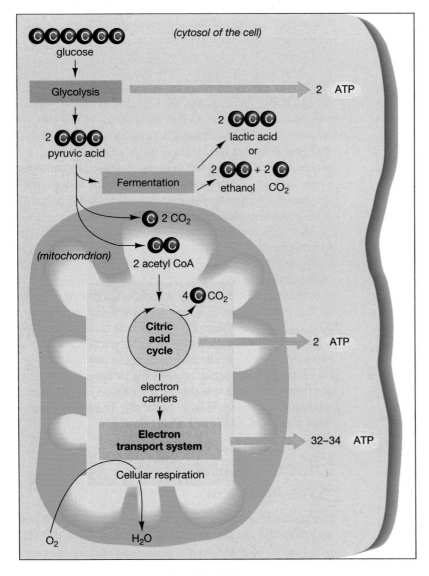

Figure 8-1   A summary of glucose metabolism

Refer to this diagram as we progress through the reactions of glycolysis (in the cell cytosol) and cellular respiration (in the mitochondria). Note that the breakdown of glucose occurs in stages, with various amounts of energy harvested as ATP along the way. The vast majority of the ATP is produced in the mitochondria, justifying their nickname, "powerhouse of the cell."

feres with its production, such as lack of oxygen or the poison cyanide, quickly results in death.

Glycolysis

Glycolysis (in Greek, "to break apart a sweet") is a complex sequence of reactions in the cytosol of a cell in which a molecule of glucose is broken down into two molecules of pyruvic acid. This breakdown results in a small energy gain of two molecules of ATP and two molecules of the electron carrier NADH. Reduced to its essentials, glycolysis consists of two major steps: glucose activation and energy harvest (Fig. 8-2).

Glucose Must Be Activated by ATP before Its Energy Can Be Harvested

In glucose activation, a molecule of glucose undergoes two enzyme-catalyzed reactions, each of which uses ATP energy. These reactions convert a relatively stable glucose molecule into a highly reactive molecule of fructose diphosphate. Fructose is a sugar molecule similar (but not identical) to glucose; diphosphate refers to the two phosphate groups acquired from the ATP molecules. Forming fructose diphosphate costs the cell two ATP molecules. But this initial consumption of energy is necessary to ultimately produce much greater energy returns.

In the energy harvest steps, fructose diphosphate splits apart into two three-carbon molecules of phosphoglyceraldehyde (PGAL; in Chapter 7 we encountered PGAL in the C$_3$ cycle of photosynthesis). The two PGAL molecules then go through a series of reactions that culminate in the production of two molecules of pyruvic acid, one from each PGAL. Two of these reactions are coupled to ATP synthesis, generating two ATP molecules per PGAL, for a total of four ATPs. Because two ATPs were used to activate the glucose molecule in the first place, *there is a net gain of only two ATPs per glucose molecule.* At another step along the way from PGAL to pyruvic acid, two high-

energy electrons and a hydrogen ion are added to the "empty" electron carrier NAD$^+$ to make the "energized" carrier NADH (note that these molecules are slightly different from NADP$^+$ and NADPH encountered in the light-dependent reactions of photosynthesis). Two PGAL molecules are produced per glucose molecule, so two NADH carrier molecules are formed when PGAL is converted to pyruvic acid.

❖ S U M M A R Y

GLYCOLYSIS

Each molecule of glucose is broken down to two molecules of pyruvic acid. During these reactions, two ATP molecules and two NADH electron carriers are formed.

Carrier molecules such as NAD$^+$ capture energy by accepting high-energy electrons, and they can carry these electrons to sites where their energy is used to form ATP. One major difference between anaerobic and aerobic glucose breakdown is the use that is made of these high-energy electrons. In the absence of oxygen, pyruvic acid acts as the electron acceptor during fermentation, producing molecules (ethanol or lactic acid) that the cell cannot use. During cellular respiration, oxygen becomes the electron acceptor, allowing the pyruvic acid to be fully broken down and its energy harvested as ATP.

Fermentation

The electrons carried in NADH are highly energetic, but their energy can be used to synthesize ATP only when oxygen is available (see "Cellular Respiration," below). Many organisms (particularly microorganisms) thrive in the guts of animals, deep in soil, in sediments underlying lakes and oceans, or in bogs and marshes where oxygen is nearly or entirely absent. As described below, even some

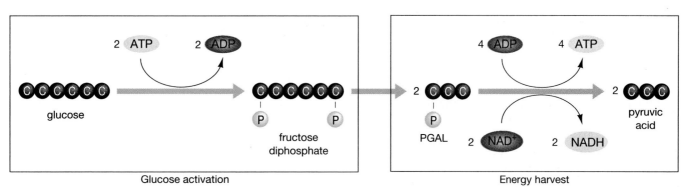

Figure 8-2 The essentials of glycolysis

(1) Glucose activation: The energy of two ATP molecules is used to convert glucose to the highly reactive fructose diphosphate. Fructose diphosphate splits into two smaller, but still reactive, molecules of phosphoglyceraldehyde (PGAL). (2) Energy harvest: Both PGAL molecules go through a series of reactions that generate four ATP and two NADH molecules. Therefore, glycolysis results in a net harvest of two ATP and two NADH molecules per glucose molecule.

Some Cells Ferment Pyruvic Acid to Lactic Acid

When you exercise vigorously, your muscles need lots of ATP as an energy source (Fig. 8-3a). Even though cellular respiration generates much more ATP than glycolysis alone, *cellular respiration is limited by organismal respiration,* that is, by the ability of the organism to provide oxygen (by breathing, for example). While you exercise vigorously, such as by sprinting to class after oversleeping, you may not be able to get enough air into your lungs and enough oxygen out of this air, into your blood, and delivered to your muscles to keep cellular respiration going fast enough. When deprived of oxygen, your muscles do not immediately stop working. Instead, glycolysis continues for a while, providing its meager two ATP molecules per glucose and generating both pyruvic acid and NADH.

To regenerate NAD⁺, muscle cells convert pyruvic acid molecules to lactic acid, using electrons and hydrogen ions from NADH:

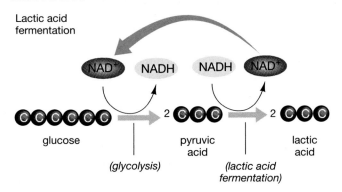

The regenerated NAD⁺ molecules produced by this conversion again are available to accept electrons during glycolysis, and energy production can continue. Lactic acid, however, is toxic in high concentrations. Soon, it causes intense discomfort and fatigue, causing you to stop or at least slow down. As you rest in the classroom, breathing rapidly after your sprint, oxygen once more becomes available and the lactic acid is reconverted to pyruvic acid. Interestingly enough, the conversion from lactic acid to pyruvic acid occurs not in the muscle cells, which lack the necessary enzymes, but in the liver. This pyruvic acid is then broken down through cellular respiration to carbon dioxide and water.

Various microorganisms also use lactic acid fermentation (some ferment even when oxygen is present, others are poisoned by oxygen), including the bacteria that produce yogurt, sour cream, and cheese. As you may know, acids taste sour, and lactic acid contributes to the distinctive tastes of these foods.

Other Cells Ferment Pyruvic Acid to Alcohol

Many microorganisms use another process to regenerate NAD⁺ under anaerobic conditions: alcoholic fermentation. This series of reactions produces ethanol and CO_2 (rather than lactic acid) from pyruvic acid, using hydrogen ions and electrons from NADH:

Figure 8-3   Fermentation

(a) A sprinter at the end of a race. The runner's respiratory and circulatory systems cannot supply oxygen to her leg muscles fast enough to keep up with the demand for energy, so glycolysis and lactic acid fermentation must provide the energy. Panting after the race brings in the oxygen needed to remove the lactic acid through cellular respiration. (b) Champagne spurts out of a newly opened bottle, driven out by CO_2 formed by fermenting yeast and trapped in the bottle by the cork.

of our own body cells must cope without oxygen for brief periods. In anaerobic conditions (under which life, and probably glycolysis, evolved), NADH production is *not* a method of energy capture; it is actually a way of getting rid of hydrogen ions and electrons produced during the metabolism of glucose to pyruvic acid. But this disposal method poses a problem for the cell because NAD⁺ is used up as it accepts electrons and hydrogen ions in becoming NADH. Without a way to regenerate NAD⁺ and to dispose of the electrons and hydrogen ions, glycolysis would have to stop once the supply of NAD⁺ was exhausted.

Fermentation solves this problem by enabling pyruvic acid to act as the final acceptor of electrons and hydrogen ions from NADH. NAD⁺ is thereby regenerated for use in further glycolysis. There are two main types of fermentation; one converts pyruvic acid to lactic acid, and the other converts it to carbon dioxide and ethanol.

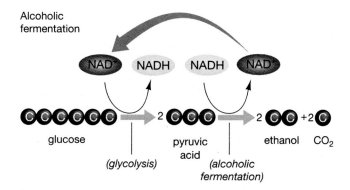

Alcoholic
fermentation

glucose   (glycolysis)   pyruvic
acid   (alcoholic
fermentation)   ethanol CO_2

Alcoholic fermentation is somewhat risky for a microbe because the alcohol it generates is poisonous. If the alcohol concentration in its environment becomes too great, the microbe will die. Most yeasts (single-celled fungi), for example, die if the alcohol concentration exceeds 12%, although certain strains developed for wine production do not succumb until the alcohol concentration reaches 18%. (Under natural conditions, of course, yeasts are unlikely to find themselves confined, as they are in a wine bottle, in a small volume of water, with lots of sugar for energy production and no oxygen—selective pressure for alcohol tolerance was probably very slight before humans developed a taste for wine!)

Sparkling wines, such as champagne, are bottled while the yeasts are still alive and fermenting, trapping both the alcohol and the CO_2. When the cork is removed, the pressurized CO_2 is released, sometimes rather explosively (Fig. 8-3b). We add baker's yeast to bread dough so that the CO_2 it produces will make the bread rise; the alcohol the yeast generates evaporates while the bread is baking.

Cellular Respiration

Cellular respiration is a series of reactions in which the pyruvic acid produced by glycolysis is broken down to carbon dioxide and water and large amounts of ATP are produced. The final reactions of cellular respiration require oxygen because oxygen acts as the final acceptor of electrons.

In eukaryotic cells, cellular respiration occurs in the mitochondria. You may recall from Chapter 5 that a mitochondrion has two membranes that produce two compartments: an inner compartment enclosed by the inner membrane and containing the fluid **matrix**, and an **intermembrane compartment** between the two membranes (Fig. 8-4). Reactions involving enzymes in the matrix, electron transport proteins in the inner membrane, and the movement of hydrogen ions through ATP-synthesizing proteins in the inner membrane yield most of the ATP produced during cellular respiration.

Figure 8-4 summarizes the main events of cellular respiration.

Step 1: Glycolysis in the cell cytosol produces two molecules of pyruvic acid per molecule of glucose.

Step 2: The pyruvic acids are transported across both mitochondrial membranes into the matrix.

Step 3: Each pyruvic acid is split into CO_2 and a two-carbon acetyl group, which enters the citric acid cycle (also called the Krebs cycle). The citric acid cycle releases the remaining carbons as CO_2, produces one ATP, and donates energetic electrons to several electron-carrier molecules.

Step 4: The electron carriers donate their energetic electrons to the electron transport system of the inner membrane. The energy of the electrons transports H^+ from the matrix to the intermembrane compartment. At the end of the system, the electrons combine with O_2 and H^+ to form H_2O.

Step 5: In chemiosmosis, the hydrogen ion gradient created by the electron transport system discharges through ATP-synthesizing enzymes in the inner membrane, and the energy is used to produce large amounts of ATP.

Step 6: The ATP is transported back out of the mitochondrion into the cytosol, available to drive needed reactions in the rest of the cell.

We have already discussed glycolysis; now let's look a little more closely at the processes of cellular respiration in the mitochondria.

Pyruvic Acid Is Transported into the Mitochondrial Matrix

The outer membrane of mitochondria contains many large pores that are very permeable to pyruvic acid. Facilitated diffusion through proteins in the inner membrane allows pyruvic acid to pass down its concentration gradient from the cytosol, where it is synthesized as the end product of glycolysis, to the mitochondrial matrix, where it is used in cellular respiration.

Pyruvic Acid Is Broken Down by Reactions in the Mitochondrial Matrix

In the matrix, pyruvic acid reacts with a molecule called coenzyme A (Fig. 8-5, left). Each pyruvic acid is split into CO_2 and a two-carbon molecule called an acetyl group, which immediately attaches to coenzyme A to form the acetyl–coenzyme A complex (acetyl CoA for short). During this reaction, two energetic electrons and a hydrogen ion are transferred to NAD^+, forming NADH.

The two acetyl CoA molecules then enter a cyclic pathway known as either the **Krebs cycle**, named after its discoverer Hans Krebs, or the **citric acid cycle**, named after the first product in the reaction sequence, citric acid (Fig. 8-5, right). Each acetyl CoA briefly combines with a molecule of oxaloacetic acid. The two-carbon acetyl group is donated to the four-carbon oxaloacetic acid (which, as you may remember, also functions in the carbon fixation stage of the C_4 pathway) to form the six-carbon citric acid. Coenzyme A is released once again (coenzyme A, like an

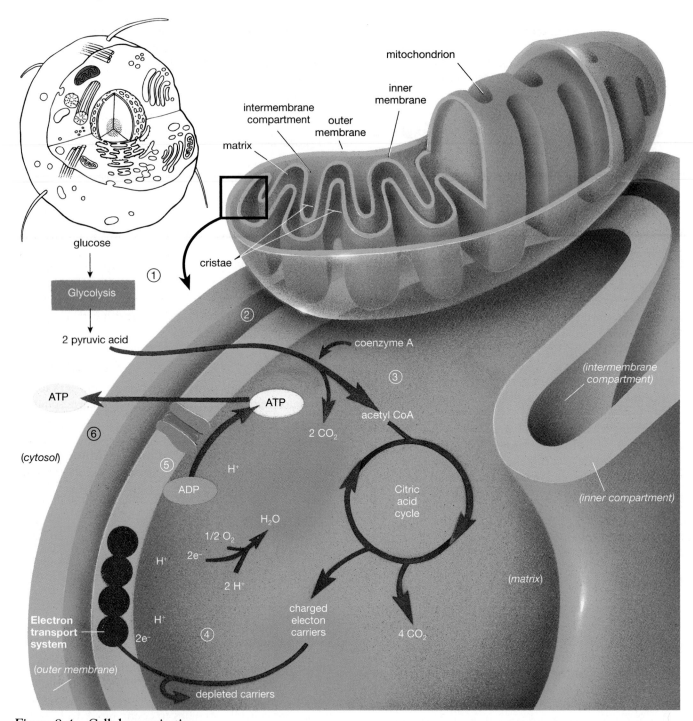

Figure 8-4   **Cellular respiration**

Cellular respiration takes place in the mitochondria, whose structure, like that of a chloroplast, reflects the compartmentalized reactions that occur there. The inner membrane separates the inner compartment, containing the soluble enzymes of the matrix, from the intermembrane compartment (between the inner and outer membranes). The lower part of the diagram summarizes the six essential steps in glucose metabolism, from the initial glycolysis in the cytosol to the final transport of ATP out of the mitochondrion and into the cytosol.

enzyme, is not permanently altered during these reactions, and is reused many times). Mitochondrial enzymes then lead each citric acid through a number of rearrangements that regenerate the oxaloacetic acid, give off two CO_2 molecules, and capture most of the energy of the acetyl group as one ATP and four electron carriers, one $FADH_2$ (flavin adenine dinucleotide) and three NADH.

▓ S U M M A R Y

THE MITOCHONDRIAL MATRIX REACTIONS

Synthesis of acetyl CoA produces one CO_2 and one NADH per pyruvic acid. The citric acid cycle produces two CO_2, one ATP, three NADH, and one $FADH_2$ per acetyl CoA. Therefore, at the conclusion of the matrix reactions, the two pyruvic acids produced from a single glucose molecule have been completely broken down by adding oxygen to form six CO_2 molecules. In the process, two ATPs, eight NADH, and two $FADH_2$ electron carriers have been produced.

Energetic Electrons Are Carried to Electron Transport Systems in the Inner Mitochondrial Membrane

At this point the cell has gained only four ATP molecules from the original glucose molecule: two during glycolysis and two during the citric acid cycle. The cell has, however, captured many energetic electrons in carrier molecules: 2 NADH during glycolysis and 8 more NADH and 2 $FADH_2$ from the matrix reactions, for a total of 10 NADH and 2 $FADH_2$. The carriers deposit their electrons in **electron transport systems** located in the inner mitochondrial membrane (Fig. 8-6). These electron transport systems are similar in function to those embedded in the thylakoid membrane of chloroplasts. The energetic electrons move from molecule to molecule along the transport systems. Energy released by the electrons during these transfers is used to pump hydrogen ions from the matrix, across the inner membrane, into the intermembrane compartment (see the section on chemiosmosis, below).

Finally, at the end of the electron transport system, oxygen and hydrogen ions accept the energetically depleted electrons: Two electrons, one oxygen atom, and two hydrogen ions combine to form water. This step clears out the transport system, leaving it ready to run more electrons through. Without oxygen, the electrons would "pile up" in the transport system, with the result that the hy-

drogen ions would not be pumped across the inner membrane. The hydrogen ion gradient would soon dissipate and ATP synthesis would stop.

Chemiosmosis Captures Energy Stored in a Hydrogen Ion Gradient and Produces ATP

Hydrogen ion pumping across the inner membrane generates a large concentration gradient; that is, a high concentration of hydrogen ions in the intermembrane compartment and a low concentration in the matrix. The inner membrane is impermeable to hydrogen ions except at protein pores that are part of ATP-synthesizing enzymes. In the process known as **chemiosmosis**, hydrogen ions move down their concentration gradient from the intermembrane compartment to the matrix through these ATP-synthesizing enzymes. The flow of hydrogen ions provides the energy to synthesize 32 to 34 molecules of ATP from ADP and inorganic phosphate in the matrix.

ATP Is Transported from the Mitochondrial Matrix and Diffuses into the Cell Cytosol

The ATP that was synthesized from ADP and inorganic phosphate in the matrix during chemiosmosis is transported across the inner membrane from the matrix to the intermembrane compartment. It then diffuses out of the mitochondrion to the cytosol through the outer membrane, which is very permeable to ATP. These ATP molecules provide most of the energy needed by the cell. ADP simultaneously diffuses from the cytosol across the outer membrane and is transported across the inner membrane to the matrix, replenishing the supply of ADP.

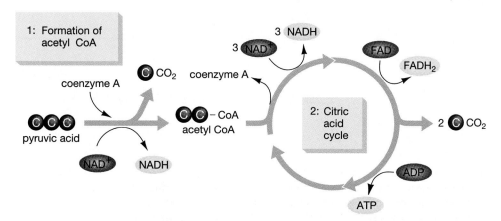

Figure 8-5 **The metabolic reactions occurring in the mitochondrial matrix**

The essentials of the metabolic reactions occurring in the mitochondrial matrix. Pyruvic acid reacts with coenzyme A to form CO_2 and acetyl CoA. During this reaction, an energetic electron is added to NAD^+ to form NADH. When acetyl CoA enters the citric acid cycle, the acetyl group combines with oxaloacetic acid to form citric acid, and coenzyme A is released. One turn through the reactions of the cycle produces three molecules of NADH, one of $FADH_2$, two of CO_2, and one of ATP for each acetyl CoA. Because each glucose molecule yields two pyruvic acids, the total energy harvest per glucose molecule in the matrix is two ATP, eight NADH, and two $FADH_2$.

HEALTH WATCH:

Metabolic Transformations—Why You Can Get Fat by Eating Sugar

As you know, humans do not live by glucose alone. Nor does the typical diet contain exactly the required amounts of each nutrient. Accordingly, the cells of the human body seethe with biochemical reactions, synthesizing one amino acid from another, making fats from carbohydrates, and channeling surplus organic molecules of all types into energy storage or release. Let's look at two examples of these metabolic transformations: producing ATP from fats and proteins and synthesizing fats from sugars.

Metabolizing Fats and Proteins

Even the leanest people have some fat in their bodies. During fasting or starvation, the body mobilizes these fat reserves for ATP synthesis, because even the bare maintenance of life requires a continuous supply of ATP, and seeking out new food sources demands even more energy. Fat metabolism flows directly into the pathways of glucose metabolism.

Chapter 3 illustrated the structure of a fat: three fatty acids connected to a glycerol backbone. In fat metabolism, the bonds between the fatty acids and glycerol are hydrolyzed (broken into subunits by the addition of water). The glycerol part of a fat, after activation by ATP, feeds directly into the middle of the glycolysis pathway (Fig. E8-1). The fatty acids are transported into the mitochondria, where enzymes in the inner membrane and matrix chop them up into acetyl groups. These attach to coenzyme A to form acetyl CoA, which enters the citric acid cycle.

In cases of severe starvation or of feeding almost exclusively on protein, amino acids too can be used to produce energy. First, they are converted to pyruvic acid, acetyl CoA, or the compounds of the citric acid cycle. These molecules then proceed through the remaining stages of cellular respiration, yielding various amounts of ATP depending on their point of entry into the pathway.

Synthesizing Fat from Sugar

The body has not only developed ways of coping with fasting or starvation; it has also evolved strategies for coping with situations in which food intake exceeds current energy needs. Just as fats can be funneled into the glucose metabolic pathway for energy production, the sugars and starches in cornflakes or candy bars can be converted into fats for energy storage. Complex sugars, such as starches and sucrose, are first hydrolyzed into their monosaccharide subunits (see Chapter 3). The monosaccharides are broken down to pyruvic acid and converted to acetyl CoA. If the cell needs ATP, the acetyl CoA will enter the citric acid cycle. If the cell has plenty of ATP, acetyl CoA will be used to make fatty acids by a series of reactions that are essentially the reverse of fatty acid breakdown. In humans, the liver synthesizes fatty acids, but fat

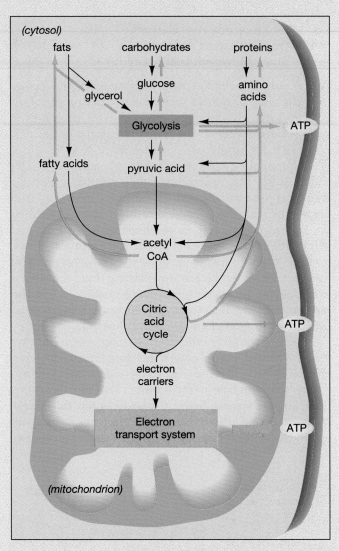

Figure E8-1   How other nutrients yield energy

Fats, carbohydrates such as starches, and proteins can all be broken down metabolically into molecules that enter glycolysis or the citric acid cycle, where they are used to generate ATP.

storage is relegated to fat cells, with their all too familiar distribution in the body, particularly around the waist and hips.

Energy use, fat storage, and eating are usually precisely balanced. Where the balance point lies, however, varies from person to person. Some people seem to eat nearly continuously without ever storing much fat, probably because their metabolic rates are high or adjust upward to compensate for excess food intake. Other people crave high-calorie foods even when they have a lot of fat already stored. As every chronic dieter knows, many people seem to have bodies that are "biologically set" to have large fat deposits. Constant restraint can slim most people down, but as soon as the discipline is relaxed, the fat once again accumulates. Before the fashionably thin become too smug about their svelte condition, they should remember that overeating during times of easy food availability is highly adaptive behavior from an evolutionary point of view. If hard times come, the plump may easily (if hungrily) survive while the trim succumb to starvation. Only recently have many societies enjoyed regular agricultural surpluses, in which the drive to eat can lead to obesity.

▓ SUMMARY

ELECTRON TRANSPORT AND CHEMIOSMOSIS

Electrons from the electron carriers NADH and $FADH_2$ enter the electron transport system of the inner mitochondrial membrane. Here their energy is used to generate a hydrogen ion gradient. Movement of hydrogen ions down their gradient through the pores of ATP-synthesizing enzymes drives the synthesis of 32 to 34 molecules of ATP. At the end of the electron transport system, two electrons combine with one oxygen atom and two hydrogen ions to form water.

Cellular Metabolism Influences the Way Entire Organisms Function

Many students feel that the details of cellular metabolism are hard to learn and don't really help them to understand the living world around them. Have you ever read a murder mystery and wondered how cyanide could kill a person almost instantly? Cyanide reacts with one of the proteins in the electron transport system, immediately blocking the movement of electrons through the system and bringing cellular respiration to a screeching halt. Under more normal conditions, metabolic processes within individual cells have enormous impacts on the functioning of entire organisms. To take just two familiar examples, let's consider the migration of ruby-throated hummingbirds and track events in the Olympics.

Hummingbird Migration Requires Efficient Energy Storage and Use

As the days shorten in August, many birds in North America prepare to migrate to Central and South America, thereby escaping the cold weather and food shortages of the northern winter. For some, migration is a very dangerous undertaking. Ruby-throated hummingbirds, for example, fly across the Gulf of Mexico from the southeastern United States to Mexico and Central America. There is nowhere to stop and no food on a journey of over 1000 kilometers (621 miles) of open sea.

This trek presents real difficulties for a hummingbird. With its short wings, a hummer couldn't fly at all if it weighed too much. Yet flying 1000 kilometers requires a lot of energy that must be stored in the bird's body. Hummers solve this dilemma with a twofold strategy of storing the highest-energy molecules possible—fat—and extracting the maximum usable energy during flight.

A ruby-throated hummingbird weighs 2 to 3 grams (0.11 to 0.16 ounces) before putting on weight for migration. It adds as much as 2 grams of fat in late summer, nearly doubling its weight. As you recall from Chapter 3, fats contain over twice as much energy per unit weight as do proteins or carbohydrates. If a hummer had to store 4 or 5 grams (0.21 to 0.26 ounces) of glycogen or protein, it would be too heavy to lift off.

Even so, the hummer must still squeeze every ATP molecule possible out of each fat molecule. The hummer that just makes it to Guatemala on 2 grams of fat using cellular respiration would collapse almost within sight of the Gulf Coast if it used lactic acid fermentation instead. The

Figure 8-6   **The electron transport system of mitochondria**
① The electron carrier molecules NADH and $FADH_2$ deposit their energetic electrons with the carriers of the transport system located in the inner membrane. ② The electrons move from carrier to carrier within the transport system. Some of their energy is used to pump hydrogen ions across the inner membrane from the matrix into the intermembrane compartment. This movement out of the inner compartment creates a hydrogen ion gradient that can be used to drive ATP synthesis. ③ At the end of the electron transport system, the energy-depleted electrons combine with oxygen and hydrogen ions in the matrix to form water.

cells of a hummingbird's flight muscles are packed with mitochondria, so each cell is capable of producing large quantities of ATP.

Furthermore, the hummingbird's respiratory system is exquisitely designed to extract oxygen out of the air even while exhaling, enabling the lungs to provide a constant supply of oxygen to the cells. Therefore, even during strenuous flight, cellular respiration never falters for lack of oxygen.

Human Runners Face Similar Challenges

Humans, like hummingbirds, must regulate energy reserves and energy use. Why is the average speed of the 5000-meter run in the Olympics slower than that of the 100-meter dash? It's not because the distance runners could not, at any given time, run faster; the finishing kick at the end of the race testifies to that. The reason is that at top speed, their leg muscles consume ATP faster than their lungs can extract oxygen from the air to keep cellular respiration going. Glycolysis and lactic acid fermentation can keep the muscles functional for a short time, but soon the toxic effects of lactic acid buildup cause fatigue and cramps.

Although runners can do a 100-meter dash anaerobically, doing a 5000-meter run anaerobically is out of the question. Distance runners must therefore pace themselves, using cellular respiration to power their muscles for most of the race and saving the anaerobic sprint for the finish.

Marathon runners face somewhat the same dilemma that hummingbirds do. Nobody ever saw a fat marathoner: It's too much work to lug a lot of weight around for 26 miles. Nevertheless, a marathon may require 3000 kilocalories of stored energy, even with the most energy-efficient process, cellular respiration, supplying nearly all the ATP. Marathoners train by running 50 or 100 miles a week, not so much to build up their leg muscles (which are usually pretty stringy) as to build up the capacity of their respiratory and circulatory systems to deliver enough oxygen to their muscles. An efficient transport of oxygen to the cells is necessary to sustain the cellular respiration that such vigorous exercise demands.

As you can see, life on Earth depends on efficiently obtaining, storing, and using energy. Your understanding of the principles of cellular metabolism should enable you to fully appreciate the energy-related adaptations of living organisms.

☒ S U M M A R Y   O F   K E Y   C O N C E P T S

Glucose Metabolism: An Overview
Cells produce usable energy by breaking down glucose to lower-energy compounds and capturing some of the released energy as ATP. In glycolysis, glucose is metabolized in the cytosol to two molecules of pyruvic acid, generating two ATP molecules. In the absence of oxygen, pyruvic acid is converted by fermentation to lactic acid or ethanol and CO_2. If oxygen is available, the pyruvic acids are metabolized to CO_2 and H_2O through cellular respiration in the mitochondria, generating much more ATP.

Figure 8-7 and Table 8-1 summarize the locations, major mechanisms, and overall energy harvest for the complete metabolism of glucose from glycolysis through cellular respiration.

Glycolysis
During glycolysis, a molecule of glucose is activated by adding phosphates from two ATP molecules, to form fructose diphosphate. The fructose diphosphate then undergoes a series of reactions that break it down into two molecules of pyruvic acid. These reactions produce four ATP molecules and two NADH electron carriers. Because two ATPs were used in the activation steps, the net yield from glycolysis is two ATPs and two NADHs. Glycolysis, in addition to providing a small yield of ATP, uses up NAD^+ to produce NADH. Once the cell's supply of NAD^+ is consumed, glycolysis must stop. NADH may be regenerated by fermentation, with no additional ATP gain, or by cellular respiration, which also produces additional ATP.

Fermentation
If oxygen is absent, the pyruvic acids undergo either lactic acid fermentation, producing lactic acid, or alcoholic fermentation, producing ethanol and CO_2. In both cases, no new ATP is formed. However, both types of fermentation regenerate NAD^+ from NADH, thus replenishing the cell's supply of NAD^+.

Cellular Respiration
If oxygen is available, cellular respiration can occur. The pyruvic acids are transported into the matrix of the mitochondria. In the matrix, each pyruvic acid reacts with coenzyme A to form acetyl CoA plus CO_2. One NADH is also formed at this step. The two-carbon acetyl group of acetyl CoA enters the citric acid cycle, which releases the remaining two carbons as CO_2. One ATP, three NADHs, and one $FADH_2$ are also formed for each acetyl group that goes through the cycle. At this point, each glucose molecule has produced 4 ATPs (two from glycolysis and one from each acetyl CoA through the citric acid cycle), 10 NADHs (two from glycolysis, one from each pyruvic acid during the formation of acetyl CoA, and three from each acetyl CoA during the citric acid cycle), and 2 $FADH_2$s (one from each acetyl CoA during the citric acid cycle).

The NADHs and $FADH_2$s deposit their energetic electrons in the electron transport system embedded in the inner mitochondrial membrane. The energy of the electrons is used to pump hydrogen ions across the inner membrane from the matrix to the intermembrane compartment. At the end of the electron transport system, the

Figure 8-7   **A summary of the energy harvest from the complete metabolism of one glucose molecule**

Glycolysis and the citric acid cycle each produce two ATP molecules. The reactions within the mitochondrial matrix produce eight NADH molecules and two $FADH_2$ molecules. By donating its electrons to the electron transport system, each NADH molecule yields three ATP molecules, for a total of 24 ATPs. Each $FADH_2$ molecule yields two ATP molecules, for a total of 4 ATPs. The electrons from the two NADH molecules produced in the cytoplasm during glycolysis must be transported into the mitochondrion to reach the electron transport system. In heart and liver cells, this transport is "free"; in most cells, transport costs one ATP per NADH. The two "glycolytic NADH" molecules therefore yield either 4 or 6 ATP molecules, depending on the cell. Therefore, the energy harvest from electron transport is 32 to 34 ATPs. Including 2 ATPs from glycolysis and 2 ATPs from the citric acid cycle, the total energy yield from glucose metabolism is 36 to 38 ATPs.

depleted electrons combine with hydrogen ions and oxygen to form water. During chemiosmosis, the hydrogen ion gradient created by the electron transport system is used to produce ATP, as the hydrogen ions diffuse back across the inner membrane through pores in ATP-synthesizing enzymes. Electron transport and chemiosmosis yield 32 to 34 additional ATPs, for a net yield of 36 to 38 ATPs per glucose molecule.

TABLE 8-1 ⁝⁝ *Summary of Glycolysis and Cellular Respiration*

Process	Location	Reactions	Electron Carriers Formed	ATP Yield
Glycolysis	Cytosol	Glucose broken down to two pyruvic acids	2 NADH	2 ATP
Acetyl CoA formation	Matrix of mitochondrion	Each pyruvic acid combined with coenzyme A to form acetyl CoA and CO_2	2 NADH	
Citric acid cycle	Matrix of mitochondrion	Acetyl group of acetyl CoA metabolized to two CO_2	6 NADH, 2 $FADH_2$	2 ATP
Electron transport	Inner membrane, intermembrane compartment	Energy of electrons from NADH, $FADH_2$ used to pump H^+ into intermembrane compartment, H^+ gradient used to synthesize ATP: three ATP per NADH, two ATP per $FADH_2$		32–34 ATP*

*Glycolysis produces two NADH molecules in the cytosol of the cell. Unlike the NADH and $FADH_2$ molecules generated in the matrix of the mitochondrion, the electrons from these two NADH molecules must be transported into the matrix before they can enter the electron transport system of the inner membrane. In most eukaryotic cells, the energy of one ATP molecule is used to transport the electrons from one NADH molecule into the matrix. Thus, the two "glycolytic NADH" molecules net only two ATPs, not the usual three, during electron transport. The heart and liver cells of mammals, however, use a different transport system, one that does not consume ATP. In these cells, then, the "glycolytic NADH" molecules net three ATPs apiece, just as the "mitochondrial NADH" molecules do.

⁝⁝ **K E Y   T E R M S**

THINKING THROUGH THE CONCEPTS

Multiple Choice

1. Where does glycolysis occur?
   a. cytoplasm
   b. matrix of mitochondria
   c. inner membrane of mitochondria
   d. outer membrane of mitochondria
   e. chloroplast stroma
2. Where does respiratory electron transport occur?
   a. cytoplasm
   b. matrix of mitochondria
   c. inner membrane of mitochondria
   d. outer membrane of mitochondria
   e. chloroplast stroma
3. What are the economically important products of the fermentation of grape juice by yeast that are essential for the manufacture of champagne?
   a. lactic acid and NAD^+
   b. ATP and CO_2
   c. ATP and ethanol
   d. CO_2 and ethanol
   e. only the French know
4. The majority of ATP produced in aerobic respiration comes from
   a. glycolysis
   b. the citric acid cycle
   c. the electron transport system
   d. fermentation
   e. reverse osmosis
5. The process that converts glucose (6 C) into two molecules of pyruvate (3 C) is
   a. glycolysis
   b. fermentation
   c. the citric acid cycle
   d. respiratory electron transport
   e. the Calvin-Benson cycle
6. The process that converts pyruvate (3 C) into three molecules of CO_2 is
   a. glycolysis
   b. fermentation
   c. the citric acid cycle
   d. respiratory electron transport
   e. the Calvin-Benson cycle

Review Questions

1. Starting with glucose ($C_6H_{12}O_6$) write the overall reactions for (a) aerobic respiration and (b) fermentation in yeast.
2. Draw a labeled diagram of a mitochondrion and explain how its structure is related to its function.
3. What role do the following play in respiratory metabolism: (a) glycolysis, (b) mitochondrial matrix, (c) inner membrane of mitochondria, (d) fermentation, and (e) NAD^+?
4. Outline the major steps in (a) aerobic and (b) anaerobic respiration, indicating the sites of ATP production. What is the overall energy harvest (in terms of ATP molecules generated per glucose molecule) for each?
5. Describe the citric acid cycle. In what form is most of the energy harvested?
6. Describe the mitochondrial electron transport system and the process of chemiosmosis.
7. Why is oxygen necessary for cellular respiration to occur?

APPLYING THE CONCEPTS

1. In detective novels, "the odor of bitter almonds" is always the telltale clue to murder by cyanide poisoning. Cyanide works by attacking the enzyme that transfers electrons from respiratory electron transport to O_2. Why is it not possible for the victim to survive using anaerobic respiration? Why is cyanide poisoning almost immediately fatal?
2. Some species of bacteria that live at the surface of sediment on the bottom of lakes are facultative anaerobes that are capable of either aerobic or anaerobic respiration. How will their metabolism change during the summer when the deep water becomes anoxic (deoxygenated)? If the bacteria continue to grow at the same rate will glycolysis increase, decrease, or remain the same after the lake becomes anoxic? Explain why.
3. Dumping large amounts of raw sewage into rivers or lakes often leads to massive fish kills, although sewage itself is not toxic to fish. Similar fish kills also occur in shallow lakes that become covered in ice during the winter. What kills the fish? How might you reduce fish mortality after the accidental release of raw sewage into a small pond containing large bass?

FOR MORE INFORMATION

McCarty, R. E. "H+-ATPases in Oxidative and Photosynthetic Phosphorylation." *BioScience*, January 1985. A description of the structure and function of the ATP-synthesizing enzymes in mitochondria and chloroplasts.

Nelson, M., Burgess, T. L., Alling, A., Alverez-Romo, N., Dempster, W. F., Walford, R. L., and Allen, J. P. "Using a Closed Ecological System to Study Earth's Biosphere." *BioScience*, 1993. An artificial ecosystem allows scientists to learn more about how natural ecosystems function.

NET WATCH

On-line resources for this chapter are on the World Wide Web at:
http://www.prenhall.com/~audesirk (click on the table of contents link and then select Chapter 8).

Unit II
Inheritance

A human white blood cell (lymphocyte) undergoes mitosis in this false-color transmission electron micrograph. The two daughter cells, each of which contains a full set of genetic material (pink), are almost separated, but are still connected by a strand of cytoplasm. This is late telophase; the nuclear envelopes are about to re-form.

9 Cellular Reproduction

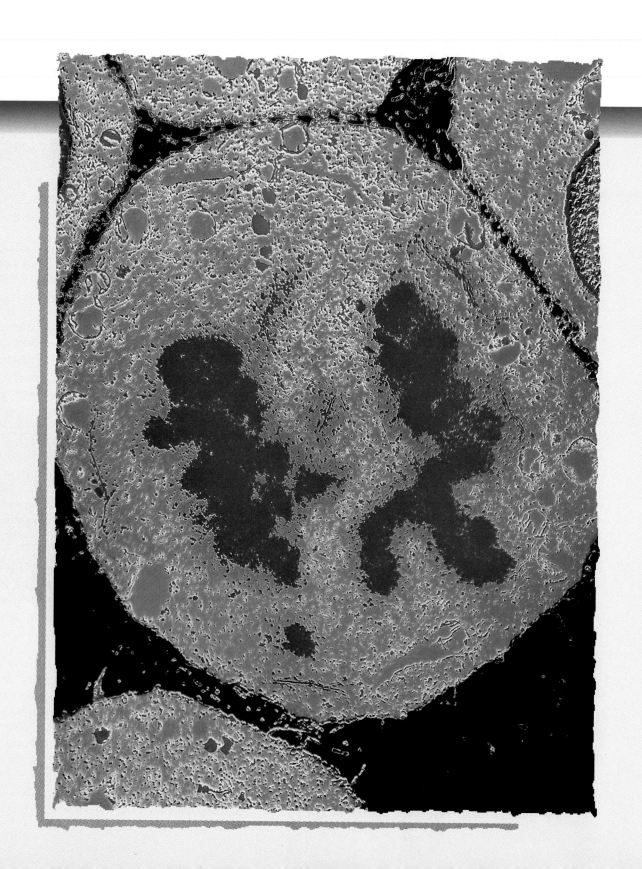

> *"All cells come from cells."*
>
> RUDOLF VIRCHOW
> German physician and early proponent of the cell theory (1858)

A human cancer cell in anaphase of cell division. Two sets of chromosomes are seen in blue in this false-color transmission electron micrograph.

"All cells come from cells." With these words, Rudolf Virchow captured the crucial importance of cellular reproduction for both unicellular and multicellular organisms. Because all living organisms consist of one or more cells, and because all cells are descended from preexisting cells, cellular reproduction is absolutely essential for the continued existence of life on Earth.

How do cells reproduce? Although there are many variations, most cells reproduce according to the simple repeating phrases: enlarge, divide in two, enlarge, divide in two, and so on. This process, appropriately called **cell division,** produces two daughter cells that, if all goes well, have exactly the same genetic information as the parent cell, carried in chromosomes that are exact copies of the chromosomes of the parent cell. Each daughter cell also inherits about half of the parent cell's cytoplasm, including a full complement of organelles. Each round of growth and cell division is called a **cell cycle.**

In unicellular organisms such as bacteria or *Amoeba,* the cell cycle is synonymous with the **life cycle,** which consists of the events in the life of an organism from one generation to the next. In multicellular organisms, the cell cycle is usually only part of the life cycle. Multicellular organisms begin life as a fertilized egg. Repeated cell divisions produce the hundreds to trillions of cells that make up the adult organism. To complete the life cycle, most multicellular organisms produce a new fertilized egg, through the process of **sexual reproduction.** Cells in the reproductive organs of the adult organism undergo a specialized kind of cell division, called **meiosis,** which we will describe in Chapter 12. In animals, the daughter cells of meiosis are sex cells—sperm or eggs. When a sperm fertilizes an egg, the life cycle starts over again.

In this unit, we will study the mechanisms of inheritance: how a cell transmits copies of its chromosomes to its daughter cells; the structure and function of the molecules of DNA contained in those chromosomes that carry the genetic information; how multicellular parents transmit chromosomes to their offspring during meiosis; and how the function of DNA and the distributions of chromosomes during meiosis produce the observable patterns of inheritance that we see in life on Earth. We will begin, in this chapter, with cell cycles and cell division. We will see that cells have evolved elaborate mechanisms that ensure that each daughter cell inherits all the materials it needs to continue the flow of life.

The Essentials of Cellular Reproduction

When a cell divides, it must transmit to its offspring cells two essential requirements for life (Fig. 9-1): hereditary information to direct life processes, and materials in the cytoplasm that the offspring need to survive and to utilize their hereditary information.

Cell Division Transmits a Complete Set of Hereditary Information to Each Daughter Cell

The hereditary information of all living cells is **deoxyribonucleic acid (DNA)**. Like many large biological molecules, a molecule of DNA consists of a long chain of smaller subunits (see Chapter 3). DNA has four different types of subunits, called nucleotides. Segments of DNA a few hundred to many thousand nucleotides long are the units of inheritance—the genes—that carry the genetic information to produce specific traits. The sequence of nucleotides in a gene encodes information for the synthesis of the RNA and protein molecules that are needed to build a cell and carry out its metabolic activities. We will see in Chapters 10 through 13 how DNA encodes genetic instructions and how a cell regulates which genes it uses at any given time.

For any cell to survive, it must have a *complete set of genetic instructions*. Therefore, when a cell divides, it cannot simply split its set of genes in half and give each daughter cell half a set. Rather, the cell must first *duplicate its*

DNA, much like making a photocopy of an instruction manual. Each daughter cell then receives a complete "DNA manual" containing all the genes.

Cell Division Transmits Essential Cytoplasmic Materials to Each Daughter Cell

Like the blueprints for a house, the instructions encoded by DNA are useless without materials to work with. Each newly formed cell must receive the molecules it needs to read its genetic instructions and to keep it alive long enough to acquire new materials from the environment and to process them into new cellular components. Furthermore, as we described in Chapter 5, a cell cannot synthesize either mitochondria or chloroplasts from scratch: These organelles arise only by the division of previously existing mitochondria and chloroplasts. Usually, when a cell divides, its cytoplasm is divided about equally between the two daughter cells. This simple mechanism normally provides both daughter cells with all the organelles, nutrients, enzymes, and other molecules they need.

The Activities of a Cell from One Cell Division to the Next Constitute the Cell Cycle

Newly formed cells usually acquire nutrients from their environment, synthesize more of their own materials, and grow larger. After a variable amount of time, depending on the organism, the type of cell, and the nutrients available, the cell divides. This general description applies to both eukaryotic and prokaryotic cells. Prokaryotic cells are structurally and functionally very different from eukaryotic cells, however, and accordingly their cell cycle differs in many respects. We will therefore discuss prokaryotic and eukaryotic cycles separately.

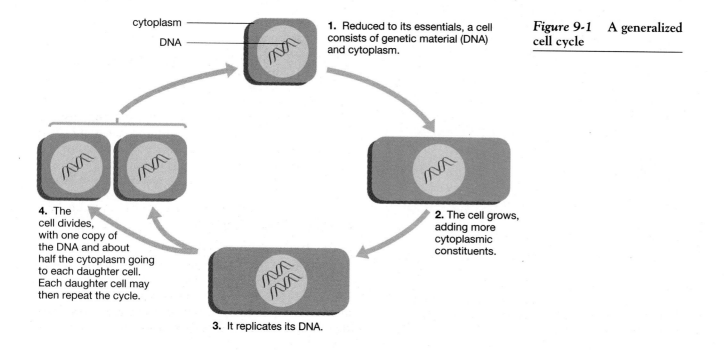

cytoplasm

DNA

1. Reduced to its essentials, a cell consists of genetic material (DNA) and cytoplasm.

2. The cell grows, adding more cytoplasmic constituents.

3. It replicates its DNA.

4. The cell divides, with one copy of the DNA and about half the cytoplasm going to each daughter cell. Each daughter cell may then repeat the cycle.

Figure 9-1   **A generalized cell cycle**

Prokaryotic and Eukaryotic Chromosomes Compared

The DNA of all living cells is packaged into one or more structures called **chromosomes** (see Chapter 5). A single, circular chromosome about a millimeter or two in circumference contains all the DNA of a prokaryotic cell (Fig. 9-2). One point on the chromosome is attached to the plasma membrane of the cell. A prokaryotic chromosome usually has relatively little protein associated with the DNA.

The chromosomes of eukaryotic cells are quite different. First, eukaryotic cells contain much more DNA than prokaryotic cells do. Second, in contrast to the single chromosome of a prokaryotic cell, a eukaryotic cell packages its enormous amounts of DNA, along with roughly equal amounts of protein, into many separate chromosomes. Even dividing up the DNA this way, eukaryotic chromosomes are often much longer than prokaryotic chromosomes. The chromosome of a bacterium is a millimeter or two in circumference, whereas the average human chromosome is over 40 millimeters long. (The total length of all the 46 chromosomes in a human cell is about *2000* millimeters, longer than the height of the average human.) Finally, the prokaryotic chromosome is not separated from the cytoplasm of the cell, whereas eukaryotic chromosomes are contained within a membrane-bound nucleus.

Differences in number, size, and location of chromosomes dictate considerable differences in the mechanisms of prokaryotic and eukaryotic cell division. Therefore, we will describe the prokaryotic cell cycle first. Then we will examine the eukaryotic chromosome in more detail, concentrating on those aspects of chromosome structure that

Figure 9-2   The bacterial chromosome

This bacterial cell (yellow central oblong) is about 1 × 3 micrometers in size. The cell was ruptured to free its chromosome. If you look carefully, you will see that there are no free ends to the chromosome: It is a single thin strand. If stretched out, the chromosome would be a circular molecule of DNA 1 or 2 millimeters in circumference.

are important in cell division. Finally, we will explore the eukaryotic cell cycle.

The Prokaryotic Cell Cycle

Under favorable conditions, the cell cycle of many prokaryotes proceeds rapidly. The common intestinal bacterium *Escherichia coli*, for example, can complete its cell cycle in 30 minutes or less. During most of this time, the cell absorbs nutrients from the environment, grows, and replicates its DNA. As soon as the DNA is replicated, the bacterium divides; in fact, in some bacteria the cell begins dividing even before DNA replication is quite finished.

With only a single chromosome and no separate nucleus, bacterial cells divide by **binary fission** (Fig. 9-3):

1. Before DNA replication, one point on the chromosome is attached to the plasma membrane.
2. The chromosome replicates (by a process that we will describe in Chapter 10). The resulting pair of identical chromosomes attach to the plasma membrane at nearby, but distinct, points.
3. The cell elongates, and new plasma membrane is added between the two attachment points, pushing them apart.
4. As the two chromosomes move toward opposite ends of the cell, the plasma membrane around the middle of the cell grows inward.
5. Finally, two new daughter cells are formed.

Each daughter cell thus receives one of the replicated chromosomes and about half the cytoplasm. With a complete set of genes and enough materials to work with, the two daughter cells begin to grow, starting the cycle over again.

The Eukaryotic Chromosome

The complicated events of eukaryotic cell division are largely an evolutionary solution for sorting out a large number of long chromosomes. To help you understand the mechanisms of eukaryotic cell division, we will begin by taking a closer look at the structure of the eukaryotic chromosome.

The Eukaryotic Chromosome Consists of a Single Strand of DNA Complexed with Proteins

Under the light microscope, interphase nuclei appear as large, darkly staining bodies without much internal structure. Early microscopists named the material in the nucleus **chromatin,** from the Greek word for "color," because chromatin is stained by some commonly used dyes. During cell division, dark threadlike structures form where

cell wall

plasma membrane

circular chromosome

attachment point

Figure 9-3   **Prokaryotic cells divide by binary fission**

① The circular chromosome is attached at one point to the cell membrane. ② The chromosome is replicated. The two copies are attached to the membrane at nearby points. ③ The cell elongates; new plasma membrane is added between the attachment points. ④ The plasma membrane grows inward at the middle of the cell. ⑤ The parent cell has divided into two daughter cells. (Inset) Cross section of a prokaryotic cell undergoing binary fission at a stage similar to ④.

the nucleus used to be. The microscopists called these threads chromosomes, meaning "colored bodies." With the advent of the electron microscope and advances in molecular biology, we now know that chromatin and chromosomes are actually the same thing (DNA and its associated proteins), but in different stages of condensation.

Each eukaryotic chromosome is a single long molecule of DNA, continuous from end to end, complexed with special proteins called histones. During most of its life (that is, interphase), a cell must be able to read the information contained in the chromosomes; it can do so only when the chromosomes are extended (see Chapter 11). In this extended state, individual chromosomes are too thin to be visible in light microscopes, or even in ordinary electron microscopes. During cell division, however, the chromosomes must be sorted out and moved into the daughter nuclei. Just as sewing thread is easier to organize when it is wound onto spools, sorting and transporting are easier if the chromosomes are condensed and shortened (Fig. 9-4).

The terms used by biologists to describe chromosomes during cell division often trouble beginning students. But you will find it difficult to understand cell division without understanding the terminology, so follow carefully. For simplicity, the drawings below illustrate chromosomes only in the condensed state.

Most chromosomes consist of two arms that extend out from a specialized region called the **centromere** (meaning "middle body"):

unreplicated chromosome

centromere

Before a cell divides, it duplicates its chromosomes. The two copies remain attached at their centromeres. As long as they are attached to one another, the copies are called sister **chromatids**. Each chromatid is a single DNA molecule identical to the DNA of the original chromosome before replication. The whole structure (two sister chromatids attached at their centromeres) is still considered to be a single chromosome:

replicated chromosome

sister chromatids

Figure 9-4 therefore shows several single chromosomes, each consisting of two sister chromatids.

During cell division, the two sister chromatids separate, with each chromatid becoming an independent daughter chromosome:

independent daughter chromosomes, each in unreplicated state

Eukaryotic Chromosomes Often Occur in Pairs Containing Similar Genetic Information

The chromosomes of each species have characteristic shapes, sizes, and staining patterns (Fig. 9-5). The nonreproductive cells of many organisms contain *pairs of chro-*

sister chromatids   centromere

Figure 9-4   **Human chromosomes during cell division**

In this scanning electron micrograph, the DNA and associated protein of the chromosomes, which are thin and extended during interphase, have coiled up into thick, short structures. The chromosomes were replicated before condensation; the two strands, called chromatids, are attached at the centromere. To get a sense of the how condensed they really are, look at the fuzzy texture of the chromosomes: Each strand of "texture" is a loop of chromosome, about as thin as the whole chromosome would look during interphase. The condensed chromosomes are about 5 to 20 micrometers long; the same chromosomes during interphase would relax into thin strands about 10,000 to 40,000 micrometers long.

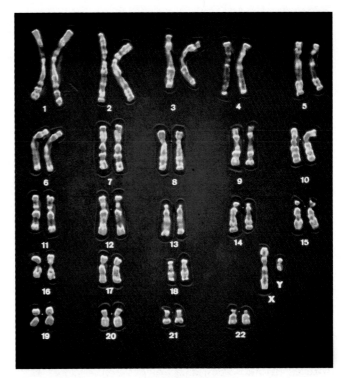

Figure 9-5   **A human male karyotype**

A karyotype is produced by staining and photographing the chromosomes of a dividing cell. Pictures of the individual chromosomes are cut out and arranged in descending order of size. Note that the chromosomes occur in pairs (homologues), which are similar in both size and staining patterns and have similar genetic material. If these chromosomes were from a female, there would be two X chromosomes. In humans, the Y chromosome is much smaller than the X chromosome.

mosomes that are the same length and that stain in the same pattern. Breeding and biochemical experiments show that the chromosomes of each pair also have similar, although usually not identical, genetic content. Therefore, the members of a pair are called **homologues,** meaning "to say the same thing" in Greek. Cells with pairs of homologous chromosomes are called **diploid.** The 46 chromosomes of the nonreproductive cells of human beings occur as 23 pairs of homologues.

Not all cells have pairs of homologous chromosomes. At some point in the life cycle of all sexually reproducing organisms, a special type of cell division called meiosis produces cells that have only one copy of each type of chromosome (see Chapter 12): These cells are called **haploid.** In animals, these haploid cells are the sex cells, or **gametes** (sperm and egg). Gametes cannot live independently; during sexual reproduction, male and female gametes fuse to re-form a diploid cell, the fertilized egg, or **zygote.** In plants, the haploid cells produced by meiosis are not gametes. They undergo ordinary cell division to produce multicellular, haploid organisms. Certain cells of these haploid organisms later become gametes and fuse to form a zygote. In both plants and animals one of each pair of chromosomes in

the diploid zygote is inherited from each parent. We will see how this happens in Chapter 12.

In biological shorthand, the number of different types of chromosomes is termed the haploid number and is designated n. For humans, $n = 23$; for other organisms, n ranges from 1 to several hundred. Diploid cells, with two homologues of each type, contain $2n$ chromosomes. Some organisms have more than two homologues of each type of chromosome in their cells and are called **polyploid.** Polyploid cells may be $3n$ (triploid), $4n$ (tetraploid), and so on.

The Eukaryotic Cell Cycle

The eukaryotic cell cycle consists of two major phases (Fig. 9-6). The first phase, **interphase,** is the period between cell divisions, during which the cell acquires nutrients from its environment, grows, and replicates its chromosomes. During the second phase, cell division (also called the M phase), one copy of each chromosome and approximately half the cytoplasm are parceled out into each of the two daughter cells.

Growth, Replication of Chromosomes, and Most Cell Functions Occur during Interphase

Most of the life of a cell (usually 90% or more) is spent in interphase. Many years ago, biologists studying the cell cycle noted that DNA synthesis, which occurs as a part of the replication of the chromosomes, is not continuous throughout interphase. Rather, interphase begins with a period in which DNA is not synthesized, then DNA synthesis occurs, and then a second period follows in which DNA is not synthesized. Cell biologists therefore divided interphase into three subphases, called G_1 (the "first gap" in DNA synthesis), S (when DNA synthesis occurs), and G_2 (the "second gap" in DNA synthesis; see Fig. 9-6).

The period that occurs after the most recent cell division and before duplication of the chromosomes is the G_1 phase. Although the designation as "first gap" in DNA synthesis implies that the cell is merely resting before getting on with the important business of DNA synthesis, this is not the case. In fact, most of the growth and activity of a cell occur during the G_1 phase. The cell acquires nutrients from its environment, carries out its specialized functions (for example, hormone synthesis and secretion), and grows.

Late in the G_1 phase, at a time called the restriction point, the cell undergoes a kind of "internal evaluation" of its ability to complete the cell cycle and produce two viable daughter cells. If the evaluation is negative, the cell does not go on to divide. If the evaluation is positive, the cell becomes committed to DNA replication and cell division. The commitment to cell division is much like launching a rocket: Once the "go-ahead" signal has been given, the cell cannot turn back and return to the G_1 phase.

Chromosome replication defines the S, or **synthesis, phase**, because this is the only time DNA synthesis normally occurs. Each chromosome is duplicated—just once. During the S phase, animal cells also duplicate their centrioles.

The period after DNA synthesis, but before the next cell division, is the G_2 **phase**. The cell is already committed to cell division before entering the G_2 phase. Most of G_2 seems to be spent in synthesizing molecules other than DNA that are required for cell division.

Cells Vary in the Duration of Interphase

Many types of mammalian cells progress steadily through interphase, spending about 5 hours in the G_1 stage, 7 hours replicating their DNA during the S phase, and 3 hours in G_2 preparing for division. Although cell division itself usually takes about an hour, certain cell types have very short cell cycles, whereas others may go for weeks or even an entire lifetime without dividing. These differences in cell cycle length usually arise from differences in the duration of G_1. The early cell divisions of an animal embryo, for instance, occur in rapid succession with virtually no G_1

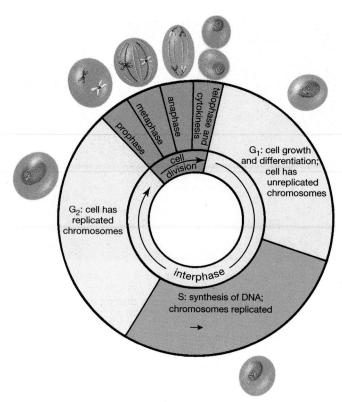

Figure 9-6   The eukaryotic cell cycle
The eukaryotic cell cycle consists of two major phases, interphase and cell division. Each is divided into several subphases, as described in the text.

phase at all and therefore almost no growth between divisions. In contrast, nerve cells in the adult mammalian brain no longer divide but remain in the G_1 phase for life.

Cell Division Consists of Nuclear Division and Cytoplasmic Division

During nuclear division, identical, complete copies of all the chromosomes are packaged into two new nuclei. Nuclear division is called **mitosis**, from the Greek word for "thread," because the chromosomes appear as threadlike structures at this time (see Fig. 9-7). During cytoplasmic division, **cytokinesis**, the cytoplasm is split into two daughter cells, with each cell receiving one of the newly formed nuclei and (usually) roughly equal amounts of cytoplasm. Therefore, the two daughter cells are essentially identical both genetically and cytoplasmically. They are also genetically identical to the parent cell.

Although most cells adhere to this scheme, nuclear division and cytoplasmic division are potentially independent events. Some cells, including some fungi and the skeletal muscle cells of vertebrates, undergo mitosis without cytokinesis. This process produces single cells with many nuclei.

As we mentioned earlier, a multicellular organism grows from a single fertilized egg by repeated cell divisions. Therefore, with a few exceptions, such as sex cells and the

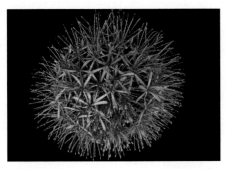

(a) The flower of the African blood lily.

(b) Interphase in a cell of the endosperm (a food-storage organ in the seed) before mitosis begins. The chromosomes are in the thin, extended state and appear as an amorphous mass of stained material in light micrographs such as this one. The microtubules during interphase form a vaguely star-shaped array leading from the nucleus to all parts of the cell.

(c) Early prophase: The chromosomes have condensed into thick, separate strands.

(d) Late prophase: The chromosomes are attached to microtubules of the almost fully developed spindle.

(e) Metaphase: The spindle has moved the chromosomes to the equator of the cell.

(f) Anaphase: The spindle is moving one set of chromosomes to each pole of the cell.

(g) Telophase: The chromosomes have been gathered into two clusters, one at the site of each future nucleus.

(h) Resumption of interphase: The chromosomes are relaxing again into their extended state. The spindle is disappearing and the microtubules are rearranging into the interphase array.

Figure 9-7   Mitosis in a plant cell

Mitosis as seen in the developing seed of the African blood lily, *Haemanthus katherinae*. In all of the micrographs, the chromosomes are stained bluish purple, and the microtubules of the spindle are stained pink to red. Compare these micrographs with the drawings of mitosis in an animal cell shown in Figure 9-8. (Each micrograph here is of a different single cell that has been fixed and stained at a particular stage of mitosis. The cells therefore have different sizes and shapes.)

(a) Late interphase: The chromosomes have been replicated but remain elongated and relaxed within the nucleus. The centrioles, within the microtubule organizing center, have also been replicated.

(b) Early prophase: The chromosomes condense and shorten, becoming visible with a light microscope. The centrioles begin to move apart, and the spindle apparatus begins to form.

(c) Late prophase: The nuclear envelope breaks down and the spindle microtubules invade the nuclear region. Some spindle microtubules (the kinetochore microtubules) attach to each chromosome at its kinetochore. Other microtubules (the polar microtubules) extend from each pole to the cell's equator, where their ends overlap.

(d) Metaphase: The spindle is fully formed. The kinetochore microtubules move the chromosomes to the cell's equator.

Figure 9-8   **Mitosis and cytokinesis in an animal cell**

cells of the immune system (see Chapter 24), all cells of a multicellular organism are genetically identical. As incredible as it seems, a brain cell and a liver cell have the same genes. Their differences in structure and function occur because of differences in which genes are used during the life of the cell.

Much of biology, reduced to its essentials, is the study of cellular activities during interphase, which result in such diverse phenomena as photosynthesis, muscle movement, and thought. Most of the rest of this text will be devoted to those topics. Here, though, we focus on the events of cell division.

Mitosis

As we outlined earlier, eukaryotic cell division consists of mitosis (nuclear division) and cytokinesis (cytoplasmic division). These two processes usually occur together, but, in some cells, mitosis may occur without cytokinesis. Before we begin our discussion of mitosis, remember that the chromosomes are duplicated during the S phase of interphase. Therefore, when mitosis begins, each chromosome already consists of two sister chromatids attached to one another at the centromere.

For convenience, mitosis is divided into four phases: prophase, metaphase, anaphase, and telophase (Figs. 9-7 and 9-8). As with most biological processes, however, these phases are not really discrete events. Rather, they form a continuum, each phase merging into the next.

During Prophase, the Chromosomes Condense and the Spindle Apparatus Forms and Attaches to the Chromosomes

The first phase of mitosis is called **prophase** (meaning "the stage before" in Greek). Prophase develops from the G_2 phase of interphase, as three major events occur: chromosome condensation, assembly of the spindle apparatus, and capture of the chromosomes by the spindle.

During Early Prophase, the Chromosomes Condense and the Nucleolus Disappears

The chromosomes have already been duplicated during interphase, and sister chromatids remain attached to each

chromosomes decondensing

nuclear envelope re-forming

(e) Anaphase: Chromatids separate at the centromere, becoming independent chromosomes. The kinetochore microtubules shorten, pulling one former chromatid from each chromosome toward each pole of the cell. At about the same time, the polar microtubules slide past one another, moving poles farther apart.
(f) Telophase: One complete set of chromosomes has reached each pole of the cell. The chromosomes relax into their extended state, the spindle disappears, and the nuclear envelopes begin to re-form.

(g) Cytokinesis: Usually simultaneously with the end of telophase, the cytoplasm is divided along the equator of the parent cell, with each daughter cell receiving one nucleus and about half the original cytoplasm.
(h) Interphase of daughter cells: The daughter cells return to interphase. The spindle disappears, the nuclear envelope completely re-forms, the chromosomes finish decondensing, and the nucleolus reappears.

other at their centromeres. During interphase, the chromosomes are extended, allowing the information in their DNA to be used by the cell (Fig. 9-8a). During prophase, the chromosomes coil and condense, becoming visible in a light microscope (Fig. 9-8b). As the chromosomes condense, the nucleolus disappears. You may recall from Chapter 5 that the nucleolus contains an aggregation of the parts of several chromosomes that are involved in ribosome synthesis. As the chromosomes condense and separate from one another, this aggregation breaks up.

The Spindle Apparatus Consists of an Array of Microtubules That Will Separate the Sister Chromatids during Anaphase

Near the end of prophase, after the chromosomes are condensed, the **spindle apparatus** begins to form. The spindle apparatus is a football-shaped arrangement of microtubules that arises in the following way.

During interphase, a cell contains an array of microtubules radiating outward from the **microtubule organizing center** near the nucleus (see Chapter 5). During prophase, the interphase microtubules disintegrate into

individual subunits of the protein tubulin. The microtubule organizing center splits in two, and each daughter center begins to form a new array of microtubules from the tubulin subunits. Gradually the two microtubule organizing centers move to opposite sides of the nucleus (the poles of the cell). The microtubules extending from each pole almost surround the nucleus. This new microtubule array is called the spindle apparatus because of its resemblance to the thread wound around the spindle of a spinning wheel—fat in the middle, and tapering more or less to a point on either end.

Although the operation of the spindle during mitosis is similar in all eukaryotic cells, the spindles of animal and plant cells differ in two details. First, in animal cells, each microtubule organizing center contains a pair of **centrioles** (Fig. 9-9). During interphase, the two centrioles separate and a new centriole develops near the base of each "parent" centriole, resulting in a microtubule organizing center containing two pairs of centrioles. When the microtubule organizing center splits, each daughter center contains one pair of centrioles. Second, in animal cells, spindle microtubules extend not only across the nucleus toward the other pole, but also outward in all directions from the microtubule organizing center toward the plasma membrane, forming a star-shaped **aster.** Plant cells, in contrast, do not have centrioles and consequently do not form asters.

The Spindle Microtubules Attach to the Chromosomes at Their Centromeres

When the spindle is fully formed, the nuclear envelope abruptly disintegrates. The double membrane of the envelope breaks up into vesicles that resemble pieces of endoplasmic reticulum. The disruption of the nuclear envelope allows the microtubules of the spindle to invade the nuclear region (see Fig. 9-8c). Some of these microtubules will attach to the chromosomes and some will interact with one another.

Each sister chromatid has a structure called a **kinetochore** located at the centromere (Fig. 9-10). Therefore, each replicated chromosome has two kinetochores. Each kinetochore grabs some of the passing microtubules (now called **kinetochore microtubules**) as they invade the nuclear region. Although the exact mechanism is still a matter of investigation, the kinetochore of one sister chromatid of each chromosome captures microtubules leading to one pole, while the kinetochore of the other sister chromatid captures microtubules leading to the opposite pole. The sister chromatids of each chromosome therefore face opposite poles of the spindle (see Fig. 9-10). When the sister chromatids separate later in mitosis, the kinetochore microtubules will pull the newly independent chromosomes to opposite poles.

Other microtubules do not attach to chromosomes but retain free ends that overlap along the cell's equator. As we will see, these **polar microtubules** will push the two spindle poles apart.

During Metaphase, the Chromosomes Are Aligned along the Equator of the Cell

During **metaphase** (the "middle stage"), the kinetochore microtubules running to opposite spindle poles engage in a tug-of-war, each pulling toward its own pole. Apparently, long kinetochore microtubules pull harder than short ones do. Therefore, if a chromosome is farther from one pole than from the other pole, it is dragged toward the more distant pole. It is also possible that, at this stage, chromosomes are pushed away from the spindle poles; the closer a chromosome is to a spindle pole, the harder it is pushed away. By being pulled toward the more distant pole or being pushed away from the closer pole, or both, each chromosome ends up aligned along the cell's equator, midway between the spindle poles. Metaphase ends when all the chromosomes have lined up along the equator (see Fig. 9-8d.).

During Anaphase, Sister Chromatids Separate and Are Pulled to Opposite Poles of the Cell

At the beginning of **anaphase** (see Fig. 9-8e), the centromere of each chromosome divides, and the sister chromatids separate into two independent daughter chromosomes. The

(a)

centrioles

nucleus

(b)

10 micrometers

aster   chromosomes

Figure 9-9   Details of the spindle apparatus in animal cells

(a) An electron micrograph of the microtubule organizing center of an animal cell. The centrioles were replicated just before cell division. The centrioles of each pair lie at right angles to each other. During mitosis, the two pairs migrate to opposite sides of the nuclear region (see Fig. 9-8b).
(b) Asters in embryonic whitefish cells. The asters consist of microtubules that extend both in toward the equator of the cell and out toward the plasma membrane. The centrioles, which are not visible at this magnification, lie in the center of the asters.

kinetochores now perform some fancy molecular moves; protein "motors" in the kinetochores, similar to those that cause bending of cilia and flagella, pull the kinetochores (and their attached chromosomes) poleward along the kinetochore microtubules (Fig. 9-11a). Simultaneously, the microtubules disassemble inside the kinetochores so that the microtubules shorten at about the same rate as the kinetochores move toward the poles.

An analogy might help you to understand these processes: Think of the kinetochores as a pair of sailors, pulling a small boat in which they are standing (the chromosome) to shore (the spindle pole) by pulling on a rope (the kinetochore microtubules) attached to the shore. One sailor (the motor) pulls the rope in, hand over hand, moving the boat to the shore. To keep from filling up the boat with rope, the other sailor cuts off the end of the rope as fast as the "motor sailor" pulls it in and throws the pieces overboard (into the cytoplasm).

At the same time as the kinetochores are towing their chromosomes toward the poles, the polar microtubules grow longer and push the spindle poles apart (Fig. 9-11b).

Note that one sister chromatid of each chromosome moves toward each pole of the cell. Because the sister chromatids are identical copies of the original chromosomes, each of the two clusters of chromosomes that form contains one copy of every chromosome.

The usefulness and accuracy of the mechanisms for sorting and separating chromosomes do not depend on the numbers of chromosomes in a cell. Mitosis works equally well in cells with only a handful of chromosomes, such as those of fruit flies (eight), and in those containing hundreds of chromosomes, such as the cells of horsetails and some ferns. Similarly, the same mechanisms are employed in haploid, diploid, and polyploid cells.

During Telophase, Nuclear Envelopes Re-form around Each Group of Chromosomes

When the chromosomes reach the poles of the spindle, **telophase** (the "end stage") has begun (see Fig. 9-8f). The spindle disintegrates. Vesicles that formed when the old nuclear envelope broke up during late prophase coalesce around each group of chromosomes, forming two new nuclear envelopes. The chromosomes relax into their extended state once again, and the nucleoli reappear. In most cells, cytokinesis occurs during telophase, enclosing each daughter nucleus into a separate cell.

Cytokinesis

In most cells, the division of the cytoplasm into nearly equal halves begins during telophase. In animal cells, microfilaments composed of the proteins actin and myosin form rings around the equator of the cell, surrounding the remnants of the spindle (Fig. 9-12). The microfilaments are attached to the plasma membrane. During cytokinesis, the rings contract and pull in the equator of the cell, much like pulling the drawstring around the waist of a pair of sweatpants. Eventually the "waist" contracts down to nothing, dividing the cytoplasm into two new daughter cells.

Cytokinesis in plant cells is quite different, perhaps because the stiff cell wall makes it impossible to divide one cell

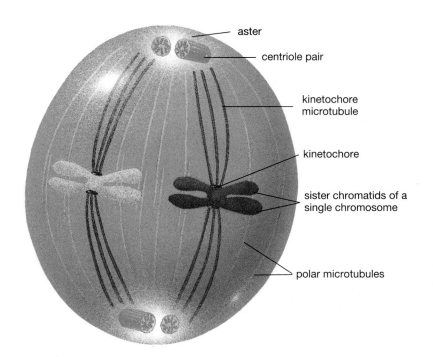

aster

centriole pair

kinetochore microtubule

kinetochore

sister chromatids of a single chromosome

polar microtubules

Figure 9-10   **A diagram of the spindle apparatus in an animal cell**

Note that there are three sets of spindle microtubules. First, the aster consists of microtubules radiating out from the pair of centrioles at the pole toward the plasma membrane (see Fig. 9-9b). Second, kinetochore microtubules run from one pole to one kinetochore of each chromosome. The two kinetochores of a single chromosome attach to kinetochore microtubules that lead to opposite poles of the cell. Finally, polar microtubules extend from each pole to, and slightly beyond, the equator of the cell. Polar microtubules overlap at the equator.

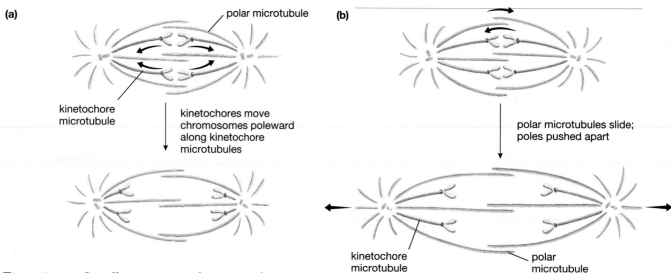

Figure 9-11   Spindle movements during anaphase

These movements occur just after the separation of chromatids to form individual daughter chromosomes. **(a)** Movement of chromosomes toward the poles: Kinetochores of the newly formed daughter chromosomes move poleward along the kinetochore microtubules, while the microtubules simultaneously shorten. The kinetochores remain attached to the microtubules even as the microtubules shorten. **(b)** Separation of the poles: At their sites of overlap along the equator of the cell, polar microtubules slide past one another, probably by a mechanism similar to the bending of cilia or flagella (see Fig. 5-22). This mechanism pushes the poles of the spindle farther apart.

into two by pinching at the waist. Instead, the Golgi complex buds off carbohydrate-filled vesicles that line up along the equator of the cell between the two nuclei (Fig. 9-13). The vesicles fuse together, producing a structure called the **cell plate,** shaped much like a hot-water bottle filled with sticky carbohydrates. When enough vesicles have fused, the edges of the cell plate merge with the original plasma membrane around the circumference of the cell. Seen as if you were facing the dividing cell in Figure 9-13, the left membrane of the cell plate becomes part of the plasma membrane of the left daughter cell, while the right membrane of the cell plate becomes part of the plasma membrane of the right

daughter cell. The carbohydrate formerly contained in the vesicles remains between the plasma membranes as the middle lamella of the cell wall (see Chapter 6).

Cell Division and Asexual Reproduction

Through cell division and subsequent diversification of cells, a single cell, usually a fertilized egg, generates the multicellular bodies of plants and animals. Cell division is

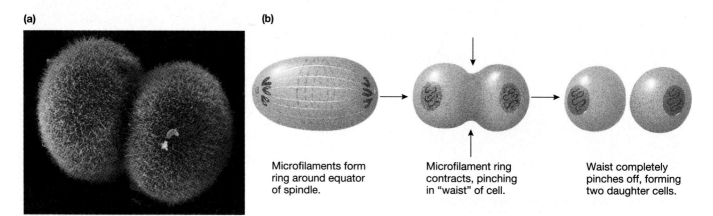

Microfilaments form ring around equator of spindle.

Microfilament ring contracts, pinching in "waist" of cell.

Waist completely pinches off, forming two daughter cells.

Figure 9-12   Cytokinesis in animal cells

(a) A ring of microfilaments just beneath the plasma membrane contracts around the equator of the cell, pinching it in two. **(b)** The mechanism of cytokinesis in animal cells.

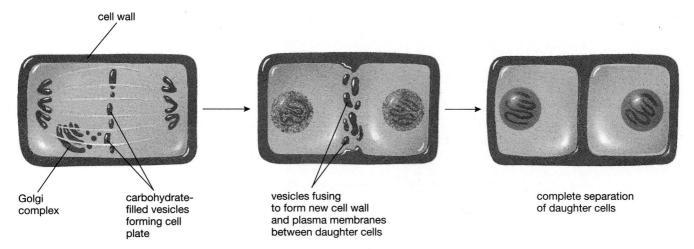

cell wall

Golgi
complex

carbohydrate-
filled vesicles
forming cell
plate

vesicles fusing
to form new cell wall
and plasma membranes
between daughter cells

complete separation
of daughter cells

Figure 9-13   **Cytokinesis in plant cells**

The mechanism of cell plate formation and cytokinesis in plant cells. Carbohydrate-filled
vesicles produced by the Golgi complex congregate at the equator of the cell, forming the cell
plate. The vesicles will fuse to form the two plasma membranes separating the daughter cells,
while their carbohydrate contents form the middle lamella.

also the basis of **asexual reproduction,** in which offspring
are formed from a single parent without the necessity of
uniting male and female gametes. This is the usual mode
of reproduction for many unicellular organisms, such as
Paramecium, Euglena, and yeasts (Fig. 9-14a).

Many multicellular organisms can also reproduce asexually. Small replicas of the parent grow through cell division. A *Hydra,* for example, can reproduce by growing
a miniature new *Hydra* as a bud (Fig. 9-14b). Eventual-

ly the bud separates from its parent, going off to live independently. Because mitosis produces genetically identical cells, these offspring are genetically identical to
their parents.

Many plants reproduce both asexually and sexually.
The beautiful aspen groves of Colorado, Utah, and New
Mexico (Fig. 9-15) develop asexually from shoots growing
up from the root system of a single parent tree. The entire
grove, although seeming to be a population of separate

(a)

(b)

1 micrometer

Figure 9-14   **Modes of asexual reproduction**

(a) In unicellular microorganisms, such as the protist *Tetrahymena* shown here, cell division
produces two new, independent organisms. **(b)** *Hydra,* a freshwater relative of jellyfish, grows
a miniature replica of itself (a bud) protruding from its side. When fully developed, the bud
breaks off and assumes independent life.

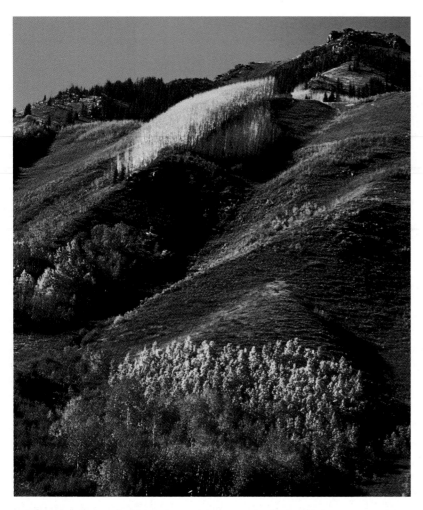

Figure 9-15   **The trees in aspen groves are genetically identical**

Each tree grows as a shoot from the roots of a single ancestral tree. This photo shows three separate groves near Aspen, Colorado. The genetic identity within a grove and genetic difference between groves are shown by their leaves in fall. One entire grove is still green, the second has turned bright gold, and the third has already lost its leaves.

trees to the admiring visitor, may actually be considered to be a single individual, with its multiple trunks interconnected by a common root system. Recently, biologists at the University of Colorado calculated that the single largest organism yet discovered on Earth is probably a huge aspen grove in Utah, covering 106 acres and including about 47,000 trunks with a total mass of perhaps almost 6 million kilograms (13 million pounds). Finally, although individual trunks may age and die, the grove lives on: Some groves are thought to be hundreds of thousands of years old. If you plant an aspen tree in your backyard, who knows what you may be starting!

�острSUMMARY OF KEY CONCEPTS

The Essentials of Cellular Reproduction
Every cell must begin life with (1) a complete set of genetic information (DNA) and (2) the materials necessary to survive and use the genetic information. Cellular reproduction involves replication of the DNA, with one copy packaged into each daughter cell, and cytoplasmic division, with about half the parental cytoplasm passing into each daughter cell.

Prokaryotic and Eukaryotic Chromosomes Compared
The DNA of a prokaryotic cell is contained in a single, circular chromosome, not separated from the cytoplasm of the cell. Eukaryotic cells contain much more DNA than prokaryotic cells do, packaged in many separate chromosomes that are contained within a membrane-bound nucleus.

The Prokaryotic Cell Cycle
The DNA of a prokaryotic cell is in the form of a single circular chromosome, attached at one point to the cell membrane. When a prokaryotic cell reproduces, it duplicates the chromosome, moves the two resulting copies to opposite ends of the cell, and divides the cell in two.

The Eukaryotic Chromosome

The chromosomes of eukaryotic cells consist of both DNA and protein. During interphase, the chromosomes are in an extended form, accessible for use in reading their genetic instructions. During cell division, the chromosomes condense to form short, thick structures. At some point in all eukaryotic life cycles, cells contain pairs of chromosomes called homologues. Homologues have virtually identical appearance and similar genetic information. Cells with pairs of homologous chromosomes are diploid. Those with only a single member of each pair are haploid, and those with more than two copies of each homologous chromosome are polyploid.

The Eukaryotic Cell Cycle

The eukaryotic cell cycle consists of two major phases, interphase and cell division. During interphase, the cell takes in nutrients, grows, and duplicates its chromosomes. Cell division consists of two processes, mitosis (nuclear division) and cytokinesis (cytoplasmic division). Mitosis parcels out one copy of each chromosome into two separate nuclei, and cytokinesis subsequently encloses each nucleus in a separate cell.

Mitosis

The chromosomes are duplicated during interphase, prior to mitosis. The two identical copies, called chromatids, are attached to one another at the centromere. During mitosis, the chromatids separate, one going to each daughter nucleus. The two daughter nuclei thus receive identical genetic information. Mitosis consists of four phases:

Prophase: The chromosomes condense and the spindle microtubules form. The nuclear membrane breaks apart into vesicles. The spindle microtubules invade the nuclear region. Some microtubules, now called kinetochore microtubules, attach to the chromosomes at kinetochores located at the centromeres. Other microtubules, called polar microtubules, extend from the spindle poles to the equator and overlap at the equator with polar microtubules extending from the opposite pole.

Metaphase: The chromosomes move to the equator of the cell.

Anaphase: The two chromatids of each chromosome separate; each former chromatid is now considered to be an individual chromosome. The spindle microtubules move the newly independent chromosomes to opposite poles of the cell.

Telophase: The chromosomes relax into their extended state, and nuclear envelopes re-form around each new daughter nucleus.

Cytokinesis

Mitosis may occur without cytokinesis, for example in mammalian muscle cells and many fungi. In most cells, however, cytokinesis roughly coincides with the end of telophase. Cytokinesis usually divides the cytoplasm into approximately equal halves. In animal cells, the plasma membrane is pinched in along the equator by a ring of microfilaments, separating the cytoplasm into two daughter cells, each containing one of the daughter nuclei. In plant cells, new plasma membrane forms along the equator by fusion of vesicles produced by the Golgi complex.

Cell Division and Asexual Reproduction

Some organisms reproduce asexually—that is, without the fusion of gametes. Asexual reproduction usually occurs by the growth of a new organism from the body of a single parent organism, accomplished by repeated cell divisions and differentiation of daughter cells. Therefore, the offspring are genetically identical to the parent.

✖ K E Y   T E R M S

�֎ T H I N K I N G   T H R O U G H   T H E   C O N C E P T S

Multiple Choice

1. Since most somatic (body) cells of a fruit fly have 8 chromosomes each, the zygote would have
   a. 2 chromosomes
   b. 4 chromosomes
   c. 8 chromosomes
   d. 16 chromosomes
   e. 32 chromosomes

2. At which stage of mitosis are chromosomes arranged along a plane at the midline of the cell?
   a. anaphase   b. telophase
   c. metaphase   d. G_2
   e. interphase

3. A diploid cell contains in its nucleus
   a. an even number of chromosomes
   b. an odd number of chromosomes
   c. one copy of each homologue
   d. either an even or an odd number of chromosomes
   e. two sister chromatids of each chromosome during G_1

4. How many chromosomes are in a human sperm cell?
   a. 8   b. 16
   c. 23   d. 46
   e. 92

5. Synthesis of new DNA occurs during
   a. prophase
   b. interphase
   c. mitosis
   d. cytokinesis
   e. formation of the cell plate

6. What is the attachment site on the chromosome for pulling chromosomes apart during mitosis?
   a. the cell plate
   b. the aster
   c. the centriole
   d. the kinetochore
   e. none of the above

Review Questions

1. Define a cell cycle and a life cycle. What is the role of the cell cycle in a bacterium? In a human?

2. Diagram and describe the prokaryotic cell cycle.

3. Diagram and describe the eukaryotic cell cycle. Name the various phases and briefly describe the events that occur during each.

4. Define mitosis and cytokinesis. Do these two processes always occur together?

5. Diagram the stages of mitosis. How does mitosis ensure that each daughter nucleus receives a complete set of chromosomes?

6. Define the following terms: homologous chromosome; centromere; kinetochore; chromatid; diploid; haploid; polyploid.

7. Diagram the spindle apparatus. How do the two types of spindle microtubules operate during anaphase?

8. Describe and compare the process of cytokinesis in animal and plant cells.

9. Compare the roles of polar and kinetochore microtubules during mitotic cell division.

✖ A P P L Y I N G   T H E   C O N C E P T S

1. Neurons in the adult human central nervous system and heart muscle cells are postmitotic. In other words, they remain in interphase. By contrast, cells lining the lumen of the small intestine divide frequently. Discuss this in terms of why damage to the former (brain in stroke and heart in heart attack) is so dangerous. What do you think might happen to tissues like the intestinal lining if some disorder or drug blocked mitoses in all cells of the body?

2. Do you think that each type of cell in your body steadily progresses through one cell division after another, or do you think that the rate of division is likely to vary from time to time? Explain and justify your answer in terms of some common events in human life: growing up as a child; healing a wound in your skin; menstruation; replacing blood donated to the Red Cross; or the unfortunate case of cancer.

3. A drug called colchicine inhibits the polymerization of microtubules. What specific effects do you think colchicine

would have on cell division? What effects would it have on the daughter cell(s)?

4. Cancer cells divide out of control. Chemotherapy and radiation therapy may interfere with cell division in other cells, with consequences such as loss of hair, and loss of the gastrointestinal lining, causing severe nausea. Are such therapies justified, in your opinion? If a doctor recommended such treatment to you, how might your answer differ depending on your age or on the severity of the cancer?

5. Suppose you were a researcher using tissue from a deceased person who had donated his body to science, and you discovered a chromosomal abnormality that might affect the health of surviving offspring. Should you be allowed (or even required) to communicate this finding to relatives of the deceased?

✖ FOR MORE INFORMATION

Glover, D. M., Gonzalez, C., and Raff, J. W. "The Centrosome." *Scientific American*, June 1993. The structure and function of the centrioles that are essential to the motility of chromosomes during anaphase are presented in this up-to-date paper.

Grant, M. C. "The Trembling Giant." *Discover*, October 1993. Aspen groves are really single individuals; huge, slowly spreading from the roots of the original parent tree, and potentially almost immortal.

Mazia, D. "The Cell Cycle." *Scientific American*, January 1974. A clearly illustrated description of the essentials of cell division.

McIntosh, J. R., and McDonald, K. L. "The Mitotic Spindle." *Scientific American*, October 1989. Beautiful micrographs and lucid drawings describe the structure and function of the spindle apparatus.

Murray, A. W., and Kirschner, M. W. "What Controls the Cell Cycle." *Scientific American*, March 1991. The synthesis, processing, and degradation of a group of proteins seems to control the progression of a cell through the stages of the cell cycle.

NET WATCH

On-line resources for this chapter are on the World Wide Web at:
http://www.prenhall.com/~audesirk (click on the underlined table of contents link and then select Chapter 9).

10 DNA:
The Molecule of Heredity

> *"... a structure this pretty just had to exist."*
>
> JAMES WATSON in *The Double Helix*

A computer-generated model of the structure of DNA. In the words of James Watson, "a structure this pretty just had to exist."

As you know, all living organisms on Earth are composed of cells. Each cell in existence today is the offspring of a previously existing cell, produced when the parent cell divided into two daughter cells. Furthermore, each daughter cell is, with extremely rare exceptions, genetically identical to its parent cell. The genetic information that guides the life of each cell is contained in its chromosomes. As we saw in Chapter 9, during mitosis, an exquisite, highly organized dance of the chromosomes is executed, which ensures that each daughter cell receives a full set of chromosomes. This information still leaves important questions unanswered: How is genetic information contained in the chromosomes? How are the chromosomes duplicated so precisely that each daughter cell receives a full set of chromosomes that are exact copies of the chromosomes of the parent cell?

Just 50 years ago, no one knew that **deoxyribonucleic acid, or DNA** (see the chapter opening art), is the molecule of heredity in all forms of life on Earth. Even after DNA was discovered and found to be part of chromosomes, it was not immediately recognized as the molecule that carries genetic information. Finally, when DNA was accepted as the molecule of heredity, questions still remained about how it encodes genetic information and how that information is faithfully copied when chromosomes are duplicated. These issues occupied many of the best minds of the mid-twentieth century. In this chapter, we will follow those minds along the paths of discovery that led to our present understanding of the structure and function of DNA. Along the way, we hope you will gain insight not only into the molecular nature of DNA but also into how scientists work and how they unraveled part of the complex fabric of inheritance. In Chapter 11, we will investigate the mechanisms whereby the genetic information carried by DNA is used by a cell to produce the structures and functions necessary for continued life.

The Composition of Chromosomes

In the early 1870s, the biochemist Friedrich Miescher extracted a previously unknown chemical substance from nuclei, an acidic material with an unusually high phosphorus content. Because of its location in the nucleus and its acidic properties, this material came to be known as nucleic acid. In the early 1900s, the discovery that chromosomes are the carriers of genetic information focused on their chemical composition. Biochemists found that a eukaryotic chromosome is composed only of protein and a specific kind of nucleic acid, deoxyribonucleic acid (DNA). One of these substances, therefore, must carry the cell's hereditary blueprint, but which one?

DNA Is Composed of Four Types of Nucleotides

DNA seemed to be a particularly simple molecule, composed of just four kinds of subunits, called **nucleotides,** strung together in a long chain. As we described in Chapter 3, each nucleotide consists of three parts: a phosphate group; deoxyribose, a five-carbon sugar; and a nitrogen-containing **base:**

Biochemists have assigned numbers to the carbon and nitrogen atoms that make up the skeletons of the sugar and the base (the numbers for the sugar are given as 1' ["one prime"], 2', etc., to distinguish them from the numbers for the base).

The four different DNA nucleotides have the same phosphate and sugar but different bases. The bases come in two types: **pyrimidines,** thymine (abbreviated T) and cytosine (C), which consist of a single ring; and the two-ringed **purines,** adenine (A) and guanine (G):

Nucleotides with pyrimidine bases:

Nucleotides with purine bases:

In a single strand of DNA, the phosphate group of one nucleotide bonds to the sugar of another. This arrangement forms a long strand consisting of a "backbone" of sugars and phosphates, with the bases protruding from the backbone:

Single strand of DNA:

The phosphate is attached to the 5′ carbon (the numbering system for the carbon atoms of the sugars comes in handy here). To form a single strand of DNA, the phosphate of one nucleotide undergoes a condensation reaction with the hydroxyl group attached to the 3′ carbon of the next nucleotide, thus bonding the phosphate and sugar. This bonding arrangement orients all nucleotides in the strand in the same direction. Notice also that the DNA strand has a "free" sugar on one end (the 3′ end of the strand) and a "free" phosphate on the opposite end (the 5′ end of the strand).

Is the Genetic Material DNA or Protein?

Because DNA has only four subunits, each quite similar to the others, many biologists were skeptical that DNA could contain the information needed to construct all the molecules of complex organisms. Proteins, in contrast, are constructed of about 20 different amino acids. Intuitively, it would seem that having a greater number of building blocks offers much more opportunity for carrying the enormous amount of genetic information that each chromosome obviously contains. Besides that, most of the substance of living cells, aside from water, is protein or is synthesized through the action of enzymes, which are also proteins. An attractive hypothesis, therefore, was that the proteins of chromosomes are templates for the proteins of the rest of the cell, much like molds from which copies could be made.

Gradually, biologists accumulated information that indicated that proteins might not be the genetic material. For example, chromosomes in the sperm of some fish contain only DNA and one protein, called protamine. Like other animals, a fish receives half its genes from its male parent, through the male's sperm. Now, protamine is a very simple protein, mostly composed of a single amino acid. If protamine *were* the genetic material in fish sperm, how could it possibly serve as a template for all the other, more complex, proteins of the whole fish? Despite contradictory evidence such as this, many scientists continued to support the protein-as-gene hypothesis. Finally, two experiments with microorganisms showed that DNA must be the hereditary material.

DNA: The Hereditary Molecule

Throughout the history of biology, major advances have been made possible by using the "right" organism for a particular experiment. Often, a certain organism is especially suitable for a certain experiment because it is simple in structure or reproduces very rapidly. Such simple, rapidly reproducing organisms were the keys to proving that DNA, not protein, is the chemical of heredity.

Bacterial Transformation Indicates That the Hereditary Molecule of Bacteria Is DNA

In the mid-1920s, the bacteriologist Frederick Griffith tried to develop a vaccine to protect people against a bacterium, *Streptococcus pneumoniae*, that causes pneumonia.

He never produced a successful vaccine, so in one sense his experiments were failures. As it turned out, however, he discovered an intriguing phenomenon, which he called **transformation,** that eventually played a major role in the rise of molecular genetics.

Griffith found two varieties of *Streptococcus*. In one variety, each bacterium has a polysaccharide capsule; in the other form, the bacteria are naked. Griffith hypothesized that the capsule might affect the ability of the two varieties to cause disease. He ran four experiments with the two varieties of *Streptococcus* (Fig. 10-1).

1. **Injection with live encapsulated bacteria.** When he injected mice with encapsulated bacteria, the mice contracted pneumonia and died (Fig. 10-1a). Autopsy showed that the blood of infected mice contained hordes of encapsulated bacteria. Later it was found that when *Streptococcus* bacteria invade a mammalian host—for example, a mouse—the host's white blood cells try to ingest the invading microbes. Encapsulated bacteria seem to be "slippery," preventing the white blood cells from getting a grip on the bacterial surface. Consequently, white blood cells cannot ingest encapsulated bacteria very well. Encapsulated bacteria therefore multiply in the host and cause disease.
2. **Injection with live naked bacteria.** If mice were injected with the naked variety of *Streptococcus*, they remained healthy (Fig. 10-1b). No *Streptococcus* could be found in the mice's blood. Later discoveries showed that naked bacteria are easily destroyed by white blood cells, so naked bacteria do not cause disease.
3. **Injection with heat-killed encapsulated bacteria.** Griffith found that heating encapsulated *Streptococcus* to a high temperature killed the bacteria. Not surprisingly, dead bacteria did not cause pneumonia, and the mice remained free of living bacteria (Fig. 10-1c).
4. **Injection with a mixture of dead encapsulated bacteria and live naked bacteria.** Griffith injected mice with a mixture of heat-killed encapsulated bacteria and live naked bacteria (Fig. 10-1d). Although neither of these caused pneumonia when injected alone, the mixture did cause pneumonia. What's more, the infected mice teemed with live encapsulated bacteria, and these encapsulated bacteria bred true; that is, all of their offspring also were encapsulated.

A few years later, in 1933, J. L. Alloway obtained the same results as Griffith's fourth experiment without any mice at all. He killed encapsulated bacteria, filtered out the larger bacterial parts, and mixed the resulting solution with living naked bacteria. The mixture produced live, true-breeding, encapsulated bacteria.

What did Alloway's experiments mean? A likely hypothesis was that the living bacteria had acquired molecules of genetic information from the dead bacteria. The molecules encoded the instructions for making capsules, thereby *transforming* the formerly naked bacteria into the encapsu-

(a) If mice are injected with the encapsulated form of *Streptococcus*, they contract pneumonia and die.

inject encapsulated bacteria

mouse contracts pneumonia and dies

(b) If the mice are injected with the naked form, their white blood cells destroy the invading bacteria and the mice do not develop pneumonia.

inject naked bacteria

mouse remains healthy

(c) Heating kills encapsulated bacteria so that they cannot cause pneumonia.

inject heat-killed, encapsulated bacteria

mouse remains healthy

(d) If heat-killed encapsulated bacteria are mixed with live naked bacteria, the resulting mixture now can cause pneumonia in mice. Blood samples taken from these mice contain live, encapsulated bacteria.

mix heat-killed, encapsulated bacteria with live naked bacteria

inject mixture into mouse

mouse contracts pneumonia and dies

live encapsulated bacteria found in mouse

Figure 10-1   **The discovery of transformation**

Griffith's discovery of transformation in pneumonia bacteria.

lated form. As it turns out, transformation is an effective evolutionary strategy because it allows bacteria to acquire new genes, which might help them to colonize new hosts.

The Transforming Molecule Is DNA

In 1944, three researchers at Rockefeller University—Oswald Avery, Colin MacLeod, and Maclyn McCarty—discovered that the transforming molecule is DNA. They isolated DNA from encapsulated bacteria, mixed it with live naked bacteria, and produced live encapsulated bacteria. To show that transformation was caused by DNA, and not by small quantities of protein contaminating the DNA, they treated different samples with protein-destroying enzymes. These enzymes did not affect the transforming ability of

the DNA samples; however, treating samples with DNA-destroying enzymes prevented transformation.

Avery, MacLeod, and McCarty deduced that DNA is the molecule responsible for transformation and that a live bacterium can take up DNA from its environment and incorporate this DNA into its own chromosome (Fig. 10-2). When DNA carrying the genes for capsules becomes part of the chromosome of another bacterium, the recipient becomes capable of synthesizing capsules. On the basis of this evidence, they concluded that DNA is the genetic material of bacteria.

Despite this evidence, many biologists remained unconvinced that DNA was the universal hereditary molecule. Perhaps, some thought, DNA induces a mutation in naked bacteria, changing a (protein) gene that does not di-

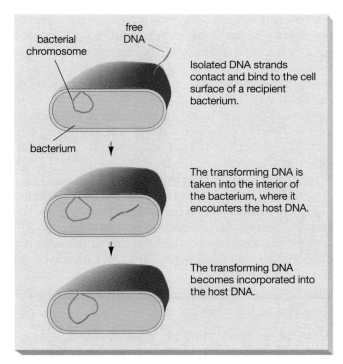

Figure 10-2   Bacterial transformation by free DNA

Transformation is not the "invasion" of a bacterium by free DNA. Rather, the recipient actively acquires the DNA and incorporates it into its chromosome. Transformation allows bacteria to acquire new genes and confers a selective advantage on the recipient bacteria. For example, transformation aids in the spread of antibiotic resistance from one type of bacterium to another.

rect capsule synthesis into one that does. Or perhaps DNA is the hereditary molecule of bacteria, but not of other organisms. A second experiment convinced most of the skeptics. This experiment used an even simpler system, a virus that infects bacteria.

The Hereditary Molecule of Bacteriophages Is Also DNA

Certain viruses infect only bacteria and are called **bacteriophages,** meaning "bacteria eaters" in Greek (Fig. 10-3). A bacteriophage (phage for short) depends on its host bacterium for every aspect of its life cycle (Fig. 10-4). When a phage encounters a bacterium, it attaches to the bacterial cell wall and injects its genetic material into the bacterium. The rest of the phage remains outside the bacterium. The phage genes subvert the bacterial metabolism into producing more phages. Finally, the genetic material of the phage directs the synthesis of an enzyme that ruptures the bacterium, liberating the newly manufactured phages.

Even though many bacteriophages have elaborate structures (see Fig. 10-3), they are chemically very simple, being composed only of DNA and protein. The question was, Which molecule is the phage genetic material?

In a stunning series of experiments published in 1952, Alfred Hershey and Martha Chase used the chemical simplicity of bacteriophages to deduce that the genetic material of phages is DNA (Fig. 10-5). Chemically, DNA and protein both contain carbon, oxygen, hydrogen, and

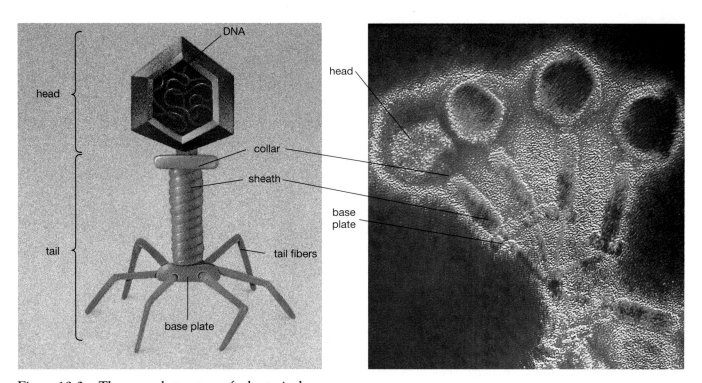

Figure 10-3   The general structure of a bacteriophage

The phage consists of a head region composed of a protein coat, containing a single DNA molecule, attached at the collar to a protein tail region (the sheath, base plate, and tail fibers) responsible for attachment to its host bacterium and injection of the genetic material.

1. A phage encounters a suitable host bacterium.

2. The phage's tail fibers chemically recognize the bacterial cell wall and attach to it.

3. The tail sheath contracts, penetrating the bacterium with the phage's central core. The phage injects its genetic material (red) into the bacterium.

4. The phage's genetic material has entered the bacterium, while its empty outer coat remains outside.

5. The phage genes use the bacterium's own metabolism to synthesize new phage parts.

6 . The phage parts self-assemble into complete bacteriophages.

7. Finally, a phage gene directs the synthesis of an enzyme that ruptures the bacterium, freeing the completed offspring phage.

Figure 10-4   **The life cycle of bacteriophages**

nitrogen. However, DNA also contains phosphorus but not sulfur, whereas protein contains sulfur (in the amino acids methionine and cysteine) but not phosphorus. Hershey and Chase forced one population of phages to synthesize DNA using radioactive phosphorus, thereby "labeling" their DNA. Another population was forced to synthesize protein using radioactive sulfur. When bacteria were infected by phages containing radioactively labeled protein, radioactivity did not appear inside the bacteria. Offspring phages were not radioactive either (Fig. 10-5a). When bacteria were infected by phages containing radioactive DNA, the radioactivity was subsequently found inside the bacteria. When the bacteria burst, some of the offspring phages also had radioactive DNA (Fig. 10-5b). These experiments showed that the genetic material that phages inject into their hosts is DNA, not protein.

The Sequence of Bases in DNA Can Encode Vast Amounts of Information

Although the experiments of Avery, MacLeod, and Mc-Carty with bacterial transformation and those of Hershey and Chase with bacteriophage replication seemed to demonstrate that DNA must be the genetic material, you may still be wondering, "Can just four different types of bases in DNA encode all of the information needed to produce thousands of proteins, each with various combinations of 20 amino acids?" The answer is yes. The four different types of bases can be arranged in any linear order along a strand of DNA. Each sequence of bases represents a unique set of genetic instructions, like a biological Morse code (which, after all, can produce all of the English language with only two symbols). A stretch of DNA just 10 nucleotides long can exist in over a million different possible sequences of the four bases. Because an "average" chromosome has millions (in bacteria) to billions (plants and animals) of nucleotides, DNA molecules can encode a staggering amount of information.

The Structure of DNA

Even before the publication of Hershey and Chase's experiments in 1952, the demonstration that the bacterial transforming factor was DNA by Avery and his colleagues had stimulated a burst of research on the chemical nature of DNA. Alfred Mirsky studied the amount of DNA in the cells of various tissues of several organisms. He found that, although the quantity of DNA varies among species, it is constant in each nucleus of a given species no matter what tissue the nucleus comes from. Gametes are a notable exception in that they have half as much DNA as the other cells of the body. When we discuss meiosis and gamete formation in animals in Chapter 12, you will see why this is expected if DNA is the hereditary material.

What seemed equally significant, if only someone could figure out what it meant, were the results of Erwin Char-

(a)

(b)

Figure 10-5   **The Hershey-Chase experiment**

To determine whether DNA or protein is the genetic material of bacteriophages, Hershey and Chase grew one population of phages in a medium containing radioactive sulfur, which radioactively labels the phage proteins. Another population of phages was grown in radioactive phosphorus, which labels the phage DNA. **(a)** The fate of protein labeled with radioactive sulfur (red): Phages attach to the cell walls of unlabeled bacteria and inject their genetic material. Whirling the mixture in a blender breaks off the phage heads and sheaths left on the outside of the bacteria. The radioactivity is found outside the bacteria, in the phage coats, but the bacteria, containing the phage genetic material, are not radioactive. New phages resulting from the infection are also not radioactive. The phage genetic material, therefore, is not protein. **(b)** The fate of DNA labeled with radioactive phosphorus (red): When the phage-bacteria combinations are broken apart in a blender, the radioactivity is found inside the bacteria, while the phage pieces are unlabeled. Some of the phages synthesized inside the infected bacteria are labeled. The genetic material must be DNA, carrying the radioactivity inside the bacteria and to offspring phages.

gaff's studies. Chargaff analyzed the amounts of the four nucleotides of DNA in several species and found a curious consistency. Although the amounts of each of the four bases vary considerably from species to species, *the DNA of any given species contains equal amounts of adenine and thymine and equal amounts of cytosine and guanine.*

Let us summarize the data compiled up to 1952:

1. DNA is the molecule of heredity.
2. All cells in the body, gametes excepted, contain the same amount of DNA.

3. The amount of cytosine is the same as the amount of guanine; the amount of thymine is the same as the amount of adenine.

Several major questions still remained unanswered: How does DNA encode genetic information? How is it duplicated before mitosis, so that each daughter cell receives exactly the same genetic information? Biologists in the early 1950s agreed that the secrets of DNA function, and therefore of heredity itself, could be found only by understanding the structure of the molecule.

Finding out the structure of any biological molecule is no simple task. Even the most powerful electron microscopes cannot reveal the structure of molecules in atomic detail. To study the structure of DNA, Maurice Wilkins and Rosalind Franklin turned to X-ray diffraction. They bombarded crystals of purified DNA with X-rays and photographed the resulting diffraction patterns (Fig. 10-6). As you can see, the X-ray pattern of DNA does not provide a direct picture of the structure of the molecule. However, the diffraction pattern suggested to Wilkins and Franklin that a DNA molecule is a **helix** (that is, twisted like a corkscrew), has a uniform diameter of 2 nanometers, and consists of subunits, each of which is separated from neighboring subunits by 0.34 nanometer (a nanometer is a billionth of a meter). One full turn of the DNA helix occurs every 3.4 nanometers. Finally, the X-ray picture suggests that the sugar-phosphate "backbone" of the molecule (see Fig. 10-6a) is on the outside of the helix, while the bases are on the inside. This last point is also supported by the fact that DNA is acidic; phosphate groups are acidic, and the bases, of course, are basic, so logically one would expect the phosphate groups to be on the outside.

The DNA of Chromosomes Is a Double Helix

The chemical and X-ray diffraction data were not nearly enough information with which to work out the structure of DNA. Some good guesses were also needed (see the box "Scientific Inquiry: The Discovery of the Double Helix"). Combining a knowledge of how complex organic molecules bond with one another with an intuition that "important biological objects come in pairs," James Watson and Francis Crick proposed that the DNA molecule consists of two strands, twisted about each other into a **double helix,** much like a ladder twisted lengthwise into a corkscrew shape (Fig. 10-7). The sugar-phosphate backbones of the two DNA strands are on the outside of the double helix, like the uprights of the ladder. Notice that the sugar-phosphate uprights run in opposite directions, so that one upright has its free sugar "foot" (3′ end) on one end of the DNA molecule and the other has its free sugar "foot" on the opposite end. The bases are packed into the middle, paired up to form the rungs of the ladder.

Watson and Crick proposed that each rung is composed of a purine and a pyrimidine, held together by hydrogen bonds. If adenine pairs with thymine, and guanine pairs with cytosine, then hydrogen bonds can form between the bases, holding the two halves of the rungs together:

Adenine and thymine are held together by two hydrogen bonds; cytosine and guanine are held together by three.

Figure 10-6 **The X-ray diffraction pattern of DNA, taken by Rosalind Franklin**

(a) The "cross" formed of dark spots is characteristic of helical molecules such as DNA. Measurements of various aspects of the pattern indicate the dimensions of the DNA helix; for example, the distance between spots in the cross corresponds to the distance between turns of the helix. (b) Rosalind Franklin, before her untimely death in 1958 at age 37, published about 40 scientific papers.

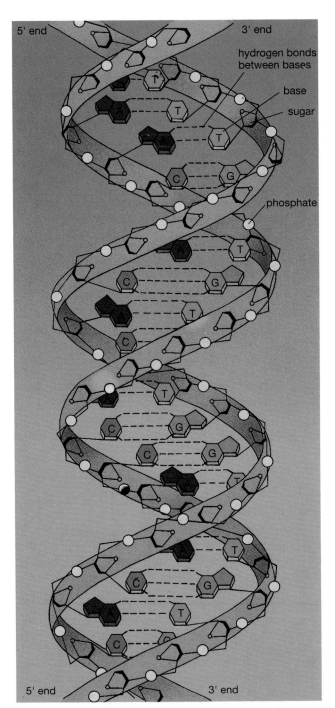

5' end   **3' end**

hydrogen bonds between bases

base

sugar

phosphate

5' end   **3' end**

Figure 10-7   **The Watson-Crick model of DNA**

In this model, two strands of DNA are wound about each other in a double helix, like a ladder twisted about its lengthwise axis. Complementary bases pair together in the center of the helix. Three hydrogen bonds hold guanine to cytosine; two hold adenine to thymine. In this and the following figure, the "arrowheads" (with the oxygen at the point of the arrow) represent the sugars, and the yellow circles the phosphate groups. The two DNA strands run in opposite directions: Note that the free sugar (the 3' end of the sugar-phosphate backbone) is at the bottom of one strand, while the phosphate (the 5' end of the backbone) is at the bottom of the other strand. (The arrowheads point in the 3' to 5' direction for each strand.) As Figures 10-8 and 10-9 show, the directionality of the strands is important during DNA replication.

Adenine-thymine and guanine-cytosine pairs make sense of Chargaff's data, that the amount of adenine in DNA equals the amount of thymine and that the amount of guanine equals the amount of cytosine. In nucleic acids, bases that pair together by hydrogen bonds are called **complementary base pairs**. In DNA, adenine is complementary to thymine, and guanine is complementary to cytosine. This is called the base-pairing rule.

The structure of DNA was solved. Although further data would be needed to confirm its details, "a structure this pretty just had to exist," as Watson later put it. Watson and Crick published their double helix model for DNA in 1953, a model that changed all of genetics and practically all of biology in just a few years. In 1962, Watson, Crick, and Wilkins shared the Nobel Prize in recognition of their revolutionary work in deciphering the structure of DNA. (Rosalind Franklin, whose work was equally deserving of the Prize, died in 1958. Nobel Prizes are not awarded after someone's death, so her contributions sometimes do not receive the recognition they deserve.)

What remains for us to find out is how the structure *works*: How are identical copies formed during DNA replication, and how does the sequence of nucleotides in DNA direct synthesis of cellular components? In the remainder of this chapter and in the following chapter, we will explore how the Watson-Crick model of DNA helps provide answers to those questions.

DNA Replication: The Key to Constancy

In classically understated scientific style, Watson and Crick included the following comment in their paper describing the double helix: "It has not escaped our notice that the specific [base] pairing we have postulated immediately suggests a possible copying mechanism for the genetic material." In other words, the base-pairing rule offers a simple hypothesis for the replication of the DNA of chromosomes before mitosis and meiosis: The sequence of bases in one strand of the double helix accurately predicts the sequence of bases in the other, complementary strand. A chromosome could be replicated by separating the two DNA strands and synthesizing new strands using nucleotides with bases complementary to the parental strands. Each new chromosome would then consist of one parental strand and one newly synthesized, complementary daughter strand.

▨   S U M M A R Y

DNA REPLICATION

There are three fundamental steps to DNA replication in all living cells (Fig. 10-8).

1. The two DNA strands of the double helix of a parental chromosome unwind and separate.

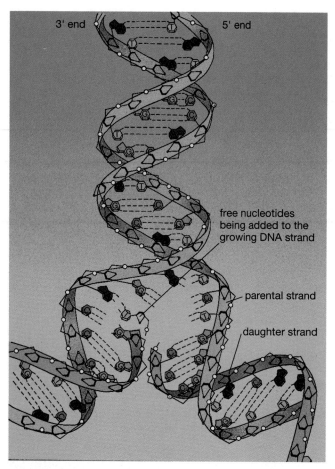

3' end   5' end

free nucleotides
being added to the
growing DNA strand

parental strand

daughter strand

Figure 10-8   **Replication of DNA**
The sugar-phosphate backbones of the parental DNA strands
are colored light tan and the backbones of the daughter DNA
strands are colored pink. The parental DNA strands unwind,
and the hydrogen bonds between complementary bases are
broken. DNA polymerase enzymes move along the two DNA
strands in opposite directions, traveling in the 3' to 5' direction (the direction that the "oxygen arrowheads" are point-
ing). The polymerase enzymes recognize the nucleotides in
the parental DNA strands, pair each one with a complementary free nucleotide, and connect the new nucleotides together into daughter strands. This semiconservative replication
process produces two new double helices of DNA, each composed of one parental strand and one daughter strand that is
an exact copy of the other parental strand. Therefore, each
new double helix is an exact copy of the parental DNA.

2. Each parental strand is used as a template for the formation of a complementary daughter strand of DNA.
The daughter strand is formed by connecting nucleotides together in an order that is dictated by the
nucleotide sequence of the parental strand: The base
of each nucleotide added to the daughter strand is
complementary to the base of the nucleotide in the
corresponding location of the parental DNA.

3. Finally, one parental DNA strand and its newly synthesized, complementary daughter strand wind together into

one double helix, while the other parental strand and
its complementary daughter strand wind together into a
second double helix. Because each new double helix
consists of one intact parental DNA strand and one
newly synthesized complementary strand, the process of
DNA replication is called **semiconservative replication**
(one half of the parental DNA is "conserved," or inherited, by each daughter cell).

We will now examine these steps in more detail.

The Steps of DNA Replication Are Catalyzed by the Coordinated Action of Several Enzymes

In eukaryotic cells, the overall sequence of events of DNA
replication is the following (Fig. 10-9):

DNA Replication Begins with the Unwinding of Small Sections of the Parental Double Helix

The first step in DNA synthesis is to separate the two
DNA strands in one or more places, thereby unwinding
the double helix (Fig. 10-9a). An enzyme appropriately
named DNA helicase ("an enzyme that breaks apart the
helix") wiggles its way between the two strands. According to one model, the helicase enzyme then "walks" along

Figure 10-9   **Molecular mechanisms of DNA
replication**

(a) The enzyme DNA helicase separates the two DNA
strands, unwinding a small portion of the double helix.
(b) Two DNA polymerase molecules attach to the separated
parental strands. DNA polymerase can synthesize new DNA
strands only by proceeding in the 3' to 5' direction on the
sugar-phosphate backbones. Therefore, the "branch point"
polymerase attaches near the helicase and proceeds away from
the helicase, while the "end" polymerase attaches at the other
end of the unwound DNA region and proceeds toward the
helicase (which is also the direction of future unwinding).
(c) The two DNA polymerase molecules match up free nucleotides with the parental DNA strands, using the base-pairing
rules, and synthesize daughter DNA strands (see Fig. 10-8).
(d) The helicase advances, unwinding more parental DNA.
The end DNA polymerase continues to synthesize additional
daughter DNA, adding new nucleotides to the end of the
strand already synthesized in part (c). The original branch
point DNA polymerase, however, has run into the end of the
unwound DNA and leaves the DNA, having synthesized only
a short daughter DNA segment. Meanwhile, a new DNA
polymerase molecule attaches at the new branch point and
proceeds away from the helicase toward the first daughter
DNA segment. **(e)** When the new branch point polymerase
has synthesized daughter DNA that reaches to the end of the
first daughter DNA segment, it leaves the DNA. Another
enzyme, DNA ligase, attaches the second daughter DNA
segment to the first daughter DNA segment.

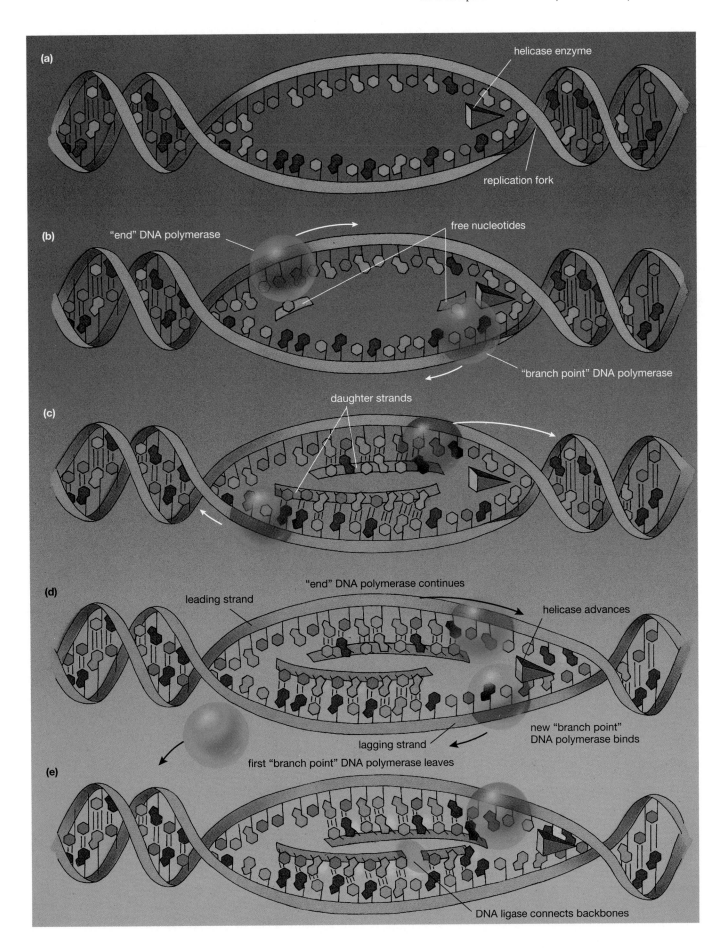

(a) helicase enzyme

replication fork

(b) "end" DNA polymerase

free nucleotides

"branch point" DNA polymerase

(c) daughter strands

(d) "end" DNA polymerase continues

leading strand

helicase advances

new "branch point" DNA polymerase binds

lagging strand

first "branch point" DNA polymerase leaves

(e) DNA ligase connects backbones

one strand, nudging the other strand out of the way as it goes. The result is that the two DNA strands separate, which exposes their bases. The junction between the separated strands and the remaining double helix is called the **replication fork.**

Daughter DNA Strands Are Synthesized That Are Complementary to Each Parental DNA Strand

DNA polymerase molecules then bind to the unwound strands, one to each strand. DNA polymerase performs a dual function. First, DNA polymerase recognizes bases exposed in a parental strand and matches them up with free nucleotides that have complementary bases. Individual nucleotides of all four types are synthesized by the cell before replication and are available within the nucleus. Second, DNA polymerase bonds together the sugars and phosphates of the free complementary nucleotides to form the daughter strand (Fig. 10-9b). In other words, when the DNA polymerase is in contact with an adenine in the parental strand, it adds thymine to the growing daughter DNA strand; when it encounters cytosine in the parental strand, it adds guanine to the daughter strand, and so on.

As far as DNA polymerase molecules are concerned, the parental DNA strands are one-way streets: DNA polymerase can travel in only one direction on a DNA strand, in the 3' to 5' direction (follow the "oxygen arrowheads" that point toward the 5' end of a DNA strand in Fig. 10-8). This means that the two DNA polymerase molecules, one on each strand, move in opposite directions. Their movement in opposite directions dictates the third step in DNA synthesis.

The Daughter DNA Pieces Are Joined into Continuous Strands

Let's look a little more closely at the process so far. Suppose that DNA helicase has wiggled its way into the helix and separated about 100 bases, forming a replication fork. Two DNA polymerase molecules then enter the fork. Now, the DNA polymerases will travel in opposite directions on the two strands of DNA, each moving toward the 5' end of its own strand. Let's assume that the two DNA polymerase molecules settle down on opposite ends of the separated DNA, one at the "end" of the fork, one at the "branch point." The end DNA polymerase builds a new daughter strand by moving along its parental strand, now called the leading strand, toward the branch point, and thus *toward the helicase.* Therefore, as the helicase continues to separate the parental DNA strands, the end polymerase simply follows along on the leading strand, synthesizing one long, continuous complementary daughter strand as it goes (Fig. 10-9c, d).

However, the polymerase that settled on the other strand, called the lagging strand, can't follow the helicase. Because it also moves in the 5' direction, it moves *away from the helicase* as it travels and soon reaches a "dead end" where the DNA is still intact (Fig. 10-9c). Therefore, as the helicase continues to separate the parental strands, more lagging strand immediately behind the enzyme is exposed, which the original polymerase can't reach (Fig. 10-9d). Soon, however, a new DNA polymerase molecule attaches to the lagging strand close behind the helicase, becoming a new branch point polymerase. Like the first branch point polymerase, it moves away from the helicase. What happens when this second branch point polymerase reaches the place where the first polymerase started? The two short DNA strands synthesized by the two DNA polymerases must be joined up to make a continuous DNA strand. An enzyme called DNA ligase bonds the two strands together (Fig. 10-9e). This process is repeated many times (perhaps 10 million or so for a human chromosome), until the entire lagging strand has been replicated. Because the original replication fork is seldom exactly at the end of the parental double helix, even the leading strand must be synthesized in several pieces and must also be stitched together by DNA ligase.

Base Pairing and Proofreading Produce Almost Error-Free Replication of DNA

A high degree of accuracy of DNA replication is provided by hydrogen bonding between correctly matched base pairs. Nevertheless, DNA replication is not perfect. Partly because of the speed of replication (50 to 500 nucleotides per second) and partly because of spontaneous chemical flip-flops in the bases, DNA polymerase occasionally incorporates incorrectly matched bases, perhaps one mistake for every 10,000 base pairs. In mammalian cells, however, the completed DNA strands contain only about one mistake for every billion base pairs. This phenomenal accuracy is ensured by several enzymes, including DNA polymerase itself, that "proofread" each daughter strand as it is synthesized. The one-way directionality of DNA structure allows the proofreading enzymes to recognize the parental strand, running in one direction, as the "right stuff," and to correct any mismatches by changing the daughter strand, which runs in the other direction.

During prophase of mitosis, each chromatid consists of a double helix of DNA, composed of one of the original strands of the parent DNA plus one new strand that is an exact copy of the other original strand. When sister chromatids separate at anaphase and are parceled out to the daughter cells, each cell receives an exact copy of each of the parental chromosomes. Thus, if there are no mistakes in the whole process, the integrity of the genetic information will be maintained from cell division to cell division and from parent to offspring.

The Discovery of the Double Helix

In the early 1950s, many biologists realized that the key to understanding inheritance lay in the structure of DNA. They also knew that whoever deduced the correct structure of DNA would receive recognition from fellow biologists, fame in the popular press, and very possibly the Nobel Prize. Less obvious were the best methods to use and who would be the person to do it.

The betting favorite in the race to discover the structure of DNA had to be Linus Pauling of Caltech. Pauling probably knew more about the chemistry of large organic molecules than any person alive, and he had hit upon the idea that accurate models could aid in deducing molecular structure. Like Rosalind Franklin and Maurice Wilkins, he was an expert in X-ray diffraction techniques. Finally, he was almost frighteningly brilliant. In 1950, he demonstrated these traits by showing that many proteins were coiled into single-stranded helices (see Chapter 3). Pauling, however, had two main handicaps. First, for years he had concentrated on protein research, and therefore he had little data about DNA. Second, he was active in the peace movement. During the early 1950s, some government officials, including Senator Joseph McCarthy, considered such activity to be potentially subversive and possibly dangerous to national security. This latter handicap may have proved decisive, as we shall see.

The second most likely competitors were Wilkins and Franklin, the English scientists who had set out to determine the structure of DNA by the most direct procedure, the careful study of the X-ray diffraction patterns of DNA. They were the only scientists who had really good data about the general shape of the DNA molecule. Unfortunately for Wilkins and Franklin, their methodical approach was also slow.

This left the door open for the eventual discoverers of the double helix, James Watson and Francis Crick, two young scientists with neither Pauling's tremendous understanding of chemical bonds nor Franklin and Wilkins's expertise in X-ray analysis. They did have three crucial advantages: the knowledge that models could be enormously helpful in studying molecular structure, a lesson learned from Pauling's work on proteins; access to the X-ray data; and a driving ambition to be first.

Watson and Crick did no experiments in the ordinary sense of the word; rather, they spent their time thinking about DNA, trying to construct a molecular model that made sense and fit the data. Because they were in England and because Wilkins was very open about his and Franklin's data, Watson and Crick were familiar with all the X-ray information relating to DNA. This information was just what Pauling lacked. Because of his presumed subversive tendencies, the U.S. State Department refused to issue Pauling a passport to leave the country, and consequently he could not attend meetings at which Wilkins presented the X-ray data, nor could he visit England to talk with Franklin and Wilkins directly.

Watson and Crick knew that Pauling was working on DNA structure and were terrified that he would beat them to it. In *The Double Helix*, Watson recounts his belief that, if Pauling could have seen the X-ray pictures, "in a week at most, Linus would have the structure."

About this time, you might be thinking, "But wait just a minute! That's not fair. If the goal of science is to advance knowledge, then everybody should have access to all the data. If Pauling was the best, he should have discovered the double helix first." Perhaps so. But science is an activity of scientists, who, after all, are people too. Virtually all scientists want to see the advancement and benefit of humanity, but each individual also wants to be the one responsible for the advancement and to receive the credit and the glory. We should not overlook the fact that the ambition to be first helps to inspire the intense concentration, the sleepless nights, and the long days in the laboratory that ultimately produce results.

At any rate, Pauling remained in the dark about the correct X-ray pictures of DNA and was beaten to the correct structure (Fig. E10-1). When Watson and Crick discovered the base-pairing rules that were the key to DNA structure, Watson sent off a letter about it to Max Delbruck, a friend and adviser at Caltech. He asked Delbruck not to reveal the contents of the letter to Pauling until their structure was formally published. Delbruck, perhaps more of a model scientist, firmly believed that scientific discoveries belong in the public domain and promptly told Pauling all about it. With the class of a great scientist and a great person, Pauling graciously congratulated Watson and Crick on their dazzling solution to the structure. The race was over.

Figure E10-1 DNA

James Watson and Francis Crick with a model of the structure of DNA.

✖ S U M M A R Y   O F   K E Y   C O N C E P T S

The Composition of Chromosomes

Eukaryotic chromosomes are composed of DNA and protein. DNA is the molecule that carries genetic information. It is composed of subunits called nucleotides, linked together into long strands. Each nucleotide consists of a phosphate group, the five-carbon sugar deoxyribose, and a nitrogen-containing base. Four different bases occur in DNA: adenine, guanine, thymine, and cytosine.

DNA: The Hereditary Molecule

Transformation in bacteria and infection of bacteria by bacteriophages provided the first evidence that DNA is the hereditary molecule. Griffith discovered that living bacteria can acquire genetic material from dead bacteria, transforming the live bacteria. Avery and his colleagues showed that the genes taken up by living bacteria during transformation are composed of DNA.

A bacteriophage consists solely of DNA and protein. When it infects a bacterium, a phage injects its genetic material into the bacterium, where it directs the synthesis of more phages. Hershey and Chase demonstrated that phages inject DNA, but not protein, into bacteria. Therefore, DNA must be the genetic material of phages.

The Structure of DNA

The DNA of chromosomes is composed of two strands wound about one another in a double helix. The sugars and phosphates that link one nucleotide to the next form the backbone on each side of the double helix, while the bases from each strand pair up in the middle of the helix. Only specific pairs of bases, called complementary base pairs, can link together in the helix, held by hydrogen bonds: adenine with thymine and guanine with cytosine.

DNA Replication: The Key to Constancy

When a chromosome is replicated before mitosis or meiosis, the two DNA strands of each double helix unwind. DNA polymerase enzymes move along each strand, linking up free nucleotides into new DNA strands. The sequence of nucleotides in each newly formed strand is complementary to the sequence on a parental strand. As a result, two double helices are synthesized, each consisting of one parental DNA strand plus one newly synthesized, complementary strand that is an exact copy of the other parental strand. The two daughter DNA molecules are therefore duplicates of the parental DNA molecule.

✖ K E Y   T E R M S

bacteriophage   p. 171
base   p. 168
complementary base pair   p. 175
deoxyribonucleic acid (DNA)   p. 167
DNA polymerase   p. 178

double helix   p. 174
helix   p. 174
nucleotide   p. 168
purine   p. 168
pyrimidine   p. 168

replication fork   p. 178
semiconservative replication   p. 176
transformation   p. 169

✖ T H I N K I N G   T H R O U G H   T H E   C O N C E P T S

Multiple Choice

1. If you wanted to label amino acids but not DNA, which of the following radioactive isotopes would you use?
   a. ^{18}F b. ^{35}S c. ^{14}C
   d. ^{3}H e. ^{32}P

2. Whose pioneering work on the gene used the bacteriophage to show that DNA was the part of this virus that went into the bacterial cell for replication?
   a. Watson and Crick   b. Avery
   c. Hershey and Chase   d. Griffith
   e. Beadle and Tatum

3. Knowing what you now know about RNA, DNA, and DNA replication from this chapter and mitosis from the last chapter, sister chromatids
   a. each have one single strand of the two complementary DNA strands
   b. each have a fully formed double helix of DNA
   c. are the same as the homologous chromosomes of a diploid cell
   d. are present in bacteriophages, bacteria, and eukaryotes
   e. none of the above

4. How many different possible base sequences are there in a nucleotide chain 3 bases long?
   a. 1   b. 3   c. 9
   d. 64   e. more than 64

5. Chargaff, in analyzing DNA, found that the amount of
   a. cytosine equaled that of guanine
   b. cytosine equaled that of thymine
   c. cytosine equaled that of adenine
   d. each nucleotide was unrelated to any other
   e. each nucleotide was equal to each other

6. Semiconservative replication refers to the fact that
   a. genetic information is contained in the transforming molecule
   b. only the DNA of the bacteriophage enters the bacterial cell
   c. certain bases pair with specific bases
   d. one DNA strand remains while a new one is made
   e. point mutations sometimes interfere with accurate replication

Review Questions

1. Draw the structure of a nucleotide. Which parts are identical in all nucleotides, and which can vary?
2. Name the four types of nitrogen-containing bases found in DNA. Which bases are complementary to one another? How are they held together in the double helix of DNA?
3. What is transformation? In Griffith's experiments, why were the transformed bacteria able to synthesize capsules?
4. Diagram the life cycle of a bacteriophage.
5. Describe the structure of DNA in a chromosome. Where are the bases, sugars, and phosphates in the structure?
6. Describe the process of DNA replication.

✖ A P P L Y I N G   T H E   C O N C E P T S

1. Diagram and describe the Hershey-Chase experiment. Suppose they had found that *both radioactively labeled DNA and radioactively labeled protein* were injected into the bacteria. What would they have been able to conclude from this? (Note that if the bacteriophage chromosome was like the eukaryotic chromosome, that is, if it contained both DNA and protein, this is exactly what would have happened.)

2. A preview question for Chapter 11: Genetic information is encoded in the sequence of nucleotides in DNA. Let's suppose that the nucleotide sequence on one strand of a double helix encodes the information needed to synthesize a hemoglobin molecule. Do you think that the sequence of nucleotides on the other strand of the double helix would also encode useful information? Why or why not? An analogy to think about: Suppose that English were a "complementary language," with letters at opposite ends of the alphabet complementary to one another (for example, A is complementary to Z, B to Y, C to X, etc.). Would the complementary sentence to "All the world's a stage" make sense?

3. Suppose mammalian cells are grown and divide many times in a culture medium containing thymine made radioactive with a form of hydrogen (^{3}H). The cells are then removed from the radioactive medium and allowed to replicate several times in normal culture medium. Daughter chromosomes are tested each generation to determine whether they contain radioactive thymine. While in the radioactive medium, every strand of DNA of each chromosome has radioactive thymine. Assuming that each chromosome contains a single long double-stranded molecule of DNA, predict the radioactive status of daughter chromosomes after one, two, and three rounds of cell division in normal medium. Explain how your predictions are consistent with the Watson-Crick explanation of semiconservative DNA replication.

4. During the Inquisition, the church forced Galileo to disavow some of his scientific statements. Today, scientific advances are being made at an astounding rate, and nowhere is this more evident than in our understanding of the biology of heredity. Using DNA as a starting point, do you think there are limits to the knowledge people should acquire?

5. Historians debate the degree to which key people actually changed the course of history. From this chapter's historical perspective, discuss whether the pace or actual progress of knowledge is influenced by the individual scientists involved.

✖ F O R   M O R E   I N F O R M A T I O N

Crick, F. H. C. "The Structure of the Hereditary Material." *Scientific American*, October 1954. It is entertaining to consider a paper on one of science's most fundamental results while the finding was still new and before the author won the Nobel Prize for Physiology and Medicine.

Holliday, R. "A Different Kind of Inheritance." *Scientific American*, June 1989. Methyl groups attached to DNA may alter its use in a cell. What's more, patterns of DNA methylation may be passed down from parent to offspring.

Judson, H. F. *The Eighth Day of Creation: Makers of the Revolution in Biology.* New York: Simon and Schuster, 1979. A remarkably readable history of the beginnings of molecular biology.

Kimura, M. "The Neutral Theory of Molecular Evolution." *Scientific American*, November 1979. Kimura persuasively argues that some evolutionary changes can occur without selection.

Mirsky, A. E. "The Discovery of DNA." *Scientific American*, June 1968. The early history of DNA research.

Radman, M., and Wagner, R. "The High Fidelity of DNA Duplication." *Scientific American*, August 1988. Faithful duplication of chromosomes requires both reasonably accurate initial replication of DNA sequences and final proofreading.

Rennie, J. "DNA's New Twists." *Scientific American*, March 1993. Relevant to this and several subsequent chapters, this simple review covers some of the interesting modern contradictions that go beyond the standard textbook dogma.

Watson, J. D. *The Double Helix.* New York: Atheneum Publishers, 1968. If you still believe the Hollywood images, that scientists are either maniacs or cold-blooded, logical machines, be sure to read this book. Although scarcely models for the behavior of future scientists, Watson and Crick are certainly human enough!

N E T   W A T C H

On-line resources for this chapter are on the World Wide Web at:
http://www.prenhall.com/~audesirk (click on the <u>table of contents</u> link and then select Chapter 10).

11 Gene Expression and Regulation

*Protein synthesis in action: The large beads are ribosomes,
traveling along a central strand of messenger RNA. The
smaller beaded strands emerging from the larger beads are
the beginnings of proteins. The molecules involved in protein
synthesis are described in this chapter.*

*S*olving the structure of the double helix of DNA and discovering how DNA is replicated during cell division is only the first step in understanding the molecular basis of inheritance. In this chapter, we will take the next two steps.

First, how does a cell use its DNA to create the cell itself? You may recall from previous chapters that most of the organic molecules of a cell are proteins or are synthesized through the actions of protein enzymes. Therefore, to go from DNA to cell structure and function means synthesizing the appropriate proteins. How does DNA direct protein synthesis and function in a cell? We will see that the sequence of bases in DNA is a code that is translated into the sequence of amino acids in proteins. Genes are, in fact, a library containing the information needed to construct the proteins of a cell.

Second, how does an organism regulate the use of its DNA in the appropriate manner during its development and subsequent interactions with its environment? For example, if you begin life as a fertilized egg, and if all the cells of your body are derived from that egg through mitosis, then it follows that all the cells of your body contain the same genes. How, then, can different cells of the body have different structures and functions? Why do hair follicle cells synthesize hair proteins and not hormones? Further, does the environment influence how the genes are used? If so, how? The answer is that, within certain limits, cells can regulate which genes are used and which are not, depending on the function of the cell and the environment in which it lives.

The Relationship between Genes and Proteins

Long before anything was known about the molecular structure of genes, biologists tried to find out how genes work by studying how the functioning of an organism was affected by changes in its genes. In the early 1900s, the English physician Archibald Garrod studied the inheritance of human metabolic disorders. Garrod knew that the human body is a chemical caldron, churning with reactions that convert molecules of food into other molecules the body needs. These biochemical conversions proceed in steps, with each step catalyzed by a specific enzyme. The

Figure 11-1   **Enzymes catalyze biochemical conversions**

Some of the pathways of phenylalanine and tyrosine metabolism in humans. The arrows represent enzyme-catalyzed chemical reaction(s) that produce a product (for example, tyrosine) from a substrate (for example, phenylalanine). If the enzyme that converts phenylalanine to tyrosine is defective, phenylalanine and phenylpyruvic acid build up, damaging developing brain cells. Tyrosine is a precursor for several other essential compounds, including thyroid hormone, the nervous system chemicals epinephrine and norepinephrine, and the pigment melanin. Hereditary disorders caused by single-gene mutations that lead to defective enzymes are shown.

metabolism of the amino acids phenylalanine and tyrosine is an instructive example (Fig. 11-1).

Dietary proteins usually contain both amino acids. Once in the body, the amino acids are used as building blocks for bodily proteins and for other functions, each catalyzed by a specific set of enzymes. If a single enzyme is defective, it may cause one or both of two types of effects. First, the product of the reaction catalyzed by that enzyme will not be synthesized. For example, one of the pathways that tyrosine may take in the human body is to be converted into the pigment melanin, which colors skin, hair, and the iris of the eye. Defects in the enzymes that produce melanin from tyrosine will result in a lack of melanin, and the affected person will be an albino, with extremely pale skin, white hair, and usually pink eyes.

Second, if a substance is not converted to a given product, either the substance itself or products of alternative pathways the substance can take are likely to accumulate in the body. In phenylketonuria (PKU), the enzyme that converts phenylalanine to tyrosine is defective. Therefore, phenylalanine and related substances called phenylketones accumulate. (*Phenylketonuria* means that phenylketones are found in the urine: hence, PKU.) In infants, high phenylalanine levels injure developing brain cells, causing severe mental retardation.

Garrod catalogued several of these "inborn errors of metabolism" and deduced, from their patterns of inheritance, that each defective enzyme was probably caused by a rare, defective version of a single gene. Given the technology of the time, this hypothesis was impossible to test in humans.

Most Genes Encode the Information for the Synthesis of a Protein

The common red bread mold, *Neurospora crassa*, proved to be an ideal organism for studying the relationship between genes and enzymes. Although we commonly see it on stale bread, *Neurospora* is an extremely independent organism that can synthesize almost all the organic compounds it needs. It can grow on a minimal medium containing an energy source, such as sucrose, a few minerals to supply essential elements, such as nitrogen and phosphorus, and a single vitamin, biotin.

Besides being easy to raise, *Neurospora* is genetically ideal as well. For most of its life cycle, *Neurospora* is haploid, with just one copy of each chromosome and, therefore, with just one copy of each gene. Most plants and animals, in contrast, are diploid, with two copies of each chromosome and thus two copies of each gene. Consequently, it may be difficult to tell if a plant or animal carries a defective copy of a gene, because there may be a normal copy that masks the effects of the defective copy. (An analogy might help: Let's imagine two workers who are both supposed to be making widgets, but only one actually works, while the other one goofs off. If the productive worker makes enough widgets, the boss may never know that anything is wrong. The same thing can happen with genes; under some circumstances, the "boss" [your body] may never suffer from a defective copy of a gene. We will see a lot more about interactions between the two copies of each gene in the next chapter.) Because *Neurospora* is haploid, the effects of a defective gene cannot be masked, there being no other copy of that gene.

Geneticists George Beadle and Edward Tatum bombarded *Neurospora* with X-rays, which cause mutations in DNA. They produced mutant molds that lost the ability to synthesize some substances (Fig. 11-2). One particularly interesting mutant couldn't grow on minimal medium unless the amino acid arginine was added. Beadle and Tatum therefore concluded that the mutant must have lost the ability to synthesize arginine. By starting with minimal medium and adding one precursor molecule in the pathway of arginine synthesis at a time, Beadle and Tatum found that the mutant lacked only the enzyme that catalyzed one specific step in arginine synthesis. Genetic analysis showed that the mutant differed from normal molds in a single gene. From this and many other experiments, they concluded that a gene encodes the information needed for the synthesis of a specific enzyme. This conclusion became known as the "one-gene, one-enzyme" hypothesis. Garrod had reached the same conclusion on the basis of his studies of human metabolic disorders, but he couldn't prove it.

Geneticists have since learned that not all genes encode the information needed by a cell to produce enzymes. Some genes carry information for the synthesis of structural proteins, such as collagen in skin and keratin in hair, or hormones, such as insulin. For a few genes, the final product isn't protein at all, but ribonucleic acid (RNA), such as the RNA of ribosomes. Nevertheless, the ultimate cellular products encoded by most genes are usually proteins or parts of proteins. In general, then, each gene encodes the information for a single protein; this generalization might be called the **one-gene, one-protein hypothesis**.

The Sequence of Bases in DNA Codes for the Sequence of Amino Acids in Proteins

Knowing the relationship between genes and DNA, we can reinterpret the conclusions of Beadle and Tatum's experiments: A gene is a segment of DNA that contains the information needed to synthesize a protein. Different genes have different base sequences, and different proteins have different amino acid sequences. Therefore, the sequence of bases in DNA must encode the sequence of amino acids in a protein. How does a cell convert the message of its DNA into proteins?

Synthesizing Proteins from the Instructions in DNA

As you know, the DNA of eukaryotic cells is located in the nucleus. Protein synthesis, on the other hand, occurs on ribosomes in the cytoplasm. Therefore, DNA cannot directly guide protein synthesis. There must be an intermediate molecule that carries the information from DNA

(a) Mold spores are bombarded with X-rays to induce mutations. They are germinated and grown on an enriched medium. When these molds reproduce, their spores are collected.

(b) Spores are germinated individually in enriched medium containing all the amino acids. A piece of each resulting mold is placed on minimal medium lacking amino acids. If the mold still grows on minimal medium (left), then no mutation has occurred and this mold is discarded. If the mold cannot grow on minimal medium (right), a mutation has occurred and the experiments are continued.

+alanine   +arginine   +cysteine   +leucine   +glycine   +lysine, etc.

(c) Pieces of mutated mold are placed in tubes containing minimal medium plus one amino acid. The mutant can grow only if arginine is added. Therefore, the mutant cannot synthesize arginine.

minimal medium   +ornithine   +citrulline   +arginine

(d) Mutants are tested in media to which precursors in the synthesis pathway of arginine have been added. Because the mold can grow if supplied with citrulline or arginine, but not ornithine, the mutation must have ruined a single enzyme, the one that normally catalyzes the conversion of ornithine to citrulline.

***Figure 11-2 Beadle and Tatum's experiments with the mold* Neurospora**

These experiments showed that a mutation of a single gene may produce a single defective enzyme.

in the nucleus to the ribosomes in the cytoplasm. This molecule is **ribonucleic acid, or RNA**.

RNA is similar to DNA but differs in three respects: (1) RNA is usually single stranded; (2) RNA has a different type of sugar in its backbone—ribose instead of deoxyribose; and (3) the base thymine in DNA is replaced by uracil in RNA:

There are three functional types of RNA in a cell; all are involved in translating genetic information in DNA to the amino acid sequence of proteins. These types are messenger RNA, transfer RNA, and ribosomal RNA. We will examine their functions in more detail shortly.

Genetic Information Flows in a Cell from DNA to RNA to Protein

Information flows from DNA to proteins in a two-step process (Fig. 11-3).

1. In **transcription**, the information contained in the DNA of a specific gene is copied into RNA. All three types of RNA are transcribed from DNA. The sequence of bases in one particular type of RNA, appropriately called **messenger RNA (mRNA),** carries information from the nucleus to the ribosomes about the sequence of amino acids in the protein to be manufactured.
2. In **translation**, the two other types of RNA, **transfer RNA (tRNA)** and **ribosomal RNA (rRNA),** convert the information of messenger RNA into the correct amino acids and help to synthesize the protein.

To understand the molecular mechanisms of this information flow from DNA to protein, geneticists first had to break the language barrier: How does the language of base sequences in DNA and messenger RNA translate into the language of amino acid sequences in proteins? This translation relies on a "dictionary" called the genetic code.

The Genetic Code

We have used the word *code* several times to refer to the information stored in DNA and ultimately translated into the amino acid sequence of proteins. This **genetic code** is conceptually similar to Morse code: One set of symbols (bases in nucleic acids, dots and dashes in Morse code) can be translated into another set of symbols (amino acids in proteins, letters of the alphabet). The question is, What combinations of bases stand for which amino acids?

The Genetic Code Uses Three Bases to Specify Each Amino Acid

There are four different bases in DNA: adenine (A), cytosine (C), guanine (G), and thymine (T). There are also four bases in RNA: adenine, cytosine, guanine, and uracil (U). But there are 20 different amino acids in proteins, so the bases cannot serve as a one-to-one code for amino acids. There are simply not enough of them. Perhaps, though, the genetic code might use a short sequence of bases to encode each amino acid, just as Morse code uses a short sequence of dots and dashes to encode the letters of the alphabet. If a sequence of two bases codes for an amino acid, then there will be 16 (4 × 4) possible combinations of bases. This isn't enough either. Three bases per amino acid, however, gives 64 (4 × 4 × 4) possible com-

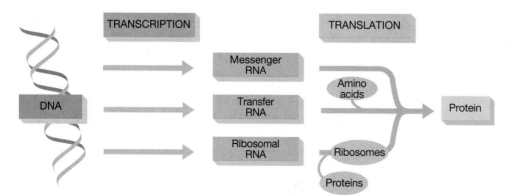

Figure 11-3   **Information flow from DNA to RNA to protein**

During transcription, a gene in DNA provides the information for the synthesis of RNA. During translation, the information in messenger RNA is used to specify the sequence of amino acids in a protein. Transfer RNA and ribosomal RNA are essential for decoding the information in messenger RNA and forming chemical bonds between amino acids to make the protein.

binations, which is more than enough. Under the assumption that nature operates as economically as possible, biologists hypothesized that the genetic code must be triplet: Three bases specify one amino acid. In 1961, Francis Crick and three co-workers demonstrated that this hypothesis is correct (see "Scientific Inquiry: Cracking the Genetic Code.")

A few requirements must be met for any language to be understood. The users must know what the words mean, where words start and stop, and where sentences start and stop. The Crick experiments demonstrated that the "words" of the genetic code are all three bases long and that a set of three bases "means" one amino acid. Shortly after this discovery, researchers began to decipher the code. They ground up bacteria and isolated the components needed to synthesize proteins. To this mixture, they added artificial messenger RNA, which allowed them to control what "words" were to be transcribed. Researchers could then see which amino acids were incorporated into the resulting proteins. For example, an RNA composed entirely of uracil (UUUUUU…) directed the mixture to synthesize a protein composed solely of phenylalanine. Therefore, the triplet specifying phenylalanine must be UUU. Because the genetic code was deciphered using these artificial RNAs, the code is usually written as the base triplets in messenger RNA that code for each amino acid (Table 11-1). These messenger RNA triplets are called **codons**.

What about punctuation? How does the cell recognize where codons start and stop and where codes for entire proteins, which are composed of many amino acids, start and stop? Research showed that the codon AUG signals

"start"—that is, the beginning of a protein **(start codon)**. Three codons, UAG, UAA, and UGA, signal "stop"—the end of a protein **(stop codon)**. Now, if all the codons have three bases, and the beginning and end of a protein are specified, then punctuation between codons is unnecessary. To see why this is so, consider what would happen if English used only three-letter words: A sentence such as THEMANSAWTHECAT would be perfectly understandable, even without spaces between the words, as long as the reader knew where the sentence started and stopped. The genetic code doesn't need, and doesn't have, punctuation between codons.

Though there are only 20 amino acids for which to code, there are 60 codons, in addition to the start and stop codons. All 60 codons are used in the genetic code. The genetic code is thus highly redundant, or **degenerate**. In other words, a single amino acid may be specified by several codons. For example, six different codons all code for arginine (see Table 11-1). Even though the code is redundant, however, it is not ambiguous: Each codon specifies one, and only one, amino acid.

RNA: Intermediary in Protein Synthesis

Protein synthesis requires all three types of RNA: messenger RNA, transfer RNA, and ribosomal RNA. In the following sections, we will take a closer look at how these RNA molecules are synthesized, how they differ, and how they function in protein synthesis.

TABLE 11-1 ⋮ *The Genetic Code (Codons of mRNA)*

First Base	Second Base								Third Base
	U		C		A		G		
U	UUU	Phenylalanine	UCU	Serine	UAU	Tyrosine	UGU	Cysteine	U
	UUC	Phenylalanine	UCC	Serine	UAC	Tyrosine	UGC	Cysteine	C
	UUA	Leucine	UCA	Serine	UAA	Stop	UGA	Stop	A
	UUG	Leucine	UCG	Serine	UAG	Stop	UGG	Tryptophan	G
C	CUU	Leucine	CCU	Proline	CAU	Histidine	CGU	Arginine	U
	CUC	Leucine	CCC	Proline	CAC	Histidine	CGC	Arginine	C
	CUA	Leucine	CCA	Proline	CAA	Glutamine	CGA	Arginine	A
	CUG	Leucine	CCG	Proline	CAG	Glutamine	CGG	Arginine	G
A	AUU	Isoleucine	ACU	Threonine	AAU	Asparagine	AGU	Serine	U
	AUC	Isoleucine	ACC	Threonine	AAC	Asparagine	AGC	Serine	C
	AUA	Isoleucine	ACA	Threonine	AAA	Lysine	AGA	Arginine	A
	AUG	Start (Methionine)	ACG	Threonine	AAG	Lysine	AGG	Arginine	G
G	GUU	Valine	GCU	Alanine	GAU	Aspartic Acid	GGU	Glycine	U
	GUC	Valine	GCC	Alanine	GAC	Aspartic Acid	GGC	Glycine	C
	GUA	Valine	GCA	Alanine	GAA	Glutamic Acid	GGA	Glycine	A
	GUG	Valine	GCG	Alanine	GAG	Glutamic Acid	GGG	Glycine	G

S C I E N T I F I C   I N Q U I R Y

Cracking the Genetic Code

The hypothesis that three bases in DNA code for one amino acid in protein is an attractive and logical possibility. Fewer than three bases can't unambiguously code for all 20 amino acids; more than three are superfluous. But how could you prove that nature really uses a triplet code?

As it did in the Hershey-Chase experiments (see Chapter 10), the simplicity of the bacteriophage proved invaluable in deciphering a fundamental principle of genetics. Francis Crick and his co-workers exposed bacteriophages to a dye called acridine, which causes insertion mutations: One, two, or three nucleotides were inserted into the DNA molecule at random places near the beginning of a particular gene. Crick and his colleagues found that inserting one or two nucleotides into the DNA causes the synthesis of defective enzymes that prevent the phages from reproducing in their host bacteria. Inserting three nucleotides, however, sometimes produces phages that synthesize normal or nearly normal enzymes.

Crick concluded that during RNA synthesis the DNA of a gene is "read" in a linear order, starting at the beginning. Each set of three bases makes up a "word"; that is, three bases in DNA encode a single amino acid. The code must specify the beginning and end of a gene, but within a gene there are no spaces or punctuation between words. How did he arrive at these conclusions?

To understand Crick's reasoning, let's suppose that we have a gene with this repetitive sequence of bases:

T	A	G	T	A	G	T	A	G	T	A	G	T	A	G	T	A	G	T	A	G	T	A	G

If DNA always has three-letter words, then this base sequence spells out the English word *tag* over and over again. You don't need spaces to separate the words if they all have three and only three letters.

Suppose that the acridine dye inserts another thymine somewhere near the beginning of the gene. The gene now reads:

first insertion   incorrect "reading"

Protein synthesis is like a computer: garbage in, garbage out. From the point of the insertion to the end of the gene, the gene now reads GTA, GTA, GTA,…, which is nonsense.

Inserting another T nearby still results in nonsense, as most of the gene now reads AGT, AGT, AGT:

second insertion   incorrect "reading"

A third insertion, however, results in most of the gene reading TAG, TAG, TAG again:

third insertion   remainder of gene reads correctly

How does this scenario explain Crick's results? One insertion near the beginning of the bacteriophage gene causes the rest of the triplets to encode the wrong amino acids, so the enzyme specified by that gene completely malfunctions. The second insertion still leaves most of the triplets calling for the wrong amino acids. If there were three insertions near the start of the gene, however, then the first few amino acids would be wrong, but all the triplets beyond the third insertion would be correct once again. Most of the enzyme would therefore be synthesized correctly. If the incorrect parts of the enzyme were relatively unimportant to its overall function, it may work well enough to allow the phages to reproduce.

Only a triplet code can account for the fact that three insertions, and not one or two, restore near-normal enzyme function. Further, only a code without punctuation or spaces between words will be confused by any of the insertions. Finally, a code without punctuation between words can work only if the start and stop points of protein synthesis are clearly marked. All three conclusions have been found to be correct, as we describe in the text.

Transcription Produces RNA Molecules That Are Complementary Copies of One Strand of DNA

All RNA molecules are synthesized using molecules of DNA as a template. As we saw earlier in this chapter, RNA nucleotides are chemically very similar to DNA nucleotides. Because the two "languages" are so much alike, RNA synthesis has been named transcription, meaning "the process of copying over."

Transcription of DNA into RNA is restricted in two major ways. First, in any cell, transcription normally copies only the DNA of selected genes into RNA. For example, hair follicle cells transcribe the DNA encoding keratin proteins, insulin-secreting cells of the pancreas transcribe the genes encoding insulin, and so on. Some genes, such as the genes for ribosomal RNA, are transcribed hundreds or thousands of times during interphase of most cells.

Second, when transcription of these selected genes is needed, transcription normally copies only one strand of DNA into RNA. This happens because the useful information of any given gene normally resides on only one strand of the DNA double helix. Why? Remember, the two strands of DNA are *complementary*, not *identical*. If the sequence of bases on the strand on which the gene lies codes for a sequence of amino acids that forms a functional protein, the complementary strand will have a different sequence of bases, which would probably not code for a useful protein. The DNA strand that actually contains the gene, and is transcribed into RNA, is called the **template strand** of the DNA, because it is the template from which the complementary RNA strand is made. A chromosome, which is one long DNA molecule, contains many genes. One strand may be the template strand of some genes, while the other strand may be the template strand for other genes.

With these constraints in mind, we can view transcription as a three-step process: initiation, elongation of the RNA molecule, and termination (Fig. 11-4). These three steps correspond to the three major parts of most genes, in both eukaryotes and prokaryotes: a promoter at the beginning of the gene; the "body" of the gene, which for most genes consists of the DNA bases that actually code for amino acids in the protein to be synthesized; and a termination signal at the end of the gene.

RNA Synthesis Begins at the Promoter of a Gene

RNA synthesis is carried out by an enzyme called **RNA polymerase.** The first step in transcription is for RNA polymerase to locate the beginning of the gene, causing initiation of transcription. The **promoter** region of a gene is a short sequence of DNA bases located just "upstream" (in the 3′ direction) of the body of the gene. RNA polymerase recognizes the promoter base sequence as marking the beginning of a gene and binds to the DNA at that site (Fig. 11-4a). Although RNA polymerase is shown occupying only a few bases in Figure 11-4, it is a fairly large protein that occupies at least 50 bases, including both the promoter and the first dozen or so bases in the body of the gene.

RNA Synthesis Proceeds from the Promoter to the End of the Gene

Once the RNA polymerase has bound to the promoter site, it changes shape, forcing the DNA double helix to open up at the beginning of the body of the gene. RNA polymerase then moves along the template strand of the DNA. RNA polymerase travels in the 3′ to 5′ direction along the template strand, just as DNA polymerase does (refer to Figs. 10-8 and 10-9). During elongation, using free RNA nucleotides present in the nucleus, RNA polymerase synthesizes a single strand of RNA that is complementary to the DNA of the template strand (Fig. 11-4b). The same base-pairing rules are used for RNA synthesis as for DNA replication, except that adenine in DNA is paired with uracil in RNA. The DNA-to-RNA base-pairing rules thus become:

Base in DNA	Complementary Base in RNA
adenine	uracil
cytosine	guanine
guanine	cytosine
thymine	adenine

Although RNA polymerase adds RNA nucleotides to the growing RNA strand according to base pairing with the DNA nucleotides of the gene, this base pairing does not persist. After about 10 nucleotides have been added to the growing RNA chain, the beginning of the RNA molecule separates from the DNA (Fig. 11-4c). As the RNA continues to elongate, it forms a long "tail" drifting away from the DNA, as can be seen in electron micrographs (Fig. 11-5).

At the End of RNA Synthesis, the RNA Molecule Separates from the DNA of the Gene

RNA polymerase continues along the template strand until it reaches the termination signal, a sequence of DNA bases that trigger two events (see Fig. 11-4d). First, the RNA molecule completely separates from both the DNA and the RNA polymerase. Second, the RNA polymerase leaves the template strand of the DNA. These events terminate transcription.

Three Types of RNA Cooperate in Protein Synthesis

Messenger RNA (mRNA) carries the code for the amino acid sequences of proteins from the genes in DNA to the ribosomes, the actual sites of protein synthesis. In contrast, ribosomal RNA (rRNA) and transfer RNA (tRNA) do not carry information to be translated into protein. Instead, these RNA molecules are the final products of certain genes and thus are an exception to the generalization that genes code for proteins.

RNA polymerase

beginning of gene   DNA

(b)

template strand

3′ end of gene   direction of transcription

5′ end of gene

RNA

(c)   growing RNA molecule

(d)   RNA polymerase leaves DNA

RNA molecule released

DNA rewinds

Figure 11-4 **A diagram of RNA transcription**

(a) The enzyme RNA polymerase binds to the promoter region of DNA near the beginning of a gene (initiation). **(b)** The DNA double helix unwinds. The RNA polymerase travels along one of the DNA strands (the template strand), catalyzing the formation of a continuous strand of RNA from free RNA nucleotides (elongation). The nucleotides incorporated into the growing RNA strand are complementary to the nucleotides in the template strand of DNA. **(c)** The RNA polymerase continues to the end of the gene. **(d)** At the end of the gene, the RNA polymerase leaves the DNA. The DNA rewinds, and the RNA molecule is released (termination).

The Sequence of Bases in Messenger RNA Encodes the Sequence of Amino Acids in a Protein

Messenger RNA is a long, single-stranded molecule that includes the codons that will be translated into the amino acid sequence of a protein. In prokaryotes, mRNA is directly transcribed from the DNA of a gene, and translation into proteins often begins even before transcription is complete. In eukaryotes, things are a bit more complicated, because the RNA transcribed from DNA contains more nucleotides than will ultimately be translated into protein. We will examine the formation of eukaryotic mRNA in more detail a bit later in this chapter.

The bottom line in eukaryotic cells, however, is that mRNA molecules are synthesized in the nucleus and enter the cytoplasm through the pores in the nuclear envelope. In the cytoplasm, mRNA binds to ribosomes, where the codons of mRNA are translated into the language of amino acids in proteins. (You might think of mRNA as a "molecular photocopy" of the DNA of the gene. The gene itself remains safely stored in the nucleus, like a valuable document in a library, while copies are sent to the cytoplasm to be used in protein synthesis.)

Ribosomal RNA Forms an Important Part of the Protein-Synthesizing Machinery of a Ribosome

Ribosomes are composites of ribosomal RNA and a variety of proteins. Each ribosome is composed of two subunits (Fig. 11-6). In eukaryotic cells, the small subunit consists of one molecule of rRNA and about 30 proteins. It recognizes and binds mRNA and part of tRNA. The large ribosomal subunit consists of three rRNA molecules and 45 to 50 proteins. It contains an enzymatic region that catalyzes the addition of amino acids to the growing protein chain and two sites (designated P and A) that bind to tRNA. It is thought that rRNA plays an important part, perhaps the major part, both in recognizing mRNA and in catalyzing the formation of peptide bonds between amino acids of the protein.

Transfer RNA Molecules Decode the Sequence of Bases of mRNA into the Amino Acid Sequence of a Protein

Transfer RNA molecules bind amino acids and deliver them to the ribosome, where they are incorporated into protein chains. There are many different types of tRNAs, at least one type for each amino acid. Transfer RNAs are like "code books," the only molecules in the cell that can decipher the codons of mRNA and translate them into the amino acids of proteins. Transfer RNAs are complex molecules, twisted about into a shape something like a three-leaf clover with a stem (Fig. 11-7a). For our purposes, the stem and the central leaf are the important parts. Enzymes in the cytoplasm recognize each specific tRNA molecule and attach the correct amino acid to the stem (Fig. 11-7b). The energy of ATP is used to form the tRNA–amino acid bond. Some of the ATP energy is stored in the tRNA–amino acid bond, and this energy will be used to forge the peptide bond when the amino acid is added to a growing protein molecule.

The outside bend of the central tRNA leaf bears three exposed bases, called the **anticodon,** that actually decipher the mRNA code: The anticodon of each tRNA is complementary to the codon of mRNA that specifies the amino acid to which that tRNA is attached. For example, the mRNA codon GUA is complementary to the anticodon CAU of a tRNA bearing the amino acid valine (Fig. 11-7).

Protein Synthesis

Now that we have introduced all the actors involved in protein synthesis, let's look at the actual events. Once the amino acids have linked up with their appropriate transfer RNAs (Fig. 11-7b), protein synthesis can be considered to occur in two stages. First, during transcription, messenger RNA is transcribed from the DNA template of the genes in the nucleus (see Fig. 11-4). The mRNA travels to a ribosome in the cytoplasm. Second, during translation, the ribosome binds mRNA and the appropriate tRNAs. On the ribosome, the codons of mRNA are translated into the amino acid sequence of a protein.

We described transcription earlier in this chapter; here we will examine the second stage, translation (Fig. 11-8). Like transcription, translation has three steps: initiation of protein synthesis, elongation of the protein chain, and termination.

Protein Synthesis Begins When the Initiator tRNA and the Start Codon of mRNA Bind to a Ribosome

As you'll remember, the genetic code signals where protein synthesis is to begin with the start codon AUG on mRNA (see Table 11-1). In eukaryotic cells, the first step in translation is the binding of several protein "initiation factors" and a tRNA bearing the complementary "start anticodon" UAC (sometimes called the "initiator" tRNA) to the small subunit of a ribosome (Fig. 11-8a). (Unless they are actively synthesizing proteins, the two subunits of a ribosome remain apart.) The small subunit then binds to an mRNA molecule

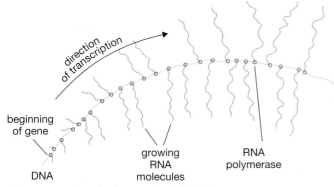

Figure 11-5   A photograph of RNA transcription

This electron micrograph shows the progress of RNA transcription in the egg of the African clawed toad (Xenopus). As shown in the diagram, in each "Christmas tree" structure, the central "trunk" is DNA and the "branches" are RNA molecules. A series of RNA polymerase molecules are traveling down the DNA, synthesizing RNA as they go. The beginning of the gene is on the left. Therefore, the short RNA molecules on the left have just begun to be synthesized, while the long RNA molecules on the right are almost finished.

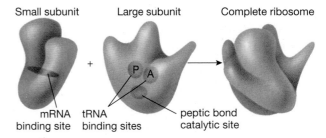

Figure 11-6   Ribosome structure

A ribosome has two subunits, each composed of protein and rRNA. The small subunit binds messenger RNA. The large subunit has three functional sites. Two, called the P and A sites, bind tRNA, while the third catalyzes the formation of the peptide bond between amino acids of the growing protein.

Figure 11-7   Transfer RNA

(a) Transfer RNA is a single RNA strand, folded back upon itself into loops like a three-leaf clover growing from a single stem. The loops are held together by hydrogen bonds between complementary bases. The central "leaf" of the tRNA clover bears three bases called the anticodon, which base pair with complementary bases in mRNA during protein synthesis (see Fig. 11-8). The anticodons differ among different tRNA molecules. The "stem" binds the amino acid encoded by the anticodon bases of the leaf. All tRNA stems end with the bases CCA. **(b)** Each type of tRNA binds to a specific amino acid. Enzymes in the cytoplasm, one for each amino acid, catalyze the bond between each tRNA and its corresponding amino acid, using the energy of ATP. Some of the ATP energy is stored in the tRNA–amino acid bond and will be used to drive the formation of the peptide bond when the amino acid is used in protein synthesis.

and moves along it until the start codon is encountered. At this point, the UAC anticodon on the initiator tRNA base pairs with the AUG of the start codon (Fig. 11-8b). The large ribosomal subunit then attaches to the small subunit. As it does so, the initiator tRNA simultaneously binds to the P site on the large subunit (Fig. 11-8c). The ribosome is now fully assembled and ready to begin translation.

Protein Synthesis Proceeds One Amino Acid at a Time

The completed ribosome is large enough to encompass two codons' worth of mRNA: Initially, this includes the start codon plus the codon coding for the next amino acid

in the protein to be synthesized (GUU in our example). The anticodon of a tRNA–amino acid complex (valine in our example) recognizes the second mRNA codon and moves into the A site on the large subunit (Fig. 11-8d). The two amino acids carried by the two tRNAs now lie next to one another. The catalytic site on the large subunit breaks the bond holding the "start" amino acid (methionine) to its tRNA and uses the released energy to form a peptide bond between the methionine and the valine carried by the second tRNA. At the end of this step, the initiator tRNA is now "empty," while the second tRNA bears a short, two-amino acid, protein chain (Fig. 11-8e). (Incidentally, because the start codon specifies the amino acid methionine, all protein chains are synthesized with

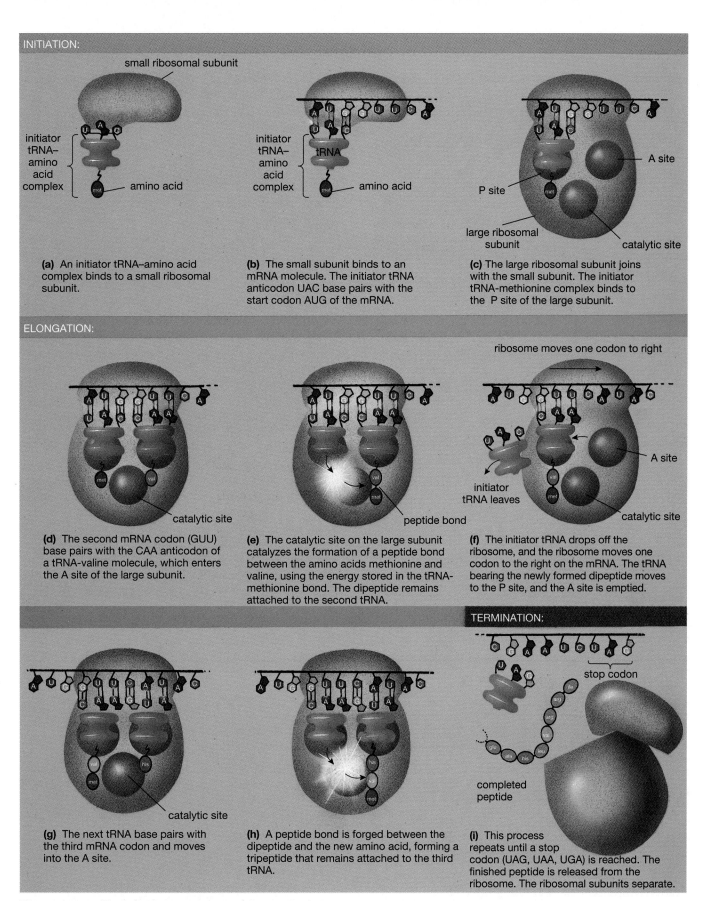

Figure 11-8 Translation: protein synthesis

Protein synthesis is the translation of the sequence of bases in mRNA to the sequence of amino acids in the encoded protein.

methionine at the beginning. In most proteins, however, the methionine is soon removed.)

At this point, the empty initiator tRNA drops off the ribosome, and the ribosome shifts to the next codon on the mRNA molecule (Fig. 11-8f). The tRNA holding the growing protein chain shifts too, from the A site to the P site on the ribosome. A new tRNA–amino acid complex binds to the emptied A site (Fig. 11-8g). The catalytic site on the large subunit breaks the bond between the dipeptide and its tRNA and links the dipeptide with the amino acid (histidine) in the A site (Fig. 11-8h). The "empty" tRNA in the P site leaves the ribosome, the ribosome shifts over another codon, and the process repeats.

Protein Synthesis Terminates When a Stop Codon Is Reached on mRNA

Near the end of the mRNA, a stop codon is reached. No tRNA recognizes a stop codon. Instead, "termination factors" cut the finished protein chain off the last tRNA, releasing it from the ribosome (Fig. 11-8i).

Messenger and Transfer RNAs Form a Decoding Chain from the Sequence of Bases in DNA to the Sequence of Amino Acids in Protein

We can now understand how a cell decodes the genetic information stored in its DNA to synthesize a protein (Fig. 11-9).

1. The DNA is organized into genes that are dozens to thousands of bases long. You might think of a gene as a sentence within the genetic information manual of the cell.
2. The "words" that make up the "gene sentences" are groups of three bases.
3. A codon of mRNA consists of three bases that are complementary to the three bases of a DNA "word."

4. An anticodon of tRNA, in turn, is complementary to a specific codon of mRNA.
5. The tRNA bears a specific amino acid, which is attached to the tRNA by enzymes that can "read" the anticodon.

This decoding chain, from bases in DNA to codon of mRNA to anticodon of tRNA to amino acid, results in the incorporation of the correct amino acid in the growing protein.

Mutations in DNA and Their Effects

So far, both in this chapter and in Chapter 10, we have emphasized the precision and fidelity of genes and their functions: DNA molecules are replicated exactly before the cell divides; mRNA molecules are exact complementary copies of the DNA of a gene; and the information in mRNA is used to produce a protein with exactly the correct amino acid sequence. Of course, no living thing is perfect. Mistakes can be made in any of these processes. A single faulty copy of mRNA or a single defective protein molecule doesn't usually affect a cell very much, because there are many correct molecules in the cell at the same time, carrying out the proper cellular functions. A single faulty copy of a gene, however, is extremely serious, because a cell may have only one (in haploid cells) or two (in diploid cells) copies of the gene and may synthesize many, or only, defective proteins as a result.

Changes in DNA are called **mutations.** Because the information in a gene is encoded in the specific sequence of bases, a mutation is a change in the sequence of bases. How can the base sequence change? One way for a mutation to occur is through a mistake in base pairing during replication (see Chapter 10). A few base-pairing mistakes

Figure 11-9   From DNA to protein

A summary of the chain of events by which a DNA sequence is decoded to produce a protein.

occur spontaneously, in the best of circumstances: Even with proofreading, replication of several billion bases results in some mistakes—not many, but a few. Certain chemicals (such as the aflatoxins synthesized by some molds that live on grain and peanuts) and some types of radiation (such as X-rays) increase errors of base pairing during replication or even induce changes in DNA composition between replications.

Random changes in DNA composition are unlikely to code for improvements in the functioning of the gene products, much as typing random words in the midst of a script of *Hamlet* will be unlikely to improve on Shakespeare's work. Therefore, cells have evolved mechanisms to monitor DNA and repair damaged regions (Fig. 11-10a). Nevertheless, mistakes do occur.

Point Mutations Are Changes in a Single Base

In a **point mutation,** a pair of bases becomes incorrectly matched (Fig. 11-10b). Repair enzymes recognize the mismatch, cut out the incorrect nucleotide, and replace it with a nucleotide bearing a complementary base. Occasionally, however, the enzymes replace the correct nucleotide instead of the incorrect one. This mistake produces a complementary pair of bases at that location once again but not the same pair that the chromosome originally had.

Insertion and Deletion Mutations Result from Addition or Removal of Nucleotides

As their names suggest, an **insertion** occurs when new nucleotide pairs are inserted in the midst of a gene (Fig. 11-11a), while a **deletion** occurs when nucleotide pairs are removed from a gene (Fig. 11-11b).

Mutations Differ in Their Effects on Protein Structure and Function

If a mutation occurs in cells whose offspring become gametes (sperm or eggs), it may be passed on to future generations. But how does a change in base sequence affect the organism that inherits the mutated DNA? As the box "Scientific Inquiry: Cracking the Genetic Code" points out, deletions and insertions can have catastrophic effects on a gene, because all the codons following the deletion or insertion will be misread. The protein synthesized from such misread directions is almost certain to be nonfunctional.

Four categories of effects may result from point mutations (Table 11-2). As a concrete example, let's consider possible mutations of the DNA sequence CTC, which codes for glutamic acid.

1. A mutation might not change the amino acid sequence of the encoded protein. Remember that the genetic code is degenerate, so one amino acid may be encoded by several different codons. If a mutation changes CTC to CTT, the new triplet still codes for

(a)

(b)

Figure 11-10   **The fate of mismatched base pairs**

Repair of damaged or mismatched regions of DNA sometimes results in restoration of the original base sequence and sometimes results in mutations. In this diagram and in Figure 11-11, the sugar-phosphate backbone of DNA is represented by solid lines, and the bases are represented by the letters A, C, G, and T. **(a)** Correction of mismatched bases. During DNA replication, a mistake in base pairing occurs, resulting in an incorrect A-C pair instead of the proper A-T pair. Repair enzymes in the daughter cell recognize the mismatch, cut out the incorrect base in the daughter DNA, and substitute the correct complementary base (T). No mutation has occurred. **(b)** Point mutation. With the same replication error as in part (a), if the repair enzymes cut out the mismatched base on the parental strand instead of on the daughter strand, then the new base pair differs from the original. A mutation has occurred.

glutamic acid. Therefore, the protein synthesized from the mutated gene remains the same.

2. A mutation might code for an amino acid that is functionally equivalent to the original amino acid. Many proteins have large "background" regions whose exact amino acid sequence is relatively unimportant. For example, in hemoglobin, the amino acids on the outside of the protein must be hydrophilic to keep the protein dissolved in the cyto-

(a)

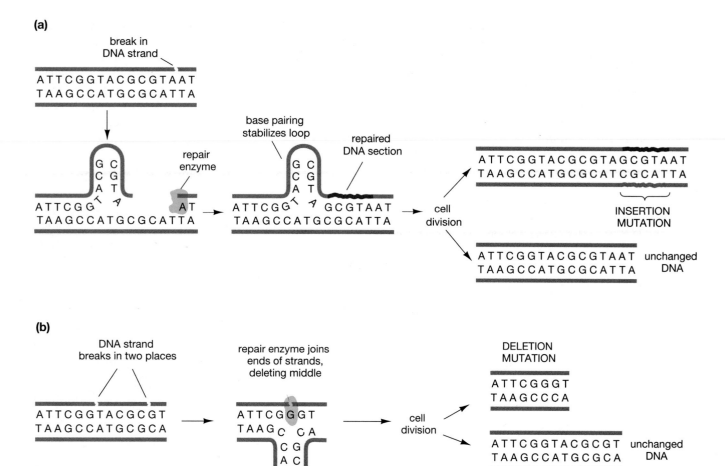

(b)

Figure 11-11   **Insertion and deletion**

(a) Insertion. Radiation or chemical agents cause a break in one of the DNA strands, and the freed DNA end loops out. If the free loop has certain sequences of bases, base pairing may stabilize the loop. Repair enzymes fill in the resulting gap with new nucleotides, making one DNA strand (with the loop) longer than the other (intact) strand. When DNA replication occurs, each strand is used as a template for synthesis of a new complementary strand. The long strand thus carries an insertion mutation to one of the daughter cells. **(b)** Deletion. If a DNA strand breaks in two places, a repair enzyme may stitch together the remaining pieces without filling in the missing nucleotides. When the DNA is replicated, one daughter cell receives the deletion mutation.

TABLE 11-2 ▪ *Examples of Functional Outcomes of Single Substitutions in the Glutamic Acid Codon of DNA*					
	DNA	*mRNA*	*Amino Acid*	*Properties*	*Functional Effect*
Original sequence	CTC	GAG	Glutamic acid	Hydrophilic, acidic	—
Mutation 1	CTT	GAA	Glutamic acid	Hydrophilic, acidic	Neutral
Mutation 2	CTA	GAU	Aspartic acid	Hydrophilic, acidic	Neutral
Mutation 3	CAC	GUG	Valine	Hydrophobic, neutral	Lose water solubility; possibly catastrophic
Mutation 4	ATC	UAG	Stop codon	Ends translation	Synthesize only part of protein; catastrophic

plasm of red blood cells. Exactly *which* hydrophilic amino acids are on the outside doesn't seem to matter too much. A mutation from CTC to CTA, replacing glutamic acid (hydrophilic) with aspartic acid (also hydrophilic), probably wouldn't affect the solubility of hemoglobin. Mutations that do not detectably change the function of the encoded protein are called **neutral mutations**.

3. A mutation might encode for a functionally different amino acid. A mutation from CTC to CAC replaces glutamic acid (hydrophilic) with valine (hydrophobic). This substitution, which is the genetic defect in sickle-cell anemia (see Chapter 15), causes hemoglobin molecules to stick together, clumping up and distorting the shape of the red blood cells. This is a potentially fatal mutation.

4. A mutation might produce a stop codon. An inappropriate stop codon will cut short the translation of mRNA before the protein is finished. This defect is almost certainly catastrophic to protein function; if the protein is essential to life, as hemoglobin is, the mutation will be lethal to the organism.

Mutations Provide the Raw Material for Evolution

Although most mutations are harmful or neutral, mutations are essential for evolution, because ultimately all genetic variation originates as these random changes in DNA sequence. Natural selection tests new sequences in the crucible of competition for survival and reproduction. Occasionally, a mutation proves beneficial in the organism's interactions with its environment. The mutant sequence may spread throughout the population and become common, as its possessors outcompete rivals bearing the original, unmutated DNA sequence.

Gene Regulation

Knowing how proteins are synthesized and the likely effects of gene mutations on protein structure and function does not provide a complete understanding of how an organism's genes are used to produce its structures and behaviors. For example, most of the cells of your body have the same DNA but don't use all the DNA all the time. Individual cells express (produce proteins encoded by) only a small fraction of their genes, those that are appropriate to the function of that particular cell type. As we mentioned earlier in the chapter, this constraint restricts the transcription of much of a cell's DNA. Muscle cells, for example, synthesize the contractile proteins actin and myosin but not insulin or hair proteins. Gene expression also changes over time, depending on the needs of the body from moment to moment. Both an organism's developmental stage and its environment can

affect which genes are transcribed. Therefore, understanding gene function requires an understanding of how genes are regulated.

The use of genetic information by a cell is a multistep process, beginning with the transcription of DNA and often ending with an enzyme's catalyzing a needed reaction (Fig. 11-12). Regulation can occur at any of these steps.

1. The Rate of Transcription of Individual Genes Can Be Regulated

The rate of transcription of specific genes depends on the type of cell, its stage in the cell cycle, and the metabolic activity of both the cell and the whole organism. Some genes are never transcribed in certain types of cells; for example, the gene for insulin is not transcribed in mus-

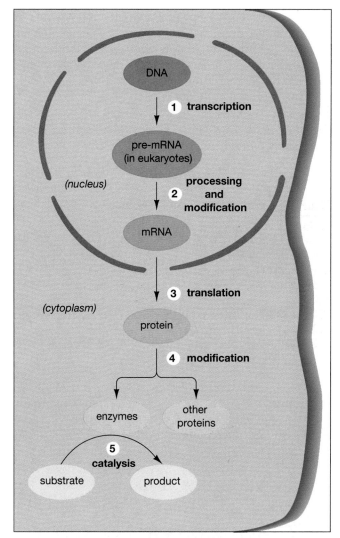

Figure 11-12 **An overview of information flow in a cell**

This simplified diagram shows the series of steps from DNA to protein to chemical reactions catalyzed by enzymes. Information flow can be regulated at any step.

cle cells. Transcription of other genes is turned on or off on demand.

2. RNA Molecules Transcribed from DNA Can Be Processed into Different Final mRNAs

As we will discuss shortly, the genes of eukaryotic cells are much larger than the final mRNA used to synthesize a protein. The genes are first transcribed into very long "pre-messenger RNA" molecules. Differential processing of these pre-mRNA molecules can produce different types of final "true mRNA," which are therefore translated into different proteins. Messenger RNA molecules must also be modified, for example by the addition or removal of nucleotides at the ends, before they can be used by the cell.

3. Messenger RNAs Can Be Translated at Different Rates

Messenger RNAs vary in their stability and the rate at which they are translated into protein. Some mRNAs are extremely stable and therefore offer the possibility of repeated translation; others are rapidly degraded. Further, a cell may block translation of certain mRNAs, depending on its metabolic requirements.

4. Proteins Might Require Modification before They Can Carry Out Their Functions in a Cell

Many proteins must be modified before they become active. For instance, the protein-digesting enzymes produced by cells of your stomach wall and pancreas are initially synthesized in an inactive form (this form keeps the enzymes from digesting the very cells that produce them). After these inactive forms are secreted into the digestive tract, portions of the enzymes are then snipped out to unveil the active site.

5. The Rate of Activity of Enzymes Can Be Regulated

Enzyme activity is often controlled by competitive or allosteric inhibition, as discussed in Chapter 4.

These methods of regulating gene activity are all important and are probably used to some extent by virtually all eukaryotic cells. We will restrict our discussion, however, to the first two steps: the transcription of DNA to RNA in prokaryotic and eukaryotic cells and the processing of pre-mRNA to true mRNA in eukaryotic cells.

Gene Regulation in Prokaryotes

Prokaryotic DNA is often organized in coherent packages called **operons,** in which the genes for related functions lie close to one another (Fig. 11-13). An operon

consists of four parts: (1) a **regulatory gene,** which controls the timing or rate of transcription of other genes; (2) a promoter, which RNA polymerase recognizes as the place to start transcribing; (3) an **operator,** which governs access of RNA polymerase to the promoter; and (4) the **structural genes,** which actually encode for the related enzymes or structural proteins (Fig. 11-13a). Whole operons are regulated as units, so that functionally related enzymes are synthesized simultaneously when the need arises. Prokaryotic operons are regulated in different ways, depending on the functions they control. Some operons synthesize enzymes that are needed by the cell just about all the time, such as the enzymes that synthesize amino acids. These operons are usually transcribed continuously, except under unusual circumstances when the bacterium encounters a vast surplus of a particular amino acid. Other operons synthesize enzymes that are needed only occasionally, for instance to digest a relatively rare food substance. They are transcribed only when the bacterium encounters the rare food.

As an example of the latter type of operon, consider the common intestinal bacteria *Escherichia coli.* These bacteria must live on whatever types of nutrients their host eats, and they can synthesize a variety of enzymes to metabolize a potentially wide variety of foods. The genes that code for most of these enzymes are transcribed only when the enzymes are needed. The enzymes that metabolize lactose, the principal sugar in milk, are a case in point. The **lactose operon** contains three structural genes, each coding for an enzyme that aids in lactose metabolism (Fig. 11-13a).

The lactose operon is shut off, or **repressed,** unless specifically activated by the presence of lactose. The regulatory gene of the lactose operon directs synthesis of a protein, called a **repressor protein,** that binds to the operator site. RNA polymerase, although still able to bind to the promotor, cannot get past the repressor protein and transcribe the structural genes. Consequently, the lactose-metabolizing enzymes are not synthesized (Fig. 11-13b).

When *E. coli* colonize the intestines of a newborn mammal, however, they find themselves bathed in a sea of lactose whenever the host nurses from its mother. Lactose molecules enter the bacteria and bind to the repressor proteins, changing their shape (Fig. 11-13c). The lactose-repressor combination cannot attach to the operator site. Therefore, when RNA polymerase binds to the promoter of the lactose operon, it can transcribe the structural genes. Lactose-metabolizing enzymes are synthesized, allowing the bacterium to use lactose as an energy source.

After the young mammal is weaned, it usually never consumes milk again. The intestinal bacteria no longer encounter lactose, the repressor proteins are free to bind to the operator, and the genes for lactose metabolism are shut down.

Structure of the lactose operon

codes for repressor protein

operator: repressor protein binds here

promoter: RNA polymerase binds here

structural genes that code for enzymes of lactose metabolism

(a) The lactose operon consists of a regulatory gene, a promoter, an operator, and three structural genes that code for enzymes involved in lactose metabolism. The regulatory gene codes for a protein, called a repressor, which can bind to the operator site under certain circumstances.

Lactose absent

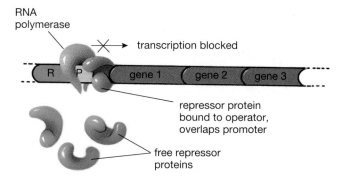

RNA polymerase

transcription blocked

repressor protein bound to operator, overlaps promoter

free repressor proteins

(b) When lactose is not present, repressor proteins bind to the operator of the lactose operon. When RNA polymerase binds to the promotor, the repressor protein blocks access to the structural genes, which therefore cannot be transcribed.

Lactose present

RNA polymerase binds to promoter, transcribes structural genes

lactose bound to repressor proteins

lactose-metabolizing enzymes synthesized

(c) When lactose is present, it binds to the repressor protein. The lactose-repressor complex cannot bind to the operator, so RNA polymerase has free access to the promoter. The RNA polymerase transcribes the three structural genes coding for the lactose-metabolizing enzymes.

Figure 11-13 **Structure and regulation of the lactose operon of *Escherichia coli***

Gene Regulation in Eukaryotes

Gene regulation is quite different in eukaryotes. Not only are genes for related functions sometimes found on entirely different chromosomes, but even many individual genes are split up on the chromosome. As a result, transcription and its regulation are more complex.

Eukaryotic Genes Consist of DNA Segments That Code for the Amino Acid Sequence of Proteins Interrupted by Noncoding DNA Segments

In the 1970s, molecular geneticists discovered that most eukaryotic structural genes have much more DNA than is needed to encode the amino acids of proteins. Each gene consists of two or more base sequences that encode the protein, interrupted by other base sequences that are not translated into protein. The coding segments are called **exons** because they are **ex**pressed in protein. The noncoding segments are called **introns** because they **int**ervene between the exons (Fig. 11-14a).

Each eukaryotic gene has its own promoter. A nearby region of the chromosome, called the **enhancer,** regulates binding of RNA polymerase to the promoter. When specific regulatory proteins bind to the enhancer, they facilitate the binding of RNA polymerase to the promoter, thus enhancing transcription.

When a eukaryotic gene is transcribed, a very long molecule of RNA is synthesized, starting before the first exon and ending after the last exon (Fig. 11-14b). The resulting RNA contains many more nucleotides than the true codons for the amino acids of the encoded protein. Two major steps convert this RNA molecule into true mRNA. First, RNA nucleotides are added at the beginning (the cap) and the end (the tail) of the molecule. Second, enzymes in the nucleus precisely cut the molecule apart, splice together the sections that code for the protein, and discard the rest.

Why are eukaryotic genes split up like this? There appear to be at least two functions served by gene fragmentation. The first function is to produce multiple proteins from a single gene. In rats, there is a gene that is transcribed in both the thyroid and the brain. In the thyroid, one splicing arrangement results in the synthesis of a hormone called calcitonin, which helps to regulate calcium concentrations in the blood. In the brain, another splicing arrangement results in the synthesis of a peptide that probably is used as a messenger molecule for communication among brain cells.

The second function is more speculative but has some good experimental evidence in its support: Fragmented genes may provide a quick and efficient way for eukaryotes to evolve new proteins with new functions.

(a) Eukaryotic Gene Structure

A typical eukaryotic gene consists of sequences of DNA called exons, which code for the amino acids of a protein (red), and intervening sequences called introns (purple), which do not. At least two control regions, the promoter (yellow) and the enhancer (pink), regulate the transcription of eukaryotic genes.

(b) RNA Synthesis and Processing in Eukaryotes

RNA polymerase transcribes both the exons and introns, producing a long RNA molecule. More nucleotides are added at the beginning and end of this initial RNA transcript. Enzymes then cut out the RNA introns and splice together the exons to form the true mRNA, which moves out of the nucleus to be translated on the ribosomes.

Figure 11-14 **Eukaryotic gene structure and the processing of RNA**

Eukaryotic Cells May Regulate the Transcription of Individual Genes, Large Parts of Chromosomes, or Entire Chromosomes

As do prokaryotes, eukaryotes regulate the rate of transcription of genes. Transcriptional regulation can operate on three levels: the individual gene, large parts of chromosomes, or entire chromosomes.

Regulatory Proteins Alter the Rate of Transcription of Individual Genes

Some of the best-known examples of transcriptional regulation at the level of the individual gene are the cellular actions of steroid hormones (Chapter 3). The stimulation by estrogen of albumin (egg white) synthesis in female birds (Fig. 11-15) is one example of this type of regulation at work. Being lipid soluble, steroid hormones readily penetrate cell membranes and enter the interiors of cells. During the breeding season, estrogen is secreted into the bloodstream by the bird's ovaries and enters the cells of the oviduct. The estrogen binds to receptor proteins in the cytoplasm. The estrogen-protein complex enters the nucleus, where it binds to DNA, probably near the enhancer for the albumin gene. Proximity to the enhancer makes it easier for RNA polymerase to contact the promoter of the albumin gene. Rapid transcription occurs and albumin is syn-

thesized. Similar activation of genes by steroid hormones occurs in other animals, including humans.

Condensed Regions of Chromosomes Normally Are Not Transcribed

Certain parts of chromosomes are in a highly condensed, compact state, in which the DNA seems to be inaccessible to RNA polymerase. Some of these regions are structural parts of chromosomes that don't contain genes. For example, condensed DNA is usually found at the centromeres that hold sister chromatids together during cell division. In other cases, DNA can change from the condensed state to a looser configuration that allows genes to be transcribed, depending on the stage in the life cycle of the animal or the type of cell (Fig. 11-16).

Entire Chromosomes May Be Inactivated, Thereby Preventing Transcription

In rare cases, an entire chromosome may be condensed and inaccessible to RNA polymerase. This occurs with one of the X chromosomes of female mammals. As you probably know, female mammals have two homologous X chromosomes. But only one X chromosome is available for transcription in any given cell. The other entire X chromosome is condensed into a tight mass. In the light

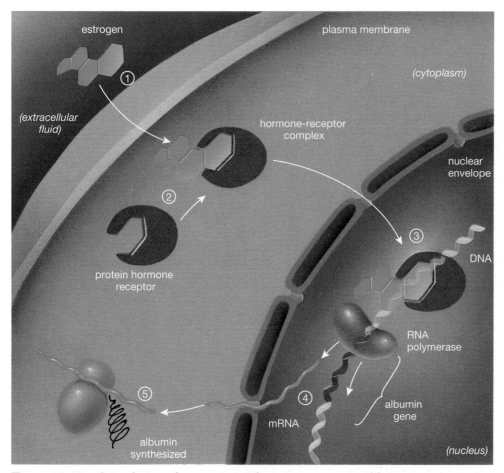

Figure 11-15   Stimulation of transcription by estrogen, a steroid hormone

① Estrogen diffuses through the plasma membrane into the cytoplasm. ② Estrogen combines with a receptor protein. ③ The hormone-receptor complex enters the nucleus and binds to the enhancer of the albumin gene. ④ RNA polymerase transcribes the albumin gene. ⑤ The mRNA leaves the nucleus and is translated into albumin.

Figure 11-16   A giant chromosome of a midge (a small fly)

Most of the chromosome is in a tight, condensed state, but a few regions have puffed out in a much looser configuration. The purple stain shows the RNA, which is transcribed mostly from the loose, puffed regions of DNA. Changing patterns of puffing and condensation of the chromosomes reflect different genes turning on and off during development.

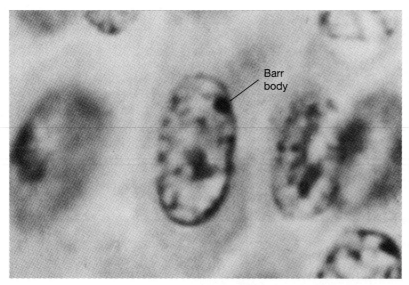

Figure 11-17   Nuclei of human cells, stained to show Barr bodies

In general, human nuclei have one fewer Barr body than the number of X chromosomes in the nucleus. Each nucleus from a normal XX woman has one Barr body. Nuclei from men (XY) lack Barr bodies.

microscope, the **inactivated** X chromosome shows up as a dark spot in the nucleus, called a **Barr body** (Fig. 11-17), named after its discoverer, Murray Barr. Apparently, both X chromosomes are in the "loose" state in fertilized eggs. After a few cell divisions, one or the other condenses and forms a Barr body. (The cells in the ovaries that will give rise to eggs are an exception; here, both X chromosomes are in the loose, uncondensed state.) Which X chromosome is inactivated in any given cell is random, but all its daughter cells will then have the same condensed chromosome. As a result, female mammals (including human females) are mosaics: patches of cells with one X chromosome active are interspersed with patches in which the other X chromosome is active.

✖ S U M M A R Y   O F   K E Y   C O N C E P T S

The Relationship between Genes and Proteins
Genes are segments of DNA on chromosomes. The ultimate cellular product encoded by a gene is usually a protein. Therefore, with a few exceptions (principally transfer RNA and ribosomal RNA), the specific base sequence of a gene encodes the amino acid sequence of a protein or a part of a protein.

Synthesizing Proteins from the Instructions in DNA
Ribonucleic acid (RNA) is a single-stranded nucleic acid involved in transcribing and translating genetic information in DNA to amino acid sequences. Information flows from DNA to proteins in a two-step process. (1) *Transcription*: The information contained in the DNA of a gene is copied into messenger RNA. (2) *Translation*: The sequence of bases in mRNA provides the information needed by transfer RNA and ribosomal RNA to synthesize a protein with the amino acid sequence specified by the base sequence of the DNA of the gene.

The Genetic Code
The sequence of bases in mRNA carries the genetic code for the amino acid sequence in a protein. Sequences of three bases in mRNA, called codons, specify the amino acids of the protein. There are also start and stop codons that signal the beginning and end of protein synthesis.

RNA: Intermediary in Protein Synthesis
Synthesizing proteins from the information in DNA requires RNA molecules as intermediates. RNA is transcribed from one DNA strand by the enzyme RNA polymerase. RNA polymerase recognizes the promoter region of DNA

at the beginning of a gene. Starting there, RNA polymerase uses free ribose nucleotides to synthesize an RNA strand that is complementary to the DNA of the gene.

There are three types of RNA. The sequence of bases in messenger RNA (mRNA) carries the information needed to determine the amino acid sequence of a protein. Ribosomal RNA (rRNA) and proteins form ribosomes. Ribosomes consist of large and small subunits. The small subunit recognizes mRNA. The large subunit bears two binding sites for transfer tRNA (tRNA) as well as a catalytic site that forges the peptide bond between amino acids as a protein is synthesized. Each tRNA binds a specific amino acid and transports it to a ribosome. A set of three bases in tRNA, called the anticodon, is complementary to the codon in mRNA that specifies the amino acid to which that tRNA is bound.

Protein Synthesis

Protein synthesis occurs in the following sequence:

1. Messenger RNA is transcribed from a gene. The mRNA leaves the nucleus and travels to a ribosome.
2. Two codons of mRNA bind to the small subunit of the ribosome. The first codon is the start codon, which signals where protein synthesis is to begin.
3. Transfer RNAs, carrying their amino acids, move to the mRNA. The anticodons of two tRNA molecules base pair with the two codons of mRNA, and the tRNAs bind to the large ribosomal subunit.
4. The large subunit of the ribosome catalyzes the formation of a peptide bond between the amino acids carried by the two tRNA molecules. The "first" amino acid detaches from its tRNA. The chain of two amino acids remains attached to the "second" tRNA.
5. The first tRNA leaves the ribosome. The ribosome moves one codon over on the mRNA. A third tRNA, with its attached amino acid, base pairs with the third codon on mRNA. A new peptide bond is formed between the amino acid of the third tRNA and the dipeptide still attached to the second tRNA.
6. This process continues until a stop codon is reached, whereupon the mRNA and the newly formed protein leave the ribosome.

Mutations in DNA and Their Effects

A mutation is a change in the sequence of bases in a gene. Mutations are caused by mistakes in base pairing during replication, by chemical agents, and by certain kinds of radiation. Common gene mutations include point mutations, insertions, and deletions. Although mutations are usually neutral or harmful, occasionally a mutation will promote better adaptation to the environment and thus will be favored by natural selection.

Gene Regulation

Which genes are transcribed in a cell at any given time is regulated by the function of the cell, the developmental stage of the organism, and the environment. In prokaryotic cells, DNA is often organized into operons. Whole operons are regulated as units. Access of RNA polymerase to the promoter of a gene may be either prevented or enhanced by other molecules in the cell, including nutrients and hormones.

Eukaryotic genes consist of exons (sequences of bases that encode the information for the amino acid sequence of a protein) and introns (bases that separate exons and do not themselves encode amino acid sequences). Therefore RNA transcribed from eukaryotic genes must be processed to form a true mRNA, by cutting out the introns and splicing together the exons. Regulatory proteins may alter the transcription rate of individual genes. Large parts of chromosomes may also be rendered inaccessible to RNA polymerase by changes in DNA structure.

⚙ K E Y   T E R M S

anticodon   p. 191
Barr body   p. 202
codon   p. 187
degeneracy   p. 187
deletion   p. 195
enhancer   p. 199
exon   p. 199
genetic code   p. 186 .
inactivation   p. 202
insertion   p. 195
intron   p. 199
lactose operon   p. 198

messenger RNA (mRNA)   p. 186
mutation   p. 194
neutral mutation   p. 197
one-gene, one-protein hypothesis   p. 185
operator   p. 198
operon   p. 198
point mutation   p. 195
promoter   p. 189
regulatory gene   p. 198
repression   p. 198
repressor protein   p. 198

ribonucleic acid (RNA)   p. 185
ribosomal RNA (rRNA)   p. 186
ribosome   p. 190
RNA polymerase   p. 189
start codon   p. 187
stop codon   p. 187
structural gene   p. 198
template strand   p. 189
transcription   p. 186
transfer RNA (tRNA)   p. 186
translation   p. 186

✖ T H I N K I N G   T H R O U G H   T H E   C O N C E P T S

Multiple Choice

1. *Anticodon* is the term applied to
   a. the list of amino acids corresponding to the genetic code
   b. the concept that multiple codons sometimes code for a single amino acid
   c. the part of the tRNA interacting with the codon
   d. the several three-nucleotide stretches which code for "stop"
   e. the control mechanism of prokaryotic genes

2. It is possible to take the cytoplasmic mRNA and reverse transcribe it into a complementary DNA (called cDNA). In contrast with genomic DNA (in the cell's nucleus) from a eukaryotic cell, the cDNA derived from the same gene
   a. would likely be shorter   b. would contain introns
   c. would have histones bound to it
   d. would have condensed chromatin
   e. would be identical

3. If ^{3}H (tritiated) thymidine were taken up by cells, what would that probably indicate?
   a. cells are dividing
   b. a high level of transcription
   c. a high level of translation
   d. the cell is in G_1 and is actively synthesizing mRNA
   e. introns are being excised in the processing of RNA

4. Eukaryotic DNA
   a. takes part directly in protein synthesis by leaving the nucleus and being translated on the ribosome
   b. takes part indirectly in protein synthesis; the DNA itself stays in the nucleus
   c. has nothing to do with protein synthesis; it is only involved in cell division
   d. is involved in protein synthesis, which takes place in the nucleus
   e. codes for mRNA but not tRNA or rRNA

5. A single-stranded molecule that is assembled using DNA as a template and contains the information for assembly of a specific amino acid sequence in the cytoplasm is
   a. messenger RNA   b. complementary DNA
   c. transfer RNA   d. intron DNA
   e. ribosomal RNA

6. Making a protein from a template of messenger RNA
   a. is catalyzed by DNA polymerase
   b. is catalyzed by RNA polymerase
   c. is called translation
   d. is called transcription
   e. utilizes the RNA exactly as it is copied from the eukaryotic DNA template

Review Questions

1. What are the three types of RNA? What are their functions? Where and how are they synthesized?

2. Compare the chromosomal DNA from a specific eukaryotic gene and the mRNA derived from it with respect to the major differences.

3. Define the following terms: genetic code, codon, anticodon. What is the relationship between the bases in DNA, the codons of mRNA, and the anticodons of tRNA? What does it mean to say that the genetic code is degenerate?

4. Diagram and describe protein synthesis.

5. In general, all cells in your body have the same genes. How is it that different cells in your body differ? What is the name of the mechanism by which different cell-specific forms of one protein can be coded for by one "gene?" What starting molecule is converted to what final molecule? What components are pieced together to make the final molecule? What components are chopped out?

6. How would a radioactively labeled precursor of uracil (^{3}H-uridine) be used to localize sites of active gene transcription? Specifically, what molecule would take up the ^{3}H-uridine?

7. What is an operon? Are operons found in prokaryotes, eukaryotes, or both?

8. Describe the process of gene regulation in prokaryotes, using the lactose operon as an example.

9. Diagram the structure of a eukaryotic gene, including both the internal structure of the gene and the nearby control regions of the chromosome.

10. How is mRNA formed from a eukaryotic gene?

11. How do steroid hormones regulate eukaryotic genes?

12. Define mutation, and give one example of how a mutation might occur. Describe four functional consequences of point mutations. Would you expect most mutations to be beneficial or harmful? Explain your answer.

13. What amino acid has the same coding sequence as the start codon?

14. How would the genetic material on the two complementary DNA strands differ in terms of coding for proteins?

✖ A P P L Y I N G   T H E   C O N C E P T S

1. The same genetic code is used in most organisms, including bacteria and people. Suppose that the gene that encodes a human protein, insulin for example, were inserted into a bacterium. If the bacterium transcribes this gene into messenger RNA, would you expect that the amino acid sequence of the protein would be the same as the protein synthesized in a human being?
2. If you look at the genetic code (Table 11-1), you will notice that most of the codons that specify the same amino acid have the same first two bases and differ only in the third base. Now, most mutations that cause a change in a protein are harmful. If you sequenced the DNA for the hemoglobin gene from 100 people, in which codon position do you think you would find the most differences? Why?
3. If a single base pair deletion occurred toward the beginning of the coding sequence of a gene, why might transcription of the whole gene be interrupted by a stop codon?
4. As you have learned in this chapter, many factors influence gene expression. Use of hormones such as anabolic steroids or growth hormones among athletes has created quite a controversy recently. Hormones certainly affect gene expression, but, in the broadest sense, so do vitamins and food. What do you think are appropriate boundary lines for the use of hormones? Should athletes take growth hormone? Should a child be given growth hormone if he is unlikely to achieve normal height without it?
5. "Basic" science research refers to the quest for knowledge without any immediate practical application, sometimes using simple organisms such as harmless bacteria. In contrast, "applied" scientific research often has a specific goal, such as a cure for a disease. Given limitations in funding, it is often suggested that applied science be given priority. Thinking over the information in the last few chapters, argue for or against equal levels of funding for basic and applied scientific research.

✖ F O R   M O R E   I N F O R M A T I O N

Beadle, G. W. "The Genes of Men and Molds." *Scientific American*, September 1948. A classic paper by the winner of the 1958 Nobel Prize on the classic experiments on *Neurospora* referenced in this chapter.

Beardsley, T. "Smart Genes." *Scientific American*, August 1991. How genes are regulated is just as important as which genes are on the chromosomes. New findings are beginning to outline how gene expression is regulated during development.

Chambron, P. "Split Genes." *Scientific American*, May 1981. The segmented nature of eukaryotic genes is described.

Cohen, J. S., and Hogan, M. E. "The New Genetic Medicines." *Scientific American*, December 1994. Can pieces of DNA be administered which tie up the expression of targeted genes for the control of disease?

Crick, F. H. C. "The Genetic Code." *Scientific American*, October 1962. The determination of the triplet nature of the genetic code.

Crick, F. H. C. "The Genetic Code: III." *Scientific American*, October 1966. The genetic code is completely solved.

Grunstein, M. "Histones as Regulators of Genes." *Scientific American*, October 1992. Histones are proteins found associated with DNA in eukaryotic chromosomes. Once thought to be merely a scaffold for DNA, they are actually important in gene regulation.

Nirenberg, M. W. "The Genetic Code: II." *Scientific American*, March 1963. Nirenberg deciphered much of the genetic code. Here he describes some of those experiments.

Steitz, J. A. "Snurps." *Scientific American*, June 1988. "Snurps" are small nuclear ribonucleoproteins, which snip introns out of eukaryotic pre–messenger RNA. Snurps are probably examples in which the RNA part of the molecule is the real catalyst.

Tjian, R. "Molecular Machines That Control Genes." *Scientific American*, February 1995. Complexes of proteins regulate which genes are transcribed in a cell, and therefore help to determine the cell's structure and function.

N E T   W A T C H

On-line resources for this chapter are on the World Wide Web at:
http://www.prenhall.com/~audesirk (click on the table of contents link and then select Chapter 11).

12 Gene Exchange, Meiosis, and Eukaryotic Life Cycles

Male elephant seals fight for control of a harem, a group of females with which one of them will breed.

In Chapters 9 through 11, we discussed several important aspects of inheritance. First, DNA molecules carry the hereditary information of all living organisms; the sequences of bases in specific stretches of DNA, the genes, encode the information needed to specify the amino acid sequences of proteins. Second, chromosomes are long strands of DNA, which, in eukaryotic organisms, are found in the nucleus of each cell. Third, the cells of most plants and animals are diploid; that is, they have pairs of chromosomes that have similar, though usually not identical, DNA sequences. Finally, when cells divide, each daughter cell receives a full set of chromosomes containing one copy of all the chromosomes of the parent cell.

In multicellular organisms, cell division is the basis of asexual reproduction (see Chapter 9): groves of aspens that are clones of an ancestral parent or a new *Hydra* budding off from its parent. Asexual reproduction through mitosis seems to be so simple and efficient. In contrast, **sexual reproduction**, in which DNA from two parents is combined in a single offspring, is such a nuisance. Consider male elephant seals, fighting to obtain a harem (see the opening photograph). These immense beasts develop long, rubbery snouts as displays to show off in front of rival males. They bash each other with their massive necks and bite each other until their blood stains the beach. A defeated male may never reproduce at all, but victory is not pure bliss, either. The victor often spends so much energy defending his harem that, by the end of the breeding season, he is drained of all reserves, a mockery of his former magnificence. Many do not survive the winter.

Why not just reproduce asexually, as aspens and *Hydra* can? Or perhaps we should word the question the other way around—why, if asexual reproduction is so effective, do aspens nevertheless flower and set seed, reproducing sexually as well as asexually? Offspring produced by asexual reproduction are usually genetically identical to their parents. This duplication may be an advantage if the parent organism is well adapted to its environment, if that environment never changes, and if the organism and its offspring never move to a new location. But suppose the environment changes—the climate grows warmer or new predators appear. Neither the parent nor its asexually reproduced offspring may be well adapted any longer, and neither may survive. On the other hand, suppose that the offspring vary somewhat, both from the parent and from one another, because the genetic material from two unique individuals is combined. Perhaps some of them would be better adapted to the changing environment.

In this chapter, we will consider some of the advantages and disadvantages of sexual reproduction from evolutionary and genetic perspectives. We will see that a combination of overall constancy of genetic information together with limited genetic variability may have been the selective pressure behind the evolution of sexual reproduction, a specialized type of cell division called meiosis, and diploid cells with paired chromosomes. We will then examine the mechanisms of meiosis, whereby diploid cells with paired chromosomes give rise to haploid cells with unpaired chromosomes. Finally, we will see how meiosis, sexual reproduction, and mitosis interact in the life cycles of eukaryotic organisms.

Genetic Variability, Genetic Exchange, and the Evolution of Sexual Reproduction

Mutations in DNA Are the Ultimate Source of Genetic Variability

Although we have emphasized the fidelity of DNA replication and proofreading, we have also noted that changes in base sequence *do* occur. These changes in DNA are **mutations**. Often we think of mutations in a negative light, focusing on genetic disorders caused by defective genes, such as Huntington's disease or muscular dystrophy, or on science fiction fantasies, such as gigantic mutant ants destroying cities in old movies. Mutations, however, are the foundation of evolution. Without mutations in DNA, the only life on Earth would probably be extremely primitive single cells floating around in the primordial soup. Bacteria are different from bison, and you are different from your relatives, because of differences in DNA sequence that originally arose as mutations.

The Effects of a Mutation Depend on the Nature of the Mutation, the Organism in Which It Occurs, and the Environment in Which the Organism Lives

What are the effects of a mutation on a cell or an entire organism? As we described in Chapter 11, depending on the exact change in base sequence, the proteins synthesized from the directions encoded in mutated DNA may be completely unchanged, subtly altered (for better or for worse), or even rendered completely nonfunctional. Whether these effects are helpful or harmful to the organism, or matter at all, depends on many factors. One of the clearest examples of the ambiguous effects of mutations and the interaction between mutations and the environment is the development of antibiotic resistance in microbes. Many bacteria are resistant to some common antibiotics. In many cases, the resistant bacteria synthesize enzymes that break down the antibiotic; in other cases, they have altered cellular constituents, such as ribosome structure. In general, synthesizing antibiotic-metabolizing enzymes is a drain on a bacterium's energy and nutrient reserves; altered cellular components usually do not function quite as well as the original "wild-type" components do. Mutations causing altered cellular components or synthesis of antibiotic-metabolizing enzymes can therefore be harmful or beneficial, *depending on the presence or absence of the antibiotic in the environment.* Without antibiotics around, the mutations are harmful to the bacteria; when people use antibiotics, the mutations become beneficial.

Genetic Exchange May Combine Useful Mutations

Let's suppose that two different useful mutations occur in two separate organisms of the same species. It would clearly be useful for the two mutations to be combined in a single organism, which would then be genetically superior in that environment. How can this be accomplished? Asexual reproduction, of course, cannot combine the mutations. Every offspring of the organism having one of the mutations would have that same mutation, and every offspring of the organism having the second mutation would have just that mutation. It is possible that the second mutation might occur later in an organism, or its asexually reproduced offspring, that already had one of the useful mutations. But most mutations are rare, so the odds against the organism's having both are enormous. (For example, if the likelihood of each useful mutation occurring separately is one in a million, then the likelihood of both occurring in a single organism is one in a trillion.)

Enter genetic exchange. If organisms with the individual mutations could exchange some DNA, they might produce an organism with both mutations. Probably all living organisms have some mechanism for genetic exchange between individuals (Fig. 12-1). Genetic exchange is the principle underlying sexual reproduction. The potential advantages of combining useful mutations provide the evolutionary rationale for the development of the complex processes that create genetic exchange.

Eukaryotic Organisms Use a Specialized Cell Division Process Called Meiosis to Combine Genetic Material from Two Separate Parents in a Single Offspring

Most modern eukaryotic cells are either **haploid**, containing a single copy of each type of chromosome, or **diploid**, containing pairs of chromosomes. The two members of each pair of a single type of chromosome are called **homologues**, from a Greek word meaning "to say the

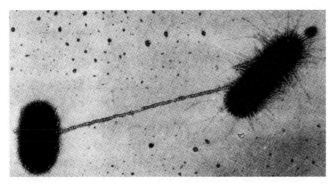

Figure 12-1   **Bacterial conjugation**

Bacteria exchange genetic material by a process called conjugation. Two bacterial cells become temporarily connected by a hollow tube, the pilus, through which DNA passes from one cell to the other.

same thing." Actually, homologous chromosomes have similar, but not identical, genetic material. As we will see later in this chapter and in Chapter 13, the fact that homologues are *not* exactly the same has extremely important consequences for inheritance. Haploid cells include sperm and eggs, of course, but also the "body cells" of organisms such as the bread mold *Neurospora*, the unicellular alga *Chlamydomonas*, and part of the plant life cycle called the gametophyte stage (discussed later in this chapter). Diploid cells include the body cells of most animals and the part of the plant life cycle called the sporophyte stage.

The first eukaryotic cells to evolve, about 1 to 1.5 billion years ago, were probably haploid. Relatively early on, single-celled eukaryotic organisms evolved two events that allowed them to shuffle and recombine genetic information. First, two haploid (parental) cells would fuse, resulting in a temporary, large, diploid cell. In this diploid cell, there would be two copies of each chromosome, one copy contributed by each of the parental cells. Second, this diploid cell would then soon undergo a type of cell division, called meiosis, that yields haploid daughter cells, each containing one copy of each type of chromosome—one of each of the pairs of homologues.

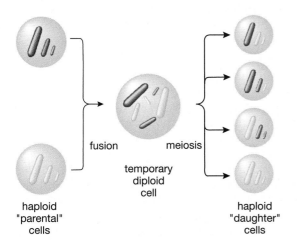

fusion   meiosis

temporary
diploid
cell

haploid
"parental"
cells

haploid
"daughter"
cells

We'll get to the actual mechanisms of meiosis shortly; for now just note that a single daughter cell formed from meiosis might contain some chromosomes from each parental cell, and perhaps combine two or more useful mutations that would allow it to survive and reproduce more successfully in its environment. This form of sexual reproduction is still used today by *Neurospora* and *Chlamydomonas* (see Fig. 12-8).

"Permanent" Diploidy Protects against Some of the Harmful Effects of Mutations

Going from haploid cells to a temporary diploid cell and immediately back to haploid cells may allow a single organism to combine two or more beneficial mutations. But, as you know, few mutations are beneficial; in fact, most are neutral or harmful. Let us use the analogy from Chapter 11: Randomly changing the amino acid sequence of a protein will usually make the protein function less effectively, just as randomly inserting words into Shakespeare's *Hamlet* will probably make the play worse. If a haploid cell had just one chromosome containing one gene encoding a vital protein, what would happen if that gene mutated and the resulting protein became nonfunctional? That's right: The cell would die or at least limp along much less successfully. But what would happen if that mutation occurred in a diploid cell with two homologous chromosomes, each bearing a copy of the vital gene? Now the same mutation, but in just one of the homologous chromosomes, would be much less harmful. There would still be some normal protein synthesized from the directions in the unmutated gene copy. Under the right conditions, for example if the protein is an enzyme not needed in huge quantities, a cell with one normal copy may function just as well as a cell with two normal copies (see Chapter 13).

A great advantage of diploidy is that it minimizes the potential negative effects of mutations. A diploid cell can have quite a few harmful mutations without suffering serious effects, whereas the same number of harmful mutations would almost certainly kill a haploid cell. On the other hand, a diploid cell might not benefit quite as much from a useful mutation, because it would produce both the original protein and the new, improved version, whereas a haploid cell would produce only the improved protein. Given the high frequency of harmful mutations compared with beneficial mutations, however, diploidy is probably advantageous.

Although we will never know for sure, diploidy probably evolved when a "temporary" diploid cell that formed from the fusion of two haploid cells delayed meiosis for a while, remaining diploid for a significant portion of its life cycle. Eventually, some of these diploid cells underwent some mitotic divisions as well, producing diploid daughter cells. From there, it is a relatively short evolutionary leap to multicellular, diploid, sexually reproducing organisms in which meiosis serves to produce haploid **gametes** (that is, sperm or eggs) that cannot live independently. Sperm fuses with egg to yield a diploid cell, the fertilized egg, or **zygote** (see Fig. 12-9), which, through repeated mitotic cell divisions, gives rise to a new, multicellular organism.

Meiosis and Sexual Reproduction

Although the details vary widely from organism to organism, there are three features typical of sexual reproduction in almost all multicellular eukaryotes.

1. Organisms engaging in sexual reproduction have diploid cells with pairs of homologous chromosomes at some stage in their life cycle.
2. At some point, the homologues are separated from one another through meiosis, which produces haploid cells. In animals, these haploid cells are gametes. In plants and many fungi, the haploid cells are spores that undergo mitosis to produce a multicellular, haploid body. Certain cells of this haploid form then differentiate into gametes at a later time (this life cycle is discussed later in this chapter).
3. The haploid gametes fuse to form a diploid cell once again (the zygote), with one copy of each homologous chromosome donated by each parent. This fusion reestablishes pairs of homologous chromosomes.

Meiosis Separates Homologous Chromosomes in a Diploid Cell to Produce Haploid Daughter Cells Containing One Copy of Each Type of Chromosome

The key to sexual reproduction in eukaryotic cells is **meiosis**, the production of haploid cells with unpaired chromosomes. The word *meiosis* comes from a Greek word meaning "to diminish," and that is just what meiosis does: It reduces the number of chromosomes by half. Meiosis is not, however, a random division. In meiosis, each daughter cell receives one member from each pair of homologous chromosomes.

Because meiosis evolved from mitosis, many of the structures and events in meiosis are similar or identical to those of mitosis. Therefore, you may want to review the essentials of mitosis in Chapter 9 before proceeding.

Let's begin with an overview of the primary events of meiosis (Fig. 12-2). Meiosis involves two nuclear divisions, termed **meiosis I** and **meiosis II**. The overall process of meiosis proceeds as follows:

1. All the chromosomes are replicated during interphase before meiosis I, with sister chromatids attached at the centromere of each chromosome.

2. During meiosis I, homologous chromosomes are separated; each daughter nucleus I receives one homologue of each pair of chromosomes. The sister chromatids do *not* separate during meiosis I, so at the completion of meiosis I each chromosome still consists of two sister chromatids. Remember that sister chromatids joined at a single centromere are *not* independent chromosomes (see Chapter 9). Therefore, because each daughter nucleus contains one homologue of each pair of chromosomes, the daughter nuclei are considered to be haploid. The chromosomes *do not replicate again* between meiosis I and meiosis II.
3. In meiosis II, the sister chromatids split into two independent daughter chromosomes, with one going into each daughter nucleus II.

Therefore, the reduction in the number of chromosomes from diploid to haploid occurs during meiosis I. Meiosis II is essentially the same as mitosis in a haploid cell, with the number of chromosomes remaining the same. Each nuclear division is usually accompanied by cytokinesis (division of the cytoplasm). Therefore, a cell that undergoes both meiosis I and meiosis II produces a total of four haploid cells.

Further, in most cases homologous chromosomes in the parental diploid cell have similar, but not identical, base sequences in their DNA. Therefore, because the haploid daughter cells produced by meiosis receive one homologue of each pair of chromosomes, the daughter cells are genetically similar, but not identical, to one another.

The Mechanisms of Meiosis

The phases of meiosis are given the same names as the roughly equivalent phases in mitosis, although, as we shall see, there are important differences in meiosis I (Fig. 12-3).

Meiosis I

In multicellular organisms, meiosis occurs only in a relatively few, specialized cells in the reproductive organs. Meiosis begins in these cells with the replication of the chromosomes. (Because one function of meiosis is to *reduce* the number of chromosomes from diploid to haploid, you may think that to begin by duplicating the chromosomes is an unnecessary step. However, the cellular mechanisms of meiosis evolved from those of mitosis, and the replication of chromosomes is probably leftover evolutionary baggage. As we will see many times throughout this text, evolution is not an engineer; it's more like a tinker.) As in mitosis, the sister chromatids of each chromosome are attached to one another at the centromere. From this point on, meiosis I differs tremendously from mitosis.

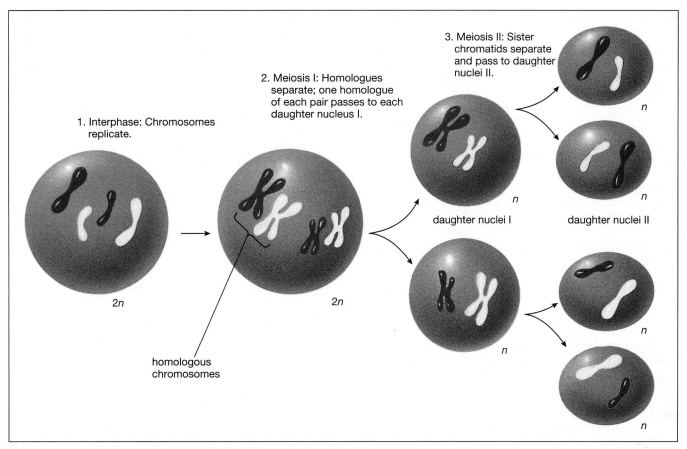

Figure 12-2   **An overview of meiosis**

Meiosis consists of two nuclear divisions: meiosis I and meiosis II. Together, the two divisions separate homologous chromosomes and produce haploid nuclei that each contain one member of each pair of homologues. The number of different types of chromosomes in a cell is designated by the letter *n*. Haploid cells, with only one copy of each type of chromosome, are indicated by *n* (1 × *n*). Diploid cells, with two copies of each type, are indicated by 2*n*.

During Prophase I, Homologous Chromosomes Pair Up and Exchange DNA

During **prophase I**, homologous chromosomes line up side by side, exchange segments of DNA, and condense (Fig. 12-4). In some way that is not yet understood, homologous chromosomes find one another early in prophase I of meiosis (step 1 in Fig. 12-4). (To make it easier to keep track of the homologues, we'll call one homologue the "maternal chromosome" and the other the "paternal chromosome," because, as we will see, one was originally inherited from the organism's mother and the other was inherited from the organism's father.) The ends of the maternal and paternal chromosomes attach to the nuclear envelope right next to each other, and the two chromosomes are brought together, side by side, much like zipping the zipper on a jacket (step 2). Protein strands join the homologous chromosomes so that they match up exactly along their entire length. Then large enzyme complexes assemble at random

places along the paired chromosomes (step 3). The enzymes snip apart the chromosomes and graft the broken ends together again. However, *the enzymes almost always graft a maternal "stump" onto a paternal "stem," and vice versa* (step 4). Thus, the maternal and paternal chromosomes now intertwine, forming crosses, or **chiasmata** (singular, **chiasma**). Eventually the enzyme complexes leave the chromosomes, and the protein zippers that held homologues together disappear. The chromosomes coil and condense, becoming visible with a light microscope, with chiasmata connecting homologues (step 5 in Fig. 12-4, and Fig. 12-5, p. 214).

Chiasmata serve two functions. First, the maternal and paternal chromosomes have swapped some segments of their DNA, an event called **crossing over**. If, as is likely, they had slightly different DNA, then neither chromosome is quite the same as it was before the chiasmata formed. We will discuss the genetic consequences of crossing over in the next chapter. The second function of chi-

MEIOSIS I

paired homologous chromosomes   recombined chromosomes

chiasma   spindle

PROPHASE I   METAPHASE I   ANAPHASE I   TELOPHASE I

Prophase I. Chromosomes thicken and condense. Homologous chromosomes come together in pairs, and chiasmata occur as chromatids of homologues exchange parts. The nuclear envelope disintegrates and the spindle forms.

Metaphase I. Paired homologous chromosomes line up along the equator of the cell. One homologue of each pair faces each pole of the cell, with both chromatids of a given homologue becoming attached to microtubules leading to the same pole.

Anaphase I. Homologues separate, one member of each pair going to each pole of the cell. Sister chromatids do not separate.

Telophase I. Two clusters of chromosomes have formed, each containing one member of each pair of homologues. The daughter nuclei are therefore haploid. Cytokinesis often occurs during telophase I. There is usually little or no interphase between meiosis I and meiosis II.

MEIOSIS II

PROPHASE II   METAPHASE II   ANAPHASE II   TELOPHASE II   FOUR HAPLOID DAUGHTER CELLS

Prophase II. If chromosomes have relaxed after telophase I, they recondense. Spindles re-form, and the spindle microtubules attach to the sister chromatids.

Metaphase II. Chromosomes line up along the equator, with sister chromatids of each chromosome attached to microtubules leading to opposite poles of the cell.

Anaphase II. Chromatids separate into independent daughter chromosomes, one former chromatid moving toward each pole.

Telophase II. Chromosomes finish moving to opposite poles.

Four haploid cells. Cytokinesis results in four haploid cells, each containing one member of each pair of homologous chromosomes.

Figure 12-3   The details of meiosis

Meiosis is the process by which the homologous chromosomes of a diploid cell are separated. Each of the four resulting haploid daughter cells contains one member of each pair of homologous chromosomes. In these diagrams, two pairs of homologous chromosomes are shown, large and small. The yellow chromosomes are from one parent (for example, father), and the violet chromosomes are from the other parent.

asmata is to hold the homologous chromosomes together during their attachment to the spindle microtubules. As we will see, this continued pairing is essential.

As in mitosis, the spindle begins to assemble outside the nucleus during prophase I. Near the end of prophase I, the nuclear envelope breaks up and the spindle microtubules capture the chromosomes by attaching to their kinetochores.

During Metaphase I, Paired Homologous Chromosomes Move to the Equator of the Cell

During **metaphase I**, the spindle microtubules move the chromosomes to the equator of the cell (see Fig. 12-3). Unlike in mitosis, where *individual chromosomes* line up along the equator, during metaphase I of meiosis *homologous pairs of chromosomes* line up along the equator. The

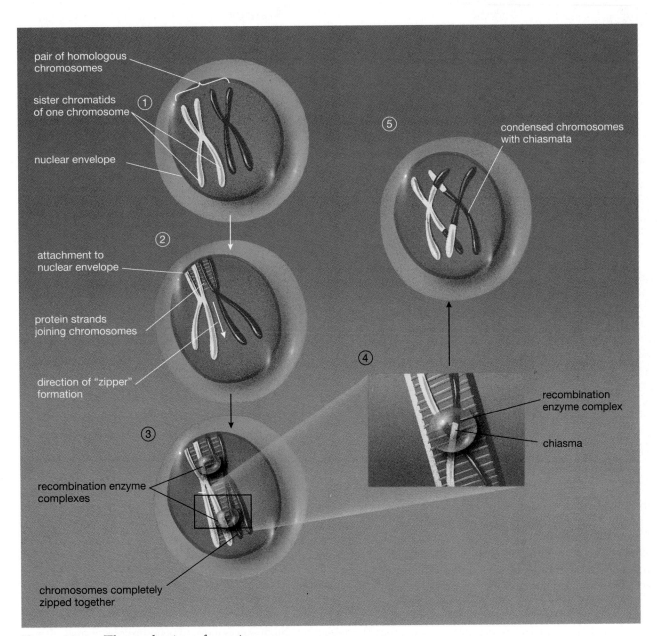

Figure 12-4   **The mechanism of crossing over**

See text for details. ① Homologous chromosomes pair up side by side. ② One end of each chromosome binds to the nuclear envelope. Protein strands "zip" homologous chromosomes together. ③ Homologous chromosomes are fully joined by protein strands. Recombination enzymes bind to the chromosomes. ④ Recombination enzymes snip chromatids apart and reattach the chromatids. Chiasmata are formed when one end of a chromatid of a paternal chromosome (yellow) is attached to the other end of a chromatid of a maternal chromosome (violet). ⑤ The protein strands and recombination enzymes leave as the chromosomes condense. The chiasmata remain as locations where homologous chromosomes are twisted around each other, helping to hold homologues together.

chiasmata

chiasmata

Figure 12-5   Chiasmata

This figure shows chiasmata in homologous chromosomes during prophase I in a salamander testis. Chiasmata are sites of DNA exchange between homologues. In this drawing, for example, the segment of yellow chromosome between the chiasmata will become part of the purple chromosome, and the purple segment below it will join the yellow chromosome.

two homologues of each pair are attached to kinetochore microtubules leading to opposite poles of the spindle.

Because the attachment of the chromosomes to the spindle is really the key to understanding the movement of chromosomes during meiosis I, let's compare the spindle attachments of mitosis with those of meiosis I (Fig. 12-6). In mitosis, the two sister chromatids of each chromosome attach to kinetochore microtubules leading toward opposite poles of the cell (Fig. 12-6a). Therefore, when the centromeres split during anaphase, the former sister chromatids are pulled to opposite poles. In meiosis I, both chromatids of one homologue attach to kinetochore microtubules extending toward one pole of the cell, while both chromatids of its paired homologue attach to kinetochore microtubules leading to the opposite pole (Fig. 12-6b). The kinetochore microtubules will thus separate the paired homologous chromosomes, but not pull apart the sister chromatids of any single chromosome.

Which member of a pair of chromosomes faces which pole of the cell is random. For some pairs, the maternal chromosome may face "north," while for other pairs, the maternal chromosome may face "south." We will look at the genetic consequences of this randomness, also called *independent assortment*, later in this chapter.

During Anaphase I, Homologous Chromosomes Separate

In **anaphase I**, the paired homologous chromosomes separate from one another and move to opposite poles of the cell (see Fig. 12-3). Anaphase I begins when the chiasmata holding homologous chromosomes together suddenly loosen up. The kinetochore microtubules tug the homologues apart, with the chiasmata apparently sliding along to the ends of the chromosomes, eventually letting the two homologues separate. The centromeres holding sister chromatids together, however, do *not* split, so the sister chromatids remain attached to one another. Therefore, one chromosome of each homologous pair (still consisting of two sister chromatids) moves to one pole of the dividing cell while its homologue moves to the other pole. This step is referred to as the "segregation" of the homologues.

At the end of anaphase I, there are two clusters of chromosomes. Each cluster contains one member of each pair of homologous chromosomes and is therefore haploid.

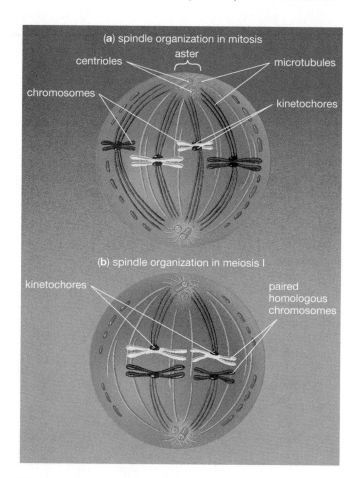

Figure 12-6   A comparison of the spindles formed during mitosis and meiosis I

(a) Homologous chromosomes are not paired. The kinetochores of sister chromatids are attached to kinetochore microtubules that lead to opposite poles. Therefore, when the sister chromatids separate during anaphase, the newly independent daughter chromosomes move to opposite poles of the cell. **(b)** Homologous chromosomes are paired. The two kinetochores of the sister chromatids of a single chromosome are attached to kinetochore microtubules that lead to the same pole. Both kinetochores of the sister chromatids of the paired homologous chromosomes are attached to kinetochore microtubules that lead to the opposite pole. Therefore, during anaphase I, sister chromatids of each chromosome remain together, moving to the same pole, but homologous chromosomes separate and move to opposite poles.

During Telophase I, Two Haploid Clusters of Chromosomes Are Formed

In **telophase I**, the spindle disappears. In many cases, nuclear envelopes may not re-form; particularly in plants, cytokinesis may not occur either. Meiosis I is usually followed immediately by meiosis II, with little or no intervening interphase. It is important to remember that *the chromosomes do not replicate between meiosis I and meiosis II*. Frequently, the chromosomes remain in their condensed, shortened state, so prophase II can be distinguished from telophase I only by the reappearance of the spindle.

Meiosis II

In meiosis II, the sister chromatids of each chromosome separate. Meiosis II is easy to follow, because it is virtually identical to mitosis in a haploid cell (see Fig. 12-3). During **prophase II** the spindle re-forms. The chromosomes attach to microtubules as they did in mitosis: *The two sister chromatids of each chromosome attach to microtubules that extend to opposite poles of the cell.* **Metaphase II** is marked by completion of the spindle apparatus and migration of the chromosomes to the equator of the cell. During **anaphase II**, the centromeres holding sister chromatids together split, and the chromatids separate from one another. The kinetochore microtubules pull each former chromatid, now an independent chromosome, to opposite poles of the cell. **Telophase II** and cytokinesis conclude meiosis II, as nuclear envelopes re-form, the chromosomes relax into their extended state, and the cytoplasm is divided. Because both daughter cells produced in meiosis I undergo meiosis II, a total of four haploid cells are formed by the end of meiosis.

Mitosis, Meiosis, and Eukaryotic Life Cycles

The **life cycles** of almost all eukaryotic organisms have a common overall pattern (Fig. 12-7). First, two haploid cells fuse during the process of **fertilization**, bringing together genes from different parent organisms and endowing the resulting diploid cell with new gene combinations. Second, at some point in the life cycle, meiosis occurs, recreating haploid cells. Third, at some time in the life cycle, mitosis

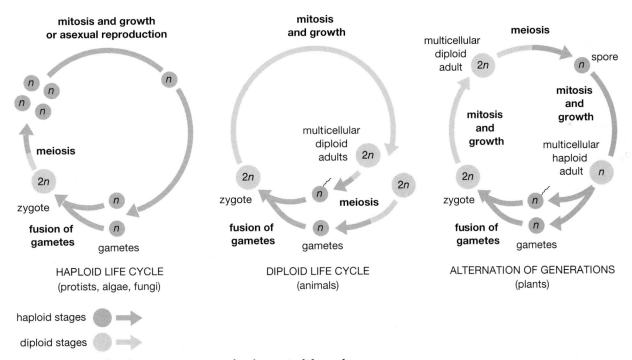

Figure 12-7   **The three major types of eukaryotic life cycles**

The lengths of the arrows correspond roughly to the proportion of the life cycle spent in each stage. **(a)** Haploid life cycle (many protists, algae, and fungi): Most of the life cycle is spent with haploid cells. Mitosis of haploid cells results in either asexual reproduction of unicellular organisms or growth of multicellular organisms. At some point, specialized reproductive haploid cells fuse, and the resulting diploid cell almost immediately undergoes meiosis. **(b)** Diploid life cycle (animals): Most of the life cycle is spent with diploid cells. When haploid cells are produced, they fuse to form a diploid zygote. Mitosis of the zygote produces multicellular bodies composed of diploid cells. **(c)** Alternation of generations (plants): Diploid and haploid stages are of roughly equal duration. Cells of the diploid stage undergo meiosis to form spores. The spores then undergo mitosis to produce a multicellular haploid body. Later, specialization of haploid cells produces gametes, which fuse to form a zygote. Through mitosis, the zygote develops into a multicellular diploid form.

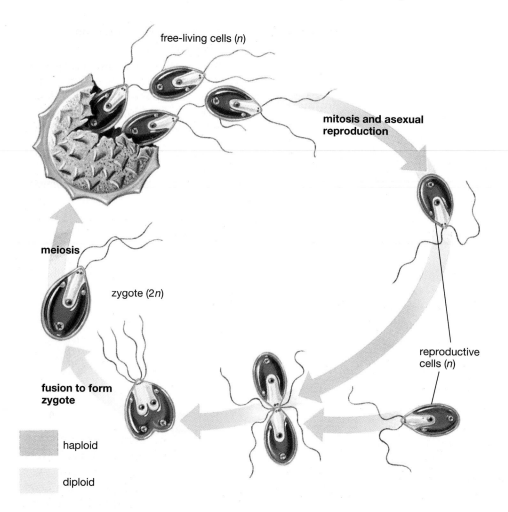

free-living cells (n)

mitosis and asexual
reproduction

meiosis

zygote (2n)

reproductive
cells (n)

fusion to form
zygote

haploid

diploid

**Figure 12-8   The life cycle
of the unicellular alga
*Chlamydomonas***

Chlamydomonas multiplies asexually by mitosis of haploid cells.
When nutrients are scarce, specialized haploid cells (usually of
different strains) fuse to form a
diploid cell. Meiosis then immediately produces four haploid cells,
usually with different genetic
composition than either of the
original parental strains.

of either haploid or diploid cells, or both, results in the
growth of multicellular bodies and/or asexual reproduction.

The seemingly vast differences between the life cycles
of, say, ferns and humans are due to variations in three aspects: (1) the interval between meiosis and the fusion of
haploid cells, (2) at what points in the life cycle mitosis
and meiosis occur, and (3) the relative proportions of the
life cycle spent in the diploid and haploid states. These
aspects of life cycles are interrelated, and we can conve-
niently label life cycles according to the relative dominance of diploid and haploid stages.

In Haploid Life Cycles, the Majority of the Cycle Consists of Haploid Cells

Some eukaryotes, such as many fungi and the unicellular
alga *Chlamydomonas*, spend most of their life cycles in the
haploid state, with single copies of each type of chromosome
(Fig. 12-8). Asexual reproduction by mitosis produces a population of identical, haploid cells. Under certain environmental conditions, specialized "sexual" haploid cells are produced. Two of these sexual haploid cells fuse, forming a
diploid cell. This cell immediately undergoes meiosis, producing haploid cells again. In organisms with a haploid life
cycle, mitosis never occurs in diploid cells. As we mentioned
at the beginning of this chapter, this life cycle is probably

fairly similar to the life cycle of eukaryotic organisms in
which meiosis and sexual reproduction first evolved.

In Diploid Life Cycles, the Majority of the Cycle Consists of Diploid Cells

Most animals have life cycles that are just the reverse of the
Chlamydomonas cycle. Virtually the entire animal life cycle
is spent in the diploid state (Fig. 12-9). Haploid gametes
(sperm in males and eggs in females) are formed through
meiosis. These fuse to form a diploid fertilized egg, the zygote. Growth and development of the zygote into the adult
organism is a result of mitosis in diploid cells.

In Alternation-of-Generations Life Cycles, There Are Both Diploid and Haploid Multicellular Stages

The life cycles of most plants include both multicellular
diploid and multicellular haploid body forms. In the typical pattern (Fig. 12-10), a multicellular diploid body gives
rise to haploid cells, called **spores**, through meiosis. The
spores then undergo mitosis to create a multicellular haploid stage (the "haploid generation"). At some point, certain haploid cells differentiate into haploid gametes. Two
gametes then may fuse to form a diploid zygote. The zygote
grows by mitosis into a diploid multicellular body (the

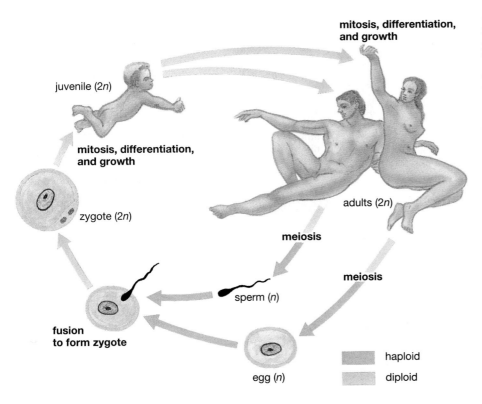

Figure 12-9 **The human life cycle**

Through meiosis, the two sexes produce different types of gametes—sperm in males and eggs in females—that fuse to form a diploid zygote. Mitosis and specialization of cells produce an embryo, child, and ultimately a sexually mature adult. The haploid state lasts only a few hours to a few days; the diploid state may survive for a century.

"diploid generation"). This life cycle is called **alternation of generations** because a multicellular diploid generation alternates with a multicellular haploid generation.

In "primitive" plants, such as ferns, both the haploid and diploid stages are free-living, independent plants. The flowering plants, however, have reduced the haploid stage to a minimum, represented only by the pollen grain and a small cluster of cells in the ovary of the flower.

The Roles of Meiosis and Sexual Reproduction in Producing Genetic Variability

As we outlined in the introduction to this chapter, genetic variability among organisms is essential for survival and reproduction in a changing environment, and therefore for evolution. In the long run of millions of years, mutation of DNA is the ultimate basis for genetic variability; on shorter time scales, down to just a single generation, shuffling of homologous chromosomes, crossing over, and fusion of gametes from separate parents during sexual reproduction provide most of the genetic variability among members of a species.

Shuffling Homologues Creates Novel Combinations of Chromosomes

Maternal and paternal homologues are randomly shuffled out into the daughter cells of meiosis I. Remember, at metaphase I the paired homologues line up at the cell's equator. For each pair of homologues, the paternal chro-

mosome faces one pole and the maternal chromosome faces the opposite pole, but which homologue faces which pole is random. Let's consider an organism with three pairs of homologous chromosomes, large, medium, and small. Further, let's color-code the paternal homologue of each pair yellow, the maternal homologue violet. The alignment of chromosomes at metaphase I can assume four configurations:

Anaphase I can therefore produce eight possible sets of chromosomes ($2^3 = 8$):

Each of these chromosome clusters will then undergo meiosis II to produce two gametes. Therefore, our hypothetical parent organism could produce gametes with eight different chromosome sets.

mature
sporophyte (2n)

meiosis

spores (n)

**mitosis,
differentiation,
and growth**

**mitosis,
differentiation,
and growth**

young
sporophyte (2n)

egg (n)

young
gametophyte (n)

zygote (2n)

**fusion to
form zygote**

mature
gametophyte (n)

sperm (n)

haploid

diploid

*Figure 12-10   Alternation of
generations*

The life cycle of a fern is
representative of alternation of
generations. The diploid stage is
called the *sporophyte*; this is the
fern usually seen in the woods.
Specialized cells in the sporophyte
undergo meiosis, producing
haploid spores. Spores differ from
gametes in that spores undergo
mitosis and grow into a
multicellular haploid body (the
gametophyte). In ferns, the haploid
stage is inconspicuous, a small
plant growing in moist places on
the forest floor. Sperm and eggs are
produced during the haploid stage,
and they fuse to make a diploid
zygote. Through mitosis, the
zygote develops into a new
multicellular diploid body, the
sporophyte.

Crossing Over Creates Chromosomes with Novel Combinations of Genes

Crossing over produces chromosomes that differ from those of either parent. Consequently, offspring may receive new combinations of genes that may never before have occurred on single chromosomes. Each time our hypothetical parent produces gametes, its homologous chromosomes cross over in new and different places. Therefore, it can produce many more than eight different types of gametes. *Because of crossing over, the parent probably never produces any two gametes that are completely identical!*

Fusion of Gametes Adds Further Genetic Variability to the Offspring

Finally, two gametes, possibly genetically unique in all the world, fuse to form a diploid offspring. As an example of the resulting variability, let's look at the potential for genetic variability in human beings. Humans have 23 pairs of homologous chromosomes. A single person, therefore, can theoretically produce 2^{23} = about 8 million different gametes, just on the basis of random separation of the homologues. Fusion of gametes from just two different people could produce 8 million × 8 million = 64 trillion genetically different children! When you add in the variability produced by crossing over, is it any wonder that, with the exception of identical twins, there is truly no one just like you?

�֍ S U M M A R Y   O F   K E Y   C O N C E P T S

Genetic Variability, Genetic Exchange, and the Evolution of Sexual Reproduction
For evolution to occur, there must be genetic differences among organisms. These genetic differences originate as

mutations. Multiple mutations that arise in separate members of a species may be combined in single individuals through genetic exchange, most often through sexual reproduction.

Meiosis and Sexual Reproduction

Sexual reproduction usually combines genetic contributions from two parental organisms. At some point in the life cycle of all sexually reproducing eukaryotic species, diploid cells undergo meiosis, which separates homologous chromosomes and produces haploid cells with only one homologue from each pair. These haploid cells or their descendants fuse to form diploid cells that receive one homologue of each pair from each parent, reestablishing pairs of homologous chromosomes.

The Mechanisms of Meiosis

During meiosis, a diploid cell undergoes two specialized cell divisions, meiosis I and meiosis II, to produce four haploid daughter cells. During interphase before meiosis, chromosomes are replicated. The cell then proceeds into meiosis I.

Meiosis I: During prophase I, homologous chromosomes, each consisting of two chromatids, pair up and exchange parts by crossing over. During metaphase I, paired homologues move together to the equator of the cell, with the two members of each pair facing opposite ends of the cell. Homologous chromosomes separate during anaphase I, and two nuclei form during telophase I. Each daughter nucleus receives only one member of each pair of homologues and is therefore haploid. However, sister chromatids do *not* separate during meiosis I, so each chromosome is still duplicated, consisting of two sister chromatids attached at the centromere. Chromosomes do *not* duplicate again between meiosis I and meiosis II.

Meiosis II: Meiosis II occurs in both of the daughter nuclei and resembles mitosis in a haploid cell. The chromosomes move to the equator of the cell during metaphase II. The two chromatids of each chromosome separate and are moved to opposite ends of the cell during anaphase II. This second division produces four haploid nuclei. Cytokinesis usually occurs during or shortly after telophase II, producing four haploid cells.

Meiosis separates pairs of homologous chromosomes in diploid cells to form unpaired chromosomes in haploid cells. Homologous chromosomes contain similar, but usually slightly different, genetic information. The separation of homologues during meiosis I is random: Either member of each pair of homologues ("maternal" or "paternal") may end up in either daughter nucleus. Therefore, the haploid cells produced by meiosis usually differ in their genetic composition.

Figure 12-11 compares mitosis with meiosis. The illustrations are arranged so that comparable stages are aligned. Note that there are no stages in mitosis that are comparable to meiosis I. Meiosis II is essentially identical to mitosis in a haploid cell.

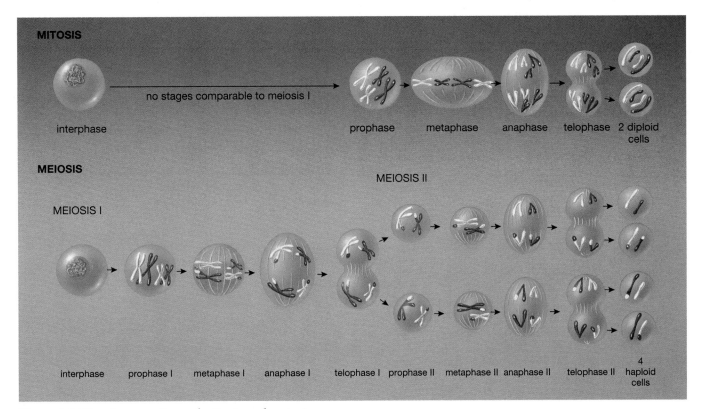

Figure 12-11 **A comparison of mitosis and meiosis**

In these diagrams, comparable phases of mitosis and meiosis are aligned. In both, chromosomes are replicated during interphase. Meiosis I, with pairing of homologous chromosomes, formation of chiasmata and exchange of chromosome parts, and separation of homologues to form haploid daughter nuclei, has no counterpart in mitosis. Meiosis II, however, is similar to mitosis occurring in a haploid cell.

Mitosis, Meiosis, and Eukaryotic Life Cycles

Most eukaryotic life cycles are similar in three respects: (1) Sexual reproduction combines haploid gametes to form a diploid cell; (2) at some point in the life cycle, diploid cells undergo meiosis to produce haploid cells once again; and (3) at some time in the life cycle, mitosis of either the haploid or diploid cells, or both, results in the growth of multicellular bodies and/or sexual reproduction. Three general types of life cycle can be distinguished on the basis of the proportion of the life cycle spent in diploid or haploid stages: (1) haploid cycles (mostly microbes and fungi), in which most of the life cycle is spent in the haploid state; (2) diploid cycles (animals), in which most of the life cycle is spent in the diploid state, with meiosis immediately preceding gamete formation; and (3) alternation of generations (plants), in which there are both haploid and diploid multicellular stages.

The Roles of Meiosis and Sexual Reproduction in Producing Genetic Variability

From one generation to the next, shuffling of homologous chromosomes, crossing over, and fusion of gametes from separate parents during sexual reproduction provides most of the genetic variability among members of a species. The random shuffling of homologous maternal and paternal chromosomes creates completely new chromosome combinations. Crossing over creates chromosomes with novel combinations of genes that may never before have occurred on single chromosomes. Because of crossing over, the parent probably never produces any two gametes that are completely identical. Finally fusion of two genetically unique gametes adds further genetic variability to the offspring. Fusion of gametes from just two different people could produce 64 trillion genetically different children!

✜ KEY TERMS

alternation of generations   p. 217
anaphase I, II   pp. 214, 215
chiasma   p. 211
crossing over   p. 211
diploid   p. 208
fertilization   p. 215

gamete   p. 210
haploid   p. 208
homologue   p. 209
life cycle   p. 215
meiosis I, II   p. 210
metaphase I, II   pp. 213, 215

mutation   p. 208
prophase I, II   pp. 211, 215
sexual reproduction   p. 207
spore   p. 216
telophase I, II   p. 215
zygote   p. 210

✜ THINKING THROUGH THE CONCEPTS

Multiple Choice

1. Which statement is most correct?
   a. All mutations are harmful.
   b. Mitosis and meiosis both add to genetic diversity.
   c. Crossing over helps each gamete get a different set of genes in meiosis.
   d. Mitosis always makes diploid daughter cells while meiosis always produces gametes.
   e. The only haploid cells are gametes.
2. When does crossing over occur?
   a. during translation   b. during fertilization
   c. during mitosis   d. during mutation
   e. when homologues have aligned and replicated
3. When do homologous chromosomes pair up?
   a. only in mitosis   b. only in meiosis I
   c. only in meiosis II   d. in both mitosis and meiosis
   e. in neither mitosis nor meiosis
4. Two homologues
   a. would be represented by maternal vs. paternal copies of the same chromosome
   b. would represent the two single strands of DNA that wrap into the double helix
   c. are still joined at the centromere between meiosis I and meiosis II
   d. are both present in haploid gametes
   e. none of the above

5. Curiously, there is no crossing over of any chromosome in the male fruit fly *Drosophila*, which has 4 pairs of chromosomes. How many different combinations of maternal vs. paternal chromosomes are possible in a male fruit fly's sperm?
   a. 2   b. 4
   c. 8   d. 16
   e. many more than the above
6. During alternation of generations
   a. the zygote produces a gametophyte
   b. the sporophyte is haploid
   c. spores give rise to gametophytes
   d. sporophytes produce sperm and eggs
   e. the gametophyte is diploid

Review Questions

1. Diagram the events of meiosis. At what stage do haploid nuclei first appear?
2. Describe the process of homologue pairing and crossing over. At what stage of meiosis does this occur? Name two functions of chiasmata.
3. In what ways are mitosis and meiosis similar? In what ways are they different?
4. Draw a generalized diagram of each of the three types of life cycles of eukaryotic organisms, and give an example of an organism using each type. Label the haploid and diploid stages, and indicate at which point in the cycle meiosis occurs.

5. What are probable advantages and disadvantages of sexual reproduction? Of diploidy?
6. Describe how meiosis provides for genetic variability. If an animal had a haploid number of 2 (no sex chromosomes), how many genetically different types of gametes could it produce (assume no crossing over)? If it had a haploid number of 5?
7. If there were no crossing over in human meiosis, how many different combinations of chromosomes (i.e., paternal vs. maternal centromeres, given this assumption) would be possible in the gametes (sperm and ova)? Express your answer as a number and an exponent. What does crossing over do to the number of possible different combinations of genes?

✖ APPLYING THE CONCEPTS

1. Crossing over apparently occurs at random locations along a pair of homologous chromosomes. Would you expect crossing over to recombine two genes more often if they are close to one another on the chromosome or if they are on opposite ends of the chromosome? Why?
2. In humans, meiosis is the first step in the formation of sperm and eggs. Normally, meiosis separates homologous chromosomes during meiosis I and separates sister chromatids during meiosis II. Occasionally, however, something goes awry. Suppose that during meiosis I, *both members of one pair of homologous chromosomes* moved to the same pole of the cell. If meiosis II occurred normally, what would be the chromosome complement of the resulting gametes? Suppose now that meiosis I occurred normally, but both sister chromatids of one chromosome migrated to the same pole during meiosis II. What would be the chromosome complement of resulting gametes? One relatively common genetic defect is Down syndrome, otherwise known as trisomy 21 (three copies of chromosome 21). Can abnormalities of meiosis explain trisomy 21? How?
3. In the bodies of humans and most other animal species, both mitosis and meiosis occur. Discuss the similar and different functions of mitosis and meiosis in our bodies and why it is important that both processes occur.
4. Some animal species can reproduce either asexually or sexually, depending on the state of the environment. Asexual reproduction tends to occur in stable, favorable environments, whereas sexual reproduction increases in unstable and/or unfavorable circumstances. Discuss the possible advantages or disadvantages this behavior might have on survival of the species in an evolutionary sense, or on survival of individuals in the population.
5. Some species, such as the cheetah, have had their populations reduced to such a small size that their genetic diversity has been reduced. Why might a lack of genetic diversity threaten a species? What can people do to help maintain genetic diversity, especially in light of captive breeding programs for endangered species in zoos? Could a lack of genetic diversity threaten a small, isolated, ethnically distinct human population?
6. Farm animals with desirable traits are sometimes produced by removing eggs from the female and fertilizing them in a dish using sperm from a male with desirable traits. After the zygote has undergone several divisions, the individual cells can be separated, returned to the same or different females, and they will develop into genetically identical animals. Human sperm banks already carry sperm from men with particular traits that a woman may choose. Human *in vitro* fertilization (see the Scientific Inquiry box in Chapter 27) could theoretically be used to create identical twins, triplets, etc. Argue for or against such human "cloning."

✖ FOR MORE INFORMATION

Amabile-Cuevas, C. F., and Chicurel, M. E. "Horizontal Gene Transfer." *American Scientist,* 81: 332–341, 1993. "Vertical" transfer of genetic material traverses generations as in sexual reproduction. More examples are being uncovered of transfer of genetic material horizontally, from one organism to another.

Kinoshita, J. "Swap Meet." *Discover,* April 1991. A discussion of hypotheses about the evolution of sex.

Lyttle, T. W. "Cheaters Sometimes Prosper: Distortion of Mendelian Segregation by Meiotic Drive." *Trends in Genetics* 9, June 1993. A gene usually perpetuates itself by survival and reproduction of the organism which expresses its coded protein. In a strikingly more direct survival of the fittest, some genes kill the gametes which do not possess the genes.

Stahl, F. "Genetic Recombination." *Scientific American,* February 1987. A look at the mechanisms of crossing over.

NET WATCH

On-line resources for this chapter are on the World Wide Web at:
http://www.prenhall.com/~audesirk (click on the <u>table of contents</u> link and then select Chapter 12).

13 Patterns of Inheritance

"In the mingling of the seed, sometimes the woman, with sudden force, overpowers the man, and then the children, born of maternal seed, will resemble more the mother; but if from paternal seed, the father. The children you see resembling both their parents, having the features of both, have been created from father's body and mother's blood, when the seeds course through the bodies excited by Venus, in harmony of mutual passion, breathing as one, with neither conquering and neither being conquered."

LUCRETIUS in *On the Nature of Things* (96–55 B.C.)

Inheritance provides for both similarity and difference. These animals are both tigers because of the vast majority of identical genes that they inherited from common ancestors; they differ in color because of differences in a single gene.

How can two brown-eyed parents produce a blue-eyed child? If you breed two black Labrador retrievers, why may the offspring be yellow, brown, or black (Fig. 13-1)? Although you may not realize it yet, you have already come a long way toward understanding the answers to questions such as these. First, you have learned that DNA molecules carry the hereditary information of all living organisms; the sequences of nucleotides in specific stretches of DNA, the **genes**, encode the information needed to specify the amino acid sequences of proteins. Second, DNA is physically part of the chromosomes, which, in eukaryotic organisms, are found in the nucleus of each cell. Third, in the cells of a diploid organism, chromosomes occur in pairs, called homologous chromosomes, that are similar in appearance and have the same genes. But the nucleotide sequences of the genes may differ on homologous chromosomes because of mutations in DNA that may have arisen in that organism or its ancestors. Fourth, during meiosis, homologous chromosomes separate, one member of each pair entering each haploid daughter cell. Therefore, when gametes are formed, either directly after meiosis or later during the organism's life cycle, each gamete contains one member of each pair of chromosomes. During sexual reproduction, each offspring receives one member of each pair of homologous chromosomes from each of its two parents.

What does this process have to do with blue-eyed children and brown dogs? Quite a bit. If two homologous chromosomes have different nucleotide sequences in some of their genes, then each homologue will direct the synthesis of slightly different proteins. Those protein differences can have profound, visible effects in an organism. For example, as we will see in Chapter 15, different eye colors in people are caused by differences in the amount and distribution of pigment molecules that are synthesized by protein enzymes. Small genetic differences produce enzyme differences, which in turn alter the amount and distribution of pigments in the iris, yielding blue, green, brown, or black eyes. Therefore, slightly different nucleotide sequences of genes on homologous chromosomes, separation of homologues during meiosis, and re-pairing of homologous chromosomes from two parents can produce offspring with genetic differences, enzyme differences, and consequently differences in appearance or physiology.

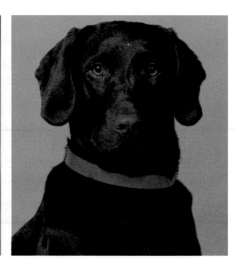

Figure 13-1   **Inheritance of color**

Labrador retrievers come in yellow (left), brown (middle), or black (right). Color is determined by interactions between two sets of genes.

In this chapter, we will examine the patterns of inheritance commonly observed in diploid organisms, starting with the work of Gregor Mendel.

Gregor Mendel and the Origin of Genetics

Gregor Mendel (Fig. 13-2) was a monk in the monastery of St. Thomas in Brünn (now Brno, in the Czech Republic). Before settling down in the monastery, Mendel tried his hand at several pursuits, including health care and teaching. In an effort to earn his teaching certificate, Mendel attended the University of Vienna for two years, where he studied botany and mathematics, among other subjects. This training proved crucial to his later experiments, which were the foundation for the modern science of genetics. At St. Thomas, in the mid-1800s, Mendel carried out both his monastic duties and a groundbreaking series of experiments on inheritance in the common edible pea. Although Mendel worked without knowledge of genes or chromosomes, it will be easier to follow his experiments after a brief look at the genetic differences that may occur on homologous chromosomes.

The specific physical location of a given gene on a chromosome is called its **locus** (Fig. 13-3). Homologous chromosomes carry the same genes, located at the same loci. Although the nucleotide sequence at a given gene locus is always *similar* on homologous chromosomes, it may not be *identical*. Different nucleotide sequences at the same gene locus on two homologous chromosomes produce alternative forms of the gene, called **alleles**. The human ABO blood types, for example, are produced by different alleles of the same gene (see Fig. 13-16).

If two homologous chromosomes in an organism have the same allele at a given gene locus, the organism is said to be **homozygous** for that gene locus. (*Homozygous* comes

Figure 13-2   **Gregor Mendel**

A portrait of Gregor Mendel, painted about the time of his pioneering genetics experiments. Mendel was born in Austria in 1822 and died in 1884.

from Greek words meaning "same pair.") If two homologous chromosomes have different alleles at a given gene locus, the organism is **heterozygous** ("different pair") at that

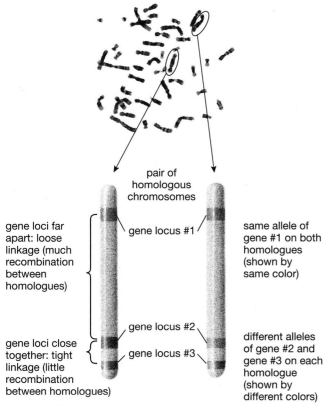

Figure 13-3   The relationships among genes, alleles, and chromosomes

The drawing illustrates a pair of homologous chromosomes. A given gene is a segment of DNA at the same location (locus) on both homologous chromosomes. A chromosome carries many genes; each gene occupies a different locus on the chromosome. Differences in nucleotide sequences at the same gene locus (depicted by different colors) produce different alleles of the gene.

locus. You will recall that, during meiosis, homologous chromosomes are separated. When gametes are formed, each gamete receives one member of each pair of homologous chromosomes. Therefore, an organism homozygous at a particular gene locus produces gametes that all contain the same allele. An organism heterozygous at the same gene locus produces gametes of two kinds; about half of the gametes contain one allele and half contain the other allele.

These facts, combined with some careful breeding experiments and a little "genetic bookkeeping," can be used to explain, and indeed to predict, the patterns of inheritance of many traits in organisms as diverse as pea plants and human beings. Interestingly, both the patterns of inheritance and many essential facts about genes, alleles, and the distribution of alleles in gametes and zygotes during sexual reproduction were deduced by Gregor Mendel before DNA, chromosomes, or meiosis had been discovered. Because you know much more about cellular and molecular genetics than Mendel did, you should be able to anticipate Mendel's results even before we present them in the text. However, because Mendel's experi-

ments are a succinct, elegant example of science in action, let's follow Mendel along his paths of discovery.

Mendel was not the first person to study inheritance. However, as Mendel himself wrote:

Whoever surveys the work in this field will come to the conviction that among the numerous experiments, not one has been carried out to an extent or in a manner that would make it possible to determine the number of different forms in which hybrid progeny appear, permit classification of these forms in each generation with certainty, and ascertain their numerical interrelationships. . . . [H]owever, this seems to be the one correct way of finally reaching the solution to a question whose significance for the evolutionary history of organic forms must not be underestimated."

Doing It Right: The Secrets of Mendel's Success

Why did Mendel succeed where others before him had failed? There are three key steps to any successful experiment in biology: (1) choosing the right organism to work with, (2) designing and performing the experiment correctly, and (3) analyzing the data properly. Mendel was the first geneticist to complete all three steps.

Mendel's choice of the edible pea as an experimental subject was excellent. The structure of the pea flower normally prevents pollen from entering from the outside—that is, from another flower (Fig. 13-4). Instead, pea flowers usually self-pollinate. As a result, the egg cells in each flower are fertilized by sperm from the pollen of the same flower (this is called **self-fertilization**). Even in Mendel's time, commercial seed dealers sold many different types of peas that were **true-breeding**. In true-breeding plants, all the offspring produced through self-fertilization are essentially identical to the parent plant.

Figure 13-4   The flower of the edible pea

In the intact flower (left), the lower petals form a container enclosing the reproductive structures, the stamens (male) and carpel (female). Pollen normally cannot enter the flower from outside, and consequently peas usually self-fertilize. Plant breeders, however, can pull apart the petals (right) and remove the stamens so that self-fertilization cannot occur. Dusting the carpels with pollen of their choice results in controlled cross-fertilization.

Although peas normally self-fertilize, plant breeders can also mate two different plants by hand. By carefully picking the flowers apart, plant breeders can collect pollen from one plant and transfer it to a flower on another plant. The sperm from the "foreign" pollen will then fertilize the egg cells of the recipient flower (**cross-fertilization**). In this way, one can mate two different true-breeding plants and see what types of offspring they produce.

Mendel began by working with only one trait at a time, and he chose traits that had unmistakably different forms, such as white versus purple flowers (Fig. 13-5). Earlier scientists had often crossed plants that differed in several traits or that had different forms of a trait that might be confused with one another. Not surprisingly, the results were difficult to interpret. Studying single traits, with clearly different alternative forms, allowed Mendel to see through to the underlying principles of inheritance.

Perhaps most important, Mendel counted the numbers of offspring bearing the various traits and critically analyzed the numbers. The use of statistics as a tool for understanding underlying principles has since become an extremely important practice in biology.

	DOMINANT TRAIT	RECESSIVE TRAIT
Seed shape	smooth	wrinkled
Seed color	yellow	green
Pod shape	inflated	constricted
Pod color	green	yellow
Flower color	purple	white
Flower location	at leaf junctions	at tips of branches
Plant size	tall (1.8 to 2 meters)	dwarf (0.2 to 0.4 meters)

Figure 13-5   **Traits of Mendel's peas**
Traits of pea plants used by Mendel in his studies of plant inheritance.

Inheritance of Single Traits: The Law of Segregation

Mendel started as simply as possible. He raised varieties of pea plants that were true-breeding for different forms of a single trait and cross-fertilized them. Such an experiment, involving organisms that differ in only one trait, is called a **monohybrid cross**. (In this usage, **hybrid** refers not to the offspring of parents of two different species, such as a mule, but to offspring of parents of the same species that differ in one or more inherited traits, such as hybrid corn.) Mendel saved the resulting hybrid seeds and grew them the next year.

In one of these experiments, Mendel cross-fertilized a white-flowered pea with a purple-flowered one. (In modern terminology, this was the **parental generation**, denoted by the letter **P**.) When he grew the resulting seeds, he found that all the first-generation offspring (the "first filial," or **F$_1$ generation**) produced purple flowers:

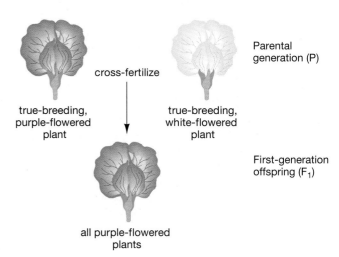

What happened to the white color? The flowers of the hybrids were every bit as purple as the flowers of their purple parent. White seemed to have disappeared completely in the F$_1$ offspring.

Mendel allowed the F$_1$ flowers to self-fertilize, collected the seeds, and planted them the next spring. The second generation (**F$_2$**) had some plants with flowers of each color:

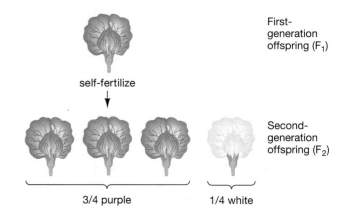

Overall, about three-fourths of the plants had purple flowers and one-fourth had white flowers. The exact numbers were 705 purple and 224 white, or a ratio of 3.15 purple to 1 white.

Mendel allowed the F_2 plants to self-fertilize and produce yet a third (F_3) generation. He found that all the white-flowered F_2 plants produced white-flowered offspring; that is, they bred true. This result was true for as many generations as he had time and patience to raise: White-flowered parents always gave rise to white-flowered offspring. About one-third of the purple-flowered F_2 plants were also true-breeding. The remaining two-thirds were hybrids and produced both purple- and white-flowered offspring, again in the ratio of 3 to 1. Therefore, the F_2 generation included ¼ true-breeding purple plants, ½ hybrid purple, and ¼ true-breeding white.

The Inheritance of Dominant and Recessive Alleles on Homologous Chromosomes Can Explain the Results of Mendel's Monohybrid Crosses

Mendel formed a five-part hypothesis to explain these results.

1. Each trait is determined by pairs of discrete physical units (what modern geneticists call genes). Each individual organism has two genes that together control the expression of a given trait (for example, two genes for flower color). As you will no doubt have guessed, the two genes are located on homologous chromosomes. There may be two or more alternative forms (alleles) of a gene. White-flowered peas, for example, have a different allele of the gene controlling flower color than true-breeding purple-flowered peas do. Different alleles of the same gene are located at the same locus on a chromosome but have different nucleotide sequences.

2. Pairs of genes separate from each other during gamete formation. This part is Mendel's **law of segregation**: Each gamete receives only one of an organism's pair of genes. When a sperm fertilizes an egg, the resulting offspring receives one gene from the father and one from the mother. Now apply your knowledge of meiosis; homologous chromosomes separate during anaphase I of meiosis. When gametes form, each receives one member of each pair of homologous chromosomes, and therefore one member of each pair of genes.

3. Which member of a pair of genes becomes included in a gamete is determined by chance. If we assign letters to each member of a pair of genes, say G1 and G2, then each gamete is just as likely to receive a G1 as a G2. The gamete will not receive both, and there will not be a preference for one member of the pair. This randomness occurs because separation of pairs of homologous chromosomes during meiosis is random.

4. When two alternative forms of a gene are present, one (the **dominant** allele) may mask the expression of the other (the **recessive** allele). However, although the dominant allele of a gene masks the *expression* of the recessive allele, it doesn't alter the *physical nature* of the recessive allele, which can be passed unchanged into the individual's gametes. In Mendel's experiments with flower color, the allele for purple flowers is dominant to the allele for white flowers. At the end of this chapter we will discuss the molecular basis of dominance.

5. True-breeding organisms have two of the same alleles for the trait under study (that is, they are homozygous); hybrids have two different alleles (that is, they are heterozygous). A homozygous organism, with only one type of allele, can produce only one type of gamete:

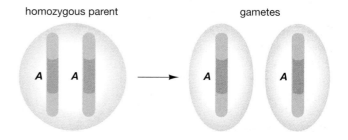

A heterozygous organism, with two different alleles, produces equal numbers of gametes with each of the two alleles:

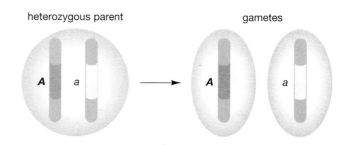

Let's see how Mendel's hypothesis explains the results of his experiments with flower color. Using letters to represent the different alleles, we will assign the uppercase letter P to the allele for purple (the dominant allele is usually represented by a capital letter) and the lowercase letter p to the allele for white (recessive). A homozygous purple-flowered plant has two alleles for purple (PP); a white-flowered plant has two alleles for white (pp). A PP plant produces sperm and eggs all of which carry the P allele; a pp plant produces sperm and eggs all of which carry the p allele:

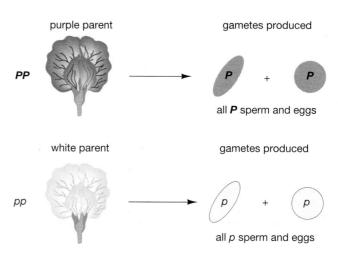

The F_1 hybrid offspring are produced when P sperm fertilize p eggs or when p sperm fertilize P eggs. In either case, the F_1 offspring are Pp. Because P is dominant to p, all the offspring are purple:

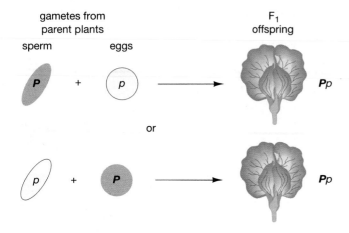

All gametes produced by a heterozygous Pp plant have an equal chance of receiving either the P allele or the p allele (that is, the plant produces equal numbers of P and p sperm and equal numbers of P and p eggs). When a Pp plant self-fertilizes, the two types of sperm have an equal chance of fertilizing each type of egg:

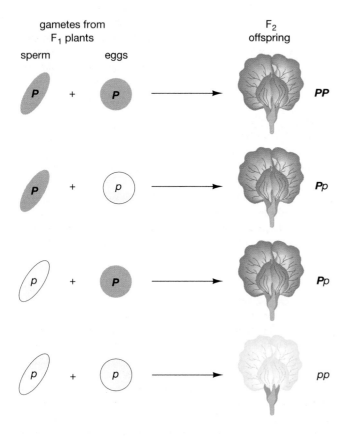

Therefore, three types of offspring can be produced: PP, Pp, and pp. These occur in the approximate proportions of $\frac{1}{4}$ PP, $\frac{1}{2}$ Pp, and $\frac{1}{4}$ pp.

The actual combination of alleles carried by an organism (for example, PP or Pp) is its **genotype**. An organism's traits, including its outward appearance, behavior, digestive enzymes, blood type, or any other observable feature, determine its **phenotype**. As we have seen, plants with genotypes PP and Pp both bear purple flowers. Thus, even though they have different genotypes, they have the same phenotype. Therefore, the F_2 generation consists of three genotypes ($\frac{1}{4}$ PP, $\frac{1}{2}$ Pp, and $\frac{1}{4}$ pp) but only two phenotypes ($\frac{3}{4}$ purple and $\frac{1}{4}$ white).

Figures 13-6 and 13-7 present two methods of "genetic bookkeeping" for determining the expected proportions of offspring in a monohybrid cross. The first method uses a diagram called a Punnett square (Fig. 13-6), named after a famous geneticist of the early 1900s. The second method relies on probability theory (Fig. 13-7). The two methods yield the same results, so you can use whichever seems easier for you. Whichever method you choose, there is one crucial point to keep in mind: *These calculations give only the most probable proportions of offspring of different genotypes and phenotypes.* In a real experiment, one would expect only that the offspring will occur in *approximately* the predicted proportions. In the F_2 generation of Mendel's monohybrid cross, note that he didn't obtain *exactly* $\frac{3}{4}$ purple-flowered and $\frac{1}{4}$ white-flowered plants.

Mendel's Hypothesis Can Be Used to Predict the Outcome of New Types of Single-Trait Crosses

You probably have recognized that Mendel used the scientific method discussed in Chapter 1. He observed the results of his experiments and then formed a hypothesis. As you know, the scientific method has a third step: to use the hypothesis to predict the results of other experiments and to see if those experiments support or refute the hypothesis. Mendel did just that. If the hybrid F_1 flowers have one allele for purple and one for white (Pp), then we should be able to predict the outcome of cross-fertilizing these Pp plants with homozygous recessive (pp) white plants. (Can you?) As Figure 13-8a shows, Mendel's hypothesis predicts that there will be equal numbers of Pp (purple) and pp (white) offspring. This is precisely what happened.

This type of experiment has practical uses, too. By just looking at an organism that has the dominant phenotype, one usually cannot tell whether it is homozygous or heterozygous. Cross-fertilization of a phenotypically dominant individual with a homozygous recessive individual is called a **test cross**, because it can be used to test whether the dominant parent is homozygous or heterozygous. When crossed with a homozygous recessive, a homozygous dominant produces all phenotypically dominant offspring (Fig. 13-8b), whereas a heterozygous dominant yields offspring with both dominant and recessive phenotypes, in a 1:1 (one-to-one) ratio (Fig. 13-8a).

Inheritance of Multiple Traits: Independent Assortment

Having determined the mode of inheritance of single traits, Mendel then turned to the more complex question of multiple traits. He began with a **dihybrid cross**; that is, he cross-bred plants that differed in two traits, for example seed color (yellow or green) and seed shape (smooth or wrinkled). If he crossed a plant that was homozygous for smooth yellow seeds with one that was homozygous for wrinkled green seeds, the F_1 offspring all bore smooth yellow seeds. This result was no great surprise, because from separate monohybrid crosses of each of these traits he al-

ready knew that smooth (S) is dominant to wrinkled (s), and that yellow (Y) is dominant to green (y) (see Fig. 13-5). The F_1 offspring, therefore, are all genotypically $SsYy$. Allowing these F_1 plants to self-fertilize, Mendel found that the F_2 generation consisted of 315 smooth yellow seeds, 101 wrinkled yellow seeds, 108 smooth green seeds, and 32 wrinkled green seeds. These numbers work out to roughly $9/16$ smooth yellow seeds, $3/16$ wrinkled yellow seeds, $3/16$ smooth green seeds, and $1/16$ wrinkled green seeds, or a ratio of 9:3:3:1. The F_2 generations produced from other dihybrid crosses had similar phenotypic ratios.

Mendel realized that these ratios could be explained if he assumed that the genes for seed color and seed shape are inherited independently of each other and do not influ-

Figure 13-6   The Punnett square

The Punnett square method of determining likely genotypes and phenotypes of offspring. The Punnett square is intuitive and relatively nonmathematical, and many students prefer it to the probability theory. The example shown here analyzes the inheritance of flower color in peas, based on a cross of F_1 hybrids. Using the Punnett square requires the following six steps.

(1) Assign letters to represent the different alleles; dominant alleles are usually upper case (P), recessive alleles lower case (p). **Be sure that you can distinguish upper and lower case letters**: for example, W and w or S and s may be difficult to tell apart, especially in the heat of a test.

(2) Assuming that the genotypes of the parents are known, determine the types of equally likely, genetically different gametes that can be produced by the male and female parents. In the case of the gene pair here, $1/2$ of the gametes should receive the P allele, and $1/2$ should receive the p allele. (Of course, if a parent is homozygous, then all of its gametes will receive the same allele.)

(3) Draw the Punnett square. The number of rows in the square is the same as the number of different types of sperm, and the number of columns in the square is the same as the number of different types of eggs. Label the rows with the types of sperm, the columns with the types of eggs.

(4) The expected offspring are the "cells" of the square. Each offspring results from the fertilization of the egg in its column by the sperm in its row. Fill out the offspring genotypes systematically, for example, first using row 1 sperm to fertilize all types of eggs, then row 2 sperm to fertilize all types of eggs, and so on.

(5) Examine the offspring cells and add up the number of offspring with the same genotypes.

(6) **Remember that the numbers generated by a Punnett square do not mean that every mating will yield exactly four offspring.** Convert the numbers of offspring genotypes to expected fractions of offspring genotypes by dividing the number of each offspring genotype by the total number of offspring cells. If desired, add together those genotypes that produce the same phenotypes (PP and Pp in this example), to yield the expected fractions of offspring with each phenotype. Remember also that the fractions of different offspring generated by the square are only *expected* fractions. Real matings would be expected only to yield *approximately* these fractions of different offspring.

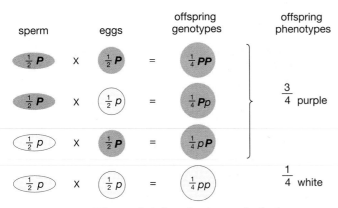

sperm		eggs		offspring genotypes	offspring phenotypes
$\frac{1}{2}$ P	X	$\frac{1}{2}$ P	=	$\frac{1}{4}$ PP	
$\frac{1}{2}$ P	X	$\frac{1}{2}$ p	=	$\frac{1}{4}$ Pp	$\frac{3}{4}$ purple
$\frac{1}{2}$ p	X	$\frac{1}{2}$ P	=	$\frac{1}{4}$ pP	
$\frac{1}{2}$ p	X	$\frac{1}{2}$ p	=	$\frac{1}{4}$ pp	$\frac{1}{4}$ white

Figure 13-7   **The probability theory method of determining likely genotypes and phenotypes of offspring**

The example is the F_1 cross illustrated in the Punnett square method in Figure 13-6. In probability theory, the probability that two independent events will occur together is the product (multiplication) of their independent probabilities. In this cross, each type of gamete (P or p) is equally likely to occur; that is, each has a probability of ½. Further, each type of sperm is equally likely to fertilize each type of egg. In terms of probability theory, the first event (which allele is in a particular sperm) does not depend on the second event (which allele is in the egg fertilized by that sperm). We can therefore obtain the probable ratio of offspring by multiplying the probability of obtaining each type of sperm by the probability of obtaining each type of egg.

To use probability theory to analyze genetic crosses, perform the following four steps:

(1) Assign letters to the alleles (the same as step 1 of the Punnett square method).

(2) Determine the types of equally likely, genetically different gametes that can be produced by the parents (the same as step 2 of the Punnett square method).

(3) Set up a multiplication table as shown below: ½ of the sperm are P, and ½ of the eggs are P; multiply these together to obtain ¼ PP offspring. Continue filling in the table; be sure that all types of sperm have the "opportunity" to fertilize all types of eggs. For a "genotype" table such as this one, the number of rows in the table (4) should equal the number of different types of sperm (2) multiplied by the number of different types of eggs (2).

(4) If desired, add together the genotypes that produce the same phenotypes (the same as step 6 of the Punnett square method).

ence each other during gamete formation (Fig. 13-9). (Imagine flipping two coins, a dime and a nickel. Whether the dime comes up heads or tails does not affect which side of the nickel comes up.) Thus the outcome for each trait could be regarded as a simple monohybrid cross, in which a 3:1 ratio of offspring would be expected. The laws of probability state that the independent combination of two 3:1 ratios yields a 9:3:3:1 ratio (Fig. 13-10b), and we can see from Mendel's results that this is just what happened. There were 423 smooth seeds (of either color) to 133 wrinkled ones (3.18:1) and 416 yellow seeds (of either shape) to 140 green ones (2.97:1). The Punnett square shows how two 3:1 ratios combine to form an overall 9:3:3:1 ratio (Fig. 13-10a).

The independence in the inheritance of two or more distinct traits is called the **law of independent assortment**: The distribution of alleles for one trait into the gametes does not affect the distribution of alleles for other traits. Think back to the events of meiosis in Chapter 12: How do you think that independent assortment of alleles for two different genes might occur? That's right: Independent assortment will occur if the gene loci are on different chromosomes. When paired homologous chromosomes line up at the equator of the cell during metaphase I, the orientation of the members of each pair is random, and the orientation of one pair of homologues does not influence the orientation of other pairs. In other words, if the paternal homologue of chromosome 1 faces the "north pole" of the cell, this orientation does not influence whether the paternal homologue of chromosome 2 faces north or south. Therefore, when the homologues separate during anaphase I, whether a given daughter cell receives the paternal homologue of chromosome 1 has no effect on the likelihood that it will receive the paternal or maternal homologue of chromosome 2.

In an Unprepared World, Genius May Go Unrecognized

In 1865, Gregor Mendel presented his theories of inheritance to the Brünn Society for the Study of Natural Science, and they were published the next year. It did *not* mark the beginning of genetics. In fact, it didn't make any impression at all on the biology of his time. Mendel's experiments, which eventually spawned one of the most important theories in all of biology, simply vanished from the scientific scene. Apparently, very few biologists read his paper, and those who did probably couldn't understand it.

It was not until 1900 that three biologists—Carl Correns, Hugo de Vries, and Erich Tschermak—working independently and knowing nothing of Mendel's work, rediscovered the principles of inheritance. No doubt to their intense disappointment, when they searched the scientific literature before publishing their results, they found that Mendel had scooped them over 30 years before.

Inheritance Patterns of Genes Located on the Same Chromosome

Gregor Mendel didn't know anything about the physical nature of genes or that genes are parts of chromosomes, because chromosomes weren't even discovered until long after his experiments were concluded. Once chromosomes were seen in microscopes, and their distribution during mitosis and meiosis was deduced, it became obvious that they were the vehicles of inheritance. It also became obvious that there are many more traits, and therefore many more genes, than there are chromosomes. The facts that genes are parts of chromosomes and that each chromosome bears many genes have important implications for inheritance.

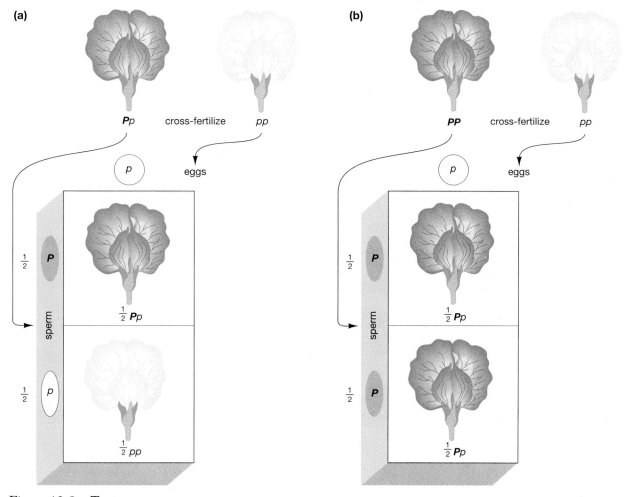

Figure 13-8 **Test crosses**

Test crosses involve breeding a homozygous recessive organism (in these examples, *pp* plants with white flowers) with a phenotypically dominant organism (here, plants with purple flowers). If the phenotypically dominant parent is heterozygous *Pp* **(a)**, then half the offspring will be heterozygous purple and half white; if the parent is homozygous *PP* **(b)**, then all the offspring will be heterozygous purple.

Genes on the Same Chromosome Tend to Be Inherited Together

If *chromosomes* assort independently during meiosis I, then only genes located on *different chromosomes* will assort independently. Genes on the *same chromosome* are said to be **linked**, because they do not assort independently but tend to be inherited together. One of the first pairs of linked genes to be discovered were those for flower color and pollen grain shape in the sweet pea (a different type than the edible pea). Purple flower color (*P*) is dominant over red (*p*), and long pollen shape (*L*) is dominant over round (*l*). When a homozygous purple-flowered, long-pollen pea is crossed with a homozygous red-flowered, round-pollen pea, all the F₁ offspring have purple flowers and long pollen. However, the genes for flower color and for pollen shape are carried on the same chromosome, so the two genes tend to assort together during meiosis and to be inherited together. Thus, the phenotypes of the F₂ generation do not occur in a 9:3:3:1 ratio, as Mendel's law of independent assortment would predict. Instead, the F₂ generation has about ¾ purple-flowered, long-pollen plants and ¼ red-flowered, round-pollen plants (Fig. 13-11).

Crossing Over May Separate Linked Genes

There is one thing wrong with this tidy scheme: Genes on the same chromosome do not *always* stay together. In the sweet pea cross just described, for example, often the F₂ generation will include a few purple-flowered, round-pollen plants and red-flowered, long-pollen plants. How can this be?

As you know, during prophase I of meiosis, the nonsister chromatids of homologous chromosomes intertwine, producing X-shaped figures called **chiasmata** (see Fig. 12-5). At chiasmata, parts of nonsister chromatids may exchange with each other, a process called **crossing over**. In fact, there is almost always at least one exchange between homologues during any meiosis. Exchange of corresponding segments of DNA during crossing over forms new combinations of alleles on both homologous chromosomes. Therefore, when homologous chromosomes separate at anaphase I, the chro-

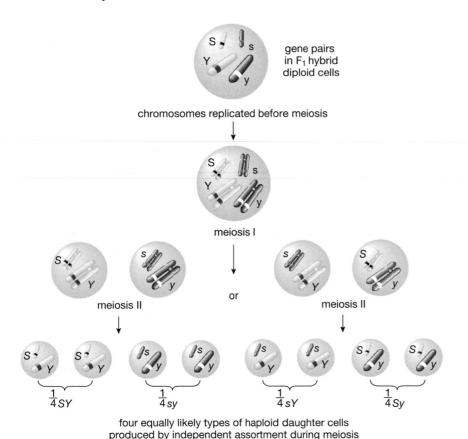

gene pairs
in F₁ hybrid
diploid cells

chromosomes replicated before meiosis

meiosis I

or

meiosis II   meiosis II

$\frac{1}{4}SY$ $\frac{1}{4}sy$ $\frac{1}{4}sY$ $\frac{1}{4}Sy$

four equally likely types of haploid daughter cells
produced by independent assortment during meiosis

Figure 13-9   Independent assortment of chromosomes

Independent assortment of alleles for two different genes arises from the mechanisms of chromosome movements during meiosis (see Chapter 12). At the end of meiosis I, two equally likely chromosome groupings are possible: SY and sy or Sy and sY. At the end of meiosis II, the SY/sy grouping yields haploid cells, half of which are SY and half of which are sy. The Sy/sY grouping yields half Sy and half sY daughter cells. Because meiosis occurs in many reproductive cells in a plant, the two combinations are equally likely to occur, on the average. Therefore, an F₁ plant would produce gametes in the predicted proportions ¼ SY, ¼ sy, ¼ Sy, and ¼ sY.

(a)

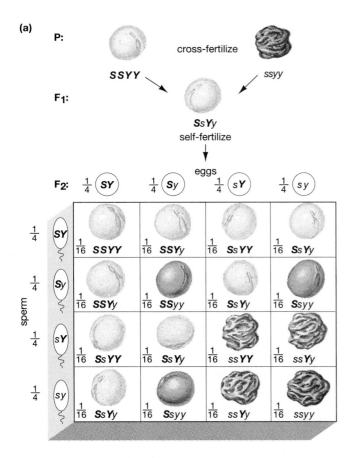

P: cross-fertilize

SSYY   ssyy

F₁:

SsYy
self-fertilize

eggs

F₂: ¼ (SY)   ¼ (Sy)   ¼ (sY)   ¼ (sy)

sperm

¼ (SY) $\frac{1}{16}$ SSYY $\frac{1}{16}$ SSYy $\frac{1}{16}$ SsYY $\frac{1}{16}$ SsYy

¼ (Sy) $\frac{1}{16}$ SSYy $\frac{1}{16}$ SSyy $\frac{1}{16}$ SsYy $\frac{1}{16}$ Ssyy

¼ (sY) $\frac{1}{16}$ SsYY $\frac{1}{16}$ SsYy $\frac{1}{16}$ ssYY $\frac{1}{16}$ ssYy

¼ (sy) $\frac{1}{16}$ SsYy $\frac{1}{16}$ Ssyy $\frac{1}{16}$ ssYy $\frac{1}{16}$ ssyy

(b)

seed shape		seed color			F₂ offspring
$\frac{3}{4}$ smooth	x	$\frac{3}{4}$ yellow	=	$\frac{9}{16}$	smooth yellow
$\frac{3}{4}$ smooth	x	$\frac{1}{4}$ green	=	$\frac{3}{16}$	smooth green
$\frac{1}{4}$ wrinkled	x	$\frac{3}{4}$ yellow	=	$\frac{3}{16}$	wrinkled yellow
$\frac{1}{4}$ wrinkled	x	$\frac{1}{4}$ green	=	$\frac{1}{16}$	wrinkled green

Figure 13-10   Calculating the expected offspring in a dihybrid cross

(a) Punnett square analysis. One of the original parents is homozygous for smooth (SS) yellow (YY) seeds, and the other original parent is homozygous for wrinkled (ss) green (yy) seeds. The F₁ offspring therefore all have smooth yellow seeds (SsYy). If we allow the F₁ to self-fertilize, we obtain the offspring shown within the square. There are now 16 squares in the analysis, but the method is the same as in Figure 13-6. The expected proportions of offspring are ⁹⁄₁₆ smooth yellow, ³⁄₁₆ smooth green, ³⁄₁₆ wrinkled yellow, and ¹⁄₁₆ wrinkled green, or 9:3:3:1. Note also that the Punnett square predicts 12 smooth to 4 wrinkled seeds and 12 yellow to 4 green seeds (both 3:1 ratios), just as would be expected from monohybrid crosses if we ignore the second trait. **(b)** Analysis by probability theory. *Remember that independent assortment means that each trait assorts independently of the others.* We know from monohybrid crosses that three-quarters of the F₂ offspring will be smooth and one-quarter will be wrinkled and that three-quarters will be yellow and one-quarter will be green. Multiplying these independent probabilities produces the expected F₂ offspring. These are identical to the offspring ratios generated by the Punnett square.

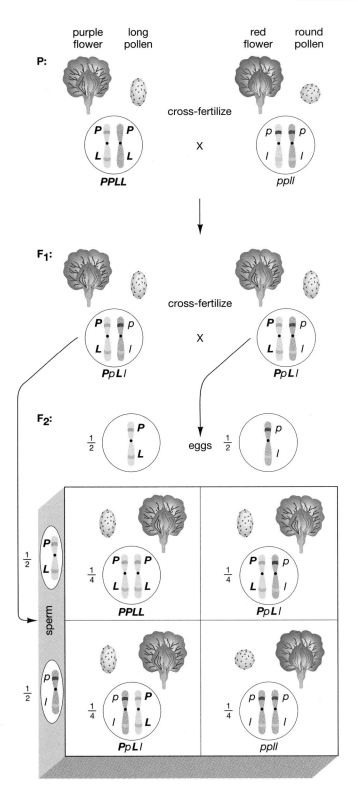

mosomes that each haploid daughter cell receives are different from those of the parent cell.

Crossing over during meiosis explains why new combinations of traits occurred in the sweet pea cross. At the beginning of prophase I, all the F₁ peas had this pair of homologous chromosomes:

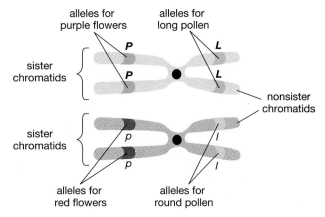

In a few reproductive cells, crossing over occurred between the locations of the genes for flower color and pollen shape, so that two nonsister chromatids exchanged alleles for flower color:

At anaphase I the separated homologous chromosomes had this gene composition:

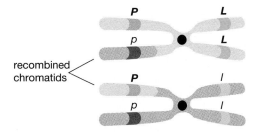

Four different types of chromosomes were distributed to the haploid daughter cells during meiosis II:

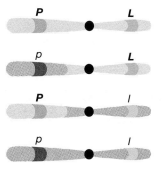

Figure 13-11 **Linkage**

Inheritance of linked genes for pollen grain shape and flower color in sweet peas. Because these genes are found on the same chromosome, the configurations *PL* and *pl* in the parental generation tend to be preserved in all subsequent generations. The F₂ offspring show a phenotypic ratio of 3 purple, long to 1 red, round rather than the 9:3:3:1 ratio one would expect from independently assorted genes.

Therefore, some gametes were produced with each of the four chromosome configurations: *PL* and *pl* (the original parental types) and *Pl* and *pL* (recombined chromosomes).

In most of the reproductive cells, crossing over between these specific pairs of alleles did not occur. Therefore, most of the F₂ offspring received chromosomes with the *PL* and *pl* allele combinations. However, fusion of sperm and eggs that formed from cells in which crossing over *did* occur gave rise to a few plants with purple flowers and round pollen and a few plants with red flowers and long pollen.

This situation is **genetic recombination**: generating new combinations of alleles by the exchange of DNA between homologous chromosomes. In most organisms, **sexual recombination** also occurs, in which offspring receive one homologous chromosome from each of two, genetically different parents. Together, genetic and sexual recombination provide great genetic variability among organisms.

The Frequency of Crossing Over between Linked Genes Can Be Used to Map the Location of Genes on a Chromosome

Crossing over occurs more often between genes that are far apart on a chromosome than between genes that are close together. In fact, if two genes are close enough together on a chromosome, crossing over almost never occurs between them; in contrast, if two genes are at opposite ends of a long chromosome, crossing over almost always occurs between them. Intermediate distances between genes result in intermediate amounts of crossing over. By swapping alleles between homologous chromosomes, crossing over produces offspring with recombined traits. Therefore, the proportion of offspring with recombined traits is a reflection of the spacing of genes along a chromosome.

This idea has been used to map chromosomes. The distance between two genes on a chromosome is defined as the percentage of recombination observed during a cross between one organism homozygous recessive for both genes and another organism heterozygous for both genes. (If no crossing over occurred, what would be the expected ratio of genotypes and phenotypes? Hint: This is a "two-gene" test cross.) If two genes recombine 10% of the time, they are placed 10 "map units" apart on the chromosome. In some organisms, such as the fruit fly *Drosophila*, it has been possible to correlate this type of recombination map with the physical appearance of certain chromosomes and locate the actual sites of genes (Fig. 13-12).

Inheritance of Sex and Sex-Linked Genes

To say that the two members of a pair of homologous chromosomes carry the same genes is a bit of an oversimplification. In mammals and many insects, males have the same number of chromosomes as females do, but one "pair," the

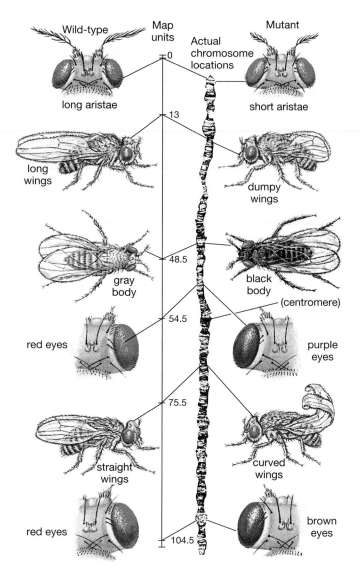

Figure 13-12   A chromosome map in the fruit fly

A map of chromosome 2 in the fruit fly, comparing the distances between genes obtained by recombination analysis with the physical locations of genes on the chromosomes. The order of the genes is the same in both maps, but the distances between genes are not. Crossing over apparently is hindered in some regions of the chromosome, especially around the centromere, distorting the recombination map.

sex chromosomes, are very different in appearance and genetic composition. Females have two identical sex chromosomes, called **X chromosomes**, and males have one X chromosome and one **Y chromosome**. Although the Y chromosome usually carries far fewer genes than the X chromosome does, small parts of the sex chromosomes are homologous, so X and Y chromosomes pair up during prophase of meiosis I and separate during anaphase I. The other chromosomes, which occur in pairs of identical appearance in both males and females, are called **autosomes**. Numbers of chromosomes vary tremendously, but there is always only one pair of sex chromosomes: For example, the

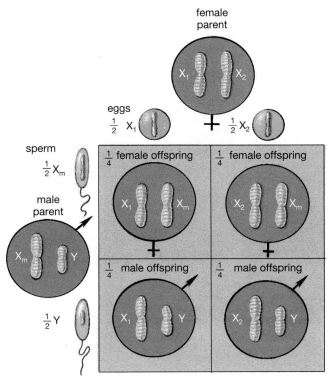

Figure 13-13   **Sex determination**

Sex determination in species such as humans and fruit flies, in which males carry two dissimilar sex chromosomes (XY) and females carry two similar sex chromosomes (XX). Only the distribution of sex chromosomes is illustrated. Male offspring receive the Y chromosome from the father; female offspring receive the father's X chromosome (labeled X_m in the drawing). The mother passes one of her X chromosomes (X_1 or X_2 in the drawing) to both male and female offspring.

fruit fly *Drosophila* has four pairs of chromosomes (three pairs of autosomes), humans have 23 pairs (22 pairs of autosomes), and dogs have 39 pairs (38 pairs of autosomes).

For organisms in which males are XY and females are XX, the sex of an offspring is determined by which sex chromosome is in the sperm that fertilizes the egg (Fig. 13-13). During sperm formation, the sex chromosomes segregate, and each sperm receives either the X or the Y chromosome (plus one member of each pair of autosomes). The sex chromosomes also segregate during egg formation, but because females have two X chromosomes, every egg receives one X chromosome. Therefore, an offspring is male if an egg is fertilized by a Y-bearing sperm, or female if an egg is fertilized by an X-bearing sperm.

Sex Linkage Is a Special Case of Linkage between Genes on the Same Chromosome

Genes that are found on one sex chromosome but not on the other are called **sex-linked**. In principle, the X chromosome might carry genes that are not found on the Y chromosome, and vice versa. However, in many animals, the Y chromosome carries relatively few genes other than

those determining maleness, whereas the X chromosome bears many genes that have nothing to do with specifically female traits. The human X chromosome, for example, contains genes for color vision, blood clotting, and certain structural proteins in muscles; those genes have no counterpart on the Y chromosome. Therefore, because they have two X chromosomes, females can be either homozygous or heterozygous for genes on the X chromosome. Normal dominant versus recessive relationships among alleles will be expressed. Males, in contrast, must fully express whatever alleles they have on their single X chromosome, whether those alleles are otherwise dominant or recessive.

How does sex linkage affect inheritance? Let's look at the first example of sex linkage to be discovered, the inheritance of eye color in the fruit fly *Drosophila*. Because of their small size, rapid reproductive rate, ease of rearing, and small number of chromosomes, *Drosophila* have been favored subjects for genetics studies for more than a century. Over the years, many mutations have been discovered in laboratory-bred *Drosophila* that produce significant phenotypic differences compared with wild-type *Drosophila*. Wild-type *Drosophila* have bright red eyes (Fig. 13-14). In the early 1900s, researchers in the laboratory of Thomas Hunt Morgan at Columbia University discovered a male fly with white eyes. This white-eyed male was mated to a virgin red-eyed female. The resulting offspring were all red-eyed flies (Fig. 13-14a), indicating that white eye color (*w*) is probably recessive to red (*W*). The F_2 generation, however, was a surprise. As expected, the total ratio of red- to white-eyed flies was about 3:1; however, *there were nearly equal numbers of red-eyed males and white-eyed males, but no females with white eyes at all* (Fig. 13-14b)! A test cross of the F_1 red-eyed females and the original white-eyed male yielded roughly equal numbers of red-eyed and white-eyed males and females.

From these data, could you figure out the mode of inheritance of eye color? Morgan made the brilliant hypothesis that *the gene for eye color must be located on the X chromosome, while the Y chromosome has no corresponding gene:*

In the F_1 generation, both male and female offspring received an X chromosome, with its *W* allele for red eyes, from their mother. The F_1 males received a Y chromosome from their father, with no allele for eye color at all, so the males had a *WY* genotype (the use of *Y* here indicates that the Y chromosome does not contain the gene for eye color). The F_1 females received the father's X chromosome with its *w* allele, so the females had a *Ww* genotype. Thus, all male and female F_1 offspring had red eyes.

Figure 13-14   Sex-linked inheritance

Morgan's interpretation of the results of sex-linked inheritance of white eye color in fruit flies. The gene for eye color is carried on the X chromosome; there is no corresponding gene on the Y chromosome. The normal red eyes allele (W) is dominant to the mutant allele for white eyes (w). **(a)** The F_1 generation. Female offspring receive the w allele on their father's X chromosome, but phenotypically this is masked by the dominant W allele on their mother's X chromosome. Male offspring receive the W allele from their mother and the Y chromosome with no eye color gene from their father. Therefore all the females are Ww and all the males are WY. Both males and females have red eyes. **(b)** The F_2 generation. All F_1 males carry the W allele, so all their F_2 female offspring receive the W allele as well. Therefore, the F_2 females are all red eyed. The F_1 females are all heterozygous Ww. Consequently, half their F_2 sons receive the W allele and half receive the w allele. The F_2 males also receive a Y chromosome from their fathers. Therefore half the F_2 males are WY (red eyes) and half are wY (white eyes).

Crossing two F₁ flies resulted in an F₂ generation with the chromosome distribution shown in Figure 13-14b. All the F₂ females received one X chromosome from their F₁ male parent with its *W* (red) allele. They therefore had red eyes. All the F₂ males, on the other hand, inherited their single X chromosome from their F₁ female parent. Because the F₁ females were heterozygous for eye color (*Ww*), the F₂ males had a 50-50 chance of receiving either an X chromosome with the *W* allele or one with the *w* allele. *With no corresponding gene on the Y chromosome, the F₂ males displayed the phenotype determined by the allele on the X chromosome.* Therefore, half the F₂ males had red eyes and half had white eyes.

Before you read on, try to answer this question: How much crossing over would you expect between X and Y chromosomes? If you concluded "very little" or "none," you're right. Most of the X and Y chromosomes are not homologous, so the normal mechanisms of crossing over do not apply.

Variations on the Mendelian Theme

In our discussion of patterns of inheritance thus far, we have made some major simplifying assumptions: that each trait is completely controlled by a single gene, that there are only two possible alleles of each gene, and that one allele is completely dominant to the other, recessive, allele. Most traits, however, are influenced in more varied and subtle ways.

In Incomplete Dominance, the Phenotype of Heterozygotes Is Intermediate between the Phenotypes of the Homozygotes

In his pea experiments, Mendel encountered a particularly simple situation: Heterozygotes and homozygous dominants had the same phenotype. This is often not the case. In snapdragons, for example, crossing homozygous red-flowered plants (*RR*) with homozygous white-flowered ones (*R´R´*) does not produce red-flowered F₁ hybrids. Instead, the F₁ flowers are pink. When the heterozygous phenotype is intermediate between the two homozygous phenotypes, the pattern of inheritance is called **incomplete dominance**. This apparent blending of the phenotype is not, however, the result of any alteration of the underlying alleles. The F₂ generation shows that the alleles for flower color have not changed—in homozygous flowers of the F₂ generation, the red and white colors are as strong as ever (Fig. 13-15). The F₂ offspring include about ¼ red, ½ pink, and ¼ white flowers. This distribution corresponds to the genotypic ratio of ¼ *RR*:½ *RR´*:¼ *R´R´*. *We will discuss the molecular basis for incomplete dominance later in this chapter.*

There May Be Multiple Alleles of a Single Gene

A single diploid organism can have at most only two different alleles for a given gene. Alleles, however, arise

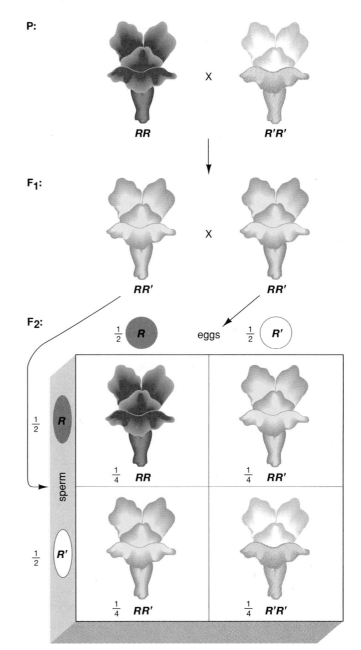

Figure 13-15   Incomplete dominance

Incomplete dominance in the inheritance of flower color in snapdragons. In cases of incomplete dominance, we will use capital letters for both alleles (here *R* and *R´*, "R prime") rather than upper- and lowercase letters. Hybrids (*RR´*) have pink flowers, and the homozygotes are red (*RR*) or white (*R´R´*). Because heterozygotes can be distinguished from homozygous dominants, the distribution of phenotypes in the F₂ generation (¼ red to ½ pink to ¼ white) is the same as the distribution of genotypes (¼ *RR* to ½ *RR´* to ¼ *R´R´*).

through mutation, and the genes of different organisms may have different mutations, each producing a new allele. If we could sample all the individuals of a species, we would often find several, sometimes even dozens, of alleles for every gene. The gene for eye color in fruit flies, for example, has many alleles, each recessive to normal red eyes and producing various shades of white, yellow, or pink when homozygous.

The ABO blood types in humans constitute a familiar system of such **multiple alleles**. The blood types A, B, AB, or O arise as a result of three different alleles of a single gene (usually designated I^A, I^B, and i). This gene directs the synthesis of glycoprotein "identification markers" that protrude from the surfaces of red blood cells. Alleles I^A and I^B direct the synthesis of glycoproteins A and B, respectively, whereas allele i produces no glycoproteins at all (Fig. 13-16). Individual humans may have one of six genotypes: I^AI^A, I^BI^B, I^AI^B, I^Ai, I^Bi, or ii. Alleles I^A and I^B are dominant to i. Therefore, individuals with genotypes

I^AI^A or I^Ai have type A glycoproteins on their red blood cells and have type A blood. Those with genotypes I^BI^B or I^Bi synthesize type B glycoproteins and have type B blood. Homozygous recessive ii individuals lack these glycoproteins and have type O blood. However, alleles I^A and I^B are **codominant** to one another—that is, *both are phenotypically detectable in heterozygotes*. I^AI^B individuals have red blood cells with both A and B glycoproteins and have type AB blood.

These red blood cell glycoproteins may react with antibodies in the blood plasma. If a patient with type B blood receives a transfusion of type A blood, the anti-A antibodies in the patient's serum cause the type A blood cells to clump (Fig. 13-16b).

(a)

Genotype	Blood Type	Red Blood Cells	Plasma Antibodies
$I^A I^A$ $I^A i$	A	A glycoprotein	anti-**B**
$I^B I^B$ $I^B i$	B	B glycoprotein	anti-**A**
$I^A I^B$	AB	A glycoprotein and B glycoprotein	none
$i\ i$	O	neither glycoprotein	anti-**A** anti-**B**

(b)

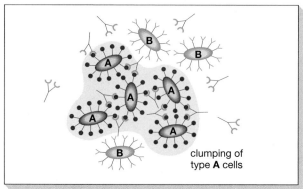

clumping of type **A** cells

Medical effects of blood transfusions			
Donor Type	Recipient Type	Effect on Recipient	Permissible Blood Donation?
A	**A**	–	Yes
	B	Clumping	No
	AB	–	Yes
	O	Clumping	No
B	**A**	Clumping	No
	B	–	Yes
	AB	–	Yes
	O	Clumping	No
AB (universal recipient)	**A**	Clumping	No
	B	Clumping	No
	AB	–	Yes
	O	Clumping	No
O (universal donor)	**A**	–	Yes
	B	–	Yes
	AB	–	Yes
	O	–	Yes

(a) Red blood cell glycoproteins and plasma antibodies found in individuals with each blood type. Type A individuals have the A glycoprotein on their red blood cell surfaces. Their plasma contains antibodies only against B glycoproteins (anti-B antibodies). Therefore, their plasma antibodies do not bind to their own red blood cell glycoproteins, and their own cells do not clump. Type B individuals have B glycoproteins and anti-A antibodies. Type AB blood contains red blood cells with both the A and B glycoproteins, but neither antibody. Type O individuals have both anti-A and anti-B antibodies but no reactive glycoproteins on their red blood cells. **(b)** The reaction between type A blood cells transfused into type B blood. Type B blood contains many anti-A antibodies. Because each antibody has two sites that can bind the A glycoprotein, the A cells become clumped, held together by the anti-A antibodies. These clumps can become large clots, with serious medical consequences. The permissible donors for each blood type are shown in the table.

Figure 13-16   **Human ABO blood group reactions**

Glycoproteins on the surfaces of the red blood cells determine blood type. If antibodies in the blood plasma bind to red blood cell glycoproteins, the red blood cells clump together. Each antibody can bind to only one specific glycoprotein; for example, anti-A antibody binds only to glycoprotein A and not to glycoprotein B.

Many Traits Are Influenced by Several Genes

If you look around your class, you are likely to see people of varied heights, skin colors, and body build, to take just a few obvious traits. Traits such as these are not governed by single genes but are influenced by interactions among two or more genes (plus, of course, interactions with the environment).

Continuous Variation in a Trait May Be Caused by Contributions from Several Genes

For many traits, such as skin color, there are more than just two or three phenotypes. There may be several phenotypes, or even seemingly continuous variation that cannot be split up into convenient, easily defined categories. In many of these cases, two or more genes may produce functionally equivalent contributions to the trait (e.g., roughly equal amounts of skin pigment), a situation called **polygenic inheritance**.

The simplest polygenic inheritance occurs when two gene pairs code for the same trait. In wheat, for example, there are two gene pairs for kernel color, which we might designate 1 and 2. Each gene has two alleles, R and R'. The R allele directs the synthesis of one "unit" of red pigment in the kernel, while the R' allele causes no pigment synthesis at all. If only gene pair 1 were present, then the inheritance of kernel color would follow simple incomplete dominance: R_1R_1 = red; $R_1R'_1$ = pink; $R'_1R'_1$ = white. But because kernel color is controlled by gene pairs 1 and 2, the color becomes more finely graded in intensity, depending on the number of R alleles of both genes. Kernels with the genotype $R_1R_1R_2R_2$ synthesize four units of pigment and therefore are dark red. Kernels with the genotype $R'_1R'_1R'_2R'_2$ produce no pigment and are white. Intermediate numbers of R alleles yield intermediate intensities of red.

As you can well imagine, the more genes that contribute to a single trait, the greater the number of categories of the trait, with increasingly fine gradation between categories. Continuing our example of color intensity, if three pairs of genes are involved, we would have seven phenotypic categories; with four pairs of genes, nine categories, and so on. With more than three pairs of genes, differences between phenotypes are small, and it is extremely difficult to classify the phenotypes reliably.

Polygenic inheritance leads to small, "quantitative" differences in traits (such as shades of color in wheat kernels) rather than the large "qualitative" differences that are caused by single-gene inheritance (such as the all-or-none, purple-or-white color of pea flowers). Therefore, polygenic inheritance is often referred to as **quantitative inheritance**.

The Action of One Gene May Modify the Action of Another Gene

In polygenic inheritance, several genes have functionally similar actions in controlling a trait. In **epistasis**, the expression of one gene depends on, or is modified by, the expression of another gene. One of the best understood cases of epistasis occurs in the inheritance of hair color in virtually all mammals, including, for example, the mouse (Fig. 13-17a). One gene controls the *synthesis* of melanin, the pigment in hair. The dominant allele of this gene (M) allows the synthesis of melanin, but the recessive allele (m) does not. A second gene controls the *distribution* of pigment in the hair. Agouti, as the normal fur of the wild house mouse is called, has individual hairs that are black with a yellow tip, giving the mouse an overall brownish gray appearance. Agouti (A) is dominant to plain black fur (a).

These two genes control hair color in the following way. If a mouse has the MM or Mm genotype, melanin will be produced. The fur will then be agouti if the mouse is AA or Aa for the melanin-distribution gene; the fur will be black if the mouse is aa. If a mouse is mm, then it cannot produce melanin. All mice with the mm genotype therefore have white hair and are albino. In an albino, the melanin-distribution gene cannot affect the phenotype, because there isn't any melanin to distribute. Therefore, the *melanin-synthesis gene* controls the expression of the *melanin-distribution gene*. A cross of two agouti mice can result in a variety of offspring, as the genes for melanin production and distribution assort independently (Fig. 13-17b).

The expression of almost all genes is influenced to some extent by other genes. To take an obvious, but usually overlooked, example, no genes can be expressed in an adult organism unless the genes that direct development operate properly. An organism is a cohesive, coordinated whole, and all its genes influence its ultimate anatomy, physiology, and behavior.

A Single Gene May Have Multiple Effects on Phenotype

We have just seen that a single phenotype may require the interaction of several genes. The reverse is also true: Single genes may have multiple phenotypic effects, a phenomenon called **pleiotropy**. Take the case of albino mice, for example. Albino mice lack pigment not only in their fur but also in their eyes. (Their eyes are bright pink because the blood circulating through capillaries inside their eyes shows through the transparent iris.) Without any eye pigment, they are extraordinarily sensitive to light. Albino mice can't see well in even fairly dim light, and normal daylight rapidly destroys the receptors in their eyes, causing blindness. Consequently, the single gene for pigment production actually can have multiple phenotypic effects: white fur, pink eyes, and blindness.

The Environment Influences the Expression of All Genes

An organism is not just the sum of its genes. Both the genotype and the environment in which an organism lives profoundly affect its phenotype, even for simple physical traits.

(a)

Alleles of Gene #1	Action	Alleles of Gene #2	Action	Final Phenotype
MM or **Mm**	produces melanin pigment	**AA** or **Aa**	distributes melanin in agouti pattern	agouti fur
		aa	distributes melanin as solid black	black fur
mm	produces no melanin	**AA** or **Aa**	would distribute melanin in agouti pattern	no melanin to distribute; white fur (albino)
		aa	would distribute melanin as solid black	

Figure 13-17   Gene interactions

Two genes interact in the inheritance of fur color in mice.
(a) To produce any melanin pigment at all, a mouse must have at least one dominant allele M for melanin production (MM or Mm genotypes). If the melanin-distribution gene is present, the mouse's fur will be agouti (AA or Aa genotypes) or black (aa genotype). If a mouse is homozygous recessive mm, then it produces no melanin and will have white fur regardless of its genotype for the melanin-distribution gene.
(b) If two mice that are heterozygous for both genes are mated, their offspring will not have a 9:3:3:1 ratio of phenotypes because of the interaction between the genes. In this cross, the phenotypic ratio of offspring is 9 agouti to 3 black to 4 albino. Other types of crosses (for example, a heterozygote mated with a homozygous recessive) yield other modifications of the 9:3:3:1 ratio.

(b)

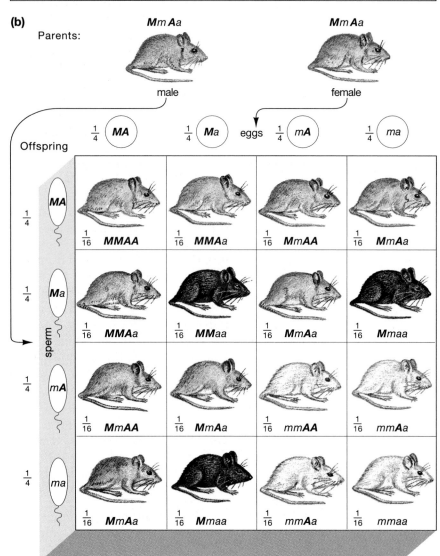

Parents: MmAa (male)   MmAa (female)

Offspring

***Figure 13-18*   Environmental influence on genotype**

A simple case of interaction between genotype and environment is the expression of the gene for black fur in the Himalayan rabbit. Cool areas (nose, ears, and feet) allow expression of the gene for black fur.

A striking example of environmental effects on gene action occurs in the Himalayan rabbit, which, like the Siamese cat, has pale body fur but black ears, nose, tail, and feet (Fig. 13-18). The Himalayan rabbit actually has the genotype for black fur all over its body. The enzyme that produces the black pigment, however, is temperature sensitive; above about 34° C (93° F), the enzyme is inactive. At the temperatures typical of rabbit hutches, extremities such as the ears and feet are cooler than the rest of the body, and black pigment can be produced there. The main body surface is warmer than 34° C, so this fur is pale.

Most environmental influences are more complicated and subtle. The interactions between complex genetic systems and varied environmental conditions can create a continuum of phenotypes that defies analysis into genetic and environmental components. This is particularly true of human characteristics. Genetic analysis of all but the simplest human traits is notoriously difficult: The human generation time is long, the number of offspring per couple is small, and in any case one can't very well kidnap a few thousand people and keep them for genetics experiments. Add to these factors the countless subtle ways in which people respond to their environments, and you can see that a precise determination of the genetic bases of complex traits such as intelligence or musical ability is, at best, extraordinarily difficult.

The Molecular Basis of Mendelian Genetics

You have now learned about the molecular basis of inheritance, the flow of information from DNA to protein, and the patterns of inheritance that can be observed in the traits of whole organisms. What is the connection among these facets of genetics? Although the precise molecular processes underlying most traits are still unknown, it has become clear that the various patterns of inheritance are outgrowths of the transcription and translation into protein of information stored in DNA. Here, we will focus on a few simple cases in which the molecular basis of relationships among alleles is fairly clear.

A Dominant Allele Usually Codes for a Functional Enzyme; a Recessive Allele Usually Codes for a Nonfunctional Enzyme

Consider the situation in which a single gene controls the expression of a single trait. There are two different alleles of the gene, each producing a different form of the trait, and one allele is completely dominant to the other allele. To see how such a system might work on the molecular level, let's look at the inheritance of body fat color in rabbits.

In domestic rabbits, body fat is usually white, but rabbits that are homozygous for a recessive allele have yellow fat. Why? As you know, rabbits eat plants. Most plants contain a yellow pigment called xanthophyll, which is fat soluble. If not broken down, xanthophyll dissolves in the rabbit's fat, coloring it yellow. The xanthophyll-metabolizing allele directs the synthesis of an enzyme that breaks down xanthophyll to a colorless compound. Rabbits that are homozygous for the xanthophyll-metabolizing allele synthesize more than enough functional enzyme to break down all of the xanthophyll in their diet, so these rabbits have white fat. The yellow-fat allele is a mutation that renders the xanthophyll-metabolizing enzyme nonfunctional. In rabbits with one xanthophyll-metabolizing allele and one mutant allele, the xanthophyll-metabolizing allele directs the synthesis of enough functional enzyme to break down dietary xanthophyll completely. Therefore, the body fat of heterozygotes will still be white. However, in rabbits homozygous for the mutant allele, there are no functional xanthophyll-metabolizing enzymes. The xanthophyll dissolves in the fat, coloring it yellow.

In general, dominant alleles direct the synthesis of functional enzymes and recessive alleles direct the synthesis of nonfunctional enzymes. An organism with one dominant allele synthesizes enough enzyme to produce a phenotype that is indistinguishable from the phenotype of organisms with two dominant alleles. This is an example of the protective effect of diploidy mentioned in Chapter 12: A functional allele on one homologous chromosome may completely compensate for a defective allele on the other homologue.

What about incomplete dominance? The simplest form of incomplete dominance is very similar to complete dominance: One allele directs the synthesis of a functional enzyme, and the other allele directs the synthesis of a nonfunctional enzyme. The difference between complete and incomplete dominance results from the difference in effects of one versus two copies of the functional allele. In incomplete dominance, the amount of enzyme synthesized under the direction of a single functional allele is not enough to catalyze all the reactions that would be needed to produce a dominant phenotype. For example, the red allele in snapdragon flowers (see Fig. 13-15) directs the synthesis of an enzyme that produces a red pigment. If a plant has two red alleles, its flowers produce a lot of red pigment and are colored red. If a plant has only one red allele, its flowers produce less red pigment and are colored pink.

Codominance Is Caused by Two or More Alleles, Each of Which Codes for a Different, Functional Enzyme

Mutations may occur anywhere in a gene. Therefore, different organisms may have quite different mutations in the same gene. These different mutations may all produce slightly different, though still functional, proteins, giving rise to multiple alleles of the same gene. In some circumstances, each of these multiple alleles may be detectable phenotypically, as is the case with the human ABO blood groups (see Fig. 13-16). The I^A and I^B alleles both code for slightly different, functional enzymes. The I^B enzyme attaches the monosaccharide galactose to a protein on red blood cell membranes, whereas the I^A enzyme attaches a slightly different compound, galactosamine. The i allele is a mutation producing a nonfunctional enzyme that can't attach anything to the red blood cell proteins. People with $I^B I^B$ or $I^B i$ genotypes have red blood cell proteins with galactose attached and have type B blood. Those with $I^A I^A$ or $I^A i$ genotypes have red blood cell proteins with galactosamine attached and have type A blood. The $I^A I^B$ genotype results in red blood cells with some galactose-bearing proteins and some galactosamine-bearing proteins (type AB blood). Finally, the red blood cells of ii individuals bear neither compound (type O blood). The I^A and I^B alleles are considered codominant because the phenotypes produced by both the I^A and I^B alleles can be detected in people with type AB blood.

Reflections on Genetic Diversity and Human Welfare

The science of genetics, which began in a monastery over a century ago, is critically important to humankind today. Genetics plays a role in many human activities, from possible new sources of energy to inherited human diseases, from paternity lawsuits to crop improvement. Research into crop and livestock improvement represents some of the greatest successes, as well as challenges, for geneticists. Virtually all the corn grown in the United States today, for example, is hybrid corn, the result of crossing parent strains that are themselves usually not very good as a crop. Faced with high prices, rising populations, and increasing demand for food, agricultural geneticists are constantly try-

ing to develop strains of plants or animals that produce more food per unit input of energy, labor, and money.

Scientists take three principal approaches to improving crops and livestock. First, breeders can search for individual plants or animals with the desired characteristics and use only those organisms as breeding stock. The individuals selected for breeding could arise from mutations or from chance recombinations of preexisting alleles. This approach relies on "good" alleles appearing spontaneously and on our ability to recognize them. Second, molecular geneticists can try to create superior genes in the laboratory or transplant them from one species to another. Gene transplantation is rapidly approaching commercial use, but creating superior genes is not yet feasible. The third approach is to look for desirable traits in wild populations of the same or closely related species. Breeders might then cross-breed or otherwise incorporate the desired genetic information into livestock or crop plants. This last approach shows great promise and, in fact, is how many crops and livestock breeds were developed in the first place, thousands of years ago. Wheat, for example, was a cross between varieties of wild grass, aided by irregularities in meiosis that resulted in a polyploid plant producing large edible kernels.

Today, our reserves of wild genes are diminishing at a frightening rate, largely because of the pressures of human population and development. An estimated 8 to 10 million species of organisms exist on Earth, and only about a million and a half of these have even been identified. What's worse, many, perhaps most, of the unidentified species may become extinct before science ever has a chance to study them. Large tracts of wilderness, both in the United States and in other countries, are diminishing rapidly. When these ecosystems go, countless species and varieties go with them.

Rare species and local varieties of otherwise widespread species have been important in crop development in the past and promise to be just as important in the future. In the 1960s, for example, plant geneticists cross-bred commercial strawberries with a wild variety growing in Cottonwood Canyon, Utah. The result: strawberries that produce fruit year-round. More recently, Florida breeders have developed a heat-tolerant blueberry bush by cross-breeding domestic blueberries with the rabbit-eye blueberry of the South. Geneticists at Oregon State University are developing a wildflower called meadowfoam into a source of high-quality commercial oil for the pharmaceutical and electronics industries. One of the most useful meadowfoam species is found in a 6-square-mile area near Medford, Oregon, and nowhere else in the world.

This is an important, but little recognized, reason to preserve wilderness: to preserve the genes of the plants and animals that live in them. This is also an important goal of efforts to save endangered species, because once a species becomes extinct, its genes are lost forever. Genes, with all their various alleles, have evolved over hundreds of millions of years and represent one of our most valuable and irreplaceable natural resources.

■ S U M M A R Y   O F   K E Y   C O N C E P T S

Gregor Mendel and the Origin of Genetics

Homologous chromosomes carry the same genes located at the same loci, but the genes at a particular locus can exist in alternate forms called alleles. An organism whose homologous chromosomes carry the same allele at a given locus is homozygous for that particular gene. If the alleles differ, the organism is heterozygous. Gregor Mendel deduced many principles of inheritance in the mid-1800s, before the discovery of DNA, genes, chromosomes, or meiosis. He did this by choosing an appropriate experimental subject, designing his experiments carefully, and analyzing his data accurately.

Inheritance of Single Traits: The Law of Segregation

Breeding experiments that deal with the inheritance of a single trait at a time are called monohybrid crosses. The inheritance of each individual trait is determined by discrete physical units called genes. Each organism possesses a pair of similar, but not necessarily identical, genes that influence each trait. However, only one of each pair of genes is included in any individual gamete (the law of segregation). An offspring formed by the fusion of two gametes therefore receives pairs of genes, one of each pair inherited from each parent. Each gene may exist in alternative forms called alleles. Different alleles cause different forms of the trait (for example, purple or white flowers). If an individual possesses two different alleles of the same gene, one allele, called dominant, may completely mask the expression of the other, recessive allele. (For example, a pea with alleles both for purple and white flower color will have purple flowers.) In individuals with two different alleles of the same gene, which allele is included in any given gamete is determined by chance. Therefore, we can predict the relative proportions of offspring through the laws of probability.

The physical appearance of an organism (its phenotype) may not always be a reliable indicator of its alleles (the genotype) because of the masking of recessive alleles by dominant alleles. Organisms with two dominant alleles (homozygous dominant) have the same phenotype as organisms with one dominant and one recessive allele (heterozygous).

Inheritance of Multiple Traits: Independent Assortment

Genes are parts of chromosomes. The F_2 generation from a dihybrid cross (parental organisms differing in two traits) may have two fundamentally different outcomes:

If the genes for the two traits are on different chromosomes, then the F_2 offspring will appear in four different phenotypes, resulting from the independent assortment of the chromosomes (and hence the alleles) during meiosis. This is Mendel's law of independent assortment.

Inheritance Patterns of Genes Located on the Same Chromosome

If the genes are found on the same chromosome, they are said to be linked. In the absence of crossing over, the F_2 offspring will express only the two parental phenotypes. If crossing over does occur, then some of the F_2 offspring will express recombined phenotypes. The frequency of crossing over depends on the distance between the genes on the chromosome: the farther apart the genes, the higher the frequency of crossing over.

Inheritance of Sex and Sex-Linked Genes

Sex linkage is a special and easily observed case of linkage of traits on the same chromosome. In many animals, females have two X chromosomes, whereas males have one X and one Y chromosome, with many fewer genes. Consequently, males have only one copy of most of the X chromosome genes, and recessive traits are more likely to be phenotypically expressed in males.

Variations on the Mendelian Theme

Not all inheritance follows the simple dominant-recessive pattern.

1. In incomplete dominance, heterozygotes have a phenotype intermediate between the two homozygous phenotypes.
2. Codominant alleles, such as those determining blood type, are both phenotypically detectable in heterozygotes. The heterozygotes are not intermediate in phenotype between the two parents, but have separately distinguishable features of both parental types.
3. Many traits are determined by several genes with similar actions, a phenomenon called polygenic inheritance. Traits that appear to exist in a continuum of finely graded forms are often determined polygenically.
4. The expression of some genes may depend on, or be modified by, the expression of other genes (epistasis).
5. Many genes have multiple phenotypic effects (pleiotropy).
6. The environment plays at least some role in the phenotypic expression of all traits.

The Molecular Basis of Mendelian Genetics

The observed patterns of inheritance in whole organisms arise from and can be explained on the molecular level. Different alleles of a given gene are slightly different nucleotide sequences of DNA that produce differences in proteins. A dominant allele usually codes for a functional enzyme, whereas a recessive allele usually codes for a nonfunctional enzyme. An organism with a single, completely dominant allele synthesizes enough enzyme to produce a phenotype indistinguishable from the phenotype of organisms with two dominant alleles. Codominant alleles both code for functional, but slightly different, enzymes that produce distinguishably different phenotypes.

KEY TERMS

allele   p. 224
autosome   p. 234
chiasma   p. 231
codominance   p. 238
cross-fertilization   p. 226
crossing over   p. 231
dihybrid cross   p. 229
dominant   p. 227
epistasis   p. 239
F_1 generation   p. 226
F_2 generation   p. 226
F_3 generation   p. 226
gene   p. 223

genetic recombination   p. 234
genotype   p. 228
heterozygous   p. 224
homozygous   p. 224
hybrid   p. 226
incomplete dominance   p. 237
law of independent assortment   p. 230
law of segregation   p. 227
linkage   p. 231
locus   p. 224
monohybrid cross   p. 226
multiple alleles   p. 238
parental generation   p. 226

phenotype   p. 228
pleiotropy   p. 239
polygenic inheritance   p. 239
quantitative inheritance   p. 239
recessive   p. 227
self-fertilization   p. 225
sex chromosome   p. 234
sex linkage   p. 235
sexual recombination   p. 234
test cross   p. 228
true-breeding   p. 225
X chromosome   p. 234
Y chromosome   p. 234

THINKING THROUGH THE CONCEPTS

Multiple Choice

1. An organism is described as *Rr* : red. The *Rr* is the organism's [A]; red is the organism's [B]; and the organism is [C].
   a. [A] phenotype; [B] genotype; [C] degenerate
   b. [A] karyotype; [B] hybrid; [C] recessive
   c. [A] genotype; [B] phenotype; [C] heterozygous
   d. [A] gamete; [B] linkage; [C] pleiotropic
   e. [A] zygote; [B] epistasis; [C] homozygous

2. The 9:3:3:1 ratio is a ratio of
   a. phenotypes in a test cross
   b. phenotypes in a monohybrid cross
   c. phenotypes in a dihybrid cross
   d. genotypes in a monohybrid cross
   e. genotypes in a dihybrid cross

3. A test cross is used to
   a. demonstrate whether two genes are linked
   b. test whether an organism expressing the dominant phenotype is homozygous or heterozygous
   c. show whether inheritance is polygenic or Mendelian
   d. determine whether a trait is sex-linked or autosomal
   e. find out whether the trait is characterized by incomplete dominance

4. Which of the following statements rewords Mendel's first law in terms of what you know about genetics and meiosis?
   a. Genes are arranged on sex chromosomes or autosomes.
   b. If two genes are on two different chromosomes, then how alleles of one gene distribute into gametes does not influence how alleles of the other gene distribute.
   c. If two genes are on the same chromosome, they are linked.
   d. Two alleles of a gene do not mix with each other while they are in a heterozygous organism; they can appear unaltered in subsequent generations.
   e. There would be no crossing over between X and Y chromosomes.

5. A man's lawyer tells him that he cannot use blood type in a paternity suit against him because the child could, in fact, be his according to blood type. Which of the following is the only possible combination supporting this hypothetical circumstance? (Answers in the order: mother:father:child)
   a. A:B:O   b. A:O:B

   c. AB:A:O   d. AB:O:AB
   e. B:O:A

6. A heterozygous red-eyed female *Drosophila* mated with a white-eyed male would produce
   a. red-eyed females and white-eyed males in the F_1
   b. white-eyed females and red-eyed males in the F_1
   c. half red- and half white-eyed females and all white-eyed males in the F_1
   d. all white-eyed females and half red- and half white-eyed males in the F_1
   e. half red- and half white-eyed females as well as males in the F_1

Review Questions

1. Define the following terms: gene, allele, dominant, recessive, monohybrid, dihybrid, true-breeding, homozygous, heterozygous, cross-fertilization, self-fertilization.

2. Explain the meaning of Mendel's law of segregation and law of independent assortment. Under what circumstances does the law of independent assortment apply? When is it violated?

3. Explain why genes located on one chromosome are linked during inheritance. Why do linked genes sometimes separate during meiosis?

4. Explain why human skin color does not occur in just two forms, e.g., black and white.

5. What is sex linkage? In mammals, which sex would be most likely to show recessive sex-linked traits?

6. What is the difference between a phenotype and a genotype? Does knowledge of an organism's phenotype always allow you to determine the genotype? What type of experiment would you perform to determine the genotype of a phenotypically dominant individual?

7. Define polygenic inheritance, epistasis, and pleiotropy. Describe an example of each.

8. If smooth versus wrinkled and yellow versus green pea seeds had happened to be close to each other on the same chromosome, what would Mendel have found?

9. In the white body with black extremities of the Himalayan rabbit, where does the enzyme work? Where doesn't it work? Can you think of an explanation for this?

✖ APPLYING THE CONCEPTS

1. There are some groups of insects in which females are XX and males are XO (the Y chromosome has not evolved in these species). In some other groups of animals, including moths, butterflies, and birds, females are ZW and males are ZZ for their sex chromosomes. Regarding chromosomal sex determination, discuss several basic ways in which these groups differ from mammals, in which females are XX and males are XY.

2. There are many hypotheses, but as yet no definitive answers, to the question, "Why does the Y chromosome carry so few genes?" Discuss the disadvantages of a Y chromosome lacking most of the genes carried on the X chromosome in the light of your knowledge of dominance and recessiveness, and the protective effect of diploidy.

3. Sometimes the term *gene* is used rather casually. Compare and contrast use of the terms *allele* and *locus* as alternatives to *gene*.

4. As you read in this chapter, a person with type B blood has antibodies that would cause blood cells of a type A transfusion to clump. Type A people have antibodies that attack type B glycoproteins, type O people have antibodies that attack both glycoproteins, and type AB people do not have antibodies that attack the blood group glycoproteins. Using this information along with the information in the chapter on which glycoproteins are in which blood group, explain why AB individuals are referred to as "universal recipients" in terms of blood transfusions and why people with type O blood are "universal donors."

5. Mendel's numbers seemed almost too perfect to be real; some believe he may have cheated a bit on his data. Perhaps he continued to collect it until the numbers matched his predicted ratios, then stopped. Recently, there has been much publicity over violations of scientific ethics, including researchers' plagiarizing others' work, using other scientists' methods to develop lucrative patents, or just plain fabricating data. How important an issue is this for society? What are the boundary lines of ethical scientific behavior? How should the scientific community or society "police" scientists? What punishments would be appropriate for violators of scientific ethics?

6. Although American society has often been described as a "melting pot," people often engage in "assortative mating," in which they marry others of similar height, socioeconomic status, race, and IQ. Discuss the consequences to society of assortative mating among humans. Would society be better off if people mated more randomly? Discuss why or why not.

✖ GENETICS PROBLEMS

(*Note: An extensive group of genetics problems, with answers, can be found in the Study Guide.*)

1. In certain cattle, hair color can be red (homozygous *RR*), white (homozygous *R'R'*), or roan (a mixture of red and white hairs; heterozygous *RR'*).
   a. When a red bull is mated to a white cow, what genotypes and phenotypes of offspring could be obtained?
   b. If one of these offspring were mated to a white cow, what genotypes and phenotypes of offspring could be produced? In what proportion?

2. The palomino horse is golden in color. Unfortunately for horse fanciers, palominos do not breed true. In a series of matings between palominos, the following offspring were obtained:

    65 palominos, 32 cream-colored,
    34 chestnut (reddish brown)

   What is the probable mode of inheritance of palomino coloration?

3. In the edible pea, tall (*T*) is dominant to short (*t*), and green pods (*G*) are dominant to yellow pods (*g*). List the types of gametes and offspring that would be produced in the following crosses:
   a. *TtGg* × *TtGg*
   b. *TtGg* × *TTGG*
   c. *TtGg* × *Ttgg*

4. In tomatoes, round fruit (*R*) is dominant to long fruit (*r*), and smooth skin (*S*) is dominant to fuzzy skin (*s*). A true-breeding round, smooth tomato (*RRSS*) was cross-bred with a true-breeding long, fuzzy tomato (*rrss*). All the F$_1$ offspring were round and smooth (*RrSs*). When these F$_1$ plants were bred, the following F$_2$ generation was obtained:

    Round, smooth: 43
    Long, fuzzy: 13

   Are the genes for skin texture and fruit shape likely to be on the same or on different chromosomes? Explain your answer.

5. In the tomatoes of question 4, an F$_1$ offspring (*RrSs*) was mated with a homozygous recessive (*rrss*). The following offspring were obtained:

    Round, smooth: 583
    Long, fuzzy: 602
    Round, fuzzy: 21
    Long, smooth: 16

   What is the most likely explanation for this distribution of phenotypes?

6. In humans, hair color is controlled by two interacting genes. The same pigment, melanin, is present in both brown-haired and blond-haired people, but brown hair has much more of it. Brown hair (*B*) is dominant to blond (*b*). Whether any melanin can be synthesized at all depends on another gene. The dominant form (*M*) allows melanin synthesis; the recessive form (*m*) prevents melanin synthesis. Homozygous recessives *mm* are albino. What will be the expected proportions of phenotypes in the children of the following parents?
   a. *BBMM* × *BbMm*
   b. *BbMm* × *BbMm*
   c. *BbMm* × *bbmm*

7. In humans, one of the genes determining color vision is located on the X chromosome. The dominant form (*C*) pro-

duces normal color vision; red-green color blindness (c) is recessive. If a man with normal color vision marries a colorblind woman, what is the probability of their having a colorblind son? A colorblind daughter?

8. In the couple described in question 7, the woman gives birth to a colorblind daughter. The husband sues for a divorce, on the grounds of adultery. Will his case stand up in court? Explain your answer.

✖ A N S W E R S   T O   G E N E T I C S   P R O B L E M S

1. a. A red bull (RR) is mated to a white cow ($R´R´$). The bull will produce all R sperm, while the cow will produce all $R´$ eggs. All the offspring will be $RR´$, and will have roan hair (codominance).

   b. A roan bull ($RR´$) is mated to a white cow ($R´R´$). The bull produces half R and half $R´$ sperm, while the cow produces $R´$ eggs. Using the Punnett square method:

	eggs
	$R´$
sperm R	$RR´$
$R´$	$R´R´$

Using probabilities:

sperm	egg	offspring
½ R	$R´$	½ $RR´$
½ $R´$	$R´$	½ $R´R´$

   The predicted offspring will be ½ $RR´$ (roan) and ½ $R´R´$ (white).

2. The offspring occur in three types, classifiable as dark (chestnut), light (cream), and intermediate (palomino). This distribution suggests incomplete dominance, with the alleles for chestnut (C) combining with the allele for cream ($C´$) to produce palomino heterozygotes ($CC´$). We can test this hypothesis by examining the offspring numbers. There are approximately ¼ chestnut (CC), ½ palomino ($CC´$), and ¼ cream ($C´C´$). If palominos are heterozygotes, we would expect the cross $CC´ \times CC´$ to yield ¼ CC, ½ $CC´$, and ¼ $C´C´$. Our hypothesis is supported.

3. a. $TtGg \times TtGg$. This is a "standard" dihybrid cross. Both parents produce TG, Tg, tG, and tg gametes. The expected proportions of offspring are ⁹⁄₁₆ tall green, ³⁄₁₆ tall yellow, ³⁄₁₆ short green, ¹⁄₁₆ short yellow.

   b. $TtGg \times TTGG$. In this cross, the heterozygous parent produces TG, Tg, tG, and tg gametes. However, the homozygous dominant parent can only produce TG gametes. Therefore, all offspring will receive at least one T allele for tallness and one G allele for green pods, and thus all the offspring will be tall with green pods.

   c. $TtGg \times Ttgg$. The second parent will produce two types of gametes, Tg and tg. Using a Punnett square:

	eggs	
	Tg	tg
sperm TG	$TTGg$	$TtGg$
Tg	$TTgg$	$Ttgg$
tG	$TtGg$	$ttGg$
tg	$Ttgg$	$ttgg$

   The expected proportions of offspring are ⅜ tall green, ⅜ tall yellow, ⅛ short green, ⅛ short yellow.

4. If the genes are on separate chromosomes—that is, assort independently—then this would be a typical dihybrid cross

with expected offspring of all four types (approximately ⁹⁄₁₆ round smooth, ³⁄₁₆ round fuzzy, ³⁄₁₆ long smooth, and ¹⁄₁₆ long fuzzy). However, only the parental combinations show up in the F₂ offspring, indicating that the genes are on the same chromosome.

5. The genes are on the same chromosome and are quite close together. On rare occasions, crossing over occurs between the two genes, producing recombination of the alleles.

6. a. $BBMM$ (brown) $\times BbMm$ (brown). The first parent can produce only BM gametes, so all offspring will receive at least one dominant allele for each gene. Therefore, all offspring will have brown hair.

   b. $BbMm$ (brown) $\times BbMm$ (brown). Both parents can produce four types of gametes: BM, Bm, bM, and bm. Filling in the Punnett square:

	eggs			
	BM	Bm	bM	bm
sperm BM	$BBMM$	$BBMm$	$BbMM$	$BbMm$
Bm	$BBMm$	$BBmm$	$BbMm$	$Bbmm$
bM	$BbMM$	$BbMm$	$bbMM$	$bbMm$
bm	$BbMm$	$Bbmm$	$bbMm$	$bbmm$

   Remembering that all mm offspring are albino, the expected proportions are ⁹⁄₁₆ brown-haired, ³⁄₁₆ blond-haired, ⁴⁄₁₆ albino.

   c. $BbMm$ (brown) $\times bbmm$ (albino):

	eggs
	bm
sperm BM	$BbMm$
Bm	$Bbmm$
bM	$bbMm$
bm	$bbmm$

   The expected proportions of offspring are ¼ brown-haired, ¼ blond-haired, ½ albino.

7. A man with normal color vision is CY (remember, the Y chromosome does not have the gene for color vision). His colorblind wife is cc. Their expected offspring will be:

	eggs
	c
sperm C	Cc
Y	cY

   We therefore expect that all the daughters will have normal color vision, while all the sons will be colorblind.

8. The husband should win his case. All his daughters must receive one X chromosome, with the C allele, from him, and therefore should have normal color vision. If his wife gives birth to a colorblind daughter, her husband cannot be the father (unless there was a new mutation for color blindness in his sperm line, which is very unlikely).

✖ FOR MORE INFORMATION

Benzer, S. "Genetic Dissection of Behavior." *Scientific American*, December 1973. Not only physical traits, but behaviors too are under the influence of genes. The fruit fly once again proves a useful model system in which to study inheritance.

Crow, J. F. "Genes That Violate Mendel's Rules." *Scientific American*, February 1979. A close look at genetic recombination and its role in evolution.

Fisher, R. A. "Has Mendel's Work Been Rediscovered?" *Annals of Science*, 1:115–137, 1936. Reprinted on pages 139–172 of Stern and Sherwood's *The Origin of Genetics*, a famous geneticist and statistician argues that Mendel's ratios are too good to be true. Was Mendel's approach to science high on insight and low on honesty in data collection?

Hoyt, E. "Wild Relatives." *Living Wilderness*, Summer 1990. Undisturbed ecosystems are a valuable gene bank with great potential for improving crop plants.

Sapienza, C. "Parental Imprinting of Genes." *Scientific American*, October 1990. It is not quite true that all genes are equal, regardless of whether they have been inherited from mother or father. In some cases, which parent a gene comes from greatly alters its expression in the offspring.

Stern, C., and Sherwood, E. R. *The Origin of Genetics: A Mendel Source Book*. San Francisco: W. H. Freeman, 1966. There is no substitute for the real thing, in this case a translation of Mendel's original paper to the Brünn Society.

Wright, S. "Mendel's Ratios." Also in Stern and Sherwood's *The Origin of Genetics: A Mendel Source Book*, pages 173–175, another famous geneticist takes issue with Fisher's accusation against Mendel.

NET WATCH

On-line resources for this chapter are on the World Wide Web at:
http://www.prenhall.com/~audesirk (click on the table of contents link and then select Chapter 13).

14 Molecular Genetics and Biotechnology

"In the year 6565, ain't gonna need no husband, won't need no wife.

You'll pick your son, pick your daughter too

From the bottom of a long glass tube. Woh-woh"

From the song *"In the Year 2525"* by RICK EVANS

A plant that glows in the dark! Molecular genetic techniques have been used to insert a gene from a firefly coding for the enzyme luciferase into this tobacco plant. This enzyme breaks down the chemical luciferin, releasing light in the process. This tobacco plant has been provided with water containing luciferin. The glow provides evidence that the plant's cell are expressing the firefly gene.

Biotechnology—virtually every week the newspaper features articles about the potential health benefits of some new medical advance in the works, based on biotechnology. Almost every state and large city in the United States tries frantically to lure "biotech" companies that promise high-paying, low-pollution jobs. Biotechnology often seems to be portrayed as both the medical and the economic miracle of our times. Yet furious debate rages about whether milk from cows treated with a recombinant growth hormone should be sold in supermarkets. What exactly *is* biotechnology? Is it boon or bane or both?

Biotechnology is the manipulation of the molecular basis of inheritance by methods collectively called **recombinant DNA technology**. Biotechnology is usually practiced to achieve one or more of three goals:

1. To understand more about the processes of inheritance and gene expression
2. To provide better understanding and treatment of various diseases, particularly genetic disorders
3. To generate economic benefits, including improved plants and animals for agriculture and efficient production of valuable biological molecules

The features of biotechnology that provide both vast promise and potential threat are the specificity with which biotechnology can, or will soon be able to, direct genetic changes; the speed with which genetic changes can be made; and the ability to transfer genetic material between species.

This chapter has three major themes. First, we look at a few important recombinant DNA technologies, briefly examining some of the methods that molecular biologists use to manipulate genes. Second, we discuss a few of the applications, both real and potential, of biotechnology. Third, we explore some of the practical and ethical issues that the use of biotechnology may raise.

Before we begin our discussion of human-directed biotechnology, however, it will be useful to provide a bit of background on what we might call "natural biotechnology." Let us start, then, with a survey of some naturally occurring methods of DNA recombination.

DNA Recombination in Nature

DNA recombination, whether by natural events or human intervention, involves two distinct processes: (1) changing the nucleotide composition of the DNA of a single cell, a few cells, or an entire organism; (2) selecting valuable DNA combinations.

DNA Recombination Occurs Naturally through Processes Such as Sexual Reproduction, Bacterial Transformation, and Viral Infection

You are already familiar with two major methods of "recombinant DNA," although we called them "genetic recombination" and "sexual recombination." Genetic recombination occurs through crossing over during meiosis I, when genes from a maternal chromosome and a paternal chromosome produce a chromosome with a new combination of alleles that may never have occurred before. In sexual recombination, chromosomes from two separate organisms combine to produce offspring that often are genetically unique. These natural recombinations, of course, usually occur within a single species. Many people have a tendency to consider single-species recombination as "natural" and therefore good, whereas they see between-species recombinations performed in the lab as "unnatural" and possibly intrinsically bad. However, between-species recombinations occur in nature, too.

Transformation May Combine DNA from Different Bacterial Species

Bacteria employ several methods of recombination that allow gene transfer between unrelated species. As you know, transformation allows bacteria to pick up free DNA from the environment (see Chapter 10). Sometimes, the free DNA acquired during transformation is part of the chromosome of another bacterium, including DNA from a bacterium of another species.

Transformation may also occur when bacteria pick up tiny circles of DNA called **plasmids** (Fig. 14-1). Plasmids, which range in size from about 1000 to 100,000 nucleotides, are self-replicating "parasites" normally found in the cytoplasm of many bacteria. A single bacterium may contain dozens or even hundreds of copies of a plasmid. Although the bacterium's "own" chromosome contains all the genes that it normally needs for survival, the genes carried by plasmids may be useful, too. For example, some plasmids contain genes that code for enzymes that digest certain antibiotics, such as penicillin. In environments where exposure to antibiotics is high, such as hospitals, these plasmids obviously are extremely valuable to their bacterial hosts. When a plasmid-containing bacterium dies, its plasmids may be liberated into the environment and transform other living bacteria. Plasmids may also be exchanged between living bacteria. The acquisition or exchange of plasmids is probably the most common form of bacterial transformation.

Figure 14-1   A bacterial plasmid

These tiny rings of DNA, only a few micrometers in diameter, have become important tools in recombinant DNA research.

Viruses May Transfer DNA among Eukaryotic Species

Recent evidence suggests that viruses sometimes transfer genes among eukaryotic organisms. The DNA of certain viruses can insert itself into a chromosome of its eukaryotic host and exist there quietly for days, months, or even years. Then, perhaps in response to a stimulus that is stressful to the host, the viral DNA leaves the chromosome, occasionally taking a bit of the eukaryotic DNA along with it. The viral DNA then takes over the host cell metabolism, replicates itself, and directs the synthesis of new viruses. The offspring viruses thus may include some host DNA. If one of these offspring infects a new host of a different species, it may insert itself, along with the piece of DNA from the former host, into a chromosome of its new host. In this way, the new host may acquire some genes that originally belonged to an unrelated species.

Recombination Provides Raw Material for Evolution

Recombination, by whatever mechanism, changes the genetic makeup of organisms. As is the case with mutations, natural recombination is random and undirected—a bacterium, for example, does not deliberately take up a plasmid that contains an antibiotic-resistance gene so that it can flourish in hospitalized patients. Naturally occurring DNA recombinations are tested, just as mutations are, by natural selection. Probably most recombinations prove to be harmful or neutral. A few, however, prove helpful in a particular environment. The organisms with the new DNA combinations thrive and pass on the new combinations to their offspring.

There Are Both Similarities and Differences between Recombination in Nature and in the Laboratory

Before we begin our discussion of recombinant DNA technologies, let's compare natural and laboratory DNA recombination.

1. Both natural and laboratory DNA recombination involve exchange of DNA between organisms, including between species.
2. Naturally occurring DNA recombinations are relatively random and undirected; in general, specific genes are not preferentially moved, and there is no "goal" that drives DNA movements. In laboratory DNA recombinations, specific pieces of DNA are moved between deliberately chosen organisms to achieve a specific goal.
3. The "usefulness" of naturally occurring DNA recombinations is determined by natural selection. Human interests determine the usefulness of laboratory DNA recombinations.

Recombinant DNA Technology

Within the past few years, the technologies and applications of recombinant DNA have mushroomed. Rather than attempting to summarize all the technologies, we will follow the sequence of procedures that might be used to solve a particular problem or produce a specific product. Although many other techniques are also commonly used, one logical sequence of activities would be the following: (1) produce a "DNA library" of an organism; (2) identify individual genes of interest; (3) produce a copy or (preferably) many copies of the gene; and (4) insert the gene into the desired organism and regulate the expression of the gene in a useful way.

A DNA Library Consists of DNA from a Particular Organism Inserted into Bacterial Plasmids

The first task in recombinant DNA technology is to produce a **DNA library**—a readily accessible, easily duplicable assemblage of all the DNA of a particular organism. Why build a DNA library? Think about your college library. It wouldn't be very useful if all the books were scattered about in random piles on the floor. Only consistent filing and a user-friendly cataloguing system allow you to find the books you need. To a geneticist, the chromosomes in a living cell are much like books in a pile: All the genes of the organism are there, but they are incredibly difficult to find and study. (The cell, of course, has its own "filing system," but geneticists are only beginning to decipher it.) A DNA library organizes the DNA in a way that researchers can use. As we will explain below, a DNA library also allows "molecular photocopying" of the DNA, so that researchers can easily obtain the thousands to billions of gene copies that they need for their experiments. Restriction enzymes, plasmids, and bacteria provide mol-

ecular geneticists with the filing systems, catalogues, and shelf space they need for their DNA libraries.

Restriction Enzymes Cut DNA at Specific Nucleotide Sequences

Many bacteria produce **restriction enzymes** that sever DNA at particular nucleotide sequences. In nature, restriction enzymes defend bacteria against bacteriophage invasion by cutting apart the phage DNA. The host bacteria protect their own DNA against being cut by the enzymes, probably by attaching methyl groups to some of the DNA bases. Restriction enzymes thus "restrict" phage infections to the types of phage whose DNA is not cut apart by the enzymes.

As shown in Figure 14-2, many restriction enzymes sever palindromic sections of DNA, which read the same in one direction on one strand as they read in the reverse direction on the other strand (a palindrome in English is a word that reads the same forward and backward, such as *madam*). Furthermore, the DNA is cut between the same two bases on the two strands, in this case between guanine and adenine. For the enzyme illustrated in Figure 14-2, this cut results in two pieces of DNA, one with a single-stranded end reading TTAA and one with a single-stranded end reading AATT. Complementary DNA regions like these, even from different organisms, can pair up, held together by hydrogen bonds between the bases. If the appropriate DNA repair enzymes are present, the two pieces can be rejoined.

Figure 14-2 Restriction enzymes sever double-stranded DNA at specific nucleotide sequences

Many of the most useful restriction enzymes, such as the one called Eco RI illustrated here, sever DNA at palindromes, where the nucleotide sequence reads the same on one DNA strand (GAATTC) as it does in the opposite direction on the other strand. When a double-stranded DNA molecule is cut by Eco RI, two short, single-stranded "sticky ends" remain: AATT on one strand and TTAA on the other. Because these sticky ends are complementary, they can base pair, temporarily holding these strands, or any strands with the same sticky-end sequences, together again (see Fig. 14-3).

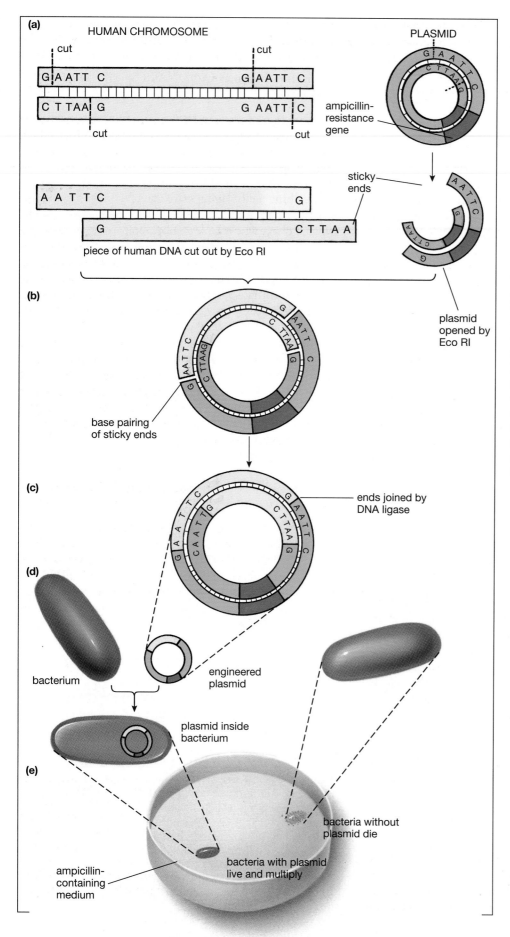

(a) HUMAN CHROMOSOME   PLASMID

cut   cut

G A ATT C   G AATT C

C T TAA G   G A ATT C

cut   cut

ampicillin-resistance gene

sticky ends

A A T T C   G

G   C T T A A

piece of human DNA cut out by Eco RI

plasmid opened by Eco RI

(b)

base pairing of sticky ends

(c)

ends joined by DNA ligase

(d)

bacterium

engineered plasmid

plasmid inside bacterium

(e)

bacteria without plasmid die

ampicillin-containing medium

bacteria with plasmid live and multiply

Figure 14-3   **The use of restriction enzymes in building a human DNA library**

(a) Human chromosomes and bacterial plasmids (with the ampicillin-resistance gene) are both cut with the Eco RI restriction enzyme. Both types of DNA therefore have AATT and TTAA sticky ends. **(b)** The severed human and plasmid DNA molecules are mixed. Plasmid-human combinations are temporarily held together by complementary base pairing of sticky ends. **(c)** DNA ligase bonds the backbones of the plasmid-human combinations together. **(d)** The new plasmid-human rings are mixed with bacteria that have been treated with calcium salts to make the bacteria permeable to DNA. Some bacteria take up a plasmid-human ring. **(e)** Bacteria are placed on culture dishes filled with medium containing the antibiotic ampicillin. Bacteria without plasmids succumb to the ampicillin; those with plasmids survive. Plasmids within the bacteria constitute a human DNA library.

Many restriction enzymes have been isolated from various species of bacteria. Each enzyme cuts DNA apart at a different nucleotide sequence. The specificity and variety of restriction enzymes have enabled molecular geneticists to identify and isolate segments of DNA from many organisms, including humans.

Restriction Enzymes Are Used to Insert DNA into Plasmids to Build a DNA Library

Suppose that DNA is isolated from a human source, say white blood cells, and is cut apart with the restriction enzyme Eco RI (Fig. 14-3a). Everywhere that the GAATTC-CTTAAG pairing occurs, the human DNA will be severed, leaving single-stranded AATT and TTAA "sticky ends" protruding. The next step is to isolate many copies of a bacterial plasmid that includes an easily identifiable "marker gene," such as a gene that confers resistance to an antibiotic called ampicillin (we'll see why this marker gene is useful in a minute). The plasmids are then exposed to the same restriction enzyme. The plasmids, too, will be cut open, with AATT and TTAA sticky ends protruding.

Now the human DNA fragments and opened plasmids are mixed together. The sticky ends will hydrogen-bond, forming human-human, plasmid-plasmid, and, most important, plasmid-human DNA combinations (Fig. 14-3b). DNA ligase enzymes are then added. DNA ligase bonds the sugar-phosphate backbones together, inserting human DNA into plasmids (Fig. 14-3c). Ideally, each plasmid receives only a relatively small piece of human DNA. Millions or billions of plasmids collectively would incorporate DNA from the entire human genome.

The new rings of plasmid-plus-human DNA are mixed with bacteria treated with calcium salts to make them permeable to DNA (Fig. 14-3d). The bacteria take up the plasmid-human DNA. Usually, 100 to 1000 times more bacteria than plasmids are used, so that no individual bacterium ends up with more than one plasmid-human DNA molecule. Of course, this procedure also ensures that most of the bacteria don't have any plasmid-human DNA combinations at all. Here's where the ampicillin-resistance marker gene becomes useful (Fig. 14-3e). The bacteria are placed on a culture dish containing medium with ampicillin. The antibiotic kills off all the bacteria that have not taken up a plasmid. The result: a population of bacteria, all with a plasmid, some with recombined plasmid-human DNA. This constitutes our human DNA library.

The Genes of Interest Must Be Identified in the DNA Library

The human **genome**, consisting of all the genes in an individual, contains about 6 billion nucleotides (3 billion pairs). The average protein has 300 to 400 amino acids, so at three nucleotides per amino acid, only about a thou-sand nucleotides encode the information for a typical protein. Identifying or finding a particular gene within the human genome, then, would appear to be a search for a very small needle within a very large haystack. Fortunately, there are several "tricks" that can be used to identify or locate genes.

It is fairly straightforward to find a gene if you first know the amino acid sequence of its encoded protein (Fig. 14-4). For example, the amino acid sequences of most of the major human hormones are known. From the amino acid sequence, one can then work backward through the genetic code to determine likely DNA sequences (remember, however, that a single amino acid may be encoded by several nucleotide sequences, which complicates things a little). Machines are now commercially available that can synthesize DNA chains from nucleotide sequences typed on a keyboard. The synthesizers cannot make really long chains, but they can accurately synthesize a chain a few dozen nucleotides in length. We'll see what to do with this artificial DNA shortly.

Another way of identifying a gene is to find a cell that you know synthesizes vast amounts of a particular protein. Immature red blood cells, for example, synthesize lots of hemoglobin. That means that they have lots of messenger RNA (mRNA) for hemoglobin. This mRNA is complementary to the DNA of the hemoglobin gene and can also serve to locate the gene.

Gene Probes Are Used to Search the DNA Library for Bacteria Containing Plasmids Bearing the Desired Gene

So far, our DNA library is pretty useless, because it lacks a card file or computer index that would tell us which bacterium contains the plasmid with the gene we want to study. This is where synthetic DNA or isolated mRNA, labeled with radioactive isotopes, comes in. Figure 14-5 explains how these "probes" are used to locate bacteria containing specific genes.

Figure 14-4   Synthetic DNA can be used to locate a gene

If the amino acid sequence of a protein is known, the likely nucleotide sequence of the DNA that encodes the protein can be deduced. The diagram shows DNA sequences that might code for a peptide with the amino acid sequence tryptophan-phenylalanine-methionine-lysine (trp-phe-met-lys). Note that, because phenylalanine and lysine may each be encoded by two DNA triplets, there are four possible DNA sequences for this single peptide.

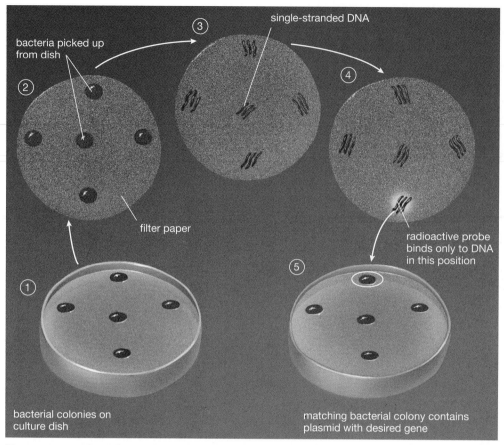

Figure 14-5   **Searching a human DNA library for a specific gene**

① Bacteria from the DNA library are sparsely spread out onto a culture dish. Each bacterium multiplies into a visible colony. If few enough bacteria were originally put in the dish, then each colony should consist of the descendants of a single bacterium, containing in turn a single type of plasmid-human combination. ② A sheet of special filter paper is pressed onto the culture dish. It picks up a few bacteria from each colony, faithfully preserving the colony positions. The original culture dish is saved. ③ The filter paper is placed in a basic solution. This solution breaks open the bacteria, freeing the plasmids, and separates the double-stranded DNA of the plasmids into single strands. ④ The paper is now bathed in solution of neutral pH containing a radioactive synthetic mRNA or DNA "probe" (yellow). The probe hydrogen-bonds only to plasmid DNA that is complementary to the nucleotide sequence of the probe; that is, to plasmids that contain the human gene complementary to the probe. ⑤ The locations of radioactivity on the paper are matched with bacterial colonies on the culture dish. Colonies in the same position consist of bacteria containing plasmids with the desired gene. Samples of these bacteria are cultured in their own dishes.

Selected DNA Sequences in the Library Can Be Amplified

Now that we have located the gene in the library, we can amplify it by making millions or billions of copies for further use, a process called **cloning**. One merely picks bacteria from the appropriate colonies located during the search and grows them in appropriate culture conditions. Biotechnology firms often use huge vats to produce many kilograms of bacteria, all containing a specific human gene (Fig. 14-6).

Another method for making copies of specific stretches of DNA, not only from DNA libraries, but even from single cells, is called the **polymerase chain reaction (PCR)**. In the polymerase chain reaction, billions of copies of selected genes can be made. The polymerase chain reaction has become so important in genetic research and biotechnology that *Science* magazine named the key enzyme involved in the procedure, a DNA polymerase from a specialized microbe, the "Molecule of the Year" in 1989.

Amplified Genes Can Be Used for Research, Industrial, Medical, or Agricultural Purposes

At this point, procedures vary greatly, depending on what one wants to do with the gene. Further, the technologies tend to be quite complex and differ for each application. Therefore, we will discuss only a couple of possibilities, and generally we will describe only the applications, not the methodologies.

Amplified Genes Can Be Used to Produce Commercially Useful Amounts of Valuable Proteins

One prominent application of recombinant DNA is to produce large amounts of human proteins for medical uses. The genes for human and bovine (cow) growth hormone, insulin, blood clotting factors, and enzymes that dissolve blood clots have all been inserted into bacteria or eukaryotic cells for commercial production. The recombinant cells are grown in huge vats. The desired product is extracted from the culture medium, or the cells themselves, and purified.

Why is this process useful? Take the case of human growth hormone. Growth hormone, secreted by the pituitary gland, governs bodily growth, protein metabolism, and perhaps aging. A small percentage of the population does not produce enough growth hormone, remains very short, and ages prematurely. Formerly, the only source of human growth hormone was human pituitaries. Tens of thousands of pituitaries were removed from corpses and processed to purify the growth hormone, which would then be injected into growth hormone–deficient children. This approach, of course, was very expensive. To make matters worse, in 1985 a few recipients died from a rare infection, Creutzfeldt-Jakob disease that had contaminated the hormone preparation.

Meanwhile, a biotechnology company, appropriately called Genentech, inserted the gene for human growth hormone into bacteria and in late 1985 began to sell bacterially produced hormone. Because no human body parts are involved, no human diseases can be transmitted by the Genentech hormone. Similarly, blood clotting factors produced through recombinant DNA technology are now given to hemophiliacs rather than blood transfusions or blood clotting factors isolated from donated human blood. (Unfortunately, this development came too late to save hundreds of hemophiliacs from contracting AIDS from contaminated blood products; see Chapter 24.)

Viral Proteins May Be Synthesized for Use as Vaccines

Presently, vaccines are effective mostly against viral diseases, such as smallpox or measles. The vaccine itself usually consists of killed or weakened virus or of a strain of virus that is genetically incapable of causing disease. The killed, weakened, or genetically defective virus, although not disease causing, is recognized by the immune system. This recognition triggers the production of defenses that would protect against later encounters with disease-causing strains of the same virus (see Chapter 24).

A safer and potentially more effective vaccine might be synthesized using recombinant DNA techniques. The immune system usually recognizes not a whole virus, but one or more proteins on the viral surface. Recombinant DNA technologies might be used to find the genes for the appropriate viral proteins, insert them into bacteria, and produce huge quantities of the protein. A vaccine would then consist of pure protein and would be completely safe, because no living viruses would ever have been involved in its production. Recombinant protein vaccines would be especially attractive in the case of the HIV virus, which causes AIDS, because the virus is so lethal; in fact, one vaccine composed of weakened virus appears to have the potential to cause AIDS, at least in certain people. In late 1994, the World Health Organization decided to begin large-scale clinical trials of two AIDS vaccines based on fragments of one of the HIV surface proteins.

Genes May Be Inserted into or Deleted from Free-Living Organisms

Genetic engineers are inserting and deleting genes from bacteria that they intend to release into the outside world. For example, several laboratories are working to produce bacteria that selectively metabolize toxic substances, from

Figure 14-6   A gene-copying "factory"

A fermenting apparatus at a biotechnology company raises huge numbers of bacteria, each containing a plasmid that incorporates a valuable human gene.

crude oil to hazardous industrial wastes. In the 1980s, one of the first intentional releases of bioengineered bacteria—the so-called ice-minus bacteria—raised a storm of controversy, even though a gene had been cut out of, rather than engineered into, the bacteria (see "Reflections on the Ethics of Biotechnology").

Genes can also be inserted into eukaryotic cells. Under the appropriate circumstances, entire multicellular organisms can be genetically engineered to contain foreign genes. One drawback to this scenario is that many valuable traits may be polygenic—the desirable phenotype may be due to the interaction of several, perhaps many, genes, some of which may not yet be known.

Many genetically engineered plants have already been developed, for instance plants resistant to herbicides. Herbicides, more commonly called weed-killers at the local greenhouse, are used to kill undesirable plants that compete with flowers or crops for water, nutrients, and light. Unfortunately, most herbicides have the potential to harm desirable plants, too, so only limited amounts can be used. Biotechnology can potentially solve this dilemma. A gene that confers resistance to a common herbicide, glyphosate, has been identified and inserted into several types of crop plants. Once inserted, the glyphosate-resistance gene allows the crops to withstand heavier applications of herbicide. Higher concentrations of herbicide kill off more weeds that otherwise compete with the crops. Whether this application is beneficial or not depends on one's views about the safety of herbicides and how likely it is that the genes for herbicide resistance might be inadvertently transferred to other plants, including weeds (see the "Reflections" section).

In other examples, the FlavrSavr tomato, developed by the biotechnology company Calgene, has been engineered to resist rotting. It can therefore be allowed to ripen longer on the vine, perhaps at last producing a supermarket tomato that actually tastes like a tomato. French researchers at the Moet-Chandon champagne company have engineered grape vines to resist the grape fan-leaf virus. Genes for virus resistance have also been inserted into cantaloupe, cucumber, papaya, squash, potato, tomato, and tobacco.

One of the most valuable genetic modifications of plants would be the insertion of genes for nitrogen fixation into common crops such as corn or wheat. With a few exceptions, such as soybeans and alfalfa, our major crop plants cannot use atmospheric nitrogen as a nitrogen source. Consequently, they require nitrogen fertilizers, which are synthesized and spread using the energy of fossil fuels. Rain washes some of the fertilizer off into nearby streams and lakes, causing pollution. Further, the expense of fertilizers is burdensome to farmers, especially in the Third World. Several types of bacteria can extract nitrogen from the atmosphere and capture it in usable form, a process called nitrogen fixation (see Chapter 30). If the genes for nitrogen fixation could be inserted into wheat and corn, farmers could grow much more food at much lower cost. Especially in poor countries, an improvement in food supplies would have a tremendous impact on human health and well-being.

Recombinant DNA Techniques Can Be Used to Diagnose Genetic Disorders

A variety of human diseases are inherited, including sickle-cell anemia, Tay-Sachs disease, Huntington's disease, and muscular dystrophy. So far, there are no cures for these genetic disorders. Carriers (heterozygotes for recessive disorders who show no significant symptoms) or affected fetuses or both can sometimes be identified using recombinant DNA technology. Some of the techniques and the ethical concerns that they raise are discussed in Chapter 15.

Amplified Genes May Be Used to Treat Genetic Disorders

A few children are born each year with severe combined immune deficiency, in which the affected person has virtually no defense against disease (see Chapter 24). In about 25% of these children, the failure of the immune system can be traced to abnormal alleles of a single gene. In the late 1980s, R. Michael Blaese of the National Cancer Institute and W. French Anderson of the National Institutes of Health inserted normal alleles of this gene into white blood cells and obtained reasonably normal gene expression. At 12:52 P.M., September 14, 1990, the first clinical trial of human gene therapy began: A 4-year-old girl received a transfusion of her white blood cells that were genetically engineered to contain the normal allele. The treatment worked; today she has a functioning immune system and attends school. Unfortunately, so far only 2 out of 10 patients have responded to the treatment.

In 1994, experimental trials for cystic fibrosis, a usually fatal inherited lung disease, were begun with a "nasal spray" containing viruses engineered to infect lung cells and provide them with a crucial protein that helps prevent fluid buildup in the lungs.

As these examples illustrate, the most likely candidates for genetic engineering in humans are defects involving discrete structures, such as lungs or bone marrow, that are normally the only, or the most important, active sites of transcription of certain genes. These therapies would, of course, offer effective treatments only for the affected individual: The patient's reproductive cells would not be "fixed," so he or she could still pass the genetic defect on to future generations. It is likely to be many years before genetic engineers are in a position to modify the genetic composition of human sperm or eggs. Whether society will allow them to do so is also an open question, which we discuss in the "Reflections" section at the end of the chapter.

Locating Genes

The applications of biotechnology that we have discussed so far all rely on some prior knowledge: The desired genes are already known by either their protein products, their mRNA, or, in a few instances, at least part of their DNA sequences. Unfortunately, geneticists do not know much, if anything, about the molecular nature of most genes or their protein products. How can such information be obtained?

You already know something about methods for locating genes: for example, linkage analysis, which uses the frequency of crossing over to map chromosomes (see Chapter 13). To locate a gene by linkage analysis, one calculates the incidence of recombination between two traits during sexual reproduction. If recombination *never* occurs between the two traits, then either the genes controlling those traits lie very close together on a chromosome or, in fact, the traits are controlled by the same gene. If the location of the gene controlling one trait is already known, then it should be easy to find the other gene.

In principle, to locate an unidentified gene in a DNA library, you could make a probe for a known marker gene that linkage analysis tells you is close to the unidentified gene. Somewhere nearby should be the gene that you're interested in.

Using ordinary linkage analysis to find a gene in a DNA library, however, has two problems. First, geneticists usually know the location of relatively few genes on any given chromosome, and eukaryotic chromosomes are very large: The average human chromosome, for example, contains about 60 million nucleotides. Therefore, if linkage analysis reveals that an unknown gene is "close" to a known gene, say, only 10% of the chromosome's length away, that means that the few thousand nucleotides that make up the unknown gene might lie anywhere within about 6 million nucleotides. That information isn't necessarily very useful. A second, related problem with linkage analysis is the relative scarcity of genes in a eukaryotic genome. Although there are tens of thousands of genes in most eukaryotic genomes, these constitute less than 5% of the genome: The rest is introns, regulatory sequences, or vast amounts of other, apparently nonfunctional, DNA. Therefore, classical linkage analysis usually cannot be expected to locate an unknown gene more closely than within a few hundred thousand nucleotides, at best. That still isn't very useful. What geneticists need are a lot more markers, a lot closer together.

Restriction Enzymes Can Be Used to Provide Markers on a Chromosome

You may recall that restriction enzymes cut DNA only at specific nucleotide sequences. Further, there are a host of known restriction enzymes, and each cuts DNA at a different sequence of nucleotides. Let's suppose that a pair of homologous chromosomes is exposed to a given restriction enzyme (Fig. 14-7). The DNA is cut up into

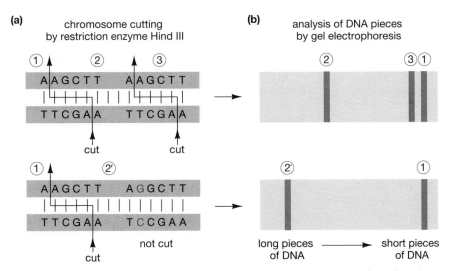

Figure 14-7   **The nature and detection of restriction fragment length polymorphisms**

(a) Two homologous chromosomes differ in a single base substitution. A restriction enzyme, Hind III, cuts DNA at AAGCTT/TTCGAA palindromes. The upper chromosome has two of these sequences and is cut into three pieces, two short (numbered 1 and 3 on the illustration) and one medium-long (number 2). The lower chromosome, because of the base substitution (a G-C pair substituted for an A-T pair), has only one sequence cut by Hind III. Therefore, the lower chromosome is cut into only two pieces, one short (same as number 1 on the upper chromosome) and one very long (numbered 2´). **(b)** The DNA pieces produced by Hind III are separated by the technique of gel electrophoresis. DNA migrates through the gel, pushed along by an electrical field. Small pieces (1 and 3) slip through the gel more easily than large pieces (2 and 2´) can and migrate farther to the right. The size and number of DNA pieces differ for the two chromosomes.

pieces called **restriction fragments** (fragments of DNA produced by restriction enzymes). These fragments can be separated by size, using a method known as gel electrophoresis. Would the pattern of sizes be the same for both homologous chromosomes? Perhaps, or perhaps not, depending on the nucleotide sequences of the chromosomes and the recognition sequence of the enzyme. In the example diagrammed in Figure 14-7, the two homologous chromosomes have different nucleotide sequences near one end. The restriction enzyme Hind III can cut apart the upper chromosome in two places, but the lower chromosome in only one place (Fig. 14-7a). Gel electrophoresis therefore shows three pieces of DNA from the upper chromosome and two pieces of DNA from the lower chromosome (Fig. 14-7b). These differences in the size of the DNA pieces are called **restriction fragment length polymorphisms**, or **RFLPs** for short (*polymorphism* means "many forms" in Greek).

Restriction Fragment Length Polymorphisms Can Be Used to Locate a Gene

Gene location by RFLP analysis is conceptually similar to locating a gene by crossing over. In RFLP analysis of a human genetic disorder, researchers look for a restriction enzyme "cut site" that is always, or almost always, inherited along with the disorder. The cut site must therefore be within, or very close to, the gene causing the disorder. In the late 1980s, RFLP analysis was used to locate and sequence the gene for cystic fibrosis.

It is easy to understand why a unique restriction enzyme cut site might be located within a defective allele: If the altered nucleotide sequence that causes the genetic defect is also the one recognized and cut by the restriction enzyme, then the enzyme will cut the defective allele but not the normal allele. But why should a unique cut site be located *close to* a defective gene? To understand this, you must consider the inheritance of genetic defects and the nature of linkage. Defective alleles for inherited genetic disorders are usually quite rare. Therefore, all affected individuals may well be descended from a single, though distant, ancestor. Suppose that this person, strictly by chance, had a second altered nucleotide sequence on the same chromosome. How he or she got it, by inheritance or mutation, doesn't matter. If this second altered nucleotide sequence is close enough to the locus of the defective allele, then there will be little or no crossing over between the two sites.

This second nucleotide change can be used to find the locus of the defective allele if one more condition is met: The "normal" and "altered" nucleotide sequences must be affected differently by a restriction enzyme. Either the normal sequence must be cut by the enzyme while the altered sequence is not cut, or vice versa. Given the vast number of restriction enzymes, a "cut–no cut" difference can almost always be found relatively close to any given defective allele, although it may take years of work to find one close enough to be useful for DNA sequencing (described shortly).

Other Uses of RFLPs

Locating a gene is not the only application of RFLPs. For example, RFLP analysis can also be used for prenatal diagnosis of certain genetic defects, if the nucleotide sequence or at least the location of the gene is known. In Chapter 15, we describe how RFLPs are used to diagnose sickle-cell anemia. A variation on standard RFLP analysis, popularly known as DNA fingerprinting, is used increasingly in forensic medicine to help convict or clear a suspect; a class of RFLPs from blood or semen samples left at the scene of a crime is compared with RFLPs from the suspect's tissues. DNA fingerprinting promises to play a crucial role in criminal cases (see "Scientific Inquiry: DNA Fingerprinting in Forensics—The Case of the Telltale Palo Verde"). The methodology and some other applications of RFLP analysis are explained in a fascinating *Scientific American* article, "Chromosome Mapping with DNA Markers" (see "For More Information").

Sequencing Genes

Both basic research and medical and industrial applications of biotechnology are greatly enhanced if the sequence of nucleotides in a gene is known. During the 1980s, automated machines were developed that allow fairly large genes (up to a few tens of thousands of nucleotides long) to be sequenced (Fig. 14-8). The process isn't cheap, and even 50,000 nucleotides are less than a tenth of a percent of the 60 million nucleotides in an average human chromosome, which is why precise localization of genes is still very important.

Why is knowing the nucleotide sequence of a gene useful? For one thing, the nucleotide sequence determines the amino acid sequence of the encoded protein. In many instances, it is now easier to locate and sequence a gene than it is to isolate, purify, and sequence a protein. In recent years, biochemists have frequently "forecast" the amino acid sequence of a protein, its location within a cell (membrane-bound or dissolved in the cytoplasm), and even some of its probable functions, solely on the basis of the nucleotide sequence of a gene. The gene for cystic fibrosis was sequenced in late 1989. The gene sequence was the starting point for pinpointing the underlying genetic defect.

Further, if one knows the nucleotide sequence of a gene and the nucleotide sequence that is cut by a variety of restriction enzymes, one may be able to design a rapid, definitive test for genetic disorders—for example, the DNA of affected people will not be cut by a certain restriction enzyme but the DNA of everyone else will be cut. This type of test is now the basis for prenatal diagnosis of sickle-cell anemia.

In one of the most remarkable instances of cooperation and collaboration in modern biology, a huge group of scientists is undertaking the ultimate sequencing task—to sequence the entire human genome. The rationale, approach, and methodology of the Human Genome Project are explored in Chapter 15.

Reflections on the Ethics of Biotechnology

In many respects, biotechnology presents a microcosm of the benefits and dilemmas that science and technology have always offered to humankind. There are few fundamental advances in science that do not have both bright and dark sides—the first spear allowed hunters to nourish their families better, but it also provided a more effective means of killing their fellow humans. There are groups in society who would greatly restrict the uses of biotechnology and who might prefer that the techniques of recombinant DNA had never been developed. Others feel that the benefits greatly outweigh the liabilities and that any potential for harm can rather easily be overcome by fairly simple legal means. Many others take no stand on biotechnology in general but have strong opinions about specific applications. Here, we will explore three aspects of biotechnology that together represent many of the ethical issues that surround this emerging field. In the next chapter, we will examine some ethical concerns in the application of biotechnology to medicine.

Some Uses of Biotechnology Have Social Impacts

Perhaps the easiest ethical issue surrounding biotechnology is the objection that some applications may be socially undesirable. For example, in the late 1980s, a biotech firm began to mass-produce bovine growth hormone by recombinant DNA techniques. Growth hormone, when injected into cows, enhances milk production. Some people are vigorously opposed to this practice, for three main reasons. First, will the recombinant hormone change the composition of the milk? Experts convened by the National Institutes of Health and the Food and Drug Administration have concluded that the changes are minor and pose no threat to human health. Second, increased milk production is likely to lead to increased incidence of udder infections in cows. This effect might not only be harmful to the cows but might also lead to increased use of antibiotics, which might in turn lead to increased antibiotic resistance among microbes. Third, there is already a milk glut in both America and Europe. Many dairy farmers oppose the use of recombinant growth hormone because they fear that greater milk production per cow will drive prices down even further and put some small farmers out of business.

These concerns are real. However, they do not have any direct bearing on biotechnology in and of itself. Millions of cattle and other mammals are slaughtered for meat every year; suppose that a cheap way was found to extract growth hormone from their pituitary glands. The same arguments would arise without biotechnology's being part of the issue at all. In instances such as this, society may wish to intervene in *a specific use of biotechnology* for reasons of public health, economic issues, or social justice. The outcome, not the methodology, is the heart of the matter.

Bioengineered Organisms May Be Released into the Environment

Although bioengineered *products* are relatively simple to deal with, bioengineered *organisms* are more problematic. Two examples illustrate the potential difficulties.

Many crop plants are quite sensitive to frost: oranges and tomatoes, to name just two. Ice formation in the air, in water, or on plant surfaces is greatly facilitated by small particles of dust or other matter that act as "seeds" for ice-crystal formation. In the absence of "seed particles," ice crystals do not form even at temperatures somewhat below freezing. One of the primary sources of ice seeding on plants is a common bacterium, *Pseudomonas syringae*. A specific protein on the surface of the bacterium promotes ice formation. Therefore, an obvious use of genetic engineering would be to remove

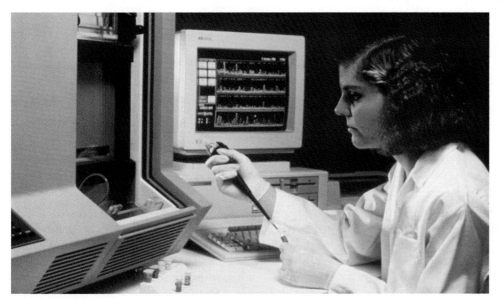

Figure 14-8 **A commercially available DNA sequencing machine**

Machines such as this, only a dream a couple of decades ago, are routine equipment in many molecular biology labs.

S C I E N T I F I C   I N Q U I R Y

DNA Fingerprinting in Forensics—The Case of the Telltale Palo Verde

The polymerase chain reaction (PCR) procedure has caused a revolution in many fields of biology, but nowhere has the impact of this technique been felt more than in forensics, the collection of information to be used as evidence in court proceedings. The technique that is proving most useful has been nicknamed "DNA fingerprinting," because it may generate patterns of DNA fragments unique to each individual. This procedure involves the amplification of a set of DNA fragments of unknown sequence, a set that is likely to differ from one individual to the next. When these fragments are amplified and separated from one another on a gel, the pattern of bands is referred to as a DNA fingerprint. In theory, it represents a characteristic "snapshot" of the unique genome of an individual, allowing it to be distinguished from that of every other individual in the population. The quantity of DNA needed to analyze using PCR is exceedingly small; this amount of DNA can easily be obtained from a single spot of dried blood, from the cells clinging to the base of a hair shaft, from skin fragments under a victim's fingernails, or other small tissue samples collected at a crime scene.

Here is an example of the forensic use of DNA fingerprints that did not involve the suspect's DNA, but rather that of plants at the crime scene. On the night of May 2, 1992, a Phoenix woman was strangled and her body was dumped near an abandoned factory in the desert. Police discovered a pager near the body, making its owner the prime suspect. When questioned, the man who owned the pager claimed that the woman must have stolen the pager from his pickup truck, since he had never been near the factory. A search of the pickup revealed two seed pods from a palo verde tree (Fig. E14-1), several of which grew near the factory. Could DNA fingerprinting demonstrate that the seed pods found in the suspect's truck had fallen from the tree near where the body was found? If it could, this evidence would place the suspect at the crime scene.

Homicide investigators turned to Dr. Timothy Helentjaris, then at the University of Arizona, in nearby Tucson. In his research on the evolution of maize and other important crop plants, he had been generating DNA fingerprints of parts of the maize genome. Helentjaris knew that any match between a seed pod and a tree would have significance only if he could first demonstrate that palo verdes show considerable genetic variation from one tree to the next. Otherwise, it would be impossible to unambiguously link a given pod to any particular tree. Fortunately, he found considerable differences in DNA between individual plants, allowing for discriminatory fingerprints.

Helentjaris then was given the two seed pods from the suspect's truck along with pods collected by investigators from 12 different palo verde trees near the factory. The investigators knew which of the 12 trees was near the body but did not tell Helentjaris. He extracted DNA from seeds taken from each pod and performed the PCR reaction, producing 10 to 15 distinct bands from each pod. The result of this experiment was unmistakable: The pattern from the pod found in the truck exactly matched the pattern from only one of the 12 trees, the one nearest the body. In an important additional test, Helentjaris found that this pattern was also different from that of 18 pods collected from trees at random sites around Phoenix. This result was admitted as evidence in the trial, the first time a DNA fingerprint of a plant had been used in a court case. This information was crucial in demolishing the suspect's alibi, and at the end of the 5-week trial the suspect was found guilty of first-degree murder.

Figure E14-1   The telltale palo verde
DNA fingerprints from the pods (inset) of the palo verde tree helped convict a murderer.

the gene encoding this protein from the bacterium and to try to replace the "wild-type" bacterium on crops with the new "ice-minus" bacterium. The genetic engineering part of this scenario was accomplished in the mid-1980s, but there were several years of controversy and litigation before field trials were permitted (Fig. 14-9).

What was the controversy about? What if the ice-minus strain, once released into the environment, drifted beyond the fields onto which it was sprayed? What if it became established in the wild and became widespread? Susceptibility to frost is a major determining factor in the distribution of plants across the planet—might ice-minus bacteria trigger massive changes in vegetation? Like the ice on leaves, raindrops also usually form around "seeds" of dust or debris in the atmosphere. What if wild-type bacteria contribute significantly to normal rain seeding—might droughts become frequent if ice-minus bacteria became the norm? Genetic engineers responded that the altered bacteria were actually less fit than the wild type and that a minuscule number of bacteria were to be released, compared with the vast hordes of wild-type bacteria in the environment. Therefore, they argued, the chances were infinitesimally small that the ice-minus bacteria could actually become established in the wild. At least so far, the genetic engineers appear to be right: A test field was sprayed in 1987, and no ecological effects seem to have occurred.

A similar controversy is brewing about herbicide-resistant plants. Some people are opposed to herbicide-resistant plants because they feel that extensive herbicide use is hazardous to human health and the environment, and herbicide-resistant crops would almost certainly lead to greatly increased herbicide use. Alternatively, what if herbicide-resistance genes were to be transferred from crops to weeds, for example by a plant virus? Such "lateral transfers" of plant genes may be possible. If such a transfer occurred, herbicides might ultimately become completely useless.

Ice-minus bacteria and herbicide-resistant plants are examples of larger issues surrounding releases of genetically engineered organisms into the environment: People are not omniscient; people make mistakes. Can we be sure that genetically released organisms, or the modified genes they carry, will not cause ecological harm? If we can determine only that the risk is "small," then how small is small enough?

Genetic engineers correctly assert that humans have practiced crude genetic engineering for millenia, breeding plants and animals with desired properties. Modern biotechnology is merely a faster, more precise version of standard agricultural practice. What's more, various forms of genetic recombination occur all the time in nature. Why should modern genetic engineering be singled out as a special threat?

Some ecologists reply that ecological disasters of varying magnitude do indeed happen: the invasion of pastureland in the western United States by cheatgrass and knapweed, the displacement of bluebirds by sparrows, the evolution of antibiotic-resistant bacteria, and the mutation of a monkey virus into the AIDS virus, to name a few modern examples. (On the other hand, bioengineering may offer the possibility of *correcting* one ecological disaster of major proportions, the destruction of American chestnut trees by chestnut blight; see "Earth Watch: Biotechnology Attacks the Chestnut Blight.") Genetically engineered organisms, such as the ice-minus bacteria, are likely to be genetically inferior to their wild-type counterparts, and therefore be outcompeted, but there may remain some risk. Although generally supportive of properly controlled releases of genetically engineered organisms, in 1989 a committee of prominent ecologists wrote the following warning in the journal *Ecology*: "The direct effects of self-replicating introduced organisms may not necessar-

ily decrease with time or with distance from the point of introduction. The absence of an immediate negative effect does not ensure that no effect will ever occur."

Biotechnology May Eventually Allow Alterations in Human Genomes

The greatest concern of many observers is that genetic engineering offers the potential to change the human genome. As we pointed out, it is not yet possible to eliminate, say, insulin-dependent diabetes by cutting the defective gene out of fertilized eggs and inserting a functional one. But suppose that it does become possible (few molecular geneticists would be willing to bet against it). Then what? Could humankind agree on what constitutes a "bad" gene? Even if we could, would we be right? Probably most people could agree on diabetes, cystic fibrosis, and muscular dystrophy. But what about more subtle changes? What about genes that might alter personality? Or length of life? Or predisposition to take risks? Are we wise enough to direct our own evolution—or, if we develop the capability, are we wise enough to refrain?

Figure 14-9   **Field tests of "ice-minus" bacteria**

Although the bacteria are completely harmless to humans, this worker spraying "ice-minus" bacteria onto a strawberry patch was required to wear protective clothing, because the bacteria were legally considered to be a pesticide that had not yet been safety-tested on people. The "frostbusters" logo, although just a publicity gimmick, was accurate: Treated plants showed about 80% less frost damage than nearby, untreated control plants.

E A R T H   W A T C H

Biotechnology Attacks the Chestnut Blight

Scarcely a century ago, the towering American chestnut tree made up a major portion of the eastern hardwood forest (Fig. E14-2). Today the splendor of the chestnuts is barely a memory; few notice the struggling sprouts emerging from the roots of the former giants. The American chestnut has been all but driven to extinction by the ravages of a fungus called the chestnut blight, inadvertently imported from Asia around 1900 on Asian chestnuts. However, there is hope that the American chestnut will reappear, thanks to chance and the research techniques of biotechnology.

In 1950, an Italian researcher noted that some European chestnuts, also attacked by chestnut blight, were recovering—had the fungus lost its virulence? In the 1960s, plant pathologists found that some strains of chestnut blight had been weakened, apparently by a new viral infection that left particles containing double-stranded RNA (dsRNA, which encodes genetic information in many fungal viruses) floating throughout the fungal cytoplasm. In an unknown way, the viral dsRNA decreases the virulence of the fungus, allowing the tree to fight it successfully (Fig. E14-2b). By spreading the viral dsRNA infection to virulent chestnut blight fungi, foresters are gradually controlling the blight in Europe. Unfortunately, the European blight's virus is not very useful against the strains of chestnut blight fungus found in the United States.

Although viral-infected strains of chestnut blight also occur in the United States, they do not successfully control the blight. Therefore, the tools of biotechnology are being applied toward weakening the parasitic fungus. In 1994, researchers at the Roche Institute of Molecular Biology in New Jersey found a way to introduce both viral dsRNA and a DNA "gene" encoding the dsRNA into chestnut blight fungi in such a way that the RNA or DNA or both are transferred from one fungus to another. Ultimately, they hope to produce a large supply of weakened fungus that can be introduced into the wild, where it would infect the deadly fungal variety with the viral nucleic acids, weakening the fungus so that it would no longer be able to overcome the defenses of the chestnut tree.

Biotechnology may eventually be used, not only to weaken the fungus, but to strengthen the chestnut. Asian chestnuts are not killed by the chestnut blight fungus, and there is hope that the gene that confers fungal resistance in the Asian chestnut may be identified and introduced into "germlings" of the American chestnut, producing a blight-resistant strain. Although you will never see the chestnut forests that delighted your great-great-grandparents, resistant chestnut trees infected by weakened fungi may yet shade the homes of your children.

(a)

(b)

Figure E14-2   **Chestnut blight**

(a) A healthy American chestnut, which escaped the blight because it had been transplanted far west of the normal range of the chestnut before the invasion of the blight fungus. (b) Chestnut blight fungus has produced a lesion on this chestnut tree. The lesioned area was treated with infected fungus, and the tree is walling off the infection with a growth of new tissue.

SUMMARY OF KEY CONCEPTS

DNA Recombination in Nature

There are many naturally occurring forms of DNA recombination, where the genetic makeup of an organism incorporates DNA from two or more other organisms. These include genetic recombination (crossing over), sexual recombination, transformation (in bacteria), and viral transfer of DNA between eukaryotic species. Bacterial transformation often involves exchange of circular pieces of DNA called plasmids, which are found in the cytoplasm of many bacteria. Bacteria may take up plasmids that have been released by the destruction of other bacteria, thereby acquiring plasmid DNA from different species.

Recombinant DNA Technology

Many common uses of recombinant DNA technology include the following steps: (1) production of a DNA library, (2) identification of the desired gene, (3) production of copies of the desired gene, and (4) insertion and regulation of the gene in specific organisms.

A DNA library is a complete set of all the DNA of a particular organism, usually cloned in plasmids. DNA libraries are produced through the use of restriction enzymes that cut DNA at specific nucleotide sequences, leaving single-stranded "sticky ends" at each side of the cut. A restriction enzyme is used to cut DNA of both source genome and plasmid. The two sets of DNA fragments are mixed together; the single-stranded sticky ends cause plasmid and source genome DNA to base pair. DNA repair enzymes are then used to bond the source genome DNA to the plasmid DNA. Bacteria then take up the plasmids and may be stored indefinitely as a DNA library.

Genes may be identified through a number of techniques. If the amino acid sequence of a protein is known, then artificial DNA may be synthesized that corresponds to the DNA sequence encoding all or part of the amino acid sequence. Alternatively, mRNA may be isolated from a source that is rich in the specific mRNA transcribed from the desired gene. In either case, radioactive mRNA or synthetic DNA can be used to highlight the desired DNA region from a DNA library by base pairing with the gene.

Copies of a specific gene may be produced in bacteria from the DNA library by culturing those bacteria that contain plasmids with the desired gene. Alternatively, the DNA containing the gene may be amplified using the polymerase chain reaction.

Uses of cloned genes include (1) producing large quantities of specific proteins; (2) synthesizing vaccines; (3) diagnosing genetic disorders; (4) modifying DNA in intact organisms, which may be bacteria, agricultural plants, or animals, and releasing the organisms into the environment for specific uses; and (5) modifying the DNA of cells of specific organs in human beings to correct genetic defects.

Locating Genes

Restriction fragment length polymorphisms (RFLPs) may be used to locate genes in chromosomes. Restriction enzymes cut DNA at specific nucleotide sequences. If the same chromosomes from two or more sources are cut with the same restriction enzymes, differences in nucleotide sequences will yield differences in restriction fragment lengths. Inheritance of phenotypic traits, such as genetic diseases, can be compared with the differences in RFLPs to locate genes quite precisely in chromosomes.

Sequencing Genes

DNA nucleotide sequences can be determined by automated processes in the laboratory. The nucleotide sequence of a gene predicts the amino acid sequence of the encoded protein. Knowledge of the amino acid sequence of the encoded protein can reveal the function and location of the protein and may assist in understanding genetic disorders and devising treatments for them.

Reflections on the Ethics of Biotechnology

Biotechnology has generated ethical controversies regarding its proper uses. These controversies mostly deal with the risks involved in the introduction of genetically engineered organisms into the environment and the potential for making significant changes in the human genome.

KEY TERMS

cloning   p. 254
DNA library   p. 251
genome   p. 253
plasmid   p. 250

polymerase chain reaction (PCR)   p. 254
recombinant DNA technology   p. 249
restriction enzyme   p. 251

restriction fragment   p. 258
restriction fragment length polymorphisms (RFLPs)   p. 258

✖ T H I N K I N G   T H R O U G H   T H E   C O N C E P T S

Multiple Choice

1. During recombinant DNA techniques, how are the bacterial cells that take up the plasmids isolated from those that do not?
   a. by screening for restriction fragment length polymorphisms
   b. by using antibiotic resistance plasmid genes and antibiotic-containing medium
   c. each of the plasmids must be sequenced
   d. by use of the polymerase chain reaction
   e. this is where you use mRNA or information on the protein sequence

2. How do you help ensure that each bacterium in a library has only one gene-containing plasmid in it?
   a. Genes with more than one plasmid do not survive the antibiotic.
   b. You need to screen using a radioactive RNA probe.
   c. It is a matter of numbers and probability—far more bacteria than plasmids are mixed together.
   d. Only a small amount of calcium salt is used to facilitate the incorporation of the plasmid into the bacterial cell.
   e. This is why it is so important to spread the bacteria sparsely onto the culture dish.

3. Knowing the DNA coding sequence of a gene would give you [A]; knowing the amino acid sequence of the coded protein would give you [B].
   a. [A] approximate information about the amino acid sequence; [B] approximate information about the DNA sequence
   b. [A] approximate information about the amino acid sequence; [B] precise information about the DNA sequence
   c. [A] precise information about the amino acid sequence; [B] approximate information about the DNA sequence
   d. [A] precise information about the amino acid sequence; [B] precise information about the DNA sequence
   e. none of the above

4. How are RFLPs detected?
   a. by looking at the chromosome in the microscope
   b. by doing a standard Mendelian cross
   c. by observing DNA of different lengths on a gel
   d. by seeing with which plasmids they will combine
   e. by amplifying the DNA using PCR

5. How would you identify the bacterium in a library if you knew the sequence of the protein it coded for?
   a. You would put radioactive protein on the petri dish.
   b. You would put a synthetic radioactive nucleotide probe, designed from the protein sequence and the genetic code, onto the petri dish.
   c. Because of the antibiotic, only the one of interest would survive.
   d. You would check to see whether a protein of the desired sequence is being synthesized by each bacterium.
   e. You would use DNA polymerase.

Review Questions

1. Describe three natural forms of genetic recombination, and discuss the similarities and differences between recombinant DNA technology and these natural forms of recombination.

2. What is a plasmid? How are plasmids involved in transformation in bacteria?

3. What is a restriction enzyme? How can restriction enzymes be used to splice a piece of human DNA within a plasmid?

4. What is a DNA library? Briefly describe the steps involved in creating a DNA library of a mouse.

5. What is a restriction fragment length polymorphism? Describe the use of RFLPs in locating a gene on a human chromosome.

✖ A P P L Y I N G   T H E   C O N C E P T S

1. If you knew the entire sequence of a protein, what strategy would you use to design a short nucleotide sequence most likely to be complementary to the corresponding sequence in the actual gene?

2. In cloning, what two tricks (one at the level of plasmid insertion and one at the level of bacterial plating and growing) must be used to ensure that each bacterium (to grow into a colony) has one and only one inserted gene? Make the simplifying assumption that the restriction nuclease used chops into small enough fragments to have just one gene per fragment.

3. Traditionally, a "marker" in a genetic cross is a phenotype such as eye color. Now, it is possible to locate genes more directly by reference to an RFLP "marker" right in the DNA. Contrast the techniques and circumstances that would apply to the utilization of these different approaches.

4. The library described in this chapter, prepared from the nuclear DNA, is sometimes called a "genomic library." It is also possible to prepare a "cDNA library" made from cDNA, the complementary DNA copied from the mRNAs expressed in a particular tissue or structure (like *Drosophila* head). How do you think such a cDNA library would differ from the genomic library?

5. Let's say that you are a citizen member of a National Institutes of Health advisory panel on the uses of recombinant DNA. Discuss the ethical issues that surround the release of bioengineered organisms into the environment. What levels of benefits and risks do you think would justify such a release?

6. Do you think that using recombinant DNA technologies to change the genetic composition of a human egg cell is ever justified? If so, what restrictions should be placed on such a use?

�֎ FOR MORE INFORMATION

Barton, J. H. "Patenting Life." *Scientific American*, March 1991. Some of the legal and economic issues surrounding the new biotechnological advances are discussed.

Baskin, Y. "DNA Unlimited." *Discover*, July 1990. A clear, succinct description of the methods and applications of the polymerase chain reaction.

Capecchi, M. R. "Targeted Gene Replacement." *Scientific American*, March 1994. Changing the genetic makeup of mice provides a new way to understand the function of genes.

Elmer-Dewitt, P. "The Genetic Revolution." *Time*, January 17, 1994. A description of the science and ethics of biotechnology, written for a nonscience audience.

Gasser, C. S., and Fraley, R. T. "Transgenic Crops." *Scientific American*, June 1992. More and more commercially valuable plants are being engineered to resist pests while growing and spoilage after ripening.

Mestel, R. "It's a Tall, Tall World." *Discover*, January 1994. Should short, but otherwise healthy, children be given recombinant growth hormone, just to make them taller?

Mullis, K. B. "The Unusual Origin of the Polymerase Chain Reaction." *Scientific American*, April 1990. Mullis conceived the idea of PCR while on a drive through the mountains of California.

Paabo, S. "Ancient DNA." *Scientific American*, November 1993. The scientific concept underlying the science fiction in the film *Jurassic Park* is not so far-fetched, but, in actuality, it is applied to reconstructing evolutionary history rather than to recreating the dinosaur.

Roberts, L. "Ethical Questions Haunt New Genetic Technologies." *Science*, and "Ecologists Wary about Environmental Releases," *Science*, both March 1990. These two articles provide a brief primer on the debates concerning the social and ecological issues surrounding the uses of recombinant DNA technology.

Verma, I. M. "Gene Therapy." *Scientific American*, November 1990. A brief discussion of the promises and pitfalls of gene therapy.

White, R., and Lalovel, J.-M. "Chromosome Mapping with DNA Markers." *Scientific American*, February 1988. The authors describe the use of RFLPs to locate genes on human chromosomes.

NET WATCH

On-line resources for this chapter are on the World Wide Web at:
http://www.prenhall.com/~audesirk (click on the <u>table of contents</u> link and then select Chapter 14).

15 Human Genetics

*Nature and nurture: These twins are genetically identical, but
have also grown up in extremely similar environmental situations.
The interplay of both heredity and environment shapes the
characteristics of all organisms, but especially of people.*

Human genetics is similar to the genetics of peas or
fruit flies, except that it is more interesting to most people, and vastly more complex. First, humans have long life
spans and few children per couple. Second, people choose
their own mates, so crosses that might be genetically informative may not occur very often. Finally, humans interact extensively with their environment, an environment that is incredibly diverse from culture to culture and
even within a single culture. These interactions often obscure underlying genetic patterns. This is particularly true
for personality traits, in which learning and experience
play such a major role.

Nevertheless, human genetics is a flourishing field of
study, and a great deal is known about the inheritance of
human traits. From the viewpoint of genetics, two factors
partially compensate for the difficulties humans present
as experimental subjects: the enormous number of human
beings on Earth and the extensive documentation of
human families, ranging from names written down in family Bibles to the records of ancestry of royalty. Given
enough time, ingenuity, and patience, geneticists often
can find records of matings that help them determine how
particular human traits are inherited, especially physical
features and genetic diseases.

Methods in Human Genetics

Because experimental crosses are out of the question,
human geneticists search medical, historical, and family records to find crosses that have already been made
voluntarily. Records extending across several generations can be arranged in the form of family **pedigrees**, diagrams that show the genetic relationships among a set
of individuals, such as those shown in Figure 15-1. Careful analysis of pedigrees shows that certain traits, such as
an unattached earlobe, are inherited as simple dominants, whereas other traits, such as albinism, are inherited as recessives.

Since the mid-1960s, great strides have been made in
understanding gene function on the molecular level. For
instance, geneticists now know the molecular basis of several inherited diseases, such as sickle-cell anemia and cys-

(a) A pedigree for a dominant trait

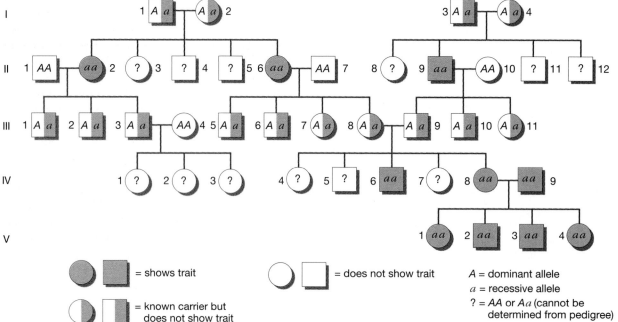

Figure 15-1   **Representative family pedigrees**

In pedigrees, squares usually represent males and circles females. Generations are indicated by the Roman numerals on the left side of the pedigree. **(a)** A pedigree for a dominant trait. Remember that all individuals bearing even one dominant allele will express the trait. Therefore, the trait will appear in every generation, in both males and females, and no child will possess the trait who does not have at least one parent with the trait. However, because heterozygotes show dominant traits, not all the children of phenotypically dominant parents will necessarily show the trait—that is, they may be homozygous recessive (for example, generation III, individuals 2 and 4). In this hypothetical pedigree, the genotypes of the individuals are given; in real pedigrees, genotypes must be deduced from inheritance patterns and cannot always be determined with certainty. **(b)** A pedigree for a recessive trait. Only homozygous recessives will show the trait. People with only one recessive allele will have the dominant phenotype. Parents with the dominant phenotype but who have a child who shows the recessive trait must themselves be heterozygotes. Therefore, the recessive phenotype may not appear in every generation (for example, generation III, families composed of individuals 5 through 8 and 9 through 11); the recessive allele may remain "hidden" in heterozygotes for several generations until two heterozygotes happen to marry (generation III, individuals 8 and 9) and have a homozygous child (generation IV, individuals 6 and 8). Two parents who both show the recessive phenotype (IV, 8 and 9) will have all homozygous recessive children (V, 1–4). In a pedigree for a rare trait, it is usually safe to assume that all individuals are homozygous normal (II, 1, 7, and 10) unless there is positive evidence of possession of the recessive allele (for example, an affected child).

tic fibrosis. Genetic engineering promises to increase our ability to predict genetic diseases and perhaps even to cure them (see Chapter 14 and "Scientific Inquiry: Collecting Fetal Samples and Analyzing Them Using Biotechnology" in this chapter).

Single-Gene Inheritance

Many "normal" human traits, such as freckles, long eyelashes, cleft chin, the ability to roll one's tongue (Fig. 15-2), are inherited in a simple Mendelian fashion; that is, they each appear to be controlled by a single gene with a dominant and recessive allele. In this chapter, rather than merely cataloguing inherited human traits, we will concentrate on a few examples of medically important genetic diseases and defects and explore their consequences for individuals and society.

Most Human Genetic Disorders Are Determined by Recessive Alleles

The human body depends on the integrated action of hundreds of enzymes and other proteins. A mutation in the gene coding for one of these enzymes almost always impairs or destroys enzyme function. As we described in the case of fat color in rabbits (Chapter 13), however, the presence of one normal allele often generates enough functional enzymes so that heterozygotes are phenotypically indistinguishable from homozygous normals. Therefore, normal alleles usually are dominant and mu-

tant alleles are recessive, although there are exceptions, such as polydactyly (more than five fingers on each hand), that are caused by dominant alleles. Most genetic diseases that are caused by lack of an essential enzyme are inherited as recessives.

If the metabolic pathway controlled by a particular gene is essential to survival, then individuals who are homozygous recessive will die. This outcome was especially certain before medical care became widely available in recent decades. In evolutionary terms, defective alleles are selected against. Consequently, even heterozygous **carriers** of serious genetic diseases (people who are phenotypically normal but can pass on their defective allele to their children) are usually rare. An unrelated man and woman will seldom both possess the same defective allele and produce a homozygous child. Related couples, however, especially first cousins or closer, have inherited some of their genes from recent common ancestors. Therefore, they are much more likely to carry the same allele and have an affected child (Fig. 15-3).

By their very nature, genetic diseases are an integral part of an affected person. They can neither be prevented with vaccines nor cured with antibiotics. Some, such as diabetes, can be treated, with varying degrees of success. As we pointed out in Chapter 14, genetic engineering offers the promise of a permanent genetic cure in certain cases. In instances in which the expression of the defective genes is localized to certain accessible organs, such as the liver or bone marrow, it may be possible to extract some of the patient's cells, insert the proper genes, and reimplant them in the patient. This procedure has been done for a few children suffering from severe combined

(a)

(b)

Figure 15-2   **Single-gene human traits**

These are controlled by dominant and recessive alleles of single genes. **(a)** Cleft chin seen in actor Kirk Douglas (left) and his son, Michael Douglas (right). **(b)** Tongue-rolling ability, demonstrated by Heather Audesirk (her father, Gerald Audesirk, also has this ability).

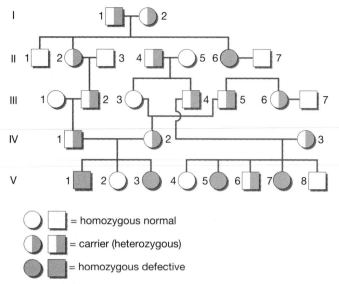

Figure 15-3 **Relatives may carry the same alleles**

Marriage between relatives increases the chance of genetic disorders. When related people marry, they often carry some of the same alleles, inherited from their common ancestors, and the likelihood that they will both carry the same defective recessive allele is much higher than normal. As a result, marriages between cousins or even closer relations are the cause of a disproportionate number of recessive diseases. In this family, marriages between cousins were common, including those between III, 3 and 5; III, 4 and IV, 3; and IV, 1 and 2.

immune deficiency (see Chapter 14). However, the genetic defect still remains in the reproductive cells that give rise to sperm or eggs, and the affected people can pass on the genes to their children.

For the present, then, the only way to prevent genetic diseases is to prevent the birth of affected babies. Reducing the incidence of recessive genetic diseases would require either identifying heterozygous carriers and preventing reproduction by couples who are both carriers or screening fetuses and aborting those that are homozygous recessive. Both courses of action are often both scientifically difficult and ethically questionable.

By definition, carriers of recessive genetic diseases cannot be identified by casual observation; if they can be identified at all, expensive medical procedures are usually required. With a few exceptions, it is impractical to screen the entire population of any country for a genetic disease. Screening tests can be useful, however, when the defective alleles are found almost exclusively within a readily identifiable group. Tay-Sachs disease and sickle-cell anemia are recessive genetic diseases, largely restricted to certain segments of the population, for which both carrier identification and prenatal diagnosis are possible. These two diseases illustrate the procedures used to diagnose recessive diseases and some possible courses of action. The practical and ethical dilemmas implicit in the diagnosis and treatment of genetic diseases are explored in the last section of the chapter.

Tay-Sachs Disease

Tay-Sachs disease occurs when nerve cells fail to synthesize an enzyme involved in the breakdown of lipids. Lipids accumulate in the brain of a child who is homozygous for the Tay-Sachs allele, causing progressive mental retardation, blindness, and failure of motor control (Fig. 15-4a). There is no cure, and death occurs in early childhood.

The Tay-Sachs allele is very rare in the United States population as a whole; only about one in 400 Americans is a carrier. However, about one in 30 American Jews is a carrier, so there is about one Tay-Sachs child in 3600 Jewish births. Heterozygotes for Tay-Sachs disease can easily be detected by a blood test (Fig. 15-4b), and many Jewish prospective parents take this test. If both husband and wife are identified as carriers, they have three choices: (1) forego reproduction entirely; (2) take a 25% chance of having an affected child who will face suffering and a certain early death; or (3) through **amniocentesis** or **chorionic villus sampling** determine whether each fetus is affected and abort homozygous recessives. The techniques of amniocentesis and chorionic villus sampling are described in "Scientific Inquiry: Collecting Fetal Samples and Analyzing Them Using Biotechnology." Genetic screening followed by abortion has reduced the incidence of Tay-Sachs disease by a factor of 20 since the 1970s.

Sickle-Cell Anemia

As with Tay-Sachs disease, almost all the carriers of **sickle-cell anemia** belong to one ethnic group, in this case Africans and African Americans. A single nucleotide substitution in DNA (adenine for thymine) causes the amino acid valine to be substituted for glutamic acid at one position on the outside of the hemoglobin molecule (see Chapter 11). Glutamic acid is highly charged and hydrophilic ("water loving"), whereas valine is neutral and hydrophobic ("water fearing"). Glutamic acid is crucially important in keeping hemoglobin molecules dissolved in the cytoplasm of the red blood cells. Substitution of valine at this position causes the hemoglobin molecules to clump together, forcing the red blood cells into the sickle-like shape that gives the disease its name (Fig. 15-5). During exercise or stress, the sickled cells break and clog the capillaries, cutting off circulation. In some instances, this lack of circulation may cause fatal strokes or heart attacks. Heterozygotes have about half normal and half abnormal hemoglobin, which you might expect would lead to partial sickle-cell symptoms. But heterozygotes usually have few sickled cells and show no symptoms whatsoever (in fact, many world-class black athletes are heterozygotes).

In the United States, about 8% of the African-American population is heterozygous; in parts of Africa, 15% to 20% of the population may be heterozygous. Given how severe the disease can be to homozygotes, why hasn't natural selection eliminated the sickle-cell allele? The prevalence of the sickle-cell allele is probably an evolutionary

(a)

(b)

Figure 15-4   Tay-Sachs disease

(a) The progression of Tay-Sachs disease. Evelyn, beautiful and seemingly bright, happy, and normal for the first few months after birth, never managed simple tasks such as sitting upright (notice that at 9 months she is using her hands for balance). As the disease progressed, she became unable to respond to sounds or other stimuli and suffered from recurring infectious diseases. She died shortly before her sixth birthday. **(b)** The Tay-Sachs allele codes for a nonfunctional enzyme. Heterozygotes may appear phenotypically normal, but, because they have only one normal allele, they often have only about half as much of the normal enzyme as homozygous dominants do. In Tay-Sachs, this difference in the concentration of normal enzymes in the blood can be used to detect heterozygotes. This blood test is the basis for genetic counseling that can spare parents from having a child with what Evelyn's parents have called "this cruel . . . disease."

(a)   (b)

Figure 15-5   Sickle-cell anemia

Sickle-cell anemia is caused by abnormal hemoglobin molecules that clump together, especially in low oxygen concentrations such as occur in capillaries during exercise. The clumps force the red blood cell out of its normal disc shape **(a)** into a longer, sickle shape **(b)**. The sickled cells are more fragile than normal red blood cells, making them likely to break or clog in capillaries. Tissues "downstream" from such a clogged capillary do not receive oxygen or have their wastes removed. This effect can cause pain, and, if it occurs in the brain or heart, serious injury.

compromise: People who are heterozygous for the sickle-cell allele enjoy some protection against malaria. In Africa, where malaria is still all too common, this protection may make the difference between life and death. In Chapter 17 we will explore how the dual selective effects of diminished hemoglobin function and protection against malaria may have favored the preservation of both the normal and sickle-cell alleles in African populations.

Heterozygous carriers of the sickle-cell allele can be detected by a blood test, but until fairly recently there was no way to find out if a fetus is homozygous or heterozygous. Recombinant DNA techniques have now been devised that can distinguish chromosomes with the normal hemoglobin allele from those with the sickle-cell allele (see "Scientific Inquiry: Collecting Fetal Samples and Analyzing Them Using Biotechnology"). Analyzing fetal cells collected by amniocentesis or chorionic villus sampling now allows medical geneticists to diagnose sickle-cell anemia in fetuses.

A Few Human Genetic Disorders Are Determined by Dominant Alleles

Many physical traits are inherited as dominants, including cleft chin and freckles. But, for two reasons, few people have serious genetic diseases caused by dominant alleles. First, as we have already explained, when a mutation strikes a normal allele, it usually produces a nonfunctional, recessive allele. Second, everyone bearing a dominant defective allele will develop the disease: There can be no

phenotypically normal carriers. Before the advent of modern medicine, if the disease was serious, these people died without reproducing and did not pass on their defective allele to future generations.

An exception to this rule is Huntington's disease (Fig. 15-6). An incurable disease, Huntington's causes a slow, progressive deterioration of parts of the brain, resulting in loss of motor coordination, flailing movements, personality disturbances, and eventual death. Huntington's disease is particularly insidious because symptoms usually do not occur until 30 to 50 years of age. Therefore, a person usually has already had children before he or she ever suffers the first symptoms.

Because everyone with even one defective dominant allele shows the defective trait (if they live long enough), dominant diseases could be virtually eliminated in a single generation if all the affected people chose not to reproduce (except, of course, for new mutations, which are usually very rare). In the case of Huntington's, however, few people know for sure if they have the disease until after they have had children. Therefore, to eliminate Huntington's, all people with a parent suffering from the disease would themselves have to forego having children. Since half of these people would be homozygous normal and could not pass on the Huntington's gene, the decision not to reproduce would have to be based on statistics. For some, the choice is clear; for others, the dilemma is excruciating.

In 1983, painstaking pedigree analysis, restriction fragment length polymorphism (RFLP) analysis (see Chapter 14), and a lot of luck combined to localize the Huntington's gene to a relatively small part of chromosome 4 (see "Sci-

(a)   (b)

Figure 15-6   **Huntington's disease**

The public became aware of Huntington's disease when folk singer Woody Guthrie **(a)** developed symptoms in the 1950s. The reason for the behavioral and motor deficits that Huntington's patients develop is appallingly clear when we compare magnetic resonance images of the brains of a living patient with Huntington's disease **(b, left)** and a normal person of the same age **(b, right)**. Notice that the total area of brain tissue (light color) is much smaller and the darker fluid-filled spaces around it and inside it (see arrows) are much larger in the Huntington's victim's brain.

entific Inquiry: In Search of the Huntington's Gene"). No one knows what the protein encoded by the normal Huntington's gene is, or what it does in the brain, much less what's wrong with the protein encoded by the defective allele. Finding the gene and studying the proteins produced by the normal and defective alleles might enable researchers to devise therapies that slow or stop the progress of the disease. After 10 frustrating years of work, geneticists finally isolated the Huntington's gene in 1993. The next steps, finding out what the Huntington's gene does in healthy individuals and what goes wrong in afflicted people, are just beginning.

RFLP analysis has also succeeded in producing a test for Huntington's disease. Although the procedure is not infallible, geneticists can now sample a person's DNA and predict whether he or she has inherited the Huntington's allele *before any symptoms appear*. At first glance, this is a wonderful medical advance, which could result in the elimination of this devastating disease. For the individuals involved, however, the knowledge will be gained at enormous personal cost. What would *you* do if you were told that in 20 or 30 years you would develop Huntington's? Some people do not want to know.

Many Human Traits Are Either Sex-Linked or Sex-Influenced

As we described in Chapter 13, the X chromosome bears many genes that have no counterpart on the Y chromosome. With one X and one Y chromosome, males are effectively haploid for X-chromosome genes; that is, all alleles of X-chromosome genes are always expressed in men, a phenomenon called **sex-linked**, or **X-linked, inheritance**.

A son receives his X chromosome from his mother and his Y chromosome from his father. A man must therefore inherit X-chromosome genes from his mother, and he can pass them on only to his daughters. For rare, recessive alleles of X-chromosome genes, there is usually a striking pattern of inheritance (Fig. 15-7). Recessive traits appear most frequently in males and typically skip generations, with an affected male passing the trait on to a phenotypically normal, carrier daughter who in turn bears affected sons. The most familiar genetic defects due to recessive alleles of X-chromosome genes are red-green color blindness (Fig. 15-7), **hemophilia** (Fig. 15-8), and one form of muscular dystrophy, in which muscles of afflicted young males degenerate.

Each sex also has its own set of **sex-influenced traits** (in addition to the obvious ones of breast and genitalia development) that occur more commonly or more strongly in that sex but are not coded by genes on the sex chromosomes. The majority of sex-influenced traits affect males more often than females and seem to be enhanced by male sex hormones, such as testosterone. A familiar example is baldness, which appears as if it were dominant in men (heterozygotes become bald) but recessive in women (heterozygotes retain their hair). The hormonal

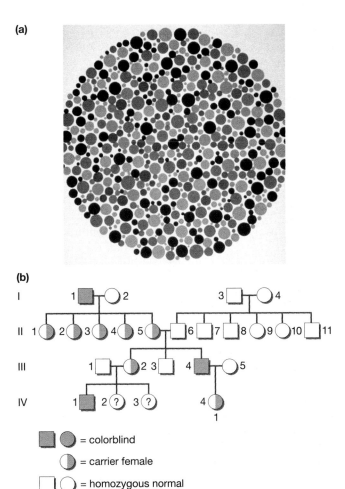

Figure 15-7   Color blindness

(a) This figure, called an Ishihara chart after its inventor, distinguishes between red-deficient and green-deficient vision in people with red-green color blindness (which is actually only color deficiency). People with normal color vision see the number 96, while those with red-deficient or green-deficient vision see either a 9 or a 6 depending on the type of deficiency. (b) Part of the family tree of Gerald Audesirk (III, 4), showing sex-linked inheritance of red-green color blindness. Audesirk and his maternal grandfather (I, 1) are both color deficient, but neither his mother and her sisters, nor any relative of his father, shows the defect. This pattern of skipping generations, a more common occurrence in males, and transmission from affected male to carrier female to affected male is typical of X-linked recessive traits.

connection in baldness is readily apparent in people with abnormal levels of sex hormones. Castrated men almost never become bald, but if castrated men with the allele for baldness are given high doses of testosterone, they lose their hair. Some women with tumors causing enlargement of the adrenal glands (sites of testosterone synthesis) produce large amounts of testosterone and may become bald. Another sex-influenced trait that is 20 times more common in men than in women is gout, which causes painful deposits of uric acid in joints.

SCIENTIFIC INQUIRY

In Search of the Huntington's Gene

As a young boy in 1860, George Huntington encountered an emaciated mother and daughter, both uncontrollably twitching, grimacing, twisting, and bowing: "I stared in wonderment, almost in fear," he later wrote. Huntington went on to study medicine at Columbia University and spent the rest of his life investigating this disease, now called Huntington's disease (HD) in his honor. Huntington showed that HD occurs in about half the children of an affected parent, evidence that it is inherited as a dominant allele of a gene.

Years later, in 1967, tragedy struck the family of Nancy Wexler, a young woman of 22, when her mother was diagnosed with HD. In response, Nancy's father, Milton Wexler, a clinical psychologist, founded the Hereditary Disease Foundation in 1968, hoping to find a cure or an effective treatment for the disease that eventually would kill his wife. Nancy Wexler was profoundly affected by her mother's suffering, and especially by her progressive loss of mental function. Trained as a clinical psychologist like her father, Nancy Wexler decided to devote her life to finding a cure for HD and now serves as president of the Hereditary Disease Foundation. She taught herself genetics by attending lectures, workshops, and seminars on the subject, by having many informal discussions with geneticists, and by reading extensively. During one of the seminars, Wexler learned that in several villages near Lake Maracaibo in Venezuela, an unusually high percentage of the villagers died of HD. Nancy visited these villages and found that about one in five persons there were stricken with the disease (Fig. E15-1).

By 1981, Wexler had begun assembling a large family tree showing the relationships of this huge, genetically interconnected Venezuelan population, where families with 13 or more children are common. Over 10,000 individuals have now been added to this extensive pedigree. Wexler also collected blood samples from these people, shipping them to the laboratory of Dr. James Guisella at Massachusetts General Hospital in Boston. In Guisella's lab, molecular genetic technology was used to identify a region of DNA near the tip of human chromosome 4 as the location of the gene for HD. Comparing the genetic in-

Figure E15-1   Tracking the gene for Huntington's disease

Nancy Wexler is shown here with one of the many inhabitants of the Lake Maracaibo region of Venezuela who suffer from or may carry the gene for Huntington's disease.

formation in the Venezuelan blood samples with the relationships depicted in the pedigree, Guisella's group devised tests that very accurately determine whether a person has inherited the gene for HD from an affected parent or is free of the "genetic time bomb," as Wexler describes the Huntington's gene.

Ironically, although Nancy Wexler was instrumental in the development of the screening test, she has decided not to find out whether she has inherited her mother's Huntington's gene. As she explains, each individual must make a deeply personal decision as to whether the pain of knowing the gene is present outweighs the joy of knowing the gene is absent, or whether living with the anxiety of not knowing is preferable. The availability of this test is of special importance to an individual who may have inherited the Huntington's gene and wants to have children. Genetic screening coupled with the voluntary decision not to risk passing the gene to the next generation could rapidly eliminate this tragic condition.

Is the Y chromosome totally devoid of genes other than those involved in determining maleness? No, but it turns out to be quite difficult to prove Y-linked inheritance. You might think that Y-linked traits should be easy to demonstrate, because *all* the males and *only* the males descended from a given affected male ancestor should show the trait. In practice, the situation is complicated by the influence of male sex hormones. Most traits that once were thought to be due to Y-chromosome genes are actually due to autosomal genes whose expression is strongly influenced by testosterone levels.

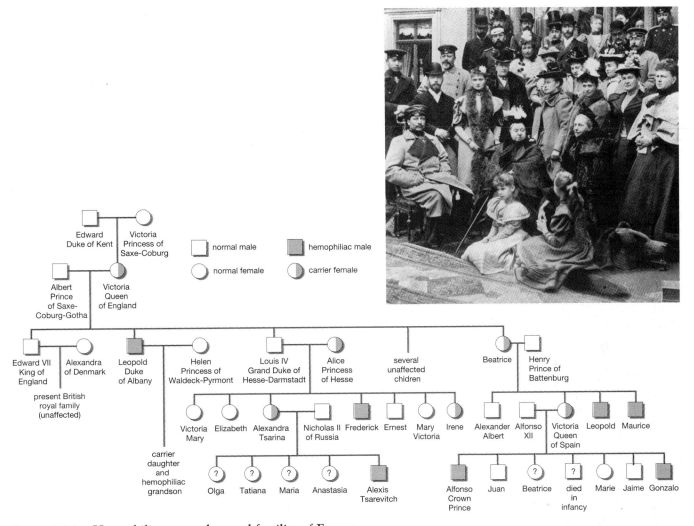

Figure 15-8   **Hemophilia among the royal families of Europe**

Hemophilia results from a deficiency of one of the substances causing blood clotting, a deficiency inherited as a recessive allele on the X chromosome. Affected males do not bleed to death from the first scratch or bruise; most hemophiliacs can clot off a minor wound. As a result, males often survive to pass on their allele to their daughters. Homozygous hemophiliac women are rare, but some are known. Surprisingly enough, they do not usually die from menstruation or even from bleeding following childbirth. In both of these instances, blood flow is stopped not by clotting, but by muscular contraction, shutting off circulation to the uterine wall.

The most famous genetic pedigree involves the transmission of sex-linked hemophilia from Queen Victoria of England (seated in right center, with crown) to her offspring, and eventually to virtually every royal house in Europe. Because all of Victoria's ancestors were free of hemophilia, the hemophilia allele must have arisen as a mutation either in Victoria herself when she was an embryo or in one of her parents.

Extensive intermarriage among royalty, who, after all, are not supposed to marry commoners, spread Victoria's hemophilia allele throughout Europe. Her most famous hemophiliac descendant was great-grandson Alexis, Tsarevitch (crown prince) of Russia. Bleeding episodes in the only son of the Tsar naturally distressed his parents greatly. The Tsarina Alexandra (Victoria's granddaughter) believed that the monk Rasputin, and no one else, could control Alexis's bleeding. Rasputin may actually have been able to do this through hypnosis, perhaps causing Alexis to cut off circulation to bleeding areas by muscular contraction. Although there were many causes underlying the Russian Revolution, which led to the rise of the Communist regime, the influence that Rasputin had over the imperial family may have contributed to the downfall of the Tsar. In any event, hemophilia was not to be the cause of Alexis's death; he was killed with the rest of his family by the Bolsheviks in 1918.

Complex Inheritance

Many human characteristics are not inherited in a simple either/or fashion. Even traits that are controlled by single genes, such as Huntington's disease, are often influenced by other genes and by environmental factors. For example, although everyone who has at least one allele for Huntington's contracts the disease, the age at which the first symptoms appear is extremely variable: Some people are affected in their teens; others do not develop symptoms until their fifties or even sixties.

Many Human Traits Are Influenced by Many Genes

Many, perhaps most, human traits are determined by the interaction of two or more genes. Examples of such polygenic inheritance in humans are the colors of eyes and skin.

The color of the iris in human eyes varies from very pale blue through green to almost black. Nevertheless, there are no blue, green, or black pigments in the human iris. Eye colors are actually caused by the distribution of a single yellowish brown pigment, melanin, the same pigment that colors skin and hair. The iris contains two layers of pigment, one at the back and one at the front. If there is little or no pigment at the front, the iris appears blue. The blue color arises from the scattering of light in the front layers, viewed against the dark background of melanin in the rear, just as the sky appears blue because of light scattered by the air, seen against the black background of space. In people with greater amounts of melanin in the front layers, the iris color may be green (blue plus yellowish brown), brown, or almost black. At least two, and probably more, genes direct the synthesis of melanin in the front of the iris, with each gene having two alleles showing incomplete dominance. The simplest scheme of two genes can create five shades of eye colors (Fig. 15-9).

Skin color is another case of different amounts of melanin. People don't really have white, yellow, red, or black skin: All are various shades of brown, with a pinkish tint from surface blood vessels showing through the paler tones. Skin color is inherited through the action of at least three pairs of genes, each with incompletely dominant pairs of alleles. As with eye color, polygenic inheritance explains both a more or less continuous gradation of skin colors and the occasional offspring whose skin color differs considerably from that of either parent.

The Environment Influences the Expression of Genes

Human physical traits are also affected by the environment. This fact is obvious to any Caucasian who spends much time in the summer sun: Skin color can be darkened by exposure to sunlight. At the same time, a person's genotype limits the range of environmental effects on phenotype. How fast we run, how high we jump, or how well we sing can all be greatly improved by training, but no amount of practice would ever enable most of us to soar above the rim like Michael Jordan.

Environmental factors influence the expression of every trait in all organisms, including behavioral traits. Humans, however, with our uniquely developed capacity for personality development, social interactions, and learning, are behaviorally much more flexible than any other animal. So great is the role of learning in human personality development that there are those who maintain that the human mind is almost infinitely changeable and that there are no genetic differences among people in personality or intelligence. However, increasing evidence indicates that at least some mental traits are influenced by both environmental factors and inherited tendencies. Some people are probably genetically predisposed to the development of certain mental diseases, such as schizophrenia, although environmental events may determine whether symptoms develop. Some of these disorders have been traced to imbalances in important chemicals in the brain. Affected people probably bear alleles coding for defective versions of the enzymes that regulate the synthesis, breakdown, and use of these chemicals.

Intelligence, too, has both genetic and environmental components. Dozens of studies have compared IQ levels in people of varying degrees of relatedness. Even when they have been separated at birth and raised in different environments, identical twins make similar scores on IQ tests, although not as similar as twins who have been raised together. Brothers and sisters who are not twins differ more than twins but are still fairly similar, while parents and children are less so. Thus, the more genetically related two people are, the more similar their IQ scores, indicating very strongly that intelligence, or at least whatever it is that IQ tests measure, is partially genetic. At the same time, unrelated people who have been raised together as children (for example, adoptees) show more similarity on IQ tests than unrelated people reared apart. Therefore, it is fair to say that *both heredity and environment play major roles in the development of intelligence*, and probably other personality traits as well.

Chromosomal Inheritance

Like other diploid organisms, humans possess two copies of most genes, one on each of two homologous chromosomes. This rule is not true, of course, for sex-linked genes in men, whose Y chromosome lacks most of the genes borne on the X chromosome. It is also not true for people who inherit abnormal numbers of chromosomes. In Chapter 12 we examined the intricate mechanisms of meiosis, which act to ensure that each sperm and each egg receive

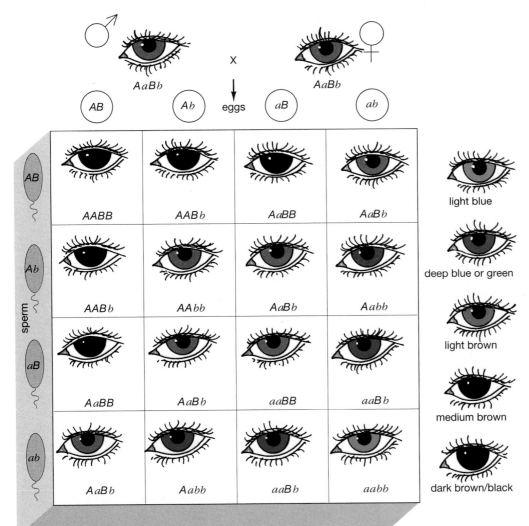

Figure 15-9   **Human eye color**

At least two separate genes, each with two incompletely dominant alleles, govern human eye color. A man and a woman, each heterozygous for both genes, could have children with five different eye colors, ranging from light blue (no dominant alleles) through light brown (two dominants) to almost black (all four alleles dominant).

light blue

deep blue or green

light brown

medium brown

dark brown/black

one homologous chromosome of each pair. Not surprisingly, this elaborate dance of the chromosomes occasionally misses a step, resulting in gametes that have too many or too few chromosomes. Such errors in meiosis, called **nondisjunction,** can affect the distribution of both sex chromosomes and autosomes. Most of the embryos that arise from the fusion of gametes with abnormal chromosome complements spontaneously abort, accounting for 20% to 50% of all miscarriages, but some survive.

Some Genetic Disorders Are Caused by Abnormal Numbers of Sex Chromosomes

Nondisjunction of the sex chromosomes occurs in both men and women. Nondisjunction in men produces sperm that are O (lacking any sex chromosome), XX, YY, or XY instead of the normal X or Y. In women, nondisjunction produces O or XX eggs instead of the normal X. When normal gametes fuse with these defective sperm or eggs, the zygotes have abnormal numbers of sex chromosomes (Table 15-1). The most common abnormalities are XO, XXX, XXY, and

XYY. (Genes on the X chromosome are absolutely essential to survival, and embryos with no X chromosome always spontaneously abort very early in development.)

Turner Syndrome (XO)

About one in every 5000 phenotypically female babies has only one X chromosome. At puberty, an XO female fails to menstruate or develop normal secondary sexual characteristics, such as breasts. These **Turner syndrome** women are sterile, usually short in stature, and often have folds of skin around their necks; under microscopic examination, their nuclei are shown to lack Barr bodies (see Fig. 11-17). Mentally, they are usually normal, except that they are frequently weak in mathematics and spatial perception. The differences between XO and XX women suggest that the theory of X chromosome inactivation in females (Chapter 11), which holds that in a normal female one X chromosome is condensed and "inactive," is oversimplified. Some genes on the inactivated X chromosome must be functional in XX females, preventing the Turner syndrome traits.

TABLE 15-1 ▓ *Effects of Nondisjunction of the Sex Chromosomes during Meiosis*			
Nondisjunction in Father			
Sex Chromosomes of Defective Sperm	*Sex Chromosomes of Normal Egg*	*Sex Chromosomes of Offspring*	*Phenotype*
O	X	XO	Female
XX	X	XXX	Female
YY	X	XYY	Male
XY	X	XXY	Male
Nondisjunction in Mother			
Sex Chromosomes of Normal Sperm	*Sex Chromosomes of Defective Egg*	*Sex Chromosomes of Offspring*	*Phenotype*
X	O	XO	Female
Y	O	YO	Dies as embryo
X	XX	XXX	Female
Y	XX	XXY	Male

Trisomy X (XXX)

About 1 in 1000 women have three copies (trisomy) of X chromosomes. These women usually have no detectable defects at all, except for a relatively high incidence of below-normal intelligence. Unlike women with Turner syndrome, **trisomy X** women are fertile, and, interestingly enough, almost always bear normal XX and XY children. Some mechanism, presently unknown, must operate during meiosis to prevent the extra X chromosome from being included in the egg.

Klinefelter Syndrome (XXY)

About one male in 1000 is born with two X and one Y chromosomes. At puberty, these men show mixed secondary sexual characteristics, including partial breast development, broadening of the hips, and small testes. Men with **Klinefelter syndrome** are always sterile but usually not impotent. As is common in people with abnormal chromosome numbers, XXY males have an increased incidence of mental deficiency; in fact, about 1% of all people institutionalized for mental retardation are XXY males.

XYY Males

The last common type of sex chromosome abnormality is XYY, occurring in about one male in 1000. You might expect that having an extra Y chromosome, which presumably has few genes, would not make very much difference, and in most cases it doesn't seem to. However, XYY males may be affected in two ways: below-average intelligence and above-average height (about two-thirds of XYY males are over 6 feet tall, compared with the average male height of 5'9"). There is some debate about whether XYY males are geneti-

cally predisposed to violence. For instance, several studies have shown that a higher-than-expected percentage of men in prison are XYY. In several countries men accused of murder attempted to use their XYY constitution as a defense, like the insanity plea. They were not acquitted. The juries were probably right: Only a tiny percentage of XYY males ever commit any sort of crime, so an extra Y chromosome certainly doesn't force anyone into a life of violence.

In Humans, the Y Chromosome Determines Maleness

Studies of men and women with abnormal numbers of sex chromosomes lead to the inescapable conclusion that the Y chromosome determines maleness in humans. Simply having only one X chromosome, as both XY males and XO females do, does not automatically lead to maleness. In most respects, including external genitalia, XO individuals are clearly female. On the other hand, having a Y chromosome produces the male phenotype, no matter how many X chromosomes are present. Even the rare XXXY or XXXXY person is male. In computer terminology, femaleness is the "default" condition for human sex; explicit instructions encoded on the Y chromosome are required to produce a male. In fact, geneticists have found a specific gene on the Y chromosome that causes both the development of male sex organs and the regression of female sex organs during fetal development.

Some Genetic Disorders Are Caused by Abnormal Numbers of Autosomes

Nondisjunction of the autosomes may also occur, producing eggs or sperm with a missing autosome or two copies of an autosome. Fusion with a normal gamete (one

(a)

(b)

Figure 15-10   **Trisomy 21, or Down syndrome**

(a) Karyotype from a Down syndrome child reveals three copies of chromosome 21. **(b)** Children with abnormal numbers of chromosomes are almost always both physically and mentally defective, in ways that are characteristic of the particular chromosome abnormality. These girls have the typical relaxed mouth and distinctively shaped eyes usually seen in cases of Down syndrome.

copy of each autosome) leads to an embryo with either one or three copies of the affected autosome. With only one copy of any of the autosomes, the embryo aborts so early in development that the woman never knows she was pregnant at all. The presence of three copies of an autosome (trisomy) usually also causes a spontaneous abortion, but often later in pregnancy. Sometimes, however, trisomic babies are born, especially those with three copies of chromosomes 13, 18, or 21. Of these, trisomy 21 is the most common.

Trisomy 21, or Down Syndrome

In about one of every 900 births, the child has an extra copy of the twenty-first chromosome. Children with **trisomy 21**, also called **Down syndrome**, have several distinctive physical characteristics, including weak muscle tone, a small mouth held partially open because it cannot accommodate the tongue, and distinctively shaped eyelids (Fig. 15-10). Much more serious defects include low resistance to infectious diseases, heart malformations, and mental retardation so severe that only about 1 in 25 ever learns to read and only 1 in 50 learns to write.

The frequency of nondisjunction is influenced by the age of the parents, especially the mother (Fig. 15-11). As we will see in Chapter 27, meiosis begins in a woman's ovaries while she is still a fetus in her mother's womb, but it is suspended during prophase of meiosis I. When

Figure 15-11   **Down syndrome frequency increases with maternal age**

This graph shows that nondisjunction of chromosome 21, producing children with Down syndrome, greatly increases after the mother reaches age 35. Geneticists have found that this relation is true even for women whose husbands are much younger.

a woman matures, meiosis resumes, a few cells at a time, during the monthly menstrual cycle. Therefore, a 40-year-old woman produces eggs that have been in meiosis for 40 years. Errors in chromosome distribution are more likely in old cells, perhaps just from aging itself, or perhaps from exposure to radiation, toxic chemicals, or viruses. In men, new sperm-producing cells are constantly being produced in the testes, and meiosis takes only a couple of weeks. Therefore, nondisjunction does not increase with age as rapidly in men as in women. Nevertheless, nondisjunction in sperm accounts for about 25% of the cases of Down syndrome, and there is a small age effect.

In the past two decades, it has become increasingly common for couples to delay having children, so that both husband and wife can establish careers. As a result, many women bear children when they are in their late 30s or early 40s, and in general they have husbands of similar age. Delaying pregnancy inevitably leads to more trisomic fetuses than if the same couples had reproduced earlier. Fetal trisomy can be diagnosed by examining the chromosomes of fetal cells collected through amniocentesis or chorionic villus sampling. Some couples may then choose a therapeutic abortion to prevent the birth of a trisomic child.

The Human Genome Project

Beginning with a few small meetings in 1985 and 1986, and now racing along with multimillion dollar budgets, the Human Genome Project is probably science's most ambitious response to the command of the ancient Oracle at Delphi: "Know thyself." The answers to a host of questions about humankind reside in the nucleotide sequences of our genes. Further, humanity is heir to about 4000 genetic disorders. If we knew the nucleotide sequence of the entire human genome, then, in theory at least, we could devise rapid screening and prenatal diagnosis of many genetic disorders, design better therapies for many diseases, and understand much more about the workings of the human brain.

Ever since DNA sequencing was first developed by Alan Maxam, Walter Gilbert, and Frederick Sanger, molecular geneticists have pondered the potential for sequencing the human genome, but the time, effort, and cost seemed daunting—$3 billion was the estimate in 1986. Recent developments in RFLP analysis, automated DNA sequencing, and polymerase chain reaction (PCR) technologies, however, have combined to make the project now seem quite reasonable. Although many meetings and much research went on before, the Human Genome Project officially got under way in 1988.

The project has many ultimate goals, but there are three of foremost importance:

1. Produce a genetic map. The locations of at least 3000 genes or other markers (such as restriction enzyme cut sites) will be sought. Remember, linkage maps locate genes by the relative frequency of crossing over (see Chapter 13). Linkage maps can be used in pedigree analysis to help locate other genes that are known only from phenotypic effects. This portion of the project is now complete.

2. Generate a physical map. Each human chromosome will be cut apart with restriction enzymes. Unique DNA nucleotide sequences will be sought in each piece, with the unique sequences no farther apart than about 100,000 nucleotides. These unique nucleotide sequences can then be used as starting points for biochemically "walking" along the chromosomes, sequencing all the DNA or looking for previously unmapped genes. This part of the project is well underway.

3. Sequence the human genome. A complete set of haploid human chromosomes contains about 6 billion nucleotides (3 billion pairs on two complementary strands). This is an ambitious and expensive undertaking, but advocates point out that, if the entire genome sequence were known, and someone later pinpointed the gene locus of a genetic disorder, then one could obtain the gene sequence merely by looking it up in a computer. It will probably be at least 10 more years before the entire genome is sequenced.

There is, of course, no one human genome—yours is different from everyone else's. By definition, the nucleotide sequence for a defective allele differs from that of the normal allele, and there is considerable variety among normal, functioning alleles, too. Therefore, at least for the DNA sequences that code for genes, it will be important to sequence the DNA from a fairly large sample of people.

The Human Genome Project has its detractors, mostly because of its cost. If the overall science budget remains constant, then the $50 million to $100 million slated to be spent each year on the Human Genome Project must mean decreased funding for other projects. Supporters, however, point to increased understanding of normal functioning, human evolution, and human genetic disorders that the project would produce. Certainly, the Human Genome Project has brought together many of the best molecular geneticists and is fostering cooperation in a field in which fierce competition is the norm.

Finally, the human genome is not the only one of interest. Plans are already being made to sequence the genomes of the bacterium *Escherichia coli*, the fruit fly *Drosophila melanogaster*, the roundworm *Caenorhabditis elegans*, the domestic dog, and at least one plant. These sequences would be used in fields as diverse as microbiology, developmental biology, veterinary medicine, and agriculture.

Medical Genetics Poses New Medical and Ethical Dilemmas

Genetics has opened up new vistas of understanding human nature. Analyses of family pedigrees, biochemical tests, amniocentesis, and recombinant DNA technologies have enabled geneticists to identify prospective parents who are carriers for some disorders, such as sickle-cell anemia and Tay-Sachs disease, and to detect fetuses who are homozygous for these and other disorders. These new powers demand a new set of decisions, both ethical and economic, by individuals, physicians, and society. Let us examine these new choices in one specific case, phenylketonuria.

Phenylketonuria Illustrates a Biomedical Dilemma

As you learned in Chapter 11, people with **phenylketonuria (PKU)** are homozygous for a recessive allele coding for a defective version of the enzyme that converts the amino acid phenylalanine to tyrosine (see Fig. 11-1). As is usual with recessive diseases, most homozygous children result from the mating of two heterozygous (and therefore phenotypically normal) parents. The level of phenylalanine in the developing fetus is regulated by its pregnant mother, so homozygous recessive infants are born with normal brain development. Once on their own, however, the infants' phenylalanine levels increase rapidly, damaging the brain. Fortunately, homozygous infants can be detected as early as 4 or 5 days after birth by a simple blood test that costs just a few dollars. Treatment consists of maintaining these children on a diet low in phenylalanine for at least the first 5 or 6 years of life, when most brain development occurs. Most naturally occurring proteins contain phenylalanine, so the low-phenylalanine diet is quite expensive and inconvenient.

This is not the end of the story, however. Women with PKU who were treated in infancy have normal mental development, but they are still deficient in phenylalanine metabolism and therefore have an elevated concentration of phenylalanine in their blood. Children of such women, fathered by homozygous normal men, will be heterozygous and should be phenotypically normal. The problem is that the phenylalanine levels of an otherwise normal fetus are raised by the high levels of its phenylketonuric mother. High phenylalanine levels during pregnancy, even just the first few months, cause mental retardation. The only protection for the fetus is if the mother adheres to the low-phenylalanine diet.

Thus, treating phenylketonuria in female infants brought with it new responsibilities. Many women now of child-bearing age went off their low-phenylalanine diets years ago. Many have since forgotten that they have PKU or, not understanding genetics, think that they *used* to have PKU but have now outgrown it. Others planned to resume the low-phenylalanine diet when they wanted to become pregnant but became pregnant unintentionally. Therefore, the best medical advice for phenylketonuric women is to stay on the low-phenylalanine diet until menopause.

Knowledge of Human Genetics Poses Ethical Dilemmas

Insofar as we cannot cure most inherited diseases, or foresee a cure in the near future, what are the responsibilities and options open to individuals and societies? A hundred years ago, when no one understood the nature of inherited diseases, the choices were limited and the responsibilities few. A prospective parent could not know whether he or she carried an inherited defect, could not predict the likelihood of having an affected child, and could not tell whether a child was affected until the child was born or perhaps was even several years old.

The situation has changed dramatically. Today, many people *do* know that they carry a seriously defective gene, and genetic counselors can predict their chances of having an affected child. Should such an individual refrain from reproduction? In some cases, such as Tay-Sachs disease and the chromosomal abnormalities, prenatal diagnosis can detect the inherited defect in a fetus. Should the fetus be aborted? Should screening tests, at least for infants, be encouraged or required, as is now the case for PKU and, in some states, sickle-cell anemia? Would such a requirement constitute governmental interference in the rights of the individual? What about screening tests for prospective parents, at least in high-risk groups? Should these tests be required?

What, if any, are the rights and responsibilities of society in these decisions? In many Western countries, society pays most or all of the costs of medical care for affected children, often running to hundreds of thousands of dollars each. What, if any, are the obligations of society to children who might be born with genetic defects? Suppose a phenylketonuric woman goes off the low-phenylalanine diet against the advice of her physician and then bears a child. Would she be guilty of child abuse?

Some people argue that the human species is becoming genetically "loaded" with defective alleles: Whereas in earlier times people who had the alleles for diabetes, sickle-cell anemia, and many other genetically influenced diseases would have died, modern medicine now allows them to survive and reproduce. As a result, some believe that carriers of alleles for serious, incurable diseases should not reproduce, to eliminate not only homozygous recessives who are afflicted with the disease but also future carriers. Eventually, when we become knowledgeable enough, we may find that almost *all* of us carry recessive alleles for some inherited disease or other, yet few would argue that the majority of the population is morally obligated to remain childless. In a free society, probably the best solution is to give people the most accurate information possible about their genetic constitution and that of their future children. The choices, whatever their consequences for society as a whole, almost certainly will remain with the individual.

SCIENTIFIC INQUIRY

Collecting Fetal Samples and Analyzing Them Using Biotechnology

Prenatal diagnosis of a variety of genetic disorders, including sickle-cell anemia, Tay-Sachs disease, and Down syndrome, requires samples of fetal cells or chemicals produced by the fetus. Presently, there are two main techniques used to obtain these samples: amniocentesis and chorionic villus sampling. Once samples are collected, several tests, including some using recombinant DNA techniques, can be performed that allow prenatal diagnosis of many genetic disorders.

(a)

Amniocentesis

The human fetus, like all animal embryos, develops in a watery environment. A waterproof membrane called the amnion (see Chapter 27) surrounds the fetus and holds the amniotic fluid. As it develops, the fetus sheds some of its own cells into the amniotic fluid. When a fetus is 16 weeks or older, enough amniotic fluid has accumulated and there is enough space between the fetus and the amniotic membrane for the amniotic fluid to be collected by a procedure called amniocentesis. A physician determines the position of the fetus by ultrasound scanning (Fig. E15-2a) and inserts a sterilized needle through the abdominal wall of the pregnant woman and into the amniotic fluid (Fig. E15-2b). Ten to 20 milliliters of fluid is withdrawn. Biochemical analysis may be performed on the fluid immediately. However, there are very few cells in the fluid sample. For many analyses, such as karyotyping for Down syndrome, the cells must first be grown in culture. The cells multiply in culture, and after a week or two there are usually enough cells for karyotyping or other analyses.

(b)

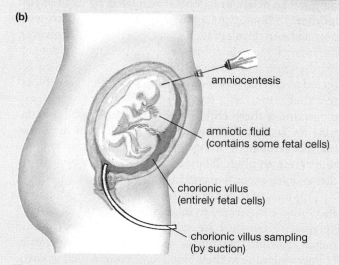

Chorionic Villus Sampling

This is a newer procedure. The chorion is a membrane, produced by the fetus, that becomes part of the placenta (see Chapter 27). The chorion produces many small projections, called villi; the loss of a few villi seems to cause no harm. In chorionic villus sampling, a physician inserts a small tube into the uterus through the vagina and suctions off a few villi for analysis (see Fig. E15-2b).

Figure E15-2   **Prenatal tissue sampling techniques**

(a) Ultrasound imaging reveals the position of the fetus. A needle can then be inserted into the amniotic sac to withdraw fluid without touching the fetus. **(b)** Two methods of obtaining fetal tissue samples.

The Two Techniques Compared

Chorionic villus sampling has two major advantages over amniocentesis. First, it can be done much earlier in pregnancy, perhaps as early as the eighth week. Early testing is especially helpful if the woman is contemplating a therapeutic abortion in case the fetus suffers from a major defect. Second, the sample contains a much higher concentration of fetal cells than amnio-

centesis can obtain, so analyses can be performed immediately rather than waiting for a week or two. However, chorionic cells seem to be more likely than amniotic fluid to have abnormal numbers of chromosomes, even in a normal fetus, which complicates karyotyping. In most large metropolitan areas, both amniocentesis and chorionic villus sampling are available, and each has advantages and drawbacks.

Analyzing the Samples

Several analyses can be performed on the fetal cells or on the amniotic fluid. Biochemical analysis is used to determine the concentration of chemicals in the amniotic fluid. For example, Tay-Sachs disease and many other metabolic disorders can be detected by the low concentration of the enzymes that normally catalyze specific metabolic pathways, or by the accumulation of precursors or by-products. Analysis of the chromosomes of the fetal cells can show if all the chromosomes are present, if there are too many or too few of some, and if there are structural abnormalities of any of the chromosomes.

Analysis of the DNA of fetal cells using recombinant DNA techniques can detect some defective alleles, such as the Tay-Sachs and sickle-cell anemia alleles. Just a few years ago, prenatal diagnosis of sickle-cell anemia or other genetic disorders required thousands or millions of copies of the DNA. Therefore, fetal cells often had to be grown in culture for as long as 2 weeks before they had multi-plied enough to be used. For sickle-cell anemia, the development of the polymerase chain reaction (PCR; see Chapter 14) has eliminated the wait.

As you may recall, PCR can produce virtually unlimited copies of a specific gene, starting with as little as one copy. The second step in prenatal diagnosis of sickle-cell anemia, therefore, is to extract the DNA from a few cells and to amplify the hemoglobin genes (both normal and sickle-cell versions) with PCR. This process takes only a few hours. The third step is to cut the PCR-amplified DNA with special restriction enzymes and sort the DNA according to size by gel electrophoresis. The hemoglobin genotype of the fetus can then easily be determined.

If the infant is homozygous for the sickle-cell allele, some therapeutic measures can be taken. In particular, regular doses of antibiotics greatly reduce bacterial infections that otherwise kill about 15% of homozygous children. Further, knowing that a child has the disorder ensures correct diagnosis and rapid treatment, should "sickling crises" occur.

✖ SUMMARY OF KEY CONCEPTS

Methods in Human Genetics

The genetics of humans is similar to the genetics of other animals, except that experimental crosses are not feasible. Analysis of family pedigrees and, more recently, molecular genetic techniques must be used to determine the mode of inheritance of human traits.

Single-Gene Inheritance

Many genetic disorders are inherited as recessives; therefore, only homozygous recessive persons show symptoms of the disease, whereas heterozygotes are phenotypically normal and usually cannot be detected by casual observation. Heterozygotes are called carriers, because the offspring of two carriers may be homozygous recessive and show the trait. Recessive genetic disorders include Tay-Sachs disease, sickle-cell anemia, and phenylketonuria.

Many normal traits, and a few diseases, are inherited as simple dominants. Both heterozygous and homozygous dominant individuals show the trait.

The sex chromosomes are alike in women (XX) but not alike in men (XY). The Y chromosome bears few genes other than those determining maleness. There-fore, women show normal dominant-recessive relationships among alleles of X chromosome genes, whereas men phenotypically display whatever allele they carry on their single X chromosome, a phenomenon called sex-linked, or X-linked, inheritance. Sex-linked conditions include red-green color blindness and hemophilia. Sex hormones help to determine the expression of sex-influenced traits. The majority of sex-influenced defects are more common in males than in females and include baldness and gout.

Complex Inheritance

Most human traits are polygenic—that is, they are influenced by the action of many genes. Examples include the color of skin and eyes and probably many personality traits. The environment affects the phenotypic expression of all genes in all organisms. This rule is particularly apparent in human beings, especially in intellectual and personality traits.

Chromosomal Inheritance

Gender in humans is determined by the presence of a Y chromosome. Individuals with one or more Y chromosomes are phenotypically male; individuals with no Y chromosome are phenotypically female. People with abnormal numbers of sex chromosomes often have mental and physical deficiencies. The most common defect is below-normal intelligence.

Abnormal numbers of autosomes usually lead to spontaneous abortion early in pregnancy. In rare instances, the fetus may survive to birth, but severe mental and physical deficiencies are always found. The likelihood of abnormal numbers of chromosomes increases with increasing age of the mother, and, to a lesser extent, the father.

The Human Genome Project
The largest single project in the history of biology, the Human Genome Project, proposes to produce (1) a detailed linkage map; (2) a precise physical map; and (3) the complete nucleotide sequence of the "average" human genome.

❋ KEY TERMS

amniocentesis   p. 270
carrier   p. 269
chorionic villus sampling   p. 270
Down syndrome   p. 279
hemophilia   p. 273
Klinefelter syndrome   p. 278

nondisjunction   p. 277
pedigree   p. 267
phenylketonuria (PKU)   p. 281
sex-influenced trait   p. 273
sex-linked inheritance   p. 273
sickle-cell anemia   p. 270

Tay-Sachs disease   p. 270
trisomy 21   p. 279
trisomy X   p. 278
Turner syndrome   p. 277
X-linked inheritance   p. 273

❋ THINKING THROUGH THE CONCEPTS

Multiple Choice

1. Which one of the following is a lethal mutation that does not express itself until late in life in men and women?
   a. Down syndrome
   b. hemophilia
   c. Huntington's disease
   d. Tay Sachs disease
   e. sickle cell anemia

2. Which is NOT true of sickle-cell anemia?
   a. It is most common in African Americans.
   b. It involves a one–amino acid change in hemoglobin.
   c. It involves red blood cells.
   d. It is lethal in heterozygotes since it is dominant.
   e. It confers resistance to malaria.

3. Which statement is NOT true?
   a. The frequency of some chromosome abnormalities goes up with maternal age.
   b. XYY individuals originate from two sperm cells' fertilizing one ovum.
   c. Sperm are created throughout life.
   d. Meiosis to produce human eggs begins before birth.
   e. Chromosomal abnormalities are different from genetic abnormalities in that the former are usually not passed to the next generation.

4. Sex-linked disorders such as color blindness and hemophilia are
   a. on the X chromosome
   b. on the autosome
   c. on the Y chromosome
   d. the same as sex-influenced traits such as baldness; they are simply more likely to be expressed in men
   e. expressed only when two chromosomes are homozygous recessive

5. The improper disjoining (nondisjunction) of chromosome 21 in human beings leads to
   a. Tay-Sachs disease
   b. crossing over
   c. Klinefelter syndrome
   d. Turner syndrome
   e. Down syndrome

6. In human beings, red color blindness is modeled as a sex-linked recessive allele. A couple has a red-blind daughter. Which of the following statements is the only one that MUST be correct on the basis of this information?
   a. The father is red blind.
   b. The mother is red blind.
   c. Both father and mother are red blind.
   d. The daughter inherited all red-blind genes from the mother.
   e. Half of the mother's sons would be red blind.

Review Questions
(The answers to questions 1 and 2 are given at the end of the question list.)

1. If the frequency of heterozygous carriers for Tay-Sachs disease is 1 in 30 American Jews, why is the frequency of homozygous recessive babies of Jewish parents 1 in 3600?

2. If one parent of a couple has Huntington's disease (assume that this parent is heterozygous), calculate the fraction of their children that would be expected to develop the disease. What if both parents were heterozygous?

3. Why are most genetic diseases inherited as recessives rather than dominants?

4. How is sex determined in humans? What is the evidence for this?

5. Define polygenic inheritance. Why could polygenic inheritance allow parents to produce offspring that are notably different in eye or skin color than either parent?

6. What is sex-influenced inheritance? What is the evidence that hormonal levels control the expression of sex-influenced traits?

7. Define nondisjunction, and describe the common syndromes caused by nondisjunction of sex chromosomes and autosomes.

8. Describe amniocentesis and chorionic villus sampling, including the advantages and disadvantages of each. What are their medical uses? What technique can be used in conjunction with amniocentesis in order to expedite production of a sufficient sample of DNA material from a gene for analysis?

9. Refer to the material in Chapter 11. What type of mutation is sickle-cell anemia?

✖ A N S W E R S   T O   Q U E S T I O N S   1   A N D   2

1. The probability of two or more independent events occurring simultaneously is the product of their individual probabilities. The probability of the husband being heterozygous is 1/30; the probability that the wife is heterozygous is also 1/30; the probability that two heterozygotes will produce a homozygous recessive child is 1/4 (see Chapter 13). Multiplying $1/30 \times 1/30 \times 1/4$ gives 1/3600.

2. Let H = the Huntington's allele (dominant) and h = the normal allele (recessive). The first set of parents would be Hh and hh. Half of their offspring would be expected to inherit Huntington's disease. If both parents are heterozygous, then the cross is $Hh \times Hh$. We would expect the offspring to be 1/4 HH, 1/2 Hh, and 1/4 hh; therefore, 3/4 would develop the disease.

✖ A P P L Y I N G   T H E   C O N C E P T S

1. *Eugenics* is the term applied to the notion that the human condition might be improved by improving the human genome. Do you think there are both good and bad sides to eugenics? What examples can you think of to back up your stand? What would an out-and-out eugenicist think of the medical advances that have ameliorated the problems of PKU?

2. Think about some of the personal, religious, and medical economic issues related to prenatal counseling and diagnosis. Would you avoid having children if you knew that both you and your spouse were heterozygous for a recessive disorder such as Tay-Sachs, which is fatal at an early age and may involve considerable suffering? What would you do if you or your spouse were pregnant and learned that your offspring, if born, would be homozygous for such a disorder? Is the situation qualitatively different for Down syndrome? (Down syndrome children have a short life expectancy [20 to 30 years] with mental retardation but have a generally pleasant disposition.) If you were heterozygous for Huntington's disease, would you want to avail yourself of the medical diagnostic tests being made available through the most recent technologies to find out? If the answer is "no" and you were a man, what would you think about your wife's potential decision to have amniocentesis or chorionic villus sampling?

3. Now that you know a bit more about human genetics and polygenic inheritance, rethink your answer to the following question from Chapter 12: Farm animals with desirable traits are sometimes produced by removing eggs from the female and fertilizing them in a dish using sperm from a male with desirable traits. After the zygote has undergone several divisions, the individual cells can be separated, returned to the same or different females, and they will develop into genetically identical animals. Human sperm banks already carry sperm from men with particular traits that a woman may choose. Human *in vitro* fertilization (see the Scientific Inquiry box in Chapter 27) could theoretically be used to create identical twins, triplets, etc. Argue for or against such human "cloning." Has your answer changed? Why or why not?

✖ F O R   M O R E   I N F O R M A T I O N

Collins, F., and Galas, D. "A New Five-Year Plan for the U.S. Human Genome Project." *Science*, 262:43–46, 1993.

Horgan, J. "Eugenics Revisited." *Scientific American*, June 1993. Determining patterns of inheritance, let alone finding suspect genes, for behavioral disorders is notoriously difficult. What's more, it isn't at all clear what one should do with such data, if they become available.

McGue, M. "Nature-Nurture and Intelligence." *Nature*, 340: 507–508, 1989. Nature works via nurture in producing the outward manifestations of intelligence.

Pines, M. "In the Shadow of Huntington's." *Science*, May 1984; Grady, D. "The Ticking of a Time Bomb in the Genes." *Discover*, June 1987; Roberts, L., "Huntington's Gene: So Near, Yet So Far." *Science*, 247:624–627, 1990; and Morell, V. "Huntington's Gene Finally Found." *Science*, 260:28-30, 1993. The story of the scientific detective work involved in the diagnosis of Huntington's disease using recombinant DNA techniques, the social dilemma it created, the frustrations of searching for the gene, and finally finding it. Particularly poignant, because one of the lead investigators may herself be a victim.

Roberts, L. "To Test or Not to Test." *Science*, 247:17-19, 1990. Some of the alleles causing cystic fibrosis cannot be detected —should the entire childbearing population be screened for carriers for this devastating disease?

Sutherland, G. R., and Richards, R. I. "Dynamic Mutations." *American Scientist*, March–April 1994. Several genetic disorders, including Huntington's disease, are characterized by large numbers of repeated nucleotide triplets.

Wivel, N. A., and Walters, L. "Germ-line Gene Modification and Disease Prevention: Some Medical and Ethical Perspectives." *Science*, 262: 533–538, 1993. If we can fix a defective allele in an egg or zygote, should we?

NET WATCH

On-line resources for this chapter are on the World Wide Web at:
http://www.prenhall.com/~audesirk (click on the <u>table of contents</u> link and then select Chapter 15).

Unit III
Evolution

Beautifully preserved large and small fossil ammonites, which are prehistoric mollusks

16 Principles of Evolution

"When on board H.M.S. Beagle, as naturalist, I was much struck with…the distribution of the inhabitants of South America, and…the geological relations of the present to the past inhabitants of that continent. These facts seemed to me to throw some light on the origin of species—that mystery of mysteries, as it has been called by one of our greatest philosophers."

CHARLES DARWIN in *On the Origin of Species by Means of Natural Selection* (1859)

A fossil trilobite, an arthropod that was common 450 million years ago

The words above introduce what is perhaps the most important work in biology. In *On the Origin of Species*, Charles Darwin proposed that over thousands of millions of years, species arise from other, preexisting species through the process of "descent with modification," or **evolution**.

Before Darwin, how species originated remained the "mystery of mysteries" for a very simple reason. Over the time span of recorded human history, let alone the life of a single human being, no new species had been recognized (although undoubtedly many new species had appeared, especially plant species). It is quite difficult to determine how something happens if no one has ever observed it happening.

Nevertheless, one of the most striking features about our world is the remarkable variety of organisms inhabiting it. Why are there dozens of types of pine trees and scores of species of warblers? To explain the origin and diversity of life on Earth, nearly all peoples of the world historically turned to hypotheses of **creationism**. The most common of these hypotheses is that a supernatural being created each type of organism individually at the beginning of the world and that all modern organisms are essentially unchanged descendants of those ancestors.

Throughout history, however, scientists have sought natural, rather than supernatural, causes to explain natural events. As we pointed out in Chapter 1, one of the fundamental principles of science is that earthly phenomena are produced by natural, earthly causes that follow natural laws. Science cannot say whether or not divine power originally established those laws, but science firmly adheres to the principle that natural events have causes that arise from the operation of natural laws.

Therefore, throughout history, scientists have sought natural causes for the origin of species. But it was only in the nineteenth century that a truly coherent theory—evolution by descent with modification, driven by natural selection—was developed. This theory was published by two British naturalists, Charles Darwin and Alfred Russell Wallace, in 1858 and today still forms the foundation of our understanding of evolution. Darwin and Wallace did not work in a vacuum. Centuries of thought and observation preceded them and influenced their ideas. Before examining the principles of evolution, let us begin, then, with a brief survey of evolutionary thought.

The History of Evolutionary Thought

What Is a Species?

Before we can study the origin of species, we must first decide what a species is. Throughout most of human history, "species" was a poorly defined concept. When using the word *species*, most Europeans meant one of the originally created "kinds" referred to in the Bible. But because no one was present at the Creation to record the criteria of the Creator, one had to distinguish among species by visible differences in structure; in fact, the word *species* is Latin for "appearance." Clearly, warblers are different from eagles and ducks. But how do biologists distinguish among species of warblers? Today, biologists define a **species** as all the populations of organisms that are capable of interbreeding among themselves under natural conditions but that usually do not interbreed with members of other species. If interbreeding with another species does occur, the hybrid offspring are usually infertile or handicapped in some way (see Chapter 17). A **population** consists of all the individual members of a particular species within a given area.

Biologists have found that differences in appearance do not always mean that two populations belong to different species. For example, field guides published in the 1970s listed the Myrtle Warbler and Audubon's Warbler (Fig. 16-1) as distinct species; they differ in range and in the color of the throat feathers. More recently, scientists decided that they are, after all, merely local varieties of the same species. The reason: Where their ranges overlap, interbreeding between these warblers occurs, and the offspring are just as vigorous as the parents.

The Philosophies of Plato and Aristotle Influenced Early Christian Beliefs

Two principal lines of thought descended from the ancient Greeks to influence later Western ideas about the nature and origin of species. The philosophy of Plato (427–347 B.C.) rested on the foundation of the "ideal Form" that exists, perfect and forever unchanging, somewhere beyond Earth in the world of Forms. In this world view, each object on Earth, living or nonliving, rock, worm, or human, is a mere temporary reflection of its ideal Form. Plato's concept of unchanging Forms greatly influenced early Christian thought and came to be embodied in the idea that every species of living thing was created by God at the beginning of time. Although minor variations may occur among individual members of a given species, each species as a whole remains unchanged, very like a Platonic Form.

Plato's student Aristotle (384–322 B.C.), one of the first great naturalists, categorized all the living things that he encountered. Aristotle thought that all organisms fit into an orderly scheme, later called the *Scala Naturae*, or Ladder of Nature. The ladder stood, so to speak, upon nonliving matter; climbed rung by rung from fungi and mosses to higher

(a)

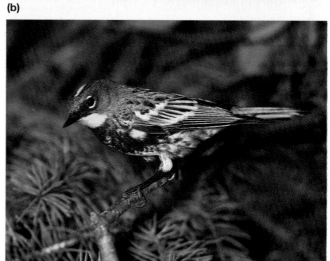

(b)

Figure 16-1   **Interbreeding blurs the distinction between species**

The Myrtle Warbler (**a**) and Audubon's Warbler (**b**) were formerly considered to be two separate species but are now considered to be merely local varieties of one widespread species.

plants, through primitive animals such as mollusks and insects; and finally culminated in human beings. Aristotle's ideas, even more than Plato's, were incorporated into Christian thought. The *Scala Naturae* was considered to be permanent and never changing: Each organism has its place on the ladder, ordained by God during creation.

Evidence Supporting Evolution Was Discovered Even before Darwin's Time

Creationism, the idea that each species was created individually by God and never changed thereafter, reigned unchallenged for nearly 2000 years. By the eighteenth century, however, new evidence arose that challenged this static view of creation.

Exploration of New Lands Revealed a Staggering Diversity of Life

As European naturalists explored the newly discovered lands of Africa, Asia, and America, they found that the diversity of living things was much greater than anyone had suspected. Some of these exotic species closely resembled one another yet also displayed variations in characteristics. This unpredicted expansion of information led some naturalists to consider that perhaps species could change after all and that some of the similar species might have developed from a common ancestor.

Fossils in Rocks Resembled Parts of Living Organims

At the same time, excavations for roads, mines, and canals revealed that rocks often occur in layers (Fig. 16-2). Sometimes, a few strangely shaped rocks were found embedded within one of these layers. These rocks, called **fossils** (from the Latin, meaning "dug up"), often resembled parts of living organisms. At first, fossils were thought to be ordinary rocks that wind, water, or people had worked into lifelike forms. As more and more fossils were discovered, however, it became obvious that they were in fact plants and animals that had died long ago and been changed into or preserved in rock (Fig. 16-3). After studying fossils carefully, William Smith (1769–1839) realized that certain fossils were always found in the same layers of rock. Further, the organization of fossils and rock layers was consistent: Fossil type A could always be found in a rock layer resting atop an older layer containing fossil B, which in turn rested atop a still older layer containing fossil C, and so on.

Fossil remains also showed a remarkable progression in form. Fossils found in the lowest (and therefore oldest) rock layers were usually very different from modern forms, with a gradual advancement to greater resemblance to modern species in younger rocks, as if there were a *Scala Naturae* stretching back in time. Many of these fossils were the remains of plants and animals that no longer lived on Earth (Fig. 16-4). Putting these facts together, scientists came to the inescapable conclusion that different types of organisms had lived at various times in the past.

But what did this newfound richness of organisms, both living and extinct, mean? Was each organism produced by a separate act of creation? If so, why? And why bother to create so many, letting thousands become extinct? Georges Louis LeClerc, known by his title Comte de Buffon (1707–1788), suggested that perhaps the original creation provided a relatively small number of founding species and that some of the modern species had been "conceived by Nature and produced by Time"—that is, they had evolved through natural processes. Most people were not convinced. First, Buffon could not provide any mechanism whereby Nature could "conceive" new species. Second, no one thought that there was Time enough for their "production."

Geology Provided Evidence That Earth Is Exceedingly Old

In the early 1700s, few scientists suspected that Earth could be more than a few thousand years old. Counting generations in the Bible's Old Testament, for example, yields a maximum age of 4000 to 6000 years. By reading the descriptions of plants and animals from ancient writers such as Aristotle, it was clear that wolves, deer, lions, and other European organisms had not changed in over 2000 years. How, then, could whole new species arise, if Earth was created only a couple of thousand years before Aristotle?

To account for a multitude of species, both extinct and modern, while preserving creationism, the French paleontologist (a scientist who studies fossils) Georges Cuvier (1769–1832) proposed the theory of **catastrophism.** Cuvier hypothesized that a vast supply of species was created in the beginning. Successive catastrophes (such as the Flood described in the Bible) produced the layers of rock and destroyed many species, fossilizing some of their remains in the process. The reduced flora and fauna of the modern world, he theorized, are the species that survived the catastrophes. However, if modern *species* have survived from an original creation, then many *individuals* of those species should have died in the ancient catastrophes. Surely some of them would have been fossilized, and even the lowest and oldest rock layers should contain at least a few fossils of present-day species. Unfortunately for Cuvier's hypothesis, they do not. Louis Agassiz (1807–1873) proposed that there was a new creation after each catastrophe and that modern species result from the most recent creation. The fossil record forced Agassiz to postulate at least 50 separate catastrophes and creations!

Alternatively, perhaps Earth *is* old enough to allow for the production of new species. Geologists James Hutton (1726–1797) and Charles Lyell (1797–1875) contemplated the forces of wind, water, earthquakes, and volcanism.

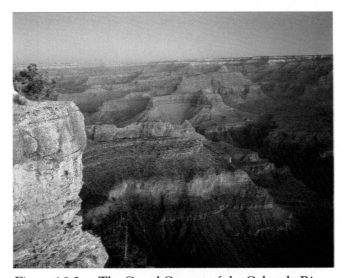

Figure 16-2   **The Grand Canyon of the Colorado River**
Layer upon layer of sedimentary rocks form the walls of the canyon, exposed in cliffs and mesas. The Grand Canyon strata cover more than a billion years of evolutionary history.

(a) The bipedal carnivore *Allosaurus* confronts a *Stegosaurus*. A sudden volcanic eruption buries both under lava and ash.

(b) Minerals from the surrounding rocks seep into their skeletons.

(c) Further eruptions bury the fossil skeletons more deeply, where they remain for tens of millions of years.

(d) Finally, erosion cuts through the overlying rock layers and exposes the fossils, where they may be found by paleontologists.

Figure 16-3   **Stages in fossilization**

Fossilization occurs when living organisms are buried beneath mud, silt, sand, or volcanic ash. This example might have occurred in Montana or Colorado.

They concluded that there was no need to invoke catastrophes to explain the findings of geology. Do not rivers in flood lay down layers of sediment? Do not lava flows produce layers of basalt? Why, then, should we assume that layers of rock are evidence of anything but ordinary natural processes, occurring repeatedly over long periods of time? This concept, called **uniformitarianism,** satisfies a scientific axiom often called Occam's Razor: The simplest explanation that fits the facts is probably correct. The implications of uniformitarianism were profound. If slow natural processes alone are enough to produce layers of rock thousands of feet thick, then Earth must be old indeed, many millions of years old. Hutton and Lyell, in fact, concluded that Earth was eternal: "No Vestige of a Beginning, no Prospect of an End," in Hutton's words. (Modern geologists estimate that Earth is about 4.5 billion years old; see Chapter 19, "Scientific Inquiry: Radiometric Dating Traces Fossils through Time.") Thus, Hutton and Lyell provided the time for evolution, but there was still no convincing mechanism.

Early Biologists Proposed Mechanisms for Evolution

One of the first to propose a mechanism for evolution was the French biologist Jean Baptiste Lamarck (1744–1829). Lamarck was impressed by the progression of forms in the fossil record. Older fossils tend to be simpler, while younger fossils are more complex and more like existing organisms. In 1801, Lamarck hypothesized that organisms evolved

(a)   **(b)**   **(c)**

Figure 16-4   **Fossils of extinct animals**

(a) Trilobites were early arthropods (relatives of spiders and crabs) that flourished for scores of millions of years. Dozens of species of fossil trilobites are found in many parts of the world, but there have been no living trilobites for hundreds of millions of years. **(b)** *Allosaurus*, a predatory dinosaur similar to the more familiar *Tyrannosaurus rex*, preyed on large and formidable prey, including *Stegosaurus* (see Fig. 16-3). **(c)** The skull of *Australopithecus boisei*, an extinct relative of human beings. Clearly unlike either today's humans or apes, this species probably has no modern descendants.

through the **inheritance of acquired characteristics:** Living organisms can modify their bodies through use or disuse of parts, and these modifications can be inherited by their offspring. (As it turns out, the first part of his hypothesis is correct to some extent, but the second part is not.) Why would organisms modify their bodies? Lamarck proposed that all organisms possess an innate drive for perfection, an urge to climb the Ladder of Nature. In his best-known example, Lamarck hypothesized that ancestral giraffes stretched their necks to feed on leaves growing high up in trees, and as a result their necks became slightly longer. Their offspring inherited these longer necks and stretched even farther to reach still higher leaves. Eventually, this process might produce modern giraffes with very long necks indeed (Fig. 16-5).

Today, Lamarck's theory seems silly: The fact that a prospective father pumps iron doesn't mean that his children will look like Arnold Schwarzenegger. Remember, though, that in Lamarck's day no one had the foggiest idea how inheritance worked. Gregor Mendel wouldn't even be born for another 20 years, and the incorporation of his principles of inheritance into mainstream biology didn't happen until the early twentieth century.

Although Lamarck's theory fell by the wayside, by the mid-1800s some biologists were beginning to realize that the fossil record and the similarities between fossil forms and modern species could be best explained if present-day species had evolved from preexisting ones. The question

remained: *But how?* In 1858, Charles Darwin and Alfred Russell Wallace independently provided convincing evidence that the driving force behind evolutionary change was natural selection.

Charles Darwin's Voyage on the *Beagle* Sowed the Seeds for His Theory of Evolution

Like many modern students, Charles Darwin excelled only in subjects that intrigued him. Although his father was a physician, Darwin was uninterested in medicine and unable to stand the sight of surgery. He eventually obtained a degree in theology from Cambridge University, although theology too was of minor interest to him. What he really liked to do was to tramp over the hills, observing plants and animals, collecting new specimens, scrutinizing their structures, and categorizing them. As he himself later put it, "I was a born naturalist." Fortunately for Darwin (and for the development of biology), some of his professors at Cambridge, most notably the botanist John Henslow, had similar interests. So constant was their companionship in field studies that Darwin was sometimes called "the man who walks with Henslow."

In 1831, when Darwin was only 22 years old (Fig. 16-6), the British government sent Her Majesty's Ship *Beagle* on a 5-year surveying expedition that took them first along the coastline of South America and then around the world. As was common on such expeditions, the *Beagle* would carry along a naturalist to observe and collect geological

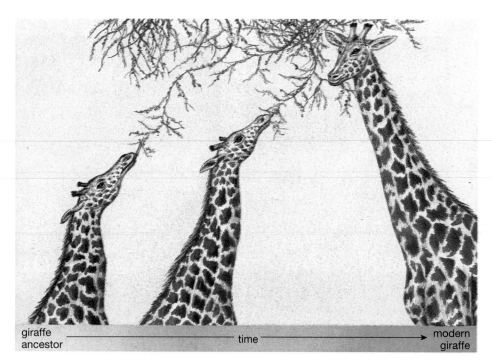

giraffe ancestor ———————— time ————————→ modern giraffe

Figure 16-5   **Lamarck's hypothesis for the evolution of the giraffe**

Time flows from left to right. Short-necked ancestral giraffes stripped the leaves from the lower branches of trees and stretched and strained to reach leaves higher up, making their necks slightly longer in the process. Their offspring inherited these longer necks, did more stretching in their turn, and passed on still longer necks to their offspring. The modern giraffe on the right would be the outcome of this continual striving to feed on tree leaves high off the ground.

and biological specimens encountered along the route. Thanks to Henslow's recommendation to the captain, Robert FitzRoy, Darwin was offered the position of naturalist aboard the *Beagle*. (When they first met, FitzRoy almost rejected Darwin for the post because of the shape of Darwin's nose! Apparently FitzRoy felt that a person's personality could be predicted by the shape of the facial features, and Darwin's nose failed to measure up, as it were. Later, Darwin expressed the opinion that FitzRoy was "afterwards well satisfied that my nose had spoken falsely.")

The *Beagle* sailed to South America, making many stops along the coast. Here Darwin observed the plants and animals of the tropics and was stunned by the diversity of species compared with Europe. Although he boarded the *Beagle* convinced of the permanence of species, his experiences soon led him to doubt this. He discovered a snake with rudimentary hind limbs, calling it "the passage by which Nature joins the lizards to the snakes." Another snake vibrates its tail like a rattlesnake, but has no rattles and therefore makes no noise. Penguins use their wings to paddle through the water rather than fly through the air. If the Creator had individually created each animal in its present form, to suit its present environment, what could be the purpose behind these makeshift arrangements?

Perhaps the most significant stopover of the voyage was the month spent on the Galapagos Islands off the northwestern coast of South America. Here Darwin found huge tortoises (*galapagos* means "tortoise" in Spanish); different islands were home to distinctively different types of tortoises. On islands without tortoises, prickly pear cactus grew with their juicy (though spiny) pads and fruits spread out over the ground. On islands where tortoises lived, the prickly pears grew substantial trunks, bearing the fleshy pads and fruits high above the reach of the voracious and tough-mouthed

tortoises (Fig. 16-7). Darwin also found several varieties of mockingbirds and finches, and, as with the tortoises, different islands supported slightly different forms. Could the differences in these organisms have arisen after they became isolated from one another on separate islands? The diversity of tortoises and birds "haunted" him for years afterward.

In 1836, Darwin returned to England after 5 years on the *Beagle* and became established as one of the foremost naturalists of his time. But constantly gnawing on his mind was the problem of the origin of species. For over 20 years, Darwin pondered the problem, collecting evidence that might provide insight into its solution. When he finally published *On the Origin of Species* in 1859, his evidence had become truly overwhelming. Although its full implications would not be realized for decades, Darwin's theory of evolution by natural selection has become a unifying concept for virtual-

Figure 16-6   **A painting of Charles Darwin as a young man**

Figure 16-7   **Tortoises act as agents of selection**
Galapagos tortoises feed on prickly pear cactuses. On islands with tortoises, a young cactus quickly grows a tall trunk, which lifts the succulent pads beyond the reach of the tortoises.

ly all of biology. Events that change the world sometimes hinge on small details, even the shape of a nose!

Darwin and Wallace Both Proposed That Evolution Occurs by Natural Selection

Although their social and educational backgrounds were very different, Darwin and Wallace were quite similar in some respects. Both had traveled extensively in the tropics and had studied the staggering variety of plants and animals living there. Both found that some species differed only in a few fairly subtle, but ecologically important, features (Fig. 16-8). Darwin and Wallace were familiar with the fossil record, which showed a trend of increasing complexity through time. Finally, both were aware of the studies of Hutton and Lyell, who proposed that Earth is extremely ancient. These facts suggested to both men that species change over time; that is, they evolve. Both sought a mechanism that might direct change over many generations, causing new species to arise.

Part of the answer came to both men from an unlikely source: the writings of an English economist and clergyman, Thomas Malthus. In his *Essay on Population*, Malthus wrote, "It may safely be pronounced, therefore, that [human] population, when unchecked, goes on doubling itself every 25 years, or increases in a geometrical ratio." Darwin and Wallace realized that a similar principle holds true for plant and animal populations. In fact, most organisms can reproduce much more rapidly than humans (consider rabbits, dandelions, and houseflies) and consequently could produce overwhelming populations in short order. Nonetheless, the world is *not* chest-deep in rabbits, dandelions, or flies: Natural populations do *not* grow "unchecked" but tend to remain approximately constant in size. Clearly, vast numbers of individuals must die in each generation, and most must not reproduce. In fact, population growth is checked

by many environmental factors, including food supply, predators, diseases, and weather (see Chapter 28).

From their experience as naturalists, Darwin and Wallace realized that the individual members of a species often differ from one another in form and function. Further, *which individuals die in each generation is not arbitrary but depends to some extent on the structures and abilities of the organisms.* This observation was the source of the theory of evolution by natural selection. As Wallace put it, "those which, year by year, survived this terrible destruction must be, on the whole, those which have some little superiority enabling them to escape each special form of death to which the great majority succumbed." Here you see the origin of the expression "survival of the fittest." That "little superiority" that confers greater fitness might be better resistance to cold, more efficient digestion, or any of hundreds of other advantages, some very subtle. Everything now fell into place. Darwin wrote: "It at once struck me that under these circumstances favorable variations would tend to be preserved, and unfavorable ones to be destroyed." If the favorable variations were inheritable, then the entire species would eventually consist of individuals possessing the favorable trait. With the continual appearance of new variations (due, as we now know, to mutations), which in turn are subject to further selection, "the result . . . would be the formation of new species. Here, then, I had at last got a theory by which to work."

In 1858, remarkably similar papers from Darwin and Wallace were presented to the Linnaean Society in London. Like Gregor Mendel's manuscript on the principles of genetics, their papers made little impact. The secretary of the society, in fact, wrote in his annual report that nothing very interesting happened that year. Fortunately, the next year Darwin published his monumental *On the Origin of Species by Means of Natural Selection*, forcing everyone to take note of the new theory.

(a)

(b) **(c)**

Figure 16-8 **Darwin's Finches, residents of the Galapagos Islands**

The Galapagos are a group of volcanic islands about 600 miles off the coast of Ecuador. Here Darwin found many species of finches, all similar to a species found on the South American mainland. Ordinarily, finches are seed-eating birds, with large bills adapted to crushing hard seeds. Apparently, thousands of years ago a finch or small flock of finches became lost during migration or were blown off course by a storm and arrived at the Galapagos. There they found few other birds. With few competitors, the finches found rich pickings of both seeds and insects. Over time, different groups of finches evolved adaptations to exploit different food sources. Modern biologists agree with Darwin that the variety of finch species on the Galapagos evolved from a common ancestor (a). Particularly significant are the differences in beaks, which vary from a slender, warblerlike bill in the insect-eating *Certhidia olivacea* (b) to the massive seed-crushing bill of *Geospiza magnirostris* (c). Beak differences are also important in species recognition. From behind, many Galapagos finches are very similar, and a male may approach a female of the wrong species. As soon as he catches sight of her beak, however, he immediately recognizes his mistake and looks elsewhere for a mate.

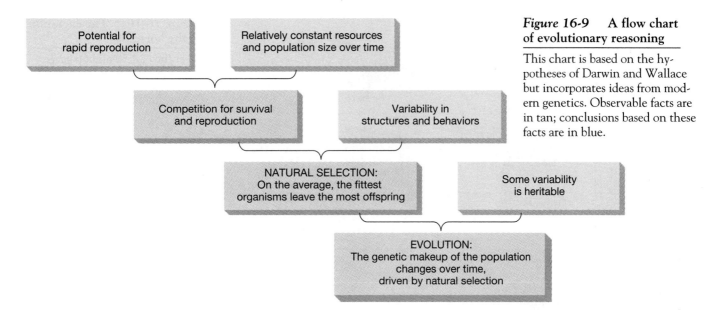

Figure 16-9   **A flow chart of evolutionary reasoning**

This chart is based on the hypotheses of Darwin and Wallace but incorporates ideas from modern genetics. Observable facts are in tan; conclusions based on these facts are in blue.

Evolutionary Theory Arises from Scientific Observations and Conclusions Based on Them

The essence of the Darwin-Wallace theory is very simple; it consists of three conclusions based on four observations. Below, we summarize their theory in modern terms (Fig. 16-9).

Observation 1: Natural populations of all organisms have the potential to increase rapidly, because organisms can produce far more offspring than are required merely to replace the parents.

Observation 2: Nevertheless, the sizes of most natural populations and the resources available to maintain them (such as food and appropriate habitat) remain relatively constant over time.

> **Conclusion 1:** Therefore, there is competition for survival and reproduction. In each generation, many individuals must die young, fail to reproduce, produce few offspring, or produce less-fit offspring that fail to survive and reproduce in their turn.

Observation 3: Individual members of a population differ from one another in their ability to obtain resources, withstand environmental extremes, escape predators, and so on.

> **Conclusion 2:** The most well-adapted (the "fittest") individuals in one generation will usually leave the most offspring. This is **natural selection**: the process by which the environment selects for those individuals whose traits best adapt them to that particular environment.

Observation 4: At least some of the variation in adaptive traits among individuals is due to genetic differences that may be passed on from parent to offspring.

> **Conclusion 3:** Over many generations, differential, or unequal, reproduction among individuals with different genetic makeup changes the overall genetic composition of the population. This is evolution through natural selection.

As you know, the principles of genetics had not yet been discovered when Darwin wrote *On the Origin of Species*. Our observation 4, therefore, was an untested assumption for Darwin and a grave weakness in his theory. Although he could not explain how inheritance operated, Darwin's theory made an important prediction that we now know is correct. According to Darwin, the variations that appear in natural populations arise purely by chance. Unlike Lamarck, he proposed no internal drives for perfection or any other mechanisms that would ensure that variations would be favorable. Molecular genetics has shown that Darwin was correct: Variations arise because of chance mutations in DNA (see Chapters 10 and 11).

How might natural selection among chance variations change the makeup of a species? In *On the Origin of Species*, Darwin proposed the following example. "Let us take the case of a wolf, which preys on various animals, securing [them] by... fleetness.... The swiftest and slimmest wolves would have the best chance of surviving, and so be preserved or selected.... Now if any slight innate change of habit or structure benefited an individual wolf, it would have the best chance of surviving and of leaving offspring. Some of its young would probably inherit the same habits or structure, and by the repetition of this process, a new variety might be formed." The same argument would apply to the wolf's prey, in which the fastest or most alert would be the most likely to avoid predation and would pass on these traits to its offspring. Over generations, the population would change as a greater percentage of its individuals acquired adaptive traits. This interplay between predator and prey is an example of **coevolution,** an important source of **adaptations,** which are traits that help organisms survive and reproduce in a particular environment: an environment that includes other organisms. Coevolution is discussed in more detail in Chapter 29.

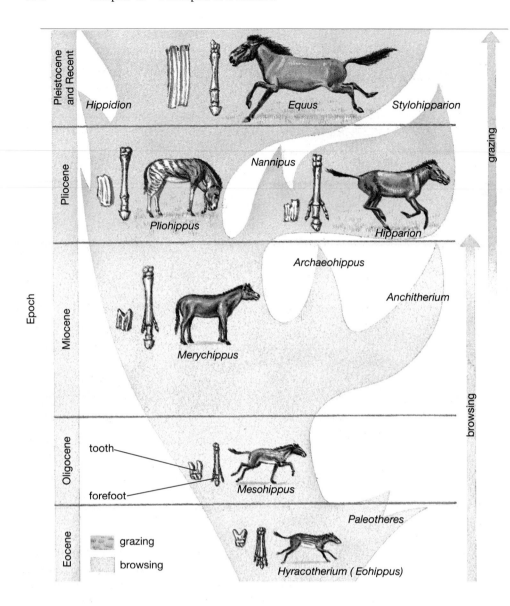

Figure 16-10   The evolution of the horse

Over the past 50 million years, horses evolved from small woodland browsers to large plains-dwelling grazers. Three major changes include size, leg anatomy, and tooth anatomy. No one is certain why horses became larger, but it may have been an anti-predator adaptation. In any case, a large body running over hard plains favored the evolution of large, hard hooves, attached by springlike, shock-absorbing joints to legs with stout bones. Finally, the teeth became larger, with more enamel. This change reflects a change in diet from the relatively soft leaves and buds of bushes to the abrasive blades of grasses, which contain silicon (glass). If a modern horse had teeth like *Hyracotherium*, its teeth would be ground away while it was still very young, leaving it to starve.

Although it is easiest to understand how natural selection would cause *changes within a species*, under the right circumstances, the same principles might produce *entirely new species*. In Chapter 18, we discuss the circumstances that give rise to new species. In the rest of this chapter, we briefly review some of the evidence for evolution.

The Evidence for Evolution

Virtually all biologists consider evolution to be a fact. Although debates still rage over the *mechanisms* of evolutionary change, exceedingly few biologists dispute that evolution occurs. Why? Because an overwhelming body of evidence permits no other conclusion.

The Fossil Record Provides Evidence of Evolutionary Change over Time

If fossils are the remains of members of species that are ancestral to modern species, then one would expect to find progressive series of fossils leading from an ancient, primitive organism, through several intermediate stages, and culminating in the modern form. (Just how fine the gradations between stages should be is currently a subject of debate among evolutionary biologists; see Chapter 18.) Probably the best-known of such series are the fossil horses (Fig. 16-10), but giraffes, elephants, and several mollusks all show a gradual evolution of body form over time, suggesting that species evolved from and replaced previous species. Certain sequences of fossil snails have such slight gradations in form between successive fossils that paleontologists cannot easily decide where one species leaves off and the next one begins.

Comparative Anatomy Provides Structural Evidence for Evolution

Through Convergent Evolution, Natural Selection Has Shaped Unrelated Organisms into Similar Forms in Similar Environments

Evolution by natural selection also predicts that, given similar environmental demands, unrelated organisms might independently evolve superficially similar struc-

(a)

(b)

Figure 16-11   **Analogous structures**

Similar selective pressures acting on unrelated animals may result in the evolution of outwardly similar structures. The wings of insects **(a)** and birds **(b)** and the sleek, streamlined shapes of seals **(c)** and penguins **(d)** are examples of such analogous structures.

(c)

(d)

tures, a process called **convergent evolution.** Such outwardly similar body parts in unrelated organisms, termed **analogous structures,** are often completely different in internal anatomy, because the parts are not derived from common ancestral structures. The wings of flies and birds and the fat-insulated, streamlined shapes of seals (mammals) and penguins (birds), are two examples of analogous structures that have arisen through convergent evolution (Fig. 16-11).

Homologous and Vestigial Structures Provide Evidence for Relatedness of Organisms Adapted to Different Environments

Modern organisms are adapted for a wide variety of habitats and life-styles. The forelimbs of birds and mammals, for example, are variously used for flying, swimming, running over several types of terrain, and grasping objects such as branches and tools. Despite this enormous diversity of function, the internal anatomy of all bird and mammal forelimbs is remarkably similar (Fig. 16-12). It is inconceivable that nearly the same bone arrangements could be ideal for such different functions, as would be expected if each animal were created separately. Such similarity is exactly what we would expect, however, if bird and

mammal forelimbs evolved from a common ancestor. Through natural selection, each has been modified to perform a particular function. Such internally similar structures are called **homologous structures,** meaning that they have a similar evolutionary origin, despite possible differences in function. Studies of comparative anatomy have long been used to determine the relationships among organisms, on the grounds that the more similar the internal structures of two species, the more closely related the species must be; that is, the more recently they must have diverged from a common ancestor.

Evolution by natural selection also helps to explain the curious circumstance of **vestigial structures.** These are structures that serve no apparent purpose, including such things as molar teeth in vampire bats (which live on a diet of blood and therefore don't chew their food) and pelvic bones in whales and certain snakes (Fig. 16-13). Both of these vestigial structures are clearly homologous to important structures found in other vertebrates (animals with a backbone). Their continued existence in animals that have no use for them is best explained as a sort of "evolutionary baggage." For example, the ancestral mammals from which whales evolved had four legs and a well-developed set of pelvic bones. Whales do not have hind legs, yet they have small pelvic and leg bones embedded in their sides. During whale evolution, there was a selective advantage to the loss of the hind legs, the better to streamline the body for movement through water. Once mutation and selection reduced the pelvic bones so that they no longer interrupted the smooth line of the body, that selective pressure (selective forces exerted by the environment) diminished. The result is the modern whale with small, useless and unused pelvic bones.

Embryological Stages of Animals Can Provide Evidence for Common Ancestry

In the early 1800s, the German embryologist Karl von Baer noted that all vertebrate embryos look quite similar to one another early in development (Fig. 16-14). Fish, turtles, chickens, mice, and humans all develop tails and gill slits early in development. Only fish go on to develop gills, and only fish, turtles, and mice retain substantial tails. Why do such diverse vertebrates have similar developmental stages? The only plausible explanation is that ancestral vertebrates possessed genes that direct the development of gills and tails. All of their descendants still retain those genes. In fish, these genes are active throughout development, resulting in gill- and tail-bearing adults. In humans and chickens, these genes are active only during early developmental stages, and the structures are lost or inconspicuous in adults.

Modern Biochemical and Genetic Analyses Reveal Relatedness among Diverse Organisms

Biochemistry and molecular biology provide striking evidence for the evolutionary relatedness of all living organisms. At the most fundamental biochemical levels, all living cells are very similar. For example, all cells have DNA as the carrier of genetic information; all use RNA, ribosomes, and approximately the same genetic code to translate that genetic information into proteins; all use roughly the same set of 20 amino acids to build proteins; and all use ATP as an intracellular energy carrier.

Evolutionary relationships among species are reflected in similarities and differences in their proteins. The amino acid sequences of a few proteins, such as hemoglobin and cytochrome c (one of the electron carriers in the mitochondrion), have been determined for many different species. The sequences are remarkably similar across a huge spectrum of species. Further, an evolutionary tree comparing the degree of differences in amino acid sequence between species closely resembles the evolutionary trees that have been deduced from anatomical and embryological studies. Relatedness among organisms can also be evaluated by examining the morphology of their chromosomes. For example, the chromosomes of chimpanzees and humans are extremely similar, showing that these species are closely related.

Within the past decade, the techniques of molecular genetics have triggered a revolution in studies of evolutionary relationships. For the first time, DNA, rather than "secondary" features such as appearance, behavior, or even proteins, can be used to investigate relatedness among organisms. Some of these methods and findings are explored in the essay "Scientific Inquiry: Modern Molecular Genetic Techniques Shed Light on Evolutionary Relationships."

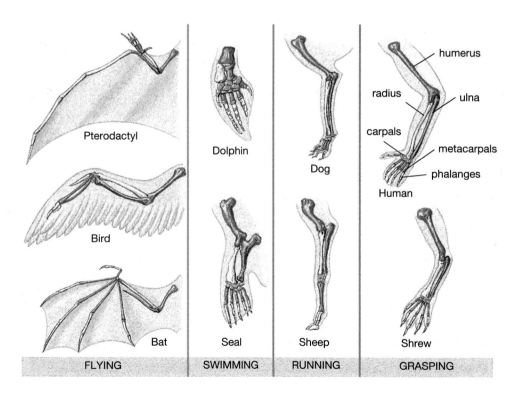

Figure 16-12 **Homologous structures**

The bones in the forelimbs of these animals are all similar, despite wide differences in function. The bones have been tinted different colors to highlight the similarities among the various species.

salamander

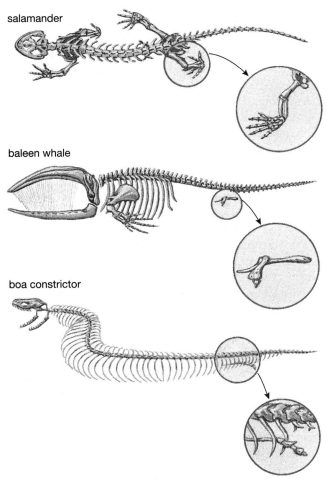

baleen whale

boa constrictor

Figure 16-13   Vestigial structures

The functional hindlimb of a salamander (top) is probably similar to the limb of the common ancestor of all amphibians, reptiles, birds, and mammals. Baleen whales (center) and boa constrictors (bottom) have no functional legs but still develop vestigial pelvic bones and even miniature leg bones buried in their sleek sides.

Artificial Selection Demonstrates That Organisms May Be Modified by Controlled Breeding

One line of evidence supporting evolution that particularly impressed Charles Darwin was **artificial selection:** breeding domestic plants and animals to produce specific desirable features. The various breeds of dogs provide a striking example of artificial selection (Fig. 16-15). Dogs descended from wolves, and even today the two will readily cross-breed. With rare exceptions, however, few modern dogs resemble wolves. Some breeds, such as the Chihuahua and the Great Dane, are so different from one another that they would be considered separate species if they were found in the wild. Interbreeding would hardly be possible without a lot of human assistance. If humans could breed such radically different dogs in a few hundred to at most a few thousand years, Darwin reasoned, it seemed quite plausible that natural selection could produce the spectrum of living organisms in hundreds of millions of years.

(a)

(b)

(c)

(d)

Figure 16-14   Embryological stages reveal evolutionary relationships

Early embryonic stages of a turtle **(a)**, mouse **(b)**, human **(c)**, and chicken **(d)**, showing strikingly similar anatomical features.

(a)

(b)

Figure 16-15   **Dog diversity illustrates artificial selection**

A comparison of the ancestral dog (the gray wolf, *Canis lupus*) (a) and various breeds of dog (b). Artificial selection by humans has caused a great divergence in size and shape in only a few thousand years. Huge differences in size would make it extremely difficult for the large dogs in the rear of the photo to interbreed with the small terriers in the foreground.

Evolution by Natural Selection Occurs Today

One problem that many people have with the theory of evolution is that they think it all happened in the distant past. Darwin himself believed this; in *On the Origin of Species* he stated: "We see nothing of these slow changes in progress, until the hand of time has marked the lapse of ages." Modern evolutionary biologists, however, maintain that evolution is *not* merely a phenomenon of the past, but that it continues today. One example of present-day

evolution has been caused by the increasing industrial pollution in the nineteenth and twentieth centuries.

Tree trunks in British woodlands usually support a lush growth of mottled gray lichens (a symbiotic association of an alga and a fungus). Before the Industrial Revolution of the nineteenth century, most peppered moths, *Biston betularia*, were white with scattered specks of black pigment. This coloration matched the color and pattern of the lichens growing on the trees. Because the moths sat quietly on the lichens during the day, predatory birds could not easily see them (Fig. 16-16a). Occasionally, mutant black individuals appeared. These black moths, extremely conspicuous against the pale lichens, were easily spotted by birds and didn't live long.

During the Industrial Revolution, Britain's growing industries began to burn huge quantities of coal for fuel. With no pollution-control technology, soot from the smokestacks soon blanketed the countryside around the mills and factories (Fig. 16-17). The lichens on the trees became covered with pollutants and died off, leaving the trunks sooty black. The pale form of the peppered moth was no longer camouflaged while sitting on the blackened trunks (see Fig. 16-16b). Increasingly, pale moths fell prey to birds. Soot-covered trees, however, provided excellent camouflage for black moths. The black moths' color became, in Wallace's words, the "little superiority enabling them to escape" predation by birds. Black moths, once rare, survived and reproduced, passing on their genes for dark pigmentation to succeeding generations. As the years passed, ever-increasing numbers of black moths appeared, until by the end of the nineteenth century, about 98% of the moths around the industrial city of Manchester were black. Fortunately, pollution-control laws have now dramatically reduced emissions of soot and other pollutants, and lichens grow once again on the trees near British cities. As predicted by the principles of evolution, the pale moths are making a comeback in these areas, and the black form is becoming increasingly rare.

Evolution has also been documented more recently, and closer to home. In Florida, homeowners were dismayed to realize that roaches were ignoring a formerly effective poison bait called Combat. Researchers discovered that the bait had acted as an agent of natural selection—roaches that liked it were consistently killed. Those that survived had inherited a rare mutation that caused them to dislike glucose, a type of sugar found in the corn syrup used as bait in Combat. By the time researchers identified the problem in the early 1990s, the formerly rare mutation had become widespread in Florida's urban roach population.

In a meadow near Carson City, Nevada, biologists have closely monitored a rare subspecies of checkerspot butterfly. In 1983, about 80% of the eggs of the butterfly were laid on a native meadow plant *Collinsia*. But over the next decade, the meadow was invaded by a weed, *Plantago*. By 1993, over 70% of the checkerspots' eggs were being laid on *Plantago*. When eggs laid on *Plantago* were hatched in

(a)   (b)

Figure 16-16 **Protective colors differ in different environments**

The two color forms of the peppered moth, resting on a lichen-covered tree trunk **(a)** and a soot-blackened trunk **(b)**. The pale form is well camouflaged on the lichen but is conspicuous on the sooty trunk. The reverse is true of the black form.

the laboratory and the butterflies reared exclusively on *Collinsia*, the mature butterflies still sought out *Plantago* to lay their eggs, just as their mothers had. The preference for *Plantago* was clearly inherited and had rapidly increased within the population during a single decade.

Figure 16-17 **Pollution alters the environment**

The countryside around the industrial city of Manchester, England, is crisscrossed by black stone fences. If you break one of the stones, you can see that they are actually a bright golden sandstone. The black outer crust is the remnant of pollution from the Industrial Revolution of over a half-century ago.

It is important to remember that black coloration of British moths, distaste for glucose of Florida cockroaches, and preference for *Plantago* of Nevada butterflies were not *produced* by the pollution, poisoned corn syrup, or the invading weed. The mutations that produced these adaptations all arise spontaneously now and then. *The variations upon which natural selection works are produced by chance mutations.* A second point to remember is that selection does not necessarily produce well-adapted species. Some other species of moths in Britain lacked black mutants, and these became extinct in industrial areas. If the right raw material isn't there (black moth mutants in this case) then selection may drive a species to extinction. Finally, some people tend to regard evolution as a mechanism for producing the *Scala Naturae*, with ever-greater degrees of perfection appearing over time. It is not. *The processes of evolution select for organisms that are best adapted to a particular environment.* Under natural conditions, a distaste for glucose would put a cockroach at a disadvantage, causing it to avoid many rich food sources. In the presence of poisoned corn syrup, however, it becomes advantageous. Natural selection does not select for the "best" in any absolute sense, but only in the context of a particular environment, which varies from place to place, and which may change over time.

Other examples of modern-day evolution include antibiotic resistance in bacteria and the appearance of the AIDS virus (HIV) (Chapter 24). Evolution is easiest to observe in rapidly reproducing, short-lived organisms such as insects or bacteria, but it occurs constantly in all the species on Earth.

Modern Molecular Genetic Techniques Shed Light on Evolutionary Relationships

Evolution deals with inherited changes over time. Because DNA is the molecule of heredity, evolutionary changes must be reflected in changes in DNA. Chapter 17 will explore a few of the genetic mechanisms that contribute to evolution. In this essay, we will consider two methods of studying DNA and how those methods are applied to determining the evolutionary relatedness of organisms.

Cross-Species DNA Hybridization
The DNA in the chromosomes of all living organisms is double stranded, with the two strands held together by hydrogen bonds between complementary base pairs. The two strands of DNA can be separated by heating; when the temperature becomes high enough, the hydrogen bonds break and the two strands separate. The amount of separation can be measured by observing the absorption of ultraviolet (UV) light. Single DNA strands absorb UV light more effectively than double strands do, so the higher the proportion of single strands, the greater the UV absorption (Fig. E16-1).

Suppose now that you had two single strands of DNA in which only half of the bases were complementary. The complementary bases would still hold the two strands together, but with only half the normal number of hydrogen bonds. If you heated this "semidouble helix," you would find that, with fewer hydrogen bonds holding the strands together, the strands would separate at a lower temperature.

In the technique called DNA-DNA hybridization, single strands of DNA from two different species are allowed to form "hybrid" double helices (the procedures are fair-

Figure E16-1   DNA-DNA hybridization
As DNA is heated, the two strands gradually separate. Separation is measured by UV light absorption; as more single strands are liberated, more UV light is absorbed. When DNA strands from two different species have been "hybridized," the lower their melting temperature, the poorer the match between them.

ly involved, but this is the final result). They are then heated up, and the temperature at which the two strands separate is measured. Separation temperatures for strands from different species will necessarily be less than those for the same species. The principle is this: The DNA of closely related species will have more-similar nucleotide sequences than does the DNA of distantly related species.

A Postscript by Charles Darwin

"It is interesting to contemplate an entangled bank, clothed with many plants of many kinds, with birds singing on the bushes, with various insects flitting about, and with worms crawling through the damp earth, and to reflect that these elaborately constructed forms . . . have all been produced by laws acting around us. These laws, taken in the highest sense, being Growth with Reproduction; Inheritance [and] Variability . . . ; a Ratio of Increase so high as to lead to a Struggle for Life, and as a con-

sequence to Natural Selection, entailing Divergence of Character and Extinction of less-improved forms . . . . There is grandeur in this view of life, with its several powers, having been originally breathed into a few forms or into one; and that, whilst this planet has gone cycling on according to the fixed law of gravity, from so simple a beginning endless forms most beautiful and most wonderful have been, and are being, evolved."

These are the concluding sentences of *On the Origin of Species* by Charles Darwin.

The more similar the nucleotide sequences, the greater the number of hydrogen bonds between them, and the higher the temperature required for the hybrid helices to separate. When the melting temperature for hybrid (two-species) DNA is very similar to that of the DNA for each single species, the two species are very similar genetically and are very closely related. Using this type of information, one can generate a tree of evolutionary relationships based on the separation temperature of the hybrid helices (Fig. E16-2).

Direct Comparison of Nucleotide Sequences

"Melting" DNA is, of course, a somewhat indirect way of measuring similarities in nucleotide sequences. With the advent of DNA sequencing machines, it is now possible to compare the nucleotide sequences of single genes from a variety of species. The number of nucleotide substitutions between genes encoding the same or similar proteins reflects the relatedness among species: the fewer the differences in nucleotide sequence, the more closely the organisms are related. One problem is that the rate of change in the nucleotide sequence may differ for different genes; for example, selection pressures may favor changes in certain proteins more than in others. Thus, comparisons of a *single* gene may not provide an accurate evolutionary tree. This is, in fact, an argument in favor of using DNA-DNA hybridization, because hybridization detects changes over millions of nucleotides, not just the few hundreds that make up any single gene. However, as gene sequencing becomes more automated and more economical, this technique will

Figure E16-2   Relatedness can be determined by DNA melting temperatures

In this study by Charles Sibley and Jon Ahlquist, double helices of DNA were made by combining single strands from various pairs of species of primate. The melting temperatures were then determined and compared with temperatures of DNA from a single species (the difference is graphed here as the decrease in melting temperature). The greater the difference in melting temperature of hybrid versus "pure" DNA, the longer ago the two species diverged from a common ancestor.

provide increasing insight into the molecular changes underlying evolution.

SUMMARY OF KEY CONCEPTS

The History of Evolutionary Thought

A species consists of all the populations of organisms that can potentially interbreed and that do not usually interbreed with members of other species. Historically, the most common explanation for the origin of species has been creationism, that a divine being created each species in its present form, and that species have not significantly changed since the creation. Evidence provided by the diversity of living things, fossils, and geology challenged this view, though no convincing mechanism for the evolution of present-day species from previous ones was proposed. Since the middle of the nineteenth century, scientists have

concluded that species originate by the operation of natural laws, as a result of changes in the genetic makeup of the populations of organisms. This process is called evolution.

Charles Darwin and Alfred Russell Wallace independently proposed the theory of evolution by natural selection. Their theory can be concisely expressed as three conclusions based on four observations. These are summarized in modern biological terms in Figure 16-9.

The Evidence for Evolution

Many lines of evidence indicate that evolution has occurred and that natural selection is the chief mechanism driving changes in the characteristics of species over time. The evidence includes:

1. Fossils of ancient organisms are simpler in form than modern organisms. Sequences of fossils have been discovered that show a graded series of changes in form. Both of these facts would be expected if modern forms evolved from older forms.

2. Organisms thought to be related through evolution from a common ancestor show many similar anatomical structures. Examples include the limbs of amphibians, reptiles, birds, and mammals. Similarly, stages in embryological development and similarities in chromosome structure, sequences of amino acids in proteins, and DNA composition all support the notion of descent of related species through evolution from common ancestors.

3. Rapid, heritable changes have been produced in domestic animals and plants by selectively breeding organisms with desired features (artificial selection). The immense variations in species produced in a few thousand years of artificial selection by humans makes it seem likely that much larger changes could be wrought by hundreds of millions of years of natural selection.

4. Evolution can be observed today. Both natural and human activities drastically change the environment over short periods of time. Significant changes in the characteristics of species have been observed in response to these environmental changes. A well-studied example is the evolution of black coloration among moths in response to the darkening of their environment by industrial pollutants.

✖ K E Y   T E R M S

adaptation   p. 297
analogous structures   p. 299
artificial selection   p. 301
catastrophism   p. 291
coevolution   p. 297
convergent evolution   p. 299

creationism   p. 289
evolution   p. 289
fossil   p. 291
homologous structures   p. 299
inheritance of acquired characteristics   p. 293

natural selection   p. 297
population   p. 290
species   p. 290
uniformitarianism   p. 292
vestigial structures   p. 299

✖ T H I N K I N G   T H R O U G H   T H E   C O N C E P T S

Multiple Choice

1. Your arm is homologous with
   a. a seal flipper
   b. an octopus tentacle
   c. a bird wing
   d. a sea star arm
   e. both a and c

2. All organisms share the same genetic code. This commonality is evidence that
   a. evolution is occurring now
   b. convergent evolution has occurred
   c. evolution occurs gradually
   d. all organisms are descended from a common ancestor
   e. life began a long time ago

3. Which of the following are fossils?
   a. pollen grains buried in the bottom of a peat bog
   b. the petrified cast of a clam's burrow
   c. the impression a clam shell made in mud, preserved in mudstone
   d. an insect leg sealed in plant resin
   e. all of the above

4. In Africa, there is a species of bird called the Yellow-throated Longclaw. It looks almost exactly like the meadowlark found in North America, but they are not closely related. This is an example of
   a. uniformitarianism
   b. coevolution
   c. gradualism
   d. vestigial structures
   e. convergent evolution

5. Which of the following are examples of vestigial structures?
   a. your tailbone
   b. nipples on male mammals
   c. sixth fingers found in some humans
   d. your kneecap
   e. none of the above

6. Which of the following would *stop* evolution by natural selection from occurring?
   a. if humans became extinct because of a disease epidemic
   b. if a thermonuclear war killed most living organisms and changed the environment drastically
   c. if ozone depletion led to increased ultraviolet radiation, which caused many new mutations
   d. if genetic recombination, sexual reproduction, and mutation stopped, so that all offspring of all organisms were exact copies of their parents
   e. all of the above

Review Questions

1. Selection acts on individuals, but only populations evolve. Explain why this is true.

2. Describe what "survival of the fittest" really means by writing a clear explanation of evolutionary fitness.

3. Distinguish between catastrophism and uniformitarianism. How did these hypotheses contribute to the development of evolutionary theory?

4. Describe Lamarck's theory of inheritance of acquired characteristics. Why is it invalid?

5. What is natural selection? Describe how natural selection might have caused differential reproduction among the ancestors of a fast-swimming predatory fish, such as a barracuda.

6. Describe how evolution occurs through the interactions among the reproductive potential of a species, the normally constant size of natural populations, variation among individuals of a species, natural selection, and inheritance.

7. What is convergent evolution? Give an example.

8. How do biochemistry and molecular genetics contribute to the evidence that evolution occurred?

✖ APPLYING THE CONCEPTS

1. Does evolution through natural selection produce "better" organisms in an absolute sense? Are we climbing the *Scala Naturae*? Defend your answer.

2. Describe and compare the views of Plato and Darwin on the existence of variation in natural populations.

3. The idea of special creation and the study of fossils have each had an impact on evolutionary thought. Discuss why one is considered scientific endeavor and the other not scientific.

4. In evolutionary terms, "success" can be defined in many different ways. What are the most successful organisms you can think of in terms of (a) persistence over time, (b) sheer numbers of individuals alive now, (c) numbers of species (for a lineage), and (d) geographical range?

5. In what sense are humans currently acting as "agents of selection" on other species? Name some organisms that are *favored* by the environmental changes humans cause.

6. Darwin and Wallace's discovery of natural selection is one of the great revolutions in scientific thought. Some scientific revolutions spill over and affect the development of philosophy and religion. Is this true of evolution? Does (or should) Darwinism affect the way humans view their place in the world?

7. In your mind, what scientific question currently represents the "mystery of mysteries"?

✖ FOR MORE INFORMATION

Altman, S. A. "The Monkey and the Fig." *American Scientist*, May–June 1989. An amusing, informative discussion of many evolutionary themes, cast as a Socratic dialogue.

Bishop, J. A., and Cook, L. M. "Moths, Melanism and Clean Air." *Scientific American*, January 1975. This followup to H. B. D. Kettlewell's 1959 work shows that cleaning up the air reverses industrial melanism in moths—that is, becoming pigmented as a consequence of industrial activities—providing graphic evidence that evolution only adapts organisms to existing environments.

Darwin, C. *On the Origin of Species by Means of Natural Selection*. Garden City, N.Y.: Doubleday, 1960 (originally published in 1859). An impressive array of evidence amassed to convince a skeptical world.

Eiseley, L. C. "Charles Darwin." *Scientific American*, February 1956. An essay on the life of Darwin, by one of his foremost American biographers. Even if you need no introduction to Darwin, read this anyway, as an introduction to Eiseley, author of many marvelous essays.

Gould, S. J. *Ever Since Darwin*, 1977; *The Panda's Thumb*, 1980; and *The Flamingo's Smile*, 1985. New York: W. W. Norton. A series of witty, imaginative, and informative essays, mostly from *Natural History* magazine. Many deal with various aspects of evolution.

Grant, P. R. "Natural Selection and Darwin's Finches." *Scientific American*, October 1991. A drought in the Galapagos Islands provides dramatic evidence for natural selection as an agent of evolutionary change.

Kettlewell, H. B. D. "Darwin's Missing Evidence." *Scientific American*, March 1959. Industrial melanism as an example of modern-day evolution.

Weiner, J. "Evolution Made Visible." *Science*, January 6, 1995. A clear summary of modern evidence for evolution in action.

NET WATCH

On-line resources for this chapter are on the World Wide Web at:
http://www.prenhall.com/~audesirk (click on the table of contents link and then select Chapter 16).

17 How Organisms Evolve

"What but the wolf's tooth whittled so fine
The fleet limbs of the antelope?
What but fear winged the birds, and hunger
Jewelled with such eyes the great goshawk's head?"

ROBINSON JEFFERS in *The Bloody Sire*

A goshawk surveys its surroundings for prey. The exceptionally keen eyes of birds of prey evolved in response to selection pressures that favored those who could spot prey from far above. As Robinson Jeffers poetically suggests, those who could not see well went hungry and produced fewer offspring.

In Chapter 16 we discussed the history of the theory of evolution and presented some of the evidence that evolution actually happens. But what processes drive evolutionary change? Is natural selection the only cause of evolution? Does evolution always occur all the time in all populations of organisms? In this chapter, we examine evolutionary processes in more detail. As we do, you will see that *evolution is an inevitable consequence of the nature of living things*. It occurs as a direct result of the chemical structure of genes and the interactions between organisms and their environment.

Evolution and the Genetics of Populations

Individual organisms live, reproduce, and die. Individuals, however, do not evolve; populations do. Evolution is the change in gene frequency that occurs in a population over time. Inheritance, therefore, is the link between the lives of individual organisms and the evolution of **populations**, which are all the individuals of a species living in a given area. We will begin our discussion of the processes of evolution by reviewing the principles of genetics as they apply to individuals and then extend those principles to the genetics of populations. You may want to refer to Unit II to refresh your memory on specific points.

Genes, Influenced by the Environment, Determine the Traits of Each Individual

Each cell of every organism contains a repository of genetic information encoded in the DNA of its chromosomes. A gene is a segment of DNA located at a particular place on a chromosome. Its sequence of nucleotides encodes the sequence of amino acids of a protein, usually an enzyme that catalyzes one particular reaction in the cell. Slightly different sequences of nucleotides at a given gene's location, called alleles, generate different forms of the same enzyme. There are usually two or more alleles of a single gene. An individual having alleles of the same type is homozygous, and an individual having alleles of different types is heterozygous. The specific alleles borne on an organism's chromosomes (its genotype), interacting with the environment, determine its physical and behavioral traits (its phenotype).

Let's illustrate these principles with an example that should be familiar to you from Unit II. A pea flower is colored purple because a chemical reaction in its petals converts a colorless molecule to a purple pigment. When we say that a pea plant has the allele for purple flowers, we mean that a particular stretch of DNA on one of its chromosomes contains a sequence of nucleotides that codes for the enzyme catalyzing this reaction. A pea with the allele for white flowers has a different sequence of nucleotides at the corresponding place on one of its chromosomes. The resulting enzyme cannot produce purple pigment. If a pea is homozygous for the white allele, its flowers produce no pigment and are white.

As we will see later, natural selection operates on the phenotype and, in doing so, either favors or selects against the particular genotype that produced it. Thus natural selection alters the gene frequencies within a population.

The Gene Pool Is the Sum of All the Genes Occurring in a Population

A branch of genetics, called **population genetics**, deals with the frequency, distribution, and inheritance of alleles in populations. Because evolution is a change in the genetic makeup of populations over generations, you will need to learn the principles of population genetics to understand the mechanisms of evolution.

In population genetics, the **gene pool** is defined as all the genes that occur in a population. It is made up of all the alleles of all the genes found in all of the individuals. Each particular gene can also be considered to have a gene pool, which consists of all the alleles of that specific gene occurring in a population. For example, in a population of 100 pea plants, the gene pool for flower color would consist of 200 alleles (peas are diploid, so there are two color alleles per plant, multiplied by 100 plants). If we could analyze the genetic composition of every plant in the population, we might find that some have alleles for white flowers, some have alleles for purple flowers, and some have both alleles. If we added up the color alleles of all the plants in the population, we could determine the relative proportions of the different alleles, a number called the **allele frequency**. Let's say that the gene pool for flower color consisted of 140 alleles for purple and 60 alleles for white. The allele frequencies would then be purple, 0.7 (70%), and white, 0.3 (30%).

Evolution Is the Change of Gene Frequencies within a Population

What does all this have to do with evolution? Quite a bit. Suppose a flower-eating cow happens upon a field of purple flowers and, being enamored of purple flowers, eats all of them before they produce seeds. Because the allele for pur-

ple flowers (P) is dominant to the allele for white (p), all the purple alleles in the entire population are in the purple-flowered plants (PP or Pp). If none of these plants reproduce, while the white-flowered plants do reproduce, then the next generation will consist entirely of white-flowered peas (pp). The allele frequency for purple will drop to 0, while the allele frequency for white will rise to 1.0 (100%). As a result of the selective eating habits of the cow, *evolution will have occurred in that field*. The gene pool of the pea population will have changed, and **natural selection**, in the form of foraging by the cow, will have caused the change.

This simple example illustrates four important points about evolution.

1. *Natural selection does not cause genetic changes in individuals*. The alleles for purple or white flower color arose spontaneously, long before the cow ever found the pea field. The cow did not cause white alleles to appear. It merely favored the differential survival of white alleles compared with purple alleles.
2. *Natural selection befalls individuals, but evolution occurs in populations*. Individual pea plants either reproduced or did not, but it was the population as a whole that evolved as its gene frequencies changed.
3. *Evolution is a change in the allele frequencies of a population, owing to differential reproduction among organisms bearing different alleles*. In evolutionary terminology, the **fitness** of an organism is measured by its reproductive success: In our example, the white flowers had greater fitness than the purple flowers, because they produced more viable offspring.
4. *Evolutionary changes are not "good" or "progressive" in any absolute sense*. The white alleles were favored only because of the dietary preferences of this particular cow; in another environment, with other predators, the white allele may well be selected against.

Mutation and the Recombination of Alleles during Sexual Reproduction Provide Sources of Variability

Your observations of the life around you have undoubtedly made you aware that even within a species, most individuals are at least slightly different from one another. The number of different individuals that can be produced from the same set of thousands of genes, considering that most genes come in multiple alleles, is staggering. This variability is produced by genetic recombination and sexual reproduction, which recombine existing alleles. But where did the different genes and alleles come from in the first place? The ultimate source of new genes and new alleles is mutation. Together, these processes provide the raw material for evolution.

Mutations Are the Source of New Genes and Alleles

Cells have efficient mechanisms that protect the integrity of their genes. Enzymes constantly scan the DNA, repairing flaws caused by radiation, chemical damage, or mistakes in copying. Nevertheless, changes in nucleotide sequence can happen. These changes are mutations, and they vary tremendously in their impact. As we explained in Chapter 11, some changes in DNA have virtually no effect on the organism; many, perhaps most, are harmful; and a few may be beneficial or may aid the organism in coping with new or changed environments.

How significant is mutation in altering the gene pool of a population? Mutations are rare, occurring once in 10,000 to 1,000,000 genes per individual in each generation. Therefore, mutation is not a major force in evolution by itself by changing gene frequencies. However, *muta-*

tions are the source of new alleles, new heritable variations upon which other evolutionary processes can work. As such, they are the foundation of evolutionary change. Without mutations there would be no diversity among life forms, and probably no life at all.

As we mentioned earlier, *mutations are not goal-directed.* A mutation does not arise as a result of, nor in anticipation of, environmental necessities (Fig. 17-1). A mutation simply happens and may in turn produce a change in the structure or function of the organism. Whether that change is helpful or harmful, now or in the future, depends on environmental conditions over which the organism

Figure 17-1   **Mutations occur spontaneously**

Experiment supporting the hypothesis that mutations occur spontaneously and not in response to specific selective pressures. **(a)** Several colonies of bacteria, each the offspring of a single individual and thus having the same genetic makeup, are grown on a solid nutrient medium in a dish. These bacteria have never been exposed to antibiotics. A piece of velvet the exact size of the dish is lightly pressed into the bacterial colonies and then touched to the surface of nutrient medium containing the antibiotic streptomycin in a second dish. Many bacteria from each original colony stick to the velvet and then come off the velvet into the second dish. Thus the exact positions of the "parent" colonies are duplicated in the second dish. **(b)** Only one daughter colony, in position #2, grows on the streptomycin-containing medium in the second dish. If mutations for streptomycin resistance were induced by the presence of streptomycin, then one would predict that any colony transplanted to the streptomycin-containing medium would develop these mutations. If, however, streptomycin resistance is an occasional spontaneous occurrence, then one would predict that the only colony grown on a normal medium that would survive is one that might already have been resistant. These alternatives are tested in **(c)**. Samples of the original colonies 1 and 2 are transferred to streptomycin-containing medium. All bacteria from colony 2 grow, but none from colony 1 grow, suggesting that the bacteria of colony 2 already possessed the mutation for streptomycin resistance prior to exposure and that the presence of streptomycin in the medium did not induce an adaptive mutation for streptomycin resistance.

has little or no control. The mutation provides *potential*; other forces, such as migration and especially natural selection, acting on that potential, may favor the spread of a mutation through the population or eliminate it.

Recombination during Sexual Reproduction Provides New Combinations of Existing Alleles

The production of new combinations of alleles occurs in three ways during sexual reproduction. During meiosis, homologous pairs of chromosomes are separated and parceled randomly into gametes. As a result of this segregation of homologues, each human (having 23 paired chromosomes) can produce 8 million different combinations of chromosomes in his or her gametes. Recombination of alleles on individual chromosomes also occurs regularly as a result of crossing over between pairs of homologous chromosomes, adding to the variability. Then, during fertilization, gametes join in random pairs. Together, these events guarantee that each fertilized egg bears a completely unique combination of alleles, or genotype. Many of these new allele combinations produce measurable differences in behavioral, anatomical, or physiological traits (new phenotypes) that in turn alter the fitness of the individual in a particular environment.

The Equilibrium Population Is a Hypothetical Population in Which Evolution Does Not Occur

It will be easier to understand the forces that cause populations to evolve if we first consider the characteristics of a population that would *not* evolve. In 1908, Godfrey H. Hardy and Wilhelm Weinberg defined an **equilibrium population** as one in which the allele frequencies and the distribution of genotypes remain constant with succeeding generations. In other words, the population remains in **genetic equilibrium**. If allele frequencies do not change, evolution does not occur. A population can remain in equilibrium only if several restrictive conditions are met:

1. *There must be no mutation.*
2. *There must be no **gene flow** between populations;* that is, there must be no net migration of alleles into the population (through immigration) or out of the population (through emigration).
3. *The population must be very large* (theoretically infinite).
4. *All mating must be random,* with no tendency for certain genotypes to mate with specific other genotypes.
5. *There must be no natural selection;* that is, all genotypes must be equally adaptive and reproduce equally well.

Under these conditions, allele frequencies within a population will remain the same indefinitely. If one or more of these conditions are violated, then allele frequencies will change: Evolution will occur.

As you might expect, few if any natural populations are truly in equilibrium. If so, then what is the importance of the Hardy-Weinberg principle? The Hardy-Weinberg conditions are useful starting points for studying the mechanisms of evolution. In the following sections, we will examine each condition, show why it is often violated by natural populations, and illustrate the consequences of its violation. In this way, you can better understand both the inevitability of evolution and the forces that drive evolutionary change.

The Mechanisms of Evolution

As the Hardy-Weinberg conditions predict, there are five major causes of evolutionary change within a population: mutation, migration, small populations, nonrandom mating, and natural selection.

Mutations Are the Ultimate Source of Genetic Variability

A population will remain in genetic equilibrium only if no mutations occur. Although mutations are not common, they are an inevitable result of the imperfections in the way DNA is copied as cells reproduce. When such a copying error occurs in a cell that produces gametes, the mutation may be passed to an offspring and enter the gene pool of a population. Over a sufficiently long time span, genetic change as a result of mutation is a certainty.

Migration Produces Gene Flow between Populations

In biology, the word *migration* has two distinct meanings. In the most familiar context, migration refers to the seasonal movement of many species between summer breeding grounds and winter refuges. In evolutionary biology, however, **migration** *is the flow of genes between populations.* Baboons, for example, live in social groupings called troops. Within each troop, all the females mate with a handful of dominant males. Juvenile males usually leave the troop. If they are lucky, they join and perhaps even become dominant in another troop. Thus the male offspring of one troop carry genes to the gene pools of other troops.

Migration has two significant effects.

1. *Gene flow spreads advantageous alleles throughout the species.* Suppose that a new allele arises in one population and that this new allele benefits the organisms that possess it. Migration can carry this new allele to other populations of the species.
2. *Gene flow helps to maintain all the organisms over a large area as one species.* If migrants constantly carry genes back and forth among populations, then the populations can never develop large differences in allele frequencies. Isolation of populations, with no gene flow to or from other populations of the same species, is a key factor in the origin of new species, as will be discussed in Chapter 18.

Small Populations Are Subject to Random Changes in Allele Frequencies

To remain in genetic equilibrium, a population must be so large that chance events have no impact on its overall genetic makeup. Disaster may befall even the fittest organism. The maple seed that falls into a pond never sprouts; the deer and elk blasted away by the eruption of Mount St. Helens left no descendants. If a population is sufficiently large, chance events are unlikely to alter the overall gene frequency, since they would be expected to interfere equally with the reproduction of organisms of all genotypes. In a small population, however, certain alleles may be carried by only a few organisms. Chance events could reduce or even entirely eliminate such alleles from the population, altering its genetic makeup.

Genetic Drift Is An Example of Random Genetic Change in Small Populations

It is much more likely that chance events will change allele frequencies in a small population than in a large population, by a process called **genetic drift**. Consider, for example, two hypothetical populations of haploid ladybugs in which the outer shell is either spotted or solid-colored, controlled by alternate alleles of a single gene. In each population, half the ladybugs are spotted and half are solid-colored (that is, the frequencies of both alleles are 0.5, or 50%), but one population has only four bugs and the other has 1000. Let us assume that each individual that survives to maturity produces two offspring identical to itself. For the population sizes to remain constant, exactly half the individuals must reproduce in each generation. Let us further assume that whether an individual ladybug reproduces is determined entirely by chance.

In the larger population, 500 bugs will be parents to the next generation. Though their survival is random, the odds against all 500 of the reproducing ladybugs being spotted are enormous. In fact, it would be extremely unlikely for even 300 parents to be spotted. In this large population, then, we would not expect a major change in allele frequencies to occur from generation to generation (Fig. 17-2). The effects of chance are minimized by the large population size. In the small population, on the other hand, only two individuals will reproduce. There is a 25% chance that both parents will be spotted (this is the same likelihood as flipping two coins and having both come up heads). If this happens, then the next generation will consist entirely of spotted ladybugs. Within a single generation, it is possible for the allele for solid-colored shell to completely disappear from the population.

Figure 17-2a illustrates two important points about genetic drift:

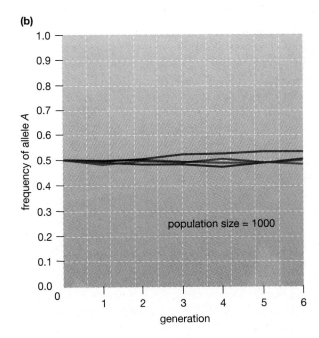

Figure 17-2   **Genetic drift**

Computer-generated graphs illustrating the effect of population size on genetic drift. In both graphs, the initial population was composed of half A and half *a* alleles, and six generations were simulated, with individuals chosen at random to contribute alleles to the next generation. Four simulations were run for each population size, producing the four lines on each graph. **(a)** With a population size of 4, one allele sometimes became "extinct" owing to chance. For example, in the top simulation run, the *a* allele became extinct by the second generation (therefore, the frequency of the A allele became 1.0). **(b)** With a population size of 1000, allele frequencies remained relatively constant.

1. *Genetic drift tends to reduce genetic variability within a small population.* In extreme cases, all members of a population may become genetically identical (Fig. 17-2a, top line).
2. *Genetic drift tends to increase genetic variability between populations.* Purely as a result of chance, separate populations may evolve extremely different allele frequencies (Fig. 17-2a, top versus bottom lines).

Two special cases of genetic drift, called the population bottleneck and founder effect, further illustrate the enormous consequences that small population size may have on the allele frequencies of a species.

A Population Bottleneck Is an Example of Genetic Drift

In a **population bottleneck**, a species undergoes a drastic reduction in population size, so that only a few individuals contribute genes to the entire future population of the species. As our ladybug example showed, population bottlenecks may cause both *differences in allele frequencies* and *reductions in genetic variability* (Fig. 17-3a). Even if the population then rebounds and the species becomes common,

these genetic effects of the bottleneck may remain for hundreds or thousands of generations.

Loss of genetic variability has been documented in the northern elephant seal and the cheetah (Fig. 17-3b). The elephant seal was hunted almost to extinction in the 1800s; by the 1890s only about 20 survived. Because elephant seals breed harem-style, with a single male mating with a stable group of females, one male may have fathered all the offspring at this extreme bottleneck point. The population today has expanded to about 30,000, but biochemical analysis shows that all northern elephant seals are genetically almost identical. Other species of seals, whose populations have historically always remained large, are much more variable. The rescue of the northern elephant seal from extinction is rightly regarded as a triumph of conservation; however, with very little genetic variation, the elephant seal has much less potential to evolve in response to environmental changes. No matter how many elephant seals there are, the species must be considered to be threatened with extinction. Cheetahs are also genetically uniform, although the reason for the bottleneck is unknown. Consequently, cheetahs too could be

(a)   original population   resulting population

time

(b)

Figure 17-3   Genetic bottlenecks reduce variability

(a) If a population is reduced to a very small number of individuals, the gene pool is reduced and a population bottleneck occurs. The recovered population shows reduced genetic and phenotypic variability, because all are offspring of the few organisms that survived the bottleneck. **(b)** Both the northern elephant seal (left) and the cheetah (right) passed through a population bottleneck in the recent past, resulting in an almost total loss of genetic diversity. As a result, the ability of these populations to adapt to changing environments is very limited.

Figure 17-4   **A human example of the founder effect**

An Amish woman with her child, who suffers from a set of genetic defects known as the Ellis-van Creveld syndrome (short arms and legs, extra fingers, occasionally heart defects). Fleeing from religious persecution, about 200 members of the Amish religion migrated from Switzerland to Pennsylvania between 1720 and 1770. Since that time, virtually all the Pennsylvania Amish moved to Lancaster County and have remained reproductively isolated from non-Amish Americans. The population increased to about 8000 by 1964. In that year, geneticist Victor McKusick surveyed the Lancaster County Amish and discovered that they had an allele frequency for Ellis-van Creveld of about 0.07, compared with a frequency of less than 0.001 in the general population. Why? One couple who immigrated in 1744 carried the allele. Inbreeding among the Amish passed the allele along to their descendants: a clear example of a founder effect. In addition, by chance the Ellis-van Creveld carriers had more children than the Amish average, further increasing the allele frequency by genetic drift. The combination of an initially high frequency in the immigrants (1 or 2 out of 200) plus genetic drift has resulted in more cases of Ellis-van Creveld syndrome from Lancaster County than from the entire rest of the world.

gravely threatened by small changes in their environment.

A special case of a population bottleneck is the **founder effect**, which occurs when isolated colonies are founded by a small number of organisms. A flock of birds, for instance, may become lost during migration or may be blown off course by a storm (this is thought to have happened in the case of Darwin's finches in the Galapagos Islands). Among humans, small groups may migrate for religious or political reasons (Fig. 17-4). Such a small group may have allele frequencies that are very different from the frequencies of the parent population because of chance inclusion of disproportionate numbers of certain alleles in the founders. If the isolation of the founders is maintained for a long period of time, a sizable new population may arise that differs greatly from the original population.

How much does genetic drift contribute to evolution? No one really knows. Only rarely are natural populations extremely small or completely cut off from gene flow from other populations. Populations occasionally do become very small, however, and it may be precisely these small populations that contribute most to major evolutionary changes. As we will see in the next chapter, biologists believe that new species often arise in small populations.

Mating within a Population Is Almost Never Random

Organisms seldom mate strictly randomly. For example, most animals have limited mobility and are most likely to mate with nearby members of their species. Further, they may choose to mate with certain individuals of their species rather than with others. The white-crowned sparrow is a case in point. Although all white-crowned sparrows sing a fundamentally similar song, each local population has its own song dialect. A female usually chooses a mate that sings the same dialect that her father sang (Fig. 17-5). Among animals, there are three common forms of nonrandom mating: harem breeding, assortative mating, and sexual selection.

In some species, such as elephant seals, baboons, and bighorn sheep, only a few males fertilize all the females. Following some sort of contest, which may involve showing off with loud sounds or flashy colors, making threatening gestures, or actual combat, only certain males succeed in gathering a harem and mating (Fig. 17-6).

Many animals mate assortatively—that is, they select mates that are similar to themselves. Humans, for example, tend to marry members of the opposite sex that are similar in height, race, intelligence, and social status.

Finally, in many mammal and bird species, mate selection is primarily the prerogative of one sex, usually the female. Males display their virtues, such as the bright plumage of a peacock (Fig. 17-7) or the rich territory of a songbird. A female evaluates the males and chooses her mate. This phenomenon, called **sexual selection**, is explored in more detail later in this chapter.

All Genotypes Are Not Equally Adaptive

Genetic equilibrium requires that all genotypes must be equally adaptive—that is, none has any selective advantage over the others. It is probably true that some alleles are adaptively neutral, so organisms possessing any of several alleles will be equally likely to survive and reproduce. However, this is clearly not true of all alleles in all environments. Any time an allele confers, in Wallace's words, "some little superiority," natural selection will favor the enhanced reproduction of the individuals possessing it.

Natural selection is not the *only* evolutionary force. As we have seen, mutation provides initial variability in her-

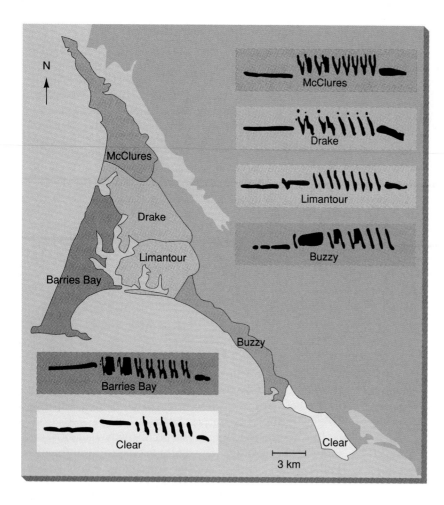

Figure 17-5   Song-dialect preference illustrates nonrandom mating

Song dialects among populations of white-crowned sparrows at Point Reyes National Seashore north of San Francisco. As analysis of the sound patterns shows, the songs are fairly similar, but both birds and human listeners can recognize the different dialects. Male birds of each population learn their local dialect while in the nest and sing it when they mature. Females preferentially mate with males that sing the dialect sung by the females' own fathers—that is, the females' own local dialect.

itable traits. The chance effects of genetic drift may change allele frequencies, even spawning new species. Further, evolutionary biologists are just now beginning to appreciate the power of random catastrophe in shaping the history of life on Earth—mass extinctions that may exterminate flourishing and floundering species alike. Nevertheless, it is natural selection that shapes the evolution of adaptations that humankind has admired for millennia. For this reason, we will examine the mechanisms of natural selection in some detail.

Natural Selection

To most people, the words *natural selection* are synonymous with the phrase *survival of the fittest*. Natural selection evokes images of wolves chasing caribou, of lions snarling angrily in competition over a zebra carcass. Natural selection, however, is not really about *survival* but about *reproduction*. It is certainly true that an organism must survive at least long enough to reproduce. In some cases, it may also be true that a longer-lived organism has more chances to reproduce. But no organism lives forever, and the only way that its genes continue into the future is

Figure 17-6   Male competition promotes nonrandom mating

Sparring contests between males result in extremely nonrandom mating among many animals, including deer, elk, seals, and many monkeys. Here, two male bighorn sheep square off against each other during the fall mating season. Although the horns are potentially deadly weapons, they are used in ritualized ways that minimize the danger of injury to either contestant.

During Predation, Both Predator and Prey Act as Agents of Selection

Although we commonly think of predation as one animal preying upon another animal, predation actually includes any situation in which one organism eats another. In some instances, coevolution between predators and prey is a sort of "biological arms race," with each side evolving new adaptations in response to "escalations" by the other. Darwin used the example of wolves and deer: Wolf predation selects against slow or incautious deer, thus leaving faster, more alert deer to reproduce and continue the species. In their turn, alert, swift deer select against slow, clumsy wolves, because such predators cannot acquire enough food.

Symbiosis Produces Adaptations between Species That Live in Intimate Association with One Another

Symbiosis is any relationship in which individuals of different species closely interact for an extended time. Examples of symbiosis are parasitism, in which one species lives and feeds on a larger species; commensalism, in which one species benefits and the other remains unharmed; and mutualism, in which both species benefit. The different types of symbiosis are described in Chapter 29. From an evolutionary perspective, symbiosis leads to the most intricate coevolutionary adaptations. Although a given predator usually preys on several species and may interact with a particular species only occasionally, partners in symbiosis often live together virtually their entire lives (Fig. 17-11). At least one of the partners, and usually both, must continually adjust to any evolutionary changes developed by the other.

Sexual Selection Sometimes Seems to Oppose Other Forms of Natural Selection

As mentioned earlier, in many species of birds and mammals and even some fish, one of the sexes, usually the female, selects the mate. Males compete for the attention of females through song, elaborate displays, the defense of large territories, or even by building elaborate structures such as that of the bowerbird. Choosing a male with a good territory is obviously advantageous, since good territories provide adequate food and shelter to raise young. However, females often also prefer elaborate "fashions" in their mates, such as bright colors and long feathers or fins that may make the male more vulnerable to predation. Why? A popular hypothesis is that structures and colors that do not serve any clear adaptive purpose actually provide the females with an outward sign of the males' fitness. Only vigorous, energetic males can survive when burdened with conspicuous coloration or large tails. Similarly, males sick or under parasitic attack may be dull and frumpy compared with healthy males. Whatever the exact selective mechanisms, it is thought that many of the elaborate structures and behaviors found only in males have

evolved through the selective pressure of female mate choice: Only the flashy males transmitted their genes to the next generation.

Darwin was so impressed with these structures that he coined the term *sexual selection* to designate the process of evolution through mate choice. Because conspicuous structures and bizarre behaviors make the males more vulnerable to predators, sexual selection often seems to work in opposition to other forms of natural selection. The trade-off between sexual selection and natural selection through predation has recently been demonstrated through observation and experiment in guppies (small freshwater fish). In streams where predation is a threat, male guppies are inconspicuous, blending with the sandy streambed. But in safer waters, male guppies show more conspicuous colors, apparently as a result of female preference for these markings. University of California biologist John Endler recently transplanted camouflage-colored male guppies from dangerous waters into safe waters. Within a year (about 20 guppy generations) their protected descendants had evolved conspicuous colors. However, nonsexual selective forces may also oppose one another; the height of a giraffe, for example, is a compromise between the advantage of reaching higher leaves for food and the disadvantage of vulnerability while drinking water (Fig.

Figure 17-11   An example of symbiosis

Several species of clownfish live in a symbiotic relationship with anemones, each species of fish favoring a particular species of anemone. The fish nestle within the stinging tentacles of the anemone, thus protected from being eaten by other fish. The clownfish evolved specialized skin secretions and behaviors, protecting it from being affected by the anemone. The fish may accidentally drop food onto the anemone once in a while, but the benefits are probably fairly one-sided.

(a)

(b)

Figure 17-12   **A compromise between opposing selection pressures**

(a) The long neck and legs of a giraffe are decided advantages in feeding on acacia leaves high up in trees. **(b)** But a giraffe has to get into an extremely awkward and vulnerable position to drink. Feeding and drinking thus place opposing selective pressures on the length of neck and legs.

17-12). In both sexual and nonsexual selection, then, some aspect of the environment (in sexual selection the "opinion" of the opposite sex, which is part of the social environment) influences reproductive success.

Kin Selection Favors Altruistic Behaviors

Evolution is often portrayed as being "red in tooth and claw." This image of bloody and vicious interaction, however, is not the complete picture. Although it is true that competitive and predatory interactions influence the evolution of most species, cooperation and even self-sacrifice can be important selective forces too. **Altruism** refers to any behavior that endangers an individual organism or reduces its reproductive success but benefits other members of its species. Altruistic behaviors are common in the animal kingdom. A mother killdeer flutters just out of reach of a predator, feigning an injured wing and luring the predator away from her nest (Fig. 17-13); female worker bees forego reproduction and devote their lives to raising the offspring of the hive queen; and young male baboons scout around the edges of the troop, even though doing so increases their danger from leopards.

You might think that altruism runs counter to natural selection: If altruism is encoded in an organism's genes, those genes are placed at risk every time the altruist performs one of its brave behaviors. But natural selection can indeed select for altruistic genes, if the altruistic individual helps relatives who possess the same alleles. This special case of natural selection is an example of **kin selection** and is explored in "Evolutionary Connections: Kin Selection and the Evolution of Altruism."

Extinction

Natural selection not only produces the fleet limbs of the antelope and the exquisite eyes of the goshawk. It may also lead to the death of all the members of a species, **extinction.** Trilobites, dinosaurs, saber-tooth cats—all are extinct, known only from fossils. Paleontologists estimate that *at least* 99.9% of all the species that ever existed are now extinct. Why? The actual cause of extinction is probably always environmental change, either in the living or the nonliving parts of the environment. Two characteristics seem to predispose a

Figure 17-13   **Altruism between mother and offspring**

A female killdeer lures a predator away from its nest by faking injury. The mother places herself in some small danger (she can always fly away if the predator comes too close) but saves her offspring from much greater danger.

E A R T H   W A T C H

Natural Selection, Genetic Diversity, and Endangered Species

Ever since the Endangered Species Act was passed in 1976, the United States has had an official policy of protecting rare species. In fact, the real goal of the act is not protection but recovery; as one U.S. Fish and Wildlife Service official put it, "the goal is to get species *off* the list." Wildlife biologists try to determine how large a population a species needs to have before it is no longer in danger of extinction from unpredictable events, such as a couple of years of drought or an epidemic of parasites. If a species reaches this critical population size, it is no longer legally "endangered" with extinction.

Does a "large enough" population (which usually is still very small by historical standards) really ensure a species' survival? From our discussion of genetic drift and population bottlenecks, you probably realize that the answer is no. If the population of a species has been reduced to the point where it is placed on the endangered species list, then it probably has lost much of its genetic diversity. As ecologist Thomas Foose aptly put it, loss of habitat and consequent reduction in population size mean that "gene pools are being converted into gene puddles." Even if the species recovers in numbers, its original gene pool has been lost. When the forces of natural selection change at some future time, the species may not have the necessary genetic variability to produce individuals adapted to the new environment, and it may become extinct.

What can be done? The best solution, of course, is to leave enough habitat of diverse types so that species never become endangered in the first place. The human population, however, has grown so large and appropriated so much of Earth's resources that this solution is not possible in many places (Fig. E17-1). For many species, the only so-lution is to preserve enough habitat so that the remaining population is large enough to retain all or most of the total genetic diversity of the species. If we are to be, as we imagine ourselves, the caretakers of the planet and not merely its ultimate consumers, then protection of other life forms and their genetic heritage will be a continuing responsibility as long as humankind exists.

Figure E17-1   Endangered by habitat destruction

This orangutan and its young, who live in the tropical rain forests of Borneo, are among the innumerable species whose continued survival is threatened by habitat destruction.

species to extinction when the environment changes: localized distribution and overspecialization. Three major changes that drive species to extinction are competition among species, novel predators or parasites, and habitat destruction.

Localized Distribution and Extreme Specialization Make Species Vulnerable in Changing Environments

Species vary widely in their range, and hence in their susceptibility to extinction. Some species, such as herring gulls, white-tailed deer, and humans, inhabit entire continents or even the whole Earth, while others, such as the Devil's Hole pupfish (Fig. 17-14), have extremely limited ranges. Obviously, if a species occurs in only a very small area, any disturbance of that area could easily result in extinction. If Devil's Hole dries up from climatic change or well drilling nearby, its pupfish will immediately vanish. Wide-ranging species, on the other hand, usually do not succumb to local environmental catastrophes.

Another factor that may make a species vulnerable to extinction is extreme specialization. Each species evolves a set of genetic adaptations in response to pressures from its particular environment. Sometimes these adaptations limit the organism to a very specialized set of environmental conditions. The Everglades kite, for example, feeds only on a certain freshwater snail (Fig. 17-15). As the swamps of the American Southeast are drained for farms and developments, the snail population shrinks. If the snail becomes extinct, the kite will surely go extinct along with it.

Figure 17-14 **Very localized distribution can endanger a species**

The Devil's Hole pupfish is found in only one spring-fed water hole in the Nevada desert. During the last glacial period, the southwestern deserts received a great deal of rainfall, forming numerous lakes and rivers. As the rainfall decreased, pupfish populations were isolated in shrinking small springs and streams. Isolated small populations and differing environmental conditions caused the ancestral pupfish species to split up into several very restricted modern species, all of which swim on the brink of extinction.

In the fossil record, such behavioral specialization is hard to recognize. Structural specializations, however, may be just as restrictive. A case in point is giantism. For poorly understood reasons, many animals evolved huge size, including certain amphibians, dinosaurs, and giant mammals, such as mammoths and ground sloths. To support their bulk, these animals must have consumed enormous amounts of food. If environmental conditions deteriorated, those giants may have been unable to find enough food and thus died out. Smaller animals that ate the same food but needed less of it survived.

Interactions with Other Organisms May Drive Species to Extinction

As described earlier, interactions such as competition, predation, and parasitism serve as forces of natural selection. In some cases, these same forces can lead to extinction, rather than adaptation.

Competition for limited resources occurs in all environments. If a species' competitors evolve superior adaptations, and it doesn't evolve fast enough to keep up, it may become extinct. A particularly striking example of extinction through competition occurred in South America 2 to 3 million years ago. For millions of years North and South America were isolated from one another, and each developed a distinctive array of animal life. When the Panamanian land bridge arose, connecting the two continents, massive migrations took place. In general, North American animals displaced their South American counterparts, and many South American species became extinct.

When formerly isolated populations encounter one another, it isn't only competitors that migrate between the areas—predators and parasites do too. With the excep-

tion of humans, who have exterminated hundreds of species, predators probably cause few extinctions. Parasites, on the other hand, can be devastating. In North America, Dutch elm disease and chestnut blight are well-known instances of introduced parasites that almost completely destroyed widespread native species. We cannot tell much about prehistoric parasite invasions, but the extinction of South American animals, mentioned above, might have been at least partly due to diseases carried south by resistant North American migrants.

Habitat Change and Destruction Are the Leading Causes of Extinction

Habitat change, both contemporary and prehistoric, is the single greatest cause of extinctions. Presently, habitat destruction due to human activities is proceeding at a frightening pace. Perhaps the most rapid extinction in the history of life will occur over the next 50 years, as tropical forests are cut for timber and to clear land for cattle and crops. As many as half the species presently on Earth may be lost because of tropical deforestation.

Prehistoric habitat alteration usually occurred over a longer time span but nevertheless had serious consequences. Climate changes, in particular, caused many extinctions. Several times, moist, warm climates gave way to drier, colder climates with more variable temperatures. Many plants and animals failed to adapt to the harsh new conditions and became extinct. One cause of climate change is continental drift (Fig. 17-16). As the continents move about over the surface of Earth, they change latitudes. Much of North America was located around the equator many millions of years ago, an area characterized by consistently warm and wet tropical weather. But drift carried the continent up into temperate and arctic regions. As a result, the tropical weather

Figure 17-15 **Extreme specialization places species at risk**

The Everglades kite feeds exclusively on the apple snail, found in swamps of the southeastern United States. Such behavioral specialization renders the kite extremely vulnerable to any environmental change that may exterminate its single species of prey.

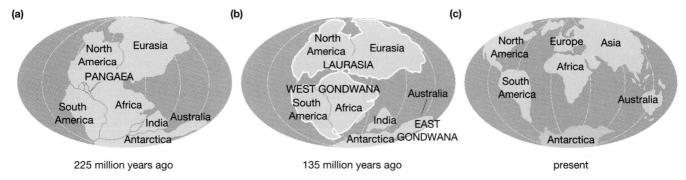

(a) 225 million years ago

(b) 135 million years ago

(c) present

Figure 17-16   **Continental drift has caused climate change**

Although slow, continental drift can cause tremendous environmental changes, as land masses are moved about on the surface of Earth. The solid surfaces of the continents slide about over a viscous, but fluid, lower layer. **(a)** About 225 million years ago, all the continents were fused together into one gigantic land mass, which geologists call Pangaea. **(b)** Gradually, Pangaea broke up into Laurasia and West and East Gondwana. **(c)** Further drift eventually resulted in the modern positions of the continents. Continental drift continues today: The Atlantic Ocean, for example, widens by a few centimeters each year.

was replaced by cooler temperatures, less rainfall, and seasonal changes.

An extreme, and very sudden, type of habitat destruction might be caused by catastrophic geological events, such as massive volcanic eruptions. Several prehistoric eruptions, which would make the Mount St. Helens explosion look like a firecracker by comparison, wiped out every living thing for dozens of miles around and probably caused global climatic changes as well.

The fossil record reveals episodes of extensive worldwide extinctions, especially among marine life (Fig. 17-17). Enormous meteorites, several kilometers in diameter, may have

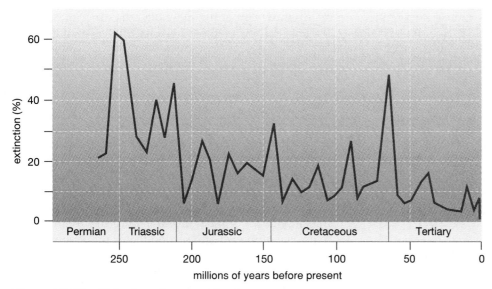

Figure 17-17   **Episodes of mass extinction**

This graph plots the percentage of genera of marine animals that have become extinct during geologic time. The higher the peak, the greater the extent of extinctions. Marine animals were chosen for this study because their fossils are abundant in sedimentary rocks and are easily dated. The large peak of extinctions near the boundary between the Cretaceous and Tertiary periods also approximately coincides with the end of the dinosaurs. Many paleontologists are convinced that the dinosaurs were going downhill for millions of years before this time, so there is hot debate over whether the Cretaceous-Tertiary extinction event (such as a meteorite impact) provided the final blow to the dinosaurs or whether the dinosaurs would have died out at that time anyway.

Figure 17-18   **Meteorites may have triggered some mass extinctions**

The Manicouagan crater in Quebec is about 45 miles in diameter. Giant impact craters such as this one often contain a central dome of rock that splashes up after the meteorite has buried itself in the ground, leaving a ring-shaped depression between the outer crater wall and the inner dome. In this satellite photo, water backed up behind a dam fills the crater ring. The Manicouagan meteorite struck a little over 200 million years ago; geologist Paul Olsen suggests that its impact triggered the mass extinction near the Triassic-Jurassic boundary (see Fig. 17-17).

hit Earth at these times. If a huge meteorite struck land, it would kick up enormous amounts of dust. The dust might be thick enough, and spread widely enough, to block out most of the sun's rays. Fires started by the impact might be widespread, adding soot to the atmosphere. Many plants would die because they couldn't photosynthesize. Many animals, all of which ultimately depend on plants for food, would also die. Smaller amounts of dust might still block out enough sunlight to cause global cooling, perhaps even triggering an ice age. Widespread extinctions would result.

Did such massive meteorite strikes really occur, and if so, would they cause extinctions? No one knows for sure, but considerable evidence points to meteorites as the causes of at least some major extinctions (Fig. 17-18). Recently, two groups of researchers have suggested that the Chicxulub crater near the Yucatan Peninsula of Mexico was the impact site of the meteorite that might have killed the dinosaurs.

E V O L U T I O N A R Y
C O N N E C T I O N S

Kin Selection and the Evolution of Altruism

Altruism is any behavior that is potentially harmful to the survival and future reproduction of an individual organism but enhances the reproductive potential of other organisms. Altruism includes worker bees' rearing the offspring of queen bees. Note that altruism does not imply conscious, voluntary decisions to engage in selfless behavior. Rather, most altruistic behaviors have a strong instinctive component; that is, many animals have altruism programmed in their genes.

From an evolutionary viewpoint, how can this be? Surely, if a mutation arose that caused altruistic behavior, and the bearers of that mutation lost their lives or failed to re-

produce because of their self-sacrificing behaviors, their "altruistic alleles" would disappear from the population. Maybe, or maybe not. To understand the evolution of altruism, we will need to introduce a new concept: **inclusive fitness**. As formulated by W. D. Hamilton, the inclusive fitness of an allele is the fitness conferred on *all* organisms that have the allele. Therefore, *if an altruist benefits related members of its own species that bear the same altruistic allele, then the altruistic allele may be favored by natural selection.*

To see how altruism might increase the inclusive fitness of an allele, let's consider the Florida scrub jay. Year-old jays usually do not mate and reproduce. Instead, these yearlings remain at their parents' nest and help out with next year's brood. Let's assume, for simplicity, that this altruistic behavior is controlled by a single "altruistic" allele and that in the distant past helper jays had the altruistic allele, and nonhelpers had another, "selfish," allele.

At least for one year, altruistic yearlings do not reproduce; some probably die from predation or accidents and never reproduce at all. How, then, can this behavior be adaptive? It all has to do with relatedness and the probability of successful reproduction. First, an animal's offspring inherit 50% of its genes (the other 50% of the offspring's genes come from the other parent). On the average, an animal also shares 50% of its genes with its siblings. Therefore, *a scrub jay is just as related to its siblings as it would be to its own offspring.* The second factor influencing scrub jay reproduction is that ideal habitat for jays is limited. Inexperienced yearling jays would probably have a hard time acquiring a good nest site and would be hard put to feed their offspring. Their best "reproductive bet," then, is to put their energy into helping their parents. Selfish yearling jays that try to nest on their own will probably contribute fewer genes to the next generation than the altruistic yearlings do. This phenomenon, whereby the actions of an individual increase the survival or reproductive success of its relatives, is called kin selection.

As this example suggests, *kin selection can favor the evolution of altruism if the altruistic behavior benefits relatives that bear the same altruistic allele*. In most cases, an animal will not know if another carries the altruism allele, but the animal must at least be able to distinguish relatives from strangers: Relatives stand a good chance of possessing the altruism allele, but you never can tell with strangers. A yearling jay that helped out at the nest of unrelated adult jays would probably waste its time and effort.

Identification of relatives isn't too hard to imagine in the case of jays and their parents. Many biologists objected to other proposed instances of altruistic behaviors, however, arguing that animals cannot evaluate degrees of relatedness. Two findings seem to address this objection. First, many social groups, including wolf packs and baboon troops, are actually family groups. Therefore, an animal would not have to identify relatives in order for its altruistic behaviors to benefit them the most. Second, many animals, including birds, monkeys, tadpoles, bees, and even tunicate larvae, can indeed identify relatives (Fig. 17-19). Given the choice between relatives and strangers, these animals preferentially associate with their relatives, *even if they were separated at birth and have never seen those relatives before*. If animals selectively form related groups, then altruistic behaviors will most likely benefit relatives. Although it is not the only mechanism, kin selection has been a powerful selective force in the evolution of altruism in many species, probably including humans.

***Figure 17-19* Cannibalistic animals don't eat close relatives**

Spadefoot toad tadpoles, found in transient water holes of the Arizona desert, are cannibalistic. Many of their prey, however, are released unharmed after being tasted briefly. Researchers have discovered that the tadpoles can indeed distinguish, and spit out, their own brothers and sisters, preferring to eat unrelated members of their own species.

▓ S U M M A R Y   O F   K E Y   C O N C E P T S

Evolution and the Genetics of Populations

The gene pool of a population is the total of all the different alleles of all the genes carried by the members of a population. The sources of genetic variability within a population are mutation, which produces new genes and alleles, and recombination during sexual reproduction. In its broadest sense, evolution is a change in the frequencies of alleles in the gene pool of a population due to enhanced reproduction by individuals bearing certain alleles.

Allele frequencies in a population will remain constant over generations only if the following conditions are met: (1) There must be no mutation; (2) there must be no gene flow, no net migration of alleles into or out of the population; (3) the population must be very large; (4) all mating must be random; (5) all genotypes must reproduce equally well (that is, no natural selection). These conditions are rarely, if ever, met in nature. Understanding why they are not met leads to an understanding of the mechanisms of evolution.

The Mechanisms of Evolution

1. Mutations are random, undirected changes in DNA composition. Although most mutations are neutral or harmful to the organism, some prove advantageous in certain environments. Mutations are rare and do not change allele frequencies very much, but they provide the raw material for evolution.

2. Migration is the flow of genes between populations. If the alleles that migrants carry are different from those in the populations from which they come or to which they migrate, migration will cause changes in allele frequencies.

3. In any population, chance events kill or prevent reproduction by some of the individuals. If the population is small, chance events may eliminate a disproportionate number of individuals bearing a particular allele, thereby greatly changing the allele frequency in the population. This change is termed genetic drift.

4. Many organisms do not mate randomly. If only certain members of a population can mate, then the next generation of organisms in the population will all be offspring of this select group, whose allele frequencies may differ from those of the population as a whole. Population bottleneck and founder effect, two types of genetic drift, illustrate the consequences small populations have on frequency of alleles.

5. The survival and reproduction of organisms are influenced by their phenotype. Because phenotype depends at least partly on genotype, natural selection will tend to favor the reproduction of certain alleles at the expense of others.

Natural Selection

Natural selection is really an issue of differential, or unequal, reproduction. Three types of natural selection are:

1. *Directional selection.* Individuals with characteristics that are different from average in one direction (for example, smaller) are favored both over average individuals and over those that differ from average in the opposite direction.
2. *Stabilizing selection.* Individuals of the "average value" for a characteristic are favored over individuals of extreme values.
3. *Disruptive selection.* Individuals of extreme characteristics are favored over individuals with average values.

Natural selection occurs as a result of the interactions of organisms with both the biotic (living) and abiotic (nonliving) parts of their environments. Within a species, sexual selection and altruism are two types of natural selection. When two or more species interact extensively so as to exert mutual selective pressures on each other for long periods of time, they both evolve in response. Such coevolution can occur as a result of any type of relationship between organisms, including competition, predation, and symbiosis.

Extinction

Two factors that contribute to the likelihood of extinction of a species are localized distribution and overspecialization. Factors that actually cause extinctions include competition among species, novel predators or parasites, and habitat destruction.

✖ KEY TERMS

adaptation   p. 320
allele frequency   p. 310
altruism   p. 322
balanced polymorphism   p. 318
coevolution   p. 320
competition   p. 320
differential reproduction   p. 317
directional selection   p. 317
disruptive selection   p. 317

equilibrium population   p. 312
extinction   p. 322
fitness   p. 310
founder effect   p. 315
gene flow   p. 312
gene pool   p. 310
genetic drift   p. 313
genetic equilibrium   p. 312
inclusive fitness   p. 326

kin selection   p. 322
migration   p. 312
natural selection   p. 310
population   p. 309
population bottleneck   p. 314
population genetics   p. 310
sexual selection   p. 315
stabilizing selection   p. 317
symbiosis   p. 321

✖ THINKING THROUGH THE CONCEPTS

Multiple Choice

1. Genetic drift is a _____ process.
   - a. random
   - b. directed
   - c. selection-driven
   - d. coevolutionary
   - e. uniformitarian

2. Most of the 700 species of fruit flies found in the Hawaiian archipelago are each restricted to a single island. One hypothesis to explain this pattern is that each species diverged after a small number of flies had colonized a new island. This mechanism is called
   - a. sexual selection
   - b. genetic equilibrium
   - c. disruptive selection
   - d. the founder effect
   - e. assortative mating

3. You are studying leaf size in a natural population of plants. The second season is particularly dry, and the following year you find that the average leaf size in the population is smaller than the year before. But the amount of overall variation is the same, and the population size hasn't changed. Also, you've done experiments that show that small leaves are better adapted to dry conditions. Which of the following has occurred?
   - a. genetic drift
   - b. directional selection
   - c. stabilizing selection
   - d. disruptive selection
   - e. the founder effect

4. You have bacteria thriving in your gastrointestinal tract. This is an example of
   - a. inclusive fitness
   - b. balanced polymorphism
   - c. symbiosis
   - d. kin selection
   - e. altruism

5. Lamarckian evolution could occur
   - a. if each gene had only one allele
   - b. if individuals had different phenotypes
   - c. if the genotype was altered by the same environmental changes that altered the phenotype
   - d. if the phenotype was altered by the environment
   - e. under none of these conditions

6. Of the following possibilities, the best way to estimate the Darwinian fitness of an organism is to measure the
   - a. size of its offspring
   - b. number of eggs it produces
   - c. number of eggs it produces over its lifetime
   - d. number of offspring it produces over its lifetime
   - e. number of offspring it produces over its lifetime that survive to breed

Review Questions

1. What is a gene pool? How would you determine the allele frequencies in a gene pool?

2. Define an equilibrium population, and outline the conditions that must be met for a population to remain in equilibrium.

3. How does population size affect the likelihood of changes in allele frequencies by chance alone? Can significant changes in allele frequencies (that is, evolution) occur because of genetic drift?

4. If you went out and measured the allele frequencies of a gene and found large differences from the proportions predicted by Hardy-Weinberg equilibrium, would that prove that natural selection is occurring in the population you are studying? Review the conditions that lead to Hardy-Weinberg equilibrium, and explain your answer.

5. People like to say that "you can't prove a negative." Study the experiment in Figure 17-1 again, and comment on what it demonstrates.

6. Describe the three types of natural selection. Which type(s) is (are) most likely to occur in stable environments and which type(s) in rapidly changing environments?

7. What is sexual selection? How is sexual selection similar to and different from other forms of natural selection?

8. Briefly describe competition, predation, symbiosis, and altruism, and give an example of each.

9. Define kin selection and inclusive fitness. Can these concepts help to explain the evolution of altruism?

✖ APPLYING THE CONCEPTS

1. In North America, the average height of human adults has been increasing steadily for decades. Is directional selection occurring? What data would you need to justify your answer?

2. Malaria is rare in North America. In populations of African Americans, what would you predict is happening to the frequency of the hemoglobin allele that leads to sickling in red blood cells? How would you go about determining if your prediction is true?

3. By the 1940s the Whooping Crane population had been reduced to under 50 individuals. Thanks to conservation measures, their numbers are now increasing. But what special evolutionary problems do Whooping Cranes have after passing through a population bottleneck?

4. In many countries, conservationists are trying to design national park systems so that "islands" of natural area (the big parks) are connected by thin "corridors" of undisturbed habitat. The idea is that this arrangement will allow animals and plants to migrate between refuges. Why would such migration be important?

5. Extinctions have occurred throughout the history of life on Earth. Why should we care if humans are causing a mass extinction event now?

6. A preview question for Chapter 18: A species is all the populations of organisms that potentially interbreed with one another but that are reproductively isolated from (cannot interbreed with) other populations. Using the five assumptions of the Hardy-Weinberg equilibrium population as a starting point, what factors do you think would be important in the splitting of a single ancestral species into two modern species?

✖ FOR MORE INFORMATION

Allison, A. C. "Sickle Cells and Evolution." *Scientific American*, August 1956. The story of the interaction between sickle-cell anemia and malaria in Africa.

Alvarez, W., and Asaro, F. "An Extraterrestrial Impact," and Courtillot, V. E. "A Volcanic Eruption." *Scientific American*, October 1990. Leading geologists debate the question, What caused the mass extinction of the dinosaurs at the end of the Cretaceous period?

Fellman, B. "To Eat or Not to Eat." *National Wildlife*, February–March 1995. How animals performing altruistic behaviors identify their relatives.

Gould, S. J. "The Evolution of Life on the Earth." *Scientific American*, October 1994. The importance of chance and catastrophe in shaping modern life.

May, R. M. "The Evolution of Ecological Systems." *Scientific American*, September 1979. Coevolution accounts for much of the structure of natural communities of plants and animals.

O'Brien, S. J., Wildt, D. E., and Bush, M. "The Cheetah in Peril." *Scientific American*, May 1986. According to molecular and immunological techniques, a population bottleneck has reduced the genetic variability of the world's cheetahs almost to zero.

Ryan, M. J. "Signals, Species, and Sexual Selection." *American Scientist*, January–February 1990. Ryan explores a variety of experiments on sexual selection, including the genetic basis of male characteristics and female choice.

Stebbins, G. L., and Ayala, F. "The Evolution of Darwinism." *Scientific American*, July 1985. A synthesis of molecular and classical evolutionary methodologies.

NET WATCH

On-line resources for this chapter are on the World Wide Web at:
http://www.prenhall.com/~audesirk (click on the table of contents link and then select Chapter 17).

18 The Origin of Species

> *"… endless forms most beautiful and most wonderful have been, and are being, evolved."*
>
> CHARLES DARWIN in *On the Origin of Species by Means of Natural Selection* (1859)

The great bird of paradise is one of almost 140 species of birds living on New Guinea, many of them unique to New Guinea and its neighboring islands. Why there are so many species, how they arose, and how they maintain reproductive isolation from other, often similar, species are major questions in evolutionary biology.

A tribe on New Guinea has 137 separate names for local birds, almost exactly the same as the number of species determined by Western ornithologists. The larvae of each of the 750 species of fig wasps on Earth can grow only in its own particular species of fig (Fig. 18-1). All three—New Guinea islanders, ornithologists, and fig wasps—perceive qualitative differences between members of different species; that is, differences in kind. Mere quantitative differences, for instance that this fig tree is a bit larger than that one, or this bird's feathers are a bit brighter blue than that one's, are usually ignored. To biologists, the major qualitative difference between species is the *ability to interbreed*. Thus, a species consists of all the populations of organisms that are potentially capable of interbreeding under natural conditions and that are reproductively isolated from (incapable of breeding with) members of other populations.

Our discussion of the mechanisms of evolution in the previous chapter dealt exclusively with quantitative changes within a population: the frequency of the sickle-cell anemia allele, the length of a giraffe's neck, or the size of a peacock's tail. At what point do these quantitative changes add up to a large enough total difference so that the population is a new species and cannot or will not interbreed with any other populations? Or is there a sudden jump, a quantum leap of speciation distinct from, and perhaps unrelated to, the accumulation of small, quantitative differences?

And what about the grand sweep of evolutionary history? Is all of evolution, from the first prokaryotic cell to the human brain, simply the accumulation of small changes over millions of years? Or are other processes at work?

These are some of the most contentious questions in biology. Virtually all biologists agree that evolution occurs and that the processes described in Chapter 17 produce quantitative changes in populations. But other questions remain. How do species arise? What is the relative importance of many small changes versus a few major changes? Is natural selection within populations a major force in generating the stunning diversity of living organisms? In this chapter, we will explore current hypotheses that try to explain the origin of species.

Figure 18-1   **Species specializations**

A female fig wasp uses a special organ at the tip of her abdomen, called an ovipositor, to inject her eggs into a fig. Each species of fig wasp is restricted to a single species of fig.

Speciation

Although we have described several evolutionary forces that lead to changes *within* species, we have not yet outlined a mechanism of **speciation**, whereby *new species* may be formed. To produce a new species, evolution must generate large enough genetic changes between populations so that either mating cannot occur or, if it does, the hybrid offspring are less fit. Speciation depends on the (1) isolation and (2) genetic divergence of two populations.

1. *Isolation of populations.* If two populations are to become sufficiently distinct, genetically, so that interbreeding is difficult or impossible, then there must be relatively little gene flow (migration) between them. If there is a great deal of gene flow, then genetic changes in one population will soon become widespread in the other as well.
2. *Genetic divergence.* It is not sufficient for two populations simply to be isolated—what if two formerly isolated populations are reunited? They will have become separate species only if, during the period of isolation, they had evolved sufficiently large genetic differences so that they cannot interbreed and produce vigorous, fertile offspring when they are reunited. If isolated populations are small, chance events may generate significant genetic differences by genetic drift (see Chapter 17). In both small and large populations, different selective pressures in separate environments may favor the evolution of large genetic differences.

Speciation has seldom been observed in the wild (with the exception of "instant speciation" in plants by poly-

ploidy, described below). However, based partly on theoretical considerations and partly on experiments and observations, evolutionary biologists have devised plausible hypotheses for the origin of new species in two different cases: **allopatric speciation** (two populations are geographically separated from one another) and **sympatric speciation** (two populations share the same geographical area) (Fig. 18-2).

At first glance, you might think that sympatric speciation violates our first principle of speciation, isolation of populations, because the speciating populations live in the same locale. However, it is *isolation from gene flow* that is crucial to speciation. Isolation from gene flow is simplest to envision if two populations are geographically separated by some kind of physical barrier, such as a river. Indeed, most cases of speciation, at least in animals, probably have occurred through such geographical isolation. As we will see below, however, even two populations living in the same area may have very little gene flow if they select different habitats within the area (for example, marshes as opposed to forests) or have differences in chromosome numbers so that they cannot form fertile hybrids. Therefore, the principle still holds: *Isolation from gene flow is the key to both allopatric and sympatric speciation.*

Allopatric Speciation Occurs in Populations That Are Physically Separated

In Greek, the word *allopatric* means "having a different fatherland." Allopatric speciation occurs when two populations become **geographically isolated** from one another; that is, they are physically separated either by distance or by an impassible barrier (Fig. 18-3). As we described in Chapter 17, migration allows gene flow, which reduces genetic differences between populations and probably eliminates any possibility that two populations could ever become reproductively isolated from one another. Subsequently, if two populations become separated by a physical barrier, or even sheer distance, little or no gene flow can occur between them. If the pressures of natural selection differ in the two locations, or if the populations are small enough for genetic drift to occur, then the two populations may accumulate large genetic differences and become separate species. Founder events, in which a few members of a species become isolated from the main body of the species, may also be important in initiating genetic differences between populations. There is some debate as to whether genetic drift or natural selection normally plays a major role in allopatric speciation. Probably the majority opinion is that different selective pressures in the two geographical locales provide the major impetus to speciation. In either case, evolutionary biologists hypothesize that geographical isolation is involved in most cases of speciation, especially in animals.

(a) Allopatric speciation

(b) Sympatric speciation

time

geographical isolation

ecological isolation

genetic divergence

genetic divergence

reproductive isolation

reproductive isolation

Figure 18-2   **Models of allopatric and sympatric speciation**

(a) Allopatric speciation. (Top) A single species (white mice) occupies a relatively homogeneous habitat. (Middle) An impassable geographical barrier (here, a river changing course) splits the habitat into two parts, separating the species into two isolated populations. Genetic drift or different selective pressures cause the two populations to diverge genetically (tan vs. white mice). (Bottom) The barrier is removed (the river changes back to its original course), and the members of the two populations can share the same habitat. If the genetic differences between the two populations have become large enough so that interbreeding cannot occur (that is, they are reproductively isolated from one another), then the two populations constitute separate species (brown vs. white mice). **(b)** Sympatric speciation. (Top) A single species occupies a homogeneous habitat. (Middle) Climate change or other factors form two distinctly different habitats that are still physically part of the same general region; that is, there are no barriers to movement between habitats. Different selective pressures in the two habitats lead to genetic divergence of organisms living in each (tan vs. white mice). (Bottom) Sufficient genetic divergence causes reproductive isolation; former occupants of the two different habitats are now separate species (brown vs. white mice).

(a)   **(b)**

Figure 18-3   Geographical isolation

Geographical isolation that prevents gene flow is usually necessary for animal populations to split into two species. The tassel-eared squirrels that live on the rims of the Grand Canyon are a classic example. In the distant past, a single population of squirrels became separated as the canyon was carved into the plateau of northwestern Arizona. The two populations are split by the treeless desert of the canyon depths. Today, the Kaibab squirrel **(a)** is confined to the north rim, and the Abert squirrel **(b)** is found on the south rim and in forests throughout northern Arizona, Utah, New Mexico, and Colorado. It is not known whether the two are genetically distinct enough to constitute separate species.

Sympatric Speciation Occurs in Populations That Live in the Same Area

Sympatric means "having the same fatherland." As the name implies, sympatric speciation refers to speciation occurring within a single population and in a single geographical area. Sympatric speciation, like allopatric speciation, requires limited gene flow. There are two likely mechanisms whereby gene flow can be reduced between members of a single population: ecological isolation and chromosomal aberrations.

Ecological Isolation Restricts Different Populations to Different Habitats within the Same Area

If the same geographical area contains two distinct types of habitats (for example, food sources, nesting places, and so on), different members of a single species may begin to specialize in one habitat or the other. If conditions are right, natural selection for habitat specialization may cause the formerly single species to split into two species. Such a split seems to be occurring, right before biologists' eyes, so to speak, in the case of the fruit fly *Rhagoletis pomonella* (Fig. 18-4).

Rhagoletis is a parasite of the American hawthorn tree, laying its eggs in the hawthorn's fruit. When the maggots hatch, they eat the fruit. About 150 years ago, entomologists (scientists who study insects) noticed that *Rhagoletis* began to infest apple trees, which were introduced into North America from Europe. Now, it appears that *Rhagoletis* is splitting into two species, based on their preference

for apples or hawthorns. There are substantial genetic differences between apple-liking flies and hawthorn-liking flies. At least some of these genetic differences, such as the timing of emergence of the adult flies (described later in this chapter), are important for survival on a particular host plant. Since apples and hawthorns are often found quite close together, and flies, after all, can fly, why don't apple-flies and hawthorn-flies interbreed and cancel out any incipient genetic differences? There appear to be at least two reasons. First, female flies usually lay their eggs in the same type of fruit in which they themselves devel-

Figure 18-4   Sympatric isolation

The fruit fly *Rhagoletis pomonella* seems to be undergoing sympatric speciation through ecological isolation on two host trees: apples and hawthorns.

oped. Males also tend to rest on the same type of fruit in which they developed. Therefore, apple-liking males are likely to encounter and mate with apple-liking females. Second, apple fruits mature 2 or 3 weeks later than hawthorn fruits, and the two types of flies emerge with a timing appropriate for their chosen host fruits. Thus the two types of flies have very little chance of meeting. There is still some interbreeding between the two types of flies, but it looks as if they are well on their way to speciation. Will they make it? Entomologist Guy Bush suggests, "Check back with me in a few thousand years."

Changes in Chromosome Number Can Cause Immediate Reproductive Isolation of a Population

In some instances, new species can arise nearly instantaneously, either through changes in chromosome configuration or number or through irregularities during meiosis. A common speciation mechanism in plants is **polyploidy**, the acquisition of multiple copies of each chromosome. As you know, most plants and animals have paired chromosomes and are described as diploid. Occasionally, especially in plants, a fertilized egg duplicates its chromosomes but doesn't divide into two daughter cells. The resulting cell thus becomes tetraploid, with four copies of each chromosome. If all of the subsequent cell divisions are normal, this tetraploid zygote will develop into a plant that consists of tetraploid cells. Tetraploid plants are usually vigorous and healthy, and many can successfully complete meiosis to form viable gametes. The gametes, however, are diploid (meiosis normally produces haploid gametes from diploid cells). Therefore, if a sperm from a tetraploid plant fertilizes an egg cell of the diploid "parental" species, the resulting offspring will be triploid. The triploid offspring may develop normally, but when this triploid offspring begins meiosis, it will be unable to pair up all of its chromosomes, because there will be an odd number (three) of each. Meiosis fails, gametes are not formed, and so the triploid hybrids are sterile. The tetraploid plant is therefore reproductively isolated from its diploid parent.

Why is speciation by polyploidy common in plants but not in animals? Many plants can either self-fertilize or reproduce asexually, or both. If a tetraploid plant self-fertilizes, then its offspring will also be tetraploid. Asexual offspring, of course, are genetically identical to the parent, and are also tetraploid. In either case, the new tetraploid plant may perpetuate itself and form a new species. Animals, on the other hand, usually cannot self-fertilize or reproduce asexually. Therefore, if an animal produced a tetraploid offspring, the offspring would have to mate with a member of the diploid parental species and would produce all triploid offspring. As we pointed out, the triploid offspring would almost certainly be sterile. Speciation by polyploidy is extremely common in plants; in fact, nearly half of all species of flowering plants are polyploid, many of them tetraploid.

TABLE 18-1 :: Mechanisms of Reproductive Isolation

Premating Isolating Mechanisms: any structure, physiological function, or behavior that prevents organisms of two different populations from mating.

1. **Geographical isolation:** the separation of two populations by a physical barrier.
2. **Ecological isolation:** lack of mating between organisms belonging to different populations that occupy distinct habitats within the same general area.
3. **Temporal isolation:** the inability of organisms to mate if they have significantly different breeding seasons.
4. **Behavioral isolation:** lack of mating between species of animals that differ substantially in courtship and mating rituals.
5. **Mechanical incompatibility:** the inability of male and female organisms to exchange gametes, usually because of incompatibility of the reproductive structures.

Postmating Isolating Mechanisms: any structure, physiological function, or developmental abnormality that prevents organisms of two different populations, once mating has occurred, from producing vigorous, fertile offspring.

1. **Gametic incompatibility:** the inability of sperm from one species to fertilize eggs of another species.
2. **Hybrid inviability:** the failure of a hybrid offspring of two different species to survive to maturity.
3. **Hybrid infertility:** reduced fertility (often complete sterility) in hybrid offspring of two different species.

Maintaining Reproductive Isolation between Species

Once a species has formed, it can remain **reproductively isolated** from other species in two ways (Table 18-1). First, members of the species usually do not mate with members of other species. This strategy has a clear adaptive value. Separate species are usually genetically different in ways that adapt them to different environments. Any individual that mates with a member of another species will probably produce unfit or sterile offspring, thereby "wasting" its genes and contributing nothing to future generations. Thus there is strong selective pressure to avoid mating between different species. Incompatibilities between species that prevent mating are called **premating isolating mechanisms**.

Sometimes premating isolation fails or has not yet evolved, and members of different species do mate. Then the second isolating mechanism comes into play. If the resulting hybrid offspring die during development, then naturally the two species are still reproductively isolated from one another. In some cases, however, viable hybrid offspring are produced. Even so, if the hybrids are less fit or infertile, the two species may still remain separate, with little or no gene flow between them. Incompatibilities that prevent the formation of vigorous, fertile hybrids between species are called **postmating isolating mechanisms**.

Premating Isolating Mechanisms Prevent Mating between Species

Mechanisms that prevent mating between different species include geographical isolation, ecological isolation, temporal isolation, behavioral isolation, and mechanical incompatibility.

Geographical Isolation Prevents Members of Different Species from Meeting One Another

Members of different species obviously cannot mate if they never get near one another. As we have already seen, geographical isolation often provides the conditions for speciation in the first place. However, we cannot determine if geographically separated populations necessarily constitute distinct species. Should the barrier separating the two populations disappear (an intervening river changes course, for example), the reunited populations might interbreed freely and not be separate species at all. If they cannot interbreed, then other mechanisms, such as different courtship rituals, must have developed during their isolation. Geographical isolation, therefore, is usually considered to be a mechanism that *allows new species to form* rather than a mechanism that *maintains reproductive isolation between different species*.

Ecological Isolation Confines Members of Different Species to Different Habitats

If two populations have different resource requirements, they may use different local habitats within the same general area, and thus exhibit **ecological isolation**. White-crowned and White-throated Sparrows, for example, have extensively overlapping ranges. The White-throated Sparrow, however, frequents dense thickets, whereas the White-crowned Sparrow inhabits fields and meadows, seldom penetrating far into dense growth. The two species may coexist within a few hundred yards of one another and yet seldom meet during breeding season. The *Rhagoletis* fruit flies discussed earlier are at least partially isolated by habitat selection for apples versus hawthorns. Although ecological isolation may slow down interbreeding, it seems unlikely that it could prevent gene flow entirely. Other mechanisms usually also contribute to interspecific isolation.

Temporal Isolation Occurs between Species That Breed at Different Times

Even if two species occupy similar habitats, they cannot mate if they have different breeding seasons, a phenomenon called **temporal** (time-related) **isolation**. Bishop pines

Figure 18-5　Temporal isolation
Bishop pines, such as these, and Monterey pines coexist in nature. In the laboratory they produce fertile hybrids. In the wild, however, they do not interbreed because they release pollen at different times of the year.

and Monterey pines coexist near Monterey on the California coast (Fig. 18-5). Viable hybrids have been produced between these two species in the laboratory. In the wild, however, they release their pollen at different times: The Monterey pine releases pollen in early spring, the bishop pine in summer. Therefore, the two species never cross-breed under natural conditions. Hawthorn-liking and apple-liking *Rhagoletis* fruit flies are also partially isolated from one another because they emerge from their host fruits and breed at slightly different times of year.

Different Courtship Rituals Create Behavioral Isolation

Among animals, the elaborate courtship colors and behaviors that so enthrall human observers have evolved not only as recognition and evaluation signals between male and female; they may also aid in distinguishing among species. These different signals and behaviors create **behavioral isolation**. The striking colors and calls of male songbirds, for example, may attract females of their own species, but females of other species treat them with the utmost indifference. Among frogs, males are often impressively indiscriminate, jumping on every female in sight, regardless of the species, when the spirit moves them. Females, however, approach only male frogs croaking the "ribbet" appropriate to their species. If they do find themselves in an unwanted embrace, they utter the "release call," which causes the male to let go. As a result, few hybrids are produced.

Mechanical Incompatibility Occurs When Physical Barriers between Species Prevent Fertilization

In rare instances, ecological, temporal, and behavioral isolating mechanisms fail, and male and female of different species attempt to mate. Among different species of animals with internal fertilization (where the sperm is deposited inside the female's reproductive tract), in some cases the male and female sexual organs simply won't fit together. Among plants, differences in flower size or structure may prevent pollen transfer between species, if, for example, the species attract different pollinators. Isolating mechanisms of this type are called **mechanical incompatibility**.

Postmating Isolating Mechanisms Prevent Production of Vigorous, Fertile Offspring

Sometimes premating isolation fails, and mating occurs between members of different species. However, if vigorous, fertile hybrids are not produced, there will still be little gene flow between the two species. Postmating isolating mechanisms include gametic incompatibility, hybrid inviability, and hybrid infertility.

Gametic Incompatibility Occurs When Sperm from One Species Are Unable to Fertilize Eggs of Another

Even though a male inseminates a female, his sperm may not fertilize her eggs, a situation called **gametic incompatibility**. For example, the fluids of the female reproductive tract may weaken or kill sperm of other species. Among plants, chemical incompatibility may prevent the germination of pollen from one species that lands on the stigma (pollen-catching structure) of the flower of another species.

Hybrid Inviability Occurs If Hybrid Offspring Survive Poorly

If fertilization does occur, the resulting hybrid may be weak or even unable to survive, a situation called **hybrid inviability**. The genetic programs directing development of the two species may be so different that hybrids abort early in development. Even if the hybrid survives, it may display behaviors that are mixtures of the two parental types. In attempting to do some things the way species A does them, and other things the way species B does, the hybrid may be hopelessly uncoordinated. Hybrids between certain species of lovebirds, for example, have great difficulty learning to carry nest materials during flight and probably could not reproduce in the wild.

Hybrid Infertility Occurs If Hybrid Offspring Are Unable to Produce Normal Sperm or Eggs

Animal hybrids, such as the mule (a cross between a horse and a donkey), are usually sterile. **Hybrid infertility** prevents hybrids from passing on their genetic material to offspring. A common reason is the failure of chromosomes to pair properly during meiosis, so that eggs and sperm never develop. Among plants that have speciated by poly-

ploidy, any offspring produced by mating between the diploid "parent" species and the tetraploid "daughter" species will be triploid and sterile.

Phyletic and Divergent Speciation

The mechanisms of speciation and reproductive isolation that we have just described all apply only to "splitting," or **divergent speciation**. In divergent speciation, two populations of a single parent species become separate species as a result of isolation from gene flow followed by genetic divergence (Fig. 18-6a). Divergent speciation is often called "true" speciation, because it is possible (in principle at least) to determine if the two presumed species are in fact reproductively isolated.

In the fossil record, however, things are not so simple. Fossils, being dead and petrified, cannot breed, so it is difficult to determine whether they are reproductively isolated from other fossils. Therefore, paleontologists make distinctions among species purely on morphological (anatomical) grounds: If two fossils are structurally as different from one another as two modern species are, then the fossils are assigned to different species. Suppose that a whole species, under massive directional selection, changes over time so that the later specimens are very different from the earlier ones (Fig. 18-6b). If the differences are large enough, it may be assumed that the later organisms could not possibly interbreed with the earlier ones and so form a separate species. In other words, over time, a single species evolves into another species. In this situation, the two species never coexist and reproductive isolation between early and late forms cannot be directly tested. This process is called **phyletic speciation**.

Owing to its incompleteness, the fossil record does not reveal whether divergent or phyletic speciation dominates evolutionary history. If divergent speciation is most common, as many evolutionary biologists hypothesize, but few enough fossils are found, then most evolutionary lineages will appear to show phyletic speciation (Fig. 18-6c). Beginning with a modern species and working backward, one may be able to reconstruct a "phyletic evolutionary history" back to an ancient ancestor. This phyletic history may, however, simply be selective sampling of a much larger, divergent history in which most of the branches left no fossils.

The Genetics of Speciation

When plants speciate by polyploidy, it is clear that a sudden, major genetic change has taken place. But what about other speciation events, whether allopatric or sympatric? There are two genetic models for speciation: the gradual accumulation of many small changes and the sudden appearance of a few major changes.

One Model of Speciation Stresses Gradual Accumulation of Many Small Changes

According to the first model, two populations, whether by genetic drift or through differential natural selection pressures, gradually accumulate many small genetic changes. Each mutation or different allele by itself has only a minor phenotypic effect. Over time, however, the accumulation of many small changes may result in reproductive isolation between the two populations, and they may become separate species.

A Second Model Stresses the Sudden Appearance of a Few Major Changes

The second model starts out with the idea that single regulatory genes, or small numbers of genes, control major developmental pathways, such as how originally similar cells become specialized to form different structures. In certain aspects of the development of animals, this is known to be true. So mutations in just a few regulatory genes might result in such significant changes in development that the mutants would immediately be reproductively isolated from their parental populations. A new species would therefore arise almost instantaneously.

Which Model Is Correct?

Because so few speciation events have been observed, and so little is known about the genetics of development, it is difficult to choose between these models. The extreme genetic similarity between certain species, for example chimps and humans, suggests to some observers that mutations in a few key regulatory genes may have been the cause of speciation. On the other hand, the "small change" model is supported by the finding that many phenotypic characteristics, including size, coloration, and various behavioral traits, are under the control of several to many genes (see Chapter 13 for a discussion of quantitative inheritance). Thus, one might expect that changes in many genes would be required to generate differences at the species level. Further, both theoretical modeling and common observation suggest that one-shot, massive genetic changes are usually harmful. This debate is not likely to be settled

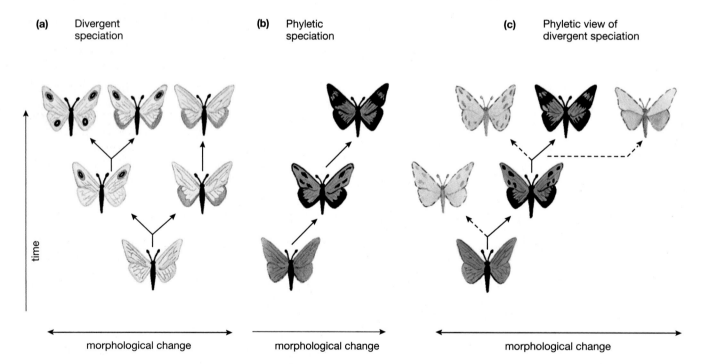

(a) Divergent speciation   **(b)** Phyletic speciation   **(c)** Phyletic view of divergent speciation

Figure 18-6   **Divergent speciation vs. phyletic speciation**

(a) In divergent speciation, a single species splits up into two or more daughter species that exist simultaneously and are reproductively isolated from one another. **(b)** In phyletic speciation, the whole population of a single species evolves over time. Sufficient genetic changes occur so that the later populations are considered to be a new species, different from earlier populations. **(c)** The fossil record is incomplete. Most species may become extinct without leaving either fossils or living descendants (dashed butterflies). Therefore, even if divergent speciation is most common, the fossil record may appear to show phyletic speciation (solid butterflies).

until molecular developmental biology comes to the rescue with a better understanding of the genetic basis of development and of developmental differences between species.

Rates of Speciation

Over the course of evolutionary history, species continually form, exist for a time, and become extinct. Geographical isolation, genetic drift, mutations, the invasion of new habitats, and coevolution all ensure that speciation never stops. The rate of speciation varies considerably over evolutionary time, however, and bursts of speciation can be seen both in the fossil record and in the distribution of modern organisms.

During Adaptive Radiation One Species Gives Rise to Many under the Differing Selection Pressures of New Habitats

Sometimes a species gives rise to many new species in a relatively short time. This process, called **adaptive radiation**, occurs when populations of a single species invade different habitats and evolve in response to the differing selective pressures in those habitats. Adaptive radiation has occurred many times and in many groups of organisms. Adaptive radiation usually results from one of two causes. First, a species may encounter a wide variety of unoccupied habitats; for example, when the ancestors of Darwin's finches colonized the Galapagos Islands, or when marsupial mammals first invaded Australia. With no competitors except other members of their own species, all the available habitats and food sources were rapidly utilized by new species that evolved from the original invaders. The ultimate in unoccupied habitats results from mass extinctions, such as those that might follow meteorite impacts (see Chapter 17). With most potential competitors exterminated by the effects of the impact, the survivors speciate prolifically, filling an array of empty habitats.

Adaptive radiation also occurs if a species develops a fundamentally new and superior **adaptation**, enabling it to displace less well adapted species from a variety of habitats. Such adaptations have apparently happened several times during evolutionary history, for example, when mammals diversified extensively at the expense of reptiles such as the dinosaurs.

The Pattern of Evolution

Although the processes that drive evolutionary change, such as genetic drift and natural selection, are well understood, there is considerable debate among evolu-

tionary biologists concerning which processes actually play the major roles in shaping evolutionary history. Three subjects of particular interest are (1) the rate and timing of evolutionary change over time; (2) the role of speciation in producing morphological change; and (3) the relative importance of natural selection versus nonselective forces in determining structures and behaviors.

In both the popular press and scientific journals, these topics often are discussed within the framework of two opposing views of evolution, termed the gradualism and punctuated equilibrium models. These two models are actually at the opposite ends of a spectrum, as we will see. In all likelihood, the evolution of real organisms usually falls somewhere closer to the middle of the spectrum, sometimes evolving mostly according to "gradualistic" rules and sometimes according to "punctuated" rules. We will briefly outline the two extreme positions and then attempt to synthesize a coherent scheme.

Gradualism Explains Speciation as the Slow, Steady Accumulation of Small Changes over Time

The extreme **gradualism** model of speciation takes the following positions:

1. Speciation occurs from the accumulation of dozens or hundreds of small genetic differences between two populations. The two populations thus become separate species in a gradual fashion, as less and less interbreeding occurs and any hybrids that are formed become less and less fit. Speciation may take hundreds of thousands or even millions of years (Fig. 18-7a).
2. Morphological change within a lineage (e.g., horses; see Fig. 16-10) occurs continuously, although the rate may change somewhat from time to time.
3. Morphological change and speciation are not closely linked. Two populations of a species may undergo considerable morphological change without necessarily becoming reproductively isolated from one another.
4. Both continual morphological changes and speciation are driven largely by natural selection among individuals, with fitter organisms leaving more offspring than less fit organisms do.

Punctuated Equilibrium Explains Speciation as Occurring Relatively Rapidly amid Long Periods of Little Change

The extreme form of the **punctuated equilibrium** model of evolution takes contrasting stands on every issue:

1. Speciation is a relatively rapid event (in the context of geological time), driven by a relatively small number of genetic changes, such as mutations in a few developmentally important regulatory genes.

The rapidity of speciation is the "punctuated" part of the model (Fig. 18-7b).

2. Species remain morphologically the same for long periods of time; this is the "equilibrium" part of the model.

3. Morphological change and speciation are very tightly linked. Virtually all morphological change occurs during the brief period of speciation, and the newly formed species then remain essentially unchanged until the next speciation event.

4. Large-scale evolutionary changes are driven by selection among species, not among individuals. Further, "species selection" often may be driven not by natural selection but by random chance.

Gradualism and Punctuated Equilibrium Models Can Both Be Applied to the Evolution of the Horse

In the evolution of the horse, the fossil record shows a progression from *Hyracotherium*, the "dawn horse," through many intermediate steps to modern *Equus* (see Fig. 16-10). Extreme gradualism would propose that fossil horses are representative samples of changes that occurred gradually throughout the evolution of the horse. As environments changed from forest to open woodlands to prairies, natural selection favored larger, faster horses with strong, shock-absorbing legs, hard hooves, and large, grinding teeth. Divergent speciation occurred from time to time as genetic changes accumulated between, for example, woodland populations and prairie populations, but phyletic specia-

tion also occurred, as whole populations changed into new species under selective pressures. During phyletic speciation, selection *between individuals within a population* drove the whole population to change, gradually becoming a new species.

One problem with this view, which provided the original stimulus for the development of the punctuated equilibrium model, is the nature of the fossil record. Every type of fossil horse is quite different from every other type; that is, the fossil record does not show a continuous accumulation of small changes, but rather a series of jumps from one fossil type to the next. Why haven't fossils been found that represent all the presumed intermediate stages? Gradualism replies that fossilization is a rare event; for every organism that becomes fossilized, and is found by modern humans, probably millions do not. Further, many thousands of years may pass during which suitable conditions for fossilization (for example, massive eruptions of volcanic ash, flash floods, buildup of sediment in shallow seas) do not occur. Therefore, many intermediate stages, perhaps even whole species, never appear in the fossil record.

The gaps in the fossil record are the starting point for the punctuated equilibrium model, championed by evolutionary biologists Stephen Jay Gould of Harvard and Niles Eldredge of the American Museum of Natural History and others. Punctuated equilibrium assumes that the gaps are not accidents: They are produced by such rapid evolutionary change that fossil remains of inter-

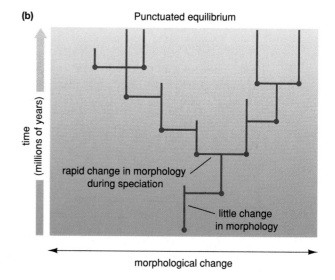

Figure 18-7   **Extreme views of gradualism vs. punctuated equilibrium**
Solid dots represent fossils considered to be separate species. Lines that do not reach the top of the graph represent extinctions. **(a)** Gradualism: Morphological change occurs gradually over time, although the rate of change may vary (more rapid during times of strong directional selection, much more slowly during times of stabilizing selection). Both phyletic and divergent speciation probably occur. **(b)** Punctuated equilibrium: Morphological change and speciation are simultaneous and very rapid, interrupted by long periods during which neither morphological change nor speciation occurs. All speciation is divergent.

mediate forms would be extremely rare. Punctuated equilibrium hypothesizes that *Hyracotherium* underwent little change for millions of years (the "equilibrium"). Small fringe populations of *Hyracotherium* split off now and then. Some evolved very rapidly, perhaps mostly by genetic drift, to become entirely new species. At some point, perhaps because of a major change in the environment, one of these new species, *Mesohippus*, quickly replaced *Hyracotherium* as the dominant form of horse. The speciation of *Mesohippus* and its replacement of *Hyracotherium* may have taken place within a few thousand years—a mere instant, in the span of geological time (the "punctuation" between equilibria). Punctuated equilibrium assumes that evolution is largely driven by this frequent, rapid divergent speciation. Selection *between species* then results in the extinction of some species and the survival of others.

Many Evolutionary Biologists Accept a Synthesis of Gradualism and Punctuated Equilibrium

As we have already cautioned, the gradualism and punctuated equilibrium models just presented are extreme positions. Although there are certainly major disagreements among evolutionary biologists, something like a majority view could be summarized as follows.

Changes Too Subtle to Be Observed between Generations Can Result in Rapid Evolution on a Geological Time Scale

The rate of evolution, as measured by morphological changes in fossils, varies greatly. In some organisms, there are indeed long periods of constancy—sharks, for example, have remained outwardly similar for scores of millions of years. Morphological constancy might arise from several causes. In a constant environment, stabilizing selection may limit the rate of evolutionary change. A rapidly fluctuating environment, with wide but frequent temperature changes, for instance, would promote equally rapid, fluctuating directional selection, with a net effect of preserving the status quo. Or even if the environment changes in a constant direction, species may simply "track" favorable habitats. During the ice ages, for example, the glaciers expanded from arctic regions, creating vast new areas of cold and severe conditions. Caribou and other cold-adapted species simply followed the vegetation that flourishes in this type of climate: south when the glaciers advanced, back north when the ice retreated. Great morphological changes were not required, because the organisms kept to the same habitat, even though its geographical location moved.

Rapid morphological changes, which may in some instances correspond with times of speciation, also occur. Genetic drift in small populations, directional selection among populations that do not track suitable habitats, coevolution owing to competition or cooperation with other

species, or sexual selection may all produce rapid evolutionary jumps.

Just how fast must evolution occur to be perceived as punctuated equilibrium? Returning to the fossil horses, let's suppose that a *Hyracotherium* 60 centimeters (24 inches) tall evolved into a *Mesohippus* 100 centimeters (40 inches) tall in 10,000 years and that the time between generations was 5 years. The height increase of 40 centimeters (16 inches) would have taken 2000 generations, or 0.2 millimeter per generation. This difference would be far too small to detect with the naked eye and is more than a hundred times less than the variability in size among modern horses, even of the same breed. To a geneticist, this rate of change doesn't seem very fast at all. In the fruit fly *Drosophila*, for example, mild selection pressure can cause evolutionary changes that are more than 100 times as fast as those required for *Hyracotherium* to evolve into *Mesohippus* in 10,000 years. In the span of geological time, however, 10,000 years is quite brief. No one would be surprised if intermediate fossils could not be found. Therefore, "punctuated" speciation would seem quite slow to a geneticist, but very fast to a geologist.

The Relationship between Speciation and Morphological Change Depends on Circumstances

Morphological change may occur whenever a small population is isolated from gene exchange with other populations and is subjected to new selective pressures. Considerable change may also occur by genetic drift or founder events or both, if the population is very small. Selection or genetic drift may also cause speciation, with the small, isolated population rapidly becoming a separate species from its larger parent population. Therefore, in these cases speciation may be linked with major morphological change.

Morphological change may occur in large populations subjected to widespread directional selection. However, if gene flow among populations is adequate, speciation is unlikely to occur. In this situation, morphological change will not be linked to speciation.

As we discussed previously, speciation could occur through the accumulation of many small genetic changes or through a few key mutations. Genetic data, although scanty, suggest that each mechanism may operate at different times in different populations. Which is more common cannot be decided with the evidence now available.

Natural Selection Occurs Both between Species and between Individuals within a Species

If two species are in direct competition for a particular resource in a particular habitat, one will probably move, evolve rapidly to use a different resource, or become extinct (see Chapters 17 and 29). In this case, species, not individuals, may be considered to be the unit of selection. However, the members of the species that "wins" in such a competition may well have attained their superior adaptations through natural selection acting on individual organisms. Further, the "winning" species may continue to evolve even after the extinction of the "losing" species, because of competition among individual members of the species. In these instances, the unit of selection is the individual. Most evolutionary biologists feel that selection occurs to some extent at both species and individual levels.

Both Natural Selection and Random Events Drive Evolution

Is natural selection the paramount force operating during evolution? Some evolutionary biologists emphasize the importance of nonselective, random events during evolution. They have two major objections to the concept that most features evolved as adaptations through natural selection. First, it is extremely difficult to prove that a particular feature, say the size of a bird's beak, was shaped by natural selection. If a biologist observes a bird with a large bill cracking a large, tough nut, it is tempting to assume that the large bill evolved because of the advantage it conferred in nutcracking. But maybe the bird acquired the large bill for other reasons, say for courtship or even through genetic drift, and eats nuts because it "accidentally" has a suitable bill. On one of the Galapagos Islands, however, zoologist Peter Grant of Princeton demonstrated quite conclusively that natural selection strongly influences beak size in a species of finch. When a prolonged wet spell altered the vegetation of the island, greatly increasing the availability of small seeds, the average bill size of breeding birds became longer and narrower over a period of just 4 years.

The second argument against the concept that natural selection is the primary force in evolution is that random forces do indeed operate in nature. In the previous chapter, we discussed the likelihood that gigantic meteorites strike the Earth from time to time, causing extinctions of many species. Species that were superbly adapted to warm, sunny climates might go extinct within a few years in the cold and gloom following the impact. The survival or extinction of species might have little to do with adaptations to their habitats, because the habitats would disappear. Smaller-scale catastrophes, such as floods, sea-level changes, or massive forest fires, could similarly cause the extinction of well-adapted but localized species.

The consensus is that both natural selection and random forces shape the evolutionary history of life on Earth. No one doubts that natural selection is the driving force behind the evolution of the magnificent structures and behaviors that we witness in the living world around us. And no one doubts that a 10-kilometer meteorite would drive many otherwise well-adapted species to extinction. Which processes dominate, over what time scales and over what fraction of species on Earth, is the real debate.

Concluding Thoughts on Evolution and Evolutionary Hypotheses

In the popular press, conflicts among evolutionary biologists are sometimes seen as conflicts about evolution itself. One occasionally reads statements to the effect that new theories are overthrowing Darwin and casting doubt on the reality of evolution. Nothing could be further from the truth. Proponents of punctuated equilibrium and gradualism, and everyone in between, unanimously agree that evolution occurred in the past and is still occurring today. The only argument is over the mechanisms of evolutionary change, their relative importance in the history of life on Earth, and which forces were most important in shaping the evolution of particular species. Meanwhile, wolves still tend to catch the slowest caribou, small populations still undergo genetic drift, habitats still change or disappear, and maybe somewhere in our galaxy another asteroid is swinging our way; evolution continues, still generating, in Darwin's words, "endless forms most beautiful."

✖ S U M M A R Y   O F   K E Y   C O N C E P T S

Speciation

A species is defined as all the populations of organisms that are potentially capable of interbreeding under natural conditions and that are reproductively isolated from other populations. Speciation, the development of new species, requires that two populations be isolated from gene flow between them and develop significant genetic divergence. Allopatric speciation occurs by geographical isolation and subsequent divergence of the separated populations through genetic drift or natural selection. Sympatric speciation occurs by ecological isolation and subsequent divergence or by rapid chromosomal changes, such as polyploidy.

Maintaining Reproductive Isolation between Species

Reproductive isolation between species may be maintained by one or more of several mechanisms, collectively called premating isolating mechanisms and postmating isolating mechanisms. Premating isolating mechanisms include geographical isolation, ecological isolation, temporal isolation, behavioral isolation, and mechanical incompatibility. Postmating isolating mechanisms include gametic incompatibility, hybrid inviability, and hybrid infertility.

Phyletic and Divergent Speciation

In divergent speciation, a single species splits into two or more subpopulations that evolve into separate species. Daughter species therefore exist simultaneously. In phyletic speciation, the entire population of a species evolves in concert over time, so that the later population is recognized as being a new species. No one knows which form of speciation dominates evolutionary history, but it is likely that phyletic speciation is a selective sampling of divergent speciation as seen through a spotty fossil record.

The Genetics of Speciation

Separate species must have large enough genetic differences to prevent interbreeding. There are two models to explain the development of genetic differences between species: the gradual accumulation of many small genetic differences and the sudden appearance of a few major differences, probably in regulatory genes. It is likely that both mechanisms occur, and each may account for the genetic divergence of some species.

Rates of Speciation

Rates of speciation vary greatly over geological time. Invasion of new, unoccupied habitats or the development of fundamentally superior adaptations may trigger extremely rapid speciation, a process called adaptive radiation.

The Pattern of Evolution

There are two extreme models for the overall pattern of evolution. The gradualism model asserts that genetic and morphological changes accumulate slowly over time, with speciation being a relatively slow process not closely linked to observable morphological change. Natural selection among individuals is the prime driving force behind evolutionary change. The punctuated equilibrium model asserts that morphological change and speciation are very rapid, simultaneous events. Speciation events are separated by long periods during which species undergo little or no morphological change. Evolutionary change is driven by natural selection among simultaneously existing species, and also by random, chance events leading to speciation and extinction. Most evolutionary biologists agree that some aspects of both models probably occur during evolution and that evolutionary history may more closely resemble either model at different times or in the lineage of different organisms.

✖ K E Y   T E R M S

adaptation   p. 339
adaptive radiation   p. 339
allopatric speciation   p. 332
behavioral isolation   p. 336
divergent speciation   p. 337
ecological isolation   p. 336
gametic incompatibility   p. 337

geographical isolation   p. 332
gradualism   p. 339
hybrid infertility   p. 337
hybrid inviability   p. 337
mechanical incompatibility   p. 337
phyletic speciation   p. 337
polyploidy   p. 335

postmating isolating mechanism   p. 336
premating isolating mechanism   p. 335
punctuated equilibrium   p. 339
reproductive isolation   p. 335
speciation   p. 332
sympatric speciation   p. 332
temporal isolation   p. 336

�֎ T H I N K I N G   T H R O U G H   T H E   C O N C E P T S

Multiple Choice

1. Many closely related species of marine invertebrates exist on either side of the isthmus of Panama. They probably resulted from:
   a. premating isolation
   b. isolation by distance
   c. postmating isolation
   d. gametic incompatibility
   e. allopatric speciation

2. Many hybrids are sterile because their chromosomes don't synapse correctly at meiosis. Why aren't polyploid plants sterile?
   a. They backcross to the parental generation.
   b. Most are triploid.
   c. They cross-pollinate.
   d. They self-fertilize, using their diploid gametes.
   e. Their eggs develop directly, without fertilization.

3. In many species of fireflies, males flash to attract females. Each species has a different flashing pattern. This is probably an example of
   a. ecological isolation   b. temporal isolation
   c. geographical isolation   d. premating isolation
   e. postmating isolation

4. In phyletic speciation, fossil species are separated by
   a. time   b. geography
   c. behavior   d. hybrid inviability
   e. hybrid infertility

5. In terms of changes in gene frequencies, founder events result in
   a. gradual accumulation of many small changes
   b. large, rapid changes
   c. polyploidy
   d. hybridization
   e. mechanical incompatibility

6. After the demise of the dinosaurs, mammals evolved rapidly into many new forms because of
   a. the founder effect
   b. a genetic bottleneck
   c. adaptive radiation
   d. geological time
   e. genetic drift

Review Questions

1. Define the following terms: species, speciation, allopatric speciation, sympatric speciation. Explain how allopatric and sympatric speciation might work, and give a hypothetical example of each.

2. Many of the oak tree species found in central and eastern North America hybridize (interbreed). Are they "true species"?

3. Review the material on the possibility of sympatric speciation in *Rhagoletis* that breed on apples or hawthorns. What types of genetic, morphological, or behavioral data would really convince you that the two forms have actually become separate species?

4. A drug called colchicine affects the mitotic spindle and prevents cell division after the chromosomes have doubled at the start of meiosis. Describe how you would use colchicine to produce a new polyploid species of your favorite garden flower.

5. List and describe the different types of premating and postmating isolating mechanisms.

6. Define punctuated equilibrium and gradualism. Look up the evolutionary history of a lineage (e.g., dinosaurs, humans) and describe how each mechanism might account for the evolution of that lineage.

7. Describe phyletic and divergent speciation. Do you think you could distinguish between these two mechanisms in the fossil record?

✖ A P P L Y I N G   T H E   C O N C E P T S

1. Do phyletic speciation and divergent speciation coincide with the gradualism and punctuated equilibrium models of evolution? Defend your answer.

2. Seedless varieties of fruits and vegetables have been created by breeders. They are triploid. Explain why they are seedless.

3. Why do you suppose there are so many endemic species on islands? (*Endemic* means "found nowhere else.") And why have the overwhelming majority of recent extinctions occurred on islands?

4. A contrarian biologist you've met says the fact that humans are pushing other species into small, isolated populations is actually good for biodiversity, because these are the conditions that theory says lead to new speciation events. Comment.

5. In southern Wisconsin there are several populations of gray squirrels (*Sciurus carolinensis*) with black fur. Design a study to determine if they are actually a separate species.

6. This chapter reviews a series of hypotheses that represent dichotomies: phyletic versus divergent speciation, punctuated equilibrium versus gradualism, species selection versus individual selection, and the importance of random events versus natural selection in the course of evolution. Think about the value of dichotomies in scientific (and political) debate. What are the advantages and disadvantages of taking extreme positions? Why do people do it?

7. It is difficult to gather data on speciation events in the past or to perform interesting experiments about the process of speciation. Does this difficulty make the study of speciation "unscientific"? Should we abandon the study of speciation?

✖ FOR MORE INFORMATION

Ayala, F. "The Mechanisms of Evolution." *Scientific American*, September 1979. Our increasing understanding of the molecular mechanisms of heredity and mutation reveals that organisms are extremely variable, providing tremendous amounts of material on which selection can operate.

Gould. S. J. "The Evolution of Life on the Earth." *Scientific American*, October 1994. The importance of chance and catastrophe in shaping modern organisms.

Gould, S. J., and Eldredge, N. "Punctuated Equilibrium Comes of Age." *Nature*, November 18, 1993. An overview of the history and evolution of the punctuated evolution hypothesis by two of its chief proponents.

Grant, P. R. "Natural Selection and Darwin's Finches." *Scientific American*, October 1991. A temporary climate change altering the food supply created selection pressures on finches that altered their average beak shape in several generations.

Smith, J. M. *Evolution Now: A Century after Darwin*. San Francisco: W. H. Freeman, 1982. The section titled "Evolution—Sudden or Gradual?" contains essays by Russell Lande and Smith himself on the "gradualist" side and by Stephen Jay Gould on the "punctuationist" side.

Stebbins, G. L., and Ayala, F. "The Evolution of Darwinism." *Scientific American*, July 1985. A synthesis of molecular and classical evolutionary methodologies.

NET WATCH

On-line resources for this chapter are on the World Wide Web at:
http://www.prenhall.com/~audesirk (click on the <u>table of contents</u> link and then select Chapter 18).

19 The History of Life on Earth

"[T]hence life was born,
Its nitrogen from ammonia, carbon from methane,
Water from the cloud and salts from the young seas . . . the cells of life
Bound themselves together into clans, a multitude of cells
To make one being—as the molecules before
Had made of many one cell.   ROBINSON JEFFERS in *The Beginning and the End*

The Great Nebula in Orion, a cloud of gases in which stars may be forming, as they did in the beginning of the Universe

Fifteen billion years ago, there was no universe as we know it: only a dense, minuscule mass in which energy and matter were melded into a single incomprehensibly ferocious state. Then the mass erupted. Particles of matter and antimatter formed, annihilating each other in great bursts of energy when they collided. So great was the energy released that astrophysicists can still detect its faint radiation, evidence to support this favored "Big-Bang" model of the formation of the universe.

According to this scenario, after the Big Bang, great clouds of matter formed. Gradually, gravity drew particle to particle, producing large aggregations. Many accumulations grew so large that their centers became dense and hot, triggering the thermonuclear reactions that fuel the light of the stars. Small, simple atoms of hydrogen fused to form helium, and these fused to give rise to still larger atoms. The nuclear forces of some stars grew so intense that they exploded, spewing their matter out into space.

Scientists theorize that our solar system originated about 5 billion years ago. At the far edge of an immense cluster of stars forming a spiral galaxy, a small cloud of matter began to condense, enriched with heavy elements provided by the self-destruction of ancient stars. The center of the cloud collapsed into the yellow dwarf star we call the sun. From the cloud of dust particles circling the newly formed sun, smaller aggregations formed the planets. Of the innermost four planets, Mercury and Venus, the two closest to the sun, became very hot, while the fourth, Mars, cooled into an eternal winter. The third settled into an orbit that would receive just the right intensity of sunlight to permit water to exist as a liquid. This became Earth, which the French oceanographer Jacques Cousteau has aptly described as the "Water Planet."

On this third planet, and so far as we know upon no other in our solar system, life evolved (Table 19-1). Although the mechanisms of evolution are known in considerable detail, they do not tell us how life first originated. Where did life come from? How did it develop into the myriad forms we see today? What is its future?

TABLE 19-1 ▓ *The History of Life on Earth*

Era	Period	Epoch	Years Ago* (millions)	Major Events
Precambrian			4600–3500	Origin of solar system and Earth.
			3500–590	Origin of first living cells; dominance of bacteria; origin of photosynthesis and evolution of oxygen atmosphere; origin of algae and soft-bodied marine invertebrates.
Paleozoic	Cambrian		590–505	Primitive marine algae flourish; origin of most marine invertebrate types.
	Ordovician		505–438	Invertebrates, especially arthropods and mollusks, dominant in sea; first fish, fungi.
	Silurian		438–408	Many fish, trilobites, mollusks in sea; first vascular plants; invasion of land by plants; invasion of land by arthropods.
	Devonian		408–360	Fishes and trilobites flourish in sea; origin of amphibians and insects.
	Carboniferous		360–286	Swamp forests of tree ferns and club mosses; dominance of amphibians; numerous insects; origin of reptiles.
	Permian		286–248	Origin of conifers; massive marine extinctions, including last of trilobites; flourishing of reptiles and decline of amphibians; continents aggregated into one land mass, Pangaea.
Mesozoic	Triassic		248–213	Origin of mammals and dinosaurs; forests of gymnosperms and tree ferns; breakup of Pangaea begins.
	Jurassic		213–144	Dominance of dinosaurs and conifers; origin of birds; continents partially separated.
	Cretaceous		144–65	Flowering plants appear and become dominant; mass extinctions of marine life and some terrestrial life, including last dinosaurs; modern continents well separated.
Cenozoic	Tertiary	Paleocene	65–54	Widespread flourishing of birds, mammals, insects, and flowering plants; drift brings continents into modern positions; mild climate at beginning of period, with extensive mountain building and cooling toward end.
		Eocene	54–37	
		Oligocene	37–24	
		Miocene	24–5	
		Pliocene	5–2	
	Quaternary	Pleistocene	2–0.01	Evolution of *Homo*; repeated glaciations in Northern Hemisphere; extinction of many giant mammals.
		Recent	0.01–present	

*From W. B. Harland, et al. *A Geologic Time Scale*. Cambridge, England: Cambridge University Press, 1982.

Origins

How and when did life first appear on Earth? Just a few centuries ago, this question would have been considered trivial. Although no one knew how life *first* arose, people thought that new living things appeared all the time, through **spontaneous generation** from both nonliving matter and other, unrelated forms of life. In 1609, a French botanist wrote, "There is a tree . . . frequently observed in Scotland. From this tree leaves are falling; upon one side they strike the water and slowly turn into fishes, upon the other they strike the land and turn into birds." Medieval writings abound with similar observations and delightful recipes for creating life— even human beings. Microorganisms were thought to arise spontaneously from broth, maggots from meat, mice from mixtures of sweaty shirts and wheat.

In 1668 the Italian physician Francesco Redi disproved the maggots-from-meat hypothesis simply by keeping flies (whose eggs hatch into maggots) away from uncontaminated meat (see Chapter 1). Then in the mid-1800s Louis Pasteur in France and John Tyndall in England disproved the broth-to-microorganism idea as well (Fig. 19-1). Although their work effectively demolished the notion of spontaneous generation, it did

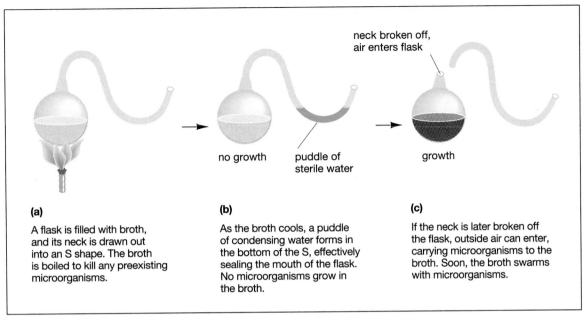

neck broken off, air enters flask

no growth   puddle of sterile water   growth

(a)

A flask is filled with broth, and its neck is drawn out into an S shape. The broth is boiled to kill any preexisting microorganisms.

(b)

As the broth cools, a puddle of condensing water forms in the bottom of the S, effectively sealing the mouth of the flask. No microorganisms grow in the broth.

(c)

If the neck is later broken off the flask, outside air can enter, carrying microorganisms to the broth. Soon, the broth swarms with microorganisms.

Figure 19-1   **Spontaneous generation refuted**

Louis Pasteur's experiment disproving the spontaneous generation of microorganisms in broth.

not address the question of how life on Earth originated in the first place.

For almost half a century the subject lay dormant. Eventually, biologists returned to the question of the origin of life and began to seek answers. In the 1920s and 1930s, Alexander Oparin in Russia and John B. S. Haldane in England noted that the oxygen-rich atmosphere that we know would not have permitted spontaneous formation of the complex organic molecules necessary for life. Oxygen reacts readily with other molecules disrupting chemical bonds, thus tending to keep molecules simple. Oparin and Haldane speculated that the atmosphere of the young Earth was very low in oxygen and rich in hydrogen in the form of hydrogen gas (H_2), methane (CH_4), and ammonia (NH_3). Given these and other conditions, which we will discuss below, Oparin and Haldane proposed that life could have arisen from nonliving matter through ordinary chemical reactions. This process is called chemical evolution, or **prebiotic evolution**: that is, evolution before life existed.

Prebiotic Evolution Was Controlled by the Early Atmosphere and Climate

The primordial Earth differed greatly from the planet we now enjoy. As rock after rock smashed into the forming planet, their energies of motion were converted into heat. Radioactive atoms decayed, releasing still more heat. Soon the rock melted, and heavier elements such as iron and nickel sank to the center of the mass, where they remain molten still because of intense heat. Gradually Earth cooled, and elements combined to form compounds of many sorts. Virtually all the oxygen combined with hydrogen to form water, with carbon to form carbon dioxide, and with heavier elements to form minerals. After millions of years, Earth cooled enough to allow water to exist as a liquid, and for thousands of years it must have rained, as water vapor condensed out of the cooling atmosphere. As the water struck the surface, it dissolved many minerals, forming a weakly salty ocean. Lightning from storms, heat from volcanoes, and intense ultraviolet light from the sun all poured energy into the young seas.

Judging from the chemical composition of the rocks formed at this time, geochemists have deduced that the primitive atmosphere probably contained substances such as carbon dioxide, methane, ammonia, hydrogen, nitrogen, hydrochloric acid, hydrogen sulfide, and water vapor. But because oxygen atoms were bound up in water, carbon dioxide, and minerals, there was virtually no free oxygen in the early atmosphere. The absence of free oxygen is an important factor in all hypotheses and experiments dealing with prebiotic evolution.

Organic Molecules Can Be Synthesized Spontaneously under Prebiotic Conditions

In 1953, inspired by the ideas of Oparin and Haldane, Stanley Miller, a graduate student, and his adviser Harold Urey of the University of Chicago set out to demonstrate prebiotic evolution in the laboratory. They mixed water, ammonia, hydrogen, and methane in a flask and provided energy with heat and electrical discharge (to simulate lightning). They found that simple organic molecules ap-

Figure 19-2   The experimental apparatus of Stanley Miller
Energy from heat and electrical discharge causes amino acids and other organic molecules to form from methane, ammonia, hydrogen, and water, all of which are thought to have been present in the atmosphere of the early Earth.

peared after just a few days (Fig. 19-2). In these and similar experiments, Miller and others have produced amino acids, short proteins, nucleotides, adenosine triphosphate (ATP), and other molecules characteristic of living things. Interestingly, the exact composition of the "atmosphere" used in these experiments is unimportant, provided that hydrogen, carbon, and nitrogen are available and that free oxygen is excluded. Similarly, a variety of energy sources, including ultraviolet light, electrical discharge, and heat, all seem about equally effective. Even though geochemists may never know exactly what the primordial atmosphere was like, it is certain that organic molecules were synthesized on the ancient Earth.

Prebiotic Conditions Would Allow Accumulation of Organic Molecules

Prebiotic synthesis would not have been very efficient or very fast; nonetheless, in a few hundred million years, large quantities of organic molecules could accumulate, especially since they didn't break down nearly as fast back then. On Earth today, most organic molecules have a short life; either they are digested by living organisms or they react

with atmospheric oxygen. Since the primeval Earth lacked both life and free oxygen, these sources of degradation were absent. However, the primordial atmosphere also lacked an ozone layer, which our present atmosphere has. The ozone layer is a region high in the atmosphere that is enriched with O_3 molecules that absorb some of the high-energy ultraviolet light of the sun before it reaches Earth. Consequently, the ultraviolet bombardment, which can break apart organic molecules, must have been fierce. Some places, however, such as those beneath rock ledges or at the bottoms of even fairly shallow seas, would have been protected from ultraviolet radiation. In these locations, organic molecules may have accumulated to relatively high concentrations.

These concentrations of organic molecules may have been crucial in the evolution of life. First, they may have provided the molecules that would form the first living organisms. Second, the chemical energy stored in these molecules would be food for the first cells. The rich accumulation of organic molecules in which scientists hypothesize that life first arose is often called the "primordial soup."

Was RNA the First Self-Reproducing Molecule?

In modern cells, proteins carry out most of the cellular functions, while DNA encodes the information the cell needs to synthesize those proteins. The old "chicken-and-egg" riddle applies here: Which came first—DNA or protein? If useful protein catalysts came first, how was the information needed to synthesize them passed from protocell to protocell? If nucleic acids came first, what would be the function of an information-storage molecule if there were no information to store?

In the 1980s, Thomas Cech of the University of Colorado and Sidney Altman of Yale offered an intriguing solution to this riddle. They found that some types of RNA molecules act as enzymes that, among other things, can cut apart RNA and synthesize more RNA molecules. During hundreds of millions of years of prebiotic chemical synthesis, RNA nucleotides may have occasionally bonded together to form short RNA chains. Let us suppose that, purely by chance, one of these RNA chains was a catalyst—dubbed a **ribozyme**—that could synthesize copies of itself from the free ribonucleotides in the surrounding waters. This first ribozyme probably wasn't very good at its job and made lots of mistakes. These mistakes, of course, were the first mutations. Like modern mutations, most undoubtedly ruined the catalytic abilities of the "daughter molecules," but a few may have been improvements. Molecular evolution could begin, as ribozymes with increased speed and accuracy of replication reproduced faster, making more and more copies of themselves.

Just as RNA-containing ribosomes are crucial to protein synthesis today, perhaps some early ribozymes began to bind amino acids and catalyze the synthesis of short proteins. Further ribozyme mutations might lead to the formation of the

(a) Organic molecules are synthesized abiotically—that is, outside of living organisms.

ribonucleotides, amino acids, lipids, simple sugars

(b) Small RNA chains form. Some of these chains, called ribozymes, can catalyze their own replication, using free ribonucleotides.

self-replicating RNA "ribozymes"

(c) Chance changes in RNA sequence allow certain ribozymes to catalyze protein synthesis from free amino acids. Proteins are more efficient as catalysts than the primitive ribozymes.

RNA ribozymes → protein enzymes

(d) Further mutations allow ribozymes to copy themselves into more stable DNA chains. DNA becomes the permanent repository of genetic information. RNA becomes the intermediary between DNA and protein.

DNA → RNA → proteins

Figure 19-3   **A hypothesis for the origin of life**

In this scenario, RNA is the first "living molecule."

first protein enzymes. At the same time, these protein-synthesizing ribozymes were vulnerable to being cut up by other ribozymes, destroying the information so slowly accumulated over millions of years of random nucleotide substitutions. Further mutations might have allowed certain ribozymes to copy themselves over into DNA molecules that would be safe from destruction by their fellow ribozymes. In this hypothesis, then, RNA occupies center stage as the first living molecule, with both DNA and proteins evolving later (Fig. 19-3). RNA gradually receded into its present role as an intermediary between DNA and the protein enzymes that carry out most of the work of modern cells.

The First Living Cells May Have Consisted of Ribozymes within Microspheres

If water containing proteins and lipids is agitated to simulate waves beating against ancient shores, hollow structures called **microspheres** are formed (Fig. 19-4). These hollow balls resemble living cells in several respects. They have a well-defined outer boundary, separating internal contents from the external solution. If the composition of the microsphere is right, a "membrane" is formed that is remarkably similar in

appearance to a real cell membrane. Under certain conditions microspheres can absorb more material from the solution ("feed"), grow, and even divide (see Fig. 19-4).

If a microsphere happened to surround the right ribozymes, something very much like a living cell would have been formed. The ribozymes and their protein products would have been protected from free-roaming ribozymes in the primordial soup. Nucleotides and amino acids might have diffused across the membrane and have been used to synthesize new RNA and protein molecules. After sufficient growth, the microsphere may have divided, with a few copies of both ribozymes and proteins becoming incorporated into each daughter microsphere. If so, the first cells would have evolved.

But Did All This Happen?

The above scenario, although plausible and supported by many research findings, is by no means certain. One of the most striking aspects of origin-of-life research is the great diversity of assumptions, experiments, and contradictory hypotheses (see the suggested reading by Horgan, "In the Beginning . . ." for a taste of the controversies). Researchers disagree as to whether life arose in quiet pools, in the sea, in moist films on the surfaces of clay or iron pyrite (fool's gold), in furiously hot deep-sea vents. A few even argue that life may have arrived on Earth from space. Can any conclusions really be drawn from the research done so far? No one really knows the answer to that question, but we can offer a few observations.

First, the experiments of Miller and others show that amino acids, nucleotides, and other organic molecules would have been formed in abundance on the primordial Earth. Second, no one imagines that prebiotic evolution would have formed any molecules as large or sophisticat-

Figure 19-4   **Do microspheres resemble the earliest cells?**

Cell-like microspheres can be formed by agitating proteins and lipids in a liquid medium. Such microspheres can take in material from the surrounding solution, grow, and even "reproduce," as these are doing.

SCIENTIFIC INQUIRY

Radiometric Dating Traces Fossils through Time

Early geologists could date rock layers and their accompanying fossils only in a *relative* way: Fossils found in deeper layers of rock were generally older than those found in shallower layers. With the discovery of radioactivity, it became possible to determine *absolute* dates, within certain limits of uncertainty. The nuclei of radioactive elements spontaneously break down, or decay, into other elements. For example, carbon-14 (usually written ^{14}C) emits an electron to become nitrogen-14 (^{14}N). Each radioactive element decays at a rate that is independent of temperature, pressure, or the chemical compound of which the element is a part. The time it takes for half the radioactive nuclei to decay at this characteristic rate is called the half-life. The half-life of ^{14}C, for example, is 5730 years.

How are radioactive elements used in determining the age of rocks? Knowing the rate of decay and measuring the proportion of decayed nuclei to undecayed nuclei, it is possible to estimate how much time has passed since these radioactive elements were trapped in rock; this process is called radiometric dating. A particularly straightforward dating technique uses the decay of potassium-40 (^{40}K) into argon-40 (^{40}Ar), which has a half-life of about 1.25 billion years. Potassium is a very reactive element and is a common constituent of volcanic rocks such as granite and basalt. Argon, on the other hand, is an unreactive gas. Let us suppose that a volcano, such as Hawaii's Kilauea volcano, erupts with a massive lava flow, covering the countryside. All the ^{40}Ar, being a gas, will bubble out of the molten lava, so when the lava solidifies into rock, it will start out with no ^{40}Ar. Potassium-40 present in the hardened lava will decay to ^{40}Ar, half the ^{40}K decaying every 1.25 billion years. The ^{40}Ar gas will be trapped in the rock. A geologist could take a sample of the rock and determine the proportion of ^{40}K to ^{40}Ar (Fig. E19-1). If the analysis finds equal amounts of the two elements, the geologist will conclude that the lava hardened 1.25 billion years ago. With appropriate care, such age estimates are quite reliable. If a fossil is found beneath a lava flow dated at, say, 500 million years, then we know that the fossil is at least that old.

Some radioactive elements, as they decay, can even give an estimate of the age of the solar system. Analysis using uranium as it decays to lead has shown that the oldest Earth rocks, meteorites, and moon rocks collected by astronauts are all about 4.5 billion years old.

Figure E19-1   **The relationship between time and the decay of radioactive ^{40}K to ^{40}Ar**

ed as, say, hemoglobin. The first ribozymes, if they existed, may have been only a dozen nucleotides long. Third, ribozymes or other primitive catalysts need not have been very efficient to have accelerated the synthesis of important molecules. The stone tools of early humans weren't as handy as a Swiss army knife, either, but they were better than fingernails. Therefore, to a biologist, the origin of life means the origin of a few small molecules with terribly inefficient catalytic activities, perhaps surrounded by a film vaguely resembling a primitive membrane. Finally, several hundred million years probably elapsed between the appearance of organic molecules and the first primitive cell-like structures, and spontaneous synthesis must have proceeded over huge areas of the Earth. Given enough time

and space, even extremely rare events may in fact happen quite often. Most biologists accept that the origin of life is probably an inevitable consequence of the working of natural laws. We should emphasize, however, that *this proposition is not proved and never will be*. Biologists investigating the origin of life have neither millions of years nor trillions of liters of reaction solutions with which to work!

The First Life

The fossil record indicates that the earliest living cells arose about 3.5 billion years ago, based on radiometric dating techniques (see "Scientific Inquiry: Radiometric Dating Traces Fossils through Time"). What were these early cells like? The first cells were prokaryotic; that is, their genetic material was not sequestered from the rest of the cell within a membrane-limited nucleus. These cells probably obtained nutrients and energy by absorbing organic molecules from the primordial soup. There wasn't any free oxygen in the atmosphere, so the cells must have metabolized the organic molecules anaerobically. You will recall from Chapter 8 that anaerobic metabolism yields only small amounts of energy.

As you probably have already recognized, the earliest cells were primitive anaerobic bacteria. As these ancestral bacteria multiplied, however, they must have eventually used up the organic molecules produced by prebiotic synthesis. Simpler molecules, such as carbon dioxide and water, were still very abundant, as was energy, in the "dilute" form of sunlight. What was lacking, then, was not *materials* or *energy itself*, but *energetic molecules*: that is, molecules in which energy is stored in chemical bonds. Eventually some cells evolved the ability to use the energy of sunlight to drive the synthesis of their own complex, high-energy molecules from simpler molecules: Photosynthesis appeared.

Several kinds of photosynthetic bacteria evolved, but the ones that proved most important in the evolution of life were the cyanobacteria, sometimes called blue-green algae. Their photosynthetic reactions converted water and carbon dioxide to organic compounds, releasing oxygen as a by-product. (You will undoubtedly breathe in some oxygen molecules today that were excreted about 2 billion years ago by a cyanobacterium!) At first, the oxygen reacted with iron atoms in Earth's crust, or surface layer, forming huge deposits of iron oxide, also called rust. After all the accessible iron turned to rust, the concentration of free oxygen in the atmosphere increased. Chemical analysis of rocks suggests that significant amounts of free oxygen appeared in the atmosphere about 2 billion years ago.

Oxygen is potentially very dangerous to living things because it reacts with organic molecules, destroying them and releasing their stored energy. The accumulation of oxygen in the atmosphere provided the selective pressure for the next great advance in the Age of Microbes: the ability to use oxygen in metabolism, channeling its destructive power through aerobic respiration to generate useful energy for the cell. Because the amount of energy available to a cell is vastly increased when oxygen is used to metabolize food molecules, aerobic cells had a significant selective advantage.

Eukaryotes Evolved Membrane-Enclosed Organelles and a Nucleus

Hordes of bacteria would offer a rich food supply to any organism that could eat them. There are no fossil records of the first predatory cells, but paleobiologists speculate that once a suitable prey population (such as these bacteria) appeared, predation would have evolved quickly. These predators would have been specialized prokaryotic cells, lacking cell walls and consequently able to engulf whole bacteria as prey. According to the most widely accepted hypothesis, these predators were otherwise quite primitive, being capable of neither photosynthesis nor aerobic metabolism. Although they could capture large food particles, namely bacteria, they metabolized them inefficiently. About 1.4 billion years ago, however, one predator probably gave rise to the first eukaryotic cell.

Eukaryotic cells differ from prokaryotic cells in many ways, but perhaps most fundamental are the membrane-bound nucleus containing the genetic material and the inclusion of organelles for energy metabolism, mitochondria and (in plants) chloroplasts. How did these organelles evolve?

Mitochondria and Chloroplasts May Have Arisen from Engulfed Bacteria

The **endosymbiotic hypothesis**, championed most forcefully by Lynn Margulis of the University of Massachusetts, proposes that cells acquired the precursors of mitochondria and chloroplasts by engulfing certain types of bacteria. These cells and the bacteria trapped inside them (*endo* means "within") gradually entered into a symbiotic relationship, a close association between different organisms over an extended time. Let us suppose that an anaerobic predatory cell captured an aerobic bacterium for food, as it often did, but for some reason failed to digest this particular prey (Fig. 19-5a). The aerobic bacterium remained alive and well. In fact, it was better off than ever, because the cytoplasm of its predator/host was chock-full of half-digested food molecules, the remnants of anaerobic metabolism. The aerobe absorbed these molecules and used oxygen to complete their metabolism, gaining enormous amounts of energy as it did so. So abundant were its food resources, and so bountiful its energy production, that the aerobe must have leaked energy, probably as ATP or similar molecules, back into its host's cytoplasm. The mitochondrion had been born. Interestingly, the amoeba *Pelomyxa palustris* may be a modern model for how this occurred. Unlike almost all other eukaryotic cells, it lacks mitochondria; however, a permanent population of aerobic bacteria carries out much the same role.

The predatory cell with its symbiotic bacteria could metabolize food aerobically, gaining a great selective advantage over its anaerobic compatriots. Soon its progeny filled the seas. One of these daughter cells managed a second feat: It captured a photosynthetic cyanobacterium and similarly failed to digest its prey (see Fig. 19-5a). The cyanobacterium flourished in its new host and gradually evolved into the first chloroplast. Some modern organisms resemble this hypothetical ancestral condition. A variety of corals, some clams, a few snails, and at least one species of *Paramecium*

(a) The probable origin of mitochondria and chloroplasts

An anaerobic, prokaryotic cell engulfs an aerobic bacterium. The bacterium is enclosed in a sac of the predatory cell's membrane. The resulting "proto-organelle" has a double membrane, one from the predator and one from the prey.

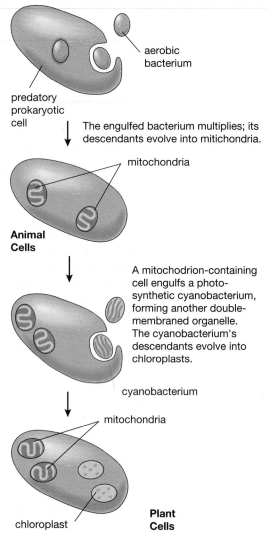

predatory prokaryotic cell

The engulfed bacterium multiplies; its descendants evolve into mitichondria.

mitochondria

Animal Cells

A mitochodrion-containing cell engulfs a photo-synthetic cyanobacterium, forming another double-membraned organelle. The cyanobacterium's descendants evolve into chloroplasts.

cyanobacterium

mitochondria

chloroplast

Plant Cells

(b) Alternative possible origins of the membrane-enclosed nucleus

1. The nucleus may have originated as an infolding of the cell membrane that came to surround the DNA.

cell membrane folds inward

nuclear envelope

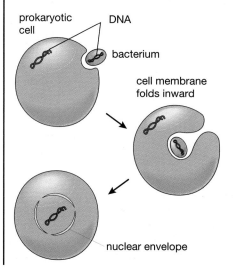

2. Or the nucleus may have originated when one bacterium engulfed another.

prokaryotic cell   DNA

bacterium

cell membrane folds inward

nuclear envelope

Figure 19-5 Possible mechanisms for the appearance of organelles in eukaryotic cells

---

harbor a permanent collection of algae in their cells (Fig. 19-6). These algae share some of their photosynthetically produced food molecules with the host cells.

Other eukaryotic organelles may have also originated through endosymbiosis. Many paleobiologists believe that cilia, flagella, centrioles, and microtubules may all have evolved from a symbiosis between a spirilla-like bacterium (a form of bacteria with an elongated corkscrew shape) and a primitive eukaryotic cell. Supporting this hypothesis was the discovery in 1990 that centrioles contain their own minute supply of DNA, which some interpret as a remnant of the DNA originally contained within the symbiotic bacterium.

The Origin of the Nucleus Is More Obscure

The evolution of the nucleus is more obscure. One possibility is that the plasma membrane folded inward, surrounding the DNA. This would create the nuclear membrane, and further infoldings could have produced the endoplasmic reticulum, which is continuous with the nuclear membrane (Fig. 19-5b, top). An alternative hypothesis is that, like many other eukaryotic organelles, the nucleus arose as a result of endosymbiosis. In this case, the engulfed bacterium took control of its host (Fig. 19-5b, bottom). However the nucleus originated, having the DNA sequestered within the nucleus seems to have con-

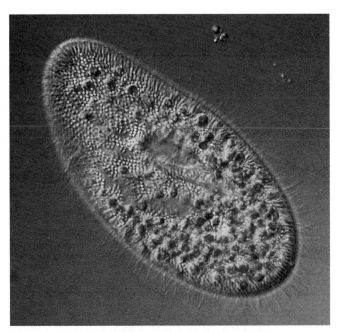

Figure 19-6   **A modern intracellular symbiosis**

Green unicellular algae of the genus *Chlorella* live within the cytoplasm of a *Paramecium*. A similar symbiotic relationship may have given rise to the ancestors of modern chloroplasts.

ferred great advantages in regulating the use of the genetic material. Today, organisms composed of eukaryotic cells are by far the most visible forms of life on Earth.

Multicellularity

Once predation had evolved, increased size became an advantage. A larger cell could more easily engulf a smaller cell, while in turn being more difficult for other predatory cells to ingest. Larger organisms can usually also move faster than small ones, making successful predation and escape more likely. But enormous single cells have problems. Oxygen and nutrients going in and waste products going out of the cell must diffuse through the plasma membrane. As we pointed out in Chapter 5, the larger a cell becomes, the less surface membrane is available per unit volume of cytoplasm. There are only two ways that an organism larger than a millimeter or so in diameter can survive. First, it can have a low metabolic rate so that it doesn't need much oxygen or produce much carbon dioxide. This seems to work for certain very large unicellular algae. Alternatively, an organism may be multicellular— that is, it may consist of many small cells packaged into a large unified body.

The fossil record reveals almost nothing about the origin of multicellularity, especially among animals. The first unicellular eukaryotic fossils occur in rocks about 1.4 billion years old, and the first signs of multicellular animals

occur as worm burrows and tracks in rocks 400 million years younger. The intervening animals almost certainly had no skeleton or other hard parts, and left few fossils. Consequently, we may never learn very much about them. Within another 500 million years, however, diverse types of multicellular animals appeared, leaving their fossilized remains in rocks in many parts of the world.

Multicellular Life in the Sea

The first multicellular organisms almost certainly evolved in the sea.

Multicellular Plants Evolved Specialized Structures That Facilitated Their Invasion of Diverse Habitats

During the Precambrian era, unicellular eukaryotic cells containing chloroplasts gave rise to the first multicellular plants. Multicellularity would have provided at least two advantages for plants. First, large, many-celled plants would have been difficult for unicellular predators to swallow. Second, specialization of cells would have conferred the potential for staying in one place in the brightly lit waters of the shoreline, as rootlike structures burrowed in sand or clutched onto rocks, while leaflike structures floated above in the sunlight. The profusion of green, brown, and red algae lining our shores today, some, such as the brown kelp, over 60 meters (200 feet) long, are the descendants of these early multicellular algae.

Multicellular Animals Evolved Specializations That Allowed Them to Capture Prey, Feed, and Escape More Efficiently

In a great burst of evolution, a wide variety of invertebrate animals (animals lacking backbones) appeared in the sea near the beginning of the Cambrian period, about 600 million years ago. For animals, one of the advantages of multicellularity is the potential for eating larger prey. The first animal capable of consuming large prey probably resembled a jellyfish. A single opening served both as a mouth, to take in food, and as an anus, to expel indigestible remains (Fig. 19-7a). A half-digested prey filling its digestive tract keeps such an animal from feeding again until it is finished with its first meal. Soon more efficient means of feeding evolved, employing a separate mouth and anus, found today in almost all animals (Fig. 19-7b). With this design, the animal can feed more or less continuously, as earthworms and sea cucumbers do today.

Coevolution of predator and prey led to increased sophistication in many kinds of animals; during the Silurian period (438 to 408 million years ago) mud-skimming, armored trilobites were preyed upon by ammonites and the chambered nautilus, which still survives almost unchanged

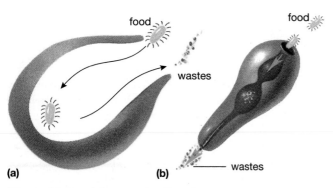

Figure 19-7   **Advances in digestive tracts**

(a) The earliest animals probably possessed a digestive system with only one opening. Feeding would have had to be periodic, because ingestion of food and expulsion of wastes would interfere with one another. **(b)** A major advance in feeding is to have separate openings for ingesting food and for ejecting wastes, allowing more or less continuous feeding.

in deep Pacific waters (Fig. 19-8). A major trend at this time was toward greater mobility. Predators often need to travel over wide areas in search of suitable prey, while speedy escape is also an advantage for prey. Locomotion is usually accomplished by the contraction of muscles that move body parts through their attachments to some sort of skeleton. Most invertebrates at this time possessed either an internal hydrostatic skeleton, much like a water-filled tube (worms), or an external skeleton covering the body (arthropods such as trilobites). Greater sensory capabilities and more sophisticated nervous systems evolved along with the ability to move more efficiently. Senses for detecting touch, chemicals, and light became highly developed. The senses were usually concentrated in the head end of the animal, along with a nervous system capable of handling the sensory information and directing appropriate behaviors.

About 500 million years ago, one group of animals evolved a new form of support for the body: an internal skeleton. For a hundred million years these were inconspicuous members of the ocean community, but by 400 million years ago some had evolved into fishes. By and large, the fishes proved to be faster than the invertebrates, with more acute senses and larger brains. Eventually, they became the dominant predators of the open seas.

The Invasion of the Land

Between the period from 600 to 400 million years ago, both plants and animals evolved significantly, but they remained cushioned by the equanimity of the sea. Life in the ocean provides buoyant support against gravity and ready access to life-sustaining water. Reproduction is also simpler in the sea. At some point in their life cycles, both plants and animals produce sex cells that must fuse to form the beginnings of a new generation. Sea-dwelling organisms usually have mobile sperm or eggs, or both, that swim to each other through the water.

In contrast, on land an organism must bear its weight against the crushing force of gravity; it must find adequate water; and to reproduce, it must ensure that its gametes, particularly the sperm, are protected from drying out. Nevertheless, the land offered great potential, particularly for plants. Water strongly absorbs light, so even in the clearest water, photosynthesis is limited to the upper few hundred meters, and usually much less. Out of the water, the sun is dazzlingly bright, permitting rapid photosynthesis. Seawater also tends to be low in certain nutrients, particularly nitrogen and phosphorus. By comparison, terrestrial soils are rich storehouses of nutrients. Finally, by this time the sea swarmed with plant-eating animals. Because the land was devoid of animal life, plants that first colonized the land would have had no predators.

Some Plants Evolved Specialized Structures That Adapted Them to Life on Dry Land

In moist soils near the shore, a few small green algae began to grow, taking advantage of the sunlight and nutrients. They didn't have large bodies to support against the force of gravity, and living right in the film of water on the soil, they could easily obtain water. About 400 million years ago, some of these algae gave rise to the first multicellular land plants. Initially simple, low-growing forms, land plants rapidly evolved solutions to two of the main difficulties of plant life on land: obtaining and conserving water and staying upright despite gravity and winds. Rootlike structures delved into the soil, mining water and minerals, while waterproof coatings on the aboveground parts reduced water loss by evaporation. Specialized cells formed tubes called vascular tissues to conduct water from roots to leaves. Extra-thick walls surrounding certain cells enabled stems to stand erect.

Primitive Land Plants Retained Swimming Sperm and Required Water to Reproduce

Reproduction out of water presented greater challenges. Like animals, plants produce sperm and eggs. Primitive marine plants have swimming sperm and sometimes swimming eggs as well. The first land plants retained swimming sperm. Consequently, they were restricted to swamps and marshes, where the sperm and eggs could be released into the water, or to areas with abundant rainfall, where the ground would occasionally be covered with water.

Conifers Encased Sperm in Pollen Grains, Allowing Them to Flourish in Dry Habitats

The reproductive strategy described above sufficed for millions of years. During the Carboniferous period, 360 to 286 million years ago, the climate was warm and moist,

(a)

(b)

(c)

(d)

Figure 19-8   **Diversity of ocean life during the Silurian period**
(a) Characteristic life of the oceans during the Silurian period, 438 to 408 million years ago. Among the most common fossils from that time are the trilobite **(b)** and its predators the ammonites **(c)** and the nautiloids. Although **(d)** illustrates a living *Nautilus*, the Silurian nautiloids were very similar in structure, showing that a successful body plan may exist virtually unchanged for hundreds of millions of years.

and great stretches of the land were covered with forests of giant tree ferns and club mosses that produced swimming sperm (see Fig. 19-9). The coal we mine today is derived from the fossilized remains of those forests. Meanwhile, some plants inhabiting drier regions had begun to evolve reproductive strategies that no longer depended on films of water. These early cone-bearing plants, called **conifers**, arose in the Permian period (286 to 246 million years ago). Their eggs were retained in the parent plant, while the sperm were encased in drought-resistant pollen grains that blew on the wind from plant to plant. Landing on a female cone near the egg, the pollen released sperm cells directly into living tissue, eliminating the need for a surface film of water. About 250 million years ago, mountains rose, swamps drained, and the moist climate dried up. Tree ferns and giant club mosses, with their swimming sperm, largely became extinct, while the conifers, which did not depend on water for reproduction, flourished and spread.

Flowering Plants Enticed Animals to Carry Pollen

About 130 million years ago, the flowering plants appeared, having evolved from a group of conifer-like plants. The initial advantage of the flowering plants seems to have been pollination by insects. The conifers are wind-pollinated, so they must produce an enormous amount of pollen because the vast majority of pollen grains fail to reach their target. Flower pollination by insects wastes far less pollen. Flowering plants also evolved other advantages, including more rapid reproduction and, in some cases, much more rapid growth. Today, flowering plants dominate the land, except in cold northern regions, where conifers still prevail.

Some Animals Evolved Specialized Structures That Adapted Them to Life on Dry Land

Soon after land plants evolved, providing potential food sources for animals, arthropods (probably early relatives of scorpions) emerged from the sea. Why arthropods? The

Figure 19-9   The swamp forest of the Carboniferous period

The treelike plants are tree ferns and giant club mosses, both now mostly extinct. Note the dragonfly at bottom center; some Carboniferous dragonflies had wingspans in excess of half a meter (about 1.5 feet)!

answer seems to be that they were **preadapted** for land life—that is, they already possessed structures, evolved under totally different selective pressures, that, purely by chance, were suited to life on land. Foremost among these was the external skeleton, or **exoskeleton**, a hard covering surrounding the body, such as the shell of a lobster or crab. Exoskeletons are both waterproof and strong enough to bear up a small animal under the stresses of gravity.

Land animals encounter another difficulty, also relatively easily solved by arthropods: breathing. To allow gas exchange, respiratory surfaces must be kept moist, and doing so is difficult in the dry air. Some arthropods, such as land crabs and spiders, evolved what amounts to an internal gill, kept moist within a waterproof sac. The insects developed tracheae, small branching tubes directly penetrating the body, with adjustable openings in the exoskeleton leading to the outside air.

For millions of years, arthropods had the land and its plants to themselves, and for tens of millions of years more, they were the dominant animals. Dragonflies with a wingspan of 70 centimeters (28 inches) flew among the Carboniferous tree ferns, while millipedes 2 meters (6.5 feet) long munched their way across the swampy forest floor. Eventually, however, the arthropods' splendid isolation came to an end.

Amphibians Evolved from Lobefin Fishes and Developed Lungs

About 400 million years ago, a group of fishes called the lobefins appeared, probably in fresh water. Lobefins had two important preadaptations to land-life: stout, fleshy fins

with which they crawled about on the bottoms of shallow, quiet waters and an outpouching of the digestive tract that could be filled with air, like a primitive lung (Fig. 19-10). The coelacanth is a lobefin that was believed to be long extinct before its discovery in 1939. In one group of lobefins, the lung evolved into a swim bladder, with which the fish could regulate their buoyancy and remain suspended in the water without active exertion. Many of these lobefins migrated back to the sea, where they evolved into the modern bony fishes. Another group of lobefins colonized very shallow ponds and streams, which shrank during droughts and whose water often became oxygen-poor. By taking air into their lungs, these lobefins could obtain oxygen anyway. Some of their descendants began to use their fins to crawl from pond to pond in search of prey or water, as some modern fish can do today (Fig. 19-11).

As the arthropods had discovered previously, the land is a rich source of food. Feeding on land and moving from pool to pool favored the evolution of fish that could stay out of water for longer periods and that could move about more effectively on land. With improvements in lungs and legs, the amphibians evolved from lobefins, first appearing in the fossil record about 350 million years ago. If an amphibian could have thought about such things, it would have thought that the Carboniferous swamp forests were heaven itself: no predators to speak of, abundant prey, and a warm, moist climate. As with the insects and millipedes, some amphibians evolved gigantic size, including salamanders over 3 meters (10 feet) long.

Despite their success, the early amphibians still lacked complete adaptation to life on land. Their lungs were

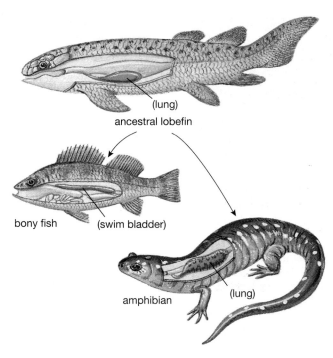

Figure 19-10   **From fish to amphibians**

A group of primitive fish, called the lobefins, gave rise both to the bony fish and the amphibians. Lobefins had a pair of lungs, which arose as outpocketings of the digestive tract. Lobefins probably used their lungs for breathing air when the pools in which they lived became stagnant, foul, and low in oxygen. In one group of lobefins, one lung evolved into a swim bladder, used by bony fishes to regulate buoyancy. The other lung disappeared. Another group of lobefins became further adapted for breathing air, evolving more complex, efficient lungs and sturdier fins for walking on land. These were the ancestors of the amphibians.

simple sacs without very much surface area, so they had to obtain some of their oxygen through their skins. Therefore, the skin had to be kept moist, restricting amphibians to swampy habitats where they wouldn't dry out. Further, amphibians deposit their sperm and eggs in water. As with the tree ferns and club mosses, when the Carboniferous climate turned dry at the beginning of the Permian period about 286 million years ago, amphibians were in trouble.

Reptiles, Which Arose from Amphibians, Evolved Several Adaptations to Dry Land

Meanwhile, just as the conifers had been evolving on the fringes of the swamp forests, so too a group of amphibians were evolving adaptations to drier conditions. These became the reptiles, which achieved four great advances over the amphibians. First, they evolved internal fertilization: The reptilian female provides the watery environment for the sperm inside her reproductive tract. Thus, sperm transfer could occur on land, without venturing back to the dangerous swamps full of fish and amphibian predators.

Figure 19-11   **A fish that walks on land**

Some modern fish resemble their lobefin ancestors in their ability to crawl about on land. The walking catfish was imported into Florida several years ago, escaped into local waters, and has spread across much of southern Florida by walking from pond to pond.

Second, they developed waterproof eggs enclosing their own supply of water for the developing embryo, again providing freedom from the swamps. Third, the ancestral reptiles developed scaly, waterproof skin. Finally, accompanying the evolution of waterproof skin came improved lungs that could provide the entire oxygen supply for an active animal. As the climate dried during the Permian period, reptiles became the dominant vertebrate land fauna, relegating the amphibians to swampy backwaters, where most remain today.

A few tens of millions of years later, the climate returned to more moist and equable conditions, providing for lush plant growth. Once again, gigantic size was selected for, as certain families of reptiles evolved into the dinosaurs (Fig. 19-12). These were among the most successful animals ever, if we consider persistence as a measure of success. They flourished for over a hundred million years, until about 65 million years ago when the last dinosaurs became extinct. No one is certain why they died, but a climate change, perhaps initiated by a gigantic meteorite impact, seems to have been the final blow (see Chapter 17). With less luxuriant plant growth, the herbivorous dinosaurs could not find enough food. Without large herbivores as prey, the carnivores were doomed too.

Even during the age of dinosaurs, many reptiles remained quite small. One major difficulty faced by small reptiles is keeping a high body temperature. Being active on land seems to require a rather warm body to maximize the efficiency of the nervous system and muscles. But a warm body loses heat to the environment unless the air is also warm. Small reptiles have a relatively large surface area through which heat is lost and a relatively small internal volume in which metabolic heat is generated. If the body is warmed metabolically when the air is cool, then an enormous amount of food must be consumed to provide sufficient energy. Apparently the food requirement is too high, for the naked-skinned small reptiles have a fairly

Figure 19-12   **A reconstruction of a Jurassic swamp**

In the foreground we see, standing in the swamp, the gigantic *Apatosaurus*, 20 meters (65 feet) long and weighing over 30 tons. On the right, a carnivorous *Allosaurus* uses its meter-long jaws to tear flesh from its prey. In the center of the background grazes the 10-ton *Stegosaurus*, an astoundingly dim-witted herbivore with a brain the size of a walnut. *Stegosaurus* had a swelling of its spinal cord in the hip region that formed a second "brain" of sorts, 20 times larger than the one inside its skull. The plates on its back probably functioned in temperature regulation. Depending on air temperature, the animal's orientation to the sun, and the amount of blood flow across the surface of the plates, they could be used as radiators to cool the blood or as solar panels to pick up warmth.

low metabolism, not enough to keep their bodies warm in cool air, especially at night when there is no sun for radiant warmth. Two groups of small reptiles independently evolved insulation that minimizes heat loss: One group evolved feathers, while another group evolved hair.

Reptiles Gave Rise to Both Birds and Mammals

In the ancestral birds, insulating feathers retained body heat. Consequently, these animals could be active in cooler habitats and during the night, when their scaly relatives became sluggish. Later, some ancestral birds developed longer, stronger feathers on their forelimbs, perhaps allowing them to glide from trees or to assist in jumping after insect prey. From this point, the evolution of flight became possible.

The hair of ancestral mammals also provided insulation. Unlike the birds, which retained the reptilian habit of laying eggs, mammals evolved live birth and the ability to feed their young with secretions of the mammary (milk-producing) glands. Because these structures do not fossilize, we may never know when the uterus, mammary glands, and hair first appeared, or what their intermediate forms looked like.

The earliest mammals were small creatures, probably living in trees and being active mostly at night. When the dinosaurs became extinct, the mammals radiated out into the vast array of modern forms. Some stayed small and nocturnal, eating mostly seeds and insects; others evolved larger size and different habits, colonizing the habitats left empty by the extinction of the dinosaurs. One group remained in the trees, giving rise to the primates.

Human Evolution

Fossils of **primates** (lemurs, monkeys, apes, and humans) are relatively rare compared with those of many other animals. There are at least three reasons. First, most primates did not live in habitats that readily preserve fossils, such as swamps and shallow lagoons. Second, until recently primates were fairly small. Therefore predators and scavengers would be more likely to break up their bones into unrecognizable fragments than they would the bones of a bison or dinosaur. Third, ancestral primates may have had small populations, thus providing less material for fossilization in the first place.

Humans are intensely interested in their own origin and evolution, especially in trying to determine what conditions led to the evolution of the gigantic human brain. This obsession has led to sweeping speculations despite an often skimpy fossil record. Therefore, while the outline of human evolution that we will present is a synthesis of current thought on the subject, it is by no means as well understood as, say, the genetic code. Paleontologists disagree about the interpretation of the fossil evidence, and many ideas may have to be revised as new fossils are found.

Primate Evolution Has Been Linked to Grasping Hands, Binocular Vision, and a Large Brain

The first ancestral primates were the insect-eating tree shrews, whose fossils are found in rocks about 80 million years old. Nimble, probably nocturnal animals, tree shrews were smaller than all but the tiniest modern primates. Over the next 50 million years, the descendants of the tree shrews evolved forms similar to the modern tarsiers, lemurs, and monkeys (Fig. 19-13). These primates stayed in the trees. Some remained nocturnal, but many become active during the day. Primates, who fed on fruits and leaves, evolved several adaptations for life in the trees.

Grasping Hands in Early Primates Allowed Both Powerful and Precise Manipulations

The tree shrews already possessed handlike paws for holding on to small branches, and the primates further refined these appendages. Most primates are much larger than tree shrews, so only relatively large tree limbs could bear their weight. Long, grasping fingers that could wrap around and hold larger limbs made life in the trees a little safer. The primate line that led to humans evolved hands with the ability to perform both delicate maneuvers using a *precision grip* (manipulating small objects, writing, sewing) and powerful actions using a *power grip* (swinging a club, thrusting with a spear).

Binocular Vision Provides Accurate Depth Perception

One of the earliest primate adaptations seems to have been large, forward-facing eyes (see Fig. 19-13). Jumping from branch to branch is risky business unless an animal can accurately judge where the next branch is located. Accurate depth perception was made possible by binocular vision provided by forward-facing eyes with overlapping fields of view. Another adaptation was color vision. We cannot, of course, tell if a fossil animal had color vision, but modern primates have excellent color vision, and it seems reasonable to assume that earlier primates did too.

Many primates feed on fruit, and color vision helps in detecting ripe fruit among a welter of green leaves.

A Large Brain Facilitated Hand-Eye Coordination and Complex Social Interactions

Primates have brains that are larger, relative to their body size, than almost all other animals. No one really knows for certain what selective forces favored the evolution of large brains. It seems reasonable, however, that controlling and coordinating binocular, color vision, rapid locomotion through trees, and dexterous movements of the hands would be facilitated by increased brain power. Most primates also have fairly complex social systems, which probably would require relatively high intelligence. If sociality promoted increased survival and reproduction, then there would have been selective pressures for the evolution of larger brains.

Hominids Evolved from Dryopithecine Primates

Between 20 and 30 million years ago, in the moist tropical forests of Africa, a group of primates called the dryopithecines diverged from the monkey line. The dryopithecines appear to have been ancestral to the **hominids** (humans and their fossil relatives) and pongids (the great apes). Around 18 million years ago, global climatic cooling began to shrink the vast expanses of forest, splitting up the woodlands into isolated islands dotted upon a sea of grassland. Diversification of habitat and isolation of small populations led to the diversification of dryopithecines. One of these gave rise to the later hominids and pongids.

(a)

(b)

(c)

Figure 19-13 **Representative primates**

(a) Tarsier, (b) lemur, and (c) lion-tail macaque monkey. Note that all have relatively flat faces, with forward-looking eyes providing binocular vision. All also have color vision and grasping hands. These features served as preadaptations for tool and weapon use by early humans.

Figure 19-14   **A possible evolutionary tree for humans**

This hypothetical family tree shows skulls and facial reconstructions of representative specimens. Although many paleontologists consider this to be the most likely human family tree, there are several alternative interpretations of the known hominid fossils, each with its strong advocates. Fossils of the earliest australopithecine, *Australopithecus ramidus*, are too fragmentary to reconstruct a skull; the photo inset is of teeth, which reveal much about the specimen's relationship to other hominids.

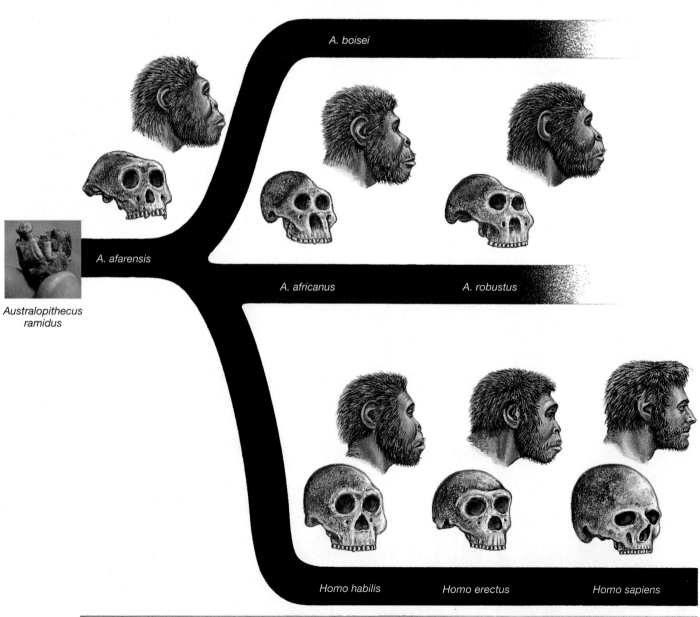

Australopithecus ramidus

A. boisei

A. afarensis

A. africanus

A. robustus

Homo habilis   Homo erectus   Homo sapiens

4   3   2   1   0

millions of years ago

Australopithecines, the First True Hominids, Could Stand and Walk Upright

A large gap ensues in the fossil record, until the first true hominids appear about 4.4 million years ago. These fossils are collectively called the australopithecines ("southern ape," after their original discovery site in southern Africa; Fig. 19-14). The brains of australopithecines were fairly large, although still much smaller than those of modern humans. Although the earliest australopithecines had legs that were shorter, for their height, than any modern humans, their knee joints allowed them to fully straighten their legs, permitting efficient bipedal (upright, two-legged) locomotion. Footprints almost 4 million years old, discovered in Tanzania by Mary Leakey, show that australopithecines could, and at least sometimes did, walk upright. Upright posture is an extremely important feature of hominids, because it freed their hands from use in walking. Later hominids thus were able to carry weapons, manipulate tools, and eventually achieve the cultural revolutions produced by modern *Homo sapiens*. What selective pressures led to the evolution of bipedalism? One interesting hypothesis is explored in "Evolutionary Connections: Why Do Humans Walk Upright?"

The oldest australopithecine species, represented by fossilized teeth, skull fragments, and arm bones, was unearthed in Ethiopia in 1993 from sediments that were dated, using radioactive isotopes, as 4.4 million years old (see "Scientific Inquiry: Radiometric Dating Traces Fossils through Time"). It was named *Australopithecus ramidus* by its discoverers (*ramid* means "root" in the local Ethiopian language). The teeth of these early fossils (although clearly hominid) have several features that are intermediate between ape and human. The second most ancient australopithecine, called *Australopithecus afarensis*, was discovered in the Afar region of Ethiopia in 1974, a partial skeleton of a 3.5-foot tall female nicknamed Lucy. Fossil remains of this species 3.9 million years old have been unearthed. Later, the *A. afarensis* line apparently gave rise to at least two distinct forms: the small, omnivorous *A. africanus* (which in size and eating habits was similar to *A. afarenis*), and the large, herbivorous *A. robustus* and *A. boisei*. Most authorities agree that either *A. afarensis* or *A. africanus* was the ancestor of modern hominids, including humans, while the other forms became extinct, without leaving descendants, about a million years ago.

Homo habilis *Was the First Hominid to Use Tools*

By about 2 million years ago, certain African populations of *A. afarensis* or *A. africanus* seem to have given rise to a new form, *Homo habilis* (see Fig. 19-14). *Homo habilis* had a considerably larger body and brain than the australopithecines. *Homo habilis* was probably the first hominid that used stone and bone tools, mostly crude weapons of various sorts (Fig. 19-15). In a relatively short time, *H. habilis* spread throughout Europe, Asia, and Africa.

Homo erectus *Had a Larger Brain and Produced More Sophisticated Tools*

About 1.8 million years ago, *Homo erectus* appeared, probably a descendant of an African population of *H. habilis*. The brain of *H. erectus* was as large as the smallest modern adult human brains (around 1000 cubic centimeters). The face was notably different from that of modern humans (see Fig. 19-14), featuring large brow ridges (bony ridges above the eyes), a slightly protruding face, and no chin (the protruding tip of the lower jaw).

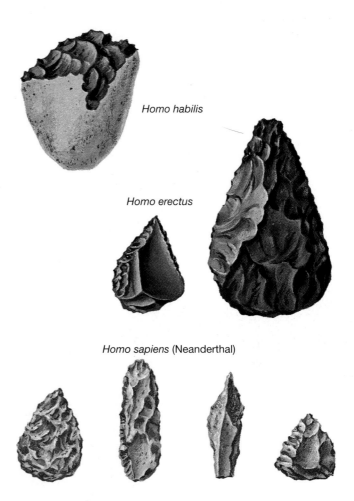

Homo habilis

Homo erectus

Homo sapiens (Neanderthal)

Figure 19-15   Representative hominid tools

Homo habilis produced only fairly crude chopping tools called hand axes, usually unchipped on one side to hold in the hand. *Homo erectus* manufactured much finer tools. The tools were often sharp all the way around the stone, so at least some of these blades were probably tied to spears rather than held in the hand. Neanderthal tools were works of art, with extremely sharp edges made by flaking off tiny bits of stone. In comparing these *H. habilis*, *H. erectus*, and *H. sapiens* weapons, note the progressive increase in the number of flakes taken off the blades and the corresponding decrease in flake size. Smaller, more numerous flakes produce a sharper blade and suggest more insight into tool making (perhaps passed down culturally from experienced tool makers to apprentices), more patience, finer control of hand movements, or perhaps all three.

Figure 19-16   Competing hypotheses for the evolution of *Homo sapiens* from *Homo erectus*

The "out of Africa" advocates hypothesize that *H. sapiens* evolved in Africa, then migrated throughout the Near East, Europe, and Asia. The "multiregional" hypothesis states that populations of *H. sapiens* evolved in many regions simultaneously from the already widespread populations of *H. erectus*.

Behaviorally, *H. erectus* seems to have been more advanced than *H. habilis*. *Homo erectus* fashioned sophisticated stone tools, ranging from hand axes used for cutting and chopping to points probably used on spears (see Fig. 19-15). These weapons suggest that *H. erectus* ate animal food, probably both hunting on their own and scavenging for the remains of prey killed by lions and other predators. *Homo erectus*, like *H. habilis*, spread throughout most of Europe and Asia. In China, evidence from 500,000 years ago suggests that in addition to sophisticated stone tools, *H. erectus* used fire, although for what purpose (to cook food, keep warm, or ward off predators) no one knows.

Homo sapiens *Evolved from* Homo erectus

The first fossils of our own species, *Homo sapiens* (*sapiens* means "intelligent"), date from about 200,000 years ago. There are currently two hotly debated hypotheses as to the origin of *H. sapiens*. Some paleoanthropologists (scientists who study human origins) believe that *H. sapiens* evolved in Africa, then migrated into the Near East, Europe, and Asia, supplanting all other hominids, including *H. erectus*. The opposing hypothesis is that populations of *H. sapiens* evolved in many regions simultaneously from the already widespread populations of *H. erectus*. According to this hypothesis, continued migrations and interbreeding among *H. erectus* populations maintained them as a single species as they gradually evolved into *H. sapiens* (Fig. 19-16).

Although the earliest fossils are fragmentary, beginning about 100,000 years ago, Neanderthals, believed by many researchers to be a variety of *H. sapiens*, left many complete fossils in northern Africa, throughout Europe, and in parts of Asia. Contrary to the popular image of the hulking, stoop-shouldered "cave man," Neanderthals were quite similar to modern humans in many ways. Although more heavily muscled, Neanderthals walked fully erect, were dexterous enough to manufacture finely crafted stone tools, and had brains, on the average, slightly larger than those of modern humans. Although many of the European fossils show heavy brow ridges and a broad, flat skull, others, particularly from areas around the eastern shores of the Mediterranean Sea, were remarkably like ourselves. Neanderthal remains show evidence of modern behaviors, too, particularly ritualistic burial ceremonies (Fig. 19-17). Neanderthal skeletons have been discovered in clearly marked burial sites surrounded with stones and often including offerings of flowers, bear skulls, and food. Altars found at other Neanderthal sites were probably used in "religious" rites associated with a bear cult.

Finally, about 90,000 years ago, fully modern humans appeared. Although the oldest specimens were discovered in Israel in 1987, these anatomically modern humans are called Cro-Magnon, after the district in France in which their remains were first discovered. Cro-Magnons had domed heads, smooth brows, and prominent chins. Their tools were precision instruments not very different from the stone tools used until recently in many parts of the world.

Where did the Cro-Magnons come from and where did the Neanderthals go? One group of paleoanthropologists believes that Cro-Magnon evolved from the modern-looking Neanderthals of the eastern Mediterranean region and spread into Europe from there. The contrary view is that Cro-Magnon evolved independently, from a separate population of *H. erectus*. Neanderthal and Cro-Magnon probably coexisted for a while, particularly in Europe. Some researchers believe that considerable interbreeding occurred between Neanderthals and Cro-Magnons and that their genes are a part of the European heritage. Others believe that the Neanderthals represent an evolutionary dead end. Whether the Neanderthals were eliminated by interbreeding, competition, or open warfare with Cro-Magnon may never be known.

Behaviorally, Cro-Magnon seems to have been similar to, if more sophisticated than, Neanderthal. Perhaps the most remarkable accomplishment of Cro-Magnon is the magnificent art left in caves such as Altamira in Spain and Lascaux in France (Fig. 19-18). Again, no one knows exactly why these paintings were made, but they attest to minds fully as human as our own.

The Evolution of Human Behavior Is Highly Speculative

Perhaps the most hotly debated subject in human evolution is the development of human behavior. Except in rare instances, such as the Neanderthal burials and the Cro-Magnon paintings, there is no direct evidence of the behavior of prehistoric hominids. A few hypotheses have been offered by biologists and anthropologists, on the basis of the fossil record and the behavior of modern humans and animals.

Figure 19-17   Neanderthal bear ceremony

Power and ferocity, and perhaps their ability to rise up on their hind legs like humans, made cave bears cult objects to Neanderthals. Some Neanderthals were buried with elaborate ceremonies, with bear skulls placed on the grave of the deceased.

Figure 19-18   **The sophistication of Cro-Magnon people**
Cave paintings by Cro-Magnon people remarkably preserved by the relatively constant underground conditions of a cave in Lascaux, France.

The fossil record shows that the development of truly huge brains occurred within the last couple of million years. What were the selective advantages of large brains? This subject continues to stir controversy among paleoanthropologists. The enlarging brain may have been selected both because it provided for improved hand-eye coordination and because it facilitated complex social interactions.

As the early hominids such as *Australopithecus* descended from the trees into the savanna, they began to walk upright. Bipedal locomotion allowed them to carry things in their hands as they walked. Hominid fossils show shoulder joints capable of powerful throwing motions and an opposable thumb (a thumb that can be placed directly against each of the other fingers) that would enhance the ability to manipulate objects. These early hominids probably could see well and judge depth accurately with binocular vision. The brain expansion that occurred in the australopithecines may have been at least partly related to integration of visual input and control of hand and arm movements.

Homo erectus, and perhaps *H. habilis* before it, was social. So, of course, are many monkeys and apes. The later hominids, however, seem to have engaged in a new type of social activity: cooperative scavenging and hunting. *Homo erectus*, in particular, seems to have hunted extremely large game, which calls for cooperation in all phases of the hunt. Some paleoanthropologists believe that selection favored larger and more powerful brains as an adaptation for success in cooperative hunting. (Lest you become too impressed with such speculation, remember that lions and wolves also hunt cooperatively, yet are not noticeably more intelligent than their relatives, such as leopards and coyotes, who do not.)

The Cultural Evolution of Humans Now Far Outpaces Biological Evolution

In recent millennia, the biological evolution of humans has been far outpaced by our **cultural evolution**: learned behaviors passed down from previous generations. Careful estimates of the human population size since our species originated reveals three major surges of population growth. Each surge was initiated by a cultural-technological revolution: the development of new forms of technology that increased the resources available to sustain human life. The first revolution was the development of tools, beginning with early hominids and extending until 10,000 years ago, when the total human population numbered 5 million. Beginning then and extending over the next 8000 years, humans grew crops and domesticated animals, and the population gradually increased by 100-fold, to about 500 million. Following this agricultural revolution, we are now 300 years into the scientific-industrial revolution, and our population has reached 5.6 billion.

Human cultural evolution, and its accompanying increases in human population, have had profound effects on the continuing biological evolution of other life forms. Our agile hands and minds have transformed much of Earth's terrestrial habitats. People have become the single overwhelming agent of natural selection. In the words of evolutionary biologist Stephen Jay Gould of Harvard University: "We have become, by the power of a glorious evolutionary accident called intelligence, the stewards of life's continuity on Earth. We did not ask for this role, but we cannot abjure it. We may not be suited for it, but here we are."

E V O L U T I O N A R Y
C O N N E C T I O N S

Why Do Humans Walk Upright?

Upright posture and bipedal locomotion, naked skin, and enlarged brains are hallmarks of humanity. Why did these features evolve in humans? The short answer is that no one knows, but British physiologist Peter Wheeler has an intriguing hypothesis: Our ancestors evolved the ability to walk upright because this posture exposes a minimum body surface to the blazing savanna sun. Wheeler constructed a small-scale model of "Lucy" (*Australopithecus afarensis*) and photographed it in bipedal and quadrupedal postures from a series of overhead locations that imitated the pathway of the sun. He found that a four-legged posture exposed the model to 60% more solar radiation than did the upright posture. Further, says Wheeler, standing upright exposes more of the body to cooler breezes a few feet above the ground. Together, these factors would increase the time the animal could spend foraging in the sun, allowing it to increase its access to food. Bipedalism may have also made possible our naked skin. Although hair is a shield against the sun in quadrupedal (four-legged) animals, in our bipedal ancestors this need was minimized, restricted to the shoulders and head. Naked skin then allowed our ancestors to dissipate even more heat. Wheeler carries the hypothesis a step further, to include the large human brain. The brain is one of the most metabolically active organs in the body and very easily damaged by overheating. Wheeler suggests that only after hominids had evolved upright posture and naked skin did they have enough heat-dissipating capacity to permit the evolution of their oversized brains.

✳ S U M M A R Y   O F   K E Y   C O N C E P T S

Origins

Before life arose, lightning, ultraviolet light, and heat formed organic molecules from water and the components of the primordial Earth's atmosphere. These molecules probably included nucleic acids, amino acids, short proteins, and lipids. By chance, some molecules of RNA may have had enzymatic properties, catalyzing the assembly of copies of themselves from nucleotides in Earth's waters. These may have been the forerunners of life. Protein-lipid microspheres enclosing these RNA molecules may have formed the first cell-like organisms.

The First Life

The first fossil cells are found in rocks about 3.5 billion years old. These cells were prokaryotes that fed by absorbing organic molecules that had been synthesized in the environment. Because there was no free oxygen in the atmosphere at this time, energy metabolism must have been anaerobic. As the cells multiplied, they depleted the organic molecules that had been formed by prebiotic synthesis. Some cells evolved the ability to synthesize their own food molecules using simple inorganic molecules and the energy of sunlight. These earliest photosynthetic cells were probably ancestors of today's cyanobacteria.

Photosynthesis releases oxygen as a by-product, and by about 2.2 billion years ago significant amounts of free oxygen had accumulated in the atmosphere. Aerobic metabolism, which generates more cellular energy than does anaerobic metabolism, probably arose about this time.

Eukaryotic cells evolved about 1.5 billion years ago. The first eukaryotic cells probably arose as symbiotic associations between predatory prokaryotic cells and bacteria. Mitochondria may have evolved from aerobic bacteria engulfed by predatory cells. Similarly, chloroplasts may have evolved from photosynthetic cyanobacteria.

Multicellularity

Multicellular organisms evolved from eukaryotic cells, first appearing about 1 billion years ago. Multicellularity offers several advantages, including increased speed of locomotion and greater size.

Multicellular Life in the Sea

The first multicellular organisms arose in the sea. In plants, increased size due to multicellularity offered some protection from predation. Specialization of cells allowed plants to anchor themselves in the nutrient-rich, well-lit waters of the shore. For animals, multicellularity allowed more efficient predation and more effective escape from predators. These in turn provided selection pressures for faster locomotion, improved senses, and greater intelligence.

The Invasion of the Land

The first land organisms were probably plants, appearing 400 to 600 million years ago. Although the land required special adaptations for support of the body, reproduction, and the acquisition, distribution, and retention of water, the land also offered abundant sunlight and protection from aquatic herbivores. Around 400 million years ago, arthropods invaded the land. Absence of predators and abundant land plants for food probably facilitated the invasion of the land by animals.

The earliest land vertebrates evolved from lobefin fishes, which had leglike fins and a primitive lung. A group of lobefins evolved into the amphibians about 350 million years ago. Reptiles evolved from amphibians, with several further adaptations for land-life: internal fertilization, waterproof eggs that could be laid on land, waterproof skin, and better lungs. Around 150 million years ago, birds and mammals evolved independently from separate groups of reptiles. Major advances included a high, constant body temperature and insulation over the body surface.

Human Evolution

One group of mammals evolved into the tree-dwelling primates. Primates show several preadaptations for human evolution: forward-facing eyes for binocular vision, color vision, grasping hands, and relatively large brains. Between 20 and 30 million years ago some primates descended from the trees; these were the ancestors of apes and humans. The australopithecines arose in Africa about 4 million years ago. These hominids walked erect, had larger brains than their forebears, and made primitive tools. One group of australopithecines evolved into true humans.

KEY TERMS

conifer   p. 357
cultural evolution   p. 366
endosymbiotic hypothesis   p. 353
exoskeleton   p. 358

hominid   p. 361
microsphere   p. 351
preadaptation   p. 358
prebiotic evolution   p. 349

primate   p. 360
ribozyme   p. 350
spontaneous generation   p. 348

THINKING THROUGH THE CONCEPTS

Multiple Choice

1. Our best estimate for the age of Earth, 4.5 billion years, is supported by
   a. the assumption of uniformitarianism
   b. the Big Bang theory
   c. gradualism
   d. punctuated equilibrium
   e. radioactive dating of the oldest rocks found

2. There was no free oxygen in the early atmosphere. Much of it was tied up in
   a. water   b. ammonia
   c. methane   d. none of the above
   e. radioactive isotopes

3. RNA became a candidate for the first information-carrying molecule when Tom Cech and Sidney Altman discovered that some RNAs can act as enzymes that
   a. degrade proteins
   b. turn light into chemical energy
   c. split water and release oxygen gas
   d. synthesize copies of themselves
   e. synthesize amino acids

4. Which three of the following observations support Lynn Margulis's endosymbiotic hypothesis for the origin of chloroplasts and mitochondria from ingested bacteria?
   a. Mitochondria are where aerobic respiration takes place.
   b. Mitochondria have their own DNA.
   c. Chloroplasts are where photosynthesis takes place.
   d. Chloroplasts have their own DNA.
   e. Bacterial plasma membranes are strikingly similar to the inner membrane of mitochondria.

5. The exoskeleton of early, marine-dwelling arthropods can be considered a preadaptation for life on land because it
   a. can support an animal's weight against the pull of gravity
   b. allows a wide diversity of body types
   c. resists drying
   d. absorbs light
   e. both a and c

6. The evolution of the shelled, amniotic (membrane-enclosed) egg was an important event in vertebrate evolution because it
   a. led to the Cambrian explosion
   b. was the first example of parents caring for their young
   c. allowed the colonization of freshwater environments
   d. freed organisms from having to lay their eggs in water
   e. allowed internal fertilization of eggs

Review Questions

1. What is the evidence that life might have originated from nonliving matter on the primordial Earth? What kind of evidence would you like to see before you would accept this hypothesis?

2. If they were so much more efficient at producing energy, why didn't the first cells with aerobic metabolism extinguish cells with only anaerobic metabolism?

3. Explain the endosymbiotic hypothesis for the origin of chloroplasts and mitochondria.

4. Name two advantages of multicellularity in plants and animals.

5. What advantages and disadvantages would terrestrial existence have had for the first plants to invade the land? For the first land animals?

6. Review the material on the evolution of flowering plants. How did flowering plants entice animals to carry pollen? (That is, what was in it for the animals?) Why do conifers "still prevail" in cold, northern regions?

7. Outline the general trends in the evolution of vertebrates, from fish to amphibians to reptiles to birds and mammals. Explain how these adaptations increased the fitness of the various groups for life on land.

8. Outline the evolution of humans from early primates. Include in your discussion such features as binocular vision, grasping hands, bipedal locomotion, social living, tool making, and brain expansion.

APPLYING THE CONCEPTS

1. What is cultural evolution? Is cultural evolution more or less rapid than biological evolution? Why?

2. Do you think that studying our ancestors can shed light on the behavior of modern humans? Why or why not?

3. A biologist would probably answer the age-old question "What is life?" by saying "the ability to self-replicate." Do you agree with this definition? If so, why? If not, how would you define life in biological terms?

4. Traditional definitions of humans have emphasized "the uniqueness of humans" because of language or tool use. But most animals can communicate with other individuals in sophisticated ways, and many vertebrates use tools to accomplish tasks. Pretend you are a biologist from Mars, and write a taxonomic description of the species *Homo sapiens*.

5. The "out of Africa" and "multiregional" hypotheses of *Homo sapiens* evolution reflect contrasting assumptions about the extent and nature of genetic divergence among human races. One assumes that races are old and highly diverged genetically; the other assumes that races are young and little diverged genetically. What data would help you determine which hypothesis is closer to the truth?

6. In biological terms, what do you think was the most significant event in the history of life? Explain your answer.

�֎ F O R   M O R E   I N F O R M A T I O N

Calvin, W. H. "The Emergence of Intelligence." *Scientific American*, October 1994. The evolution of several hallmarks of human intelligence may have been linked to the ability to co-ordinate rapid movements.

Diamond, J. "How to Speak Neanderthal." *Discover*, January 1990. Could the Neanderthals speak? How could we tell? Jared Diamond lucidly explains the arguments and methodology behind a scientific dispute.

Fenchel, T., and Finlay, B. J. "The Evolution of Life without Oxygen." *American Scientist*, January/February 1994. Clues to the origin of the first eukaryotic cells are provided by symbiotic relationships of organisms in oxygen-free environments.

Hay, R. L., and Leakey, M. D. "The Fossil Footprints of Laetoli." *Scientific American*, February 1982. The actual footprints of a hominid family were discovered by Hay and Leakey in volcanic ash 3.5 million years old.

Horgan, J. "In the Beginning . . . " *Scientific American*, February 1991. An exploration of the controversies surrounding research into the origin of life.

Morell, V. "Announcing the Birth of a Heresy." *Discover*, March 1987. Paleontologists Robert Bakker and Jack Horner speculate that dinosaurs were not the plodding beasts of monster flicks, but warm-blooded, advanced animals that may even have cared for their young.

Orgel, L. E. "The Origin of Life on Earth." *Scientific American*, October 1994. The evolution of self-reproducing molecules, of which RNA may have been the first, made life possible.

Waters, T. "Almost Human." *Discover*, May 1990. Artist John Gurche reconstructs fossil hominids by adding clay "muscle" and "skin" to replicas of hominid skulls.

 ## N E T   W A T C H

On-line resources for this chapter are on the World Wide Web at:
http://www.prenhall.com/~audesirk (click on the <u>table of contents</u> link and then select Chapter 19).

20 Taxonomy: Imposing Order on Diversity

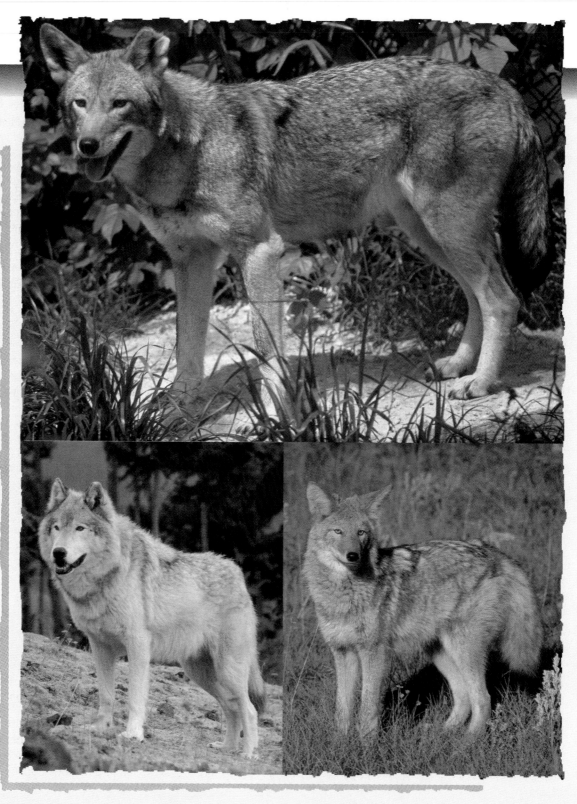

> "Systems of classification are not hat racks, objectively presented to us by nature. They are dynamic theories developed by us to express particular views about the history of organisms. Evolution has provided a set of unique species ordered by differing degrees of genealogical relationship. Taxonomy, the search for this natural order, is the fundamental science of history."
>
> STEPHEN JAY GOULD, *Natural History* (1987)

Molecular biology provides new insights into the origin of species. Mitochondrial DNA sequencing has recently shown that the red wolf (top), currently listed as an endangered species, is almost certainly a hybrid between the gray wolf (bottom left) and the coyote (bottom right) and is probably not a separate species at all.

Taxonomic Categories

Taxonomy (from the Greek word *taxis*, meaning "arrangement") is the science by which organisms are classified and placed into categories based on their structural similarities and evolutionary relationships. Taxonomic categories form a hierarchy—that is, a series of levels each more inclusive than the last. There are seven major categories: **species, genus, family, order, class, division** or **phylum** (divisions are used for plants and plant-like microorganisms; phyla for animals and animal-like microorganisms), and **kingdom**. Each category from species to kingdom is increasingly more general and includes organisms whose common ancestor was increasingly remote in its evolutionary relationship. Some examples of classifications of specific organisms are given in Table 20-1.

The **scientific name** of an organism is actually formed from the two smallest of these categories: the genus and the species. For example, the scientific name for humans, *Homo sapiens*, places us in the genus *Homo* ("man") and the species *sapiens* ("thinking") within the genus *Homo*. The genus is a category that includes very closely related organisms that do not usually interbreed. The species is a category limited to organisms that will interbreed under natural conditions. Thus, the genus *Sialia* (bluebirds) includes the Eastern Bluebird (*Sialia sialis*), the Western Bluebird (*Sialia mexicana*), and the Mountain Bluebird (*Sialia currucoides*), very similar birds that do not normally interbreed (Fig. 20-1).

Scientific names are always underlined or italicized. The first letter of the genus name is always capitalized, and the first letter of the species name is always lower-cased. These names are recognized by biologists worldwide, overcoming language barriers and allowing precise communication.

The Origins of Taxonomy

Aristotle (384–322 B.C.) was among the first to attempt to formulate a logical, standardized language for naming living things. Using characteristics such as structural complexity, behavior, and degree of development at birth, he

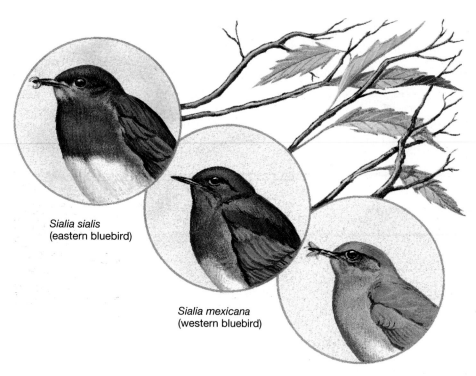

Figure 20-1   Three species of bluebird

The similarities among these species of bluebirds (genus *Sialia*) are obvious, but the species remain distinct because they do not interbreed.

Sialia sialis
(eastern bluebird)

Sialia mexicana
(western bluebird)

Sialia currucoides
(mountain bluebird)

classified about 500 organisms into 11 categories. Aristotle placed organisms into a hierarchy of categories each more inclusive than the one before it, a concept that is still used today.

Building on this foundation over 2000 years later, the Swedish naturalist Carolus Linnaeus (1707–1778) laid the groundwork for the modern classification system. He placed each organism into a series of hierarchically arranged categories based on its resemblance to other life forms, and he also introduced the scientific name based on genus and species. Nearly 100 years later, Charles Darwin (1809–1882) published *On the Origin of Species*, which added a new significance to these categories. Taxonomists then began to recognize that taxonomic categories reflect the evolution-

ary relatedness of organisms. The more categories two organisms share, the closer their evolutionary relationship.

Modern Criteria for Classification

As mentioned earlier, organisms are considered to belong to the same species if they will interbreed under natural conditions. However, this criterion can be applied only to organisms that reproduce sexually. Taxonomists must use other criteria to classify organisms that rarely engage in sexual reproduction (many bacteria and

	Human Being	Chimpanzee	Wolf	Fruit Fly	Sequoia Tree	Sunflower
Kingdom	**Animalia**	**Animalia**	**Animalia**	**Animalia**	**Plantae**	**Plantae**
Phylum	**Chordata**	**Chordata**	**Chordata**	Arthropoda	Coniferophyta	Anthophyta
Class	**Mammalia**	**Mammalia**	**Mammalia**	Insecta	Coniferosida	Dicotyledoneae
Order	**Primates**	**Primates**	Carnivora	Diptera	Coniferales	Asterales
Family	Hominidae	Pongidae	Canidae	Drosophilidae	Taxodiaceae	Asteraceae
Genus	*Homo*	*Pan*	*Canis*	*Drosophila*	*Sequoiadendron*	*Helianthus*
Species	*sapiens*	*troglodytes*	*lupus*	*melanogaster*	*giganteum*	*annuus*

TABLE 20-1 ❊ *Classification Reflects the Degree of Relatedness of Organisms**

*Boldface categories are those that are shared by more than one of the organisms classified. Genus and species are always italicized or underlined.

(a)   (b)

Figure 20-2   **Convergent evolution molds similar body forms**

External similarities between organisms in completely different classes can be the result of convergent evolution (see Chapter 16) rather than evidence for a close evolutionary relationship. **(a)** Although dolphins (like humans) are mammals, the fish, such as this shark **(b)**, that share their habitat and many of their external features are in a different taxonomic class.

some fungi and plants) and to classify all types of organisms above the species level. Some of the most important criteria are anatomy, developmental stages, and biochemical similarities.

Historically, the most important and useful of these has been anatomy. In addition to obvious similarities in external body structure (see Fig. 20-1), taxonomists look carefully at details such as skeletons and tooth structure. For example, the presence of homologous structures, such as the finger bones of dolphins, bats, seals, and humans (Fig. 20-2; see also Fig. 16-12), provides evidence of an evolutionarily distant common ancestor. To distinguish between closely related species, taxonomists may use microscopes to discern finer details—the number and shape of the "teeth" on the tonguelike radula of a mollusk, the spines on a marine worm, or the fine structure of pollen grains of a flowering plant (Fig. 20-3).

Developmental stages also provide clues to common ancestry. For example, although the "sea squirt" spends its adult life permanently attached to rocks on the ocean floor, its free-swimming larva has a nerve cord, tail, and gills, placing it in the same phylum as the vertebrates: phylum Chordata.

Modern taxonomists are also aided by sophisticated biochemical techniques. One such technique is **electrophoresis.** During this procedure, molecules (such as proteins and nucleic acids) are separated according to their electrical charge and molecular weight. The presence and relative abundance of specific molecules in various species provides a way to assess their relatedness. Another technique is **DNA hybridization,** de-

scribed in detail in Chapter 16. When heated, the double-stranded DNA separates; as the DNA cools, it reforms. If DNA from two closely related species is combined and heated, hybrids of the two DNA types form upon cooling. Only complementary sequences can hybridize. Thus, the degree of hybridization is proportional to the degree of similarity between the DNA molecules from the two species. This similarity, in turn, reflects how closely the two species are related. Modern biochemical technology also allows scientists to determine the exact sequence of nucleotides in DNA and RNA and the amino acid sequences of proteins. The greater the similarity in the sequence of the subunits in genetic material or protein between two organisms, the closer their evolutionary relationship. The **polymerase chain reaction** (which can amplify minuscule quantities of DNA, allowing it to be analyzed) permits scientists to sequence DNA from the remains of organisms that are now extinct and compare it with that of their living relatives.

The Five Kingdoms of Life

Before 1970, taxonomists classified all forms of life into two kingdoms: Animalia and Plantae. Bacteria, fungi, and photosynthetic protists were considered plants, and the protozoa were classified as animals. In 1969, Robert H. Whittaker proposed the five-kingdom classification scheme that is widely used today, and which we follow in this text.

(a)   (b)   (c)

Figure 20-3   **Microscopic structures help taxonomists classify organisms**

(a) The "teeth" on the tonguelike radula of a snail or (b) the bristles on a marine worm are potential taxonomic criteria. (c) The shape and surface features of pollen grains help identify plants, particularly from fossil deposits and sediments.

Whittaker identified two kingdoms of primarily unicellular microorganisms based on whether they showed prokaryotic or eukaryotic cellular organization. The kingdom Monera consists of generally single prokaryotic cells, whereas the kingdom Protista consists of generally single eukaryotic cells. All of the organisms in the remaining three kingdoms (Plantae, Fungi, and Animalia) are eukaryotic, and most of them are multicellular. They may be classified further on the basis of their way of acquiring nutrients. Members of the kingdom Plantae photosynthesize, and members of the kingdom Fungi secrete enzymes outside their bodies and then absorb the externally digested nutrients. In contrast, members of the kingdom Animalia ingest their food and then digest it, either within an internal cav-

ity or within individual cells. Comparisons among the five kingdoms are presented in Table 20-2.

Recently, a sixth kingdom has been proposed. Many microbiologists argue that the cell structure of the Archaebacteria is so different from that of prokaryotic and eukaryotic cells that they should occupy their own kingdom. For the sake of simplicity, however, this text retains the five-kingdom classification system. The taxonomic categories described in the following chapters are listed in Table 20-3.

During evolution, the prokaryotic Monera gave rise to the eukaryotic Protista, which show all possible modes of nutrition. Different groups of protists then evolved into the multicellular plants, animals, and fungi, as illustrated in Figure 20-4.

TABLE 20-2 ⁞⁞ *Some Characteristics of the Five Kingdoms*

Kingdom	Cell Type	Cell Number	Major Mode of Nutrition	Motility (Movement)	Cell Wall	Reproduction
Monera	Prokaryotic	Unicellular	Absorb or photosynthesize	Both motile and nonmotile	Present: peptidoglycan	Usually asexual, rarely sexual
Protista	Eukaryotic	Unicellular	Absorb, ingest, or photosynthesize	Both motile and nonmotile	Present in algal forms: varies	Both sexual and asexual
Animalia	Eukaryotic	Multicellular	Ingest	All motile at some stage	Absent	Both sexual and asexual
Fungi	Eukaryotic	Most multicellular	Absorb	Generally nonmotile	Present: chitin	Both sexual and asexual
Plantae	Eukaryotic	Multicellular	Photosynthesize	Generally nonmotile	Present: cellulose	Both sexual and asexual

TABLE 20-3 ▪ *The Classification of the Major Groups of Organisms**

Kingdom	Division/Phylum	Common Name
Monera (unicellular, prokaryotic)	Division Eubacteria	"true" bacteria
	Division Archaebacteria	"ancient" bacteria
Protista (unicellular, eukaryotic)	Division Pyrrophyta	dinoflagellates
	Division Chrysophyta	diatoms
	Division Euglenophyta	euglenoids
	Division Myxomycota	plasmodial slime molds
	Division Acrasiomycota	cellular slime modes
	Phylum Sarcomastigophora	zooflagellates, amoebae
	Phylum Apicomplexa	sporozoans
	Phylum Ciliophora	ciliates
Animalia (multicellular, eukaryotic, heterotrophic, ingest nutrients)	Phylum Porifera	sponges
	Phylum Cnidaria	hydra, anemone, jellyfish, corals
	Phylum Platyhelminthes	flatworms
	Phylum Nematoda	roundworms
	Phylum Annelida	segmented worms
	Class Oligochaeta	earthworms
	Class Polychaeta	tube worms
	Class Hirudinea	leeches
	Phylum Arthropoda	"jointed legs"
	Class Insecta	insects
	Class Arachnida	spiders, ticks
	Class Crustacea	crabs, lobsters
	Phylum Mollusca	"soft-bodied"
	Class Gastropoda	snails
	Class Pelecypoda	mussels, clams
	Class Cephalopoda	squid, octopus
	Phylum Echinodermata	sea stars, sea urchins
	Phylum Chordata	chordates
	Subphylum Vertebrata	vertebrates
	Class Agnatha	lampreys, hagfish
	Class Chondrichthyes	sharks, rays
	Class Osteichthyes	bony fish
	Class Amphibia	frogs, salamanders
	Class Reptilia	turtles, snakes, lizards
	Class Aves	birds
	Class Mammalia	mammals
Fungi (multicellular, eukaryotic, heterotrophic, absorb nutrients)	Division Zygomycota	"zygote fungi"
	Division Ascomycota	"sac fungi"
	Division Deuteromycota	"imperfect fungi"
	Division Basidiomycota	"club fingi"
	Division Oomycota	"egg fungi"
Plantae (multicellular, eukaryotic, photosynthetic)	Division Rhodophyta	red algae
	Division Phaeophyta	brown algae
	Division Chlorophyta	green algae
	Division Bryophyta	mosses
	Division Pteridophyta	ferns
	Division Coniferophyta	evergreens
	Division Anthophyta	flowering plants

*There is no agreement on taxonomic classification systems. This table lists only those taxonomic categories described in this text.

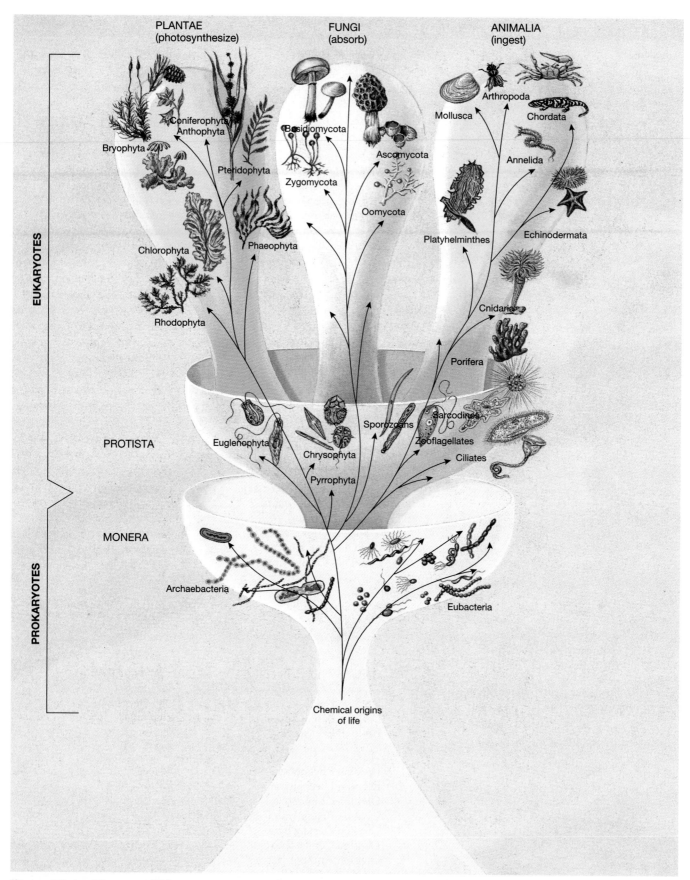

Figure 20-4   **The tree of life**

The five kingdoms and some of the major phyla or divisions within them are illustrated.

Taxonomy: An Inexact Science

As the Archaebacteria debate suggests, taxonomic categories are not the unchanging frameworks that we might wish them to be, even at the kingdom level (see "Evolutionary Connections: Classification Conundrums; or, Where Shall We Put the Algae?"). For example, in 1973 ornithologists officially declared the Baltimore Oriole and the Bullock's Oriole to be a single species, which they named the Northern Oriole. The reason: Where their ranges overlap, the two "species" of birds were interbreeding.

Another example is the red wolf from the southern United States, which is currently listed as an endangered species (see the photograph on the opening page of this chapter). Researchers recently analyzed the DNA from mitochondria taken from red wolves and found it to be identical to mitochondrial DNA from coyotes in some cases and to mitochondrial DNA from gray wolves in others. Mitochondrial DNA is inherited directly from the mother via the mitochondria present in the original egg cell. This mitochondrial DNA evidence, then, strongly suggests that red wolves are actually hybrids between gray wolves and coyotes and may not be a distinct species at all.

Asexually reproducing organisms pose a particular challenge to taxonomists, because the criterion of interbreeding cannot be used to distinguish species. For example, some taxonomists recognize 200 species of the parthenogenetic (a plant that can produce seeds without fertilization) British blackberry; others recognize only 20 species. Most unicellular organisms reproduce asexually most of the time, making their classification difficult. For example, molecular biologists recently sequenced the DNA of two different strains of the bacterium *Legionella pneumophila*, which is responsible for Legionnaire's disease, and found only a 50% correspondence between the strains. This degree of genetic dissimilarity, which occurs within a single species of bacterium, is as great as the genetic difference between mammals and fishes! The taxonomy of bacteria is especially controversial.

Taxonomic categories are continuously debated and revised as taxonomists learn more about evolutionary relationships, particularly with the application of techniques derived from molecular biology. Although the precise evolutionary relationships of many organisms continue to elude us, taxonomy is enormously helpful in ordering our thoughts and investigations into the diversity of life on Earth.

Exploring Biodiversity: How Many Species Exist?

Scientists do not know even within an order of magnitude how many species share our world. Each year, between 7000 and 10,000 new species are named, most of them insects, many from the tropical rain forests. The total number of named species is currently around 1.4 million. However, many scientists believe that 7 to 10 million species may exist, and estimates range as high as 30 million. Of all the species that have been identified, about 5% are monerans and protists. An additional 22% are plants and fungi, and the rest are animals. This distribution has little to do with the actual abundance of these organisms and a lot to do with the size of the organisms, how easy they are to classify, how accessible they are, and the number of scientists studying them. An entire new "continent" of species lies largely unexplored on the deep-sea floor at depths of 1000 meters (3250 feet) and more. From the limited samples available, scientists estimate that hundreds of thousands of unknown species may reside there.

Although about 4000 species of bacteria have been named, bacterial diversity also remains largely unexplored. Consider a recent study by Norwegian scientists who used DNA hybridization to identify the different bacteria in a small sample of forest soil. To distinguish species, they arbitrarily defined bacterial DNA as coming from separate species if it differed by at least 30% from that of any other bacterial DNA. Using this criterion, they reported over 4000 types of bacteria in their soil sample and an equal number of new forms in a sample of shallow marine sediment!

Our ignorance of the full extent of life's diversity adds a new dimension to the tragedy of the destruction of the tropical rain forests, discussed in Chapter 31. Although these forests cover only about 6% of Earth's land area, they are believed to be home to two-thirds of the world's existing species, most of which have never been studied or named. Because the forests are being destroyed so rapidly, Earth is losing many species that people will never even know existed! For example, in 1990 a new species of primate, the black-faced lion tamarin, was discovered in a small patch of dense rain forest on an island just off the east coast of Brazil (Fig. 20-5). Only 12 individuals of this squirrel-sized monkey have ever been spotted, and captive breeding may be its only hope for survival. At current rates of destruction, most of the tropical rain forests, with their undescribed wealth of life, will be gone within the next century.

 E V O L U T I O N A R Y C O N N E C T I O N S

Classification Conundrums; or, Where Shall We Put the Algae?

Classification schemes are used by biologists to place the huge number of organisms on Earth into natural groupings. Ideally these groupings are based on evolutionary relationships, which are inferred by taxonomists on the basis of the presence of shared distinguishing features. Early classification schemes placed all organisms into either the plant or the animal kingdom. Later, close examination of the unique structure of fungi and the diversity of single-

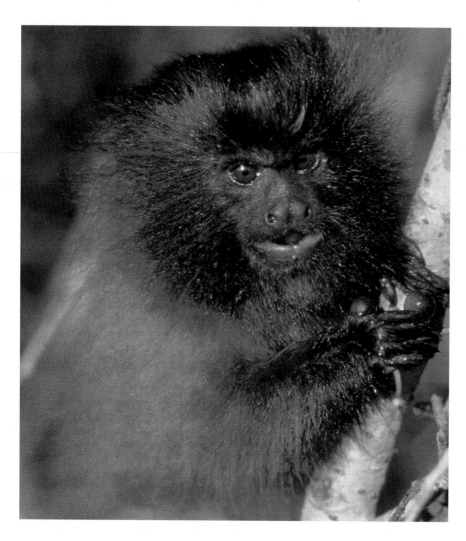

Figure 20-5   **The black-faced lion tamarin**

This rare new species of monkey was discovered in 1990 in the vanishing Brazilian rain forest.

celled organisms made it necessary to propose additional kingdoms that recognized the fundamental differences among plants, animals, fungi, and unicellular prokaryotes and eukaryotes. The current five-kingdom scheme was devised in response to this need.

The kingdom Protista, defined as comprising all single-celled eukaryotic organisms, is not a natural grouping, and scientists disagree about which organisms it should include. Plants, animals, and fungi all have close protistan relatives, and the separation of single-celled organisms from multicellular organisms is sometimes problematic. It is especially so for the algae, which have both single-celled and multicellular representatives within most smaller taxonomic groupings. Can closely related organisms be placed into separate kingdoms, Protista and Plantae, simply on the basis of multicellularity? If you look at different textbooks, you will see that the algae, photosynthetic organisms with simple reproduction, are sometimes placed entirely into Protista, and sometimes they are split between Protista and Plantae depending on whether they are single-celled or multicellular. Some taxonomists split the multicellular algae between two kingdoms, placing the multicellular brown and red algae with the protists and the multicellular green algae into the plant kingdom. These different attempts to classify closely related organisms are good ex-

amples of how difficult it is to develop standard criteria for grouping organisms, even at the kingdom level.

One approach to this problem, enthusiastically endorsed by Dr. Lynn Margulis, a biologist at the University of Massachusetts, is the creation of the kingdom Protoctista. This taxonomic category would include single-celled organisms and their close descendants (for example, the multicellular algae, but *not* the animals, fungi, and plants). Margulis describes the kingdom Protoctista as "the entire motley and unruly group of non-plant, non-animal, non-fungal organisms representative of lineages of the earliest descendants of the eukaryotes."

It is conceptually difficult to group one of the largest multicellular organisms in the world, the brown algae called giant kelps, with simple microscopic single-celled organisms. Kelps, some of which are up to 60 meters (almost 200 feet) long, possess a tissuelike level of organization that is relatively complex and can transport materials over long distances, as can the tissues of higher plants. The cells in kelps and some other algae are specialized and show division of labor. However, kelps reproduce like other algae and differently from plants. Dr. Thomas Cavalier-Smith of the University of British Columbia has proposed that brown algae merit their own kingdom (kingdom Chromista) based on ultrastructural features and molecular comparisons of all algae. So, even among the algae, there are clear differences that some scientists believe are sufficient to justify the status of a separate kingdom.

As we learn more about the relationships between organisms and refine the criteria used to classify them, classification schemes will change. As the superficially simple question "In which kingdom should we place the algae?" illustrates, the taxonomic categories in textbooks are tentative and subject to revision as we continue to discover more about life on Earth.

✖ SUMMARY OF KEY CONCEPTS

Taxonomic Categories
Taxonomy is the science by which organisms are classified and placed into hierarchical categories that reflect their structural similarities and evolutionary relationships. The seven major categories, in order of increasing inclusiveness, are species, genus, family, order, class, division or phylum, and kingdom. The scientific name of an organism is composed of its genus name and species name.

The Origins of Taxonomy
A hierarchical concept was first used by Aristotle, but Linnaeus in the mid-1700s laid the foundation for modern taxonomy. In the 1860s, evolutionary theory proposed by Darwin provided an explanation for the observed similarities and differences between organisms, and modern taxonomists attempt to classify organisms according to their evolutionary relationships.

Modern Criteria for Classification
Today, taxonomists use features such as anatomy, developmental stages, and biochemical similarities to categorize organisms. Molecular biological techniques are used to determine the sequences of nucleotides in DNA and RNA and of amino acids in proteins. Biochemical similarities among organisms are a measure of evolutionary relatedness.

The Five Kingdoms of Life
A five-kingdom classification system consisting of the kingdoms Monera, Protista, Animalia, Fungi, and Plantae is widely used (see Table 20-2 for characteristics).

Taxonomy: An Inexact Science
Taxonomic categories are controversial and subject to revision, particularly in the case of asexually reproducing species. However, taxonomy is essential for precise communication and contributes to our understanding of the origins and diversity of species.

Exploring Biodiversity: How Many Species Exist?
Although only about 1.4 million species have been named, estimates of the total number of species range up to 30 million. New species are being identified at the rate of 7000 to 10,000 each year.

✖ KEY TERMS

class   p. 371
division   p. 371
DNA hybridization   p. 373
electrophoresis   p. 373
family   p. 371

genus   p. 371
kingdom   p. 371
order   p. 371
phylum   p. 371
polymerase chain reaction   p. 373

scientific name   p. 371
species   p. 371
taxonomy   p. 371

✖ THINKING THROUGH THE CONCEPTS

Multiple Choice
1. It is possible to imagine the various levels of taxonomic classification as a kind of "family tree" for an organism. If the kingdom is analogous to the trunk of the tree, what taxonomic category would be analogous to the large limbs coming off that trunk?
   a. class   b. family
   c. order   d. phylum or division
   e. subfamily
2. The more taxonomic categories shared by two organisms, the more closely they are related in an evolutionary sense. Which biologist's work led to this insight?
   a. Aristotle   b. Darwin
   c. Linnaeus   d. Whittaker
   e. Pauling

3. Which of the following criteria could not be used to determine how closely related two different organisms are?
   a. similarities in the presence and relative abundance of specific molecules
   b. DNA hybridization
   c. the presence of homologous structures
   d. developmental stages
   e. occurrence of both organisms in the same habitat
4. Which of the following habitats appears to have the greatest number of species?
   a. the seafloor   b. deserts
   c. tropical rain forests   d. grasslands
   e. mountaintops

5. What organisms are considered plants by some taxonomists and protists by others?

a. algae
b. fungi
c. mosses
d. archaebacteria
e. diatoms

6. An organism is described to you as having many nuclei-containing cells, each surrounded by a cell wall of chitin and absorbing its food. In what kingdom would you place it?

a. Plantae
b. Protista
c. Animalia
d. Fungi
e. Monera

Review Questions

1. What contributions did Aristotle, Linnaeus, and Darwin each make to modern taxonomy?
2. What features would you study to determine whether a dolphin is more closely related to a fish or to a bear?
3. What techniques might you use to determine whether the extinct cave bear is more closely related to a grizzly bear or to a black bear?
4. Only a small fraction of the total number of species on Earth has been scientifically described. Why?
5. In England, "daddy long-legs" refers to a long-legged fly, but the same name refers to a spiderlike animal in the United States. How do scientists attempt to avoid such confusion?

✖ A P P L Y I N G   T H E   C O N C E P T S

1. There are many areas of disagreement among taxonomists, for example, whether algae belong in the kingdom Protista or the kingdom Plantae, or whether archaebacteria should be in the kingdom Monera or in an entirely new kingdom. What difference does it make whether biologists consider algae as plants or protists, or archaebacteria as monerans or something else? As Shakespeare put it, "What's in a name?"

2. The pressures created by human population growth and economic expansion place storehouses of biological diversity such as the tropics in peril. The seriousness of the situation is clear when we consider that probably only 1 out of every 20 tropical species is known to science. What sorts of arguments can you make for preserving biological diversity in poor and developing countries? Does such preservation require that these countries sacrifice economic development? Suggest some solutions to the conflict between the demand for resources and the importance of conserving diversity.

3. During major floods only the topmost branches of submerged trees may be visible above the water. If you were asked to sketch the branches below the surface of the water, based solely on the positions of the exposed tips, you would be attempting a reconstruction somewhat similar to the "family tree" by which taxonomists link various organisms according to their common ancestors (analogous to branching points). What sources of error do both exercises share? What advantages do modern taxonomists have?

4. The Florida panther, found only in the Florida Everglades, is currently classified as an endangered species, protecting it from human activities that could lead to its extinction. It has long been considered a subspecies of cougar (mountain lion), but recent mitochondrial DNA studies have shown that the Florida panther may actually be a hybrid between American and South American cougars. Should the Florida panther be protected by the Endangered Species Act?

✖ F O R   M O R E   I N F O R M A T I O N

Avise, J. C. "Nature's Family Archives." *Natural History*, March 1989. Shows how evolutionary relationships can be determined by analyzing differences in DNA contained in mitochondria.

Diamond, J. "The Search for Life on Earth." *Natural History*. A brief overview of the modern practice of taxonomy.

Gould, S. J. "What Is a Species?" *Discover*, December 1992. Discusses the difficulties of distinguishing separate species.

Lowenstein, J. M. "Molecular Approaches to the Identification of Species." *American Scientist*, 73: 541–547, 1985. A basic introduction to immunological methods of taxonomy.

Margulis, L., and Schwartz, K. *Five Kingdoms*. 2nd ed. New York: W. H. Freeman, 1988. An illustrated paperback guide to the diversity of life.

May, R. M. "How Many Species Inhabit the Earth?" *Scientific American*, October 1992. Although no one knows the answer to this question, an accurate count is crucial in an effort to manage the biological resources of this planet.

Moffett, M. W. *The High Frontier: Exploring the Tropical Rainforest Canopy*. Cambridge, MA: Harvard University Press, 1994. The tremendous diversity of life in the rainforest treetops is only now becoming known to us. This book documents the unexpected and spectacular diversity of animals in the upper reaches of this endangered habitat.

Wilson, A. C. "The Molecular Basis of Evolution." *Scientific American*, 1985. On using nucleotide and amino acid sequences to determine evolutionary relationships.

Wilson, E. O. *The Diversity of Life*. Cambridge, Mass. Harvard University Press, 1992. An outline of the processes that created the diversity of life; and a discussion of the threats to that diversity and the steps required to preserve it.

N E T   W A T C H

On-line resources for this chapter are on the World Wide Web at:
http://www.prenhall.com/~audesirk (click on the table of contents link and then select Chapter 20).

Unit IV
Human Anatomy and Physiology

Human head and torso imaged using a thermogram, a device that maps heat distribution. Blue areas are cooler, yellow areas are warmest.

21 Homeostasis and the Organization of the Animal Body

"As a physiologist I am especially impressed by the common ignorance of bodily organs and their functions. It seems to me now, as it seemed to Robert Boyle nearly 300 years ago, that it is 'highly dishonorable for a Reasonable Soul to live in so Divinely built a Mansion as the Body [he or] she resides in, altogether unacquainted with the exquisite structure of it.'"

W. B. CANNON in *The Way of An Investigator* (1945)

Through a variety of adaptations, animals are able to maintain relatively constant internal environments that differ substantially from their surroundings. This harp seal keeps warm on a Canadian ice floe with dense, waterproof fur and a thick layer of fat.

At the dawn of life, over 3.5 billion years ago, complex organic molecules arose within warm, primordial waters. But before assemblages of molecules could be considered alive, it was necessary for a boundary to form between the evolving cell and its environment. This enclosing membrane allowed the earliest life forms to maintain an internal environment distinct from their surroundings. As organisms evolved increasingly sophisticated mechanisms to regulate their internal environment, they were able to emerge from the nurturing waters. Today we find organisms thriving in extremely inhospitable regions of Earth because they are able to maintain internal environments that protect and promote the delicate processes of life. This ability is called homeostasis.

Homeostasis

In the mid-nineteenth century, the French physiologist Claude Bernard first described the "constancy of the interior milieu," the concept now known as homeostasis. The term **homeostasis** was coined in 1932 by Walter Cannon, who suggested that nearly all physiological processes operate in a manner that tends to maintain the internal constancy required for the maintenance of life. Although the term homeo*stasis* might imply a "static," or unchanging, environment, in fact, the "internal milieu" seethes with activity as the body continuously responds and adjusts to internal and external changes. The term *dynamic equilibrium* better describes this internal condition. Physical and chemical changes do occur, but only within the narrow range that cells require to function. The mechanisms by which organisms control the homeostatic conditions are called feedback systems. The continuous activation of negative feedback systems counteracts changes, while the less frequent use of positive feedback systems intensifies change for a specific purpose.

Negative Feedback Tends to Maintain Constant Internal Conditions

The most important principle governing homeostasis is **negative feedback,** in which a change in the internal environment triggers a response that tends to counteract the

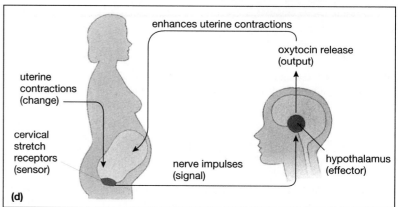

Figure 21-1   Negative and positive feedback

(a) Negative feedback controls the temperature in most homes. When the system changes (the house cools below a specific temperature), the change is sensed by a sensor with a specific set point (the thermostat, set to 68° F). The sensor sends a signal to an effector (the furnace) whose output (heat) returns the system to its set point, turning off the signal, which turns off the effector.
(b) Body temperature is both set by and sensed by neurons in the hypothalamus (the sensor with a set point). A drop in body temperature causes the hypothalamic neurons to send a signal to effectors such as muscles, which begin shivering. Shivering generates heat, which helps restore body temperature to its set point, shutting off the signal, and causing shivering to cease. (c) In positive feedback, a change in the system is sensed by a sensor, which then sends a signal to an effector. The output of the effector causes an even greater change in the system, rather than attempting to return it to previous conditions. (d) At the onset of labor, uterine contractions are sensed by receptor neurons in the cervix, which send a signal to the hypothalamus that results in secretion of a hormone (oxytocin) from the posterior pituitary gland. The hormone, in turn, stimulates more and stronger uterine contractions, ultimately leading to the expulsion of the baby and placenta from the uterus.

change and to restore and maintain the original conditions. This feedback is called "negative" because it opposes or resists the change.

A familiar example is your home thermostat (Fig. 21-1a). A drop in temperature from its setting is detected by a thermometer, which activates a switch that turns on a heating device. When the temperature is restored to the set point, the heater is turned off. Negative feedback requires a *set point* (the thermostat setting), a *sensor* (the thermometer), and an *effector* (the furnace), which accomplishes the change.

A biological example is the maintenance of your body at a temperature within about 1° of 98.6° F (37°C). The set point is established by neurons in the hypothalamus, a region of your brain that controls many homeostatic responses.

Other neurons in the hypothalamus, abdomen, spinal cord, and large veins act as sensors for body temperature and transmit this information to the hypothalamus. When your body temperature drops, the hypothalamus activates various effector mechanisms including shivering (which generates heat through muscular activity), constriction of blood supply to the skin (which reduces heat loss), and elevation of metabolic rate (which generates heat). When normal body temperature is restored, the hypothalamus switches off these temperature control mechanisms (Fig. 21-1b).

As you read the following chapters, you will find many examples of homeostatic control using negative feedback. In later chapters, you will encounter the mammalian hormones erythropoietin, ADH (antidiuretic hormone), and insulin, which regulate blood oxygen content, water balance, and blood sugar levels, respectively, all through negative feedback.

Positive Feedback Events Are Self-Limiting

A **positive feedback** system, in contrast to a negative feedback system, is one in which a change in the internal environment initiates a response that intensifies the original change (Fig. 21-1c). Instead of immediately returning conditions to a set point, this type of system encourages change under controlled circumstances to achieve a specific end. Positive feedback, as you can imagine, tends to create explosive events that must be carefully restricted. An example of positive feedback is nuclear fission, in which each of many particles that are split from one atom triggers the splitting of another atom, and so on. Controlled, the reaction supplies nuclear power. When deliberately set out of control, it produces the atomic bomb. Population growth is a familiar biological example of positive feedback, described in Chapter 28. The ecologist Paul Ehrlich coined the apt expression "population bomb" to describe unchecked population growth.

Consequently, positive feedback events in animal physiology are self-limiting and relatively infrequent. Positive feedback occurs, for example, during childbirth (see Fig. 21-1d). The early contractions of labor begin to force the baby's head against the cervix, located at the base of the uterus; this pressure causes the cervix to dilate (open). Stretch-receptive neurons in the cervix respond to this expansion by signaling the hypothalamus, which responds by triggering the release of a hormone (oxytocin) that stimulates more and stronger uterine contractions. Stronger contractions create further pressure on the cervix, which in turn prompts the release of more hormones. The feedback cycle is finally terminated by the expulsion of the baby and its placenta.

The Organization of the Animal Body

The animal body is made up of **organ systems** (for example, the digestive system and the excretory system), each made up of **organs** (the stomach and intestine of the digestive system; the kidney and urinary bladder of the excretory system). Each organ, in turn, consists of two or more of the four primary types of **tissue** (epithelial, connective, nerve, and muscle tissue). Each tissue type is composed of dozens to billions of structurally similar cells that act in concert to perform a similar function (Fig. 21-2).

Animal Tissues Are Composed of Similar Cells That Perform a Specific Function

A tissue is composed of cells that are similar in structure and designed to perform a specialized function. Tissue may also include extracellular components produced by these cells, as in the case of cartilage and bone. Here, we present a brief overview of the four major categories of

animal tissue: epithelial tissue, connective tissue, muscle tissue, and nerve tissue.

Epithelial Tissue Forms Membranes That Cover the Body and Line Its Cavities

The cells of **epithelial tissues** (also called the epithelium) form continuous sheets called **membranes.** Epithelial membranes cover the body and line all the body cavities. This means that one surface of epithelial membranes is free; that is, it faces either the outside of the body or the inside of cavities within the body. Epithelial membranes create a barrier that either resists the movement of substances across it (such as the skin) or that allows movement of specific substances across it (such as the lining of the small intestine). Epithelial membranes can serve as effective barriers because their cells are packed closely together and are connected to one another by tight junctions (see Chapter 6). No blood vessels penetrate epithelial tissue; it is nourished by diffusion from capillaries within the connective tissue that lies beneath it.

Another important property of epithelial tissues is that they are continuously lost and replaced by cell division. Consider the abuse suffered, for example, by the epithelium that lines your mouth. Scalded by coffee and scraped by corn chips, it would be destroyed within a few days if it were not continuously replacing itself. The stomach lining, abraded by food and attacked by acids and protein-digesting enzymes, is completely replaced every 2 to 3 days. Your skin's outer layer, the epidermis, is renewed about twice a month.

Some Epithelial Tissues Also Form Glands

During development, some epithelial tissues fold inward, and their cells change shape and function and form **glands**, clusters of cells that are specialized to secrete (release) substances. Glands are classified into two broad categories: exocrine glands and endocrine glands. **Exocrine glands** remain connected to the epithelium by a passageway or duct. Examples of exocrine glands are sweat and sebaceous (oil-secreting) glands, both found in the skin and derived from skin epithelium (see Fig. 21-7). Exocrine glands called salivary glands release saliva into the mouth, and still others line the stomach, where they secrete a protective layer of mucus. **Endocrine glands** become separated from the epithelium that produced them. The products of endocrine glands are mostly hormones, which are secreted into the extracellular fluid surrounding the glands and then diffuse into nearby capillaries. Endocrine glands and their hormones are covered in detail in Chapter 25.

Connective Tissues Have Diverse Structures and Functions

There are several diverse connective tissues, including the dermis of the skin, tendons, ligaments, cartilage, bone, fat, and blood. Most **connective tissues** share a common fea-

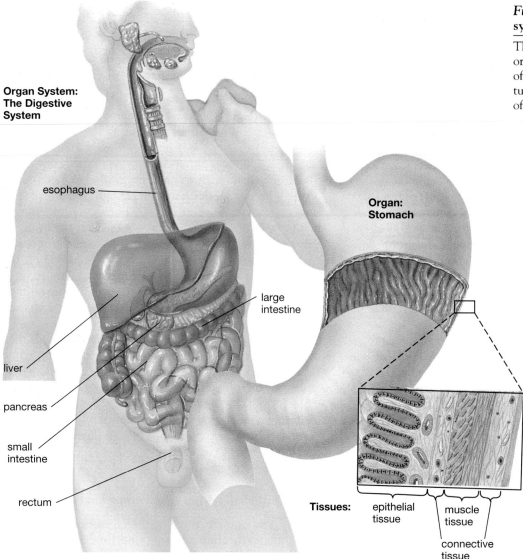

**Organ System:
The Digestive
System**

esophagus

liver

pancreas

small
intestine

rectum

large
intestine

**Organ:
Stomach**

Tissues: epithelial
tissue

muscle
tissue

connective
tissue

Figure 21-2   **Organs, organ
systems, and tissues**

The animal body contains
organ systems, each composed
of several organs, which are in
turn composed of several types
of tissue.

ture: Their cells are surrounded by large quantities of extracellular substances, often secreted by the connective
tissue cells themselves. With the exception of blood, most
connective tissue is interwoven with fibrous strands of an
extracellular protein called **collagen** secreted by the cells.
Connective tissue underlies all epithelial tissue, containing capillaries and fluid-filled spaces that nourish the epithelium. Underlying the epidermis of the skin, for example, is connective tissue, called the **dermis**, that is richly
supplied with capillaries (see Fig. 21-7).

Tendons and **ligaments** are types of connective tissue
that attach muscles to bones and bones to bones, respectively (Fig. 21-3). They contain densely packed collagen
fibers in an orderly parallel arrangement. **Cartilage** is a flex-

Figure 21-3   **The knee joint illustrates several types of
connective tissue**

Tendons connect muscles to bones; ligaments connect bones to
bones; and cartilage provides a smooth surface that allows adjacent bones to slide past one another.

quadriceps
(straightens
leg)

tendon
(to thigh
muscle)

femur

kneecap
(bone)

ligament (to
kneecap)

tibia

biceps
femoris
(bends
leg)

cartilage

ligament

fibula

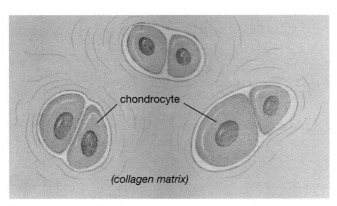

Figure 21-4   **Cartilage**

Cartilage cells (called chondrocytes) are embedded within an extracellular matrix of collagen fibers, which they secrete.

ible and resilient form of connective tissue. It consists of widely spaced cells surrounded by a thick, nonliving matrix of collagen, which they secrete (Fig. 21-4). Cartilage covers the ends of bones at joints, provides the supporting framework for the respiratory passages, supports the ear and nose, and forms shock-absorbing pads between the vertebrae. **Bone** resembles cartilage that has been hardened by deposits of calcium phosphate. During bone formation, bone cells, or **osteocytes,** become embedded in concentric layers of hardened collagen matrix. These layers surround a central canal containing a blood capillary that nourishes the bone cells. These concentric circles of bone, each surrounding a canal, are called **Haversian systems** (Fig. 21-5). The skeletal system, described later in this chapter, contains many types of connective tissue.

Blood and **lymph** are considered connective tissues because they are composed largely of extracellular fluid, the plasma and lymph, respectively. The cellular portion of blood consists of red and white blood cells and cell fragments called platelets. Blood, pumped by the heart and confined to vessels, carries dissoved oxygen and other nutrients to cells and carries away carbon dioxide and other cellular waste products. Blood distributes hormones within the body. White blood cells help destroy bacteria and other foreign invaders. Platelets aid in blood clotting. Lymph consists largely of fluid that has leaked out of blood capillaries and is carried back to the circulatory system within lymph vessels. Lymph contains some white blood cells and is responsible for transporting globules of fat from the small intestine to the blood. Blood and lymph are described in detail in Chapter 22.

Fat cells, collectively called **adipose tissue,** are specially modified to act as storage sacs for triglycerides, molecules used for long-term energy storage (see Chapter 3). Triglycerides may make up 90% of the volume of fat cells.

Muscle Tissue Has the Ability to Contract

Muscle tissue is a specialized tissue whose cells can contract (shorten) when stimulated, then relax passively. Muscle cells derive their elasticity from the arrangement of **actin** (thin) and **myosin** (thick) protein filaments, which can use energy to move relative to one another. There are three types of muscle: skeletal (striated), cardiac, and smooth (Table 21-1).

Skeletal muscle is also called **striated muscle** because its orderly arrangement of thin and thick protein filaments give it a striped appearance. Skeletal muscle is generally under voluntary, or conscious, control. As its name im-

Figure 21-5   **The structure of bone**

(Left) A long bone, such as those found in the arms and legs, showing "spongy bone" that contains bone marrow, where blood cells are formed, and hard compact bone that provides support and strength. A layer of cartilage covers the end of the bone where it contacts adjacent bones, reducing friction between them. (Right) Haversian systems, concentric circles of bone deposited around a central canal containing a blood vessel, are clearly visible in this micrograph. Bone cells, or osteocytes, are trapped within small chambers within the Haversian systems.

TABLE 21-1 ✷ *Location, Characteristics, and Functions of the Three Muscle Types*

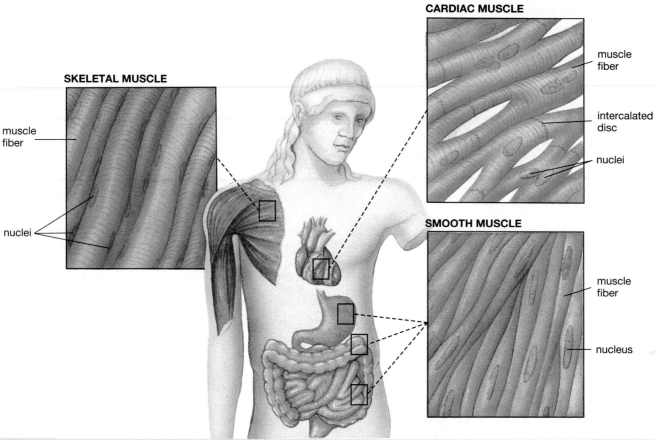

	Type of Muscle		
Property	*Smooth*	*Cardiac*	*Skeletal*
Appearance	Unstriped	Irregular stripes	Regular stripes
Cell shape	Spindle	Branched	Spindle or cylindrical
Number of nuclei	One per cell	Many per cell	Many per cell
Speed of contraction	Slow	Intermediate	Slow to rapid
Contraction caused by	Spontaneous, stretch, nervous system, hormones	Spontaneous	Nervous system
Function	Controls movement of substances through hollow organs	Pumps blood	Moves the skeleton
Voluntary control	Usually no[a]	Usually no[a]	Yes

[a]Smooth and cardiac muscle normally contract without conscious control. In some cases, however, their contractions may be initiated or modified voluntarily (for example, heart rate can be voluntarily slowed after biofeedback training, and bladder contractions are initiated consciously).

plies, its main function is to move the skeleton, as occurs when you walk or turn the pages of this text. **Cardiac muscle,** found only in the heart, also has a striated appearance. Unlike skeletal muscle, it is spontaneously active and not under conscious control. Cardiac muscle cells are interconnected by gap junctions, which allow electrical signals to spread rapidly through the heart. **Smooth muscle** appears more uniform ("smooth") under the microscope because it lacks the orderly arrangement of thick and thin filaments seen in cardiac and skeletal muscle. Smooth mus-

cle is embedded in the walls of the digestive tract, the uterus, the bladder, and the large blood vessels. It produces slow, sustained contractions that are usually involuntary.

Nerve Tissue Is Specialized to Transmit Electrical Signals

Nerve tissue makes up the brain, the spinal cord, and the nerves that travel from them to all parts of the body. Nerve tissue is composed of two types of cells: neurons and glial

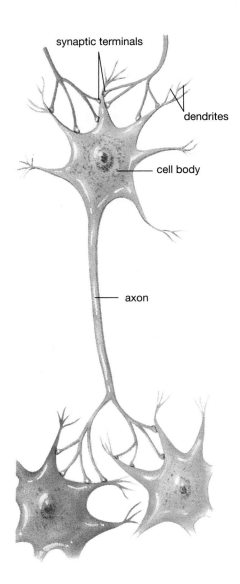

synaptic terminals

dendrites

cell body

axon

Figure 21-6   A nerve cell

A nerve cell, or neuron, has four major parts, each specialized for a specific function.

cells. **Neurons** are specialized to generate electrical signals and to conduct these signals to other neurons, muscles, or glands. A neuron has four major parts, each with a specialized function (Fig. 21-6). The **dendrites** receive signals from other neurons or from the external environment. The **cell body** directs maintenance and repair of the cell. The **axon** conducts the electrical signal to its target cell, and the **synaptic terminals** transmit the signal to the target cell at a specialized contact region called a **synapse.** **Glial cells** surround, support, and protect neurons and regulate the composition of the extracellular fluid, allowing neurons to function optimally. Neurons and the nervous system are covered in detail in Chapter 26.

Organs Include Two or More Interacting Tissue Types

Organs are formed from at least two different tissue types that function together; some organs, such as the skin, include all four of the tissue types described earlier. In this section, we examine the components and functions of a representative animal organ, the skin.

The Skin Illustrates the Properties of Organs

The **epidermis,** or outer layer of the skin, is a specialized epithelial tissue (Fig. 21-7). It is covered by a protective layer of dead cells produced by underlying living epidermal cells. These dead cells are packed with the protein **keratin,** which helps keep the skin both airtight and relatively waterproof.

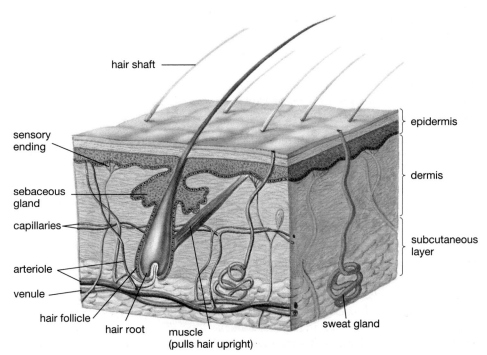

hair shaft

sensory ending

sebaceous gland

capillaries

arteriole

venule

hair follicle

hair root

muscle (pulls hair upright)

sweat gland

epidermis

dermis

subcutaneous layer

Figure 21-7   Skin

Mammalian skin, a representative organ, in cross section.

Immediately beneath the epidermis lies a layer of connective tissue, the dermis. The loosely packed cells of the dermis are permeated by arterioles (small arteries), which feed blood pumped from the heart into a dense meshwork of capillaries that nourish both the dermal and epidermal tissue. Loss of heat through the skin is precisely regulated by neurons (nerve tissue) controlling the degree of dilation (expansion) of the arterioles. When cooling is required, the arterioles dilate and flood the capillary beds with blood, thus releasing excess heat; when heat conservation is required, the arterioles supplying the skin capillaries are constricted. Lymph vessels collect and carry off extracellular fluid within the dermis. A variety of sensory nerve endings responsive to temperature, touch, pressure, vibration, and pain are scattered throughout the dermis and epidermis and provide feedback to the nervous system.

The dermis is also packed with glands derived from epithelial tissue. Glands called **hair follicles** produce hair from protein-containing secretions. Sweat glands produce watery secretions that cool the skin and excrete substances, including salts and urea. **Sebaceous glands** secrete an oily substance (sebum) that lubricates the epithelium.

In addition to the epithelial, connective, and nerve tissues already mentioned, the skin also contains muscle tissue. Tiny muscles attached to the hair follicles can cause the hairs of the skin to "stand on end" in response to signals from motor neurons. Although this reaction is useless for heat retention in humans, most mammals are able to increase the thickness of their insulating fur in cold weather by erecting the individual hairs.

The structure of the skin is, in a general sense, representative of many organs. An outer epithelium is underlain by connective tissue containing a blood supply, a nerve supply, sometimes muscle, and glandular structures derived from the epithelium. If the organ is hollow, such as the bladder or blood vessels, its interior is also lined with epithelium underlain by connective tissue. Different organs have different types and proportions of glandular, muscular, and nervous tissue.

Organ Systems Consist of Two or More Interacting Organs

Organ systems consist of two or more individual organs (sometimes located in different regions of the body) that work together, performing a common function.

The Vertebrate Skeletal System and Digestive System Are Examples of Organ Systems

The organs of the skeletal system are bones, which consist of bone tissue, a blood supply, and sometimes another form of connective tissue called bone marrow, in which blood cells are formed (see Fig. 21-5 and Chapter 22). The ends of bones are often covered with a layer of cartilage tissue, which allows adjacent bones in joints to slide past one another (see Fig. 21-3). Cartilage also forms a flexible frame-

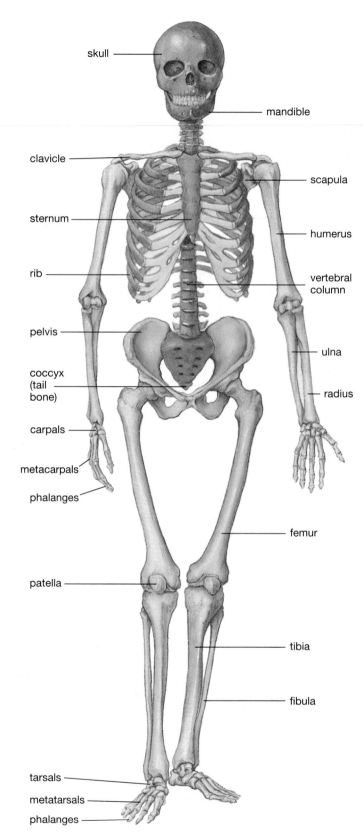

Figure 21-8   **The human skeleton**

The human skeleton, showing the axial skeleton (tinged in blue-gray) and the appendicular skeleton (bone color).

work for the ears and nose and connects the ribs to the sternum. Ligaments connect bones to one another.

The skeletal system serves a wide variety of functions, including the following:

1. The skeleton provides a rigid framework that supports the body and protects the internal organs. The central nervous system, for example, is almost completely enclosed within the skull and vertebral column, while the rib cage protects the lungs and the heart with its major blood vessels.
2. Bones produce red blood cells, white blood cells, and platelets (see Chapter 22). In adults, these cells of the circulatory system are produced by red bone marrow, found in porous areas of bone in the sternum (breastbone), ribs, upper arms and legs, and hips.
3. Bone serves as a storage site for calcium and phosphorus. Bone contains 99% of the calcium and 90% of the phosphorus in the human body. It absorbs and releases these minerals as needed, maintaining a constant concentration in the blood.
4. The skeleton even participates in sensory reception. Three tiny bones of the middle ear (hammer, anvil, and stirrup) transmit sound vibrations between the eardrum and the cochlea (see Chapter 26).

The 206 bones of the human skeleton can be placed in two categories: the axial skeleton and the appendicular skeleton. The **axial skeleton**, whose bones form the axis of the body, includes the bones of the head, vertebral column, and rib cage. The **appendicular skeleton**, whose bones form the appendages and their attachments to the axial skeleton. includes the pectoral (shoulder) and pelvic (hip) girdles and the bones of the arms, legs, hands, and feet (Fig. 21-8).

Another organ system is the digestive system, in which the mouth, esophagus, stomach, intestines, and other organs that supply digestive enzymes, such as the liver and pancreas, all function together to convert food into nutrient molecules and absorb these into the circulatory system (see Fig. 21-2 and Chapter 23). The major organ systems of the vertebrate body and their representative organs and functions are listed in Table 21-2. The structure and physiology of these organ systems (except muscular and skeletal, which we discussed here) are the subjects of the following chapters.

TABLE 21-2 ⣿ *Major Vertebrate Organ Systems*		
Organ System	*Major Structures*	*Physiological Role*
Circulatory system	Heart, vessels, blood	Transports nutrients, gases, hormones, metabolic wastes; also assists in temperature control
Respiratory system	Nose, trachea, lungs (mammals, birds, reptiles, amphibians), gills (fish and some amphibians)	Provides an area for gas exchange between the blood and the environment; allows oxygen acquisition and carbon dioxide elimination
Digestive system	Mouth, esophagus, stomach, small and large intestines, glands producing digestive secretions	Supplies the body with nutrients that provide energy and materials for growth and maintenance
Excretory system	Kidneys, ureters, bladder, urethra	Maintains homeostatic conditions within bloodstream; filters out cellular wastes, certain toxins, and excess water and nutrients
Lymphatic/immune system	Lymph, lymph nodes and vessels, white blood cells	Carries fat and excess fluids to blood; destroys invading microbes
Endocrine system	A variety of hormone-secreting glands including the hypothalamus, pituitary, thyroid, pancreas, and adrenals	Controls physiological processes, often in conjunction with the nervous system
Nervous system	Brain, spinal cord, peripheral nerves	Controls physiological processes in conjunction with the endocrine system; senses the environment, directs behavior
Muscular system	Skeletal muscle	Moves the skeleton
	Smooth muscle	Controls movement of substances through hollow organs (digestive tract, large blood vessels)
	Cardiac muscle	Initiates and implements heart contractions
Skeletal system	Bones, cartilage, tendons, ligaments	Provides support for the body, attachment sites for muscles, and protection for internal organs
Reproductive system	Male: testes, seminal vesicles, penis Female (mammal): ovaries, oviducts, uterus, vagina, mammary glands	Male: produces sperm, inseminates female Female (mammal): produces egg cells, nurtures developing offspring

HEALTH WATCH

Osteoporosis—The Hidden Crippler

As many as 20 million Americans, most of them women past menopause, are victims of a preventable but insidious form of bone loss called **osteoporosis** (literally, "porous bones"; Fig. E21-1a). Bone density increases steadily from birth through middle adulthood, reaching a peak at around age 35. From this point on, bone density begins a slow, natural decline.

Women are eight times as likely to suffer from osteoporosis as men. Why? One reason is that the bones of women are about 30% less massive than those of men to start with, so they can afford to lose less bone. Another factor is dietary calcium, which is lower in women's diets than in men's. Two-thirds of women between the ages of 18 and 30 get less than the RDA (Recommended Daily Allowance) of calcium, and women tend to consume even less calcium as they become older. As a result, when women reach the age when bone loss begins naturally, their bones may already be more fragile than they should be. Another factor unique to women is the role of the hormone estrogen. In women, estrogen stimulates bone-forming cells and helps maintain bone density. After menopause, when estrogen production drops dramatically, women lose 3% to 5% of their bone mass each year for several years. Half of all women over age 65 are estimated to have some degree of osteoporosis.

Another important factor contributing to osteoporosis is lack of weight-bearing exercise such as walking, dancing, or running. Bones thrive on moderate stress, but as people age they tend to be less active. Being inactive or bedridden (or being weightless, as astronauts have discovered) results in rapid loss of bone calcium. In contrast, exercise, even in elderly people, can reverse bone loss and even increase bone mass. Alcoholism and smoking also contribute to bone loss and osteoporosis.

Although some bone loss is normal, in osteoporosis the loss is sufficient to weaken the bones, making them vulnerable to fractures and deformities. Frequently, the vertebrae of osteoporosis victims compress, causing a hunch-backed appearance (Fig. E21-1b). In extreme cases, simple activities such as lifting a shopping bag, opening a window, or sneezing can break a bone. Of women living to age 85, nearly one-third will fracture a hip weakened by osteoporosis. As a result of hip fractures, the elderly often become much less self-sufficient and active, and one quarter of them die within 6 months of complications such as pneumonia.

Supplemental hormones (estrogen or a combination of estrogen and progesterone) may slow the progress of osteoporosis in women past menopause. Sodium fluoride, in combination with supplements of calcium and vitamin D (which is crucial to the proper metabolism of calcium), is sometimes prescribed to increase bone deposition. Such treatments, however, may occasionally have undesirable side effects. Fortunately, much of the pain, incapacitation, and expense (estimated at $3.8 billion per year) caused by fractures due to osteoporosis can be prevented. A combination of regular exercise and adequate dietary or supplemental calcium and vitamin D will assure that bone mass is as high as possible before natural, age-related losses begin and will also minimize such losses in old age. In fact, easy measures taken to prevent osteoporosis can enhance the quality of life at any age.

(a)

(b)

Figure E21-1   Osteoporosis

(a) Cross section of a normal bone (left) compared with a bone from a woman suffering from osteoporosis (right).
(b) The devastating effects of osteoporosis extend beyond the obvious deformities, such as a hunch-backed appearance. Its victims are also at high risk for bone fractures.

✖ SUMMARY OF KEY CONCEPTS

Homeostasis

Homeostasis refers to the tendency of many physiological processes to maintain an organism's internal conditions within a narrow range that permits the continuation of life. These conditions are maintained through negative feedback, in which a change triggers a response that counteracts the change and restores conditions to a set level. Temperature control as well as many hormone systems use negative feedback to maintain homeostasis. Positive feedback, in which a change initiates events that intensify the change, occurs relatively rarely and is self-limiting. For example, the uterine contractions that lead to childbirth are driven by positive feedback.

The Organization of the Animal Body

The animal body is composed of organ systems consisting of two or more organs. Organs, in turn, are made up of tissues, which are composed of cells. A tissue is a group of cells and extracellular material that forms a structural and functional unit, specialized for a specific task. Animal tissues include epithelial, connective, muscle, and nerve tissue.

Epithelial tissue forms membranous coverings over internal and external body surfaces and also gives rise to glands. Connective tissue usually contains considerable extracellular material and includes dermal tissue, bone, cartilage, tendons, ligaments, fat, and blood. Muscle tissue is specialized for movement, using sliding filaments of actin and myosin protein. There are three types of muscle tissue: skeletal, cardiac, and smooth. Nerve tissue is specialized for the generation and conduction of electrical signals.

Organs include at least two tissue types that function together. Mammalian skin is a representative organ. The epidermis, an epithelial tissue, covers and protects the dermis beneath it. The dermis contains blood and lymph vessels, a variety of glands, and tiny muscles that erect the hairs. Animal organ systems include organs that work together, performing a common function. One example is the vertebrate skeletal system; it consists of bones, which are organs formed from several types of connective tissue. The bones are held together by ligaments and are often separated by cartilage. The skeletal system has several important functions including support for the body and protection of internal organs, production of blood cells, storage of calcium and phosphorus, and even sensory reception. Other organ systems are the digestive, excretory, respiratory, circulatory, nervous, muscular, lymphatic/immune, endocrine, and reproductive systems summarized in Table 21-2.

✖ KEY TERMS

actin p. 387
adipose tissue p. 387
appendicular skeleton p. 391
axial skeleton p. 391
axon p. 389
blood p. 387
bone p. 387
cardiac muscle p. 388
cartilage p. 386
cell body p. 389
collagen p. 386
connective tissue p. 385
dendrite p. 389
dermis p. 386
endocrine gland p. 385

epidermis p. 389
epithelial tissue p. 385
exocrine gland p. 385
fat p. 387
gland p. 385
glial cells p. 389
hair follicle p. 390
Haversian system p. 387
homeostasis p. 383
keratin p. 389
ligament p. 386
lymph p. 387
membrane p. 385
myosin p. 387
negative feedback p. 383

nerve tissue p. 388
neuron p. 389
organ p. 385
organ system p. 385
osteocyte p. 387
osteoporosis p. 392
positive feedback p. 385
sebaceous gland p. 390
skeletal muscle p. 387
smooth muscle p. 388
striated muscle p. 387
synapse p. 389
synaptic terminal p. 389
tendon p. 386
tissue p. 385

✜ T H I N K I N G   T H R O U G H   T H E   C O N C E P T S

Multiple Choice

1. The skin contains
   a. epithelial tissue   b. connective tissue
   c. nerve tissue   d. muscle tissue
   e. all of the above
2. Glands that become separated from the epithelium that produced them are called _____ glands.
   a. sebaceous   b. sweat
   c. exocrine   d. endocrine
   e. saliva
3. Epithelial membranes
   a. cover the body
   b. line body cavities
   c. may create barriers that alter the movement of certain substances
   d. are continuously replaced by cell division
   e. all of the above
4. All of the following are examples of connective tissue EXCEPT
   a. tendons   b. ligaments
   c. blood   d. muscle
   e. adipose
5. Which of the following statements about muscle is true?
   a. Smooth muscle has a striated appearance.
   b. Skeletal muscle is not under conscious control.
   c. Cardiac muscle utilizes gap junctions.
   d. Smooth muscle is called voluntary muscle.
   e. Smooth muscle moves the skeleton.

6. All of the following are found in the dermis EXCEPT
   a. arteries
   b. sensory neuron endings
   c. hair follicles
   d. sebaceous glands
   e. cells packed with keratin

Review Questions

1. Define homeostasis and explain how negative feedback helps maintain it. Explain one example of homeostasis in the human body.
2. Explain positive feedback and provide one physiological example. Explain why this type of feedback is relatively rare in physiological processes.
3. Explain why body temperature cannot be maintained at *exactly* 37° C (98.6° F) at all times.
4. Describe the structure and functions of epithelial tissue.
5. What property distinguishes connective tissue from all other tissue types? List five types of connective tissue, and briefly describe the function of each type.
6. Describe the skin, a representative organ. Include the various tissues that compose it and the role of each tissue.
7. Where is collagen found in the body, and what functions does it serve?
8. List three important functions of the vertebrate skeletal system.

✜ A P P L Y I N G   T H E   C O N C E P T S

1. Why does life on land present more difficulties in maintaining homeostasis than life in water? What made it evolutionarily advantageous for organisms to colonize dry land?
2. The majority of homeostatic regulatory mechanisms in animals are "autonomic," not requiring conscious control. Discuss several reasons why this is more advantageous to the animal than conscious regulation of homeostatic controls.
3. Third-degree burns are usually painless. Skin regenerates only from the edges of these wounds. Second-degree burns regenerate from cells located at the burn edges, in hair follicles and in sweat glands. First-degree burns are painful but heal rapidly from undamaged epidermal cells. From this information, draw the depth of first-, second-, and third-degree burns on Figure 21-7.

4. A coroner dictates the following description during an autopsy: "The tissue I am looking at forms part of the fetal skeleton. The matrix appears transparent. Fibers of collagen are present but are small and evenly dispersed in the matrix. Chondrocytes appear in tiny spaces, lacunae, within the matrix. Blood vessels have not yet penetrated the matrix." What tissue is the coroner describing?
5. Imagine you are a health-care professional teaching a prenatal class for fathers. Design a real-world analogy with sensors, electrical currents, motors, etc. to illustrate feedback relationships involved in the initiation of labor that could be understood by a layperson.

⌗ FOR MORE INFORMATION

Alexander, R. "Muscles Fit for the Job." *New Scientist*, April 15, 1989. Diverse muscles adapted for different functions.

Degabriele, R. "The Physiology of the Koala." *Scientific American*, June 1980 (Offprint No. 1476). Koalas eat poisonous eucalyptus leaves and almost never drink water. How can they survive?

Huyghe, P. "No Bone Unturned." *Discover*, December 1988. The story of a remarkable forensic anthropologist who uncovers the secrets hidden in unidentified skeletons.

Merton, P. A. "How We Control the Contraction of Our Muscles." *Scientific American*, May 1972. Useful muscular activity requires precise simultaneous adjustments of the contractions of many muscles.

Moore, K. "Little Is beyond His Scope." *Discover*, March 1985. A description of intricate arthroscopic surgery on joints.

Schmidt-Nielson, K. "The Physiology of the Camel." *Scientific American*, December 1959. Do camels really store water in their humps? How long can one go without drinking?

Schmidt-Nielson, K., and Schmidt-Nielson, B. "The Desert Rat." *Scientific American*, July 1953 (Offprint No. 1050). The ultimate desert dweller, the kangaroo rat never needs to drink water.

Smith, K. K., and Kier, W. M. "Trunks, Tongues, and Tentacles: Moving with Skeletons of Muscle." *American Scientist*, January–February 1989. In certain organs of many animals, muscles provide support as well as movement.

Storey, K. B., and Storey, J. M. "Frozen and Alive." *Scientific American*, December 1990. Some animals have special adaptations that allow them to withstand freezing

Wills, Christopher. "The Skin We're In." *Discover*, November 1994. A scientist looks at the controversial issue of melanism.

NET WATCH

On-line resources for this chapter are on the World Wide Web at:

http://www.prenhall.com/~audesirk (click on the <u>table of contents</u> link and then select Chapter 21).

22 Circulation and Respiration

*Red blood cells flow into a capillary from a larger vessel in this
scanning electron micrograph. In its 4-month life span, a red
blood cell will travel over 900 miles through the circulatory
system, ferrying oxygen to cells throughout the body.*

*E*ach cell in the animal body is a tiny factory that demands a continuous influx of energy and nutrients to maintain itself. Billions of years ago, the first living cells were
nurtured by the sea in which they evolved. The sea brought
them nutrients, which diffused into the cell, and it washed
away the wastes, which had diffused out. Today, microorganisms and some simple multicellular animals still rely almost exclusively of diffusion to exchange oxygen and other
nutrients and carbon dioxide and other wastes with the
environment. As larger, more complex animals evolved,
individual cells became increasingly distant from the outside world. To avoid starving and stewing in its own wastes,
each cell needed to have both a source of nutrients and a
dumping place for wastes within diffusing distance. With
the evolution of the circulatory system, a sort of "internal
sea" was created, serving the same purpose as the sea did for
the first cells. This internal sea transports a fluid (blood)
rich in food and oxygen close to each cell and carries away
wastes produced by the cells. Humans and other vertebrates
have **closed circulatory systems**, in which blood is confined to the heart and a continuous series of vessels. Closed
circulatory systems rapidly and efficiently carry blood to
and from body tissues.

Most cells use a biochemical process called cellular respiration (see Chapter 8) to convert the energy in nutrients such as sugar into an energy form usable by the cell.
The production of energy by cellular respiration requires
a steady supply of oxygen and generates carbon dioxide as
a waste product. As natural selection produced increasingly large, active animals, specialized respiratory systems
became necessary to guarantee ample gas exchange
throughout their bodies. Respiratory systems support cellular respiration by bringing a large, moist surface into
intimate contact with both the external and the internal
environments so that oxygen and carbon dioxide may be
exchanged by diffusion. The circulatory system facilitates
this exchange by carrying carbon dioxide to the respiratory system for elimination and oxygen away for use by
the body cells. This process maintains a concentration
gradient that enhances diffusion. Humans and other vertebrates possess both circulatory and respiratory systems;
these mutually supporting adaptations are the subject of
this chapter.

The Human Circulatory System

All circulatory systems have three major parts:

1. A fluid, **blood**, that serves as a medium of transport
2. A system of channels, or **vessels**, that conduct the blood throughout the body
3. A pump, the **heart**, that keeps the blood circulating

The circulatory system has diverse roles in the vertebrates. Among the most important functions of the circulatory system are the following:

1. Transport of oxygen from the lungs to the tissues and of carbon dioxide from the tissues to the lungs
2. Distribution of nutrients from the digestive system to all body cells
3. Transport of waste products and toxic substances to the liver, where many of them are detoxified, and to the kidney for excretion
4. Distribution of hormones from the organs that produce them to the tissues on which they act
5. Regulation of body temperature, which is achieved partly by adjustments in blood flow
6. Prevention of blood loss by means of the clotting mechanism
7. Protection of the body from bacteria and viruses by circulating antibodies and white blood cells

In the following sections covering circulation we examine the three parts of the circulatory system: the heart, the blood, and the vessels, using the human as a representative vertebrate. The human circulatory system is shown in Figure 22-1. Finally, we describe the lymphatic system, which works closely with the circulatory system.

The Vertebrate Heart Consists of Muscular Chambers Whose Contraction Is Controlled by Electrical Impulses

Atria Collect Blood, and Ventricles Pump It throughout the Body

The vertebrate heart consists of muscular chambers capable of strong contractions that circulate blood through the body. The complete separation of oxygenated and deoxygenated blood, which is made possible by the four-chambered heart, assures that blood that reaches the tissues has the highest possible oxygen content.

Mammalian hearts consist of two separate pumps, each with two chambers. In each pump, an **atrium** receives and briefly stores the blood, passing it to a **ventricle** that propels it through the body (Figs. 22-2 and 22-3). One pump is for **pulmonary circulation** and consists of the right atrium and ventricle. Oxygen-depleted blood from the body empties into the right atrium through a large vein, the superior vena cava. The right atrium contracts, transferring

Figure 22-1   **The human circulatory system**

Oxygenated blood is shown in red, deoxygenated blood in blue.

the blood to the right ventricle. Contraction of the right ventricle sends the oxygen-depleted blood to the lungs via pulmonary arteries. The other pump, consisting of the left atrium and ventricle, powers **systemic circulation**. Newly oxygenated blood from the lungs enters the left atrium through pulmonary veins and is passed to the left ventricle. Strong contractions of the left ventricle, the heart's most muscular chamber, send the oxygenated blood coursing out through the aorta to the rest of the body.

The Coordinated Contractions of Atria and Ventricles Produce the Cardiac Cycle

The alternating contraction and relaxation of the heart chambers is called the **cardiac cycle**. The two atria contract in synchrony, emptying their contents into the ventricles. A fraction of a second later, the two ventricles contract simultaneously, forcing blood into arteries leaving the heart. Both chambers then relax briefly before the cycle is repeated. The period of ventricular contraction is called **systole**. The rest of the cycle, including relaxation of all the chambers followed by contraction of the atria, is called **diastole**. At normal resting heart rate, systole lasts about 0.3 second and diastole about 0.5 second. Systole and the last phase of diastole are illustrated in Figure 22-3. When blood pressure is measured (Fig. 22-4), the higher of the two readings is the systolic pressure, which is measured during ventricular contraction, and the lower reading is the diastolic pressure, measured between contractions.

Valves Maintain Directionality of Blood Flow, While Electrical Impulses Coordinate the Sequence of Contractions

Coordinating the activity of the four chambers to maintain proper blood flow through the heart and its vessels presents some challenges. First, when the ventricles contract, the blood must be directed out through the **arteries**, the large vessels that carry blood away from the heart, and not back up into the atria. Then, once blood has entered the arteries, it must be prevented from flowing back as the heart relaxes. These problems are solved by four simple one-way valves (see Fig. 22-2). Pressure in one direction opens them readily, while reverse pressure forces them tightly closed. **Atrioventricular valves** separate the atria from the ventricles; a **tricuspid** (from Latin, meaning "three-pointed") **valve** separates the right ventricle and right atrium, and a **bicuspid** ("two-pointed") **valve** lies between the left atrium and the left ventricle. Two **semilunar** (Latin, "half-moon") **valves** allow blood to enter the pulmonary artery and the aorta when the ventricles contract but prevent it from returning as the ventricles relax.

A second challenge is to create smooth, coordinated contractions of the muscle cells that make up each chamber. When they are excited, muscle cells produce electrical signals that cause the muscle cells to contract. In the heart, the individual heart muscle cells communicate directly with one another through gap junctions in their adjacent membranes (Fig. 22-5). These connecting pores allow electrical signals that cause contraction to pass freely and rapidly be-

Figure 22-2   The human heart and its valves and vessels

The right atrium receives deoxygenated blood and passes it to the right ventricle, which pumps it to the lungs. Blood returning from the lungs enters the left atrium, which passes it to the left ventricle, which pumps oxygenated blood through the rest of the body. Note the thickened walls of the left ventricle, which must pump blood over a considerably longer distance. One-way valves are located between the aorta and the left ventricle, between the pulmonary artery and the right ventricle, and between the atria and ventricles.

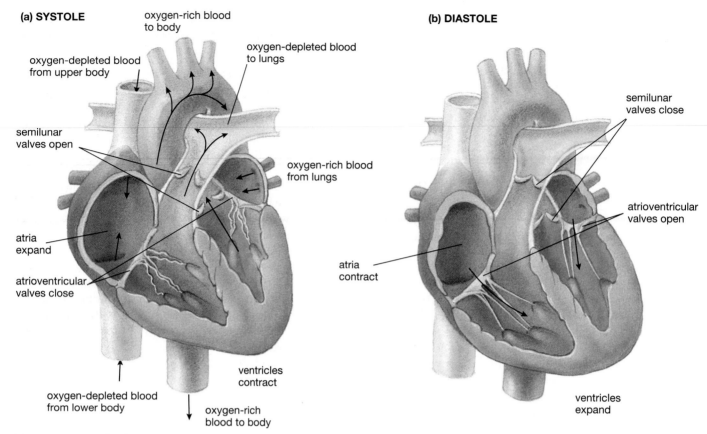

(a) SYSTOLE

oxygen-rich blood
to body

oxygen-depleted blood
to lungs

oxygen-depleted blood
from upper body

semilunar
valves open

oxygen-rich blood
from lungs

atria
expand

atrioventricular
valves close

oxygen-depleted blood
from lower body

ventricles
contract

oxygen-rich
blood to body

(b) DIASTOLE

semilunar
valves close

atrioventricular
valves open

atria
contract

ventricles
expand

Figure 22-3   **The heart valves in action**

(a) During ventricular contraction (systole), the pressure within the ventricles forces the semilunar valves open, allowing blood to flow into the aorta and pulmonary arteries. The atrioventricular valves are simultaneously pressed shut, preventing blood flow back into the atria. **(b)** As the ventricles reexpand (diastole), they would tend to draw blood back from the arteries, but this backward pressure forces the semilunar valves closed. Simultaneously, contraction of the atria forces open the atrioventricular valves, allowing blood to flow from the atria into the ventricles (arrows indicate the direction of blood flow).

tween heart cells. The contraction of the heart is initiated and coordinated by a **pacemaker**, a cluster of specialized muscle cells that produce spontaneous electrical signals at a regular rate. Although the nervous system can alter the rate of these signals, they are initiated by the pacemaker muscle cells themselves. The heart's primary pacemaker is the **sinoatrial (SA) node**, located in the wall of the right atrium (Fig. 22-6). Signals from the SA node spread rapidly through both the right and left atria, causing the atria to contract in smooth synchrony.

A final challenge is to coordinate contractions of all four chambers. The atria must contract first and empty their contents into the ventricles in order to be able to refill while the ventricles contract. Thus, there must be a delay between the contractions of the atria and those of the ventricles. From the SA node, an electrical impulse creates a wave of contraction that sweeps through the muscles of the atria until it reaches a barrier of unexcitable tissue separating the atria from the ventricles. Here the excitation is channeled through a second small mass of specialized muscle cells, the **atrioventricular (AV) node**, located in the

floor of the right atrium (see Fig. 22-6). The impulse is delayed at the AV node, postponing the ventricular contraction for about 0.1 second after contraction of the atria. This delay gives the atria time to complete the transfer of blood into the ventricles before ventricular contraction begins. From the AV node, the signal to contract spreads to the base of the two ventricles along tracts of excitable fibers. The impulse then travels rapidly from these fibers through the communicating muscle fibers, causing the ventricles to contract in unison. When the pacemaker fails to coordinate muscle contractions, uncoordinated, irregular contractions impair the heart's function, a condition known as **fibrillation**. Fibrillation of the ventricles is soon fatal because blood is not pumped out of the heart to the brain and other organs but is merely sloshed around.

The Nervous System and Hormones Influence Heart Rate

Left on its own, the SA node pacemaker would maintain a steady rhythm of about 100 beats per minute. However, the

Figure 22-4 Measuring blood pressure

Blood pressure is measured using an inflatable blood pressure cuff and a stethoscope. The cuff is placed around the upper arm and the stethoscope positioned over the artery just below the cuff. The cuff is inflated until its pressure closes off the main artery in the arm and blood ceases to flow. No pulse, therefore, can be detected below the cuff. Then the pressure is gradually reduced. When the pulse is first audible in the artery, this means that the pressure pulses created by the contracting left ventricle are just overcoming the pressure in the cuff and blood is flowing. This is the upper of the two readings: the systolic pressure. Cuff pressure is then further reduced until no pulse is audible. An inaudible pulse at this point indicates that the artery is no longer constricted and blood is flowing continuously through the artery; the pressure between ventricular contractions is just overcoming the cuff pressure. This is the lower of the two readings: the diastolic pressure. A blood pressure reading of 130 (systolic) over 80 (diastolic) is within the normal range. The numbers are in millimeters of mercury, a standard measure of pressure also used in barometers.

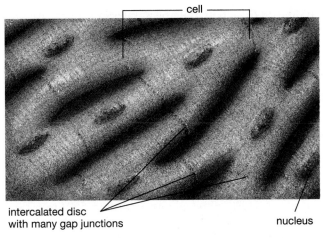

Figure 22-5 The structure of cardiac muscle

Cardiac muscle cells are branched. Adjacent cell membranes meet in folded areas called intercalated discs. These regions are densely packed with gap junctions (pores) that connect the interiors of adjacent cells. This arrangement allows direct transmission of electrical signals between the cells, coordinating their contractions.

heart rate is significantly altered by the influence of nervous impulses and hormones. In the resting person, activity of the parasympathetic nervous system (which controls body functions during periods of rest; see Chapter 26) slows the heart rate to around 70 beats per minute. When exercise or stress creates a demand for greater blood flow to the muscles, the parasympathetic influence is reduced and the sympathetic nervous system (which prepares the body for emergency action) accelerates the heart rate. Likewise, the hormone epinephrine (also known as adrenaline), increases heart rate as it mobilizes the entire body for response to threatening or unfamiliar events. When astronauts were

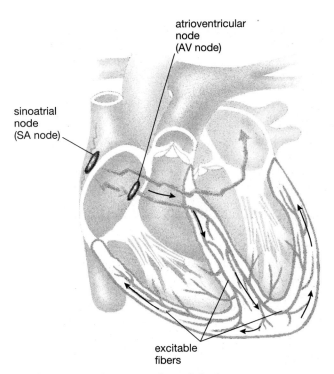

Figure 22-6 The pacemaker of the heart

A spontaneously active mass of modified muscle fibers in the right atrium called the sinoatrial (SA) node serves as the heart's pacemaker. The signal to contract spreads from the SA node through the muscle fibers of both atria, finally exciting the atrioventricular (AV) node in the lower right atrium. The AV node then transmits the signal to contract through bundles of excitable fibers that stimulate the ventricular muscle. The delay in transmission of the signals from the SA node through the AV node allows the atria to empty their contents completely into the ventricles before ventricular contraction begins.

landing on the moon, their heart rates were over 170 beats per minute, even though they were sitting still!

Blood Transports Dissolved Nutrients, Wastes, and Hormones throughout the Body

Blood is the medium in which dissolved nutrients, gases, hormones, and wastes are transported through the body. It has two major components: (1) a fluid called **plasma** and (2) specialized cells (red blood cells, white blood cells, and platelets) that are suspended in the plasma (Table 22-1). On the average, the cellular components of blood account for 40% to 45% of its volume; the other 55% to 60% is plasma. The average person has 5 to 6 liters of blood, constituting about 8% of his or her total body weight.

Plasma Is Primarily Water in Which Proteins, Salts, Nutrients, and Wastes Are Dissolved

About 90% of the straw-colored plasma is water in which a number of substances are dissolved. Those substances include proteins, hormones, nutrients (glucose, vitamins, amino acids, lipids), gases (carbon dioxide, oxygen), salts (sodium, calcium, potassium, magnesium), and wastes such as urea. Plasma proteins are the most abundant of the dissolved substances. The three major plasma proteins are (1) albumins, which help maintain the blood's osmotic pressure (which controls the flow of water across cell membranes); (2) globulins, which transport nutrients and also play a role in the immune system; and (3) fibrinogen, important in blood clotting, which is discussed later in this chapter.

Red Blood Cells Carry Oxygen from the Lungs to the Tissues

The oxygen-carrying red blood cells, also called **erythrocytes**, make up about 99% of the cells in the blood. They constitute about 40% of the total blood volume in females, and 45% in males. Each milliliter of blood contains about 5 billion erythrocytes. Proteins on the surface of red blood cells differ among individuals, thereby creating different blood types, as described in Chapter 13. The red blood cell resembles a ball of clay squeezed between thumb and forefinger (Fig. 22-7). Its biconcave (dented inward on both sides) shape provides a larger surface area than would a spherical cell of the same volume and increases the cell's ability to absorb and release oxygen through its membrane.

TABLE 22-1 Blood Cells

Cell Type		Description	Average Number Present	Major Function
Red blood cell (erythrocyte)		Biconcave disc without nucleus, about one-third hemoglobin. Approximately 8 μm[b] in diameter	5,000,000 per mm[3a]	Transports oxygen and a small amount of carbon dioxide
White blood cells (leukocytes)			7500 per mm[3]	
1. Neutrophil		About twice the size of red cells, nucleus with two to five lobes	62% of white cells	Destroys relatively small particles by phagocytosis
2. Eosinophil		About twice the size of red cells, nucleus with two lobes	2% of white cells	Inactivates inflammation-producing substances; attacks parasites
3. Basophil		About twice the size of red cells, nucleus with two lobes	Less than 1% of white cells	Releases anticoagulant to prevent blood clots and histamine, which causes inflammation
4. Monocyte		Two to three times larger than red cells, nuclear shape varies from round to lobed	3% of white cells	Gives rise to macrophage, which destroys relatively large particles by phagocytosis
5. Lymphocyte		Only slightly larger than red cell, nucleus nearly fills cell	32% of white cells	Functions in the immune response
Platelet		Cytoplasmic fragment of cells in bone marrow called megakaryocytes	250,000 per mm[3]	Important in blood clotting

[a]mm[3] = cubic millimeter.
[b]μm = micrometer.

The red color of erythrocytes is caused by the pigment **hemoglobin** (Fig. 22-8). This large, iron-containing protein accounts for about one-third the weight of the blood cell. About 97% of the oxygen carried by the blood is bound to hemoglobin. The hemoglobin binds loosely to oxygen, picking up oxygen in the capillaries of the lungs, where the concentration is high, and releasing it where the concentration is low, in other tissues of the body. After releasing its oxygen, some of the hemoglobin picks up carbon dioxide from the tissues for transport back to the lungs. The role of blood in gas exchange is discussed later in this chapter.

Red Blood Cells Have a Relatively Short Life Span

Red blood cells are formed in the marrow, the soft interior portion of certain bones, including those of the chest, upper arms, upper legs, and hips. During their development, mammalian red blood cells lose their nuclei and their ability to divide. Without the ability to synthesize cellular materials, their lives are necessarily short; each cell lives about 120 days. Every second, over 2 million red blood cells die and are replaced by new ones from the bone marrow. Dead or damaged red blood cells are removed from circulation, primarily in the liver and spleen, and broken down to release their iron. The salvaged iron is carried in the blood to the bone marrow, where it is used to make more hemoglobin and packaged into new red blood cells. Although the recycling process is efficient, small amounts of iron are excreted daily and must be replenished by the diet. Bleeding from injury or menstruation also tends to deplete iron stores.

Negative Feedback Regulates Red Blood Cell Numbers

The number of red blood cells in the blood is maintained at an adequate level through a negative feedback system

Figure 22-8   **Hemoglobin**
A molecule of hemoglobin is composed of four polypeptide chains (two pairs of similar chains), each surrounding a heme group. The heme group contains an iron atom and is the site of oxygen binding. When saturated, each hemoglobin molecule can carry four oxygen molecules (eight oxygen atoms).

that involves a hormone called **erythropoietin**. Erythropoietin is produced by the kidneys in response to oxygen deficiency. This lack of oxygen may be caused by a loss of blood, insufficient production of hemoglobin, high altitude (where less oxygen is available), or lung disease that interferes with gas exchange in the lungs. The hormone stimulates rapid production of new red blood cells by the bone marrow. When adequate oxygen levels are restored, erythropoietin production is inhibited, and the rate of red blood cell production returns to normal. This negative feedback system is diagrammed in Figure 22-9.

Blood Type Is Determined by Specific Proteins on Red Blood Cell Membranes

Blood is classified as type A, B, AB, or O depending on the presence or absence of specific proteins (designated A and B) on the cell membranes of red blood cells. Type A blood has the A protein on its red blood cells, type B has the B protein, type AB has both, and type O has neither. Also, each blood type carries antibodies in the plasma to the proteins *not* present on its own red blood cells. Thus, a person with type A blood has antibodies to the B protein. If this person is transfused with type B blood, the antibodies to the B protein attack the transfused red blood cells, causing them to clump together and block small blood vessels, sometimes with fatal results. Blood type is inherited; the genetics and properties of these major blood types are discussed in detail in Chapter 13.

Another type of protein on red blood cells is the **Rh factor**. If it is present, blood is described as Rh-positive, and if the protein is absent, the blood is Rh-negative. Unlike A or B antibodies, antibodies to the Rh protein form only after massive exposure to the protein. The first exposure—for example, transfusion of Rh-positive blood into

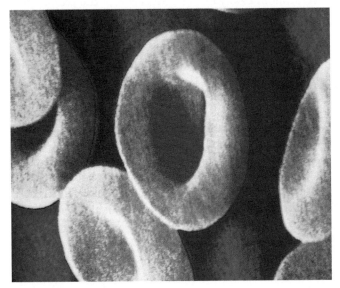

Figure 22-7   **Red blood cells**
Under the scanning electron microscope, the biconcave disc shape of red blood cells is clearly visible. These cells have been artificially colored.

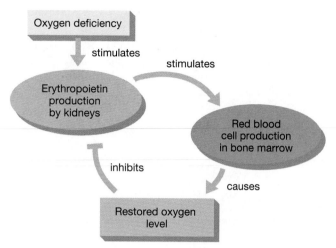

Figure 22-9   Red blood cell regulation

The production of red blood cells is regulated by a negative feedback system.

an Rh-negative individual—generally causes no ill effects but triggers the production of antibodies. Upon further exposure, the antibodies attack and destroy the Rh-positive red blood cells. If an Rh-negative woman marries an Rh-positive man, her children are likely to be Rh-positive, because Rh-positive blood is a dominant genetic trait. Her first Rh-positive child will trigger antibody production in her blood. Subsequent Rh-positive children will be in danger of being born with **erythroblastosis fetalis**, in which the mother's antibodies invade the fetus and attack its red blood cells, causing the child to be born severely anemic. Fortunately this condition can now be easily prevented by injections of a substance that prevents formation of Rh antibodies by the pregnant woman.

White Blood Cells Help Defend the Body against Disease

There are five common types of white blood cells, or **leukocytes**, which together make up less than 1% of the total cellular component of the blood. These cells, described and illustrated in Table 22-1, are distinguished from one another by their staining characteristics, size, and the shape of their nuclei. All of these are derived from cells that originate in bone marrow. Most white blood cells function in some way to protect the body against foreign invaders and use the circulatory system to travel to the site of invasion. For example, **monocytes** and **neutrophils** travel through capillaries to wounds where bacteria have gained entry, then ooze out through narrow openings in the capillary walls. After leaving the capillaries, monocytes differentiate into **macrophages** ("big-eaters"), amoebalike cells that engulf foreign particles. Macrophages and neutrophils feed on bacterial invaders or other "foreign" cells including cancer cells (Fig. 22-10). They often die in the process, and their dead bodies accumulate and contribute to the white substance called pus, seen at infection sites.

Lymphocytes, described in detail in Chapter 24, are white blood cells responsible for the production of antibodies that help provide immunity against disease. Cells that give rise to lymphocytes migrate from bone marrow to tissues of the lymphatic system such as the thymus, spleen, and lymph nodes, described later in this chapter. Least abundant are the **eosinophils** and **basophils**. Eosinophil production is stimulated by parasitic infections. Eosinophils converge on the parasitic invaders, releasing substances that kill the parasite. Basophils release substances that inhibit blood clotting as well as chemicals, such as histamine, that participate in allergic reactions and responses to tissue damage and microbial invasion.

Platelets Are Cell Fragments That Aid in Blood Clotting

Platelets are not complete cells, but fragments of large cells called **megakaryocytes**. Megakaryocytes remain in the bone marrow, pinching off membrane-enclosed pieces of their cytoplasm, called platelets (Fig. 22-11). Platelets then enter the blood and play a central role in blood clotting. Like red blood cells, platelets lack a nucleus, and their life span is even shorter: about 10 to 12 days.

Clot formation is a complex process. It starts when platelets and other factors in the plasma contact an irregular surface, such as a damaged blood vessel. Platelets tend to stick to irregular surfaces, and they may build up and plug the damaged vessel if it is narrow enough. This mechanism is supplemented by blood coagulation, or clotting, which is the most important of the body's defenses against bleeding. The ruptured surface of an injured blood vessel not only causes platelets to adhere, but it also initiates a complex sequence of events among circulating plasma proteins. These events culminate in production of the enzyme **thrombin**.

Thrombin catalyzes the conversion of the plasma protein fibrinogen into stringlike molecules called **fibrin**. Fibrin molecules adhere to one another, end to end and side to side, forming a fibrous matrix. This protein web immobilizes the fluid portion of the blood, causing it to solidify in much the same way that gelatin does as it cools. The web traps red blood cells, further increasing the density of the clot (Fig. 22-12). Platelets then adhere to the fibrous mass and send out sticky projections that attach to one another. Within half an hour, the platelets contract, pulling the mesh tighter and forcing liquid out. This action creates a denser, stronger clot (on the skin it is called a scab) and also constricts the wound, pulling the damaged surfaces closer together in a way that promotes healing.

The Blood Vessels Carry Blood to All Parts of the Body

As blood leaves the heart, it travels from arteries to arterioles to capillaries to venules to veins, which return it finally to the heart. In this section we examine each type of blood vessel in more detail.

Figure 22-10   A white blood cell attacks bacteria

An amoebalike white blood cell is seen capturing bacteria. These bacteria are *Escherichia coli*, intestinal bacteria that can cause disease if they enter the bloodstream.

Arteries and Arterioles Are Thick-Walled Vessels That Carry Blood Traveling Away from the Heart

After leaving the heart, blood first enters large vessels called arteries. These have thick walls containing smooth muscle and elastic tissue (Fig. 22-13, p. 408). With each surge of blood from the ventricles, the arteries expand slightly, like thick-walled balloons. Between heartbeats they recoil, helping to pump the blood and maintain a steady flow through the smaller vessels. Arteries branch into vessels of even smaller diameter called **arterioles**, which play a major role in determining how blood is distributed within the body, as described later.

Figure 22-11   The production of platelets

Megakaryocytes are large cells in bone marrow that constantly manufacture proteins, enzymes, and cell membrane. But instead of growing larger, megakaryocytes bud off membrane-enclosed pieces of cytoplasm, called platelets, that enter the blood and help initiate and control blood clotting. In this photomicrograph, a single megakaryocyte is seen producing dozens of platelets.

Capillaries Are Microscopic Vessels That Allow Exchange of Nutrients and Wastes between the Blood and Body Cells

The entire circulatory system is an elaborate device for providing each cell of a complex, multicellular organism with the type of exchange—diffusion—practiced by the simplest unicellular organisms. This exchange is accomplished at the level of the **capillaries**, the tiniest of all vessels. Here wastes, nutrients, gases, and hormones are exchanged between the blood and the body cells. Capillaries are finely adapted to their role of exchange. Their walls are only a single cell thick. Most nutrients, oxygen, and carbon dioxide diffuse readily through capillary cell membranes. Salts and small charged molecules (including some small proteins) move through fluid-filled spaces within the capillary cell membrane or between adjacent capillary cells (Fig. 22-14, p. 408).

Figure 22-12   Blood clotting

In response to damage to a blood vessel, threadlike proteins called fibrin form in the blood. These produce a tangled sticky mass that traps red blood cells and eventually forms a clot.

HEALTH WATCH

Cardiovascular Disorders

Cardiovascular disorders, including heart attacks, strokes, and congestive heart failure, are the leading cause of death in the United States, killing nearly 1 million Americans annually. Consider the stresses under which your circulatory system constantly operates. Your heart is expected to contract vigorously over 2.5 billion times during your lifetime without once stopping to rest. It is also expected to force blood through a series of vessels whose total length would encircle the globe twice. Add the possibility that the complex network of vessels may become constricted, weakened, or clogged for a number of reasons, and it's easy to see why the cardiovascular system is a prime candidate for malfunction.

Hidden Killers: High Blood Pressure and Atherosclerosis

High blood pressure, also called **hypertension**, is usually caused by constriction of the arterioles, which results in increased resistance to blood flow. In the majority of the 50 million Americans afflicted by this condition, the cause of this constriction is unknown. Heredity seems to play a role. For some individuals who are predisposed to hypertension, high salt intake in the diet, as well as obesity, may

aggravate it. Although normal blood pressure tends to increase somewhat with age, an approximate borderline reading for high blood pressure is 140/90.

High blood pressure often gives few warning signals, but it undermines the cardiovascular system in several insidious ways. First, it causes strain on the heart by increasing resistance to blood flow. Although the heart may enlarge in response to this added demand, its own blood supply may not increase proportionately. The heart muscle is then inadequately supplied with blood, especially during exercise. Lack of sufficient oxygen to the heart can cause chest pain called **angina pectoris**. Second, high blood pressure contributes to "hardening of the arteries," or atherosclerosis, described below. Third, high blood pressure in conjunction with hardened arteries can lead to rupture of an artery and internal bleeding. Rupture of vessels supplying the brain causes **stroke**, in which brain function is lost in the area deprived of blood and the vital oxygen and nutrients it carries.

Hypertension can be treated in several ways. Mild hypertension may be alleviated by weight reduction, exercise, and sometimes by reduction of dietary salt. Stress-reduction therapies such as relaxation techniques, meditation,

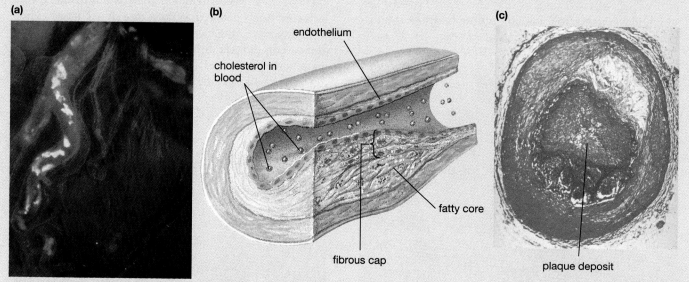

(a)   **(b)**   **(c)**

endothelium

cholesterol in blood

fatty core

fibrous cap

plaque deposit

Figure E22-1   **Plaque deposits clog arteries**

(a) In this remarkable photo of coronary arteries, plaque deposits are seen in glowing yellow. If they block a coronary artery, a heart attack will occur. **(b)** The anatomy of a plaque deposit is shown in this cross section of an artery. If the cap ruptures, a clot will form that may completely obstruct the artery, or the clot may break loose and clog a narrower artery "downstream." **(c)** This cross section of a coronary artery shows a large plaque deposit that has almost completely obstructed it. These deposits can become dislodged and be carried to a smaller artery that they may totally block, causing a heart attack or stroke.

and biofeedback may also be helpful. For more severe cases, drugs may be prescribed. These drugs include diuretics, which increase urination and reduce blood volume, drugs that reduce heart rate, and others that cause dilation (expansion) of the arteries and arterioles. Between 1970 and 1990, the number of deaths in the United States due to hypertension decreased by about half, as a result of increased public awareness, early detection, and successful treatment.

Atherosclerosis (derived from the Greek word *athero*, meaning "gruel" or "paste," and *scleros*, meaning "hard") causes a loss of elasticity in the large arteries and a thickening of the arterial walls. The thickening results from deposits called **plaques** composed of cholesterol and other fatty substances as well as calcium and fibrin. These plaques are deposited within the wall of the artery between the smooth muscle and the inner lining (Fig. E22-1). Occasionally the plaque ruptures through the lining into the interior of the vessel. This rupture stimulates the blood platelets to initiate blood clots. These clots further obstruct the artery and may completely block it. Arterial clots are responsible for the most serious consequences of atherosclerosis: heart attacks and strokes.

The American Heart Association estimates that about 1.5 million Americans suffered **heart attacks** in 1994, and about half a million people died from them. A heart attack occurs when one of the coronary arteries (arteries that supply the heart muscle itself; see Fig. E22-1) is blocked. If a blood clot suddenly breaks loose, it may be carried to a narrower part of the artery, obstructing blood flow. Deprived of nutrients and oxygen, the heart muscle supplied by the blocked artery rapidly and painfully dies. If the damaged area is small, the victim may recover, but death of large areas of heart muscle is almost instantly fatal. Although heart attacks are the major cause of death from atherosclerosis, this disease causes plaques and clots to form in arteries throughout the body. If a clot or plaque deposit obstructs an artery supplying the brain, it can cause a stroke, with results similar to those caused by a ruptured artery.

As with hypertension, the exact cause of atherosclerosis is unclear, but several factors promote it. These factors include hypertension, cigarette smoking, genetic predis-position, obesity, diabetes, a sedentary life-style, and high blood levels of a type of cholesterol bound to a carrier molecule called low-density lipoprotein (LDL). If LDL-bound cholesterol levels are too high, cholesterol may be deposited in arterial walls. In contrast, cholesterol bound to HDL (high-density lipoprotein) is metabolized or excreted and hence is often called "good" cholesterol. By exercising regularly, controlling one's weight, not smoking, and reducing dietary cholesterol and saturated fat, one can greatly reduce the risk of developing atherosclerosis. Moderation in the consumption of animal and other saturated fats is also important because recent evidence links high dietary intake of these substances with high blood cholesterol levels.

Traditional treatment of atherosclerosis includes the use of drugs to lower blood pressure and blood cholesterol levels. In extreme cases, nitroglycerin is used to dilate blood vessels and ease the pain of angina caused by constriction of the coronary arteries. Coronary bypass surgery accounts for over 400,000 operations per year in the United States. This procedure consists of bypassing an obstructed coronary artery with a piece of vein, usually removed from the patient's leg.

Although prevention is by far the most cost-effective and successful strategy, a variety of other high-technology treatments are under development to fight atherosclerosis. Blood clots are often dissolved by injecting an enzyme, streptokinase, or another drug, TPA (tissue plasminogen activator), into the coronary artery. Both work by stimulating the production of an enzyme that breaks down fibrin, the protein that binds the clot together. When performed immediately after a heart attack, this treatment can significantly increase the victim's chances of survival. Another procedure involves squashing plaques flat against the artery walls by inserting a hollow tube with a tiny balloon attached to the end into the obstructed artery. The balloon is inflated, crushing the deposits and restoring blood flow. Alternatively, a tube bearing a tiny laser or drill is manipulated up the artery to the clot. The laser vaporizes the plaques; the drill shaves them away. It is hoped that these new technologies, in conjunction with changes in diet and life-style, will significantly reduce the rate of early deaths from atherosclerosis.

The pressure within capillaries causes a continuous leakage of fluid from the blood plasma into the spaces surrounding the capillaries and tissues. This fluid is known as **the interstitial fluid** and consists primarily of water in which are dissolved nutrients, hormones, gases, wastes, and small proteins from the blood. The large plasma proteins, red blood cells, and platelets are unable to leave the capillaries because of their size, although white blood cells can ooze through the openings between capillary cells. The exchange of materials between capillary blood and nearby cells occurs through this interstitial fluid, which bathes nearly all the cells of the body.

Capillaries are so narrow that red blood cells must pass through them in single file (Fig. 22-15). Consequently, all

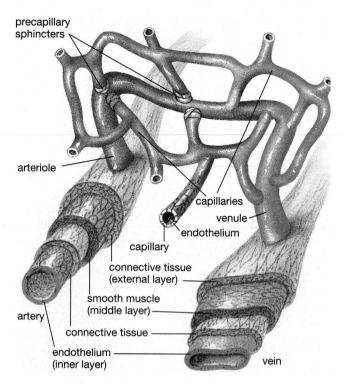

precapillary
sphincters

arteriole

capillaries

venule

endothelium

capillary

connective tissue
(external layer)

smooth muscle
(middle layer)

artery

connective tissue

endothelium
(inner layer)

vein

Figure 22-13   The structure and interconnections of the types of blood vessels

Arteries and arterioles are more muscular and maintain a greater muscular tension than veins and venules, whose walls are thinner and can stretch more. Capillaries have walls only a single cell thick, allowing movement of dissolved substances and white blood cells across the capillary wall. Oxygenated blood moves from arteries to arterioles to capillaries. Capillaries empty deoxygenated blood into venules, which empty into veins. The movement of blood from arterioles into capillaries is regulated by muscular rings called precapillary sphincters.

the blood is sure to pass very close to the capillary walls, where exchange occurs. In addition, capillaries are so numerous that no body cell is more than 100 micrometers (0.004 inch; about as thick as four pages of this book) from a capillary. These factors facilitate the exchange of materials by diffusion. It is estimated that the total length of capillaries in a human is over 80,600 kilometers (50,000 miles), enough to encircle the globe twice! The speed of blood flow drops very quickly as blood is forced through this narrow, almost endless network of capillaries, allowing more time for diffusion to occur.

Veins and Venules Carry Blood Traveling Back to the Heart

Blood from the capillaries, now carrying carbon dioxide and other cellular wastes, drains into larger vessels called **venules** that empty into still larger **veins** (see Fig. 22-13). Veins provide a low-resistance pathway by which blood can return to the heart. The walls of veins are much thinner and more expandable than those of arteries, although both contain a

layer of smooth muscle. Because blood pressure in the veins is low, the contractions of skeletal muscle during exercise and breathing must assist in the return of blood to the heart. These muscular movements squeeze the veins, forcing blood through them. When veins are compressed, you might predict that blood would be forced away from the heart as well as toward it. To prevent this, veins are equipped with one-way valves that allow blood flow only toward the heart (Fig. 22-16). When you sit or stand for long periods, the lack of muscular activity allows blood to accumulate in the veins of the lower legs. This accumulation accounts for the swollen feet often experienced by airplane passengers. Long periods of inactivity can also contribute to varicose veins, in which the valves become stretched and weakened.

If blood pressure should fall, for instance after extensive bleeding, veins can help restore it. The sympathetic nervous system (which prepares the body for emergency action) automatically stimulates contraction of the smooth muscles in the vein walls. This action decreases the internal volume of the veins and raises blood pressure, speeding up the return of blood to the heart.

Arterioles Control the Distribution of Blood Flow

The muscular walls of arterioles are under the influence of nerves, hormones, and chemicals produced by nearby tissues. They can therefore contract and relax in response to the changing needs of the tissues and organs they supply. For instance, as you read in your paperback thriller " . . . the

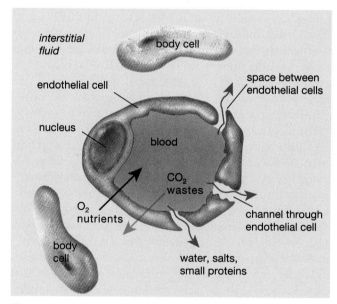

interstitial fluid

body cell

endothelial cell

space between endothelial cells

nucleus

blood

O_2 nutrients

CO_2 wastes

channel through endothelial cell

body cell

water, salts, small proteins

Figure 22-14   Capillary wall structure

The structure of capillary walls facilitates the exchange of dissolved gases, wastes, and nutrients. Capillary walls are only one endothelial cell thick. Narrow spaces separate adjacent endothelial cells, and channels penetrate through the cell membranes. Water containing salts and small proteins is forced out through the holes. Gases and most wastes and nutrients diffuse through the endothelial cell membranes along their concentration gradients.

Figure 22-15   **A capillary containing red blood cells**

Capillaries are so narrow that red blood cells must pass through them single file, thereby facilitating exchange of gases by diffusion.

blood drained from her face as she beheld the gruesome sight . . . ," keep in mind that the heroine is experiencing constriction of the arterioles that supply her skin with blood. In such threatening situations, the sympathetic nervous system is activated and stimulates the smooth muscle of the arterioles to contract. This contraction raises blood pressure overall, but selective constriction also redirects blood to the heart and muscles, where it may be needed for vigorous action, and away from the skin, where it is less essential.

On a hot summer day, however, you become flushed as the arterioles in your skin expand and bring more blood to the skin capillaries. Bringing this blood closer to the surface enables your body to dissipate excess heat to the outside and to maintain a proper internal temperature. In contrast, in extremely cold weather, fingers and toes can become frostbitten because the arterioles supplying the extremities constrict. The blood is shunted to vital organs such as the heart and brain, which cannot function properly if their temperature drops. By minimizing blood flow to the heat-radiating extremities, the body conserves heat.

The flow of blood in capillaries is regulated by tiny rings of smooth muscle called **precapillary sphincters** that surround the junctions between arterioles and capillaries (see Fig. 22-13). These sphincters open and close in response to local changes that signal the needs of nearby tissues. For example, accumulation of carbon dioxide, lactic acid, or other cellular wastes signals the need for increased blood flow to the tissues. These signals cause the precapillary sphincters as well as the muscles in nearby arterioles to relax, thus increasing blood flow through the capillaries.

The Lymphatic System

The **lymphatic system** consists of a network of lymphatic capillaries and larger vessels that empty into the circulatory system, numerous small lymph nodes, and two

additional organs: the thymus and the spleen (Fig. 22-17). Although not strictly part of the circulatory system, the lymphatic system is closely associated with it. The lymphatic system has several important functions:

1. Removal of excess fluid and dissolved substances that leak from the capillaries
2. Transport of fats from the intestine to the bloodstream
3. Defense of the body by exposing bacteria and viruses to white blood cells

Lymphatic Vessels Resemble the Veins and Capillaries of the Circulatory System

Like blood capillaries, lymph capillaries form a complex network of microscopically narrow, thin-walled vessels into which substances can move readily. In contrast to those of blood capillaries, lymph capillary walls are composed of cells with openings between them that act as one-way valves. These openings allow relatively large particles, along with fluid, to be carried into the lymph capillary. Also unlike blood capillaries, which form a continuous connected network, lymph capillaries dead-end in the body's tissues (Fig. 22-18). Materials collected by the lymph capillaries flow into larger lymph vessels, through which lymph is forced as blood is through veins. Large lymph vessels have somewhat

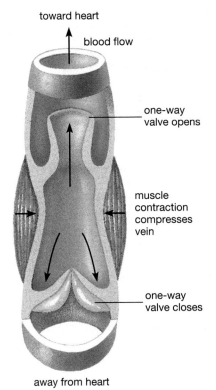

Figure 22-16   **Valves maintain directionality of blood flow in veins**

Veins and venules have one-way valves that maintain flow in the proper direction. When the vein is compressed by nearby muscles, the valves allow blood to flow toward the heart but clamp shut to prevent backflow.

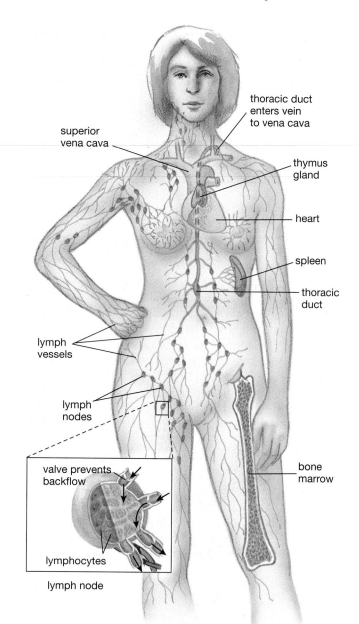

Figure 22-17   **The human lymphatic system**

This figure illustrates lymph vessels, lymph nodes, and two auxiliary lymph organs, the thymus and spleen. The lymph is returned to the circulatory system by way of the thoracic duct, which empties into the vena cava, a large vein. (Inset) A cross section of a lymph node. The node is filled with channels lined with white blood cells (lymphocytes) that attack foreign matter in the lymph.

muscular walls, but most of the impetus for lymph flow comes from the contraction of nearby muscles, such as those used in breathing and walking. As in blood veins, the direction of flow is regulated by one-way valves (Fig. 22-19).

The Lymphatic System Returns Fluids to the Blood

As described earlier, dissolved substances are exchanged between the capillaries and body cells by means of interstitial fluid (derived from blood plasma), which bathes nearly all

the cells of the body. In an average person, about 3 liters more fluid leaves the blood capillaries than is reabsorbed by them each day. One function of the lymphatic system is to return this excess fluid and its dissolved proteins and other substances to the blood. As interstitial fluid accumulates, its pressure forces the fluid through the openings in the lymph capillaries (see Fig. 22-18). The lymphatic system transports this fluid, now called **lymph**, back to the circulatory system.

The Lymphatic System Transports Fats from the Intestine to the Blood

As you will learn in Chapter 23, the small intestine is richly supplied with lymph capillaries. After absorbing digested fats, intestinal cells release fat globules into the interstitial fluid. These globules are too large to diffuse into blood capillaries but can move easily through the openings between lymph capillary cells. Once in the lymph, they are dumped into a large vein that enters the heart. After a fatty meal, these fat globules may make up 1% of the lymphatic fluid.

The Lymphatic System Helps Defend the Body against Disease

In addition to its other roles, the lymphatic system helps defend the body against foreign invaders such as bacteria and viruses. In the linings of the respiratory, di-

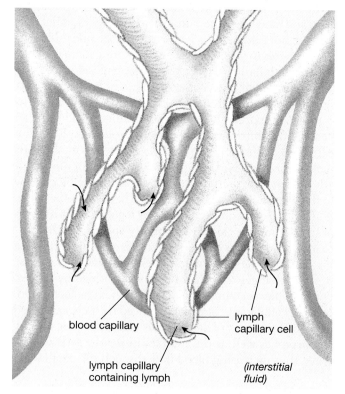

Figure 22-18   **Lymph capillaries**

Lymph capillaries end blindly in the body tissues, where pressure from the accumulation of interstitial fluid forces the fluid into the lymph capillaries through valvelike openings between lymph capillary cells.

Figure 22-19 A valve in a lymph vessel
Like blood-carrying veins, lymph vessels have internal one-way valves that direct the flow of lymph toward the large veins into which they empty.

gestive, and urinary tracts are patches of connective tissue containing large numbers of lymphocytes. The largest of these patches are the **tonsils**, located in the cavity behind the mouth. The large lymph vessels are interrupted periodically by kidney bean–shaped structures about 2.5 centimeters (1 inch) long called **lymph nodes** (see Fig. 22-17). Lymph is forced through channels within the nodes that are lined with masses of macrophages. Lymphocytes are also produced in the nodes. Both of these white blood cells recognize and destroy foreign particles, such as bacteria and viruses, and are killed in the process. The painful swelling of lymph nodes that accompanies certain diseases, such as mumps, is largely a result of the accumulation of dead lymphocytes and macrophages and of the virus-infected cells they have engulfed.

The **thymus** and the **spleen** are often considered part of the lymphatic system (see Fig. 22-17). The thymus is located beneath the breastbone slightly above the heart and is responsible for production of lymphocytes. The thymus is particularly active in infants and young children but decreases in size and importance in early adulthood. The spleen is located in the left side of the abdominal cavity, between the stomach and diaphragm. Just as the lymph nodes filter lymph, the spleen filters blood, exposing it to macrophages and lymphocytes that destroy foreign particles and aged red blood cells.

Respiratory Systems Facilitate Gas Exchange by Diffusion

Although animal respiratory systems are amazingly diverse, they all share two features that facilitate diffusion. First, the respiratory surface must remain moist, because gases must be dissolved in fluid when they enter or leave living cells. Second, the respiratory system must have a sufficiently large surface area in contact with the environment to allow adequate gas exchange by diffusion. As you may recall from Chapter 6, substances tend to diffuse from areas of high concentration to areas of low concentration: a situation described as a concentration gradient. Gas exchange in all organisms is ultimately a result of diffusion. In animals, carbon dioxide builds up and oxygen is depleted in cells as a result of their metabolic activities. This activity creates concentration gradients that favor diffusion of carbon dioxide out of cells and diffusion of oxygen into them.

Most animals have evolved specialized respiratory systems that interact closely with their circulatory systems, exchanging gases between the cells and the environment. The transfer of gases from the environment to the blood to the cells and back usually occurs in stages that alternate bulk flow with diffusion. During **bulk flow**, fluids or gases move in "bulk" through relatively large spaces, from areas of higher pressure to areas of lower pressure. Bulk flow contrasts with diffusion, in which molecules move individually from areas of high concentration to areas of lower concentration (see Chapter 6). In general, gas exchange in respiratory systems occurs in the following stages (see Fig. 22-22):

1. Air or water containing oxygen is moved across a respiratory surface by bulk flow, often facilitated by muscular breathing movements.
2. Oxygen and carbon dioxide are exchanged through the respiratory surface by diffusion. Diffusion across the respiratory membranes carries oxygen into the capillaries of the circulatory system and removes carbon dioxide.
3. Gases are transported between the respiratory system and the tissues by bulk flow of blood as it is pumped throughout the body by the heart.
4. Gases are exchanged between the tissues and the circulatory system by diffusion. At the tissues, oxygen diffuses out of the capillaries and carbon dioxide diffuses into them along their concentration gradients.

The Human Respiratory System

Human and most other terrestrial vertebrates respire using lungs. **Lungs** are chambers containing moist, delicate respiratory surfaces that are protected within the body, where water loss is minimized and the body wall provides support. The respiratory system in humans and other lung-breathing vertebrates can be divided into two parts: the **conducting portion** and the **gas-exchange portion**. The conducting portion consists of a series of passageways that carry air into the gas-exchange portion, where gas is exchanged with the blood in tiny sacs in the lungs.

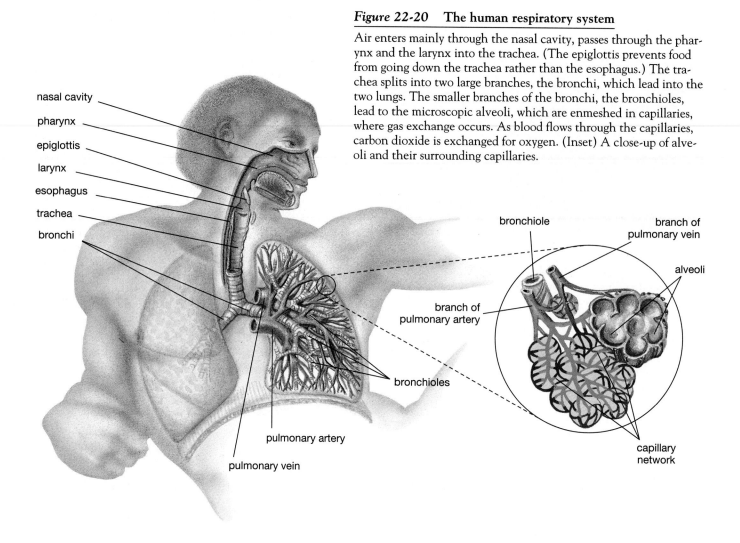

Figure 22-20   **The human respiratory system**

Air enters mainly through the nasal cavity, passes through the pharynx and the larynx into the trachea. (The epiglottis prevents food from going down the trachea rather than the esophagus.) The trachea splits into two large branches, the bronchi, which lead into the two lungs. The smaller branches of the bronchi, the bronchioles, lead to the microscopic alveoli, which are enmeshed in capillaries, where gas exchange occurs. As blood flows through the capillaries, carbon dioxide is exchanged for oxygen. (Inset) A close-up of alveoli and their surrounding capillaries.

The Conducting Portion of the Respiratory System Carries Air to the Lungs

Air enters through the nose or the mouth, passes through a common chamber, the **pharynx**, and then travels through the **larynx** (Fig. 22-20). The opening to the larynx is guarded by the epiglottis, a flap of tissue supported by cartilage. During normal breathing, the epiglottis is tilted upward, allowing air to flow freely into the larynx (see Fig. 22-20). During swallowing, the epiglottis tilts downward and covers the larynx, directing substances into the esophagus. Within the larynx are the **vocal cords**, bands of elastic tissue controlled by muscles. Muscular contractions can cause the vocal cords to partially obstruct the opening within the larynx, so that exhaled air causes them to vibrate, giving rise to the tones of speech or song. The tones can be varied in pitch by stretching the cords and articulated into words by movements of the tongue and lips.

Inhaled air continues past the larynx into the **trachea**, a flexible tube whose walls are reinforced with semicircular bands of stiff cartilage. Within the chest, the trachea splits into two large branches called **bronchi** (singular, **bronchus**), one leading to each lung. Inside the

lung, each bronchus branches repeatedly into ever-smaller tubes called **bronchioles**. These lead finally to the microscopic alveoli, tiny air sacs where gas exchange occurs (see Fig. 22-20).

During its passage through the conducting system, air is warmed and moistened. Much of the dust and bacteria it carries is trapped in mucus secreted by cells lining the passages. The mucus, with its trapped debris, is continuously swept upward toward the pharynx by cilia lining the bronchioles, bronchi, and trachea. Upon reaching the pharynx, the mucus is coughed up or swallowed. Smoking interferes with this cleansing process by paralyzing the cilia (see "Health Watch: Smoking—A Life and Breath Decision").

Gas Exchange Occurs in the Alveoli

Each lung is packed with around 1.5 to 2.5 million **alveoli** (singular, **alveolus**). These tiny (0.2-millimeter diameter) chambers give magnified lung tissue the appearance of sponge cake (see Fig. E22-3). The thin-walled alveoli provide an enormous surface area for diffusion—about 75 square meters (800 square feet), similar in size to a tennis

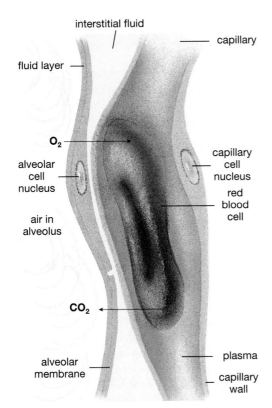

Figure 22-21   **Gas exchange between alveoli and capillaries**

The relationship between the alveolar membranes and the capillary walls facilitates the exchange of oxygen (O_2) and carbon dioxide (CO_2) by diffusion. Because the alveoli and capillaries are only one cell thick and because the cells are coated in a thin layer of water, gases dissolve and diffuse easily.

court and 80 times the total skin surface area of an adult human. The alveoli, which cluster about the end of each bronchiole like a bunch of grapes, are entirely enmeshed in capillaries (see Fig. 22-20). Because both the alveolar wall and the adjacent capillary walls are only one cell thick, the air is extremely close to the blood in the capillaries. The lung cells remain moist because they are coated by a thin layer of water lining each alveolus. Gases dissolve in this water and diffuse through the alveolar and capillary membranes (Fig. 22-21).

Blood is pumped to the lungs by the heart after circulating through the body tissues. The incoming blood surrounding the alveoli is low in oxygen (because the body cells have used it up) and high in carbon dioxide (which is released by the cells). In the alveoli, oxygen diffuses from the air, where its concentration is high, into the blood, where its concentration is low. Conversely, carbon dioxide diffuses out of the blood, where its concentration is high, into the air in the alveoli, where its concentration is lower (Fig. 22-22).

Blood from the lungs, oxygenated and purged of carbon dioxide, returns to the heart, which pumps it to the body tissues. In the tissues, the concentration of oxygen is lower than in the blood, and oxygen diffuses into the cells. Carbon dioxide, which has built up in the cells, diffuses into the blood.

Oxygen and Carbon Dioxide Are Transported Using Different Mechanisms

In the blood, oxygen binds loosely and reversibly with a large, iron-containing protein, hemoglobin, in the red blood cells, as noted earlier (Fig. 22-23, p. 416). Each hemoglobin molecule can bind up to four oxygen molecules (eight oxygen atoms). Nearly all the oxygen carried by the blood is bound to hemoglobin. By removing oxygen from solution in the plasma, hemoglobin maintains a high concentration gradient of oxygen from the air to the blood, enhancing diffusion into the blood. Thanks to hemoglobin, our blood can carry about 70 times as much oxygen as it could if the oxygen were simply dissolved in the plasma. As hemoglobin binds oxygen, the protein undergoes a slight change in shape, which alters its color. Deoxygenated blood is dark maroon-red and appears blue through the skin; oxygenated blood is a bright cherry-red. Carbon monoxide gas (CO) is highly toxic because it

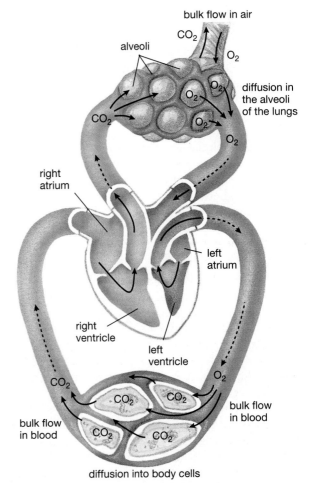

Figure 22-22   **An overview of gas exchange**

Oxygen is transported first by bulk flow as air is pumped in and out of the lungs. It moves by diffusion through the alveolar capillary walls and then is transported again by bulk flow in the bloodstream. Finally, diffusion carries oxygen into the cells of the body. Carbon dioxide follows the opposite path. Red blood is oxygenated, blue blood is deoxygenated.

HEALTH WATCH

Smoking—A Life and Breath Decision

The American Lung Association reports that an estimated 419,000 Americans die of smoking-related diseases each year: 1 out of every 5 deaths in this country. Further, the cost of health care and loss of productivity due to smoking-related ailments costs the United States about $65 billion annually. The American Cancer Society estimates that 153,000 people died of lung cancer in 1994 and that 87% of those deaths were attributable to smoking. The rest of the deaths due to smoking result from a combination of smoking-induced emphysema, chronic bronchitis, heart disease, and other cancers including cancer of the mouth, larynx, esophagus, pancreas, bladder, and kidney.

Tobacco smoke has a dramatic impact on the human respiratory tract (Fig. E22-2). As smoke is inhaled through the nose, trachea, and bronchi, toxic substances such as nicotine and sulfur dioxide paralyze the cilia lining the respiratory tract; a single cigarette can inactivate them for a full hour. Because the function of these ciliary sweepers is to remove inhaled particles, smoking inhibits them just when they are most needed. The visible portion of cigarette smoke consists of billions of microscopic carbon particles. Adhering to these particles are toxic substances, of which a dozen or more are carcinogenic (cancer causing). With the cilia out of action, the particles stick to the walls of the respiratory tract or enter the lungs. Thus, smokers encounter a higher-than-normal risk of cancer in all areas of the respiratory tract touched by smoke.

Cigarette smoke also impairs the amoebalike white blood cells (macrophages) that defend the respiratory tract by engulfing foreign particles and bacteria. Consequently, still more bacteria, dust, and smoke particles enter the lungs. In response to the irritation of cigarette smoke, the respiratory tract increases production of mucus, a third method of trapping foreign particles. But without the cilia to sweep it along, the mucus builds up and can obstruct the airways; the familiar "smoker's cough" is an attempt to expel the mucus. Microscopic smoke particles find a secure lodging place in the tiny, moist alveoli deep within the lungs. There they accumulate over the years until the lungs of a heavy smoker are literally blackened (see Fig. E22-2). The longer the delicate tissues of the lungs are exposed to the carcinogens on the trapped particles, the greater the chance that cancer will develop.

Chronic bronchitis and emphysema together kill about 85,000 persons each year in the United States, and smoking is the leading contributing factor to each of these diseases, according to the American Lung Association. Chronic bronchitis is a persistent lung infection characterized by coughing, swelling of the lining of the respiratory tract, an increase in mucus production, and a decrease in the number and activity of cilia. The result: a decrease in air flow to the alveoli. **Emphysema** occurs when toxic substances in cigarette smoke, such as nitrogen oxides and sulfur dioxide, cause the body to produce substances that reduce the elasticity and increase the brittleness of lung tissue. As the brittle alveoli rupture, the lung gradually loses its normal sponge-cake appearance and more closely resembles blackened Swiss cheese (see Fig. E22-3). The loss of the alveoli, where gas exchange occurs, leads to oxygen deprivation of all body tissues. The emphysema victim's breathing becomes labored and grows increasingly worse until death.

Figure E22-2   **Smoking causes lung cancer**

In this photo, lung cancer is visible as a large, pale mass; the lung tissue surrounding it is blackened by trapped smoke particles. Lung cancer is usually fatal.

Carbon monoxide, present in high levels in cigarette smoke, binds tenaciously to red blood cells in place of oxygen. This reduces the blood's oxygen-carrying capacity and thereby increases the work the heart must do. Chronic bronchitis and emphysema compound this problem. As a result, smokers are 70% more likely than nonsmokers to die of heart disease. The carbon monoxide in cigarette smoke may also contribute to the reproductive problems experienced by pregnant women who smoke, including increased incidence of miscarriage, lower birth weight of their babies, and learning impairment of their young children.

There is convincing evidence that "passive smoking" (which occurs when nonsmokers are forced to breathe air polluted by cigarette smoke) poses real health hazards as well. Various studies have concluded that infants whose mothers smoke are more likely to suffer from bronchitis and pneumonia than are infants whose mothers don't smoke.

Decreased lung capacity has been reported in children whose mothers smoked, and abnormally thickened and stiffened heart walls have been reported in 11- and 12-year-old boys whose parents smoked. Angina sufferers are likely to experience chest pain after sitting in a smoke-filled room. Studies have concluded that nonsmoking spouses of smokers face a 30% higher risk of both heart attack and lung cancer than do spouses of nonsmokers. The Environmental Protection Agency estimates that passive smoking is responsible for 3000 deaths of nonsmokers each year, and it has classified environmental tobacco smoke as a known human carcinogen.

There is hope for people who stop smoking before irreversible lung damage has occurred. Healing begins immediately after smoking stops, and the chances of a heart attack or lung cancer gradually drop. After 10 to 15 smoke-free years, former smokers' mortality rates approach those of people who never smoked.

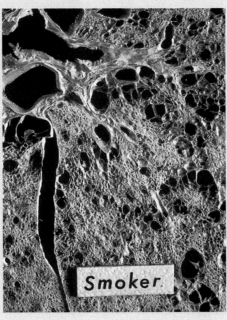

Non Smoker

Smoker

Figure E22-3 **Smoking causes emphysema**

A section through a normal lung appears almost opaque, but the lung of a smoker suffering from emphysema is full of large holes, each caused by the rupture of hundreds of alveoli as the lung tissue loses elasticity.

"fools" the hemoglobin, binding in place of oxygen and over 200 times as tenaciously. Hemoglobin containing CO is also bright red, but it is incapable of transporting oxygen. Most victims of asphyxiation have bluish lips and nail beds because their hemoglobin is deoxygenated, but the lips and nail beds of victims of carbon monoxide poisoning are brighter red than normal.

The transport of carbon dioxide (CO_2) is more complex. Most of the CO_2 carried in the blood (about 70%) reacts with water to form bicarbonate ion (HCO_3^-), which is carried in the blood plasma. About 20% of the CO_2 is returned to the lungs bound to hemoglobin that has released its O_2 to the tissues. The small amount remaining is carried dissolved in the plasma (Fig. 22-23). The production of bicarbonate ion and the binding of CO_2 to hemoglobin both reduce the concentration of dissolved CO_2 in the blood and increase the gradient for CO_2 to flow from the body cells into the blood.

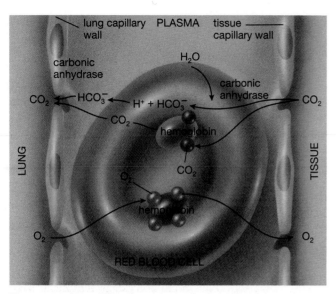

Figure 22-23   **The chemistry of gas exchange**

The chemical combination of water and carbon dioxide forming bicarbonate ion (HCO_3^-) occurs within red blood cells and is catalyzed by the enzyme carbonic anhydrase. The bicarbonate then diffuses back into the plasma. Carbonic anhydrase embedded in the capillary walls catalyzes the reverse reaction, allowing carbon dioxide (CO_2) to diffuse out of the capillaries and into the alveoli.

Air Is Inhaled Actively and Exhaled Passively

Outside the lungs, the chest cavity is airtight, bounded by neck muscles and connective tissue on top and the dome-shaped, muscular **diaphragm** on the bottom. Surrounding and protecting the lungs is the rib cage within the wall of the chest. Lining the chest and surrounding the lungs is a double layer of membranes called **pleural membranes**. These contribute to the airtight seal between the lungs and the chest wall.

Breathing occurs in two stages: **inhalation**, during which air is actively drawn into the lungs, and **exhalation**, during which it is passively expelled from the lungs. Inhalation is accomplished by making the chest cavity larger. To do this, the diaphragm muscles contract, drawing the diaphragm downward, and the rib muscles also contract, lifting the ribs up and outward (Fig. 22-24). When the chest cavity is expanded, the lungs expand with it, because a vacuum holds them tightly against the inner wall of the chest. (If the chest is punctured and air leaks in between the pleural membranes and the lung, the lung will collapse.) As the lungs expand, their increased volume creates a partial vacuum that draws air into the lungs.

Exhalation occurs automatically when the muscles causing inhalation are relaxed. The relaxed diaphragm domes upward, and the ribs fall down and inward, decreasing the size of the chest cavity and forcing air out of the lungs. More air can be forced out by contracting the abdominal muscles. After exhalation, the lungs still contain air. This air prevents the thin alveoli from collapsing

and fills the space within the conducting portion of the respiratory system. A normal breath moves only about 500 milliliters (1 pint) of air into the respiratory system. Of this, only about 350 milliliters reaches the alveoli for gas exchange. Deeper breathing during exercise causes several times this volume to be exchanged.

Breathing Rate Is Controlled by the Respiratory Center of the Brain

Breathing occurs rhythmically and automatically without conscious thought. But, unlike the heart, the muscles used in breathing are not self-activating; each contraction is stimulated by impulses from nerve cells. These impulses originate in the **respiratory center**, which is located in the brainstem just above the spinal cord. Nerve cells in the respiratory center generate cyclic bursts of impulses that cause alternating contraction and relaxation of the respiratory muscles.

The respiratory center receives input from several sources and adjusts breathing rate and volume to meet the body's changing needs. The brainstem contains receptor neurons that monitor the concentration of carbon dioxide in the blood. The respiratory rate is regulated to maintain a constant level of carbon dioxide in the blood. An elevated level of carbon dioxide (caused, for example, by an increase in cellular activity such as occurs in muscle cells during exercise) signals a need for more oxygen and causes the receptors to stimulate an increase in the rate and depth of breathing. These receptors are extremely sen-

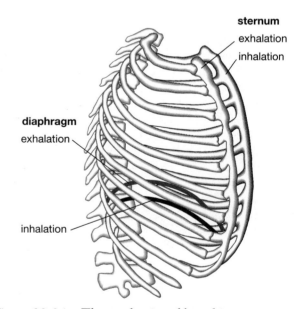

Figure 22-24   **The mechanics of breathing**

During inhalation, rhythmic nerve impulses from the brain stimulate the diaphragm muscle to contract (pulling it downward) and the muscles surrounding the ribs to contract (moving them up and outward). The result is an increase in the size of the chest cavity, causing air to rush in. Relaxation of these muscles (exhalation) allows the diaphragm to dome upward, and the rib cage to collapse, forcing air out of the lungs.

sitive; an increase in carbon dioxide of only 0.3% can double the breathing rate.

The respiratory rate is much less sensitive to changes in oxygen concentration, because even normal breathing supplies an overabundance of oxygen. But if blood oxygen levels fall drastically, receptors in the aorta and carotid arteries stimulate the respiratory center. Surprisingly, when a person begins strenuous activity, such as running, an increase in breathing rate actually *precedes* any changes in blood gas levels. Apparently, when higher brain centers activate muscles during heavy exercise, they simultaneously stimulate the respiratory center to increase breathing rate. Breathing activity is then "fine-tuned" by the receptors monitoring carbon dioxide concentrations.

✖ S U M M A R Y O F K E Y C O N C E P T S

The Human Circulatory System

Circulatory systems transport blood rich in dissolved nutrients and oxygen close to each cell, where nutrients can be released and wastes absorbed by diffusion. All circulatory systems have three major parts: a fluid, blood; a system of channels, vessels, to conduct the blood; and a pump, a heart, to circulate the blood. Circulatory systems in humans and other vertebrates transport gases, hormones, and wastes; distribute nutrients; help regulate body temperature; and defend the body against disease.

In the four-chambered heart, blood is pumped separately to the lungs and through the body, maintaining complete separation of oxygenated and deoxygenated blood. Deoxygenated blood is collected from the body in the right atrium and passed to the right ventricle, which pumps it to the lungs. Oxygenated blood from the lungs enters the left atrium, is passed to the left ventricles and pumped to the rest of the body.

The heart cycle consists of two stages: systole, during which the ventricles contract, and diastole, when the ventricles relax. The direction of blood flow is maintained by valves within the heart. The contractions of the heart are initiated and coordinated by the sinoatrial node, the heart's pacemaker. Heart rate can be modified by the nervous system and by hormones such as epinephrine.

Blood is composed of both fluid and cellular materials. The fluid plasma consists of water that contains proteins, hormones, nutrients, gases, and wastes. Red blood cells, or erythrocytes, are packed with a large iron-containing protein called hemoglobin, which carries oxygen. Their numbers are regulated by the hormone erythropoietin. Proteins on their cell membrane determine blood type. There are five types of white blood cells, or leukocytes, that fight infection (see Table 22-1). Platelets, which are fragments of megakaryocytes, are important for blood clotting.

Blood leaving the heart travels in sequence through arteries, arterioles, capillaries, venules, veins, and then back to the heart. Each vessel is specialized for its role. Elastic, muscular arteries help pump the blood along. The thin-walled capillaries are the sites of exchange of materials between the body cells and the blood. Veins provide a low-resistance path back to the heart, with one-way valves maintaining the direction of blood flow. The distribution of blood is regulated by the constriction and dilation of arterioles under the influence of the sympathetic nervous system and local factors such as the amount of carbon dioxide in the tissues. Local factors also regulate precapillary sphincters, which control blood flow in the capillaries.

The Lymphatic System

The human lymphatic system consists of lymphatic vessels resembling blood veins and capillaries, tonsils, lymph nodes, and the thymus and spleen. The lymphatic system removes excess interstitial fluid that leaks through blood capillary walls. It transports fats to the bloodstream from the intestine and fights infection by filtering the lymph through lymph nodes, where white blood cells ingest foreign invaders such as bacteria. The thymus, which is most active in young children, produces lymphocytes that function in immunity. The spleen filters blood past macrophages and lymphocytes, which remove bacteria and damaged blood cells.

Respiratory Systems Facilitate Gas Exchange by Diffusion

The transfer of gases between respiratory systems and tissues occurs in a series of stages that alternate bulk flow with diffusion. Air or water moves by bulk flow past the respiratory surface, and gases in blood are carried by bulk flow. Gases move by diffusion across membranes between the respiratory system and the capillaries and between the capillaries and the tissues.

The Human Respiratory System

The human respiratory system consists of a conducting portion and a gas-exchange portion. Air passes first through the conducting portion, consisting of the nose and mouth, pharynx, larynx, trachea, bronchi, and bronchioles, and then into the gas-exchange portion, composed of microscopic sacs called alveoli. Blood within a dense capillary network surrounding the alveoli releases carbon dioxide and absorbs oxygen from the air.

Most of the oxygen in the blood is bound to hemoglobin within red blood cells. Hemoglobin binds oxygen at concentrations typical of those in alveolar capillaries and releases it at the lower oxygen concentrations found in the body tissues. Carbon dioxide diffuses into the blood from the tissues and is transported in three ways. Most is combined with water to form bicarbonate ion. Some is transported bound to hemoglobin, and a small amount is carried as dissolved carbon dioxide.

Breathing involves actively drawing air into the lungs by contracting the diaphragm and the rib muscles, which expand the chest cavity. Relaxing these muscles causes the chest cavity to collapse, expelling the air. Respiration is controlled by nerve impulses originating in the respiratory center of the brainstem, and its rate is modified by receptors, such as those in the brainstem that monitor carbon dioxide levels in the blood.

✖ K E Y   T E R M S

alveolus   p. 412
angina pectoris   p. 406
arteriole   p. 405
artery   p. 399
atherosclerosis   p. 407
atrioventricular (AV) node   p. 400
atrioventricular valve   p. 399
atrium   p. 398
basophil   p. 404
bicuspid valve   p. 399
blood   p. 398
bronchiole   p. 412
bronchus   p. 412
bulk flow   p. 411
capillary   p. 405
cardiac cycle   p. 399
closed circulatory system   p. 397
conducting portion   p. 411
diaphragm   p. 416
diastole   p. 399
emphysema   p. 414
eosinophil   p. 404
erythroblastosis fetalis   p. 404
erythrocyte   p. 402

erythropoietin   p. 403
exhalation   p. 416
fibrillation   p. 400
fibrin   p. 404
gas-exchange portion   p. 411
heart   p. 398
heart attack   p. 407
hemoglobin   p. 403
hypertension   p. 406
inhalation   p. 416
interstitial fluid   p. 407
larynx   p. 412
leukocyte   p. 404
lung   p. 411
lymph   p. 410
lymphatic system   p. 409
lymph nodes   p. 411
lymphocyte   p. 404
macrophage   p. 404
megakaryocyte   p. 404
monocyte   p. 404
neutrophil   p. 404
pacemaker   p. 400
pharynx   p. 412

plaque   p. 407
plasma   p. 402
platelet   p. 404
pleural membrane   p. 416
precapillary sphincter   p. 409
pulmonary circulation   p. 398
respiratory center   p. 416
Rh factor   p. 403
semilunar valve   p. 399
sinoatrial (SA) node   p. 400
spleen   p. 411
stroke   p. 406
systemic circulation   p. 399
systole   p. 399
thrombin   p. 404
thymus   p. 411
tonsils   p. 411
trachea   p. 412
tricuspid valve   p. 399
vein   p. 408
ventricle   p. 398
venule   p. 408
vessel   p. 398
vocal cord   p. 412

✖ T H I N K I N G   T H R O U G H   T H E   C O N C E P T S

Multiple Choice

1. Which of the following is NOT an important function of the vertebrate circulatory system?
   a. transport of nutrients and respiratory gases
   b. regulation of body temperature
   c. protection of the body by circulating antibodies
   d. removal of waste products for excretion from the body
   e. defense against blood loss, through clotting

2. What event initiates blood clotting?
   a. contact with an irregular surface by platelets and other factors in plasma
   b. production of the enzyme thrombin
   c. conversion of fibrinogen into fibrin
   d. conversion of fibrin into fibrinogen
   e. excess flow of blood through a capillary

3. The sites of waste, nutrient, gas, and hormone exchange between the blood and body cells are the
   a. arteries   b. arterioles
   c. capillaries   d. veins
   e. all blood vessels

4. What produces systolic blood pressure?
   a. contraction of the right atrium
   b. contraction of the right ventricle
   c. contraction of the left atrium
   d. contraction of the left ventricle
   e. the pause between heartbeats

5. Which of the following is NOT a component of plasma?
   a. water   b. globulins
   c. fibrinogen   d. albumins
   e. platelets

6. Lymph most closely resembles which of the following?
   a. blood   b. urine
   c. plasma   d. interstitial fluid
   e. water

7. With which other system do specialized respiratory systems most closely interact in exchanging gases between the cells and the environment?
   a. the skin   b. the excretory system
   c. the circulatory system   d. the muscular system
   e. the nervous system

8. The gas-exchange portion of the human respiratory system is the
   a. larynx   b. trachea
   c. bronchi   d. pharynx
   e. alveoli

9. How is most of the oxygen transported in the blood?
   a. dissolved in plasma   b. bound to hemoglobin
   c. in the form of CO_2 d. as bicarbonate
   e. dissolved in water

10. Which of the following statements regarding cigarette smoking is true?
    a. Cigarette smoke damages the alveoli.
    b. Cigarette smoke decreases the blood oxygen level.
    c. Cigarette smoke causes many types of cancer.
    d. Cigarette smoke can lead to heart damage.
    e. all of the above

Review Questions

1. Trace the flow of blood through the circulatory system, starting and ending with the right atrium.
2. List three types of blood cells and describe their principal functions.
3. What are five functions of the vertebrate circulatory system?
4. In what way do veins and lymph vessels resemble one another? Describe how fluid is transported in each of these vessels.
5. Describe three important functions of the lymphatic system.
6. Distinguish among plasma, interstitial fluid, and lymph.
7. Describe veins, capillaries, and arteries, noting their similarities and differences.
8. Explain in detail what causes the vertebrate heart to beat.
9. Describe the cardiac cycle, and relate systole and diastole to the two readings taken during the measurement of blood pressure.

10. Describe how red blood cell number is regulated by a negative feedback system.
11. Describe the formation of an atherosclerotic plaque. What are the risks associated with atherosclerosis?
12. Trace the route taken by air in the vertebrate respiratory system, listing the structures through which it flows and the point where gas exchange occurs.
13. How are human respiratory movements initiated? How are they modified, and why are these controls adaptive?
14. What events occur during human inhalation? Exhalation? Which of these is always an active process?
15. Trace the pathway of an oxygen molecule in the human body starting with the nose and ending with a body cell.
16. Describe the effects of smoking on the human respiratory system.
17. Explain how bulk flow and diffusion interact to promote gas exchange between air and blood and between blood and tissues.
18. Compare carbon dioxide and oxygen transport in the blood. Include in your answer the source and destination of each.
19. Explain how the structure and arrangement of alveoli make them well suited for their role in gas exchange.

✵ A P P L Y I N G   T H E   C O N C E P T S

1. Doctors have attempted to transplant baboon hearts into humans whose hearts were failing. Discuss the implications of this operation from as many angles as you can think of.
2. Considering the prevalence of cardiovascular and respiratory diseases and the high, increasing costs of treating them, certain treatments may not be available to all who might benefit. What factors would you take into account in rationing high-cost, high-risk cardiovascular and respiratory procedures, such as heart transplants or heart-lung transplants?
3. Joe, a 45-year-old executive of a major corporation, has been diagnosed with mild hypertension. What treatments or lifestyle changes might Joe's doctor recommend? If Joe's hypertension becomes more severe, what treatments might Joe's doctor use? Should Joe be concerned about mild hypertension? Explain your answer.
4. Nicotine is a drug in tobacco that is responsible for several of the effects that smokers crave. Discuss the advantages and disadvantages of low-nicotine cigarettes.
5. Discuss why a brief exposure to carbon monoxide is much more dangerous than a brief exposure to carbon dioxide.
6. Mary, a strong-willed 3-year-old, threatens to hold her breath until she dies if she doesn't get her way. Can she carry out her threat? Explain.

✵ F O R   M O R E   I N F O R M A T I O N

Adler, T. "Fish Dishes May Catch On among Smokers." *Science News*, July 30, 1994. Discusses chronic obstructive pulmonary disease.

Houston, C. "Mountain Sickness." *Scientific American*, October 1992. The mechanisms of potentially fatal altitude sickness are explained.

Lawn, R. M. "Lipoprotein(a) in Heart Disease" *Scientific American*, June 1992. Describes the interaction of this protein with blood fats and its role in heart disease.

Perutz, M. "Hemoglobin Structure and Respiratory Transport." *Scientific American*, December 1978. How hemoglobin plays the dual role of transporting oxygen to the tissues and carbon dioxide back to the lungs.

Radetsky, P. "The Mother of All Blood Cells." *Discover*, March 1995. Discusses the discovery by Irving Weissman of a "stem cell" that gives rise to all other types of blood cells.

Zimmer, C. "The Body Electric." *Discover*, February 1993. Describes computer imaging of the heart's electrical activity.

Zucker, M. "The Functioning of Blood Platelets." *Scientific American*, June 1980. Describes the complex role of platelets in blood clotting.

N E T   W A T C H

On-line resources for this chapter are on the World Wide Web at:
http://www.prenhall.com/~audesirk (click on the table of contents link and then select Chapter 22).

23 Nutrition, Digestion and Excretion

A Brazilian toucan hungrily attacks an orange. Serrations on its long, colorful beak help it grip and cut the fruit.

In the primordial "soup" where life evolved, the earliest cells absorbed energy-rich nutrients directly through their cell membranes. But before simple life forms could invade the less hospitable environments that covered much of Earth, more complex ways of obtaining both energy and nutrients had to evolve. Over millions of years of natural selection, some organisms became photosynthetic, harvesting the energy of sunlight and the abundant carbon dioxide of the early atmosphere to produce high-energy molecules for immediate use and storage. The dead or living bodies of photosynthetic organisms, with their rich supply of stored energy, offered an abundant food source for organisms that evolved ways to digest them.

The simplest form of digestion is utilized by organisms such as fungi and bacteria, which live in environments where potential food surrounds them. These organisms secrete digestive enzymes outside their bodies, where the enzymes break down complex molecules into simple subunits that can be absorbed. More complex cells, such as those of certain unicellular aquatic organisms (for example, amoebae), take in particles of food (often smaller cells) and digest them within a vacuole into which enzymes are released. Such miniature intracellular digestive systems are also found in sponges, the simplest animals. More complex animals evolved digestive sacs and, finally, one-way digestive tracts.

Living cells have been compared to tiny factories, requiring a constant influx of raw materials and spewing out a constant stream of waste products, some of which are toxic at high levels. But living cells are delicate, requiring a precisely regulated environment. The extracellular fluid that bathes them must be close to neutral in pH, it must have precisely regulated concentrations of various salts, it must have appropriate levels of water and dissolved substances, and cellular waste products must not reach toxic levels.

The cells of the digestive system, through which most substances enter the body, are relatively unselective. Any molecule that can move into the body through the intestinal lining does so, including an excess of water, nutrients, salts, and minerals as well as nonnutritive substances such as drugs. However, even though we may load our digestive systems with pepperoni pizzas and hot fudge sundaes, our cells remain bathed in a precisely regulated solution of salts and nutrients. The burden of restoring and maintaining proper internal balance falls on the organs of homeostasis, particularly the kidneys of the urinary system.

In this chapter we discuss human nutrient requirements, the human digestive system through which we obtain those nutrients, and the excretory system, which eliminates cellular waste products and excess nutrients. Together, these systems fuel the body and help maintain the precise internal conditions necessary for survival.

Nutrition

Animal nutrients fall into five major categories: lipids, carbohydrates, proteins, minerals, and vitamins. These substances provide the body with its basic needs: (1) energy to fuel cellular metabolism and activities; (2) the chemical building blocks, such as amino acids, to construct complex molecules unique to each animal; and (3) minerals and vitamins that participate in a variety of metabolic reactions.

The Primary Sources of Energy Are Carbohydrates and Fats

Each living cell in an animal's body relies on a continuous expenditure of energy to maintain its incredible complexity and perform its specific functions. Three nutrients provide energy for animals: fats, carbohydrates, and proteins. In a "typical" American diet, fats provide about 38%, carbohydrates about 46%, and protein about 16% of the energy. These molecules are broken down during cellular respiration, and the energy derived from them is used to produce adenosine triphosphate (ATP) (see Chapter 8).

The energy in nutrients is measured in **calories**. A calorie is the amount of energy required to raise the temperature of 1 gram of water by 1 degree Celsius. The calorie content of foods is measured in units of 1000 calories (kilocalories), also known as **Calories** (with a capital C). The human body at rest burns about 1550 Calories per day (somewhat more for males and less for females). Exercise significantly boosts caloric requirements: Well-trained athletes can temporarily raise their calorie consumption from a resting rate of about 1 Calorie per minute to nearly 20 Calories per minute during vigorous exercise (Table 23-1).

Lipids Include Fats, Phospholipids, and Cholesterol

Lipids are a diverse group of molecules that generally contain long chains of carbon atoms and are insoluble in water. The principal types of lipids are fats, or triglycerides, phospholipids, and cholesterol (see Chapter 3). Fats are used primarily as a source of energy. Phospholipids are important components of cell membranes and also provide the insulating covering of neurons (nerve cells). Cholesterol is used in the synthesis of cell membranes, bile (which aids in fat breakdown), and certain hormones. Some animals can synthesize all the specialized lipids they need. Others must acquire specific types of lipid building blocks, called **essential fatty acids**, from their food. For example, humans are unable to synthesize linoleic acid, which is required for the synthesis of certain phospholipids, so we need to obtain this essential fatty acid from our diet.

Humans and most other animals store energy primarily as fat. When an animal's diet provides more energy than is expended through its metabolic activities, most of the excess carbohydrate, fat, or protein is converted to fat for storage; about 3600 Calories are stored in each pound of fat. Fats have two major advantages as energy-storage molecules. First, they are the most concentrated energy source, containing over twice the energy per unit weight of either carbohydrates or protein (about 9 Calories per gram for fats compared with about 4 per gram for proteins and carbohydrates). Second, lipids are hydrophobic, meaning they do not mix with water. Fat deposits, therefore, do not cause any extra accumulation of water in the body. For both these reasons, fats store more calories with less weight than do other molecules. Minimizing weight allows an animal to move faster (important for escaping predators and hunting prey) and to move using less energy (important when food supplies are limited).

TABLE 23-1 :: *Approximate Energy Consumed by Different Activities for a 150-Pound Person*

Activity	Calories/ min	Time to "Work Off"		
		500 Cal Cheeseburger	300 Cal Ice Cream Cone	70 Cal Apple
Cross-country skiing (5 mph)	12	42 min	25 min	6 min
Rowing (machine, vigorous)	11	45 min	27 min	6 min
Jogging (5.5 mph)	11	45 min	27 min	6 min
Bicycling (10 mph)	7	1 hr 10 min	43 min	10 min
Dancing	7	1 hr 10 min	43 min	10 min
Gardening	6.5	1 hr 17 min	46 min	11 min
Swimming (crawl, 20 yd/min)	5	1 hr 40 min	1 hr	14 min
Walking (3 mph)	3.8	2 hr 12 min	1 hr 19 min	18 min
Sitting quietly	1.7	4 hr 54 min	2 hr 56 min	41 min

Figure 23-1   **Fat provides insulation**

This walrus can withstand the icy waters of the polar seas because it is insulated with a thick layer of fat beneath the skin.

Mammals, who maintain an elevated body temperature, often make their fat deposits do double duty by providing insulation as well as storing energy. Fat is typically stored in a layer beneath the skin. Here it insulates the body, because fat conducts heat at only one-third the rate of other body tissues. Mammals who live near the North or South Pole or in cold ocean waters are particularly dependent on this insulating layer, which reduces the amount of energy they must expend to keep warm (Fig. 23-1).

Carbohydrates, Including Sugars and Starches, Are a Source of Quick Energy

Carbohydrates consist of monosaccharide and disaccharide sugars as well as longer chains of sugars called polysaccharides (see Chapter 3). Polysaccharides include starches, the principal energy-storage material of plants; glycogen, a short-term energy-storage molecule in animals; and cellulose, the major structural component of plant cell walls. During digestion, carbohydrates are broken down into sugars and absorbed. For practical purposes, body cells obtain their energy from a single sugar: glucose. Glucose can be derived from fats, amino acids, and the carbohydrates consumed in the diet.

Animals, including humans, store a carbohydrate called **glycogen** (a large, highly branched chain of glucose molecules) in the liver and muscles. Although humans can potentially store hundreds of pounds of fat, we store less than half a pound of glycogen. During exercise, such as running, the body draws on this store of glycogen as a source of quick energy. When the activity is prolonged, as in the case of a marathon runner, the stored glycogen can be totally depleted. The expression "hitting the wall" describes the extreme fatigue that distance runners may experience after exhausting their glycogen supply.

Proteins, Composed of Amino Acids, Perform a Wide Range of Functions within the Body

Each day, the body breaks down 20 to 30 grams of its own protein, which is metabolized for energy. This is replaced by protein taken in through the diet, and any excess amino acids are either broken down and used for energy or stored as fat. The breakdown of protein produces the waste product **urea**, which is filtered from the blood by the kidneys. Specialized diets in which protein is the major energy source place extra stress on the kidneys. The major role of dietary protein is as a source of amino acids to make new molecules (see Chapter 3).

Amino acids are used to synthesize certain hormones, other amino acids, some neurotransmitters (chemicals used in communication between neurons), and new proteins. These proteins have diverse roles in the body, acting as enzymes, receptors on cell membranes, oxygen transport molecules (hemoglobin), structural components (hair and nails), hormones, antibodies, and muscle proteins.

In the digestive tract, dietary protein is broken down into its amino acid subunits. Then, in the body cells, the amino acids are linked in specific sequences to form new proteins. The human liver can synthesize (from other amino acids) 9 of the 20 different amino acids used in proteins. Those that cannot be synthesized, called **essential amino acids,** must be supplied by the diet in foods such as meat, milk, eggs, corn, beans, and soybeans. Because many plant proteins are deficient in some of the essential amino acids, individuals on a vegetarian diet must include a variety of plants whose proteins together will provide all nine, or they risk protein deficiency. The nine essential amino acids and examples of how different plant proteins complement one another in providing them are illustrated in Table 23-2.

Minerals Are Elements and Small Inorganic Molecules Required by the Body

Animals require a wide variety of minerals, which are small inorganic molecules and elements (Table 23-3). Minerals must be obtained through the diet, either from food or dissolved in drinking water, because the body cannot manu-

TABLE 23-2 ⚏ *The Essential Amino Acids*

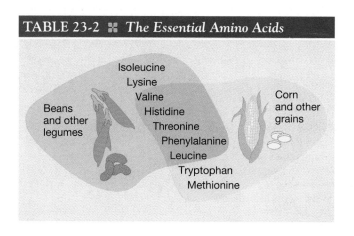

Beans and other legumes

Isoleucine
Lysine
Valine
Histidine
Threonine
Phenylalanine
Leucine
Tryptophan
Methionine

Corn and other grains

TABLE 23-3 ■ Human Water and Mineral Requirements[a]

Mineral	RDA[b] for Healthy Adult Male (milligrams)	Dietary Sources	Major Functions in Body	Deficiency Symptoms
Water	1.5 liters/day	Solid foods, liquids, drinking water	Transport of nutrients Temperature regulation Metabolic reactions	Thirst, dehydration
Calcium	800	Milk, cheese, green vegetables, legumes	Bone and tooth formation Blood clotting Nerve impulse transmission	Stunted growth Rickets, osteoporosis Convulsions
Phosphorus	800	Milk, cheese, meat, poultry, grains	Bone and tooth formation Acid-base balance	Weakness Demineralization of bone Loss of calcium
Potassium	2,500	Meats, milk, fruits	Acid-base balance Body water balance Nerve function	Muscular weakness Paralysis
Chlorine	2,000	Table salt	Formation of gastric juice Acid-base balance	Muscle cramps Apathy Reduced appetite
Sodium	2,500	Table salt	Acid-base balance Body water balance Nerve function	Muscle cramps Apathy Reduced appetite
Magnesium	350	Whole grains, green leafy vegetables	Activation of enzymes in protein synthesis	Growth failure Behavioral disturbances Weakness, spasms
Iron	10	Eggs, meats, legumes, whole grains, green vegetables	Constituent of hemoglobin and enzymes involved in energy metabolism	Iron-deficiency anemia (weakness, reduced resistance to infection)
Fluorine	2	Fluoridated water, tea, seafood	May be important in maintenance of bone structure	High frequency of tooth decay
Zinc	15	Widely distributed in foods	Constituent of enzymes involved in digestion	Growth failure Small sex glands
Iodine	0.14	Seafish and shellfish, dairy products, many vegetables, iodized salt	Constituent of thyroid hormones	Goiter
Copper Silicon Vanadium Tin Nickel Selenium Manganese	Not established (trace amounts)	Widely distributed in foods	Some unknown; some work in conjunction with enzymes	Unknown

[a]Modified from N. S. Scrimshaw and V. R. Vernon, "The Requirements of Human Nutrition," *Scientific American*, September 1976.
[b]Recommended Daily Allowance.

TABLE 23-4 ❖ *Human Vitamin Requirements*

Vitamin	*RDA for Adult (milligrams)*	*Dietary Sources*	*Major Functions in Body*	*Deficiency Symptoms*
Water soluble				
Vitamin B$_1$ (thiamin)	1.9	Milk, meat, bread	Coenzyme in metabolic reactions	Beriberi (muscle weakness, peripheral nerve changes, edema, heart failure)
Vitamin B$_2$ (riboflavin)	1.8	Widely distributed in foods	Constituent of coenzymes in energy metabolism	Reddened lips, cracks at corner of mouth, lesions of eye
Niacin	20	Liver, lean meats, grains, legumes	Constituent of two coenzymes in energy metabolism	Pellagra (skin and gastrointestinal lesions, nervous, mental disorders)
Vitamin B$_6$ (pyridoxine)	2	Meats, vegetables, whole-grain cereals	Coenzyme in amino acid metabolism	Irritability, convulsions, muscular twitching, dermatitis, kidney stones
Pantothenic acid	5–10	Milk, meat	Constituent of coenzyme A with a role in energy metabolism	Fatigue, sleep disturbances, impaired coordination
Folacin	0.4	Legumes, green vegetables, whole wheat	Coenzyme involved in nucleic and amino acid metabolism	Anemia, gastrointestinal disturbances, diarrhea, red tongue
Vitamin B$_{12}$	0.003	Meats, eggs, dairy products	Coenzyme in nucleic acid metabolism	Pernicious anemia, neurological disorders
Biotin	0.15–0.3	Legumes, vegetables, meats	Coenzymes required for fat synthesis, amino acid metabolism, and glycogen formation	Fatigue, depression, nausea, dermatitis, muscular pains
Choline	Not established Usual diet provides 500–900	Egg yolk, liver, grains, legumes	Constituent of phospholipids, precursor of the neurotransmitter acetylcholine	None reported in humans
Vitamin C (ascorbic acid)	45	Citrus fruits, tomatoes, green peppers	Maintenance of cartilage, bone, and dentine; collagen synthesis	Scurvy (degeneration of skin, teeth, blood vessels; epithelial hemorrhages)
Fat soluble				
Vitamin A (retinol)	1	Beta-carotene in green, yellow, red vegetables Retinol added to dairy products	Constituent of visual pigment Maintenance of epithelial tissues	Night blindness, permanent blindness
Vitamin D	0.01	Cod-liver oil, eggs, dairy products	Promotes bone growth and mineralization Increases calcium absorption	Rickets (bone deformities) in children; skeletal deterioration
Vitamin E (tocopherol)	15	Seeds, green leafy vegetables, margarines, shortenings	Antioxidant, prevents cellular damage	Possibly anemia
Vitamin K	0.03	Green leafy vegetables Intestinal bacteria produce	Important in blood clotting	Bleeding, internal hemorrhages

facture the inorganic substances. Required minerals include calcium, magnesium, and phosphorus, which are major constituents of bones and teeth. Others, such as sodium and potassium, are essential for muscle contraction and the conduction of nerve impulses. Iron is used in the production of hemoglobin, and iodine is found in hormones produced by the thyroid gland. In addition, trace amounts of several other minerals, including zinc, copper, and selenium, are required, often as parts of enzymes.

Vitamins Are Required in Small Amounts and Play Many Roles in Metabolism

Vitamins are a diverse group of organic compounds that animals require in small amounts. In general, vitamins cannot be synthesized by the body (or cannot be synthesized in adequate amounts) and must be obtained from food. Although our skin can manufacture some vitamin D when it is exposed to sunlight, because we spend so much time indoors most of us do not synthesize enough and must supplement it through our diet. The vitamins considered essential in human nutrition are listed in Table 23-4.

Human vitamins are often grouped into two categories, water soluble and fat soluble. Water-soluble vitamins include vitamin C and the 11 different compounds that make up the B vitamin complex. These substances dissolve in the water of the blood plasma and are excreted by the kidneys. They are not stored in the body in any appreciable amounts. Water-soluble vitamins generally work in the body in conjunction with enzymes to promote chemical reactions that supply energy or synthesize materials. Because each vitamin participates in several metabolic processes, a deficiency of a single vitamin can have wide-ranging effects (see Table 23-4).

The fat-soluble vitamins A, D, E, and K have even more varied roles. Vitamin K, for example, helps regulate blood clotting, and vitamin A is used to produce visual pigment (the light-capturing molecule in the retina of the eye). Fat-soluble vitamins can be stored in body fat and may accumulate in the body over time. For this reason, fat-soluble vitamins may be toxic if consumed in excessively high doses.

Nutritional Guidelines Help People Obtain a Balanced Diet

We are fortunate to live amid an abundance of food. However, the overwhelming diversity of food available in an American supermarket and the easy availability of "fast food" can lead to poor nutritional choices. To help people make informed choices, the U.S. government has developed recommendations and goals for the "average American," which are summarized in Table 23-5. Further help is provided by the food pyramid designed by the U.S. Department of Agriculture that illustrates the relative abundances of different food groups in an optimal diet (Fig. 23-2). A third source of information is the nutritional labeling now required on commercially packaged foods. These provide complete information on calorie, fiber, fat, sugar, and vitamin content (Fig. 23-3). Some fast-food chains also make fliers available listing nutritional information about their products.

The Challenge of Digestion

The digestive systems of humans and other animals take in, and then **digest**, that is, physically grind up and chemically break down, the complex molecules of their food into

TABLE 23-5 :: Dietary Changes Advocated by the U.S. Government		
	Percentage of Total Daily Energy Intake	
Dietary Component	*Average American Diet*	*Dietary Goals*
Carbohydrates	46	58
Lipids	38	30
Proteins	16	12

Summary of Recommendations:
1. Increase consumption of fruits, vegetables, and whole grains.
2. Decrease consumption of refined sugars.
3. Decrease consumption of fats, replace saturated with unsaturated fats.
4. Decrease consumption of animal fats by selecting lean meats, poultry, and fish.
5. Decrease consumption of high-cholesterol foods, such as butter and eggs.
6. Decrease consumption of salt and foods high in salt content.
7. Decrease caloric intake to maintain desirable weight.

Figure 23-2   **The dietary pyramid**

Recommended by the U.S. Department of Agriculture, this chart shows suggested daily servings.

simpler molecules that can be absorbed. Material that cannot be used or broken down is then expelled from the body.

Animals eat the bodies of other organisms, bodies that may resist becoming food. The plant body, for example, armors each cell with a wall of indigestible cellulose. Animal bodies may be covered with equally indigestible fur, scales, or feathers. In addition, the complex lipids, proteins, and carbohydrates found in food are not in a form that can be used directly. They must be broken down before they can be absorbed and distributed to the cell of the animal that has consumed them, where they are recombined in unique ways. Different types of animals meet the challenge of acquiring nutrients with various types of digestive tracts, each finely tuned to a unique diet and life-style. Among this diversity, however, we find certain tasks that all digestive systems must accomplish:

1. *Ingestion.* The food must be brought into the digestive tract through an opening, usually called a **mouth**.
2. *Mechanical breakdown.* The food must be physically broken down into smaller pieces. This is accomplished by gizzards or teeth as well as by the churning action of the digestive cavity itself. The particles produced

by mechanical breakdown provide a large surface area for attack by digestive enzymes.
3. *Chemical breakdown.* The particles of food must be exposed to digestive enzymes and other digestive fluids that cause large molecules to be broken down into smaller subunits.
4. *Absorption.* The small molecules must be transported out of the digestive cavity and into cells.
5. *Elimination.* Indigestible materials must be expelled from the body.

In the following section, we will describe the mechanisms by which the human digestive system accomplishes these functions.

Human Digestion

Humans and most other animals have a digestive system that is basically a tube running through the body (Fig. 23-4). The tube is divided into a series of specialized regions that process the food in an orderly sequence: first physically grinding it up, then enzymatically breaking it down, and then absorbing the small nutrient molecules into body cells.

The Mechanical and Chemical Breakdown of Food Begins in the Mouth

Both the mechanical and chemical breakdown of food begin in the mouth. In the adult human, 32 teeth of varying sizes and shapes cut and grind the food into small pieces. Human and other vertebrate teeth are specialized for the diet of the animal (Fig. 23-5). As the food is pulverized by the teeth, the first phase of chemical digestion occurs as three pairs of salivary glands (see Fig. 23-4) pour out saliva in response to the smell, feel, taste, and (if you're hungry) even the thought of food.

Saliva contains the digestive enzyme **amylase**, which begins the breakdown of starches into sugar (Table 23-6). Saliva has other functions as well. It contains a bacteria-killing enzyme and antibodies that help guard against infection. It also lubricates the food to facilitate swallowing and dissolves some food molecules such as acids and sugars, carrying them to taste buds on the tongue. The taste buds help identify the type and quality of the food.

With the help of the muscular tongue, the food is manipulated into a mass and pressed backward into the **pharynx**, a cavity connecting the mouth with the **esophagus** (Fig. 23-6a, p. 430). The pharynx also connects, via the larynx, the nose and mouth with the trachea, which conducts air to the lungs. This arrangement occasionally causes problems, as anyone who has ever choked on a piece of food can attest. Normally, however, the swallowing reflex elevates the larynx so that it meets the **epiglottis**, a flap of tissue that blocks off the respiratory passages. Food is thus directed into the esophagus (Fig. 23-6b).

NUTRITION FACTS

Serving Size	1 Cup (55g)
Servings Per Package	8

Amount Per Serving

Calories 210	Calories from Fat 0
	% Daily Value*
Total Fat 0g	0%
Saturated Fat 0g	0%
Cholesterol 0mg	0%
Sodium 20mg	1%
Total Carbohydrate 46g	15%
Dietary Fiber 6g	24%
Sugars 12g	
Protein 6g	
Vitamin A* •	Vitamin C 2%
Calcium 4% •	Iron 18%
Thiamin 38% •	

*Percent Daily Values are based on a 2,000 calorie diet. Your Daily Values may be higher or lower depending on your calorie needs:

	Calories	2,000	2,500
Total Fat	Less than	65g	80g
Saturated Fat	Less than	20g	25g
Cholesterol	Less than	300mg	300mg
Sodium	Less than	2,400mg	2,400mg
Total Carbohydrate		300g	375g
Dietary Fiber		25g	30g

Calories per gram:
Fat 9 • Carbohydrate 4 • Protein 4

Figure 23-3 **New food labels**
The U.S. government recently required more complete nutritional labeling of foods, such as in this sample. The weight in grams of various nutrients (such as fat, cholesterol, and sodium) is converted into a percentage of the recommended daily allotment, assuming a 2000-Calorie diet. At the bottom of the label, the total recommended number of grams of these nutrients is listed for a 2000- and a 2500-Calorie diet.

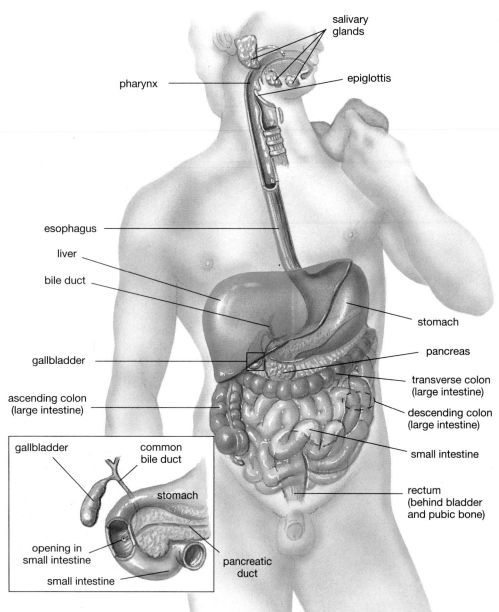

Figure 23-4   **The human digestive tract**

Some organs that produce and store digestive secretions, such as the salivary glands, liver, gallbladder, and pancreas, are included.

The Esophagus Conducts Food to the Stomach, Where Digestion Continues

The esophagus is a muscular tube that propels food from the mouth to the stomach. Circular muscles surrounding the esophagus contract in sequence above the swallowed food mass, squeezing it down toward the stomach (Fig. 23-7). This muscular action, called **peristalsis**, also occurs in the stomach and intestines, where it helps move food along the digestive tract. Peristalsis is so effective that a person can actually swallow when upside down. Mucus secreted by cells lining the esophagus helps protect it from abrasion and lubricates the food during its passage.

The **stomach** is an expandable muscular sac capable of holding from 2 to 4 liters (as much as a gallon) of food and liquids. Food is retained in the stomach by a ring of circular muscle separating the lower portion of the stomach from the upper small intestine. This muscle, called

the **pyloric sphincter**, regulates the passage of food into the small intestine, as described later. The stomach has three major functions. First, the stomach stores food and releases it gradually into the small intestine, at a rate suitable for proper digestion and absorption. Thus, the stomach enables us to eat large, infrequent meals. Carnivores carry this ability to an extreme. A lion, for instance, may consume about 18 kilograms (40 pounds) of meat at one meal, then spend the next few days quietly digesting it. A second function of the stomach is to assist in the mechanical breakdown of food. In addition to peristalsis, its muscular walls undergo a variety of contracting, churning movements that help break apart large pieces of food.

Finally, glands in the lining of the stomach secrete enzymes and other substances that facilitate the chemical breakdown of food. These include gastrin, hydrochloric acid (HCl), pepsinogen, and mucus. **Gastrin** (a hormone) stimulates secretion of hydrochloric acid by specialized stom-

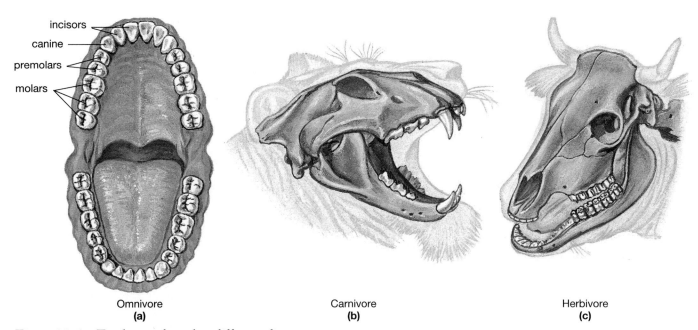

incisors
canine
premolars
molars

Omnivore
(a)

Carnivore
(b)

Herbivore
(c)

Figure 23-5   **Teeth are adapted to different diets**

(a) The upper and lower jaws of a human adult showing teeth specialized for cutting and grasping (incisors and canines) and grinding (molars and premolars). The shape of human teeth reflects our omnivorous diet, which includes both plant and animal material. **(b)** The teeth of a lion, a carnivore (meat eater), are specialized for grasping and tearing flesh and bone. **(c)** The teeth of the herbivorous (plant-eating) cow have large flat surfaces for grinding tough plant material.

ach cells. Other cells release pepsinogen, an inactive form of the protein-digesting enzyme pepsin. Pepsin is a **protease**, an enzyme that helps break proteins into shorter chains of amino acids called peptides (see Table 23-6). Pepsin must be secreted in an inactive form to prevent it from digesting the very cells that produce it. The highly acidic conditions in the stomach (pH 1 to 3), though, convert pepsinogen into pepsin, which functions best in an acidic environment.

As you may have noticed, the stomach produces all the ingredients necessary to digest itself, and indeed, this is what happens when a person develops ulcers (see "Health Watch: Ulcers—When the Digestive Tract Digests Itself"). However, cells lining the stomach normally produce a large quantity of thick mucus that coats the stomach lining and serves as a barrier to self-digestion. The protection is not perfect, however, and the cells lining

TABLE 23-6 ▪ *Digestive Secretions*

Site of Digestion	Source of Secretion	Secretion	Role in Digestion
Mouth	Salivary glands	Amylase	Breaks down starch into disaccharides
	Salivary glands	Mucus, water	Lubricates, dissolves food
Stomach	Cells lining stomach	Hydrochloric acid	Allows pepsin to work, kills bacteria, solubilizes minerals
	Cells lining stomach	Pepsin	Breaks down proteins into large peptides
	Cells lining stomach	Mucus	Protects stomach
Small intestine	Pancreas	Sodium bicarbonate	Neutralizes acidic chyme from stomach
	Pancreas	Amylase	Breaks down starch into disaccharides
	Pancreas	Peptidases	Split large peptides into small peptides
	Pancreas	Trypsin	Breaks down proteins into large peptides
	Pancreas	Chymotrypsin	Breaks down proteins into large peptides
	Pancreas	Lipase	Breaks down lipids into fatty acids and glycerol
	Liver	Bile	Emulsifies lipids
	Cells lining small intestine	Peptidases	Split small peptides into amino acids
	Cells lining small intestine	Disaccharidases	Split disaccharides into monosaccharides

(a)

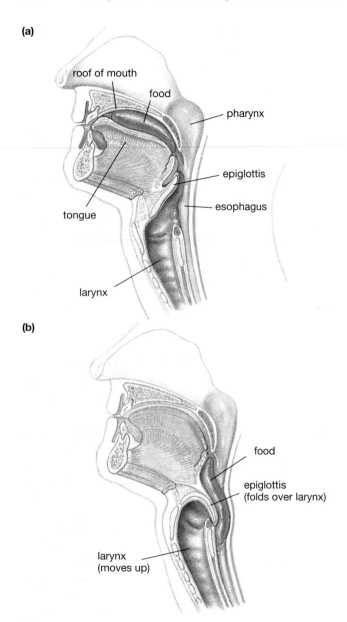

(b)

Figure 23-6   The challenge of swallowing

(a) Swallowing is complicated by the fact that both the esophagus (part of the digestive system) and the larynx (part of the respiratory system) open into the pharynx. **(b)** During swallowing, the larynx moves upward beneath a small flap of cartilage, the epiglottis. The epiglottis folds down over the larynx, sealing off the opening to the respiratory system and directing food down the esophagus.

the stomach are digested to some extent and must be replaced every few days.

Food in the stomach is gradually converted to a thick, acidic liquid called **chyme,** which consists of partially digested food and digestive secretions. Peristaltic waves then propel the chyme toward the small intestine. The sphincter at the base of the stomach allows only about a teaspoon of chyme to be expelled with each contraction, which occurs about every 20 seconds. It takes 2 to 6 hours, depending on the size of the meal, to empty the stomach completely. The continued churning movements of an empty stomach are felt as "hunger pangs."

Only a few substances, including water, some drugs, and alcohol, can enter the bloodstream through the stomach wall. Alcohol that is consumed when the stomach is empty is immediately absorbed into the bloodstream, with strong and rapid effects. Because food in the stomach slows alcohol absorption, the advice "never drink on an empty stomach" is based on sound physiological principles.

Most Digestion Occurs in the Small Intestine

The **small intestine** is a coiled, narrow tube about 3 meters (about 10 feet) long in a living person (lengths of over 20 feet are based on measurements made from cadavers in which the muscular intestinal walls have relaxed and stretched). Its diameter ranges from about 2 to 5 centimeters (1 to 2 inches) in an adult human. The small intestine has two major functions: to digest food into small molecules and to absorb these molecules, passing them to the bloodstream. The first role of the small intestine—digestion—is accomplished with the aid of digestive secretions from three sources: the liver, the pancreas, and the cells of the small intestine itself.

The Liver and Gallbladder Provide Bile, Important in Fat Breakdown

The **liver** is the largest and perhaps most versatile organ in the body. Its many functions include storage of fats and carbohydrates for energy, regulation of blood glucose levels, synthesis of blood proteins, storage of iron and certain vitamins, conversion of toxic ammonia (released by the breakdown of amino acids) into urea, and detoxification of other harmful substances such as nicotine and alcohol. The role of the liver in digestion is to produce bile, which is stored and concentrated in the **gallbladder** and released into the small intestine through the bile duct (see Fig. 23-4).

Bile is a complex mixture composed of **bile salts,** water, other salts, and cholesterol. Bile salts are synthesized in the liver from cholesterol and amino acids. Although they assist in the breakdown of lipids, bile salts are not enzymes. Rather, they act as detergents or emulsifying agents, dispersing globs of fat in the chyme into microscopic particles. These particles expose a large surface area for attack by **lipases,** lipid-digesting enzymes produced by the pancreas.

The Pancreas Supplies Several Digestive Secretions to the Small Intestine

The **pancreas** lies in the loop between the stomach and small intestine (see Fig. 23-4). It consists of two major types of cells. One type produces hormones involved in blood sugar regulation (see Chapter 25), and the other produces a digestive secretion called **pancreatic juice,** which is released into the small intestine. The role of pancreatic juice is to neutralize the acidic chyme and to digest carbohydrates, lipids, and proteins. About 1 liter (1.06 quarts) of pancreatic juice is released into the small intestine each day. It contains water, sodium bicarbonate, and several digestive enzymes (see Table 23-6). Sodium bicarbonate (the active ingredient in baking soda) neutralizes the acidic chyme in the small intestine, producing a slightly basic pH. Pancreatic digestive enzymes require this basic pH for proper functioning, in contrast to the stomach's digestive enzymes, which require an acidic pH.

The pancreatic digestive enzymes break down three major types of food: Amylase breaks down carbohydrates, lipases digest lipids, and several proteases separate proteins and peptides. The pancreatic proteases include trypsin, chymotrypsin, and carboxypeptidase. Both trypsin and chymotrypsin break proteins and peptides into shorter peptide chains. Carboxypeptidase then completes protein digestion by separating the individual amino acids from the ends of the peptide

chains. These proteases are secreted in an inactive form and become activated after they reach the small intestine.

The Intestinal Wall Completes the Digestive Process

The wall of the small intestine is studded with cells that are specialized to complete the digestive process and absorb the small molecules that result. These cells have various enzymes on their external membranes, which form the lining of the small intestine. These enzymes include proteases, which complete the breakdown of peptides into amino acids, and sucrase, lactase, and maltase, which break down disaccharides into monosaccharides (see Chapter 3). Small amounts of lipase digest lipids. Because these enzymes are actually embedded in the membranes of the cells lining the small intestine, this final phase of digestion occurs as the nutrient is being absorbed into the cell. Like the stomach, the small intestine is protected from digesting itself by large amounts of mucous secretions from specialized cells in its lining.

Most Absorption Occurs in the Small Intestine

The small intestine is not only the principal site of chemical digestion; it is also the major site of nutrient **absorption** into the blood. The small intestine has numerous foldings and projections that give it an internal surface area that is 600 times that of a smooth tube of the same length (Fig. 23-8). Covering the entire folded surface of the wall are minute, fingerlike projections called **villi** (literally, "shaggy hairs"). Villi, which range from 0.5 to 1.5 millimeters in length, give the intestinal lining a velvety appearance to the naked eye. They move gently back and forth in the chyme passing through the intestine. This movement increases their exposure to the molecules to be digested and absorbed. Further, each individual cell of the villi also bears a fringe of microscopic projections called **microvilli**. Taken together, these specializations of the small intestine wall give it a surface area of about 250 square meters, about the size of a tennis court.

Rhythmic, unsynchronized contractions of the circular muscles of the intestine, **segmentation movements**, slosh the chyme back and forth, bringing nutrients into contact with the absorptive surface of the small intestine (see Fig. 23-7). When absorption is complete, coordinated peristaltic waves conduct the leftovers into the large intestine.

Nutrients absorbed by the small intestine include water, monosaccharides, amino acids and short peptides, fatty acids produced by lipid digestion, vitamins, and minerals. The mechanisms by which this absorption occurs are varied and complex. In most cases, energy is expended to transport nutrients into the intestinal cells. (Water follows by osmosis as described in Chapter 6.) The nutrients then diffuse out of the intestinal cells into the interstitial fluid, where they then enter the bloodstream.

Each villus of the small intestine is provided with a rich supply of blood capillaries and a single lymph capillary, called a **lacteal**, to carry off the absorbed nutrients and dis-

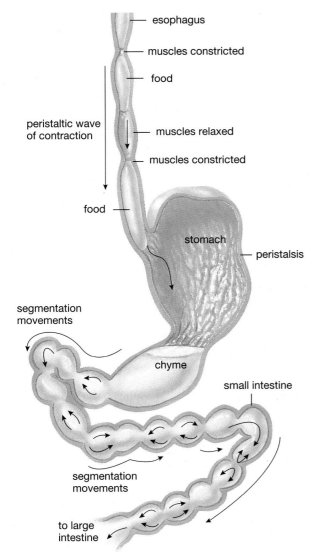

Figure 23-7 **Muscular movements of the digestive tract**
Food is propelled through the digestive system by the rhythmic peristaltic contractions of circular muscles. These contractions proceed downward through the digestive system, forcing the food along in front of them. In the stomach and small intestine, a variety of contractions, such as segmentation movement of the small intestine, help break down food particles and speed absorption.

tribute them throughout the body (see Fig. 23-8). Most of the nutrients enter the bloodstream through the capillaries, but fat subunits take a different route. After diffusing into the epithelial cells, they are resynthesized into fats, combined with other molecules, and then released as droplets into the interstitial fluid. By this means, they enter the lymph capillaries and are eventually delivered to the bloodstream when the lymph vessels empty into the veins.

Water Is Absorbed and Feces Are Formed in the Large Intestine

The **large intestine** is about 1.5 meters (5 feet) long in the adult human and about 7.5 centimeters (3 inches) in diameter, which is considerably wider than the small intestine. The large intestine has two parts: For most of its length it is called the **colon**, but its final 15 centimeters (6 inches) is called the **rectum**. Into the large intestine flow the leftovers of digestion: a mixture of water, undigested fats and proteins, and indigestible fibers such as the cell walls of vegetables and fruits. The large intestine contains a flourishing population of bacteria that live on unabsorbed nutrients. These bacteria earn their keep by synthesizing vitamin B_{12}, thiamin, riboflavin, and, most important, vitamin K, which would otherwise be deficient in a normal diet. Cells lining the large intestine absorb these vitamins, as well as leftover water and salts.

After absorption is complete, the result is the semisolid feces consisting of indigestible wastes and the dead bodies of bacteria, which account for about one-third of the dry weight of feces. The feces are transported by peristaltic movements until they reach the rectum. Expansion of this chamber stimulates the desire to defecate. Although defecation is a reflex (as any new parents can attest), it is initiated voluntarily after around the age of two.

Digestion Is Controlled by the Nervous System and Hormones

Considerable coordination is required to break down a chef salad into amino acids and peptides, sugars, fatty acids, vitamins, minerals, and indigestible cellulose. As the mouth responds to the first bite, the stomach must be warned that food is on the way. In addition, the enzymes of the stomach and small intestine require different environments for proper functioning (highly acidic in the

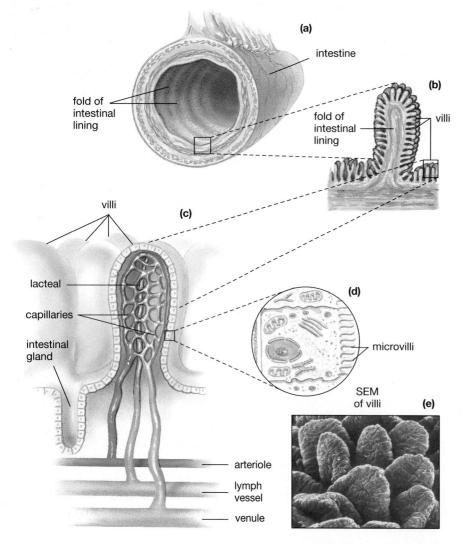

(a)

intestine

fold of intestinal lining

(b)

fold of intestinal lining

villi

villi

(c)

lacteal

capillaries

intestinal gland

(d)

microvilli

SEM of villi

(e)

arteriole

lymph vessel

venule

Figure 23-8   **The small intestine**

The surface area of the small intestine is greatly increased by **(a)** folds in the intestinal lining and **(b)** projections called villi (singular, villus) that extend from the folds, giving the lining a velvety appearance. The villi are enlarged in the scanning electron micrograph **(e)** and the diagram **(c)**, showing the blood capillaries and lymph vessels that form a network within each villus where nutrients are absorbed. **(d)** Microvilli, which are microscopic projections of individual cell membranes, further increase the surface area for nutrient absorption.

HEALTH WATCH

Ulcers—When the Digestive Tract Digests Itself

About 1 out of every 10 Americans eventually develops an ulcer. Ulcers occur when the mucus barriers of the stomach and upper small intestine break down and the inner lining and deeper layers of the digestive tract are eroded by acid and the protein-digesting enzyme pepsin. The upper small intestine is the most common site for ulcers (Fig. E23-1), because it receives the highly acidic chyme but is less protected by mucus than the stomach is.

The causes of ulcers are complex and poorly understood. A tendency to develop ulcers can be inherited. Some, but not all, ulcer sufferers secrete excess stomach acid. In some cases, decreased secretion of sodium bicarbonate or mucus by cells lining the stomach and small intestine may contribute to the disease. There is a correlation between the development of ulcers and stress, which may stimulate acid and pepsin secretion. Smoking and chronic consumption of alcohol and aspirin all may reduce the resistance of the digestive tract lining to the effects of pepsin and acids and can aggravate ulcers. In 1983, B. J. Marshall, an Australian physician, identified a bacterium in stomach tissue from ulcer patients. When his claims that the bacterium caused the ulcer were met with skepticism, he swallowed a batch of them, and soon developed ulcer symptoms. The bacterium (Helicobacter pylori) is found in the digestive tracts of nearly all ulcer victims, and antibiotics that kill these bacteria have been very effective in reducing the recurrence of ulcers. Although an estimated half of the world's population harbors these bacteria, they cause ulcers in only 10% to 20% of those infected.

Ulcers can sometimes be relieved by stress-reduction programs, eliminating smoking and drinking, and dietary changes. Medications for ulcers include antibiotics, antacids, and drugs that decrease stomach acid production.

Figure E23-1   **A bleeding ulcer in the stomach**

An illustration of a bleeding ulcer in the stomach as it would be seen through a fiber-optic viewing device called an endoscope. The stomach lining has been digested away and blood seeps through the opening.

stomach, slightly basic in the small intestine), and secretions into various parts of the digestive tract must be coordinated with the arrival of food. Not surprisingly, the secretions and activity of the digestive tract are coordinated by both nerves and hormones (Table 23-7).

The initial phase of digestion is under the control of the nervous system and involves responses to signals originating in the head. These signals include the sight, smell, taste, and sometimes the thought of food, as well as the muscular activity of chewing. In response to these stimuli,

TABLE 23-7 ■ *Some Important Digestive Hormones*

Hormone	Site of Production	Stimulus for Production	Effect
Gastrin	Stomach	Food in mouth Distension of stomach Peptides in stomach	Stimulates acid secretion by cells in stomach
Secretin	Small intestine	Acid in small intestine	Stimulates bicarbonate production by pancreas and liver; increases bile output by liver
Cholecystokinin	Small intestine	Amino acids, fatty acids in small intestine	Stimulates secretion of pancreatic enzymes and release of bile by gallbladder
Gastric inhibitory peptide	Small intestine	Fatty acids and sugars in small intestine	Inhibits stomach movements and release of stomach acid

saliva is secreted to the mouth, while nervous signals to the stomach walls initiate secretion of acid and the hormone gastrin, which stimulates further acid secretion. The concentration of acid is regulated by a negative feedback mechanism (see Chapter 21). When acid levels reach a certain point, they inhibit gastrin secretion, thereby inhibiting further acid production.

The arrival of food in the stomach triggers the second phase of digestion. Stimulation of the stomach wall causes production of a large quantity of mucus, which protects the stomach against self-digestion. The acidity of the stomach converts pepsinogen to its active form, pepsin, which begins protein digestion. However, the presence of protein in food tends to reduce the concentration of stomach acid. Thus, as protein is broken down, the acidity drops and the release of gastrin is no longer inhibited; gastrin is released again and stimulates further acid production. The cells secreting stomach acid are also activated by expansion of the stomach and by the presence of peptides produced by protein digestion.

As the liquid chyme is gradually released into the small intestine, its acidity stimulates the release of a second hormone, **secretin**, into the bloodstream by cells of the upper small intestine. Secretin causes the pancreas to pour sodium bicarbonate into the small intestine. Sodium bicarbonate neutralizes the acidity of the incoming chyme, creating an environment in which the pancreatic enzymes can function. A third hormone, **cholecystokinin**, is also produced by cells of the upper small intestine in response to the presence of chyme. This hormone stimulates the pancreas to release various digestive enzymes into the small intestine. It also stimulates the gallbladder to contract, squeezing bile through the bile duct to the small intestine. Bile assists in fat breakdown, as described earlier.

Gastric inhibitory peptide, a hormone secreted by the small intestine in response to the presence of fatty acids and sugars in chyme, inhibits acid production and peristalsis in the stomach. This inhibition slows the rate at which chyme is dumped into the small intestine, providing additional time for digestion and absorption to occur.

The Functions of the Urinary System

All of the homeostatic functions of the urinary system are performed simultaneously as blood is filtered through the kidneys. Within the kidneys, the fluid portion of the blood is collected. From this fluid, water and important nutrients are reabsorbed into the blood, while toxic substances, cellular waste products, and excess vitamins, salts, hormones, and water are left behind to be excreted as urine. The rest of the urinary system channels and stores urine until it is released from the body, a process called **excretion**. The mammalian urinary system helps maintain homeostasis in several ways. These include:

1. Regulation of the blood levels of ions such as sodium, potassium, chloride, and calcium
2. Regulation of the water content of the blood
3. Maintenance of proper pH of the blood
4. Retention of important nutrients such as glucose and amino acids in the blood
5. Secretion of hormones, such as erythropoietin, which stimulates red blood cell production (see Chapter 22)
6. Elimination of cellular waste products such as urea

Urea is a product of amino acid metabolism. As you may recall, the digestive system breaks proteins into their amino acid building blocks, which are then absorbed. When amino acids are taken into cells, some are used directly to synthesize new proteins. Others have their amino ($-NH_2$) groups removed and are then used either as a source of energy or in the synthesis of new molecules. The amino groups are released as **ammonia** (NH_3), which is very toxic. In mammals, ammonia is carried in the blood to the liver, were it is converted to urea, a far less toxic substance (Fig. 23-9). Urea is filtered from the blood by the kidneys and excreted in **urine**, a fluid consisting of water and dissolved wastes and some excess nutrients.

The Human Urinary System

Humans and other vertebrates filter their blood using **kidneys**, complex organs that are part of a larger group of structures collectively called the urinary system (Fig. 23-10). The kidneys actually produce the urine, and other portions of the system transport, store, and eliminate it. In the following sections we examine the major structures of the urinary system, tracing the pathway taken by waste products as they travel through the system. Then we explain kidney function in greater detail, and finally, we describe the role of the kidneys in homeostasis.

Human kidneys are paired, kidney bean-shaped organs located on either side of the spinal column and extending slightly above the waist. Each is approximately 13 centimeters (5 inches) long, 8 centimeters (3 inches) wide, and 2.5 centimeters (1 inch) thick. Blood carrying dissolved cellular wastes enters each kidney through a **renal artery.** After it has been filtered, the blood exits through the **renal vein** (Fig. 23-11). Urine leaves each kidney through a narrow, muscular tube called the **ureter.** Using peristaltic contractions, the ureters transport urine to the **bladder.** This hollow, muscular chamber collects and stores the urine.

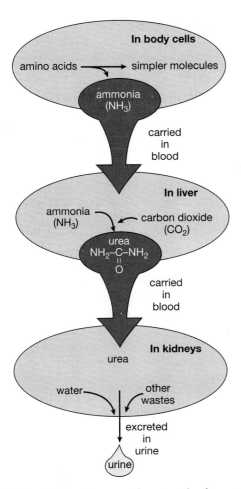

Figure 23-9   **A flow diagram showing the formation and excretion of urea**

The walls of the bladder, which contain smooth muscle, are capable of considerable expansion. Urine is retained in the bladder by two sphincter muscles located at its base, just above the junction with the urethra. When the bladder becomes distended, receptors in the walls signal its condition and trigger reflexive contractions. The sphincter nearest the bladder, the internal sphincter, is opened during this reflex. However, the lower, or external, sphincter, is under voluntary control, so the reflex can be suppressed by the brain unless bladder distension becomes acute. The average adult bladder will hold about 500 milliliters (approximately a pint) of urine, but the desire to urinate is triggered by considerably smaller accumulations. Urine completes its journey to the outside through the **urethra,** a single narrow tube about 3.8 centimeters (1.5 inches) long in the female and about 20 centimeters (8 inches) long in the male.

Urine Is Formed in the Kidneys

The kidney contains a solid outer layer where urine is formed and a hollow inner chamber called the **renal pelvis.** The renal pelvis is a branched collecting chamber that funnels urine into the ureter (see Fig. 23-11). The

outer layer of the kidney is divided into a fan-shaped inner **renal medulla** and an overlying **renal cortex.** Microscopic examination of these structures reveals an array of tiny individual filters, or **nephrons.** Over 1 million nephrons are packed into the cortex of each kidney, with many extending into the renal medulla.

The nephron has three major parts: the **glomerulus,** a dense knot of capillaries from which fluid from the blood is filtered into a surrounding cuplike structure, called **Bowman's capsule,** and finally, a long, twisted **tubule** (from the Latin, "little tube"). The tubule is further subdivided into first the **proximal tubule,** then the **loop of Henle** (which extends into the renal medulla), and finally the **distal tubule,** which leads to the **collecting duct** (Fig. 23-12). In the tubule, nutrients are selectively reabsorbed from the filtrate back into the blood, and wastes and some of the water are left behind to form urine. Additional wastes are also secreted into the tubule from the blood. Different portions of the tubule selectively modify the filtrate as it travels through (Fig. 23-13). These processes are examined in the following sections

Blood Is Filtered by the Glomerulus

Blood is conducted to each nephron by an arteriole that branches from the renal artery. Within a cup-shaped portion of the nephron—Bowman's capsule—the arteriole branches into numerous microscopic capillaries that form an in-

Figure 23-10   **The human urinary system and its blood supply**

Figure 23-11   **Cross section of a kidney**

The cross section shows the blood supply and internal structure of a kidney. The renal artery, which brings blood to the kidney, and the renal vein, which carries the filtered blood away, branch extensively within the kidney. The two are joined by a highly permeable capillary network through which substances are exchanged between the blood and the nephrons. A nephron, considerably enlarged, is drawn to show its orientation in the kidney. The renal pelvis is the branched collecting chamber that funnels urine out of the kidney.

tertwined mass, the glomerulus (see Fig. 23-12). The walls of the glomerular capillaries are extremely permeable to water and small dissolved molecules, but they prevent the movement of most large proteins, such as the albumin found in blood. Beyond the glomerulus, the capillaries reunite to form an arteriole whose diameter is smaller than that of the incoming arteriole. The differences in diameter between the incoming and outgoing arterioles create pressure within the glomerulus, driving water and many of the dissolved substances from the blood through the capillary walls. This process is called **filtration** (see Fig. 23-13), and the resulting fluid is called the **filtrate**. The watery filtrate, resembling blood plasma minus its plasma proteins, is collected in Bowman's capsule for transport through the nephron.

With the filtrate removed, the blood in the arteriole leaving the glomerulus is now very concentrated, having had much of its water removed but retaining substances too large to pass through the glomerular capillary walls, such as blood cells, large proteins, and fat droplets. Beyond the glomerulus, the arteriole branches into smaller, highly porous, capillaries. These capillaries surround the tubule, forming intimate contacts with it. At these points of contact, water and nutrients remaining in the filtrate after filtration are reab-

sorbed as the filtrate passes through the nephron and are returned to the blood; in addition, wastes remaining in the blood after filtration are passed into the filtrate for disposal.

The Blood Filtrate Is Converted to Urine in the Nephron

The blood filtrate collected in Bowman's capsule contains a mixture of both wastes and essential nutrients, including most of the blood's vital water. The nephron restores the nutrients and most of the water to the blood, while retaining wastes for elimination. This is accomplished by two processes: tubular reabsorption and tubular secretion (see Fig. 23-13).

Tubular reabsorption is the process by which cells of the tubule remove water and nutrients from the filtrate within the tubule and pass them back into the blood. Reabsorption of salts and other nutrients, such as amino acids and glucose, generally occurs by active transport (that is, cells of the tubule expend energy to transport these substances out of the tubule). These nutrients then enter adjacent capillaries by diffusion. Water follows the nutrients out of the tubule by osmosis. Wastes such as urea remain in the tubule and become more concentrated as water leaves.

Tubular secretion is the process by which wastes and excess substances that were not initially filtered out into Bowman's capsule are removed from the blood for excretion. These wastes are actively secreted *into* the tubule by

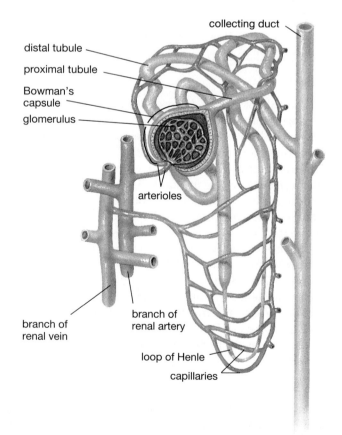

Figure 23-12   **An individual nephron and its blood supply**

① **FILTRATION:** Water, nutrients, and wastes are filtered from glomerular capillaries into Bowman's capsule of the nephron.

③ **TUBULAR SECRETION:** In the distal tubule, additional wastes are actively secreted into the tubule from the blood.

② **TUBULAR REABSORPTION:** In the proximal tubule, most water and nutrients are reabsorbed into the blood.

④ **CONCENTRATION:** In the collecting duct, additional water may leave the blood, creating urine that is more concentrated than the blood.

Figure 23-13 **Urine formation in the nephron**
A summary of events that occur during the formation of urine in the nephron.

tubule cells. Secreted substances include hydrogen and potassium ions, ammonia, and many drugs.

The Loop of Henle Allows Urine to Become Concentrated

The kidneys of mammals and birds are able to produce urine that is more concentrated (has a higher osmotic pressure) than their blood. The ability to concentrate urine is a result of the structure of both the nephron and the collecting duct into which several nephrons empty. Urine can become concentrated because there is an osmotic concentration gradient of salts and urea in the interstitial fluid surrounding the loop of Henle. This gradient is produced by the loop of Henle within the renal medulla; the longer the loop, the greater the concentration gradient. The most concentrated fluid (with the greatest amount of dissolved substances and the least amount of water), which is far more concentrated than blood, surrounds the bottom of the loop. The collecting duct passes through this osmotic gradient. As the filtrate passes through the portion of the collecting duct surrounded by the osmotically concentrated fluid, additional water leaves the filtrate by osmosis, but wastes do not. Therefore, as it moves through the collecting duct, the filtrate, now called urine, can reach an osmotic equilibrium, becoming as concentrated as the surrounding fluid. Because the rest of the excretory system does not allow water to enter or urea to escape, the urine remains concentrated.

It is important to produce concentrated urine when water is scarce and to produce dilute watery urine when there is excess water in the blood. The degree of concentration of the urine is controlled by the amount of antidiuretic hormone, to be described shortly.

The Kidneys Are Important Organs of Homeostasis

Each drop of blood in your body passes through a kidney about 350 times a day; thus, the kidney is able to fine-tune the composition of the blood helping to maintain **homeostasis**. The importance of this task is illustrated by the fact that kidney failure results in death within a short time (see "Health Watch: When the Kidneys Collapse").

The Kidneys Regulate the Water Content of the Blood

One of the most important functions of the kidney is to regulate the water content of the blood (see Fig. 23-13). Human kidneys filter 125 milliliters (about half a cup) of fluid from the blood each minute. This means that, without reabsorption of water, you would produce over 180 liters (45 gallons) of urine daily! Water reabsorption occurs passively by osmosis as the filtrate travels through the tubule and the collecting duct.

The amount of water reabsorbed into the blood is controlled by a negative feedback system (see Chapter 21) involving the amount of **antidiuretic hormone (ADH;** also called vasopressin) circulating in the blood. This hormone increases the permeability of the distal tubule and the collecting duct to water, thereby allowing more water to be reabsorbed from the urine. ADH is produced by cells in the hypothalamus and is released by the posterior pituitary gland (see Chapter 25).

ADH release is regulated by receptor cells in the hypothalamus that monitor the osmotic concentration of the blood and by receptors in the heart that monitor blood volume. For example, as the lost traveler staggers through the searing desert sun, dehydration occurs. With the loss of water, the osmotic concentration of his blood rises and

Figure 23-14   Regulation of the water content of the blood
Blood-water content is hormonally controlled by ADH through
a negative feedback process. Stimulation is indicated by a plus
sign (+), reduction by a minus sign (−). Dehydration, in
addition to stimulating ADH release, triggers the sensation of
thirst, leading to increased water intake. This response helps
restore blood volume and diminish ADH secretion.

his blood volume falls, triggering the release of more ADH
(Fig. 23-14). Release of ADH increases water reabsorption and produces urine more concentrated than the
blood. In contrast, a partygoer overindulging in beer will
experience a decrease in blood concentration and an increase in blood volume, and her receptors will cause a decrease in ADH output. Reduced ADH concentration will
make the distal tubule and collecting duct less permeable
to water. When ADH is very low, little water is reabsorbed

after the urine leaves the loop of Henle, and the urine produced will be more dilute than the blood. In extreme cases,
urine flow may exceed 1 liter (about 1 quart) per hour. As
the proper water level is restored, the increased osmotic
concentration of the blood and decreased blood volume
will stimulate increased ADH production, which will in
turn stimulate increased water reabsorption.

Mammalian Kidneys Are Adapted to Diverse Environments

In addition to regulating water balance using hormones
such as ADH, different mammals have kidneys whose
structures are adapted to the availability of water in their
natural habitats. Mammals that must conserve water do so
by producing urine that is more concentrated (has a higher osmolarity) than their blood. The degree of concentration that can be achieved is determined by the length
of the loop of Henle. The longer the loop, the higher the
salt concentration produced in the fluid surrounding it.
The higher the salt concentration, the more water moves
out of the urine by osmosis as urine passes through the
collecting duct, and the more concentrated the urine becomes. As you might predict, animals living in very dry climates have the longest loops of Henle, while those living
in watery environments have relatively short loops.

The beaver, for example, has only short-looped
nephrons and is unable to concentrate its urine to more
than twice its plasma concentration. Human kidneys have
a mixture of long- and short-looped nephrons and can
concentrate urine to about four times the plasma con-

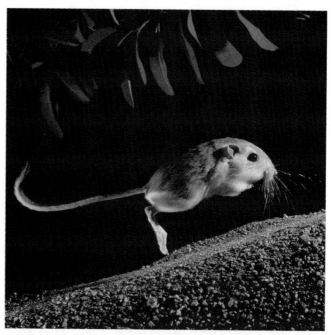

Figure 23-15   An adaptation to a dry environment
The desert kangaroo rat of the southwestern United States
can dispense with drinking partly because its long loops of
Henle allow it to produce very concentrated urine.

HEALTH WATCH

When the Kidneys Collapse

The kidney is a relatively rugged organ. Normally the need to urinate is our only evidence of its industrious activity. But if the kidneys fail, death occurs rapidly. The kidneys are vulnerable to attack from several sources. Some toxic substances can destroy kidney tubules. These include heavy metals (mercury, arsenic), organic compounds such as solvents (carbon tetrachloride), insecticides, and overdoses of some antibiotics and pain relievers. A second source of injury is through infection by bacteria, particularly intestinal bacteria that reach the kidney via the urethra. Still another source of injury is glomerulonephritis. In this disease, an abnormal immune response to an infection causes the antigen and antibody to fuse into insoluble particles, which are trapped in the glomeruli, resulting in inflammation and death of the tubules. Kidney damage is often treated using a remarkable machine: the artificial kidney (Fig. E23-2).

Figure E23-2 **A diagram of the artificial kidney**

First used in 1945, the artificial kidney operates on the simple principle of diffusion of substances from areas of high concentration to areas of low concentration across an artificial permeable membrane. The passive diffusion of substances across an artificial semipermeable membrane is called **dialysis;** when blood is filtered using this principle, the process is called **hemodialysis.** During hemodialysis, the patient's blood is diverted from the body and pumped through narrow tubing made of a cellophane membrane suspended in dialyzing fluid. This membrane has pores too small to permit the passage of blood cells and large proteins, but large enough to pass small molecules such as water, sugar, salts, amino acids, and urea. The composition of the dialyzing fluid is adjusted to have normal blood levels of these substances so that important nutrients in the blood will be retained. As a result, only molecules whose concentration is in excess of normal will diffuse into the dialyzing fluid. Urea, for example, is present in relatively high concentration in the blood and absent in the dialyzing fluid, so it diffuses out of the blood and into the dialyzing fluid.

One challenge of dialysis is to eliminate excess water. Since the blood has a higher osmolarity than the dialyzing fluid, water tends to diffuse *into* the blood from the dialyzing fluid. This process is counteracted by applying pressure to the blood, which forces some of the small water molecules out of the blood against their concentration gradient.

Roughly 500 milliliters (1 pint) of blood are in the machine at any one time, flowing at a rate of several hundred milliliters per minute. Still, the patient must remain attached to the dialysis machine for 4 to 6 hours three times a week. Although patients requiring dialysis have remained alive for up to 20 years, their lives are far from normal. Their diet and fluid intake must be carefully regulated. Even so, their blood composition fluctuates, and many more toxic substances accumulate than is normal. In spite of its drawbacks, dialysis keeps about 60,000 Americans alive. For those who are fortunate enough to find a compatible donor, kidney transplants are a far better solution; about 20,000 of these are performed yearly in the United States.

centration. The masters of urine concentration are desert rodents such as kangaroo rats, which can produce urine 14 times their plasma concentration (Fig. 23-15). Kangaroo rats (as you might predict) have only very long-looped nephrons. Because of their extraordinary ability to conserve water, they can completely dispense with drinking, relying entirely on water derived from their food.

Kidneys Help Regulate the Blood Pressure and Oxygen Content of the Blood

Two hormones produced by the kidneys are extremely important in regulating blood pressure and the blood's oxygen-carrying capacity. When blood pressure falls, the kidneys release **renin** into the bloodstream. Renin acts as an enzyme, catalyzing the formation of a second hormone, **angiotensin**, from a protein circulating in the blood. Angiotensin in turn causes arterioles to constrict, elevating blood pressure. Constriction of the arterioles carrying blood to the kidneys also reduces the rate of blood filtration, causing less water to be removed from the blood. Water retention causes an increase in blood volume and, consequently, an increase in blood pressure.

In response to low blood oxygen levels, the kidneys release a second hormone, **erythropoietin**, described in Chapter 22. Erythropoietin travels in the blood to the bone marrow, where it stimulates more rapid production of red blood cells, whose role is to transport oxygen.

Kidneys Monitor and Regulate Dissolved Substances in the Blood

As the kidney filters the blood, it monitors and regulates blood composition in order to maintain a constant internal environment. Substances regulated by the kidney, in addition to water, include nutrients such as glucose, amino acids, vitamins, urea, and a variety of ions, including sodium, potassium, chloride, and sulfate. The kidney maintains a constant blood pH by regulating the amount of hydrogen and sodium bicarbonate ions. This remarkable organ also eliminates potentially harmful substances, including some drugs, food additives, pesticides, and toxic substances from cigarette smoke, such as nicotine.

▓ S U M M A R Y   O F   K E Y   C O N C E P T S

Nutrition
Each type of animal has specific nutritional requirements. These requirements include: molecules that can be broken down to liberate energy, such as lipids, carbohydrates, and protein; chemical building blocks used to construct complex molecules, such as amino acids that can be linked together to form proteins; and minerals and vitamins that facilitate the diverse chemical reactions of metabolism.

The Challenge of Digestion
Digestive systems must accomplish five tasks: ingestion, mechanical followed by chemical breakdown of food, absorption, and elimination of wastes. Digestive systems convert the complex molecules of the bodies of other animals or plants that have been eaten into simpler molecules that can be utilized by the organism.

Human Digestion
Humans and most other animals utilize a tubular compartment with specialized chambers where food is processed in a well-defined sequence. Digestion begins in the mouth, where food is physically broken down by chewing and chemical digestion is initiated by saliva. Food is then conducted to the stomach by peristaltic waves of the esophagus. In the acidic environment of the stomach, food is churned into smaller particles, and protein digestion begins. Gradually, the liquefied food, now called chyme, is released to the small intestine. Here it is neutralized by sodium bicarbonate from the pancreas. Secretions from the pancreas, liver, and the cells of the intestine complete the breakdown of proteins, fats, and carbohydrates. In the small intestine, the simple molecular products of digestion are absorbed into the bloodstream for distribution to the body cells. The large intestine absorbs the remaining water and converts indigestible material to feces.

Digestion is regulated by the nervous system and hormones. The smell and taste of food and the action of chewing cause the secretion of saliva and production of gastrin by the stomach. Gastrin stimulates stomach acid production. As chyme enters the small intestine, three additional hormones are produced by intestinal cells: secretin, which causes sodium bicarbonate production to neutralize the acid chyme; cholecystokinin, which stimulates bile release and causes the pancreas to secrete digestive enzymes into the small intestine; and gastric inhibitory peptide, which inhibits acid production and peristalsis by the stomach. This inhibition slows the movement of food into the intestine.

The Functions of the Urinary System
The urinary system plays a crucial role in homeostasis, the maintenance of a stable internal environment. The kidneys of mammalian urinary systems regulate water and ion content of the blood and blood pH. They help retain nutrients and eliminate cellular wastes and toxic substances, and they secrete the hormone erythropoietin.

The Human Urinary System
The human urinary system consists of kidneys, ureters, bladder, and urethra. Kidneys produce urine, which is conducted by the ureters to a storage organ, the bladder. Distension of the muscular bladder wall triggers urination, during which urine passes out of the body through the urethra.

Each kidney consists of over a million individual nephrons in an outer renal cortex, with many extending into an inner renal medulla. Urine formed in the nephrons enters collecting ducts that empty into the renal pelvis, from which it is funneled into the ureter.

Each nephron is served by an arteriole that branches from the renal artery. The arteriole further branches into a mass of porous-walled capillaries called the glomerulus. Here water and dissolved substances are filtered from the blood by pressure. The filtrate is collected in the cup-shaped Bowman's capsule and conducted along the tubular portion of the nephron. During tubular reabsorption, nutrients are actively pumped out of the filtrate through the walls of the tubule. Nutrients then enter capillaries that surround the tubule, and water follows by osmosis. Some wastes remain in the filtrate, while others are actively pumped into the tubule by tubular secretion. The tubule forms the loop of Henle, which creates a salt concentration gradient surrounding it. After completing its passage through the tubule, the filtrate enters the collecting duct, which passes through the concentration gradient. Final passage of the filtrate through this gradient via the collecting duct allows concentration of the urine.

The kidneys are important organs of homeostasis. The water content of the blood is regulated by antidiuretic hormone (ADH), produced in the hypothalamus and released by the posterior pituitary gland. Low blood volume and high osmotic concentration of the blood signal dehydration and stimulate release of ADH into the bloodstream. ADH increases the permeability to water of the distal tubule and the collecting duct, allowing more water to be reabsorbed into the blood. In addition, the kidneys control blood pH, remove toxic substances, and regulate ions such as sodium, chloride, potassium, and sulfate. Excess glucose, vitamins, and amino acids are also excreted by the kidneys.

✖ KEY TERMS

absorption   p. 431
ammonia   p. 434
amylase   p. 427
angiotensin   p. 440
antidiuretic hormone (ADH)   p. 437
bile   p. 430
bile salt   p. 430
bladder   p. 434
Bowman's capsule   p. 435
calorie   p. 422
Calorie   p. 422
cholecystokinin   p. 434
chyme   p. 430
collecting duct   p. 435
colon   p. 432
dialysis   p. 439
digestion   p. 426
distal tubule   p. 435
epiglottis   p. 427
erythropoietin   p. 440
esophagus   p. 427
essential amino acid   p. 423
essential fatty acid   p. 422

excretion   p. 434
filtrate   p. 436
filtration   p. 436
gallbladder   p. 430
gastric inhibitory peptide   p. 434
gastrin   p. 428
glomerulus   p. 435
glycogen   p. 423
hemodialysis   p. 439
homeostasis   p. 437
kidney   p. 434
lacteal   p. 431
large intestine   p. 432
lipase   p. 430
liver   p. 430
loop of Henle   p. 435
microvillus   p. 431
mouth   p. 427
nephron   p. 435
pancreas   p. 430
pancreatic juice   p. 430
peristalsis   p. 428
pharynx   p. 427

protease   p. 429
proximal tubule   p. 435
pyloric sphincter   p. 428
rectum   p. 432
renal artery   p. 434
renal cortex   p. 435
renal medulla   p. 435
renal pelvis   p. 435
renal vein   p. 434
renin   p. 439
secretin   p. 434
segmentation movement   p. 431
small intestine   p. 430
stomach   p. 428
tubular reabsorption   p. 436
tubular secretion   p. 436
tubule   p. 435
urea   p. 423
ureter   p. 434
urethra   p. 435
urine   p. 434
villus   p. 431
vitamin   p. 621

�֍ T H I N K I N G   T H R O U G H   T H E   C O N C E P T S

Multiple Choice

1. An acidic mixture of partially digested food that moves from the stomach into the small intestine is called
   a. cholecystokinin
   b. bile
   c. lymph
   d. secretin
   e. chyme

2. The hormone responsible for stimulating secretion of hydrochloric acid by stomach cells is
   a. pepsin
   b. gastrin
   c. cholecystokinin
   d. insulin
   e. secretin

3. Humans lack digestive enzymes to attack chitin, a complex polysaccharide in the exoskeleton of lobster and crayfish. They also lack carbohydrases that degrade
   a. glycogen
   b. plant starch
   c. cellulose
   d. sucrose
   e. maltose

4. Which of the following is NOT characteristic of bile?
   a. It is produced in the gallbladder.
   b. It is a mixture of special salts, water, and cholesterol.
   c. It acts as a detergent or emulsifying agent.
   d. It helps expose a large surface area of lipid for attack by lipases.
   e. It works in the small intestine.

5. "Water-soluble compound that works primarily as enzyme helper" would be a good definition of
   a. vitamin C
   b. vitamin A
   c. vitamin B
   d. vitamin E
   e. vitamin D

6. Which of the following is false?
   a. Urea is more toxic than ammonia.
   b. Ammonia is converted to urea in the liver.
   c. Ammonia is produced in body cells.
   d. The fluid collected in Bowman's capsule is called the filtrate.
   e. *Tubule* means "little tube."

7. The walls of the _____ are made more or less permeable to water, depending on the need to conserve water.
   a. ureter
   b. urethra
   c. tubule
   d. collecting duct
   e. glomerulus

8. Which of the following will cause a decrease in ADH production?
   a. dehydration
   b. drinking beer
   c. an increase in osmotic pressure of blood
   d. abnormally high blood sugar
   e. strenuous exercise

9. The function of the glomerulus and Bowman's capsule of the nephron is to
   a. reabsorb water into the blood
   b. eliminate ammonia from the body
   c. reabsorb salts and amino acids
   d. filter the blood and capture the filtrate
   e. concentrate the urine

10. What determines the ability of a mammal to concentrate its urine?
    a. the number of nephrons
    b. the length of the tubules
    c. the length of the collecting duct
    d. the size of the glomerulus
    e. the length of the loop of Henle

11. Which of the following processes does NOT occur in the nephron and collecting duct?
    a. filtration
    b. elimination of urea from the body
    c. reabsorption of nutrients
    d. tubular secretion
    e. concentration of urine

Review Questions

1. List four general types of nutrients and describe the role of each in nutrition.

2. List and describe the function of the three principal secretions of the stomach.

3. List the substances secreted into the small intestine, and describe the origin and function of each.

4. Name and describe the muscular movements that usher food through the human digestive tract.

5. Name four structural or functional adaptations of the human small intestine that assure good digestion and absorption.

6. Describe protein digestion in the stomach and small intestine.

7. Explain the two major functions of excretory systems.

8. Trace a urea molecule from the bloodstream to the external environment.

9. What is the function of the loop of Henle? Of the collecting duct? Of antidiuretic hormone?

10. Describe and compare the processes of filtration, tubular reabsorption, and tubular secretion.

11. Describe the role of the kidneys as organs of homeostasis.

✖ A P P L Y I N G   T H E   C O N C E P T S

1. The food label on a soup can shows that the product contains 10 grams of protein, 4 grams of carbohydrate, and 3 grams of fat. How many Calories are found in this soup?

2. Stomach ulcers that resist antibiotic treatment are treated with several kinds of drugs. Anticholinergic drugs decrease nerve signals to the stomach walls that are produced by the sight, smell, and taste of food. Antacids neutralize stomach acid. Why would it be inadvisable to take anticholinergics and antacids together?

3. Control of the human digestive tract involves several feedback loops and messages that coordinate activity in one chamber with those taking place in subsequent chambers. List the coordinating events you discovered in this chapter, in order, beginning with tasting, chewing, and swallowing a piece of meat and ending with residue that enters the large intestine. What turns on and what shuts off each process?

4. Symbiotic protozoa in the digestive tracts of termites produce cellulase, an enzyme that breaks down cellulose, for them. In return, termites provide protozoa with food and shelter. Imagine that the human species is gradually invaded, over many generations, by symbiotic protozoa capable of producing cellulase. What evolutionary adaptive changes in body structure and function might occur simultaneously?

5. Trace a ham and cheese with lettuce sandwich through the human digestive system, discussing what happens to each part of the sandwich as it passes through each region of the digestive tract.

6. One of the common remedies for constipation (difficulty eliminating feces) is a laxative solution containing magnesium salts. In the large intestine, magnesium salts are absorbed very slowly by the intestinal wall, remaining in the intestinal tract for long periods of time. This has an effect on water movement in the large intestine. On the basis of this information, explain the laxative action of magnesium salts.

7. Some "quick weight loss" diets require ingestion of much protein-rich food and elimination of carbohydrates. Two side effects of such diets are increased thirst and increased urination. Explain the connections between the diets and the side effects.

8. Your teenage son is dying from injuries sustained in a motorcycle accident. You have been asked to allow your son's kidneys to be donated to people currently relying on dialysis. Explain what benefits these kidneys would confer on a dialysis patient. How could two dialysis patients be helped by your son?

9. Some employers require their employees to submit to urine tests before they can be employed and at random intervals during their employment. Refusal to take the test or failure to "pass" the test could be grounds for termination. What is the purpose of the urine test? What employers might find such tests necessary? How do you feel about urine tests for obtaining or keeping a job? Explain your answers.

✖ F O R   M O R E   I N F O R M A T I O N

Davenport, H. "Why the Stomach Does Not Digest Itself." *Scientific American*, January 1972. How the stomach protects itself from its own strongly acidic secretions, and what happens when the defenses fail.

Diamond, J. "Dining with the Snakes." *Discover*, April 1994. Follow food as it is consumed and digested by a snake.

Kretchmer, N. "Lactose and Lactase." *Scientific American*, April 1972. Discusses differences in tolerance to milk sugar among humans from different populations.

Livingston, E. H., and Guth, P. H. "Peptic Ulcer Disease." *American Scientist*, November–December 1992. Why and how ulcers develop.

Martin, R. J., White, B. D., and Hulsey, M. G. "The Regulation of Body Weight." *American Scientist*, November–December 1991. Describes some of the complex mechanisms that control eating, obesity, and the body's energy balance.

Mason, M. "Why Ulcers Run in Families." *Health*, September 1994. The story of the bacterium that causes ulcers.

Mayfield, E. "A Consumer's Guide to Fats." *FDA Consumer*, May 1994. Commonsense information on fat, cholesterol, diet, and health.

Moog, F. "The Lining of the Small Intestine." *Scientific American*, November 1981. A description of this intricate tissue, which is responsible for absorbing nutrients into the body.

Nuland, S. B. "The Beast in the Belly." *Discover*, February 1995. An interesting medical tale of bacterial disease, digestive enzymes, and food.

O'Brien, C. "Lucky Break for Kidney Disease Gene." *Science*, June 24, 1994. Discusses research on a dominant gene causing kidney disease.

Willett, W. C. "Diet and Health: What Should We Eat?" *Science*, April 22, 1994. Summarizes studies that suggest that diet can play a major role in the prevention of disease.

N E T   W A T C H

On-line resources for this chapter are on the World Wide Web at:
http://www.prenhall.com/~audesirk (click on the <u>table of contents</u> link and then select Chapter 23).

24 Defenses against Disease: The Immune Response

"If you wish to be well and keep well, take Braggs Vegetable Charcoal and Charcoal Biscuits. Absorbs all impurities in the stomach and bowels, effectively warding off cholera, smallpox, typhoid, and all malignant fevers. Eradicate worms in children. Sweeten the breath."

From an early 1900s newspaper ad

A macrophage, which is a type of white blood cell, sends out a pseudopod to capture a rod-shaped bacterium. The drama reproduced in this culture dish is played out many times each second in your body.

*E*very week, it seems, in newspapers, magazines, and on television, we encounter news of devastating, sometimes incurable, sometimes apparently new diseases: Lyme disease in the 1970s; AIDS in the 1980s; rodent-carried hantavirus in the Four Corners region of the American Southwest in 1993; Ebola virus outbreaks in Zaire in 1995; and new, deadlier forms of familiar old enemies, such as the bacterial infection that killed Muppets creator Jim Henson. It's enough to make you wonder how any of us survive more than a few years, at best. And the threat is real, too. Trillions of microbes lurk out there—in the air, in the water, in the foods we eat, and on the objects we touch—ready to invade our bodies. Nevertheless, most of us are healthy most of the time, not even aware of the constant combat between body and microbe. Even when an infection does occur, we almost always fend it off after a few days. How do we keep microbes out? If infected, how do we destroy the invaders?

Defenses against Microbial Invasion

The human body has three lines of defense against microbial attack: (1) external barriers that keep microbes out of the body; (2) nonspecific internal defenses that combat all invading microbes; and (3) the immune system, which directs its assault against specific microbes (Fig. 24-1).

The Skin and Mucous Membranes Form Barriers

The first, and obviously best, defense is to keep microbes out in the first place. The human body has two surfaces exposed to the environment: the **skin** and the **mucous membranes** of the digestive and respiratory tracts. These surfaces are barriers to microbial invasion.

The Intact Skin Is Both a Barrier to Entry and an Inhospitable Environment for Microbial Growth

The outer surface of the skin consists of dry, dead cells filled with horny proteins similar to those in hair and nails. Consequently, most microbes that land on the skin cannot ob-

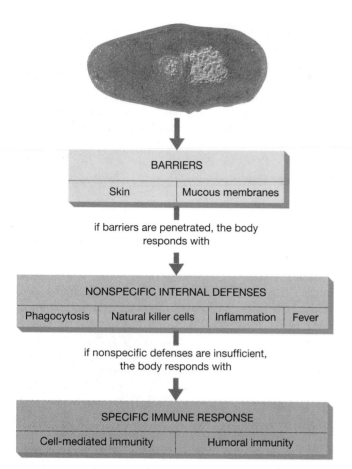

BARRIERS

| Skin | Mucous membranes |

if barriers are penetrated, the body
responds with

NONSPECIFIC INTERNAL DEFENSES

| Phagocytosis | Natural killer cells | Inflammation | Fever |

if nonspecific defenses are insufficient,
the body responds with

SPECIFIC IMMUNE RESPONSE

| Cell-mediated immunity | Humoral immunity |

Figure 24-1   **Levels of defense against infection**

these defenses, the warm, moist mucous membranes are much more vulnerable than the dry, oily skin, and many disease organisms enter the body through these membranes.

Nonspecific Internal Defenses Combat Microbes

Three nonspecific internal defenses are mustered against microbes that penetrate the skin or mucous membranes. These defenses are nonspecific because they attack a wide variety of microbes, rather than targeting specific invaders as the immune response does. First, the body has a standing army of **phagocytic cells** that destroy microbes and **natural killer cells** that destroy cells of the body that have been infected by viruses. The steady trickle of microbes that pass through the body's external barriers are mostly mopped up by these cells. Second, an injury, with its combination of tissue damage and relatively massive invasion of microbes, provokes an **inflammatory response.** The inflammatory response simultaneously recruits new members of the army of phagocytic cells and killer cells and walls off the injured area, isolating the infected tissue from the rest of the body. Third, if a population of microbes succeeds in establishing a major infection, the body often produces a **fever,** which both slows down microbial reproduction and enhances the body's own fighting abilities.

tain the water and nutrients they need. Secretions from sweat glands and sebaceous glands also cover the skin. These secretions contain acids and natural antibiotics, such as lactic acid, that inhibit the growth of bacteria and fungi. These multiple defenses make the unbroken skin an extremely effective barrier against microbial invasion.

Antimicrobial Secretions, Mucus, and Ciliary Action Defend the Mucous Membranes against Microbes

The membranes of the digestive and respiratory tracts are also well defended. First, they secrete mucus that contains antibacterial enzymes such as lysozyme, which destroys bacterial cell walls. Second, the mucus physically traps microbes that enter the body through the nose or mouth (Fig. 24-2). Cilia on the membranes sweep up the mucus, microbes and all, until it is either coughed or sneezed out of the body, or swallowed. If microbes are swallowed, they enter the stomach, where they encounter a combination of extreme acidity (pH 1 to 3) and protein-digesting enzymes, which can kill many types of microbes. Farther along in the digestive tract, the intestine is inhabited by bacteria that are harmless to the human body but secrete substances that destroy invading foreign bacteria or fungi. Despite

0.5 micrometer

Figure 24-2   **The protective function of mucus**

Mucus traps microbes and debris in the respiratory tract (a brown strand of dirt is shown caught in mucus atop the red cilia). The cilia lining the walls of the respiratory tract then sweep both mucus and foreign matter out of the body.

Phagocytic Cells and Natural Killer Cells Destroy Invading Microbes

The body contains several types of amoeboid cells that can engulf and digest microbes. The most important of these are the **macrophages** (literally, "big eaters"), white blood cells that crawl around in the extracellular fluid. Macrophages ingest microbes by phagocytosis. As we will see shortly, besides the immediate effect of destroying an individual microbe, macrophages also play a crucial role in the immune response by "presenting" parts of the microbe to other cells of the immune system.

Natural killer cells are another class of white blood cells. In general, natural killer cells do not directly attack invading microbes. Instead, natural killer cells strike at the body's own cells that have been invaded by viruses. Virus-infected cells usually bear some viral proteins on their surfaces. Natural killer cells recognize these proteins and kill the infected cell. In some way, natural killer cells also recognize and kill cancerous cells. Natural killer cells do not eat their victims; they strike from the outside. Their weapons are proteins that they secrete onto the plasma membrane of the infected or cancerous cell. Some of the proteins insert themselves into the target plasma membrane in a ring, much like making a barrel out of individual staves of wood, thereby opening up a large pore in the membrane (Fig. 24-3). Killer cells also secrete enzymes that break up some of the molecules of the target cell. Shot full of holes and chewed on by enzymes, the target cell soon dies.

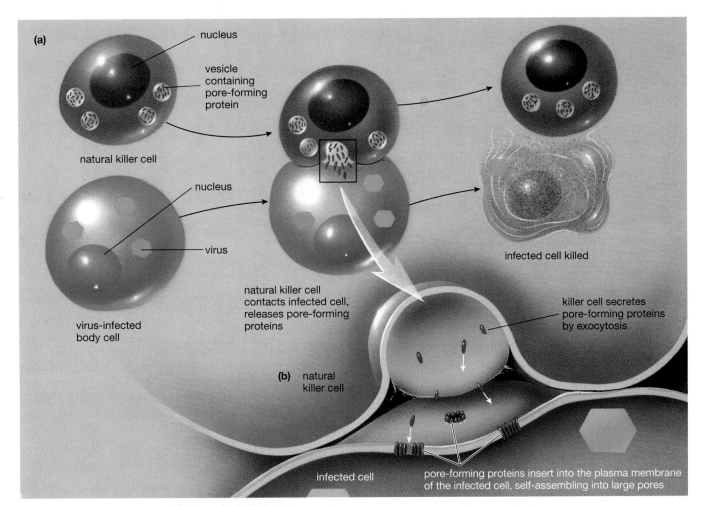

Figure 24-3   How natural killer cells kill their targets

(a) A natural killer cell contains vesicles full of pore-forming proteins. When it encounters a suitable target cell (here a virus-infected body cell), it briefly contacts the target and releases the pore-forming proteins by exocytosis. The pore-forming proteins self-assemble into large holes in the target cell's plasma membrane; the cell's contents leak out, and the cell dies.
(b) A close-up view of the killing process. Pore-forming proteins leave the killer cell and insert themselves into the target cell's plasma membrane. The proteins line up next to one another like the staves in a barrel, forming large pores. The membrane of the killer cell is not affected; apparently it contains other proteins that bind to and inactivate any pore-forming proteins that try to insert themselves into the killer's membrane.

5 micrometers

Figure 24-4   **The inflammatory response**

During an inflammatory response triggered by a cut in the skin, the cells forming capillary walls separate slightly. Macrophages and other phagocytic cells squeeze through the gaps and enter the
wounded tissue, where they engulf microbes, dirt, and debris from damaged cells. The micrograph
shows a white blood cell leaving a capillary to join the fray against bacteria that have entered a cut.

The Inflammatory Response Defends against Localized Injury

Large-scale breaches of the skin or mucous membranes, such as a cut, elicit an inflammatory response. Damaged cells release the chemical **histamine** into the wounded area. Histamine makes capillary walls leaky and relaxes the smooth muscle surrounding arterioles, leading to increased blood flow. With extra blood flowing through leaky capillaries, fluid seeps from the capillaries into the tissues around the wound. The wound becomes red, swollen, and warm (*inflammation* literally means "to set on fire"). Meanwhile, other chemicals released by injured cells initiate blood clotting (see Chapter 22), which both blocks damaged blood vessels, preventing microbes from escaping into the bloodstream, and seals off the wound from the outside world, limiting the entry of more microbes.

Still other chemicals, some released by wounded cells and others produced by the microbes themselves, attract macrophages and other phagocytic cells to the wound. The phagocytic cells squeeze out through the capillary walls, enter the wound, and engulf microbes, dirt, and tissue debris (Fig. 24-4). Unfortunately, each phagocyte can eat just so many microbes, and then it dies. The pus that collects around a wound consists largely of microbes, tissue debris, and living and dead white blood cells.

Fever Is a Response by the Body That Combats Large-Scale Infections

If enough microbes make it past the inflammatory response, they may severely infect larger areas of the body, triggering a fever. Most Americans, having been trained by advertisements and TV dramas to regard fevers as dangerous and debilitating, believe that fevers are part of a microbe's assault on the body. The facts are just the opposite: A fever is part of the body's assault on the microbes. Severe fevers *are* dangerous, even fatal, but "average" fevers of 38° or 39°C (100° to 102°F) are beneficial.

The hypothalamus, a part of the brain, contains temperature-sensing nerve cells that are the body's thermostat. Normally, the thermostat is set at about 37°C

(98.6° F). When disease organisms invade, however, the thermostat is turned up. Certain white blood cells, in responding to the infection, release hormones collectively called **endogenous pyrogens** ("self-produced fire-makers"). Pyrogens travel in the bloodstream to the hypothalamus and raise the thermostat's set point, triggering behaviors that increase body temperature: shivering, increased fat metabolism, or feeling cold so more clothing is put on. Pyrogens also cause other cells to reduce the concentration of iron in the blood.

Fever has both beneficial effects for the body's defenses and detrimental effects on the invading microbes. Many bacteria require more iron to reproduce at temperatures of 38° or 39° C than at 37° C, so fever and reduced iron in the blood combine to slow down their rate of reproduction. Simultaneously, fever increases the activity of phagocytic white blood cells that attack the bacteria, thereby producing a shorter and less serious infection.

Fever also helps fight viral infections. When certain cells of the body are invaded by viruses, they synthesize and release a protein called **interferon.** Interferon travels to other cells and increases their resistance to viral attack. Fevers increase the production of interferon. In one study, patients with colds were treated with aspirin (to reduce fever) or a placebo (an inactive substance that looks like the real drug, so the patients don't know whether they have been given the drug or not). The patients given aspirin had far more viruses in their noses and throats—and consequently sneezed and coughed out far more viruses—than the patients given the placebo did. This means that the immune systems of patients with lowered fevers weren't as effective at controlling their own infections and that these patients were much more infectious to other people.

As you can see, these defenses are all nonspecific; that is, their roles are to prevent or overcome any and all microbial invasions of the body. However, these nonspecific defenses are not impregnable. When they fail to do the job, the body then mounts a highly specific **immune response** directed against the particular organism that has successfully invaded the body.

The Immune Response

More than 2000 years ago, the Greek historian Thucydides recognized the essential features of the response of the immune system to infection. He observed that occasionally someone contracts a disease, recovers, and never catches that particular disease again—the person has become immune. With rare exceptions, however, immunity to one disease confers no protection against other diseases. Thus, the immune system attacks one type of microbe, overcomes it, and provides future protection against that microbe but no others. This is why we refer to the immune response as a specific defense against invasion.

The immune system consists of about 2 trillion white blood cells, of a type called lymphocytes. Many are distributed throughout the body in the blood and lymph; others are clustered in specific organs, particularly the thymus, lymph nodes, and spleen. The immune response results from the interactions among several types of lymphocytes and the molecules that they produce. With such a large cast of characters, the theater of the immune response is difficult to follow without a program. Table 24-1 provides a brief summary of the major actors and their roles.

A Successful Immune Response Recognizes, Overcomes, and Remembers

The immune response is provided chiefly by two types of lymphocytes, called **B cells** and **T cells** (Fig. 24-5). As do all white blood cells, B and T lymphocytes arise from precursor cells in the bone marrow (see Chapter 22). Early in embryonic development, the newly forming T cells migrate to the thymus (hence the name T cell) and differentiate. B cells differentiate in the bone marrow itself (hence the name B cell). As we shall see shortly, B and T cells play quite different roles in the immune response. However, immune responses produced by both B and T cells consist of the same three fundamental steps: (1) recognizing the invader, (2) launching a successful attack, and (3) retaining the memory of the invader to ward off future infections.

Recognition

To understand how the immune system responds to invading microbes, we must answer three related questions: (1) How do the immune cells recognize foreign molecules? (2) How can they produce specific responses to so many different molecules? (3) How do they determine that a substance is "foreign" and not "self," thereby avoiding destruction of the tissues of their own body?

Antibodies and T-Cell Receptors Recognize and Bind to Foreign Molecules, Triggering the Immune Response

The key to the immune system's ability to attack invading microbes lies in the structure and function of large proteins called **antibodies** and **T-cell receptors.** Antibodies are found either attached to the surfaces of B cells or dissolved in the plasma of the blood, where they are called immunoglobulins (often abbreviated Ig). As their name implies, T-cell receptors are attached to the surfaces of T cells. They are never secreted from the cells into the bloodstream.

An Antibody Contains Both Receptor and Effector Regions

Antibodies are Y-shaped molecules composed of four peptide chains, one large (heavy) chain and one small (light) chain on each side of the Y (Fig. 24-6). Both heavy and

TABLE 24-1 ■ *The Major Molecules and Cells of the Immune Response*

Molecules

Antigens	Large organic molecules, usually proteins, polysaccharides, or glycoproteins, that can trigger an immune response; often found on the surface of cells.
Antibodies	Proteins produced by the cells of the immune system that bind to antigens and either neutralize the antigenic molecule itself or mark cells that bear the antigens for destruction.
Major histocompatibility complex (MHC)	A set of proteins found on the surface of cells that "label" the cell as belonging to a unique individual organism.
Effector molecules	A diverse group of molecules, including histamine and the cell-destroying proteins of killer cells and complement (soluble proteins found in blood).
Regulatory molecules	Hormonelike molecules produced by cells of the immune system that regulate the immune response.

Cells

Macrophages	Phagocytic white blood cells that both destroy invading microbes and help to alert other immune cells to the invasion.
B cells	Lymphocytes that produce antibodies; when stimulated, certain of their daughter cells (*plasma cells*) secrete large quantities of antibodies into the bloodstream.
T cells	A set of lymphocytes that regulate the immune response or kill certain types of cells. *Cytotoxic T cells* destroy specific targeted cells, usually either foreign eukaryotic cells, infected body cells, or cancerous body cells. *Helper T cells* stimulate immune responses by both B cells and killer T cells. *Suppressor T cells* inhibit immune responses by other lymphocytes.
Memory cells	A subset of the offspring of B and T cells that are long-lived and provide future immunity against a second invasion by the same antigen.

light chains consist of a **constant region** that is similar in all antibodies of the same class (see Table 24-2) and a **variable region** that differs among antibodies. The combination of light and heavy chains results in an antibody with two functional parts: the "arms" and the "stem" of the Y.

The Arms of an Antibody Form Binding Sites That Recognize Specific Antigens

The variable regions that make up the arms form highly specific binding sites for large molecules. These binding sites are a lot like the active sites of enzymes (Chapter 4):

Figure 24-5   The major cells of the immune system and their roles in the immune response

Figure 24-6   Antibody structure

Antibodies are proteins composed of two pairs of peptide chains, called light and heavy chains, arranged something like the letter Y. The variable regions on the two chains form a specific binding site at the end of each arm of the Y. Different antibodies have different variable regions, forming unique binding sites. The human body synthesizes millions of distinct antibodies, each binding a different antigen.

Each binding site has a particular shape and electrical charge, so only certain molecules can fit in and bind. The binding sites at the tips of an antibody's arms are so specific that each antibody can bind just a few different types of molecules, perhaps only one.

In general, only large, complex molecules, such as proteins, polysaccharides, and glycoproteins, can bind to antibodies. These molecules are called **antigens** (short for "*anti*body response *gen*erating"). Antigens either may be attached to the surfaces of cells (for example, the body's own cells or invading microbes) or may be dissolved in

the blood or extracellular fluid (such as snake venom or a bacterial toxin). As we will discuss shortly, the binding of antigen to antibody triggers the immune response.

The Stem of an Antibody Determines Its Function

The stem of the Y-shaped antibody molecule consists of part of the constant regions of the two heavy chains. The stem determines the activity of an antibody. As we already mentioned, the constant regions are *similar* in different antibodies; they are not, however, the same. For example, the heavy chain constant regions that make up the stem of one type of antibody may function to attach the antibody to the plasma membrane of a cell; the constant regions of another type of antibody may bind to certain proteins in the blood, called **complement**, to promote destruction of microbes.

Antibodies Are Both Receptors for Foreign Molecules and Effectors That Help Destroy Invading Molecules and Microbes

Antibodies perform two very different functions in the immune response (Fig. 24-7). First, antibodies are receptors. In an antibody's role as a receptor, the stem attaches the antibody to the B cell's plasma membrane, while the two arms of the antibody protrude outward, sampling the blood and lymph for antigen. The binding of antigen to these "receptor antibodies" triggers responses in the lymphocytes bearing the antibodies.

Second, antibodies are effectors. As we will explain shortly, certain descendants of B cells (called plasma cells) secrete antibodies into the bloodstream. There the antibodies play several roles in neutralizing poisonous antigens or destroying microbes that bear antigens. Small differences in the stem—that is, the constant region—yield five classes of true antibodies (not including T-cell receptors; Table 24-2) that serve different roles in the defense of the body. The stems of some antibodies form a coating on microbes that targets the microbes for ingestion by macrophages (IgG); other stems promote secretion of the antibody into the respiratory and digestive tracts (IgA); still other stems bind to certain cells lining the respiratory and digestive tracts, which release histamine during allergic responses (IgE).

T-Cell Receptors Bind Antigen and Trigger Responses

T-cell receptors are somewhat different from antibodies in structure and function. First, they are found only on the surfaces of T cells, not in the bloodstream. Second, they consist of two peptide chains of about equal size. However, the ends of the two chains protrude out from the T cell, forming highly specific binding sites for antigen, just as the ends of the arms of an antibody molecule do. Third, T-cell receptors serve only a receptor function, recognizing antigen molecules and triggering a response in the T cell. They do not directly contribute to the destruction of invading microbes or toxic molecules.

TABLE 24-2 ⁛ *Antibodies and Their Roles*

IgM		IgM is composed of five Y-shaped antibody "monomers" held together at their stems. IgM is usually the first antibody secreted during an immune response. In the bloodstream it agglutinates antigens (because of the large number of antigen binding sites), activates complement proteins, and stimulates phagocytosis of bound microbes by macrophages.
IgG		IgG, the most common antibody in the blood, is composed of a single antibody "monomer." It activates both complement and macrophages. Special transport processes carry IgG across the placenta, where it protects the developing fetus against disease.
IgA		IgA is usually a dimer of two antibody monomers held together by their stems. An additional protein wound around the antibody stems helps IgA to be secreted from the bloodstream into saliva, tears, and mucus. Therefore, IgA is abundant on the surfaces of the respiratory and digestive tracts. It binds microbes on these surfaces, prevents them from entering the body, and allows them to be swept out of the body with mucus or other secretions. Thus, IgA provides a front line of defense for mucous membranes.
IgE		IgE, the "allergy antibody," is composed of a single antibody monomer. The IgE stem binds to histamine-releasing cells in connective tissue and to certain white blood cells, including basophils and eosinophils. Its normal function appears to be protection against parasites, which are expelled by the sneezing and coughing induced by histamine release and weakened or killed by eosinophil activation. Allergic responses are produced when harmless substances bind to IgE. Allergy "shots" work by stimulating synthesis of IgG antibodies that bind the same antigen the IgE antibodies bind. If enough IgG antibodies are present, they bind up most of the allergy antigen and keep the allergy antigens from contacting IgE.
IgD		IgD is a single antibody monomer, usually found bound to the plasma membranes of B cells. Its function is unknown; it may serve mainly as an antigen receptor.

The Immune System Can Recognize Millions of Molecules

A human being may encounter thousands of different types of invaders in a lifetime, from pollen grains and flu viruses to botulinus toxin (which causes botulism, a kind of food poisoning). The immune system recognizes and responds to this multiplicity of antigens because its cells produce millions of different antibodies and T-cell receptors, each capable of binding a different antigen.

This fact presented immunologists with a major problem: Because antibodies and T-cell receptors are proteins, and proteins are encoded by the genes, it would seem that a human being must have millions of antibody and T-cell receptor genes. But there are probably fewer than 100,000 genes in the entire human genome. How, then, can a fraction of these genes code for millions of antibodies and T-cell receptors? Two distinct but complementary mechanisms join forces to produce an enormous diversity of antibodies and T-cell receptors from a relative handful of genes. For simplicity, we will discuss only the generation of antibody diversity in B cells; similar phenomena produce a large diversity of T-cell receptors.

1. Genes encode parts of antibodies, not entire antibodies. These "antibody part" genes join together during immune cell development to form complete genes for the light and heavy chains of antibodies (Fig. 24-8). The progenitors of B cells contain a few genes for the constant regions of antibodies and several hundred genes for the variable regions (see Fig. 24-8a). During the devel-

opment of each individual B cell, the chromosomes containing the "antibody part" genes are cut, sections are removed, and the remaining pieces are joined together, bringing one variable-region gene next to one constant-region gene (see Fig. 24-8b). Each cell comes to possess one "constant + variable" light-chain gene and one "constant + variable" heavy-chain gene. Which variable-region gene winds up alongside the constant-region gene is probably random. Only the constant and variable genes that are next to one another are transcribed and ultimately provide the genetic information for antibody molecules; the rest remain unused. Therefore, each B cell produces a single type of antibody, specified by the chance recombination of variable- and constant-region genes, that is different from the antibody produced by most or all other B cells, except of course its own daughter cells (see Fig. 24-8c).

It may help you to think of antibody gene formation in terms of card playing. Each B cell is dealt a "hand" of two variable-region genes, one for the light chain and one for the heavy chain, randomly chosen from two large "decks" of genes. With each deck containing hundreds of "cards" (genes), virtually every cell will synthesize its own unique antibody.

2. The genes for certain antibody parts are incredibly prone to mutate, thereby constantly generating new antibody genes. As B cells reproduce, some of their daughter cells accumulate mutations in their antibody genes. Thus, two sister cells may produce different antibodies.

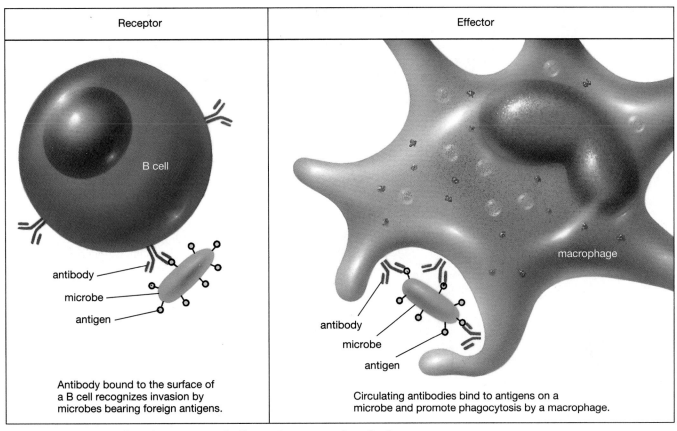

Receptor	Effector
Antibody bound to the surface of a B cell recognizes invasion by microbes bearing foreign antigens.	Circulating antibodies bind to antigens on a microbe and promote phagocytosis by a macrophage.

Figure 24-7   **Examples of receptor and effector functions of antibodies**

There are many other effector functions that are described in the text but not illustrated here.

Kmart Versus Custom Tailoring: The Immune System Does Not Design Antibodies or T-Cell Receptors Expressly to Bind Invading Antigens

The end result of mutation and gene recombination is that each B cell has its own particular antibody genes, different from those of most other B cells. At any given time, the body may contain an army of perhaps 100 million different antibodies, so antigens almost always encounter antibodies that can bind them. A key fact about the antigen-antibody match, often misunderstood by students, is this: The immune system does not "design" antibodies to fit invading antigens. Instead, the immune system randomly synthesizes millions of different antibodies. The fact that virtually every possible invading antigen binds to at least a few antibodies is due purely to the immense numbers of different antibodies present in the body.

A more familiar example may help make this clear. When the Queen of England wants a dress, tailors go to Buckingham Palace, obtain the Queen's exact measurements, and custom-sew a dress to the Queen's specifications. Most of the rest of us go to a department store and look through the racks of ready-made clothes. If a store has a large enough selection, we will probably find something that fits reasonably well and is more or less the style we want.

The immune system is like a department store: The array of antibodies are simply there, waiting. They have *not* been designed to bind to any particular antigen. When an antigen enters the body, it may randomly bump into the variable regions of many antibodies. It cannot bind to most antibodies (in our clothes analogy, these antibodies don't "fit"), but, with millions of different antibodies around, the antigen inevitably encounters one that binds it pretty well. Antigen-antibody binding triggers changes in the B cells, usually leading to the destruction of microbes bearing that antigen. We will examine these changes in more detail in a moment.

The Immune System Distinguishes "Self" from "Non-Self"

The surfaces of the body's own cells bear large proteins and polysaccharides, just as microbes do. Some of these proteins, collectively called the **major histocompatibility complex** (**MHC** for short), are unique to each individual (except identical twins, which have the same genes and hence the same MHC proteins). Because your MHC proteins are different from those of everyone else, they act as antigens in other people's bodies (this is why transplants are rejected; the recipient's immune system recognizes MHC proteins on the donor's cells as foreign and destroys the transplant). Why doesn't your own immune system respond to these "self" antigens and destroy your own cells? The key seems to be the continuous presence of the body's antigens while the immune cells mature. As an embryo develops, some differentiating immune cells do indeed produce anti-

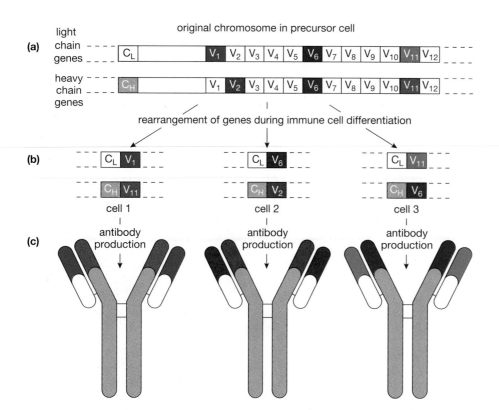

Figure 24-8 Recombination during the construction of antibody genes

(a) Each precursor cell of the immune system contains one or perhaps a few genes for the constant regions of the light and heavy chains of antibodies and many genes for the variable regions. (b) During the development of each B cell, these genes are rearranged, moving one of the variable-region genes next to a constant-region gene. Each cell thus generates a "recombined antibody gene" for each chain, which differs from the recombined antibody genes generated by other B cells. (c) Representations of the different antibodies synthesized by each B cell.

bodies or T-cell receptors that can bind the body's own proteins and polysaccharides, not only the MHC proteins, but lots of other molecules as well. However, if these *immature* immune cells contact molecules that bind to their antibodies or T-cell receptors, the immune cells are destroyed. In this way, potentially dangerous immune cells, which respond to the body's own cells as they would to foreign cells, are eliminated during immune system development. Thus the immune system distinguishes "self" from "non-self" by retaining only those immune cells that do not respond to the body's own molecules.

Attack

If the body is invaded by microbes, the immune system mounts two types of attack: B cells provide **humoral immunity,** which is provided by antibodies circulating in the bloodstream, while T cells produce **cell-mediated immunity,** as cytotoxic T cells destroy infected body cells (Fig. 24-9). The two types of responses are not as distinct as this description implies; however, the humoral and cellular responses are most easily understood when they are considered separately.

Humoral Immunity Is Produced by Antibodies in Blood

Each B cell bears a specific antibody on its surface (Fig. 24-10). When a microbial infection occurs, the antibodies borne by a few B cells will bind to antigens on the

microbe (in reality, the process is considerably more complex than this, but this is the end result). Antigen-antibody binding causes these B cells to divide rapidly (a type of T cell also helps stimulate cell division in B cells; see below. The resulting populations of cells are genetically identical to the parent B cells (a clone); they have been "selected" to multiply by the presence of particular invading antigens. The process is therefore called **clonal selection.** The daughter cells differentiate into two cell types: **plasma cells** and **memory cells.** Plasma cells become enlarged and packed with endoplasmic reticulum, churning out huge quantities of that cell type's specific antibody (Fig. 24-11). These antibodies are released into the bloodstream (hence the name "humoral" immunity; to the ancient Greeks, blood was one of the four body "humors"). Memory cells do not release antibodies. As we will see, memory cells play an important role in future immunity to this specific microbe.

Antibodies circulating in the bloodstream can affect only molecules or microbes that are exposed to them. Invaders that penetrate into the body's cells, as viruses do, are safe from antibody attack as long as they are within the cytoplasm of a body cell. Therefore, humoral immunity defends primarily against bacteria (most of which never enter the body's cells), bacterial toxins, and some fungi and protists. Viruses can be attacked during the times that they are outside of the body's cells, for example during the first stage of an infection or when they have finished replicating in one host cell, ruptured it, and have been released into the body fluids. As we will see shortly, however, cell-mediated immune responses are the primary defense against viral infections.

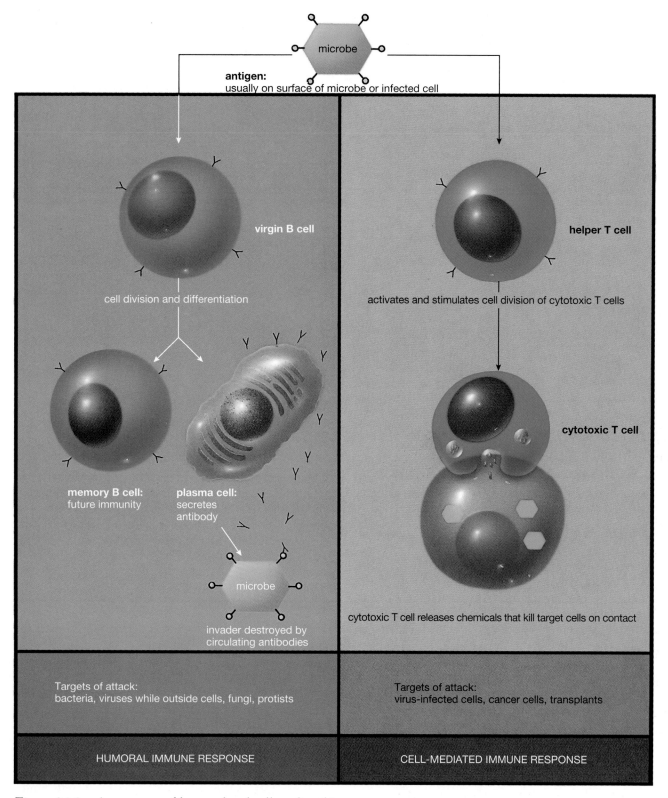

Figure 24-9   **A summary of humoral and cell-mediated immune responses**

See the text for details.

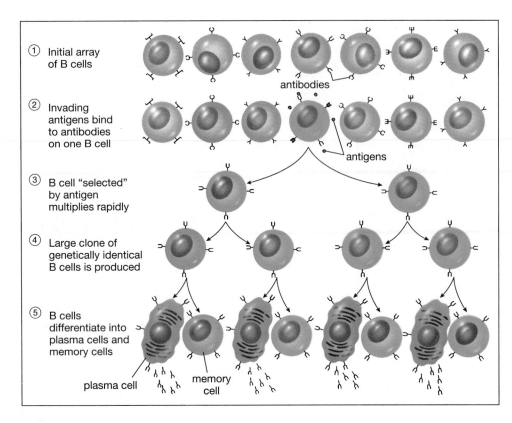

① Initial array of B cells

antibodies

② Invading antigens bind to antibodies on one B cell

antigens

③ B cell "selected" by antigen multiplies rapidly

④ Large clone of genetically identical B cells is produced

⑤ B cells differentiate into plasma cells and memory cells

plasma cell   memory cell

Figure 24-10   **Clonal selection among B cells by invading antigens**

① The body contains a large array of B cells, each of which synthesizes and bears on its surface a specific antibody, different from the antibodies borne by other B cells (see Fig. 24-8). ② When an antigen enters the bloodstream, it binds to the antibodies on one (or a few) B cells. ③ Antigen-antibody binding causes the selected B cell to multiply rapidly. ④ A clone of B cells, all synthesizing the same antibody, is produced. ⑤ The daughter B cells differentiate into plasma cells that secrete antibodies directed against the specific activating antigen or into long-lived memory cells.

Antibodies Destroy Extracellular Microbes and Molecules by Several Mechanisms

Antibodies in the blood may affect antigenic molecules, microbes bearing antigens, and infected body cells in four ways:

1. **Neutralization.** The antibody may combine with or cover up the active site of a toxic antigen such as a bacterial toxin, thereby preventing the toxin from harming the body.
2. **Promotion of phagocytosis.** The antibody may coat the surface of a microbe. The protruding antibody stems seem to identify the microbe as a target for phagocytic white blood cells to engulf.
3. **Agglutination.** Each antibody has two binding sites for antigen, one on each arm (the exception being IgM antibodies, which have 5 arms and thus 10 binding sites). These binding sites may attach to antigens on two different microbes, holding them together. As more and more antibodies link up with antigens on different microbes, the microbes clump together, or agglutinate. Agglutination seems to enhance phagocytosis.
4. **Complement reactions.** The antibody-antigen complex on the surface of an invading cell may trigger a series of reactions with blood proteins called the **complement system.** When these proteins bind to the antibody stems, they attract phagocytic white blood cells to the site, promote phagocytosis of the foreign cells, or in some instances directly destroy the invaders by creating holes in their plasma membranes, much as natural killer cells do (see Fig. 24-3).

T Cells Produce Cell-Mediated Immunity

Cell-mediated immunity is the primary defense against the body's own cells when they have become cancerous or have been infected by viruses. Cell-mediated immunity is also important in overcoming infection by fungi and protists. Several types of T cells contribute to cell-mediated immunity: helper cells, cytotoxic cells, and suppressor cells.

Three Types of T Cells Contribute to the Immune Response

Helper T cells, when their T-cell receptors bind an antigen, release chemicals that assist other immune cells in their defense of the body. These hormone-like chemicals stimulate cell division and differentiation in both B cells and cytotoxic T cells responding to the same microbial invasion. In fact, very little immune response, either cell-mediated or humoral, can occur without the boost provided by helper T cells. This is why AIDS, which destroys helper T cells, is such a deadly disease. We will discuss AIDS more thoroughly shortly.

Receptors on the surfaces of **cytotoxic T cells** bind to antigens on the surface of an infected cell. The cytotoxic T cell then releases proteins that disrupt the infected cell's plasma membrane (Fig. 24-12). In at least some instances, the cytotoxic T cell's proteins are similar or identical to those of the natural killer cells that create giant holes in the target cell's membrane. After the infection has been conquered, **suppressor T cells** seem to shut off the immune response in both B and cytotoxic T cells.

mal feeds as growth promoters. Some farm animals harbor a strain of *Salmonella* that causes a virulent form of food poisoning. These bacteria have now developed antibiotic resistance, probably as a result of continued exposure. Tetracycline is losing its effectiveness against gonorrhea, meningitis, and urinary and respiratory tract infections. An ominous new development is the emergence of strains of tuberculosis bacterium resistant to most antibiotics. Once considered vanquished in developed countries, tuberculosis is now spreading among the homeless and AIDS patients. Vigorous treatment of AIDS patients with antibiotics is believed to have hastened the emergence of these deadly new strains, which typically kill 50% of the people they infect.

Excessive use of antibiotics unnecessarily increases the exposure of bacteria to these powerful selective agents and encourages the spread of resistant strains. Antibiotics continue to be prescribed inappropriately, as in the treatment of viral infections such as colds. Researchers are working to develop new antibiotics, but drug development and testing take many years, whereas resistant bacterial strains spread rapidly. Unfortunately, most of the new drugs are more toxic to humans and far more expensive than those they replace. Clearly we must restrict use of antibiotics to situations in which they are urgently required, rather than relying on a steady flow of new ones. Awareness and restraint may yet enable our children to benefit from some of the same "miracle drugs" that protected our parents.

SUMMARY OF KEY CONCEPTS

Defenses against Microbial Invasion
The human body has three lines of defense against invasion by microbes: (1) the barriers of skin and mucous membranes; (2) nonspecific internal defenses, including phagocytosis, killing by natural killer cells, inflammation, and fever; and (3) the immune response.

The skin physically blocks the entry of microbes into the body. It is also covered with secretions from sweat and sebaceous glands that inhibit bacterial and fungal growth. The mucous membranes of the respiratory and digestive tracts secrete antibiotic substances, IgA antibodies, and mucus. Microbes are trapped in the mucus, swept up to the throat by cilia, and expelled or swallowed.

If microbes enter the body, white blood cells travel to the site of entry and engulf the invading cells. Natural killer cells secrete proteins that kill infected body cells or cancerous cells. Injuries stimulate the inflammatory response, in which chemicals are released that attract phagocytic white blood cells, increase blood flow, and make capillaries leaky. Later, blood clots wall off the injury site, preventing further invasion and spread of the microbes to other parts of the body. Fever is caused by chemicals called pyrogens released by white blood cells in response to infection. High temperatures inhibit bacterial growth and accelerate the immune response.

The Immune Response
The immune response involves two types of lymphocytes, B cells and T cells. Plasma cells, which are descendants of B cells, secrete antibodies into the bloodstream, causing humoral immunity. Cytotoxic T cells destroy some microbes, cancer cells, and virus-infected cells on contact, causing cell-mediated immunity. Helper T cells stimulate both the humoral and cell-mediated immune responses. Immune responses have three steps: recognition, attack, and memory.

Recognition
Antibodies (on B cells) and T-cell receptors (on T cells) recognize foreign molecules and trigger the immune response. Antibodies are Y-shaped proteins composed of a constant and a variable region. Antigens are molecules that generate an antibody response. Antibodies function both as receptors, detecting the presence of antigens, and as effectors, actively working to destroy antigens. Each B cell synthesizes only one type of antibody, unique to that particular cell and its progeny. The diversity of antibodies arises from gene shuffling and mutation of antibody genes during B-cell development. Each antibody has specific sites that bind one or a few types of antigen. Normally, only foreign antigens are recognized by the B cells.

Attack
Antigens bind to and activate only those B and T cells with the complementary antibodies or T-cell receptors. In humoral immunity, B cells with the proper antibodies, stimulated by the presence of particular antigens, divide rapidly to produce plasma cells that synthesize massive quantities of the antibody. The circulating antibodies destroy antigens and antigen-bearing microbes by four mechanisms: neutralization, promotion of phagocytosis by white blood cells, agglutination, and complement reactions. In cell-mediated immunity, T cells with the proper receptors bind antigen and divide rapidly. Cytotoxic T cells bind to antigens on microbes, infected cells, or cancer cells and kill the cells. Helper T cells stimulate, and suppressor T cells turn off, both the B-cell and cytotoxic-T-cell responses.

Memory
Some progeny cells of both B and T cells are long-lived memory cells. If the same antigen reappears in the bloodstream, these memory cells are immediately activated, divide rapidly, and cause an immune response that is much faster and more effective than the original response.

Medicine and the Immune Response
Antibiotics kill microbes or slow down their reproduction, thus allowing the immune system more time to respond and exterminate the invaders. Vaccinations are injections of anti-

gens from disease organisms, often the weakened or dead microbes themselves. An immune response is evoked by the antigens, providing memory and a rapid response should a real infection occur. Allergies are immune responses to normally harmless foreign substances, such as pollen or dust. Certain cells respond to the presence of these substances by releasing histamine, which causes a local inflammatory response. Some diseases are caused by defective immune responses. Autoimmune diseases arise when the immune system destroys some of the body's own cells. Immune deficiency diseases occur when the immune system cannot respond strongly enough to ward off normally minor diseases.

AIDS

AIDS (acquired immunodeficiency syndrome) is caused by one of two viruses, called human immunodeficiency

viruses 1 and 2. These viruses invade helper T cells and destroy them. Without helper T cells to stimulate the immune responses of B cells and cytotoxic T cells, the AIDS victim is extremely susceptible to a wide assortment of diseases. These infections eventually kill the patient.

Cancer

Cancer is a population of the body's cells that divides without control. Some cancers may be caused by activation of growth genes, called oncogenes, that cause cells to grow and multiply. These genes may be activated by viral infection or by mutations caused by chemicals or radiation. Other cancers may be caused by loss or inactivation of tumor-suppressor genes, which turn off or suppress cell division in normal cells.

✖ KEY TERMS

acquired immunodeficiency syndrome (AIDS)   p. 460
agglutination   p. 456
allergy   p. 458
antibody   p. 449
antigen   p. 451
autoimmune disease   p. 459
B cell   p. 449
cancer   p. 464
cell-mediated immunity   p. 454
clonal selection   p. 454
complement   p. 451
complement reaction   p. 456
complement system   p. 456
constant region   p. 450

cytotoxic T cell   p. 456
endogenous pyrogen   p. 449
fever   p. 446
helper T cell   p. 456
histamine   p. 448
humoral immunity   p. 454
immune response   p. 449
inflammatory response   p. 446
interferon   p. 449
macrophage   p. 447
major histocompatibility complex (MHC)   p. 453
memory cell   p. 454
mucous membrane   p. 445
natural killer cell   p. 446

neutralization   p. 456
oncogene   p. 465
phagocytic cell   p. 446
plasma cell   p. 454
retrovirus   p. 460
reverse transcriptase   p. 460
severe combined immune deficiency (SCID)   p. 459
skin   p. 445
suppressor T cell   p. 456
T cell   p. 449
T-cell receptor   p. 449
tumor-suppressor gene   p. 465
vaccination   p. 458
variable region   p. 450

✖ THINKING THROUGH THE CONCEPTS

Multiple Choice

1. In addition to the immune system, we are protected from disease by
   a. the skin
   b. mucous membranes
   c. natural secretions such as acids, protein-digesting enzymes, and antibiotics
   d. cilia
   e. all of the above
2. Body cells infected by viruses are destroyed mostly by
   a. IgA   b. phagocytes
   c. natural killer cells   d. histamines
   e. natural antibiotics
3. Fevers
   a. decrease interferon production
   b. decrease the concentration of iron in the blood
   c. decrease the activity of phagocytes
   d. increase the rate of reproduction of invading bacteria
   e. do all of the above
4. T and B cells are
   a. lymphocytes   b. macrophages

   c. natural killer cells   d. red blood cells
   e. phagocytes
5. During allergic responses, IgE
   a. covers invaders and marks them for phagocytosis
   b. pokes holes in cell membranes of invading parasites
   c. binds to mast cells, causing them to release histamine
   d. triggers the production of antigens by lymphocytes
   e. promotes secretion of antibodies into the digestive and respiratory tracts
6. What shuts off the immune response in T and B cells after an infection has been conquered?
   a. IgA   b. histamine
   c. pyrogens   d. natural killer cells
   e. suppressor T cells

Review Questions

1. List the three lines of defense of the human body against invading microbes. Which are nonspecific (i.e., act against all types of invaders) and which are specific (i.e., act only against a particular type of invader)? Explain your answer.

2. How do natural killer cells and cytotoxic T cells destroy their targets?

3. Describe humoral immunity and cell-mediated immunity. Include in your answer the types of immune cells involved in each, the location of antibodies and receptors that bind foreign antigens, and the mechanisms by which invading cells are destroyed.

4. How does the immune system construct so many different antibodies?

5. How does the body distinguish "self" from "non-self"?

6. Diagram the structure of an antibody. What parts bind to antigens? Why does each antibody bind only to a specific antigen?

7. What are memory cells? How do they contribute to long-lasting immunity to specific diseases?

8. What is a vaccine? How does it confer immunity to a disease?

9. Describe the causes, progression, and eventual outcome of AIDS. How do AIDS treatments work?

10. What is cancer? How do oncogenes cause cancer? How do tumor-suppressor genes prevent cancer? How can environmental factors "turn on" oncogenes to cause cancer?

✱ A P P L Y I N G   T H E   C O N C E P T S

1. The text states that smallpox has been eradicated from Earth. This is not quite true: There are smallpox stocks in two laboratories, one in the United States, one in Russia. A debate is raging about whether these stocks should be destroyed. One side argues that having smallpox around is too dangerous. The other side argues that we may be able to learn things from smallpox, answers to questions we don't even know enough to ask yet, that may help us conquer future diseases. Do you think that the smallpox stocks should be destroyed? Why or why not?

2. The box "Health Watch: Flu—The Unbeatable Bug" states that the flu virus is different each year. If that is true, what good is it to get a "flu shot" each winter?

3. Pollen, house dust, or poison ivy provokes allergic reactions in some individuals. When these allergens cannot be avoided, a "hyposensitization" procedure is usually prescribed. Hyposensitization is most likely due to an opposite process—immunization. Explain how giving a patient gradually increasing doses of poison ivy extract over a period of time helps subdue allergic dermatitis.

4. Throughout pregnancy a mother may pass IgG antibodies across the placenta. After delivery she may pass IgAs on through breast milk. The immunity provided to her offspring lasts only a few months. Explain.

✱ F O R   M O R E   I N F O R M A T I O N

Beardsley, T. "A War Not Won." *Scientific American*, January 1994. The war on cancer remains a standoff.

Caldwell, M. "The Long Shot." *Discover*, August 1993. To produce an effective vaccine against HIV turns out to be a lot more difficult than to produce one against smallpox or polio.

Goodenough, Ursula. "Deception by Pathogens." *American Scientist*, July–August 1991. Describes tactics disease organisms use to gain access to the body.

Greene, W. C. "AIDS and the Immune System." *Scientific American*, September 1993. We may know more about the AIDS virus than any other; can this knowledge produce effective treatment?

Janeway, C. A., Jr. "How the Immune System Recognizes Invaders." *Scientific American*, September 1993. All about antibodies, major histocompatibility complexes, antigen presentation and immune response.

Leder, P. "The Genetics of Antibody Diversity." *Scientific American*, May 1982. How only a few hundred genes can be used to make millions of antibodies.

Lemonick, Michael D. "The Killers All Around." *Time*, September 12, 1994. Are we losing the war against infectious diseases?

Levins, R., Awerbuch, T., Brinkmann, U., Eckardt, I., Epstein, P., Makhoul, N., de Possas, C. A., Puccia, C., Spielman, A., and Wilson, M. E. "The Emergence of New Diseases." *American Scientist*, January–February 1994. Diseases that appear to be new continue to challenge the human body and its physicians.

Lichtenstein, L. M. "Allergy and the Immune System." *Scientific American*, September 1993. Allergic responses, evolved as protection against parasites, turn against us when we respond violently to harmless pollen and foods.

Marrack, P., and Kappler, J. W. "How the Immune System Recognizes the Body." *Scientific American*, September 1993. How the immune system's potential weapons against the body are disarmed.

Nossal, G. J. V. "Life, Death, and the Immune System." *Scientific American*, September 1993. A nice introduction to the major players of the immune system.

Paul, W. E. "Infectious Diseases and the Immune System." *Scientific American*, September 1993. Microbes and the immune system engage in constant evolutionary warfare.

Steinman, L. "Autoimmune Disease." *Scientific American*, September 1993. An estimated 5 percent of Americans suffer from some type of autoimmune disorder.

Young, J. D-E., and Cohn, Z. A. "How Killer Cells Kill." *Scientific American*, January 1988. Cytotoxic T cells strike by secreting proteins that form large holes in the plasma membrane of their targets.

N E T   W A T C H

On-line resources for this chapter are on the World Wide Web at:
http://www.prenhall.com/~audesirk (click on the <u>table of contents</u> link and then select Chapter 24).

25 Chemical Control of the Human Body: The Endocrine System

*The next time you pass by an elementary school playground,
notice how young boys tend to interact almost exclusively with
other boys and girls with other girls. Hormonal changes that
begin at puberty dramatically alter the bodies, as well as the
behaviors, of teenagers.*

With the evolution of complex multicellular organisms came the need to coordinate the activities of cells in different parts of the body. Cell-to-cell communication is crucial to the control of movement, growth, reproduction, and the maintenance of homeostasis. Many different mechanisms have evolved by which cells within organisms communicate among themselves. One method is by direct contact. Molecules protruding from the surface membrane identify cells as belonging to an individual of a particular species, as parts of a unique individual organism, and as specific cell types, such as skin or liver. Surface contacts are important in the development of embryos, in which cells migrate around one another to arrive at their proper destination. Direct contact also plays a key role in defense against disease organisms, in which immune cells recognize invading foreign cells by their surface molecules (see Chapter 24).

Cells can also communicate with one another over distances. Cells of the nervous system convey electrical signals over long or short distances, then transmit those signals to other cells by releasing chemicals into the immediate vicinity of the nerve cell ending (Chapter 26). Cells of the **endocrine system**, in contrast, release chemicals into the bloodstream, where they are carried throughout the body and may have wide-ranging effects. These chemicals, or hormones, and the glands and organs that release them, are the subject of this chapter.

Hormone Structure and Function

A **hormone** is a chemical secreted by cells in one part of the body that is transported in the bloodstream to other parts of the body, where it affects particular target cells. Hormones are released by the cells of major endocrine glands and endocrine organs located throughout the body (Fig. 25-1). There are four classes of chemicals used as hormones in the animal kingdom, illustrated in Table 25-1.

Hormones Have a Variety of Chemical Structures

Peptide Hormones Include Both Short and Longer Chains of Amino Acids

Most hormones are chains of amino acids ranging from a few to over a hundred amino acids in length. Technically, short amino acid chains are called peptides, whereas longer chains are called proteins. For convenience, however, all hormones composed of amino acid chains are described as **peptide hormones.** Insulin, antidiuretic hormone, all the hormones of the hypothalamus and the anterior and posterior pituitary, as well as many other hormones are peptide hormones.

Some Hormones Consist of Modified Single Amino Acids

A few hormones are modified amino acids. The amino acid tyrosine forms the basis for the hormones epinephrine (also called adrenaline), norepinephrine (also called noradrenaline), and the thyroid hormone thyroxine.

Steroid Hormones Resemble Cholesterol in Structure

Steroid hormones, also called **steroids,** all have a chemical structure resembling cholesterol, from which most of them are synthesized. Steroid hormones are secreted by the ovaries and placenta (estrogen and progesterone), the testes (testosterone), and the adrenal cortex (aldosterone).

Prostaglandins Are Modified Fatty Acids

Nearly every type of cell in the body has been found to produce **prostaglandins.** These substances consist of two fatty acid carbon chains attached to a five-carbon ring.

Hormones Function by Binding to Specific Receptors on Target Cells

Because nearly all cells have a blood supply, once hormones enter the bloodstream, they reach nearly every cell of the body. But in order to exert their precise control, hormones must act only on certain **target cells.** Hormone specificity is determined by receptors on target cells; if a cell lacks a specific receptor for a hormone, the hormone will not affect the cell. In addition, the same hormone may have several different effects depending on the nature of the target cell it contacts. Receptors for hormones are found in two general locations on target cells: on the cell membrane and inside the cell, usually within the nucleus.

Hormones Are Regulated by Feedback Mechanisms

Animals usually regulate the release of hormones through negative feedback (see Chapter 21). During negative feedback, the secretion of a hormone causes effects in target cells that inhibit further secretion of the hormone. Neg-

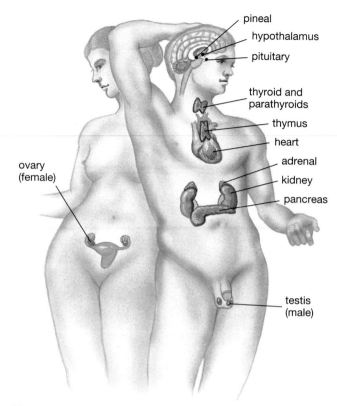

Figure 25-1   Major endocrine glands

The major endocrine glands discussed in the text are the hypothalamus-pituitary complex, the thyroid and parathyroid glands, the pancreas, the sex organs (ovaries in females, testes in males), and the adrenal glands. Other organs that secrete hormones include the pineal gland, thymus, kidneys, heart, and digestive tract.

ative feedback is an important way of maintaining homeostasis—that is, keeping conditions within the body relatively constant over time.

Most hormones exert such powerful effects on the body that it would be harmful to have too much hormone working for too long. For example, suppose you have jogged several miles on a hot, sunny day and lost a pint of water through perspiration. In response to the loss of water from your bloodstream, your pituitary gland releases antidiuretic hormone (ADH), which causes your kidneys to reabsorb more water and produce a very concentrated urine (see Chapter 23). However, if you arrive home and drink a quart of Gatorade, you will more than replace the water you lost in sweat. Continued retention of this excess water could raise blood pressure and possibly damage your heart. Negative feedback ensures that when your blood water content returns to normal, ADH secretion is turned off and your kidneys begin eliminating the excess water (see Chapter 23). Negative feedback is also at work in the control of thyroxine secretion (see Fig. 25-8).

In a few cases, positive feedback controls hormone release. As mentioned in Chapter 21, contractions of the uterus early in childbirth cause the release of oxytocin by

TABLE 25-1 ❖ *The Chemical Diversity of Vertebrate Hormones*

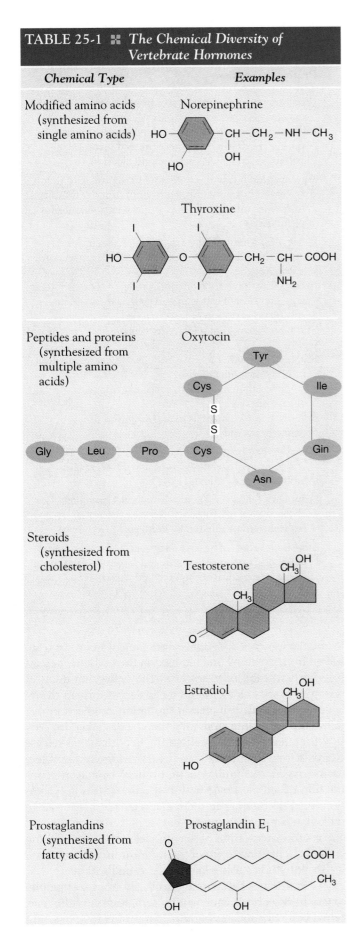

Chemical Type	Examples
Modified amino acids (synthesized from single amino acids)	Norepinephrine
	Thyroxine
Peptides and proteins (synthesized from multiple amino acids)	Oxytocin
Steroids (synthesized from cholesterol)	Testosterone
	Estradiol
Prostaglandins (synthesized from fatty acids)	Prostaglandin E_1

the posterior pituitary. The oxytocin stimulates stronger contractions of the uterus, which cause more oxytocin release, creating a positive feedback cycle. Simultaneously, oxytocin causes uterine cells to release prostaglandins that further enhance uterine contractions, another example of positive feedback. Both positive and negative feedback are involved in the interactions between mother and child during breastfeeding (see Fig. 25-4).

The Human Endocrine System

Endocrinologists (biologists who study the endocrine system) are still far from fully understanding hormonal control in humans and other animals. New hormones, or new roles for previously known hormones, are discovered nearly every year. What we might call the major endocrine glands and endocrine organs, however, have been known for many years. These are the hypothalamus-pituitary complex, the thyroid and parathyroid glands, the pancreas, the sex organs, and the adrenal glands (Fig. 25-1). Table 25-2 lists these and other glands, their major hormones, and their principal functions.

Mammals Have Both Exocrine and Endocrine Glands

Mammals, including humans, have two types of glands: exocrine glands and endocrine glands (Fig. 25-2, p. 478). **Exocrine glands** produce secretions that are released outside the body (*exo* means "out of" in Greek) or into the digestive tract (which is actually a hollow tube continuous with the outside world). Exocrine gland secretions are released through tubes or openings called **ducts**. The exocrine glands include the sweat and sebaceous (oil-producing) glands of the skin, the lacrimal (tear-producing) glands of the eye, the mammary (milk-producing) glands, as well as glands producing digestive secretions.

Endocrine glands, sometimes called ductless glands, release their hormones within the body (*endo* means "inside of"). An endocrine gland generally consists of clusters of hormone-producing cells embedded within a network of capillaries. The cells secrete their hormones into the extracellular fluid surrounding the capillaries. The hormones then enter the capillaries by diffusion and are distributed throughout the body by the bloodstream.

The Hypothalamus Controls the Secretions of the Pituitary Gland

The **hypothalamus** is a part of the brain that contains clusters of specialized nerve cells, called **neurosecretory cells**. Neurosecretory cells synthesize peptide hormones, store them, and release them when stimulated. The **pituitary** is a pea-sized gland that dangles from the hypothalamus by a stalk (Figs. 25-1 and 25-3, p. 479). Anatomically, the pituitary consists of two distinct lobes, or parts: the **posterior pituitary** and the **anterior pituitary**. The hypothalamus controls the release of hormones from both parts. The anterior

TABLE 25-2 ⁙ *Human Endocrine Glands and Hormones*

Endocrine Gland	Hormone	Type of Chemical	Principal Function
Hypothalamus (via posterior pituitary)	Antidiuretic hormone (ADH)	Peptide	Promotes reabsorption of water from kidneys; constricts arterioles
	Oxytocin	Peptide	In females, stimulates contraction of uterine muscles during childbirth, milk ejection, and maternal behaviors; in males, causes sperm ejection
Hypothalamus (to anterior pituitary)	Releasing and inhibiting hormones	Peptides	At least nine hormones; releasing hormones stimulate release of hormones from anterior pituitary; inhibiting hormones inhibit release of hormones from anterior pituitary
Anterior pituitary	Follicle-stimulating hormone (FSH)	Peptide	In females, stimulates growth of follicle, secretion of estrogen, and perhaps ovulation; in males, stimulates spermatogenesis
	Luteinizing hormone (LH)	Peptide	In females, stimulates ovulation, growth of corpus luteum, and secretion of estrogen and progesterone; in males, stimulates secretion of testosterone
	Thyroid-stimulating hormone (TSH)	Peptide	Stimulates thyroid to release thyroxine
	Growth hormone	Peptide	Stimulates growth, protein synthesis, and fat metabolism; inhibits sugar metabolism
	Adrenocorticotropic hormone (ACTH)	Peptide	Stimulates adrenal cortex to release hormones, especially glucocorticoids
	Prolactin	Peptide	Stimulates milk synthesis in and secretion from mammary glands
Thyroid	Thyroxine	Modified amino acid	Increases metabolic rate of most body cells; increases body temperature; regulates growth and development
	Calcitonin	Peptide	Inhibits release of calcium from bones
Parathyroid	Parathormone	Peptide	Stimulates release of calcium from bone; promotes absorption of calcium by intestines; promotes reabsorption of calcium by kidneys
Adrenal medulla	Epinephrine and norepinephrine (adrenaline and noradrenaline)	Modified amino acids	Increase levels of sugar and fatty acids in blood; increase metabolic rate; increase rate and force of contractions of the heart; constrict some blood vessels

pituitary is a true endocrine gland, composed of several types of hormone-secreting cells enmeshed in a network of capillaries. The posterior pituitary, on the other hand, is derived from an outgrowth of the hypothalamus.

The Posterior Pituitary Releases Hormones Produced by Cells in the Hypothalamus

The posterior pituitary contains the endings of two types of neurosecretory cells whose cell bodies are found in the hypothalamus. These neurosecretory cell endings are enmeshed in a capillary bed into which they release hormones to be carried into the bloodstream (Fig. 25-3). Two peptide hormones are synthesized in the hypothalamus and released from the posterior pituitary: **antidiuretic hormone (ADH)** and **oxytocin**.

Antidiuretic hormone, which literally means "hormone that prevents urination," helps prevent dehydration. As you learned in Chapter 23, ADH causes more water to be reabsorbed from the urine and retained in the body, by increasing the permeability to water of the collecting ducts of nephrons in the kidney. Interestingly, alcohol inhibits the release of ADH, greatly increasing urination, so a beer drinker may temporarily lose more fluid than he has taken in.

Other neurosecretory cells of the hypothalamus release oxytocin from their endings in the posterior pituitary. Oxytocin triggers the "milk ejection reflex" by causing contraction of muscle tissue within the breasts during lactation (breastfeeding). This reflex ejects milk from the saclike milk glands into the nipples (Fig. 25-4, p. 480). Oxytocin also causes contractions of the muscles of the uterus during childbirth, helping to expel the infant from the womb.

Recent studies using laboratory animals indicate that oxytocin also has behavioral effects. In rats, for example, oxytocin injections cause virgin females to exhibit maternal behavior such as building a nest, licking pups, and

TABLE 25-2 ▦ *(Continued)*

Endocrine Gland	Hormone	Type of Chemical	Principal Function
Adrenal cortex	Glucocorticoids	Steroid	Increase blood sugar; regulate sugar, lipid, and fat metabolism; anti-inflammatory effects
	Aldosterone	Steroid	Increases reabsorption of salt in kidney
	Testosterone	Steroid	Causes masculinization of body features, growth
Pancreas	Insulin	Peptide	Decreases blood glucose levels by increasing uptake of glucose into cells and converting glucose to glycogen, especially in liver; regulates fat metabolism
	Glucagon	Peptide	Converts glycogen to glucose, raising blood glucose levels
Ovaries[a]	Estrogen	Steroid	Causes development of female secondary sexual characteristics and maturation of eggs; promotes growth of uterine lining
	Progesterone	Steroid	Stimulates development of uterine lining and formation of placenta
Testes[a]	Testosterone	Steroid	Stimulates development of genitalia and male secondary sexual characteristics; stimulates spermatogenesis

Other Sources of Hormones

Digestive tract[b]	Secretin, gastrin, cholecystokinin, and others	Peptides	Control secretion of mucus, enzymes, and salts in digestive tract; regulate peristalsis
Thymus	Thymosin	Peptide	Stimulates maturation of cells of immune system
Pineal gland	Melatonin	Modified amino acid	Regulates seasonal reproductive cycles and sleep-wake cycles; may regulate onset of puberty
Kidney	Renin	Peptide	Acts on blood proteins to produce hormone (angiotensin) that regulates blood pressure
	Erythropoietin	Peptide	Stimulates red blood cell synthesis in bone marrow
Heart	Atrial natriuretic peptide (ANP)	Peptide	Increases salt and water excretion by kidneys; lowers blood pressure

[a]See Chapter 27.
[b]See Chapter 23.

retrieving pups that have strayed. Oxytocin may also have a role in male reproductive behavior. In several animals, oxytocin stimulates the contraction of muscles surrounding the tubes that conduct sperm from the testes to the penis, causing ejaculation.

The Anterior Pituitary Produces and Releases a Variety of Hormones

The anterior pituitary (see Fig. 25-3) produces six peptide hormones, four of which help regulate hormone production in other endocrine glands. Two of these, **follicle-stimulating hormone (FSH)** and **luteinizing hormone (LH)**, stimulate production of sperm and testosterone in males and eggs, estrogen, and progesterone in females. We will discuss the roles of FSH and LH in more detail in Chapter 27. **Thyroid-stimulating hormone (TSH)** stimulates the thyroid gland to release its hormones, and **ACTH**, or **adrenocorticotropic hormone** ("hormone that stimulates the adrenal cortex"), causes the release of hormones from

the adrenal cortex. We discuss the effects of thyroid and adrenal cortical hormones later in this chapter.

The remaining two hormones of the anterior pituitary, prolactin and growth hormone, do not act on other endocrine glands. **Prolactin**, in conjunction with other hormones, stimulates the development of the mammary glands (which are exocrine glands) during pregnancy. Suckling by the newborn infant then stimulates further release of prolactin, which in turn stimulates milk secretion. When the infant no longer suckles, prolactin secretion is turned off, and the ability to produce milk is lost within a few days.

Growth hormone regulates the growth of the body. Growth hormone acts on all the body's cells, increasing protein synthesis, fat utilization, and the storage of carbohydrates. During maturation, it has a stimulatory effect on bone growth, which influences the ultimate size of the adult organism. Much of the normal variation in human height is due to differences in secretion of growth hormone from the anterior pituitary. Too little growth hormone causes some cases of

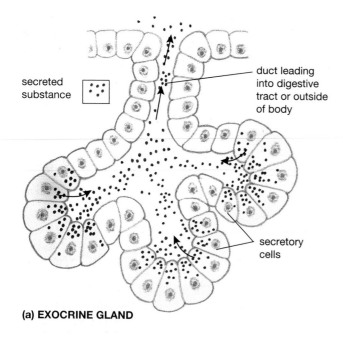

secreted
substance

duct leading
into digestive
tract or outside
of body

secretory
cells

(a) EXOCRINE GLAND

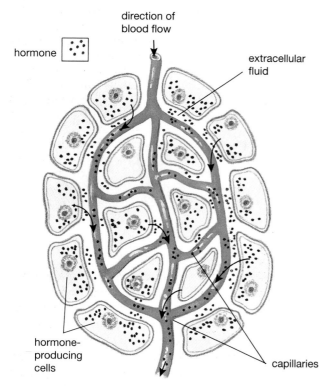

direction of
blood flow

hormone

extracellular
fluid

hormone-
producing
cells

capillaries

(b) ENDOCRINE GLAND

Figure 25-2   **The structure of exocrine and endocrine glands**

(a) The secretory cells of exocrine glands secrete substances into ducts that usually open outside the body (sweat glands, mammary glands) or into the digestive tract (pancreas, salivary glands). (b) Endocrine glands consist of hormone-producing cells embedded with a network of capillaries. The cells secrete hormones into the extracellular fluid from which they diffuse into the capillaries.

dwarfism, whereas too much can cause gigantism (Fig. 25-5). Although in adulthood many bones lose their ability to grow, growth hormone continues to be secreted throughout life, helping to regulate protein, fat, and sugar metabolism.

A major advance in the treatment of dwarfism occurred when molecular biologists successfully inserted the gene for human growth hormone into bacteria, which churn out large quantities of the substance (see Chapter 14). Previously, tiny amounts were extracted from human cadavers at great cost. Now children with underactive pituitary glands who would previously have been dwarfs can achieve normal height.

Hypothalamic Hormones Exert Control over the Anterior Pituitary

Neurosecretory cells of the hypothalamus produce at least nine peptide hormones that regulate the release of hormones from the anterior pituitary. These peptides are called **releasing hormones** or **inhibiting hormones** depending on whether they stimulate or prevent the release of pituitary hormone (Fig. 25-6). Releasing and inhibiting hormones are synthesized in nerve cells in the hypothalamus, secreted into a capillary bed in the lower portion of the hypothalamus, and travel a short distance through blood vessels down the pituitary stalk to a second capillary bed surrounding the endocrine cells of the anterior pituitary (see Fig. 25-3). There, the releasers and inhibitors diffuse out of the capillaries and influence pituitary hormone secretion.

Because the releasing and inhibiting hormones are secreted very close to the anterior pituitary, they are produced only in minute amounts. Not surprisingly, they were extremely difficult to isolate and study. Andrew Schally and Roger Guillemin, American endocrinologists who shared the Nobel Prize in 1977 for characterizing several of these hormones, used the brains of millions of sheep and pigs (obtained from slaughterhouses) to extract enough releasing hormone to analyze.

The Thyroid and Parathyroid Glands Are Located in the Neck

In the front of the neck, nestled around the larynx, lies the **thyroid gland** (Fig. 25-7a, p. 482). The four small discs of the **parathyroid glands** are embedded in the back of the thyroid.

The thyroid produces two major hormones, **thyroxine** and **calcitonin**. Thyroxine is an iodine-containing modified amino acid that raises the metabolic rate of most body cells. Calcitonin is a peptide important in calcium metabolism.

In juvenile animals, including humans, thyroxine helps regulate growth, stimulating both metabolic rate and the development of the nervous system. Cretinism, caused by undersecretion of thyroid hormone from birth, results in mentally retarded dwarfs. Fortunately, early diagnosis and thyroxine supplementation can prevent this tragedy. Precocious development in vertebrate animals is triggered by oversecretion of thyroxine. In 1912, in one of the first demons-

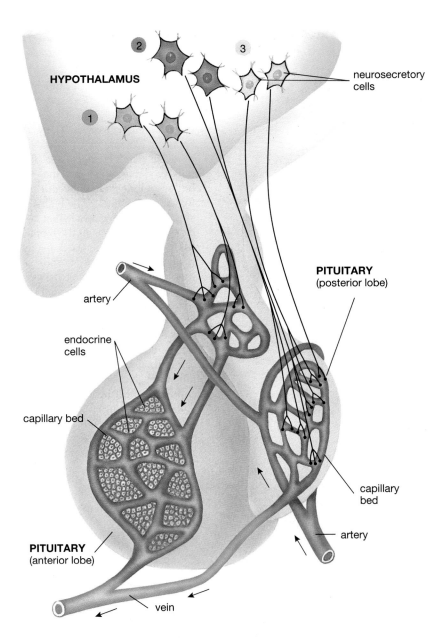

ANTERIOR LOBE	
Hormone	Target Organs
Follicle-stimulating hormone (FSH)	Gonads
Luteinizing hormone (LH)	Gonads
Thyroid-stimulating hormone (TSH)	Thyroid
Adrenocorticotropic hormone (ACTH)	Adrenal cortex
Prolactin	Mammary glands
Growth hormone	Most cells
POSTERIOR LOBE	
Hormone	Target Organs
Oxytocin	Uterus, mammary glands
Antidiuretic hormone (ADH)	Kidneys

Figure 25-3 **Anatomical relationships between the hypothalamus and pituitary**

The anterior lobe of the pituitary (left) consists of secretory cells enmeshed in a capillary bed. Release of hormones from these cells is controlled by releasing and inhibiting hormones produced by neurosecretory cells ① of the hypothalamus. These neurosecretory cells release their hormones into a capillary network directly "upstream" from the anterior pituitary (see Fig. 25-6). The posterior lobe of the pituitary (right) is an extension of the hypothalamus. Two types of neurosecretory cells, ② and ③, whose cell bodies are found in the hypothalamus, have cell endings on a capillary bed in the posterior lobe, where they release their hormones. The table shows the hormones of the anterior and posterior pituitary and their target organs.

trations of hormone action, a physiologist discovered that thyroxine can induce early metamorphosis in tadpoles (see "Evolutionary Connections: The Evolution of Hormones").

Thyroxine influences most of the cells in the body, elevating their metabolic rate. Its effects include increasing oxygen consumption and heart rate and stimulating the synthesis of enzymes that break down glucose and provide energy. In adults, an elevated metabolic rate seems to be involved in regulating body temperature and stress reactions. Exposure to cold, for example, greatly increases thyroid hormone production.

Levels of thyroxine in the bloodstream are finely tuned by negative feedback loops. Thyroxine release is stimulated by thyroid-stimulating hormone (TSH) from the anterior pituitary, which in turn is stimulated by a releasing hormone from the hypothalamus. The amount of TSH released from the pituitary is regulated by thyroxine levels in the blood (Fig. 25-8, p. 482): High concentrations of thyroxine inhibit the secretion of both the releasing hormone and TSH, thus inhibiting further release of thyroxine from the thyroid.

Iodine deficiency causes a reduction in thyroxine that stimulates dramatic growth of the thyroid (increasing the number of thyroxine-producing cells), an example of a feedback mechanism acting to restore normal hormone levels. The enlarged gland bulges from the neck, producing a condition called **goiter** (Fig. 25-7b). Goiter was once common in some regions of the United States, but widespread use of iodized salt has now all but eliminated this condition in developed countries.

Calcitonin, along with **parathormone**, the hormone secreted by the parathyroids, controls the concentration of calcium in the blood and other body fluids. Calcium is essential for many processes, including nerve and muscle function, so the calcium concentration in body fluids must be kept within narrow limits. Calcitonin and parathormone regulate calcium absorption and release by the

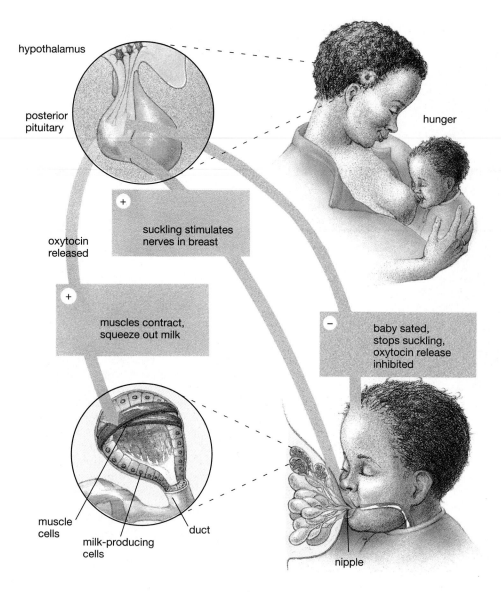

hypothalamus

posterior
pituitary

+

oxytocin
released

suckling stimulates
nerves in breast

+

muscles contract,
squeeze out milk

muscle
cells

milk-producing
cells

duct

hunger

−

baby sated,
stops suckling,
oxytocin release
inhibited

nipple

Figure 25-4   **Hormones
and breastfeeding**

The control of milk letdown by
oxytocin during breastfeeding is
regulated by positive (+) and
negative (−) feedback between
the baby and its mother. The
breast, or mammary gland, is an
exocrine gland. Here, clusters of
milk-producing cells surround
hollow bulbs, where milk collects
in lactating women. The bulbs
are surrounded by muscle that can
expel the milk through the nipple.
Milk is expelled when suckling by
the baby stimulates nerve endings
that send a signal to the mother's
hypothalamus, causing secretion
of oxytocin into the bloodstream
by the posterior pituitary. When
oxytocin reaches the muscles
surrounding the milk ducts, it
causes them to contract and expel
milk through the nipple. This cycle
continues until the infant is full
and stops suckling. With the
nipple no longer being stimulated,
oxytocin release stops, the muscles
relax, and milk flow ceases.

bones, which serve both as a skeleton and as a bank into
which calcium can be deposited or withdrawn as necessary.
In response to low blood calcium, the parathyroids release
parathormone, which causes release of calcium from bones.
The parathyroids increase in size in pregnant and lactating women, thereby enhancing parathormone output and
allowing the mother's body to meet the extra demands for
calcium imposed by the developing fetus and, later, milk
production. If blood calcium levels become too high, the
thyroid releases calcitonin, which inhibits the release of
calcium from bones.

The Pancreas Is Both an Exocrine
and an Endocrine Gland

The **pancreas**, which nestles behind the stomach, is a double gland producing both exocrine and endocrine secretions (Fig. 25-9). The exocrine portion synthesizes digestive secretions that are released into the pancreatic duct
and flow into the small intestine (see Chapter 23). The endocrine portion consists of clusters of cells, called **islet cells**,

that produce peptide hormones. One type of islet cell produces the hormone **insulin**; another type produces **glucagon**.

Insulin and glucagon work in opposite ways to regulate
carbohydrate and fat metabolism. When blood glucose
rises (for example, after a meal), insulin is released. Insulin causes most of the cells of the body to take up glucose and either metabolize it for energy or convert it to
fat or glycogen (a starchlike molecule) for storage. By far
the most important storage organ for glycogen is the liver.
When blood glucose levels drop (for example, after a person skips breakfast or runs a 10-kilometer race), glucagon
is released. Glucagon activates a liver enzyme that breaks
down glycogen, releasing glucose into the blood. It also
promotes lipid breakdown, releasing fatty acids that are
metabolized for energy. Insulin, then, reduces blood glucose levels, whereas glucagon increases them; together
they help keep blood glucose levels nearly constant.

Defects in insulin production, release, or reception by target cells result in **diabetes mellitus**, a condition in which
blood glucose levels are high and fluctuate wildly with sugar
intake. The lack of functional insulin in diabetics causes the

Figure 25-5 **When the anterior pituitary malfunctions**

An improperly functioning anterior pituitary can produce either too much or too little growth hormone. Too much can result in gigantism; too little causes dwarfism.

body to rely much more heavily on fats as an energy source, leading to high circulating levels of lipids, including cholesterol. Severe diabetes causes fat deposits in the blood vessels, resulting in high blood pressure and heart disease; diabetes is an important cause of heart attacks in the United States. The fatty deposits in small vessels can also damage the retina of the eye, leading to blindness, and the kidneys, leading to kidney failure. Insulin supplements traditionally contained insulin extracted from the pancreases of cows and pigs, obtained from slaughterhouses. Recently, however, the gene for human insulin has been inserted into bacteria, allowing the production of large quantities of human insulin, which is now commercially available.

The Sex Organs Secrete Steroid Hormones

The male testes and female ovaries are important endocrine organs. The testes secrete several steroid hormones, collectively called **androgens**. The most important of these is **testosterone**. The ovary secretes two types of steroid hormones, **estrogen** and **progesterone**. The role of the sex hormones in development, the menstrual cycle, and pregnancy is discussed in Chapters 27.

The Adrenal Glands Have Two Parts That Secrete Different Hormones

Like the pituitary and pancreas, the adrenals (Latin for "on the kidney") are two glands in one: the adrenal medulla and the adrenal cortex (Fig. 25-10, p. 484). The **adrenal**

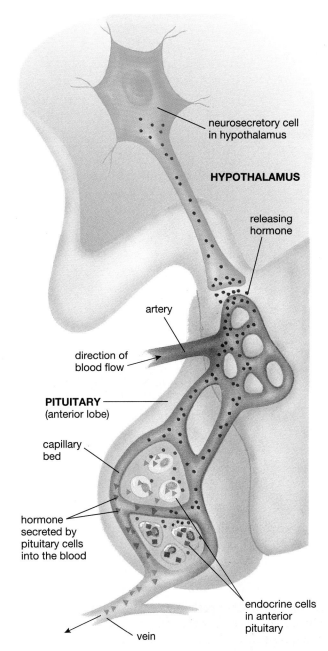

Figure 25-6 **The hypothalamus controls the anterior pituitary**

Hormone release from the anterior pituitary is under the control of releasing and inhibiting hormones from neurosecretory cells of the hypothalamus. Releasing hormones enter a capillary bed in the hypothalamus and travel downstream to capillaries in the anterior pituitary. There the releasing hormones contact the various endocrine cells of the pituitary. Only endocrine cells with matching cell membrane receptors respond to a given releasing hormone, so each releasing hormone stimulates a particular type of endocrine cell to release its hormone while leaving other types unaffected.

medulla is located in the center of each gland (*medulla* means "marrow" in Latin). It consists of secretory cells derived during development from nervous tissue, and its hormone secretion is controlled directly by the nervous system. The adrenal medulla produces two hormones, **epi-**

(a)

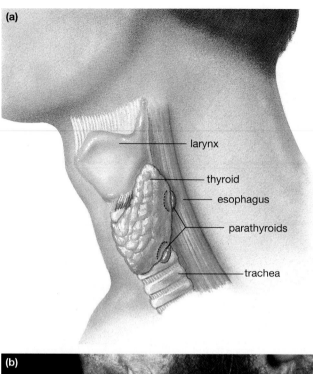

larynx

thyroid

esophagus

parathyroids

trachea

(b)

Figure 25-7   **The thyroid and parathyroid glands**

(a) The thyroid and parathyroid glands are located around the front of the larynx in the neck. **(b)** Individuals with iodine-deficient diets may suffer from goiter, a condition in which the thyroid becomes greatly enlarged.

nephrine and norepinephrine (also called **adrenaline** and **noradrenaline**), in response to stress. These hormones, which are modified amino acids, prepare the body for emergency action. They increase the heart and respiratory rates, cause blood glucose levels to rise, and direct blood flow away from the digestive tract and toward the brain and muscles. The adrenal medulla is activated by the sympathetic nervous system, which prepares the body to respond to emergencies, as described in Chapter 26.

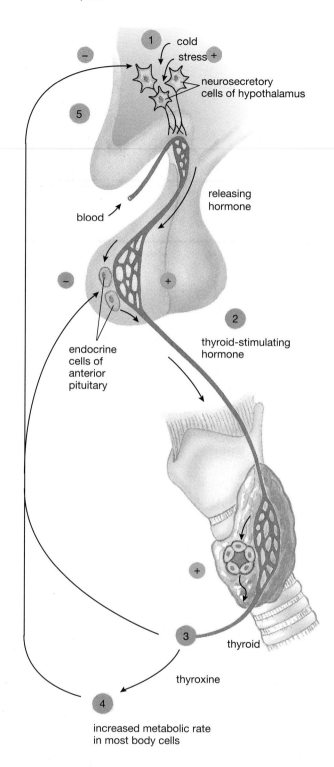

Figure 25-8   **Negative feedback in thyroid function**

Low body temperature or stress stimulates neurosecretory cells of the hypothalamus ①, whose releasing hormones trigger thyroid-stimulating hormone (TSH) release in the anterior pituitary ②. TSH then stimulates the thyroid to release thyroxine ③. Thyroxine causes increased metabolic activity in most cells of the body, generating ATP energy and heat ④. Both the raised body temperature and high thyroxine levels in the blood inhibit the releasing-hormone cells and the TSH-producing cells ⑤.

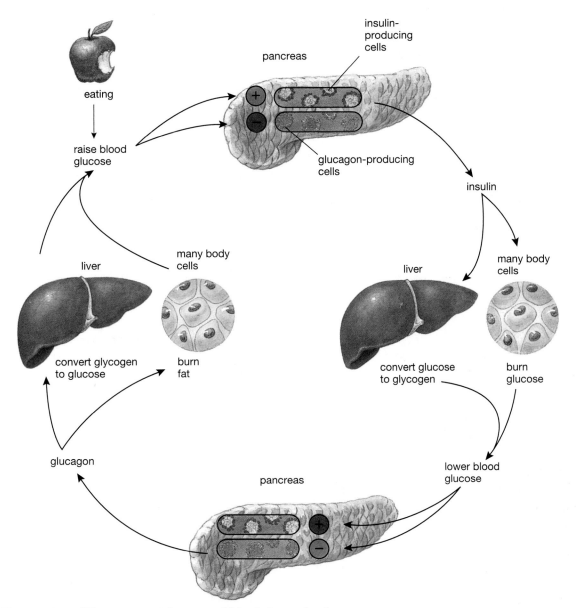

Figure 25-9   **The pancreas islets control blood glucose levels**

The pancreatic islets contain two populations of hormone-producing cells, one producing insulin, the other glucagon. These two hormones cooperate in a two-part negative feedback loop to control blood glucose concentrations. High blood glucose stimulates the insulin cells and inhibits the glucagon cells; low blood glucose stimulates the glucagon cells and inhibits the insulin cells. This dual control quickly corrects either high or low blood glucose levels.

The outer layer of the adrenal gland forms the **adrenal cortex** (*cortex* is Latin for "bark"). The cortex secretes three types of steroid hormones synthesized from cholesterol, called **glucocorticoids**. These hormones help control glucose metabolism. Glucocorticoid release is stimulated by ACTH from the anterior pituitary; ACTH release in turn is stimulated by releasing hormones produced by the hypothalamus in response to stressful stimuli, including trauma, infection, or exposure to extremes of temperature. In some respects, the glucocorticoids act similarly to glucagon, raising blood glucose levels by stimulating glucose production and promoting the use of fats instead of glucose for energy production.

You may have noticed that many different hormones are involved in glucose metabolism: thyroxine, insulin, glucagon, epinephrine, and the glucocorticoids. Why? The reason can probably be traced to a metabolic requirement of the brain. Although most body cells can produce energy from fats and proteins as well as carbohydrates, brain cells can burn only glucose. Thus, blood glucose levels cannot be allowed to fall too low, or brain cells rapidly starve, leading to unconsciousness and death.

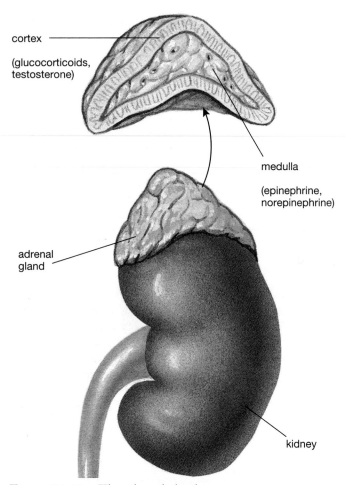

cortex

(glucocorticoids, testosterone)

medulla

(epinephrine, norepinephrine)

adrenal gland

kidney

Figure 25-10   The adrenal glands

Atop each kidney sits an adrenal gland, which is a two-part gland composed of very dissimilar cells. The outer cortex consists of ordinary endocrine cells that secrete steroid hormones. The inner medulla is derived from nervous tissue during development and secretes epinephrine and norepinephrine.

The adrenal cortex also secretes **aldosterone**, which regulates the sodium content of the blood. Sodium ions, derived from salt in the diet, are the major positive ions in blood and fluid surrounding cells. The sodium ion gradient across cell membranes (high outside, low inside) is used in many cellular events, including the production of electrical signals by nerve cells. If blood sodium falls, the adrenal cortex releases aldosterone, which causes the kidneys and sweat glands to retain sodium. Then salt and other sources of dietary sodium, combined with aldosterone-induced sodium conservation, raise blood sodium levels again, shutting off further aldosterone secretion. Too much salt increases the osmolarity of blood and extracellular fluid, drawing water out of cells by osmosis and increasing extracellular fluid and blood volume. The result is a decrease in aldosterone release that allows the kidneys to excrete more sodium.

Finally, the adrenal cortex produces the male sex hormone testosterone, although normally in much smaller amounts than the testes produce. Tumors of the adrenal medulla sometimes lead to excessive testosterone release,

causing masculinization of women. Many of the "bearded ladies" who once appeared in circus sideshows probably suffered from this condition.

Many Types of Cells Produce Prostaglandins

Unlike most other hormones, which are synthesized by a limited number of cells, prostaglandins are produced by many, perhaps all, cells of the body. They are modified fatty acids, synthesized by the cell from membrane phospholipids. Several prostaglandins are known, and probably a great many more await discovery. One prostaglandin causes arteries to constrict and stops bleeding from the umbilical cords of newborn infants. Another prostaglandin works in conjunction with oxytocin during labor, stimulating uterine contractions. Prostaglandin-soaked vaginal suppositories are currently used to induce labor. Menstrual cramps are caused by the overproduction of uterine prostaglandins, stimulating uterine contractions.

Some prostaglandins cause inflammation (such as occurs in arthritic joints) and stimulate pain receptors. Drugs such as aspirin and ibuprofen, which inhibit prostaglandin synthesis, can provide relief. Some prostaglandins expand the air passages of the lungs, and one day asthma sufferers may benefit from research into this effect. Others stimulate production of the protective mucus that lines the stomach, a potential boon for ulcer patients. Research on this diverse and potent family of compounds is still in its infancy, but it promises many health benefits in the future.

Other Endocrine Organs Include the Pineal Gland, Thymus, Kidneys, Heart, and Digestive Tract

The **pineal gland** is located between the two hemispheres of the brain, just above and behind the brainstem (see Fig. 25-1). Named for its resemblance to a pine cone, the pineal is smaller than a pea. In 1646, the philosopher René Descartes described it as "the seat of the rational soul." Since then, scientists have learned more about this organ, although many of its functions are still poorly understood. The pineal produces the hormone **melatonin**, a modified amino acid. Melatonin is secreted in a daily rhythm, which in mammals is regulated by the eyes. In some vertebrates, such as the frog, the pineal itself contains photoreceptive cells, and the skull above it is thin, so the pineal can detect sunlight and thus daylength. By responding to daylengths characteristic of different seasons, the pineal appears to regulate the seasonal reproductive cycles of many mammals. Despite years of research, the function of the human pineal and of melatonin is still unclear. One hypothesis is that the pineal and melatonin secretion influence sleep-wake cycles. Improper pineal function may contribute to the depression that some people experience during the short days of winter.

The **thymus** is located in the chest cavity behind the breastbone, or sternum (see Fig. 25-1). In addition to producing white blood cells, the thymus produces the hormone **thymosin**, which stimulates the development of specialized white blood cells (T cells) that play an important role in the immune system (see Chapter 24). The thymus is extremely large in the infant, but begins decreasing in size after puberty.

The kidney, which plays a central role in maintaining body fluid homeostasis, has recently been recognized as an important endocrine organ as well. When the oxygen content of the blood drops, the kidney produces the hormone **erythropoietin**, which increases red blood cell production, as described in Chapter 22. The kidney also produces a second hormone, **renin**, in response to low blood pressure, such as may be caused by bleeding. Renin is an enzyme that catalyzes the production of another hormone, called **angiotensin**, from proteins in the blood. Angiotensin raises blood pressure by constricting arterioles. It also stimulates aldosterone release by the adrenal cortex, causing the kidneys to retain sodium, which in turn increases blood volume.

The heart seems an unlikely endocrine organ, but in 1981, a substance extracted from heart atrial tissue was found to cause an increase in the output of salt and water by the kidneys when injected into rats. Two years later, the active substance, **atrial natriuretic peptide (ANP)**, was described and its amino acid sequence determined. This peptide is released by cells in the atria of the heart when blood volume increases. Increased blood volume causes extra distension of the heart. Atrial natriuretic peptide causes reduction of blood volume by decreasing the release of both ADH and aldosterone. The pace of modern biomedical research is astonishing: Only 5 years after the factor was first detected in heart extract, clinical trials of synthetic ANP were started to test its value in treating high blood pressure and related problems.

The stomach and small intestine produce a variety of peptide hormones that help regulate digestion. These hormones include **gastrin**, **secretin**, and **cholecystokinin**, discussed in Chapter 23.

EVOLUTIONARY CONNECTIONS

The Evolution of Hormones

Not long ago, vertebrate endocrine systems were considered unique to our phylum, and the endocrine chemicals were thought to have evolved expressly for their role in vertebrate physiology. In recent years, however, physiologists have discovered that hormones are evolutionarily ancient. Insulin, for example, is found not only in vertebrates but also in protists, fungi, and bacteria, although research has not yet determined the function of insulin in most of those organisms. Protists also manufacture ACTH, although of course they have no adrenal glands to stimulate. Yeasts have receptors for estrogen but, of course, no ovaries.

Thyroid hormones have been found in certain invertebrates, such as worms, insects, and mollusks, as well as in vertebrates. Even among vertebrates, the effects of chemically identical hormones, secreted by the same glands, may vary dramatically from organism to organism. Let's look briefly at the diverse effects that the thyroid hormone thyroxine has on several different organisms.

Some fish undergo radical physiological changes during their lifetimes. A salmon, for example, begins life in fresh water, migrates to the ocean, and finally returns to fresh water to spawn. In the stream where it hatched, fresh water tends to enter the fish's tissues by osmosis; in salt water, the fish tends to lose water, becoming dehydrated. The fish's migrations, therefore, require complete revamping of salt and water control. In salmon, one of the functions of thyroxine is to produce the metabolic changes necessary to go from life in streams to life in the ocean.

In amphibians, thyroxine has the dramatic effect of triggering metamorphosis. In 1912, in one of the first demonstrations of the action of any hormone, tadpoles were fed minced horse thyroid. As a result, they metamorphosed prematurely into miniature adult frogs (Fig. 25-11). In high mountain lakes in Mexico, where the water is deficient in the iodine needed to synthesize thyroxine, natural selection has produced one species of sala-

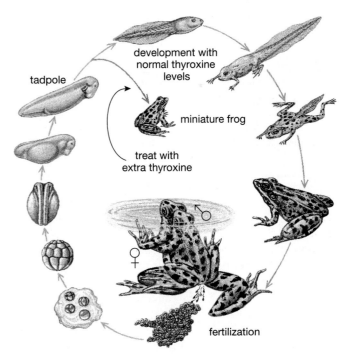

Figure 25-11   **Thyroxine controls metamorphosis in amphibians**

The life cycle of the frog begins with fertilization of the eggs (bottom); then development into an aquatic, fishlike tadpole; growth of the tadpole; and ultimately metamorphosis into an adult frog. Metamorphosis is triggered by a surge of thyroxine from the tadpole's thyroid gland. If a young tadpole is injected with extra thyroxine, it will metamorphose ahead of schedule into a miniature adult frog.

mander that has the ability to reproduce while still in its juvenile form.

Thyroxine regulates the seasonal molting of most vertebrates. From snakes to birds to the family dog, surges of thyroxine stimulate the shedding of skin, feathers, or hair. In people (who neither migrate, metamorphose, nor molt), thyroxine regulates growth and metabolism.

The use of chemicals to regulate cellular activity is extremely ancient. The diversity of life on Earth rests upon a conservative foundation: A relative handful of chemicals coordinate activities within single cells and among groups of cells. Life's diversity originated in part by changing the systems used to deliver the chemicals and by evolving new types of responses. Early in their evolution, animals developed a complement to hormonal communication that provides faster, more precise delivery of chemical messages: the nervous system. As we explain in the next chapter, the nervous system permits rapid responses to environmental stimuli, flexibility in response options, and ultimately consciousness itself.

✖ SUMMARY OF KEY CONCEPTS

Hormone Structure and Function
A hormone is a chemical secreted by cells in one part of the body that is transported in the bloodstream to other parts of the body, where it affects the activity of specific target cells.

Four types of molecules are known to act as hormones: peptides, modified amino acids, steroids, and prostaglandins.

Most hormones act on their target cells by binding to receptors. These receptors are found in two general locations on target cells: on the outer cell membrane and inside the cell, usually within the nucleus.

Hormone action is often regulated through negative feedback, a process in which the hormone causes changes that inhibit further secretion of that hormone.

The Human Endocrine System
Hormones are produced by endocrine glands, which are clusters of cells embedded within a network of capillaries. Hormones are secreted into the extracellular fluid and diffuse into the capillaries. The major endocrine glands of the human body are the hypothalamus-pituitary complex, the thyroid and parathyroid glands, the pancreas, the sex organs, and the adrenal glands. The hormones released by these glands and their actions are summarized in Table 25-2. Prostaglandins, unlike other hormones, are not secreted by discrete glands but are synthesized and released by many cells of the body. Other endocrine organs include the pineal gland, kidneys, heart, and the stomach and small intestine.

✖ KEY TERMS

adrenal cortex p. 483	goiter p. 479	pineal gland p. 484
adrenal medulla p. 481	hormone p. 473	pituitary p. 475
androgen p. 481	hypothalamus p. 475	posterior pituitary p. 475
anterior pituitary p. 475	inhibiting hormone p. 478	prostaglandin p. 474
diabetes mellitus p. 480	islet cell p. 480	releasing hormone p. 478
duct p. 475	neurosecretory cell p. 475	steroid p. 474
endocrine gland p. 475	pancreas p. 480	target cell p. 474
endocrine system p. 473	parathyroid gland p. 478	thymus p. 484
exocrine gland p. 475	peptide hormone p. 474	thyroid gland p. 478

✖ THINKING THROUGH THE CONCEPTS

Multiple Choice
1. Examples of posterior pituitary hormones are
   a. FSH and LH
   b. prolactin and parathormone
   c. secretin and cholecystokynin
   d. melatonin and prostaglandin
   e. ADH and oxytocin
2. Negative feedback to the hypothalamus controls the level of _____ in the blood.
   a. thyroxine   b. estrogen
   c. glucocorticoids   d. estrogen
   e. all of the above

3. The primary targets for FSH are cells in the
   a. hypothalamus   b. ovary
   c. thyroid   d. adrenal medulla
   e. pituitary
4. The kidney is a source of
   a. thyroxine and parathormone
   b. calcitonin and oxytocin
   c. renin and erythropoietin
   d. ANP and epinephrine
   e. glucagon and glucocorticoids

5. Hormones that are produced by many different body cells and cause a variety of localized effects are known as
   a. peptide hormones   b. parathormones
   c. releasing hormones   d. prostaglandins
   e. exocrine hormones

Review Questions

1. What are the four types of molecules used as hormones in vertebrates? Give an example of each.
2. What is the difference between an endocrine and an exocrine gland? Which type releases hormones?
3. Diagram the process of negative feedback, and give an example of negative feedback in the control of hormone action.
4. What are the major endocrine glands in the human body, and where are they located?

5. Describe the structure of the hypothalamus-pituitary complex. Which pituitary hormones are neurosecretory? What are their functions?
6. Describe how releasing hormones regulate the secretion of hormones by cells of the anterior pituitary. Name the hormones of the anterior pituitary and give one function of each.
7. Describe how the hormones of the pancreas act together to regulate the concentration of glucose in the blood.
8. Compare the adrenal cortex and medulla by answering the following questions. Where are they located within the adrenal gland? What are their embryological origins? What hormones do they produce? What organs do their hormones target? What homeostatic processes regulate blood levels of the respective hormones?

✖ APPLYING THE CONCEPTS

1. An enterprising student decides to do a science project on the effect of the thyroid gland on frog metamorphosis. She sets up three aquaria with tadpoles. She adds thyroxine to the water of one, thiouracil to a second, and nothing to the third. Thiouracil reacts with thyroxine inside tadpoles to produce an ineffective compound. Assuming the student uses appropriate physiological concentrations, predict what will happen.
2. Diabetes mellitus is common diabetes. Two other forms, adrenal and pituitary diabetes, are also characterized by high levels of blood glucose. What specific parts of regulatory cycles are disrupted in these latter forms to cause hyperglycemia?
3. Anabolic steroids, used by risk-taking athletes and body-builders, are chemically related to testosterone. They increase bone and muscle mass and do seem to improve athletic performance. But anabolic steroids can cause liver problems,

heart attacks, strokes, testicular atrophy, and personality changes in males. Females on anabolic steroids also have liver and circulatory problems. In addition, their voices deepen, their bodies develop more hair, and their menstrual cycles are disturbed. Explain how the same compound can produce testicular atrophy in males but virilism in females.
4. Some parents, interested in college sports scholarships for their children, are asking physicians to prescribe growth hormone treatments. Farmers also have an economic incentive to treat cows with growth hormone, which can now be produced in large quantities with genetic-engineering techniques. What biological and ethical problems do you foresee for parents, children, physicians, coaches, college scholarship boards, food consumers, farmers, the U.S. Food and Drug Administration, and bio-tech companies?

✖ FOR MORE INFORMATION

Atkinson, M., and MacLaren, N. "What Causes Diabetes?" *Scientific American*, July 1991.

Berridge, M. J. "The Molecular Basis of Communication within the Cell." *Scientific American*, October 1985. Both the "classical" cyclic AMP and more recently discovered second messengers convey information from cell surface receptors to DNA and cellular metabolism.

Guillemin, R., and Burgus, R. "The Hormones of the Hypothalamus." *Scientific American*, November 1972. The interaction between hypothalamus and pituitary is explored.

Sapolsky, R. "Stress in the Wild." *Scientific American*, January 1990. An interesting study of baboons showing the effects of hormones on stress responses.

Snyder, S. H. "The Molecular Basis of Communication between Cells." *Scientific American*, October 1985. Snyder describes the similarities and differences between neural and hormonal control systems in the body.

NET WATCH

On-line resources for this chapter are on the World Wide Web at:
http://www.prenhall.com/~audesirk (click on the <u>table of contents</u> link and then select Chapter 25).

26 The Nervous System and the Senses

Three neurons from the human cerebral cortex, the gray material that forms the outer layer of the brain, are seen in this scanning electron micrograph. The oval structures are cell bodies; the threadlike structures projecting out from them are dendrites and axons.

"Know thyself."

—*Inscribed above the entrance to the Temple of Apollo, home of the Oracle of Apollo at Delphi*

"Nothing is understood by the intellect which is not first perceived by the senses."

ARISTOTLE (384–322 B.C.)

As you read this page, light reflected from it bombards your eyes, and sound waves from the radio assault your ears. Your senses of smell and taste are simultaneously stimulated as you drink coffee. Touch-sensitive receptors all over your body are activated as your clothes, the chair, and the pages of this book all push at your skin. You breathe, swallow, and shift positions in your chair. Untaxed by these marvelous but ordinary activities, your mind calls upon its reading skill, directs dexterous manipulations of a pen as you take notes, stores memories of what you have read, and still has the capacity to appreciate the music from the radio.

All this activity is controlled by your brain, an organ weighing about 3 pounds (about 1.5 kilograms) encased within your skull. Composed of billions of nerve cells, the brain is connected to the rest of the body by dozens of slender cables, or nerves. In this chapter, we first compare types of communication controlled by the endocrine and nervous systems. We explore the interrelated structure and function of individual neurons and of the entire nervous system, leading up to the mystery of consciousness. We then introduce basic sensory mechanisms, followed by a more in-depth analysis of the senses of hearing, sight, taste, smell, and pain.

A Comparison of Nervous and Endocrine Communication

Although it is convenient to discuss hormonal control (Chapter 25) separately from nervous control, in some ways the two are quite similar. Both hormone-producing cells and nerve cells synthesize "messenger" chemicals that they release into extracellular spaces (Fig. 26-1). However, there are four main differences in how the nervous and endocrine systems use chemical messages. First, whereas hormone-producing and neurosecretory cells release hormones into the bloodstream, nerve cells usually release their chemical messengers (called neurotransmitters) very close to the cells they influence, often from less than a micrometer away. Second, whereas blood-borne hormones bathe millions of cells indiscriminately, a nerve cell releases its neurotransmitter onto one or a small group of specific target cells. Third, a nerve cell speeds informa-

(a) Hormone-producing cell

target cell

① ② ③

capillary   hormone

hormone travels
in bloodstream

(b) Neurosecretory cell

target cell

② ③

①

neurohormone

hormone travels
in bloodstream

(c) Nerve cell

target cells

neurotransmitter
released onto
target cells   ③

①

②

electrical signal
travels along nerve
cell process

Figure 26-1   **Three major types of "control cells"**
These all release chemicals that influence the activity of other cells of the body: **(a)** "classic" hormone-producing cells, **(b)** neurosecretory cells, and **(c)** ordinary nerve cells. Three common steps occur in each system: ① The chemical-releasing cell is stimulated. ② It sends a message to distant cells. ③ Selected target cells respond. Both classic hormone-producing cells and neurosecretory cells release their chemical messages into the bloodstream, which transports the chemicals to distant target cells. Regular nerve cells grow long cell extensions, called processes, to their target cells and release their chemical messages directly onto the target cell.

tion from one part of the body to another by means of electrical signals traveling within the cell itself, and it releases its neurotransmitter only when it reaches its target. Hormones move much more slowly and are released at varying distances from the target cell. Finally, the effects of messages sent by neurons tend to be of much shorter duration than the effects of hormones.

Yet even these differences blur as we learn more about the hormonal and nervous systems of animals. Some hormones are in fact produced and released into the bloodstream by nerve cells, such as the neurosecretory cells of the hypothalamus. Other chemicals that biologists formerly believed were strictly hormones, such as insulin,

have recently been discovered in the brain, where they are synthesized and released by neurons and act more like neurotransmitters. As you read about the nervous system in this chapter, remember that no system in your body works alone. The hormonal and nervous systems are closely coordinated in their control of bodily functions.

The Functions and Structure of Neurons

Our study of the nervous system begins with the individual nerve cell, or **neuron.** As the fundamental unit of the nervous system, each neuron must perform five functions:

1. Receive information from the internal or external environment or from other neurons
2. Integrate the information it receives and produce an appropriate output signal
3. Conduct the signal to its terminal ending, which may be some distance away
4. Transmit the signal to other nerve cells, glands, or muscles
5. Coordinate the metabolic activities that maintain the integrity of the cell

Although neurons vary enormously in structure, a "typical" vertebrate neuron has four distinct structural regions that carry out the functions above. These regions are the dendrites, the cell body, the axon, and the synaptic terminals (Fig. 26-2).

Dendrites Receive Signals from Other Neurons or the Environment

Dendrites, branched tendrils that extend outward from the cell body, are specialized to respond to signals from other neurons or from the environment. Their branched form provides a large surface area to receive these signals. Dendrites of sensory neurons have special membrane adaptations that allow them to respond to specific stimuli from the environment such as pressure, odorous molecules, light, or heat. In neurons of the brain and spinal cord, dendrites respond to the chemical neurotransmitters released by other neurons. These dendrites have protein receptors in their membranes that bind specific neurotransmitters and produce small, short-lived electrical signals as a result of this binding. These signals may be either excitatory or inhibitory, as described below.

The Cell Body Integrates Electrical Signals from the Dendrites and Maintains the Neuron

Electrical signals travel down the dendrites and converge upon the **cell body** of the neuron, which serves as an integration center. In its integrating role, the cell body adds up the various electrical signals from the dendrites and "decides" whether to produce an **action potential**. An action potential is a brief, large, positive electrical signal that the neuron will transmit to other neurons, muscles, or glands. If

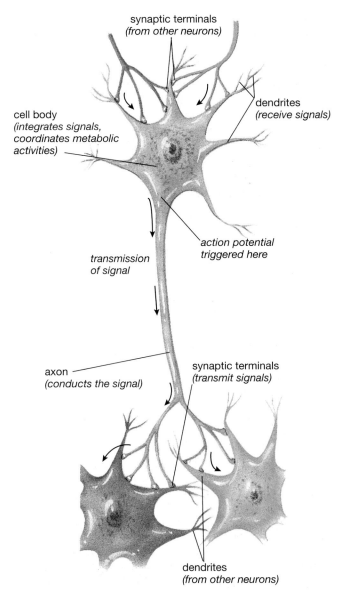

synaptic terminals
(from other neurons)

dendrites
(receive signals)

cell body
*(integrates signals,
coordinates metabolic
activities)*

*action potential
triggered here*

*transmission
of signal*

axon
(conducts the signal)

synaptic terminals
(transmit signals)

dendrites
(from other neurons)

Figure 26-2 **A nerve cell, showing its specialized parts and their functions**

spinal cord to your toes, a distance of about a meter (3 feet). Axons are distribution lines, carrying action potentials from the cell body to the synaptic terminals, located at the far end of each axon. Axons are usually bundled together into **nerves**, much like the wires in an electrical cable. However, unlike electrical power distribution cables (in which energy is lost along the way from power station to customer), the cell membranes of axons are specialized to conduct action potentials undiminished in size from the cell body to their synaptic terminals. Some axons have a fatty coating called **myelin** that increases the speed of transmission of action potentials. In vertebrates, nerves emerge from the brain and spinal cord and extend out to all regions of the body.

Synaptic Terminals Communicate with Other Neurons, Muscles, or Glands at Synapses

Signals are transmitted to other cells at **synaptic terminals**, which appear as swellings at the branched endings of axons (see Fig. 26-2). Most synaptic terminals contain a specific type of chemical, a **neurotransmitter**, that they release in response to an action potential traveling down the axon. The synaptic terminals of one neuron may communicate with a gland, a muscle, or the dendrites or cell body of a second neuron, so that the output of the first cell becomes the input to the second. In response to the signals communicated by the synaptic terminal, a gland may release hormones, a muscle may contract, or a second neuron may be stimulated.

The site at which a synaptic terminal communicates with another cell is called a **synapse**. A tiny gap, called a **synaptic cleft**, separates the synaptic terminal of the first neuron (the **presynaptic neuron**) from the second, or **postsynaptic**, **neuron** (we will use the dendrites of another neuron to illustrate the synapse; Fig. 26-3). When an action potential reaches a synaptic terminal, it triggers the synaptic terminal to release a chemical neurotransmitter into the synaptic cleft. The neurotransmitter molecules rapidly diffuse across the gap and bind to receptors in the membrane of the postsynaptic cell (see Fig. 26-3). The synapse includes the synaptic terminal of the presynaptic cell, the synaptic cleft, and the specialized membrane of the second neuron just across the synaptic cleft that contains receptors for neurotransmitters. When neurotransmitters bind these receptors, they cause small, short-lived electrical signals in the dendrite of the second neuron. As described earlier, these signals then travel to the cell body and may excite or inhibit an action potential in the second neuron.

the signals from the dendrites are sufficiently excitatory at any given moment, the neuron will generate an action potential. If enough inhibitory signals reach the cell body to cancel out the excitatory signals, an action potential is temporarily prevented.

The cell body of the neuron, containing the usual assortment of organelles, also carries on activities common to most other body cells. These include the synthesis of complex molecules such as proteins, lipids, and carbohydrates and the coordination of the metabolic activities of the cell.

The Axon Carries Electrical Signals from the Cell Body to Their Destination

In a typical neuron, a long, thin fiber, called an **axon**, extends outward from the cell body, making neurons the longest cells in the body. Single axons, for example, stretch from your

Building and Operating a Nervous System

The individual neuron uses a language of action potentials. Yet somehow this basic language allows even simple animals to perform a variety of complex behaviors. One key to the versatility of the nervous system is the presence of complex networks of neurons. These neural networks range from a few to billions of cells. As in computers, small, simple elements can perform amazing feats when connected properly.

Information Processing Requires Four Basic Operations

Before we delve into the basic anatomy of nervous systems, we should first examine the operating principles. At a minimum, a nervous system must be able to perform four operations:

1. Determine the type of stimulus
2. Signal the intensity of a stimulus
3. Integrate information from many sources
4. Initiate and direct the response

Let's examine each of these operations in more detail.

The Type of Stimulus Is Distinguished by Wiring Patterns in the Brain

The nervous system must be able to identify the type of stimulus (for example, light, touch, or sound). Action potentials are all pretty much alike; their properties tell us

nothing about the kind of stimulus that elicited them. Instead, the nervous system monitors *which* neurons are firing action potentials. Thus, your brain interprets action potentials occurring in the axons of your optic nerves (originating in the eye and traveling to a specific area of the brain) as the sensation of light, action potentials in olfactory nerves (originating in receptors in the nose and traveling to a different region of the brain) as odors, and so on. This genetic wiring may occasionally yield false information. Being poked in the eye may cause action potentials in the optic nerve, because of slight trauma. Even though the stimulus is mechanical, your brain nevertheless interprets all action potentials occurring in the optic nerve as light, and you "see stars." For the same reason, a blow to the head can make your ears "ring."

The Intensity of a Stimulus Is Coded by the Frequency of Action Potentials

Because all action potentials are of roughly the same magnitude and duration, no information about the strength, or **intensity,** of a stimulus (for example, the loudness of a

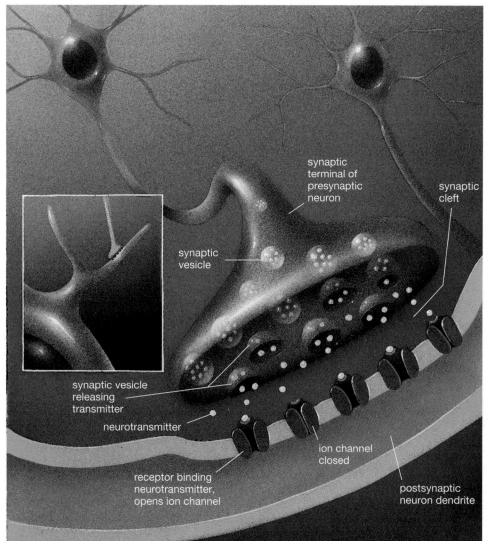

Figure 26-3   The structure and function of the synapse

The ending of the presynaptic cell contains numerous membrane-bound spheres (called *synaptic vesicles*) that contain neurotransmitter. The postsynaptic cell has membrane receptors for the transmitter. When an action potential enters the synaptic terminal of the presynaptic cell, the vesicles dump their neurotransmitter into the space, or synaptic cleft, between the neurons. The neurotransmitter diffuses rapidly across the space, binds to the postsynaptic receptors, and causes ion channels to open. Ions flow through these open channels, causing a postsynaptic potential in the postsynaptic cell.

synaptic terminal of presynaptic neuron

synaptic cleft

synaptic vesicle

synaptic vesicle releasing transmitter

neurotransmitter

ion channel closed

receptor binding neurotransmitter, opens ion channel

postsynaptic neuron dendrite

sound) can be encoded in a single action potential. Instead, intensity is coded in two other ways. First, intensity can be signaled by the frequency of action potentials in a single neuron. The more intense the stimulus, the faster the neuron fires (Fig. 26-4). Second, a nervous system usually has many neurons that can respond to the same input. Stronger stimuli tend to excite more of these neurons, whereas weaker stimuli excite fewer. Thus, intensity can also be signaled by the number of similar neurons firing at the same time (see Fig. 26-4).

The Nervous System Processes Information from Many Sources through Convergence

Your brain is continuously bombarded by sensory stimuli originating both inside and outside the body. The brain must filter all these inputs, determine which ones are important, and decide how to respond. Nervous systems integrate information much as do individual neurons, through **convergence.** In this process, many neurons funnel their signals to fewer neurons. For example, many sensory neurons may converge onto a smaller number of brain cells (Fig. 26-5). The brain cells add up the postsynaptic potentials resulting from the synaptic activity of these sensory neurons, and, depending on their relative strengths (and other internal factors such as hormones or metabolic activity), they produce appropriate outputs.

Divergence of Signals Allows Complex Responses

The output of the integrating cells is responsible for initiating activity. The actions directed by the brain may involve many parts of the body and require **divergence,** the flow of electrical signals from a relatively small number of decision-making cells onto many different neurons controlling muscle or glandular activity (see Fig. 26-5).

Neural Pathways Direct Behavior

Most behaviors are controlled by neuron-to-muscle pathways composed of four elements:

1. **Sensory neurons** that respond to a stimulus, either internal or external to the body
2. **Association neurons** that "decide" what to do, on the basis of input from many sensory neurons, stored memories, hormonal states, and other factors
3. **Motor neurons** that receive instructions from the association neurons and activate the muscles or glands
4. **Effectors**, usually muscles or glands, that perform the behavior or other response directed by the nervous system

The Simplest Behavior Is the Reflex

The simplest type of behavior is the **reflex,** a relatively involuntary movement of a body part in response to a stimulus. Examples of human reflexes include the familiar knee jerk and pain-withdrawal reflexes. The pain-withdrawal reflex (which moves a body part away from a painful stimulus) uses one neuron of each type (Fig. 26-6).

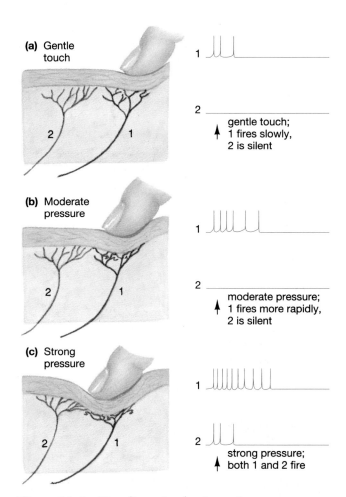

Figure 26-4   Signaling stimulus intensity

The intensity of a stimulus is signaled by the rate at which individual neurons produce action potentials and by the number of neurons firing. For example, two touch receptors may have endings in adjacent patches of skin. **(a)** A gentle touch elicits only a few action potentials and from only one of the sensory neurons. **(b)** Moderate pressure still stimulates only one receptor, but this receptor now fires faster, informing the brain that the touch is more intense than before. **(c)** Strong pressure activates both receptors, firing one very fast, and the other more slowly, thus signaling to the brain that the pressure is very intense.

Reflexes of this sort do not require the brain, although, as we know, other pathways inform the brain of pricked fingers and may in fact trigger other more complex behaviors (cursing, for example!).

Nearly all animals are capable of much more subtle and varied behavior than can be accounted for by simple reflexes. In principle, these more complex behaviors can be organized by *interconnected neural pathways* in which several types of sensory input (along with memories, hormones, and other factors) converge on a set of association neurons (see Fig. 26-5). By integrating the postsynaptic potentials from several sources, the association neurons can "decide" what to do and stimulate the motor neurons to direct appropriate activity in muscles and glands.

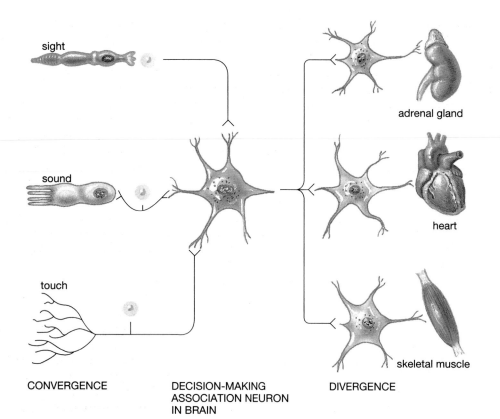

sight

sound

touch

CONVERGENCE

DECISION-MAKING
ASSOCIATION NEURON
IN BRAIN

adrenal gland

heart

skeletal muscle

DIVERGENCE

Figure 26-5   **Convergence and divergence in the nervous system**

Integration of information and initiation of coordinated action involves convergence of inputs to and divergence of outputs from neurons. In this simplified example, inputs of sight, sound, and touch (say, being attacked by a swarm of bees) all converge on a "decision-making" neuron, which is strongly stimulated as a result. Its outputs then diverge to the adrenal glands ("pump out adrenaline"), the heart ("beat faster and stronger"), and skeletal muscles ("move, legs!").

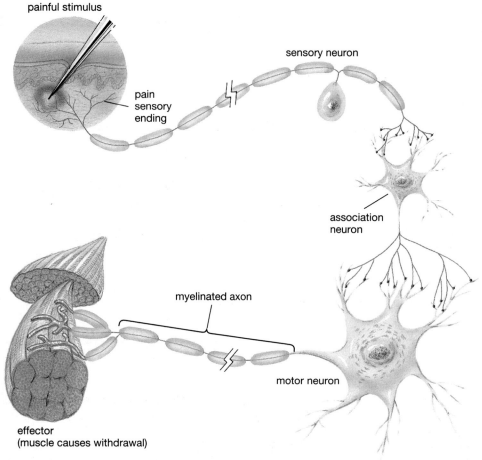

painful stimulus

pain
sensory
ending

sensory neuron

association
neuron

myelinated axon

motor neuron

effector
(muscle causes withdrawal)

Figure 26-6   **The pain-withdrawal reflex**

This simple reflex illustrates the three types of neurons and an effector (muscle). Note that the neurons vary tremendously in size and shape, a reflection of their varied functions.

The Human Nervous System

The human nervous system may be divided into two parts: central and peripheral. The **central nervous system** consists of a **brain** and a **spinal cord** that extends down the dorsal part of the torso. The **peripheral nervous system** consists of nerves connecting the central nervous system to the rest of the body (Fig. 26-7).

The Peripheral Nervous System Links the Central Nervous System to the Body

The peripheral nervous system consists of **peripheral nerves** that link the brain and spinal cord to the rest of the body, including the muscles, the sensory organs, and the organs of the digestive, respiratory, excretory, and circulatory systems. Within the peripheral nerves are axons of sensory neurons that bring sensory information *to* the central nervous system from all parts of the body. Peripheral nerves also contain the axons of motor neurons that carry signals *from* the central nervous system to the organs and muscles.

The motor portion of the peripheral nervous system can be subdivided into two parts: the **somatic nervous system** and the **autonomic nervous system.** Motor neurons of the somatic nervous system synapse on skeletal muscles and control voluntary movement. Their cell bodies are locat-

ed in the gray matter of the spinal cord (see Fig. 26-9), and their axons go directly to the muscles they control.

Motor neurons of the autonomic nervous system control involuntary responses. They synapse on the heart, smooth muscle, and glands. The autonomic nervous system is controlled both by the medulla and the hypothalamus of the brain, described later in the chapter. It consists of two divisions, the **sympathetic nervous system** and the **parasympathetic nervous system** (Fig. 26-8). The two divisions of the autonomic nervous system generally make synaptic contacts with the same organs but usually produce opposite effects.

The sympathetic nervous system acts on organs in ways that prepare the body for stressful or highly energetic activity, such as fighting, escaping, or giving a speech. During such "fight-or-flight" activities, the sympathetic nervous system curtails activity of the digestive tract, redirecting some of its blood supply to be used by the muscles of arms and legs. Heart rate accelerates. The pupils of the eyes open wider, admitting more light, and the air passages in the lungs expand, accommodating more air. The parasympathetic nervous system, in contrast, dominates during maintenance activities that can be carried on at leisure, often called "rest and rumination." Under its control, the digestive tract becomes active, heart rate slows, and air passages in the lungs constrict.

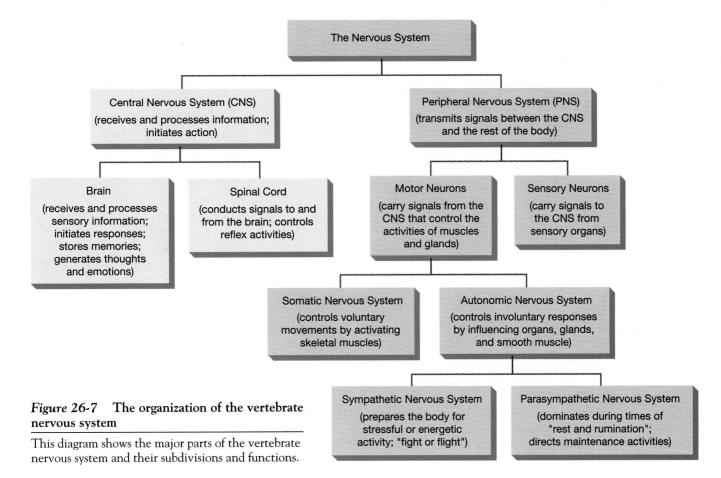

Figure 26-7 **The organization of the vertebrate nervous system**

This diagram shows the major parts of the vertebrate nervous system and their subdivisions and functions.

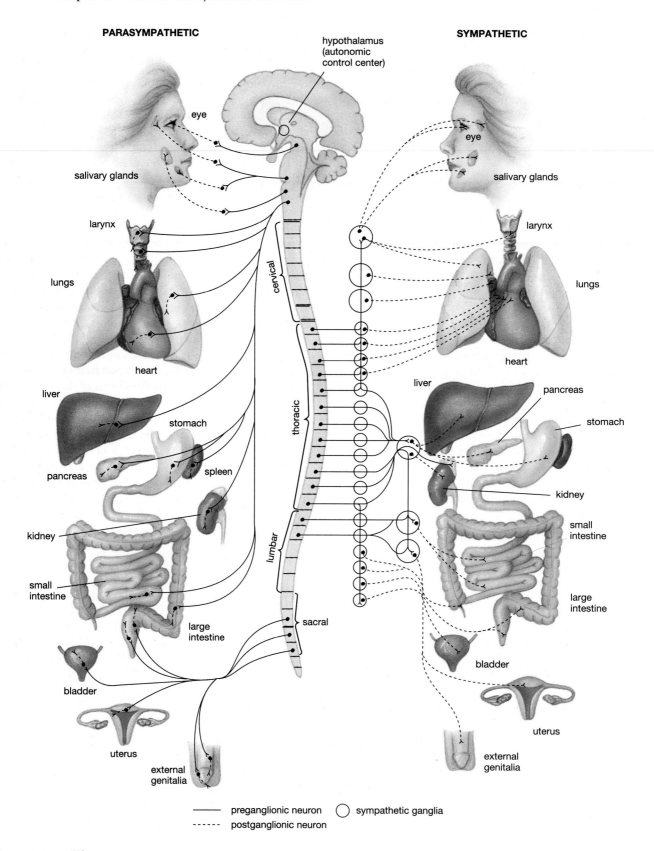

Figure 26-8   The autonomic nervous system

The autonomic nervous system has two divisions, the sympathetic and parasympathetic.
The two divisions supply nerves to many of the same organs but produce opposite effects.
Activation of the autonomic nervous system is mostly involuntary, produced by nervous outputs from the hypothalamus.

Two differences in the organization of the sympathetic and parasympathetic nervous systems are evident in Figure 26-8. First, parasympathetic axons are found in nerves that originate from two separate locations, the brain (midbrain and medulla) and the base of the spinal cord. In contrast, sympathetic axons are found in nerves that originate from the middle and lower portions of the spinal cord. Second, in both the sympathetic and parasympathetic divisions, there are two neurons that carry messages in sequence from the central nervous system to each target organ, but they synapse at different locations. In the sympathetic nervous system, the synapse occurs in ganglia (ganglia—singular, **ganglion**—are clusters of neurons) that are near the spinal cord. In the parasympathetic nervous system, the synapse occurs in smaller ganglia located at or very near each target organ.

The Central Nervous System Consists of the Spinal Cord and Brain

The central nervous system consists of the brain and spinal cord. It is the integrating portion of the nervous system, where sensory information is received and processed, thoughts are generated, and responses are directed. The central nervous system consists primarily of association neurons—somewhere between 10 and 100 billion of them!

The brain and spinal cord are protected in three ways. The first line of defense is a bony armor, consisting of the skull that surrounds the brain and the vertebral column that protects the spinal cord. Beneath the bones lies a triple layer of connective tissue called **meninges** (see Fig. 26-12). Between the layers of the meninges, a clear lymph-like liquid, the **cerebrospinal fluid,** cushions the brain and spinal cord.

The Spinal Cord Is a Cable of Axons Protected by the Backbone

The spinal cord is a neural cable about as thick as your little finger that extends from the base of the brain to the hips, protected by the bones of the vertebral column (Fig. 26-9). Between the vertebrae, nerves called dorsal roots, carrying axons of sensory neurons, and ventral roots, carrying axons of motor neurons, arise from the dorsal and ventral portions of the spinal cord, respectively; these merge to form the peripheral nerves of the spinal cord, which are part of the peripheral nervous system. In the center of the spinal cord are neuron cell bodies, which form a butterfly-shaped area of **gray matter.** These are surrounded by bundles of axons called **white matter** owing to their white insulating myelin coating (see Fig. 26-9). The spinal cord relays signals between the brain and the rest of the body, and it contains the neural circuitry for certain behaviors, including reflexes.

To illustrate some of the functions of the parts of the spinal cord, let's examine a simple spinal reflex, the pain-withdrawal reflex, which involves neurons of both the central nervous system and the peripheral nervous system (Fig. 26-10; see also Fig. 26-6). The cell bodies of the sensory neurons from the skin (in this case signaling pain) are found just outside the spinal cord in a row of ganglia. Each of these **dorsal root ganglia** is located on a spinal nerve and nestled close to the vertebral column. Both association and motor neuron cell bodies are found in the gray mat-

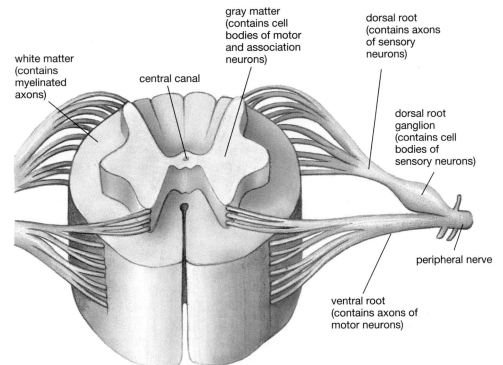

white matter (contains myelinated axons)

central canal

gray matter (contains cell bodies of motor and association neurons)

dorsal root (contains axons of sensory neurons)

dorsal root ganglion (contains cell bodies of sensory neurons)

peripheral nerve

ventral root (contains axons of motor neurons)

Figure 26-9 **The spinal cord**

The spinal cord runs from the base of the brain to the hips, protected by the vertebrae of the spine. Peripheral nerves emerge from between the vertebrae. A cross section of the spinal cord reveals an outer region of myelinated axons (white matter) traveling to and from the brain, surrounding an inner, butterfly-shaped region of dendrites and the cell bodies of association and motor neurons (gray matter). The cell bodies of the sensory neurons are located outside the cord in the dorsal root ganglion.

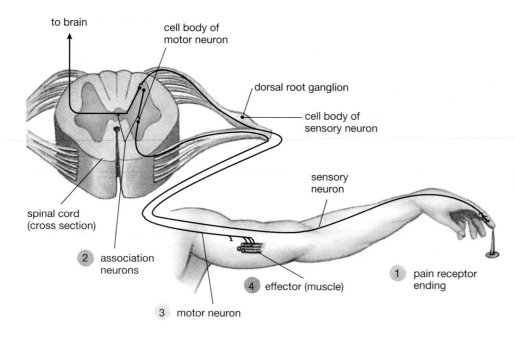

Figure 26-10   **The vertebrate pain-withdrawal reflex**

This simple reflex circuit includes one each of the four elements of a neural pathway. The sensory neuron has pain-sensitive endings in the skin ① and a long fiber leading to the spinal cord. The sensory neuron stimulates an association neuron in the spinal cord ②, which in turn stimulates a motor neuron, also in the cord. The axon of the motor neuron ③ carries action potentials to muscles ④, causing them to contract and withdraw the body part from the damaging stimulus. Note that the sensory neuron also makes a synapse on other association neurons not directly involved in the reflex, which carry signals to the brain, informing it of the danger below.

ter in the center of the spinal cord. The axons in the surrounding white matter communicate with the brain. Association neurons for the pain-withdrawal reflex, for example, not only synapse on motor neurons but also have axons extending up to the brain. Signals carried along these axons alert the brain to the painful event. The brain, in turn, sends impulses down axons in the white matter to cells in the gray matter. These signals can modify spinal reflexes. With sufficient motivation, you can suppress the pain-withdrawal reflex; to rescue a child from a burning building, for example, you could reach into the flames.

In addition to simple reflexes, the entire program for operating some fairly complex activities also resides within the spinal cord. All the neurons and interconnections needed to walk and run, for example, are found within the cord. In these cases, the role of the brain is to initiate and guide the activity of spinal neurons. The advantage of this semi-independent arrangement is probably an increase in speed and coordination, since messages do not have to travel all the way up the cord to the brain and back down again (in the case of walking) merely to swing forward one of your legs. The motor neurons of the spinal cord also control the muscles involved in conscious, voluntary activities such as eating, writing, or playing tennis. Axons of the brain cells directing these activities carry signals down the cord and stimulate the appropriate motor cells.

The Brain Consists of Several Parts Specialized for Specific Functions

All vertebrate brains have the same general structure, with major modifications corresponding to life-style and intelligence. Embryologically, the vertebrate brain begins as a

simple tube, which soon develops into three parts: the hindbrain, midbrain, and forebrain (Fig. 26-11). It is believed that in the earliest vertebrates, these three anatomical divisions were also functional divisions: The **hindbrain** governed automatic behaviors such as breathing and heart rate, the **midbrain** controlled vision, and the **forebrain** dealt largely with the sense of smell. In nonmammalian vertebrates, these three divisions remain prominent. However, in mammals, and particularly in humans, the brain regions are significantly modified. Some have been reduced in size, and others, especially the forebrain, greatly enlarged.

The Hindbrain Includes the Medulla, Pons, and Cerebellum

In humans, the hindbrain is represented by the medulla, the pons, and the cerebellum (Fig. 26-12). In both structure and function, the **medulla** is very much like an enlarged extension of the spinal cord. Like the spinal cord, the medulla has neuron cell bodies at its center, sur-

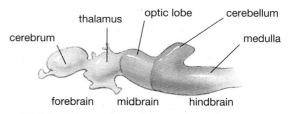

Figure 26-11   The embryonic vertebrate brain

The embryonic vertebrate brain shows three distinct regions: the forebrain, midbrain, and hindbrain. This basic structure persists in all adult brains.

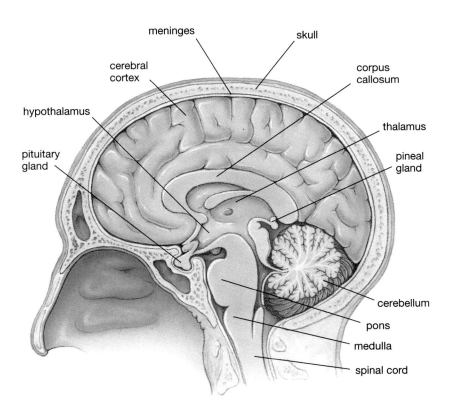

meninges

skull

cerebral cortex

corpus callosum

hypothalamus

thalamus

pituitary gland

pineal gland

cerebellum

pons

medulla

spinal cord

Figure 26-12   **The human brain**

A section taken through the midline of the human brain reveals some of its major structures.

rounded by a layer of myelin-covered axons. The medulla controls several automatic functions, such as breathing, heart rate, blood pressure, and swallowing. Certain neurons in the **pons,** located above the medulla, appear to influence transitions between sleep and wakefulness and between stages of sleep. Others influence the rate and pattern of breathing. The **cerebellum** is crucially important in coordinating movements of the body. It receives information from command centers in the higher, conscious areas of the brain that control movement and also from position sensors in muscles and joints. By comparing what the command centers ordered with information from the position sensors, the cerebellum guides smooth, accurate motions and body position. Not surprisingly, the cerebellum is largest in animals whose activities require fine coordination. It is best developed in birds, who engage in the complex activity of flight.

The Midbrain Contains the Reticular Formation

The midbrain is extremely reduced in humans, but an important relay center, the **reticular formation,** passes through it (Fig. 26-13). The neurons of the reticular formation extend all the way from the central core of the medulla, through the pons, the midbrain, and on into lower regions of the forebrain. It receives input from virtually every sense and every part of the body and from many areas of the brain as well. The reticular formation plays a role in sleep and arousal, emotion, muscle tone, and certain movements and reflexes. It filters sensory inputs before they reach the conscious regions of the brain, although the selectivity of the filtering seems to be set by

higher brain centers. Through a combination of genetically determined wiring and learning, the reticular formation "decides" which stimuli require attention. Important stimuli are forwarded to the conscious centers for processing,

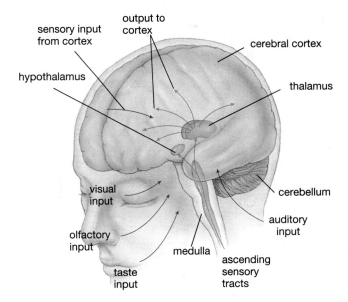

output to cortex

sensory input from cortex

cerebral cortex

hypothalamus

thalamus

cerebellum

visual input

auditory input

olfactory input

medulla

taste input

ascending sensory tracts

Figure 26-13   **The reticular formation**

The human reticular formation (shown here in blue) is a diffuse network of neurons running through the lower regions of the brain from the medulla in the hindbrain, through the midbrain, and up into the thalamus and hypothalamus of the forebrain. It receives input from most of the senses and sends outputs to many higher brain centers, filtering the sensory information that reaches the conscious brain.

and unimportant stimuli are suppressed. The fact that a mother wakens upon hearing the faint cry of her infant but sleeps through loud traffic noise outside her window testifies to the effectiveness of the reticular formation in screening inputs to the brain and to the role of learning in determining the importance of sensory stimulation.

The Forebrain Includes the Thalamus, Limbic System, and Cerebral Cortex

The forebrain, also called the **cerebrum,** can be divided into three functional parts: the thalamus, the limbic system, and the cerebral cortex. In mammals, the cerebral cortex is much enlarged compared with that of fish, amphibians, and reptiles. This trend culminates in the human cerebral cortex, which is thrown into complex folds that increase its area (see Fig. 26-12).

The Thalamus   The **thalamus** (see Figs. 26-12 and 26-13) carries sensory information to the limbic system and cerebrum. This information includes sensory input from auditory and visual pathways, from the skin, and from within the body. Inputs from the cerebellum and limbic system are also channeled through this busy thoroughfare. Very little information processing goes on in the thalamus.

The Limbic System   Anatomically, the **limbic system** is a diverse group of structures located in an arc between the thalamus and the cerebrum (Fig. 26-14). These structures work together to produce our most basic and primitive emotions, drives, and behaviors, including fear, rage, tranquility, hunger, thirst, pleasure, and sexual responses. The limbic system includes the hypothalamus, portions of the thalamus, the amygdala, and the hippocampus.

The **hypothalamus** (literally "under the thalamus") contains many different clusters of neurons. Some of these are neurosecretory cells that release hormones (see Chapter 25). Through its hormone production and neural connections, the hypothalamus acts as a major coordinating center, controlling body temperature, hunger, the menstrual cycle, water balance, and the autonomic nervous system. In addition, stimulation of specific areas of the hypothalamus elicits emotions such as rage, fear, pleasure, and sexual arousal.

The **amygdala** is believed to be responsible for the production of appropriate behavioral responses to environmental stimuli. It receives input from many sources, including the auditory and visual areas of the cerebral cortex. Different clusters of neurons in the amygdala produce sensations of pleasure, punishment, or sexual arousal when stimulated. By stimulating different portions of the amygdala, researchers can either reduce or enhance aggressive behavior. Conscious humans whose amygdalas are electrically stimulated have reported feelings of rage or fear. Recent studies have revealed that damage to the human amygdala eliminates the person's ability both to feel fear and to recognize fearful facial expressions in other people.

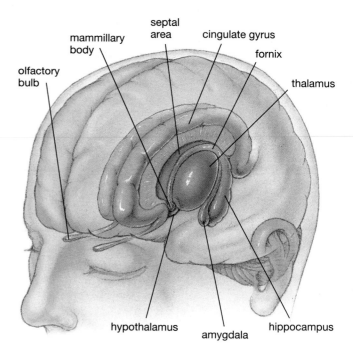

Figure 26-14   The limbic system
The limbic system extends through several brain regions. It seems to be the center of most unconscious, emotional behaviors, such as love, hate, hunger, sex, and fear.

The shape of the **hippocampus** as it curves around the thalamus inspired its name, which is derived from the Greek word meaning "sea horse." As in the amygdala and hypothalamus, behaviors that reflect a variety of emotions, including rage and sexual arousal, can be elicited by stimulating portions of the hippocampus. The hippocampus also plays an important role in the formation of long-term memory and is thus required for learning, discussed in more detail later in this chapter.

The Cerebral Cortex   In humans, by far the largest part of the brain is the **cerebral cortex,** the outer layer of the forebrain. The cerebral cortex and underlying parts of the forebrain are divided into two halves, called **cerebral hemispheres,** that communicate with each other by means of a large band of axons, the **corpus callosum.** The differences between the hemispheres is discussed a bit later in this chapter. The cerebral cortex is the most sophisticated information processing center known, and it is also the area of the brain that scientists know the least about. Roughly *50 to 100 billion* neurons are packed into this thin surface layer. To accommodate this profusion of cells, the cortex is thrown into folds, called **convolutions,** that greatly increase its area. In the cortex, cell bodies of neurons predominate, giving this outer layer of the brain a gray appearance. These neurons receive sensory information, process it, store some in memory for future use, and direct voluntary movements.

The cerebral cortex is divided into four regions based on anatomical criteria: the frontal, parietal, occipital, and temporal lobes (Fig. 26-15). Functionally, the cortex contains primary sensory areas where signals originating in sensory organs such as the eyes and ears are received and converted into subjective impressions, for example, light and sound. Nearby association areas interpret the sounds, as speech, for example, and the visual stimuli as recognizable objects. Association areas also link the stimuli with previous memories stored in the cortex and generate commands to produce speech. Primary sensory areas in the parietal lobe interpret sensations of touch originating in all parts of the body, which is "mapped" in an orderly sequence. In an adjacent region of the frontal lobe, primary motor areas generate commands for movements in corresponding areas of the body (Fig. 26-15). The association area of the frontal lobe protected by bones of the forehead seems to be involved in complex reasoning such as decision making, predicting the consequences of actions, controlling aggression (see Fig. E26-3), and planning for the future, as discussed later in the chapter. Damage to the cortex due to trauma, stroke, or a tumor results in specific deficits, such as problems with speech, difficulty reading, or the inability to sense or move specific parts of the body. Because brain cells cannot reproduce, once a brain region is destroyed it cannot be repaired or replaced, so these deficits are often permanent. Fortunately, however, in some cases diligent training can cause undamaged regions of the cortex to take control over and restore some of the lost functions.

Brain and Mind

Historically, people have always had difficulty reconciling the physical presence of a few pounds of grayish material in the skull with the range of thoughts, emotions, and memories of the human mind. This "mind-brain problem" has occupied generations of philosophers and, more recently, neurobiologists. Beginning with observations of patients with head injuries and progressing to sophisticated surgical, physiological, and biochemical experiments, the outlines of how the brain creates the mind are beginning to emerge. Here, we will be able to touch upon only a few of the more fascinating features.

The "Left Brain" and "Right Brain" Are Specialized for Different Functions

The human brain appears bilaterally symmetrical, particularly the cerebrum, which consists of two extremely similar-looking hemispheres. However, it has been known since

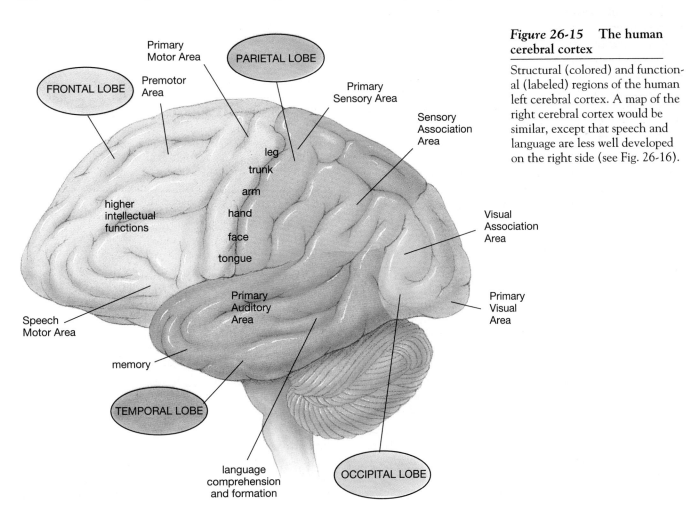

Figure 26-15   The human cerebral cortex

Structural (colored) and functional (labeled) regions of the human left cerebral cortex. A map of the right cerebral cortex would be similar, except that speech and language are less well developed on the right side (see Fig. 26-16).

the early 1900s that this symmetry does not extend to brain function. Much of what is known of the differences in hemisphere function comes from two sources: studies of accident victims with localized damage to one hemisphere and studies of patients who have had the corpus callosum (which connects the two hemispheres) severed. This surgical procedure was performed in rare cases of uncontrollable epilepsy to prevent the spread of seizures through the brain.

Studies based on selective damage to the left cerebral hemisphere had led to the belief that the right hemisphere was relatively retarded, lacking the ability to speak, write, recognize words, or reason. For example, people suffering damage to localized areas of the left hemisphere, but not the right, often became unable to speak, read, or understand spoken language. In addition, the left hemisphere for most people is superior in mathematical ability and in logical problem-solving tasks (Fig. 26-16).

Roger Sperry, of the California Institute of Technology, worked with people whose hemispheres had been sur-

gically separated by cutting the corpus callosum. In his studies, Sperry made use of the knowledge that axons within each optic nerve follow a pathway that causes the left half of each visual field to be projected on the right cerebral hemisphere, and vice versa (Fig. 26-16). Through an ingenious device that projected different images onto the left and right visual fields (thus sending different signals to each hemisphere), he and other investigators have gained more insight into the roles of the two hemispheres. If he projected an image of a nude figure onto the left visual field only, the patients would blush and smile but would claim to have seen nothing, because the image had reached only the nonverbal right side of the brain! The same figure projected onto the right visual field was readily described verbally. These experiments, begun in the 1960s and refined since then, have revealed that the right side of the brain is actually superior to the left in several areas, including musical skills, artistic ability, recognition of faces, spatial visualization, and the ability to recognize

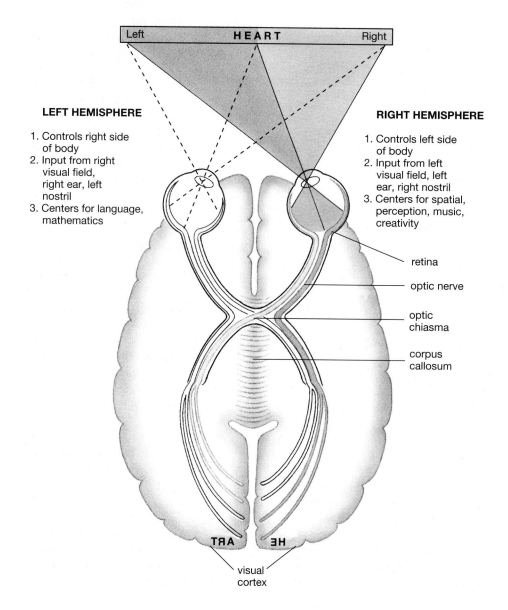

LEFT HEMISPHERE

1. Controls right side of body
2. Input from right visual field, right ear, left nostril
3. Centers for language, mathematics

RIGHT HEMISPHERE

1. Controls left side of body
2. Input from left visual field, left ear, right nostril
3. Centers for spatial, perception, music, creativity

retina

optic nerve

optic chiasma

corpus callosum

visual cortex

Figure 26-16   **Specialization of the two cerebral hemispheres**

In general, each hemisphere controls sensory and motor functions of the opposite side of the body. Further, the left side seems to predominate in rational and computational activities, whereas the right side governs creative and spatial abilities. Roger Sperry's split-brain experiments made use of the fact that images on the left half of the visual field are projected onto the right half of the retina and from there reach only the right visual cortex (orange), while the right half of the visual field is projected onto the left visual cortex (blue).

and express emotions. For his pioneering work, Sperry was awarded the Nobel Prize in 1981.

Recent experiments indicate that the left-right dichotomy is not as rigid as was once believed. Patients who have suffered a stroke that disrupted blood supply to the left hemisphere typically show symptoms such as loss of speaking ability. Frequently, however, training can partially overcome these speech or reading deficits, even though the left hemisphere itself has not recovered. This fact suggests that the right hemisphere has some latent language capabilities. Interestingly, female stroke victims recover some lost abilities more often than males, and females also have a larger corpus callosum. These findings suggest a sex difference in the degree of specialization of the two hemispheres and the extent of their interconnections. Further evidence of this difference has recently been provided by sensitive techniques that allow imaging of neural activity in the brains of normal subjects performing various mental tasks. When subjects were asked to compare word lists for rhyming words, a specific region of the left cortex of male subjects became active, but in females, similar areas in *both* left and right hemispheres were activated (further brain-imaging studies are described in "Scientific Inquiry: Scanning the Brain").

The Mechanisms of Learning and Memory Are Poorly Understood

Although theories abound as to the cellular mechanisms of learning and memory, we are a long way from understanding these phenomena. In mammals, and particularly in humans, however, we do know a fair amount about two other aspects of learning and memory: the time course of learning and some of the brain sites involved in learning, memory storage, and recall.

Memory May Be Brief or Long Lasting

Experiments show that learning occurs in two phases: an initial **working memory** followed by **long-term memory**. For example, if you look up a number in the phone book, you will probably remember the number long enough to dial but forget it promptly thereafter. This is working memory. But if you call the number frequently, eventually you will remember the number more or less permanently. This is long-term memory.

Some working memory seems to be electrical in nature, involving the repeated activity of a particular neural circuit in the brain. As long as the circuit is active, the memory stays. If the brain is distracted by other thoughts, or if electrical activity is interrupted, such as by electroconvulsive shock or by a concussion, the memory disappears and cannot be retrieved no matter how hard you try. In other cases, working memory involves temporary biochemical changes within neurons of a circuit, with the result that synaptic connections between them are strengthened.

Long-term memory, on the other hand, seems to be structural—the result, perhaps, of persistent changes in the expression of certain genes. It may require the formation of new, permanent synaptic connections between specific neurons or the permanent strengthening of existing but weak synaptic connections, for example, by increasing the area of synaptic contacts. These new or strengthened synapses last indefinitely, and the long-term memory persists unless certain brain structures are destroyed. Working memory can be converted into long-term memory, apparently by the hippocampus, which processes new memories then transfers them to the cerebral cortex for permanent storage.

Learning, Memory, and Retrieval May Be Controlled by Separate Regions of the Brain

Learning, memory, and the retrieval of memory seem to be separate phenomena, controlled by separate areas of the brain. Ample evidence shows that the hippocampus (part of the limbic system) is involved in learning. For example, intense electrical activity occurs in the hippocampus during learning. Even more striking are the results of hippocampal damage. A person whose hippocampus is destroyed retains most of his or her memories but is unable to learn anything that occurs after the loss. One victim was still unable to recall his address or find his way home after 6 years at the same residence. He could be entertained indefinitely by reading the same magazine over and over, and people whom he saw regularly required reintroduction at each encounter. People with extensive hippocampal damage can recall events momentarily, but the memory rapidly fades, as does the memory of a dream upon awakening. This phenomenon has led to the hypothesis that the hippocampus is responsible for transferring information from working into long-term memory.

Retrieval, or recall, of established long-term memories is localized in another area of the brain, the outer **temporal lobes** of the cerebral hemispheres. In a famous series of experiments in the 1940s, neurosurgeon Wilder Penfield electrically stimulated the temporal lobes of conscious patients undergoing brain surgery. The patients did not merely recall memories but felt that they were experiencing the past events right there in the operating room!

The site of storage of complex long-term memories is much less clear. The psychologist Karl Lashley spent many years training rats and subsequently damaging parts of their brains in an effort to locate the site of the memory, but failed. None of the injuries could erase a memory completely. In 1950, a frustrated Lashley wrote: "I sometimes feel, in reviewing the evidence on the localization of the memory trace, that the necessary conclusion is that learning just is not possible."

Some researchers suggest that each memory is stored in numerous distinct places in the brain. Or perhaps memories are stored like a hologram image, both everywhere and nowhere at the same time: The memory is more pre-

S C I E N T I F I C   I N Q U I R Y

Scanning the Brain

For most of human history, the brain has been seen as a "black box" whose inputs and outputs were observable but whose internal workings were inherently unknowable. New techniques, however, are providing exciting insights into brain function. The most spectacular is the PET scan (positron emission tomography, described in Chapter 2), which allows a glimpse into the working

Figure E26-1   Localization of language tasks

Changes in glucose utilization reveal different cortex regions involved in different language-related tasks, based on research by Dr. Marcus Raichle of the Washington University School of Medicine in St. Louis. The scale ranges from white (lowest) through red (highest).

Figure E26-2   Pinpointing a stroke

As a result of stroke, part of the brain dies from lack of blood flow. The damaged region can be precisely localized by the lack of neural activity to the upper left region of this brain scan.

brain. Regions of the brain that are most active have higher energy demands and utilize more glucose than inactive areas. Scientists inject the subject with radioactive glucose then monitor levels of radioactivity (which reflect glucose consumption) throughout the brain. Different levels of glucose utilization are translated into colors on cross-sectional images of the brain. By monitoring radioactivity while a specific task is per-

cise if the whole brain is intact, but each "bit" (probably several thousands of neurons) of cerebral hemisphere can store an essentially complete memory. Further research might provide definitive answers, but for now, the storage site of memories remains an unsolved mystery.

Insights on How the Brain Creates the Mind Come from Diverse Sources

Humans have always been intensely interested in the workings of their own minds. But until about 100 years ago, the mind was more appropriately a subject for philosophers than for scientists, because the tools to study the brain did not yet exist. Through the first half of the twentieth century, the

mind was treated by psychologists as a "black box" whose internal workings could be deduced only through the investigation of how past and present experiences were interpreted and influenced behavior. New discoveries, however, are rapidly changing our views of the workings of the brain.

During recent decades, we have begun to understand the neural bases of at least some psychological phenomena. Many forms of mental illness, such as schizophrenia, manic depression, and autism, once thought to be due to childhood trauma or inept parenting, are now recognized as the result of biochemical imbalances in the brain. Studies are revealing a strong heritability factor (and hence, a biological basis) for traits that were once considered entirely learned, such as shyness and alcoholism.

formed, scientists can identify parts of the brain involved in that task. Researchers can observe changes over a period of a few seconds as the brain responds to a specific odor or a visual or auditory stimulus or performs a type of thinking. Through PET scans, scientists have confirmed that different aspects of the processing of language occur in distinct areas of the cerebral cortex (Fig. E26-1). Pet scans can also precisely localize damaged portions of the brain, such as result from a stroke (Fig. E26-2). The contrast between normal and disturbed brain functioning can also be observed, for example, in some murderers and victims of Alzheimer's disease and schizophrenia (Fig. E26-3).

Figure E26-3   **Neural activity in normal and diseased brains**

PET scans can be used to compare neural activity in normal and diseased brains. These are images of sections of the brain taken so that the top of each image corresponds to the front of the brain, just behind the forehead, and the bottom to the back of the brain. Red areas are highest in metabolic activity, then yellow, then green. Blue areas show very low activity, and black areas exhibit little or no metabolic activity. **(a)** During a task requiring continuous attention, the frontal lobe (the region forming an arc across the top of the image) is strongly activated in a normal volunteer but much less active in a patient with schizophrenia. Schizophrenic individuals have difficulty focusing their attention on specific tasks. **(b)** The brain of a healthy elderly volunteer is compared with that of a victim of Alzheimer's disease, in which larger areas of blue and black indicate reduced neuronal activity in many brain regions.

A striking illustration of how the physical structure of the brain is related to personality was unwittingly provided by Phineas Gage in 1848. An explosion propelled a large metal rod through his skull, removing his left temporal lobe (Fig. 26-17). He miraculously survived, but his personality changed radically. Before the accident, Phineas was conscientious, industrious, and well liked. After his recovery, he became impetuous, profane, and incapable of planning or working toward a goal. Subsequent research has implicated the frontal lobe in emotional expression, control of aggression, and the ability to work for delayed rewards. Other sites of damage have revealed additional anatomical specializations. One patient with very localized damage to the left frontal lobe of the cerebral cortex was unable to name fruits and vegetables (although he could name everything else). Describing this patient, one science writer quipped: "Does the brain have a produce section?" Similarly, damage to certain areas of the cortex on the underside of the brain results in a selective inability to recognize faces.

In the past, much of our understanding of the human mind-brain connection came from the study of victims of brain damage such as that caused by a stroke, trauma, tumor, or surgical procedure. If the victim was cooperative, and if the case came to the attention of an interested researcher, tests were administered to define the change or loss of ability. Often, the exact extent of the damage remained unknown until revealed by autopsy.

Figure 26-17   **A famous accident**
On the basis of studies of the skull of Phineas Gage, scientists have created this computer-generated reconstruction of the path taken by the steel rod that was blown completely through his head by an explosion, extensively damaging the frontal lobe of his cerebral cortex.

Now new techniques, such as the PET and MRI scans, are permitting insight into the functioning of normal, as well as diseased, brains (see "Scientific Inquiry: Scanning the Brain"). These and increasingly sophisticated techniques of the future will create ever-larger windows into the "black box" that is the human brain. It is possible that the exact nature of consciousness will always remain obscure, but the next decades will see continued merging of the fields of psychology and neurophysiology, and a clearer understanding of how the human brain generates the human mind.

Receptor Mechanisms

The word *receptor* is used in several contexts in biology. In the most general sense, a **receptor** is a structure that changes when it is acted upon by a stimulus from its surroundings, causing a signal to be produced. All receptors are **transducers**: structures that convert signals from one form to another. The receptor may be a membrane protein that changes configuration when it binds a specific hormone or neurotransmitter, as discussed in previous chapters. Alternatively, as described in this chapter, a **sensory receptor** may be an entire cell (often a neuron) that is specialized to produce an electrical response to particular stimuli in its environment—that is, it translates sensory stimuli into the language of the nervous system. All sensory receptors produce electrical signals, but each receptor type is specialized so that it produces its signal only in response to a particular type of environmental stimulus, as discussed below. Sensory receptors are often clustered into sensory organs, such as the eye, ear, or tongue. Their electrical activity gives rise to the subjective perceptions of light, sound, and taste that we describe as our "senses."

Stimulation of a sensory receptor causes a **receptor potential,** an electrical signal whose size is proportional to the strength of the stimulus. Receptor potentials may give rise to action potentials in the receptor neurons themselves. Alternatively, receptor potentials in small receptor cells cause release of neurotransmitter onto postsynaptic neurons, which in turn produce action potentials that travel to the central nervous system. In contrast to action potentials, receptor potentials vary according to the intensity of the stimuli—the stronger the stimulus, the larger the receptor potential. Intensity is then conveyed to the nervous system by the frequency, not the size, of action potentials.

Receptor cells are named after the stimulus to which they respond, as summarized in Table 26-1. **Thermoreceptors** ("heat" receptors) respond to fluctuations in temperature. **Mechanoreceptors** respond to mechanical deformation of their cell membranes, which causes receptor potentials in one of two ways: by stretching the membrane of the receptor cell (as in receptors for touch or pressure)

TABLE 26-1 ⠿ *Vertebrate Receptor Types*			
Type of Receptor	**Specific Sensory Cell Type**	**Stimulus**	**Location**
Thermoreceptor	Free nerve ending	Heat, cold	Skin
Mechanoreceptor	Hair cell	Vibration, motion, gravity	Inner ear
	Variety of specialized endings in skin (Pacinian corpuscle, Merkel's disc)	Vibration, pressure, touch	Skin
	Free nerve endings in muscles or joints (muscle spindle, Golgi tendon organ)	Stretch	Muscles, tendons
Photoreceptor	Rod, cone	Light	Retina of eye
Chemoreceptor	Olfactory receptor	Odor (airborne molecules)	Nasal cavity
	Taste receptor	Taste (waterborne molecules)	Tongue
Pain receptor	Free nerve ending	Chemicals released by tissue injury	Widespread in body

or by bending "hairs" that project from the receptor cell membrane. Receptor cells for sound, motion, and gravity bear hairlike structures and are called **hair cells.** Bending of these hairs produces a receptor potential.

Many other receptor cells produce receptor potentials in response to stimuli that alter specific receptor proteins embedded in their cell membranes. **Photoreceptors** (receptors for light), **chemoreceptors** (receptors for chemicals, which we perceive as tastes or odors), and pain receptors possess extensive areas of membrane studded with specialized receptor proteins. When light or a chemical contacts a receptor protein specialized to respond to that stimulus, the protein changes shape. This change alters the permeability of the cell membrane, generating a receptor potential.

Sound Perception: Hearing

Sound is produced by any vibrating object—a drum, a motor, or the vocal cords of a person or animal. These vibrations are transmitted through air in the form of sound waves and are intercepted by our ears. The ears of humans and other animals are elaborate and remarkable structures that detect the direction, intensity, and pitch of sound.

The Structure of the Ear Helps Capture, Transmit, and Transduce Sound

The ear of humans and most other vertebrates consists of three parts: the outer, middle, and inner ear (Fig. 26-18). The **outer ear** consists of the **external ear** and the **audi-**

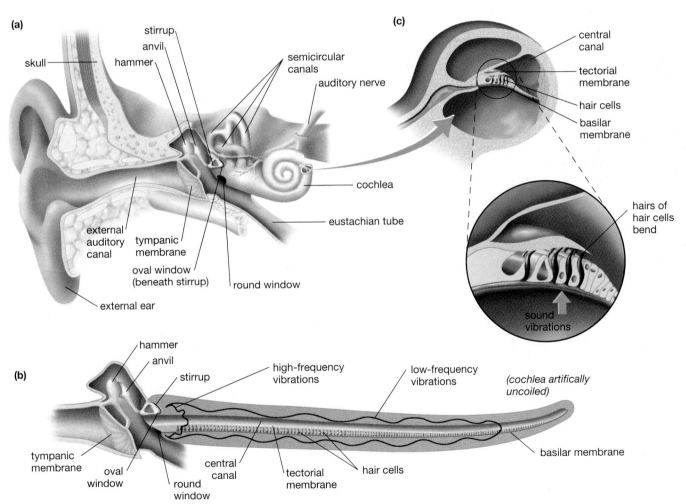

Figure 26-18 **The human ear**

The human ear contains structures that detect sound, gravity, and movement. **(a)** Sound waves enter the auditory canal and vibrate the tympanic membrane. These vibrations are transmitted through the bones of the middle ear to the membrane of the oval window connected to the fluid-filled cochlea, surrounded by thin bone. **(b)** Uncoiled, the cochlea consists of an outer tube surrounding the central canal. Vibrations of the fluid filling the cochlea cause the membranes of the central canal to vibrate, stimulating the hair cells, shown in part (c). The basilar membrane varies in width and tension along its length, causing lower-frequency vibrations (low-pitched sounds) to activate receptors toward the tip of the (uncoiled) cochlea and higher-frequency vibrations (high-pitched sounds) to activate receptors closer to the oval window. **(c)** The hairs of hair cells span the gap between the basilar and tectorial membranes in the central canal. Sound vibrations move the membranes relative to one another, bending the hairs and producing a receptor potential in the hair cells. The hair cells then cause action potentials in the auditory nerve.

tory canal. The external ear modifies sound waves with its fleshy folds. These modifications help the brain determine the location of the sound source. The air-filled auditory canal conducts the sound waves to the **middle ear,** consisting of the **tympanic membrane,** or eardrum; the **eustachian tube;** and three tiny bones called the hammer, anvil, and stirrup. The middle ear is surrounded by membranes and bone, but the eustachian tube connects the middle ear to the pharynx and equalizes the air pressure between the middle ear and the atmosphere.

Within the middle ear, sound first vibrates the tympanic membrane, which in turn vibrates the hammer, the anvil, and the stirrup. These bones transmit vibrations from the tympanic membrane to a much smaller membrane, the **oval window.** The oval window covers the opening of the **inner ear,** in which sound vibrations travel through fluid. The fluid-filled hollow bones of the inner ear form the spiral-shaped cochlea (*cochlea* means "snail" in Latin) and the vestibular apparatus (described later), where vibrations are translated into neural signals.

Sound Transduction Is Aided by the Structure of the Cochlea

The receptor cells for hearing are located within a portion of the inner ear called the **cochlea** (see Fig. 26-18a). If we mentally unroll the cochlea, in lengthwise cross section it consists of two fluid-filled tubes, an outer U-shaped canal and a straight central canal (Fig. 26-18b). The central canal contains the **basilar membrane,** on top of which are located the receptors, or hair cells. Protruding into the central canal is another membrane, the **tectorial membrane,** a gelatinous structure in which the "hairs" of the hair cells are embedded.

How do these structures allow the perception of sound? When the bones of the middle ear vibrate the oval window, it passes these vibrations to the fluid in the cochlea, which in turn vibrates the basilar membrane, causing it to move relative to the tectorial membrane. This movement bends the hairs spanning the gap between the membranes (Fig. 26-18c), producing receptor potentials in the hair cells. The hair cells then release transmitter onto neurons of the **auditory nerve.** Action potentials triggered in the auditory nerve axons travel to the brain.

The inner ear also allows us to perceive loudness (the magnitude of sound vibrations) and pitch (the frequency of sound vibrations). A weak sound causes small vibrations, which bend the hairs only slightly. This slight bending produces small receptor potentials in the hair cells and a low frequency of action potentials in the auditory nerve axons. A loud sound causes large vibrations, which cause greater bending of the hairs and a larger receptor potential. This leads to a high frequency of action potentials in the auditory nerve. Loud sounds sustained for a long time can actually damage the hairs (Fig. 26-19), resulting in hearing loss, a fate suffered by many prominent rock musicians and their fans.

The perception of pitch is a little more complex. Humans can detect vibration frequencies from about 30 vibrations per second (very low pitched) up to around 20,000 vibrations per second (very high pitched). The basilar membrane is stiff and narrow at the end near the oval window and more flexible and wider near the tip of the cochlea. This progressive change in structure causes different parts of the membrane to resonate best to particular frequencies of sound. High-frequency sound waves cause the greatest vibration toward the end of the basilar membrane near the oval window; increasingly lower frequencies produce vibration progressively farther toward the opposite end (see Fig. 26-18b). Thus, where the basilar membrane vibrates most, and consequently which receptors are stimulated most, varies with the frequency of sound. The brain interprets signals from receptors near the oval window as high-pitched sound, whereas signals from receptors farther along the uncoiled cochlea are interpreted as lower in pitch.

Figure 26-19   **Loud sound damages hair cells**

Scanning electron micrographs show the effect of intense sound on the hair cells of the inner ear. **(a)** Top view of the hairs of the hair cells in a normal guinea pig; hairs emerge from each receptor in a V-shaped pattern. **(b)** After 24-hour exposure to a sound level approached by loud rock music (2000 vibrations per second at 120 decibels), many of the hairs are damaged or missing entirely, leaving "scars." Because hair cells in humans do not regenerate, this type of hearing loss is permanent. (Scanning electron micrographs by Robert S. Preston, courtesy of Professor J. E. Hawkins, Kresge Hearing Research Institute, University of Michigan Medical School.)

Sensing Light: Photoreception

Animal vision varies in its ability to provide a sharp, accurate representation of the real world, and several types of eyes have evolved independently. All forms of vision, however, use photoreceptors. These sensory cells contain receptor molecules called **photopigments** (because they are colored) that absorb light and chemically change in the process. This chemical change alters ion channels in the receptor cell membrane, producing a receptor potential.

The Mammalian Eye Collects, Focuses, and Transduces Light Waves

Incoming light first encounters the **cornea,** a transparent covering over the front of the eyeball. Behind the cornea is a chamber filled with a watery fluid called **aqueous humor,** which provides nourishment for both the lens and cornea. The amount of light entering the eye is adjusted by the pigmented muscular tissue, the **iris,** whose circular opening, the **pupil,** can be expanded or contracted. Light passing through the pupil encounters the **lens,** a structure resembling a flattened sphere and composed of transparent protein fibers. The lens is suspended behind the pupil by ligaments and muscles that regulate its shape. Behind the lens is another, much larger chamber filled with a clear jellylike substance, the **vitreous humor,** which helps maintain the shape of the eye (Fig. 26-20a).

After passing through the vitreous humor, light reaches the **retina,** a multilayered sheet of photoreceptors and neurons, where the light energy is converted into electrical nerve impulses that are transmitted to the brain (Fig. 26-20b). The retina is richly supplied with blood vessels and contains a layer of pigment that absorbs stray light rays that escape the photoreceptors. Behind the retina is a darkly pigmented tissue, the **choroid.** The choroid's rich blood supply helps nourish the cells of the retina. Its dark pigment also absorbs stray light whose reflection inside the eyeball would interfere with clear vision. Surrounding the outer portion of the eyeball is a tough connective tissue layer, the **sclera,** visible as the white of the eye, which is continuous with the cornea.

In vertebrates (such as deer) that are most active at dusk, when little light is available, the choroid may be modified to reflect light, rather than absorb it. By reflecting light that escaped the photoreceptors during its initial passage, the choroid gives the receptors a second chance to detect it, maximizing the animal's ability to see in dim light. Reflective choroids give the eyes of these animals an eerie glow when bright light (such as from car headlights) is reflected back through the wide-open pupil.

The Adjustable Lens Allows Focusing of Both Distant and Nearby Objects

The visual image is focused most sharply on a small area of the retina called the **fovea.** Focusing is aided by the cornea, which contributes significantly to the bending of incoming light rays, producing an image of approximately the right size in the general vicinity of the retina. The shape of the cornea cannot be adjusted, however, and the lens is responsible for final, sharp focusing.

The shape of the lens can be adjusted so that it is either more rounded or more flattened when viewed from the side. This adjustment is accomplished by a circular muscle surrounding the lens. The size of the opening encircled by the muscle is largest when the muscle is relaxed. In its relaxed state, the muscle therefore applies maximum pull on the lens, which stretches and flattens it. In this configuration, the lens focuses distant objects. When focusing nearby, the muscle contracts, relaxing tension on the lens, and allowing the lens to resume a more rounded shape. The curved lens bends light rays more, allowing the eye to focus on closer objects (Fig. 26-21a).

Nearsighted people cannot focus on distant objects; farsighted people cannot focus on nearby objects. These con-

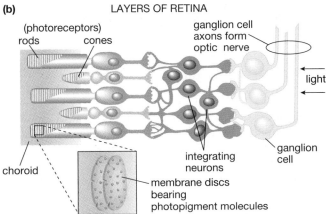

Figure 26-20   The human eye

(a) The anatomy of the human eye. **(b)** The human retina has rods and cones (photoreceptors), integrating cells, and ganglion cells. Each rod and cone bears a long extension packed with membranes in which the light-sensitive molecules are embedded.

(a) NORMAL EYE

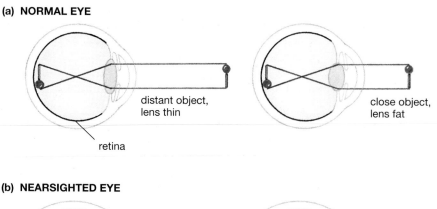

distant object, lens thin

retina

close object, lens fat

(b) NEARSIGHTED EYE

distant object focused in front of retina

long eyeball

concave lens diverges light rays, object focused on retina

(c) FARSIGHTED EYE

close object focused behind retina

short eyeball

convex lens converges light rays, object focused on retina

Figure 26-21   **Focusing in the human eye**

(a) (Left) To focus on a distant object, the lens is made thinner, causing relatively little bending of the light rays. (Right) To focus on a nearby object, the lens assumes a more nearly spherical shape, bending the light rays more sharply. **(b)** (Left) Nearsighted people usually have eyeballs that are too long. Light rays focus in front of the retina and are out of focus again by the time they strike the retina. (Right) Eyeglasses with a concave lens cause the light rays to diverge slightly so that the focal point falls on the retina. **(c)** (Left) Farsighted people have eyeballs that are too short. The lens cannot bend incoming light rays enough to focus on the retina, so the focal point falls behind the retina. (Right) Eyeglasses with a convex lens converge light rays, causing the focal point to fall on the retina.

ditions, which can be caused by abnormally long or short eyeballs, are corrected by external lenses of the appropriate shape (Fig. 26-21b, c). As humans age, the lens stiffens. By their mid-forties, most people have lost much of their ability to change the shape of the lens, and they become farsighted, requiring glasses for close work such as reading.

Light Striking the Retina Is Captured by Photoreceptors and the Signal Is Processed by Layers of Overlying Neurons

The vertebrate eye provides the sharpest vision in the animal kingdom, even though the complex, multilayered retina is "built backward" from an engineering perspective (see Fig. 26-20b). There may be some unknown advantage to the structure, but more likely, it is a good example of how a complex structure that evolves as a result of the accumulation of chance mutations may not have the same features as a structure that had been designed "from scratch" for its purpose. The photoreceptors, called **rods** and **cones** after their shapes, have their light-gathering parts farthest away from the light, at the rear of the

retina. Between the receptors and incoming light are several layers of neurons that process the signals from the photoreceptors. The outermost layer consists of **ganglion cells**, whose axons make up the **optic nerve.** The receptor potential from the photoreceptors is processed in complex ways by the retinal neurons. This processing results in an enhancement of our ability to detect edges, movement, dim light, and changes in light intensity.

The much modified signal from the photoreceptors is finally converted to action potentials carried by the ganglion cell axons in the optic nerve to the brain. Here, further processing ultimately results in the sensation of vision. Ganglion cell axons emerge from the outermost layer of the retina, then must pass back through the retina to reach the brain. The point where axons pass through the retina is called the **optic disc**, or **blind spot** (Fig. 26-22; see also Fig. 26-20a). This area lacks receptors, and objects focused here seem to disappear. You can locate your blind spot by closing your left eye and focusing on the star with your right eye. Start with the book about a foot away and gradually move it closer. The spot will disappear when the image falls on the optic disc.

Interestingly, the camera eye of mollusks such as the octopus, which evolved independently from ours, is more efficient. Its photoreceptors are located in the outermost layer of the retina, so incoming light strikes the photoreceptors immediately. This design eliminates the blind spot and increases the availability of light to the receptors.

The Photoreceptor Cells Are Called Rods and Cones

Photoreception in both rods and cones begins with absorption of light by photopigment molecules embedded in the membranes of the photoreceptors. These membranes form flattened, hollow discs in rods and are deeply folded in cones, giving them a large surface area on which to bear the photopigments (see Fig. 26-20b).

Light hitting the photopigment molecule causes the molecule to change shape. The altered shape of the molecule initiates a series of biochemical reactions inside the photoreceptor. These result in a change in the permeability of the receptor membrane to ions, producing a receptor potential in the photoreceptor cell. The receptor potential is then transmitted as described above.

Rods are far more numerous than cones (125 million rods versus 5 million cones) and dominate in the peripheral portions of the retina. Compared with cones, rods have much deeper stacks of pigment-bearing membrane (see Fig. 26-20b) and are about 100 times more sensitive to light. Studies have shown that a rod is capable of re-

sponding to a single photon, the smallest unit of light. Thus, rods are almost entirely responsible for our vision in dim light. Unlike cones, rods do not distinguish colors, so in moonlight, which is too dim to activate the cones, the world appears in shades of gray.

Although cones are found throughout the retina, they are concentrated in the fovea, where the lens focuses images most sharply (see Fig. 26-20a). The fovea, which consists entirely of densely packed cones, appears as a depression near the center of the retina because the layers of neurons normally covering the cones are pushed aside here while still retaining their synaptic connections. Because of this arrangement, light reaches the cones of the fovea with relatively little interference.

Unlike rods, cones respond to different wavelengths of light (colors). Human cones come in three varieties, each containing a slightly different photopigment. Each type of photopigment is most strongly stimulated by a particular wavelength of light, corresponding roughly to red, green, or blue. The brain distinguishes color according to the relative intensity of stimulation of different cones. For example, the sensation of yellow is caused by fairly equal stimulation of red and green cones. About 8% of all males lack normal color vision. Although described as "color-blind," they are actually only color deficient. The most common abnormality is red-green color deficiency due to a recessive allele on the X chromosome (see Chapter 15) that codes for a defective photopigment in the red cones. The altered red photopigment has about the same light-absorbing properties as the green photopigment, so the affected individual has trouble distinguishing red from green.

Not all animals have both rods and cones. Animals active almost entirely during the day (certain lizards, for example) may have all-cone retinas, whereas night-active animals (such as the ferret) or those dwelling in dimly lit habitats (such as deep-sea fishes) often have mostly rods.

Binocular Vision Allows Depth Perception

Most animals are bilaterally symmetrical and possess a pair of eyes. Among vertebrates, we find two basically different eye placements: Herbivores usually have one eye on each side of the head; predators have both eyes facing forward (Fig. 26-23). The forward-facing eyes of predators and omnivores such as humans have slightly different but extensively overlapping visual fields. This **binocular vision** allows depth perception and accurate judgment of the size and distance of an object from the eyes. These abilities are important to a cat about to pounce on a mouse or to a monkey leaping from branch to branch.

In contrast, the widely spaced eyes of herbivores have little overlap in their visual fields; accurate depth perception is sacrificed in favor of a nearly 360 degree field of view. This allows these animals, who are frequently preyed upon, to spot a predator approaching from any direction.

blind spot

fovea

Figure 26-22   **The human retina**

A photograph of a portion of the human retina, taken through the cornea and lens of a living person. The blind spot (optic disc) and fovea are visible. Blood vessels supply oxygen and nutrients; notice that they are dense in the blind spot (where they won't interfere with vision) and there are fewer near the fovea.

(a)

(b)

Figure 26-23   **Eye position reflects life-style**

There are two usual placements for the eyes—one on each side of the head or both in front. The location of the eyes is related to life-style. **(a)** Herbivorous prey animals, such as rabbits, mice, horses, and deer, tend to have eyes placed at the sides, the better to scan all around for possible predators. **(b)** Predators, such as this owl, and primates tend to have eyes in front, and both can be brought to bear on a target. Each eye gets a slightly different view of the target, so that size and distance can be judged fairly accurately.

Smell and Taste: Chemoreception

Through chemical senses, animals find food, avoid poisonous materials, and may locate homes or find mates (Fig. 26-24). Virtually all animals have chemoreceptors that sample the chemical composition of the environment. Terrestrial vertebrates have two separate chemical senses: one for airborne molecules, called smell, or **olfaction**, and one for chemicals dissolved in water or saliva, the sense of **taste**.

(a)

Figure 26-24   **Pheromones**

(a) Male moths find females not by sight, but by following airborne scents (pheromones) released by the females. These odors are sensed by receptors on the male's huge antennae, whose enormous surface area maximizes the chances of detecting the female scent. **(b)** When dogs meet, they often sniff each other near the base of the tail. Scent glands there broadcast information about sex (both type and interest in) and status.

(b)

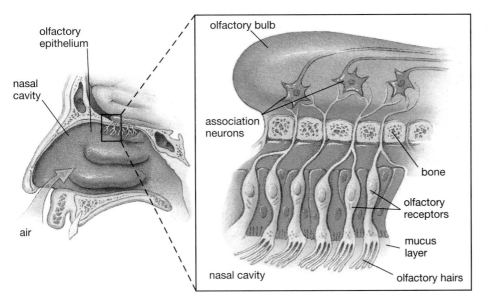

Figure 26-25   **Olfactory receptors**

The receptors for olfaction in humans are neurons that bear hairlike projections protruding into the nasal cavity. The projections are embedded in a mucus layer, in which odor molecules dissolve before contacting the receptors.

The Ability to Smell Arises from Olfactory Receptors

In humans and most other vertebrates, receptors for smell, or olfaction, are nerve cells located in a patch of mucus-covered epithelial tissue in the upper portion of each nasal cavity (Fig. 26-25). The human olfactory epithelium is small compared with that of many other mammals (dogs are an obvious example) whose sense of smell is far more acute than ours. Olfactory neurons have hairlike dendrites that protrude into the cavity and lie embedded in a layer of mucus. Odorous molecules in the air diffuse into the mucus layer and bind with receptors on the dendrites.

Recent research suggests that there may be 1000 types of receptor proteins embedded in the olfactory dendrites. Each receptor protein is specialized to bind a particular type of molecule and stimulate the olfactory neuron to send a message to the brain. Researchers hypothesize that each neuron may bear only one type of receptor protein, and the brain may interpret the electrical signals coming from different neurons as distinct odors. Some chemicals may bind to more than one receptor type, and the brain may perceive *more* than 1000 different odors through its ability to interpret signals from particular *combinations* of olfactory neurons as distinct odors.

Taste Receptors Are Found in Clusters on the Tongue

The human tongue bears about 10,000 **taste buds**, embedded in small bumps covering the surface of the tongue. Each taste bud consists of a cluster of about 60 to 80 taste receptors surrounded by supporting cells in a small pit that communicates with the mouth through a pore (Fig. 26-26). Microvilli (thin membrane projections) of taste receptor cells protrude through the pore. Dissolved chemicals enter the pore and bind to receptor molecules on the microvilli. There are only four types of taste receptors: sweet, sour, salty, and bitter. The great variety of appar-

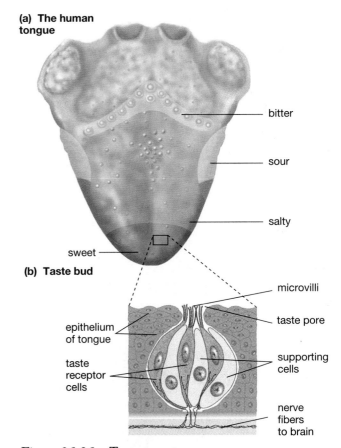

(a) The human tongue

(b) Taste bud

Figure 26-26   **Taste receptors**

(a) The human tongue bears numerous bumps in which are clustered masses of taste buds. **(b)** Each taste bud consists of supporting cells surrounding 60 to 80 taste receptor cells, whose microvilli bind the tasty molecules. Receptor cells synapse on nerve fibers that carry electrical signals to the brain.

ent tastes we perceive is the result of two mechanisms. First, a particular substance may stimulate two or more receptor types to different degrees, making it taste "salty-sweet," for example. Second and more important, material being tasted usually also gives off molecules into the air inside the mouth. These odor molecules diffuse to the olfactory receptors, which contribute an odor component to the basic flavor (remember, the mouth and nasal passages are connected).

If you don't believe that what we call taste is really mostly smell, try eating different jelly beans while holding your nose (and closing your eyes). The tastes of grape, lime, and cherry jelly beans will be indistinguishable, like sweet sticky pastes. Likewise, when you have a bad cold, notice how normally tasty foods seem bland and lose much of their appeal.

Sensing Pain

Pain is an unusual sense in that it is somewhat non-specific. Whether you burn, cut, or crush a fingertip, you will feel the same type of sensation: pain. This fact gives us a clue to the nature of pain perception: Most pain is produced by tissue damage, regardless of the

cause. Over the past few years, researchers have found that pain perception is actually a special kind of chemical sense (Fig. 26-27).

When cells are broken open by a cut or a burn, for example, their contents flow into the extracellular fluid and blood. The cell contents include enzymes that convert certain blood proteins into a chemical called **bradykinin.** Pain receptors, which are dendrites of specific sensory neurons, have receptor molecules for bradykinin. Binding of bradykinin to these receptors results in action potentials that are interpreted as pain by the brain. Because each part of the body has a separate set of pain neurons that provide input to particular brain cells, the brain knows where the pain is occurring.

Drugs that provide pain relief, such as morphine or Demerol, block synapses in the pain pathways of the brain or spinal cord. In ways that we are just beginning to understand, the brain can modulate its perception of pain through its own narcotic-like **endorphins.** In critical situations, such as combat or during escape from a fire, endorphins may allow us to function by blocking our perception of pain until the emergency is over.

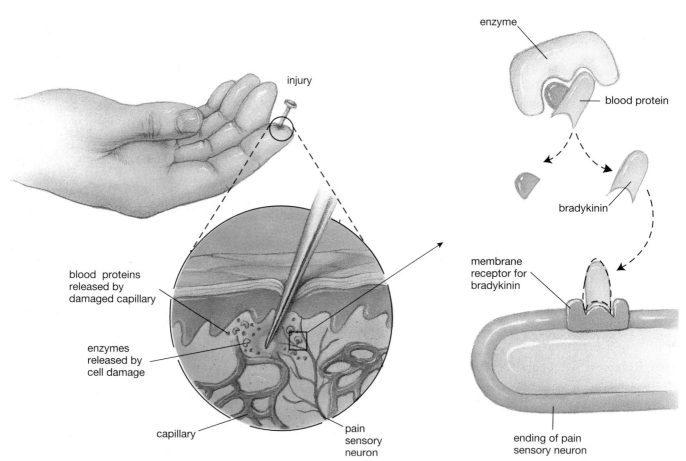

Figure 26-27   **Pain perception**

Pain perception is a specialized chemical sense. An injury, such as stabbing the finger with a needle, damages both cells and blood vessels. The cells release enzymes that convert certain blood proteins into bradykinin. Bradykinin stimulates pain-sensitive neurons by binding with membrane receptor molecules.

(a)

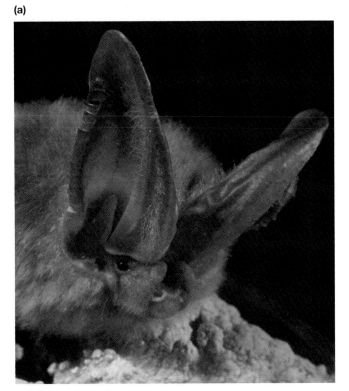

Figure 26-28   **Echolocation**

Echolocation by bats and porpoises is based on very different anatomical structures. **(a)** A long-eared bat, showing the enormous size and elaborate folds of the external ears that help it localize returning echoes. **(b)** The bottlenose porpoise focuses ultrasonic clicks using the oil-filled bulge in the front of its head.

(b)

E V O L U T I O N A R Y
C O N N E C T I O N S

Natural Selection Shapes Perception

This chapter has reviewed the "common" senses of sound, sight, smell, taste, and pain. But had this text been written by bats, it undoubtedly would have included a large section on echolocation and nearly omitted the coverage of vision! Here we review a few of the "uncommon" senses that have evolved in response to different environments.

Echolocation

Some animals who hunt in darkness or murky water have evolved a type of sonar, similar to the navigational system used by ships. These animals transmit high-frequency sound waves (higher than can be detected by the human ear) then detect the vibrations reflected from objects to produce an "auditory image" of their nearby surroundings.

This ability, called **echolocation,** is highly developed in bats and porpoises.

Bats can navigate and hunt insect prey in total darkness using echolocation. An echolocating bat emits pulses of noise at ultrasonic frequencies that bounce back from nearby objects. The intensity of this sound would make it unpleasantly loud if we could detect it. The patterns of returning sound convey accurate information about the size, shape, surface texture, and location of objects in the environment. Little brown bats can detect wires only 1 millimeter thick from a distance of 2 meters (over 6 feet). Several adaptations contribute to this remarkable sensitivity. The enormous, elaborately folded outer ears of the bat collect the returning echoes and help the bat locate their source (Fig. 26-28a). As the bat emits its cry, muscles attached to the bones of the middle ear contract briefly, reducing the bones' vibrations and preventing the bat from being deafened by its own calls. The tympanic membrane and bones of the middle ear are exceptionally light and easily vibrated by the faint returning echoes.

Porpoises produce ultrasonic clicks within their nasal passages and emit them through the front of their head (Fig. 26-28b). There, a large, flexible, oil-filled sac acts

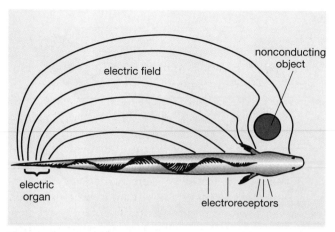

Figure 26-29   **Electroreception**

Weak electric fishes locate nearby objects by sensing distortions in their own electrical field that are produced by the objects. The fishes' fields are generated by electric organs near the tail and detected by electroreceptors in the pores of the lateral line system.

like an acoustic lens, directing the sound forward in a broad beam (for navigation) or a narrow beam (to locate prey). An echolocating porpoise can locate a pea-sized object on the floor of its tank and distinguish different species of fish. Porpoises may also use the narrowly focused beam to stun fish with a blast of sound, making them easier to capture.

Detecting Electrical Fields

Some fishes, called weak electric fishes, use electrical fields for electrolocation in much the same way that bats and porpoises use sound waves for echolocation. These fishes produce high-frequency electrical signals from an electric organ just in front of their tails, and they detect these signals using electroreceptor cells located inside the pores of their lateral lines (Fig. 26-29). Objects near the fish distort the electrical field surrounding the fish, and this distortion is detected by the electroreceptors, which send an altered pattern of action potentials to the brain. The fish is able to use this information to detect and localize nearby objects.

Recently, electroreceptors have been discovered in two mammals, the platypus and the star-nosed mole. Rather than sensing distortions in their own electrical fields, these mammals use their electroreceptors to detect electrical fields produced by other animals, such as potential prey.

Both hunt under conditions where their eyes are nearly useless. The platypus, whose bill is covered with electroreceptors (as well as sensitive mechanoreceptors) hunts for crayfish and tadpoles at night in murky freshwater ponds and streams. The star-nosed mole digs tunnels into streams, where it hunts worms in muddy water. The 22 writhing tentacles that surround the mole's nose had long puzzled zoologists. Researchers have now demonstrated that the moles can detect, and are attracted to, weak electrical fields such as those produced by earthworms, presumably by using their electroreceptor-studded noses.

Detecting Magnetic Fields

Homing pigeons are famous for their ability to fly home after being released some distance away. They can accurately locate their home roost even under cloudy skies in terrain with few landmarks. A pigeon with a small magnet strapped to its back will still home successfully in sunny weather but will lose its way under overcast skies. Apparently, pigeons can navigate either by the sun or, if the sun is hidden, by magnetic fields. In cloudy weather, the magnets confuse the pigeon's magnetic compass. How do the pigeons detect magnetic fields? Pigeons have deposits of magnetite (a magnetic iron compound) located just beneath their skulls. These deposits may act as a built-in magnet that the pigeons use to tell direction.

Eels of eastern North America and western Europe swim out of streams and rivers into the Atlantic Ocean and migrate to the Sargasso Sea (near the West Indies) to spawn. Eels probably also use magnetic fields for navigation. You may recall from high school physics that moving an electrical conductor through a magnetic field induces an electrical current in the conductor (this is how we generate electricity commercially). Seawater, with its high salt concentration, is a fairly good conductor. The currents of the Gulf Stream, which flow from the Gulf of Mexico along the east coast of the United States, provide movement through Earth's magnetic field. The Gulf Stream generates an extremely weak electrical field, roughly equivalent to the field produced by a 1-volt battery with its poles more than 7 kilometers (12 miles) apart. At first, this field seemed far too weak to be detected. But researchers discovered that eels can detect electrical fields as weak as a 1-volt battery with poles *over 5000 kilometers* (3000 miles) apart. For an eel, finding the Gulf Stream must be a piece of cake!

✖ SUMMARY OF KEY CONCEPTS

A Comparison of Nervous and Endocrine Communication

There are four main differences between hormonal and neural control: (1) Hormones are released into the blood, but nerve cells release neurotransmitters directly onto their target cells; (2) a nerve cell releases neurotransmitter onto relatively few target cells; (3) neurons use electrical signals; and (4) neural signals last a much shorter time than most hormonal messages. These distinctions, however, are not absolute, and the hormonal and nervous systems are closely coordinated.

The Functions and Structure of Neurons

Nervous systems are composed of billions of individual cells called neurons. A neuron has five major functions, which are reflected in its structure: (1) Dendrites receive information from the environment or other neurons. (2) The cell body coordinates the cell's metabolic activities. (3) The cell body also adds together electrical signals from the dendrites and from synapses on the cell body itself and "decides" whether to produce an action potential. (4) The axon conducts the action potential to its output terminal: the synapse. (5) Synaptic terminals transmit the signal to other nerve cells, glands, or muscles.

A synapse, where two neurons communicate, consists of the synaptic terminal of one neuron and a specialized region of the second neuron, separated by the synaptic cleft. Neurotransmitter, released in response to an action potential, binds to receptors on the dendrite cell membrane. This produces a brief electrical signal that either excites or inhibits an action potential in the second neuron.

Building and Operating a Nervous System

Information processing in the nervous system requires four operations. The nervous system must: (1) signal the intensity of the stimulus, (2) determine the type of stimulus, (3) integrate information, and (4) initiate and direct the response. Nervous systems signal intensity by the frequency of action potentials in single neurons and by the number of similar neurons firing at the same time. The type of stimulus is determined by which neurons produce action potentials. The nervous system collects and processes sensory information from many sources. These sensory stimuli may come together (converge) on a few neurons whose activity in response to the inputs determines action. The "decision" to act may then be transmitted to many more neurons (divergence), which direct the activity.

Neural pathways normally have four elements: sensory neurons, association neurons, motor neurons, and effectors. Overall, nervous systems consist of numerous interconnected neural pathways, which may be either diffuse or centralized.

The Human Nervous System

The nervous system of humans and other vertebrates consists of the central nervous system (brain and spinal cord) and the peripheral nervous system (nerves leading from the central nervous system to the rest of the body).

The peripheral nervous system is further subdivided into sensory and motor portions. The motor portions consist of the somatic nervous system (which controls voluntary movement) and the autonomic nervous system (directing involuntary responses). The autonomic nervous system has both a sympathetic and parasympathetic component. The sympathetic nervous system prepares the body to "fight or take flight," whereas the parasympathetic nervous system promotes maintenance activities such as digestion.

Within the central nervous system, the spinal cord contains (1) neural pathways for reflexes and certain simple behaviors; (2) motor neurons controlling voluntary muscles; and (3) axons leading to and from the brain. The brain consists of three parts—the hindbrain, midbrain, and forebrain, each further subdivided into distinct structures.

The hindbrain in humans consists of the medulla and pons, which control involuntary functions (such as breathing), and the cerebellum, which coordinates muscular activities (such as walking). In humans, the small midbrain contains the reticular formation, a filter and relay for sensory stimuli. The forebrain includes the thalamus, a sensory relay station that shuttles information to and from higher, conscious centers in the forebrain. The diverse structures of the limbic system of the forebrain are involved in emotion, learning, and the control of instinctive behaviors such as sex, feeding, and aggression. Finally, the cerebral cortex of the forebrain is the center for information processing, memory, and initiation of voluntary actions. It includes primary sensory and motor areas and association areas that analyze sensory information and plan movements.

Brain and Mind

The cerebral hemispheres are each specialized. In general, the left hemisphere is dominant in speech, reading, writing, language comprehension, mathematical ability, and logical problem solving. The right hemisphere specializes in recognizing faces and spatial relationships, artistic and musical abilities, and recognition and expression of emotions.

Memory takes two forms. Short-term memory is electrical or chemical. Long-term memory probably involves structural changes that increase the effectiveness of synapses. The hippocampus is an important site for learning and for the transfer of information into long-term memory. The temporal lobes are important for memory retrieval. The sites of memory storage are poorly understood.

Receptor Mechanisms

Receptors are transducers, converting signals from one form to another. In response to external stimuli, sensory receptor cells produce receptor potentials that in turn produce action potentials, either in the sensory neuron itself or in a postsynaptic neuron. Receptor cells are named after the stimulus to which they respond. Thermoreceptors respond to heat. Mechanoreceptors respond to mechanical deformation of the cell membrane and produce sensations of touch, stretch, hearing, or gravity. Photoreceptors, chemoreceptors, and pain receptors respond to stimuli—light, taste and odor, and tissue damage, respectively—that influence receptor molecules embedded in the receptor cell membrane.

Sound Perception: Hearing

Hearing is a modified mechanical sense specialized for reception of vibrations of the air or water. In the human ear, air vibrates the tympanic membrane, which transmits vibrations to the bones of the middle ear and then to the oval window of the fluid-filled cochlea. Within the cochlea, vibrations bend the hairs of hair cells, which are receptors located between the basilar and tectorial membranes. This bending produces receptor potentials in the hair cells that cause action potentials in the axons of the auditory nerve to the brain.

Sensing Light: Photoreception

All eyes have photoreceptor cells that contain a pigment that, upon absorption of light, causes a receptor poten-tial. In the vertebrate eye, light enters the cornea and passes through the pupil to the lens, which focuses an image on the fovea of the retina. Two types of photoreceptor, rods and cones, are found deep in the retina. They produce receptor potentials in response to light. These signals are processed through several layers of neurons in the retina and finally are translated into action potentials in ganglion cells. Ganglion cell axons form the optic nerve, which carries action potentials to the brain. Rods are more abundant and more light-sensitive than cones and provide vision in dim light. Cones, which are concentrated in the fovea, provide color vision.

Smell and Taste: Chemoreception

Terrestrial vertebrates detect chemicals in the external environment either by olfaction (for airborne sources of chemicals) or by taste (chemicals dissolved in water or saliva). Each olfactory or taste receptor cell type responds to only one or a few specific types of molecules, allowing discrimination among tastes and odors. Olfactory neurons of vertebrates are found in a tissue lining the nasal cavity. Taste receptors are found in clusters called taste buds on the tongue.

Sensing Pain

Pain is a special type of chemical sense in which sensory neurons respond to chemicals released by damaged cells that produce the substance bradykinin. Binding of bradykinin to pain receptors results in action potentials that are interpreted as pain by the brain.

�֎ KEY TERMS

✖ T H I N K I N G   T H R O U G H   T H E   C O N C E P T S

Multiple Choice

1. _____ are integration centers in neurons.
   a. dendrites   b. axons
   c. cell bodies   d. ion channels
   e. synapses

2. The role of the axon is to
   a. integrate signals from the dendrites
   b. release neurotransmitter
   c. conduct the action potential to the synaptic terminal
   d. synthesize cellular components
   e. stimulate a muscle, gland, or other neuron

3. The intensity of a stimulus can be signaled by
   a. the frequency of action potentials produced
   b. the size of action potentials produced
   c. the number of sensory neurons firing action potentials
   d. the duration of each action potential
   e. both a and c

4. Which of the following statements is FALSE?
   a. The hindbrain contains the medulla, pons and cerebellum.
   b. The thalamus is an important sensory relay structure.
   c. The cerebral cortex in humans is folded into convolutions.
   d. The hippocampus has an important role in recognizing and experiencing fear.
   e. The hypothalamus coordinates the activities of the autonomic nervous system.

5. Disturbances in body temperature, hunger perception, menstrual cycles, water balance, and sexual arousal are symptoms of defects in the
   a. cerebrum   b. thalamus
   c. medulla   d. hypothalamus
   e. cerebellum

6. Color vision is a function of _____ cells.
   a. cochlear   b. cupular
   c. corneal   d. choroid
   e. cone

7. To go from the pharynx to the middle ear, a germ must travel through the
   a. oval window   b. vestibular apparatus
   c. eustachian tube   d. cochlea
   e. auditory canal

8. The fovea is
   a. the "blind spot"
   b. a clear area in front of the pupil and iris
   c. the area of the retina where an optic nerve exits the eye
   d. a point of attachment between the retina and choroid coat
   e. a central focal region of the vertebrate retina

9. Light entering the eye and striking the choroid would travel, in order, through the
   a. lens, vitreous humor, cornea, aqueous humor, and retina
   b. retina, aqueous humor, lens, vitreous humor, and cornea
   c. cornea, aqueous humor, retina, vitreous humor, and lens
   d. cornea, aqueous humor, lens, vitreous humor, and retina
   e. lens, aqueous humor, cornea, vitreous humor, and retina

10. Pain perception is carried out by a special type of
   a. mechanoreceptor   b. proprioceptor
   c. chemoreceptor   d. magnetoreceptor
   e. thermoreceptor

Review Questions

1. List four major parts of a neuron and explain what function each part is specialized for.

2. Diagram a synapse. How are signals transmitted from one neuron to another at a synapse?

3. How does the brain perceive the intensity of a stimulus? The type of stimulus?

4. What are the four elements of a simple nervous pathway? Describe how these elements function in the human pain reflex.

5. Draw a cross section of the spinal cord. What types of neurons are located in the spinal cord? Explain why severing the cord paralyzes the body below the level where it is severed.

6. Describe the functions of the following parts of the human brain: medulla, cerebellum, reticular formation, thalamus, limbic system, cerebrum.

7. What structure connects the two cerebral hemispheres? Describe the evidence that the two hemispheres are specialized for distinct intellectual functions.

8. Distinguish between long-term and working memory.

9. What are the names of the specific transducers used for taste, vision, hearing, smell, and touch?

10. Why are we apparently able to distinguish hundreds of different flavors, when we have only four types of taste

receptors? How are we able to distinguish so many different odors?

11. Describe the structure and function of the various parts of the human ear. Do this by tracing a sound wave from the air outside the ear to the cells causing action potentials in the auditory nerve.

12. How does the structure of the inner ear allow for the perception of pitch? Of sound intensity?

13. Diagram the overall structure of the human eye. Label the cornea, iris, lens, sclera, retina, and choroid. Describe the function of each structure.

14. How does the lens change shape to allow focusing of faraway objects? What defect makes focusing on faraway objects impossible, and what is this condition called? What type of lens can be used to correct it, and how does it do it?

15. List the similarities and differences between rods and cones.

16. Distinguish between taste and olfaction.

17. Describe how pain is signaled by tissue damage.

✖ APPLYING THE CONCEPTS

1. Argue for or against the statement: "Consciousness by its nature is incomprehensible; the brain will never understand the mind."

2. In Parkinson's disease, which afflicts several million Americans, the cells that produce a neurotransmitter called dopamine degenerate in a small part of the brain important in the control of movement. Some physicians have reported improvement in patients after injecting cells taken from the same general brain region of an aborted fetus into appropriate parts of the Parkinson's victim's brain. Discuss this type of surgery from as many viewpoints as possible: ethical, financial, practical, etc. On the basis of your responses, is fetal transplant surgery the answer to Parkinson's disease?

3. If human spinal cord neuron axons were unmyelinated, would the spinal cord be larger or smaller? Would you move faster or slower? Explain.

4. Explain the statement "Your sensory perceptions are purely a creation of your brain." Discuss the implications for communicating with other human beings, with other animals, and with intelligent life from another universe.

5. Bones in the middle ear help amplify sounds. Muscles attached to these bones help dampen sounds by a reflex arc through two different cranial nerves leading to and from the brain. Where do you think the receptor neurons are located to sense loud sounds that need to be dampened? Why is your hearing more likely to be damaged from a sudden loud sound than from a sound that gradually increases in volume?

6. Discuss the biological, ethical, and social implications of regulating noise as a form of pollution in the following places: rock concerts, aerobics classes, a neighborhood that was built near an existing airport, factories.

7. Corneal transplants can help restore vision and greatly improve the quality of life for the recipient. What are some properties of the cornea that make it an excellent candidate for transplantation? Suggest some ways in which society could improve the availability of corneal and other tissue for transplantation.

✖ FOR MORE INFORMATION

Angier, N. "Storming the Wall." *Discover*, May 1990. Describes the blood-brain barrier and new methods of penetrating it with drugs.

Changeau, J-P. "Chemical Signaling in the Brain." *Scientific American*, November 1993. Acetylcholine receptors reveal information on communication between brain neurons.

Fenton, M. B., and Fullard, J. H. "Moth Hearing and the Feeding Strategies of Bats." *American Scientist*, May/June 1981. Explores evolutionary interactions between bats and moths.

Freedman, D. H. "In the Realm of the Chemical." *Discover*, June 1993.

Gutin, J. C. "Good Vibrations." *Discover*, June 1993.

Holloway, M. "Rx for Addiction." *Scientific American*, March 1991. By studying drug addiction, neurobiologists are learning more about the brain, and new ways to counteract addictions.

Koretz, J. F., and Handelman, G. H. "How the Human Eye Focuses." *Scientific American*, July 1988. Describes focusing in the human eye with an emphasis on the loss of focusing ability that occurs with age.

McKean, K. "Pain." *Discover*, October 1986. Describes the discovery of bradykinin and its role in pain perception.

Montgomery, G. "Color Perception: Seeing with the Brain." *Discover*, December 1988. Describes, in layperson's terms, how the higher centers of the brain interpret signals from the receptors to arrive at a perception of color. Beautifully illustrated.

Montgomery, G. "The Mind in Motion." *Discover*, March 1989. Describes the use of the PET scan to view the workings of the living brain.

Nathans, J. "The Genes for Color Vision." *Scientific American*, February 1989. Good discussion of the physiology and genetics of color blindness.

Parker, D. E. "The Vestibular Apparatus." *Scientific American*, November 1980. Perception of gravity and motion requires not only the utricle, saccule, and semicircular canals but also input from other senses.

Radetsky, P. "The Brainiest Cells Alive." *Discover*, April 1991. What can you learn from culturing brain cells?

Scientific American special issue on Mind and Brain, September 1992. Articles describe the development of the brain, the biochemical basis of learning, major neurological disorders, and aging effects on the brain.

Souchurek, H. "Medicine's New Vision." *National Geographic*, January 1987. Striking photography and interesting case histories highlight this readable overview of noninvasive techniques for viewing inside the body.

Zimmer, C. "The Electric Mole." *Discover*, August 1993. Describes the discovery of the electrolocation ability of the star-nosed mole.

NET WATCH

On-line resources for this chapter are on the World Wide Web at:

http://www.prenhall.com/~audesirk (click on the <u>table of contents</u> link and then select Chapter 26).

27 Human Reproduction and Development

A 4-month-old human fetus showing the head, hands, and umbilical cord, which connects the fetus to the placenta, where it obtains oxygen and nutrients from the mother's bloodstream.

Although the word *reproduction* may bring to mind images of romantic courtship and cute cuddly babies, from an evolutionary perspective, romance and the universal appeal of babies are frills that have evolved only because they further the real evolutionary goal: to pass on one's genes to another generation. Reproduction in humans and other mammals, covered in the first half of this chapter, is the process by which sperm and eggs are produced, brought together, join and develop into a new individual carrying the genes of both parents. Successful reproduction is the key to the continued existence of the species.

During reproduction, out of millions of contestants, a single sperm fuses with the egg. The two haploid nuclei—one from the egg, contributed by the mother, and one from the sperm, contributed by the father—fuse to create a diploid cell, the beginning of a new generation. How does this single cell give rise to the trillions of cells of the adult body? In the second half of this chapter, we describe the development of the human embryo from fertilization to birth.

Human Reproduction

Male and female sexes of mammals are separate. Mammals reproduce sexually, uniting **sperm** and **eggs** inside the female body, a process called **internal fertilization**. Many mammals reproduce only during certain seasons of the year and consequently produce sperm and eggs only at that time. Human reproduction is similar to that of other mammals, but it is not restricted by season. Men produce sperm more or less continuously, and women ovulate about once a month. Our discussion will focus on human reproduction.

The Male Reproductive Tract Includes the Testes and Accessory Structures

The male reproductive tract consists of the paired **gonads** (organs that produce sex cells), the **testes** (singular, **testis**) where sperm are produced, and accessory structures that store the sperm, produce secretions that activate and nourish them, and finally conduct them to the inside of the female reproductive tract (Fig. 27-1 and Table 27-1).

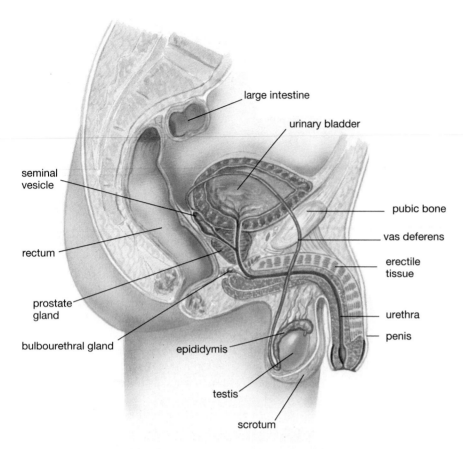

seminal vesicle

large intestine

urinary bladder

pubic bone

vas deferens

erectile tissue

urethra

penis

rectum

prostate gland

bulbourethral gland

epididymis

testis

scrotum

Figure 27-1   **The human male reproductive tract**

The male gonads, the testes, hang beneath the abdominal cavity in the scrotum. Sperm pass from the seminiferous tubules of a testis to the epididymis, and from there through the vas deferens and urethra to the tip of the penis. Along the way, fluids are added from three sets of glands: the seminal vesicles, the bulbourethral glands, and the prostate gland.

Structure	Type of Organ	Function
Testis	Gonad	Produces sperm and testosterone
Epididymis and vas deferens	Ducts	Store sperm; conduct sperm from testes to penis
Urethra	Duct	Conducts semen from vas deferens and urine from urinary bladder to the tip of the penis
Penis	External "appendage"	Deposits sperm in female reproductive tract
Seminal vesicles	Glands	Secrete fluids that contain fructose (energy source) and prostaglandins (possibly cause "upward" contractions of vagina, uterus, and oviducts, assisting sperm transport to oviducts); fluids may wash sperm out of ducts of male reproductive tract into vagina
Prostate	Gland	Secretes fluids that are basic (neutralize acidity of vagina) and contain factors that enhance sperm motility
Bulbourethral glands	Glands	Secrete mucus (may lubricate penis in vagina)

TABLE 27-1   *Structures and Functions of the Human Male Reproductive Tract*

The Testes Are the Site of Sperm Production

The testes produce both sperm and male sex hormones. The testes are located in the **scrotum,** a pouch that hangs outside the main body cavity. This location keeps the testes about 4° C cooler than the core of the body and provides the optimal temperature for sperm development. (Tight jeans may look sexy, but they push the scrotum up against the body, raising the temperature of the testes. Some researchers

think that wearing tight pants may reduce sperm counts and decrease fertility. This is not, however, a reliable means of birth control!) Coiled, hollow **seminiferous tubules,** in which sperm are produced, nearly fill each testis (Fig. 27-2a). In the spaces between the tubules are the **interstitial cells,** which synthesize the male hormone testosterone.

Just inside the wall of each seminiferous tubule lie the diploid cells, or **spermatogonia** (singular, **spermatogonium**), from which all the sperm eventually will arise, and the much

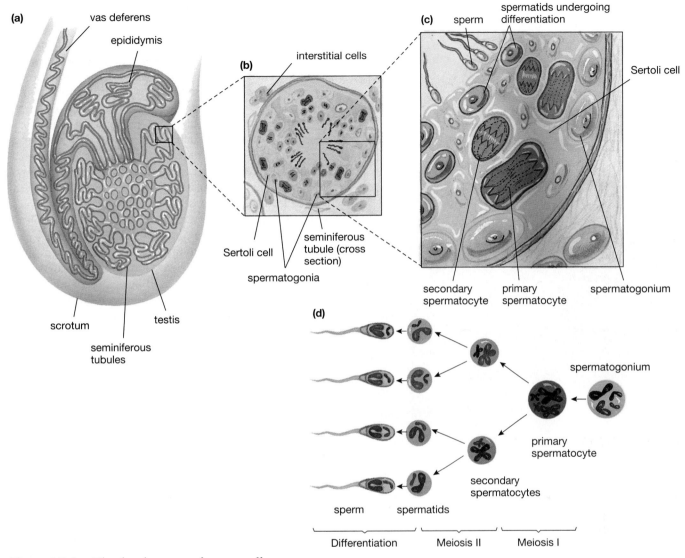

Figure 27-2   **The development of sperm cells**

(a) A lengthwise section of the testis, showing the location of the seminiferous tubules, epididymis, and vas deferens. **(b)** A cross section of a seminiferous tubule. The walls of the seminiferous tubules are lined with Sertoli cells and spermatogonia protruding into the central cavity (lumen) of the tubule. **(c)** As spermatogonia undergo meiosis, the daughter cells move inward, embedded in infoldings of the Sertoli cells. There they differentiate into sperm, drawing upon the Sertoli cells for nourishment. Mature sperm are finally freed into the lumen of the tubules for transport to the penis. Testosterone is produced by the interstitial cells found in the spaces between tubules. **(d)** Spermatogenesis is accomplished by meiotic divisions that produce haploid sperm (compare with the actual locations shown in part [c]). Although four chromosomes are shown for clarity, in humans the diploid number is 46 and the haploid number is 23.

larger **Sertoli cells** (Fig. 27-2b, c). Each time a spermatogonium divides, it can take one of two developmental paths. First, it may undergo mitosis. Mitosis ensures that the male has a steady supply of new spermatogonia throughout his life. Second, it may undergo **spermatogenesis**—that is, the production of sperm by the process of meiosis followed by differentiation (Fig. 27-2d). Spermatogenesis begins with growth and differentiation of spermatogonia into **primary spermatocytes.** These are large diploid cells that will develop into sperm. The primary spermatocytes then under-

go meiosis (see Chapter 12). At the end of meiosis I, each primary spermatocyte gives rise to two haploid **secondary spermatocytes.** Each secondary spermatocyte divides again during meiosis II to produce two **spermatids,** for a total of four spermatids per primary spermatocyte. Spermatids undergo radical rearrangements of their cellular components as they differentiate into sperm.

Sertoli cells regulate the process of spermatogenesis and nourish the developing sperm. The spermatogonia, spermatocytes, and spermatids are embedded in infoldings of

Figure 27-3   A human sperm cell

A mature sperm is a stripped-down cell equipped with only the essentials: a haploid nucleus containing the male genetic contribution to the future zygote, a lysosome (called the acrosome) containing enzymes that will digest the barriers surrounding the egg, mitochondria for energy production, and a tail (actually a long flagellum) for locomotion. The photo is a false-color electron micrograph of a human sperm.

the Sertoli cells. As spermatogenesis proceeds, they migrate up from the outermost edge of the seminiferous tubule to the central cavity of the tubule (Fig. 27-2c). The mature sperm, several hundred million a day, are finally liberated into the lumen.

A human sperm (Fig. 27-3) is unlike any other cell of the body. Most of the cytoplasm disappears, leaving a haploid nucleus nearly filling the head. Atop the nucleus lies a specialized lysosome, called the **acrosome.** The acrosome contains enzymes that will be needed to dissolve protective layers around the egg, enabling the sperm to enter and fertilize it. Behind the head is the midpiece, which is packed with mitochondria that provide the energy needed to move the tail that protrudes out the back. Whiplike movements of the tail, which is really a long flagellum, propel the sperm along inside the female reproductive tract.

At puberty, the time of rapid growth and sexual maturation, the hypothalamus releases **gonadotropin-releasing hormone (GnRH)**, which stimulates the anterior pitu-

itary to produce **luteinizing hormone (LH)** and **follicle-stimulating hormone (FSH)**. Spermatogenesis begins as a result of the interplay of LH and FSH and **testosterone** secreted by the testes (Fig. 27-4). Luteinizing hormone stimulates the interstitial cells to produce testosterone. The combination of testosterone and FSH stimulates the Sertoli cells and spermatogonia, causing spermatogenesis.

Testosterone also stimulates the development of secondary sexual characteristics (such as the growth of facial hair in males and breast development in females), maintains sexual drive, and is required for successful intercourse (a term we will use for human copulation). Sperm, however, are not involved in these functions. Therefore, if one could suppress FSH release (blocking spermatogenesis) but not LH release (thereby allowing continued testosterone production), a man would be infertile but not impotent. Efforts are under way to develop a drug to do just that, as a form of male birth control.

Accessory Structures Produce Semen and Conduct the Sperm outside the Body

The seminiferous tubules merge to form a single convoluted tube, the **epididymis** (see Fig. 27-2). The epididymis becomes the **vas deferens,** which leaves the scrotum and enters the abdominal cavity. Most of the hundreds of millions of sperm produced each day are stored in the vas def-

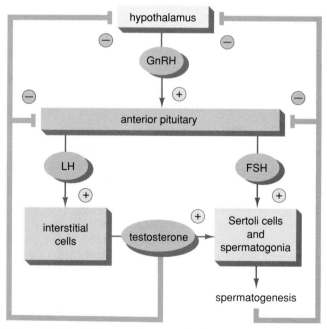

Figure 27-4   Hormonal control of spermatogenesis

Gonadotropin-releasing hormone (GnRH) from the hypothalamus stimulates the anterior pituitary to release LH and FSH. LH stimulates the interstitial cells to produce testosterone. Testosterone and FSH stimulate the Sertoli cells and the spermatogonia, causing spermatogenesis. Testosterone and chemicals produced during spermatogenesis inhibit further release of FSH and LH, forming a negative feedback loop that keeps the rate of spermatogenesis and the concentration of testosterone in the blood nearly constant. + stimulates; – inhibits.

erens and epididymis. The vas deferens joins the **urethra,** leading from the bladder to the tip of the penis. This final common path is shared, at different times, of course, by sperm (during ejaculation) and urine (during urination).

The fluid ejaculated from the penis, called **semen,** consists of sperm mixed with secretions from three glands that empty into the vas deferens or urethra: the **seminal vesicles,** the **prostate gland,** and the **bulbourethral glands.** The secretions activate swimming by the sperm, provide energy for swimming, and neutralize the acidic fluids of the vagina that normally inhibit bacterial growth (see Table 27-1).

The Female Reproductive Tract Includes Ovaries and Accessory Structures

The female reproductive tract is almost entirely contained within the abdominal cavity (Fig. 27-5 and Table 27-2). It consists of paired gonads, the **ovaries** (Fig. 27-6a), and accessory structures that accept sperm, conduct the sperm to the egg, and nourish the developing embryo.

The Ovaries Are the Site of Egg Production

The human female produces precursor egg cells, or **oogonia** (singular, **oogonium**), while still a fetus in her mother's womb, beginning the process of **oogenesis,** the formation of egg cells. The oogonia divide by mitosis and then grow into **primary oocytes.** No oogonia remain after the third month of fetal development, and no new ones form during the rest of her life. Still during the fetal stage, all the primary oocytes begin meiosis but then halt during prophase of meiosis I. At birth, the ovaries contain about 2 million primary oocytes; many die each day, until at puberty (usually 11 to 14 years of age) only about 400,000 remain. Because only a few oocytes resume meiosis during each month of a woman's reproductive span (from

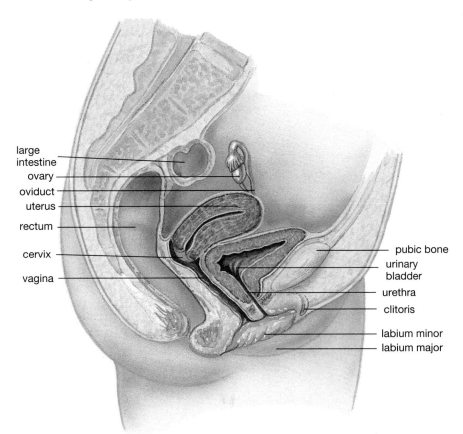

large intestine
ovary
oviduct
uterus
rectum
cervix
vagina

pubic bone
urinary bladder
urethra
clitoris
labium minor
labium major

Figure 27-5   The human female reproductive tract

Eggs are produced in the ovaries and swept by cilia into the oviduct. A male deposits sperm in the vagina, from which they move up through the cervix and uterus into the oviduct. Sperm and egg usually meet in the oviduct, where fertilization occurs. The fertilized egg attaches to the lining of the uterus, where the embryo develops.

TABLE 27-2 ▪▪ *Structures and Functions of the Human Female Reproductive Tract*		
Structure	*Type of Organ*	*Function*
Ovary	Gonad	Produces eggs, estrogen, and progesterone
Fimbria	Mouth of duct	Cilia sweep egg into oviduct
Oviduct	Duct	Conducts egg to uterus; site of fertilization
Uterus	Muscular chamber	Site of development of fetus
Cervix	Connective tissue ring	Closes off lower end of uterus, supports fetus, and prevents foreign material from entering uterus
Vagina	Large "duct"	Receptacle for semen; birth canal

(a)

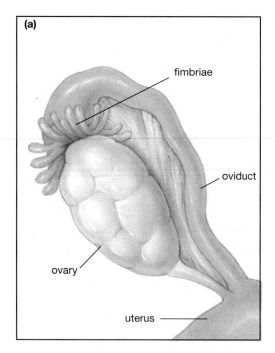

fimbriae

oviduct

ovary

uterus

Figure 27-6 **Oogenesis in the human female**

(a) External view of the ovary and oviduct. (b) The development of follicles in an ovary, portrayed in a time sequence going clockwise from the lower right. ① A primary oocyte begins development within a follicle. ②, ③ The follicle grows, providing both hormones and nourishment for the enlarging oocyte. ④ At ovulation, the secondary oocyte, or egg, bursts through the ovary wall, surrounded by some follicle cells (now called the corona radiata). The remaining follicle cells develop into the corpus luteum, which secretes hormones. If fertilization does not occur, the corpus luteum breaks down after a few days. (c) The cellular stages of oogenesis. The oogonium enlarges to form the primary oocyte. At meiosis I, almost all the cytoplasm is included in one daughter cell, the secondary oocyte. The other daughter cell is a small polar body that contains chromosomes but little cytoplasm. At meiosis II, almost all the cytoplasm of the secondary oocyte is included in the egg, and a second small polar body discards the remaining "extra" chromosomes. The first polar body sometimes also undergoes the second meiotic division. In humans, meiosis II does not occur unless the egg is fertilized.

(b)

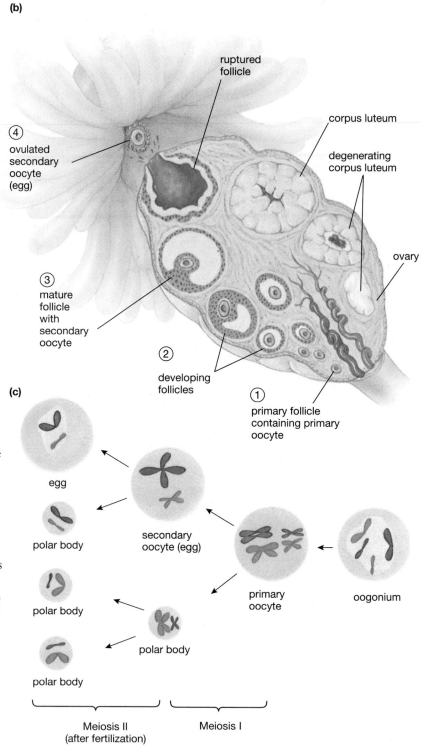

ruptured follicle

corpus luteum

degenerating corpus luteum

④ ovulated secondary oocyte (egg)

ovary

③ mature follicle with secondary oocyte

② developing follicles

① primary follicle containing primary oocyte

(c)

egg

polar body

secondary oocyte (egg)

polar body

primary oocyte

oogonium

polar body

polar body

Meiosis II (after fertilization)   Meiosis I

puberty to menopause at age 45 to 55), there is no shortage of oocytes.

Surrounding each oocyte is a layer of much smaller cells that both nourish the developing oocyte and secrete female sex hormones. Together, the oocyte and these accessory cells make up a **follicle** (Fig. 27-6b). Ap-

proximately once a month during a woman's reproductive years, she undergoes a menstrual cycle, which is described below. During the menstrual cycle, pituitary hormones stimulate development of a dozen or more follicles, although usually only one completely matures. At this time, the primary oocyte completes the first mei-

(a)

(b)

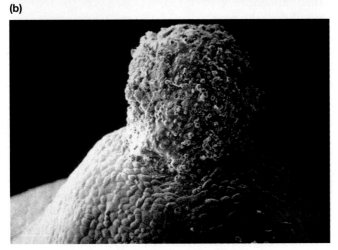

Figure 27-7   **A follicle erupts from the ovary**

The mature follicle grows so large and is filled with so much fluid that it moves to the surface of the ovary **(a)** and literally bursts through the ovary wall like a miniature volcano **(b)**. It then releases the secondary oocyte into the oviduct.

otic division (which was halted during development) to become a single **secondary oocyte** and a **polar body,** which is little more than a discarded set of chromosomes (Fig. 27-6c). Meanwhile, the small accessory cells of the follicle multiply and secrete **estrogen.** As it matures, the follicle grows, eventually erupting through the surface of the ovary and releasing the secondary oocyte (Fig. 27-7). The second meiotic division occurs, not in the ovary, but in the oviduct (the tube leading out of the ovary), and then only if the secondary oocyte is fertilized. For convenience, we will refer to the ovulated secondary oocyte as the "egg."

Some of the follicle cells accompany the egg, but most remain behind in the ovary. These cells enlarge and become glandular, forming the **corpus luteum,** which secretes both estrogen and a second hormone, **progesterone.** If fertilization does not occur, the corpus luteum breaks down a few days later.

Accessory Structures Include the Oviducts, Uterus, and Vagina

Each ovary is adjacent to, but not continuous with, an **oviduct** (sometimes called the **fallopian tube** in humans; see Fig. 27-6a). The open end of the oviduct is fringed with ciliated "fingers" called **fimbriae** that nearly surround the ovary. The cilia create a current that sweeps the egg into the mouth of the oviduct. Fertilization usually occurs in the oviduct, and the **zygote,** as the fertilized egg is now called, is swept down the oviduct by beating cilia and released into the pear-shaped **uterus,** or womb. Here it will develop for the next 9 months. The wall of the uterus has two layers that correspond to its dual functions of nourishment and childbirth. The inner lining, or **endometrium,** is richly supplied with blood vessels. (This lining will form the mother's contribution to the **placenta,** the structure that transfers oxygen, carbon dioxide, nutrients, and wastes between fetus and mother, covered later in this chapter.) The outer muscular wall of the uterus, the **myometrium,** contracts strongly during delivery, expelling the infant out into the world.

Developing follicles secrete estrogen, which stimulates the uterine lining to grow an extensive network of blood vessels and nutrient-producing glands. After **ovulation,** the release of the egg cell from the ovary, estrogen and progesterone released by the corpus luteum promote continued growth of the endometrium. Thus, if an egg is fertilized, it encounters a rich environment for growth. If the egg is not fertilized, however, the corpus luteum disintegrates, estrogen and progesterone levels fall, and the overgrown endometrium disintegrates as well. The uterus contracts, squeezing out the excess endometrial tissue (and sometimes causing menstrual cramps in the process). The resulting flow of tissue and blood is called **menstruation** (from the Latin *mensis,* meaning "month").

The outer end of the uterus is nearly closed off by a ring of connective tissue, the **cervix.** The cervix holds the developing baby in the uterus, expanding only at the onset of labor to permit passage of the child. Beyond the cervix lies the **vagina,** which opens to the outside of the body. The vagina serves both as the receptacle for the penis during intercourse and as the birth canal.

The Menstrual Cycle Is Controlled by Complex Hormonal Interactions

The human male produces sperm continuously, thereby increasing the number of potential offspring that a man can father. In contrast, it is fruitless for a woman to ovulate unless her reproductive tract is properly prepared for pregnancy. The human female reproductive system goes through a complex **menstrual cycle,** in which hormonal interactions among the hypothalamus, pituitary gland, and ovary coordinate ovulation with the preparation of the uterus to receive and nourish the fertilized egg.

You may recall from Chapter 25 that hormone release by the anterior pituitary gland is controlled by neurose-

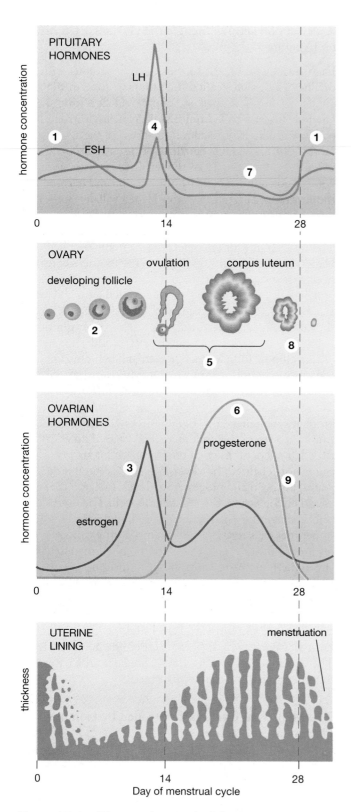

Figure 27-8   Hormonal control of the menstrual cycle

The menstrual cycle is generated by interactions among the hormones of the hypothalamus, anterior pituitary gland, and ovaries. The hormonal changes in turn drive cyclic changes in the uterine lining. The numbers on the graphs refer to the hormonal interactions discussed in the text. They zigzag back and forth among pituitary hormones, ovarian structures, and ovarian hormones, as each of these exerts effects on the others.

cretory cells in the hypothalamus. Some of these neurosecretory cells produce gonadotropin-releasing hormone (GnRH), which stimulates endocrine cells in the anterior pituitary to release FSH and LH. A key to understanding the menstrual cycle is that these neurosecretory cells spontaneously release GnRH all the time, unless actively prevented from doing so by other hormones, notably progesterone. We will begin our discussion of the menstrual cycle with the spontaneous release of GnRH.

Gonadotropin-releasing hormone stimulates the anterior pituitary to release FSH and LH (step 1 in Fig. 27-8). FSH and LH circulate in the bloodstream and initiate the development of several follicles within the ovaries. The follicle cells surrounding the developing oocyte are stimulated by FSH and LH to secrete estrogen. Under the combined influences of FSH, LH, and estrogen, the follicles grow during the next two weeks (step 2). Simultaneously, the primary oocyte within each follicle enlarges, storing both food and regulatory substances (mostly proteins and messenger RNA) that will be needed by the fertilized egg during early development. For reasons that are not completely understood, only one, or rarely two, follicles complete development each month.

As the maturing follicle enlarges, it secretes ever greater amounts of estrogen (step 3). This estrogen has three effects. First, it promotes the continued development of the follicle itself and the primary oocyte contained within it. Second, it stimulates growth of the endometrium of the uterus. Third, high levels of estrogen stimulate both the hypothalamus and pituitary, resulting in a surge of LH and FSH at about the twelfth day of the cycle (step 4).

The function of the peak in FSH concentration is still a subject of investigation, but the surge of LH has three important consequences: (1) It triggers the resumption of meiosis I in the oocyte, resulting in the formation of the secondary oocyte and the first polar body; (2) it causes the final explosive growth of the follicle, culminating in ovulation (step 5); and (3) it transforms the remnants of the follicle that remain in the ovary into the corpus luteum.

The corpus luteum secretes both estrogen and progesterone (step 6). The combination of these hormones inhibits the hypothalamus and pituitary, shutting down the release of FSH and LH (step 7), thereby preventing the development of any more follicles. Simultaneously, estrogen and progesterone stimulate further growth of the endometrium, which eventually becomes about 5 millimeters thick. (Note that the effects of progesterone and estrogen depend on the target organ. Progesterone *stimulates* the endometrium but *inhibits* hormone release from the hypothalamus and pituitary.)

In menstrual cycles in which pregnancy does not occur, the corpus luteum disintegrates about 1 week after ovulation. The corpus luteum survives only while it is stimulated by LH (or by a similar hormone released by the developing embryo, described below). However, because progesterone secreted by the corpus luteum shuts off the LH secretion that sustained it, the corpus luteum essentially "self-destructs" around the twenty-first day of the cycle (step 8). With the corpus luteum gone, estrogen and progesterone levels plummet (step 9). Deprived of stimulation by estrogen and progesterone, the endometrium of the uterus also dies, and its blood and tissue are shed, forming the menstrual flow beginning about the twenty-seventh or twenty-eighth day of the cycle. Simultaneously, the reduced progesterone level no longer inhibits the hypothalamus and pituitary, and the spontaneous release of GnRH from the hypothalamus resumes. Release of GnRH in turn stimulates release of FSH and LH (step 1), which initiates development of a new set of follicles, starting the cycle over again.

During pregnancy, the embryo itself prevents these changes from occurring. Shortly after the ball of cells formed by the dividing fertilized egg embed themselves in the endometrium, they start secreting an LH-like hormone called **chorionic gonadotropin (CG)**. This hormone travels in the bloodstream to the ovary, where it prevents breakdown of the corpus luteum. The corpus luteum continues to secrete estrogen and progesterone, and the uterine lining continues to grow, nourishing the embryo. So much CG is released by the embryo that it is excreted by the mother in her urine; in fact, most pregnancy tests use the presence of CG in a woman's urine to determine pregnancy.

Although negative feedback regulates the levels of most hormones, the hormones of the menstrual cycle are regulated by both positive and negative feedback. During the first half of the cycle, FSH and LH stimulate estrogen production by the follicles. High levels of estrogen then *stimulate* the midcycle surge of FSH and LH release (positive feedback). During the second half of the cycle, estrogen and progesterone together *inhibit* the release of FSH and LH (negative feedback). The early positive feedback causes hormone concentrations to reach high levels, and the later negative feedback shuts the system down again unless pregnancy intervenes.

Copulation Allows Internal Fertilization

Humans use internal fertilization to deposit the sperm in the moist environment of the female reproductive tract. To do this, the penis is inserted into the vagina, where sperm are released during ejaculation. The sperm swim upward in the female reproductive tract, from the vagina through the opening of the cervix into the uterus, and on up into the oviducts. If the female has ovulated within the past day or so, the sperm will meet an egg in one of the oviducts. Only one sperm can succeed in fertilizing it, starting the development of a new human being.

During Copulation, Sperm Are Deposited inside the Female Vagina

The male role in **copulation** begins with erection of the penis. Before erection, the penis is relaxed (flaccid), because the arterioles supplying it are constricted, allowing little blood flow (Fig. 27-9a). Under the dual influences of

Figure 27-9   Changes in blood flow within the penis cause erection

(a) Normally, smooth muscles encircling the arterioles leading into the penis are contracted, limiting blood flow. **(b)** During sexual excitement, these muscles relax, and blood flows into spaces within the penis. The swelling penis squeezes off the veins leaving the penis, thereby increasing the pressure produced by fluids within the penis and causing it to become elongated and firm.

psychological and physical stimulation, the arterioles dilate and blood flows into spaces in the tissue within the penis. As these tissues swell, they squeeze off the veins that drain the penis (Fig. 27-9b). Pressure builds up, causing an erection. After the penis is inserted into the vagina, movements further stimulate touch receptors on the penis, triggering ejaculation. Ejaculation occurs when muscles encircling the epididymis, vas deferens, and urethra contract, forcing semen out of the penis and into the vagina. On average, 3 or 4 milliliters of semen, containing 300 to 400 million sperm, is ejaculated. Male orgasm causes both ejaculation and a feeling of intense pleasure and release.

Similar changes occur in the female. Sexual excitement causes increased blood flow to the vagina and external parts of the reproductive tract, including the labia and clitoris (see Fig. 27-5). The clitoris, which is derived from the same embryological tissue as the penis, becomes erect. Stimulation by the penis of the male often, but not always, results in female orgasm, a series of rhythmic contractions of the vagina and uterus accompanied by sensations of pleasure and release. Female orgasm is not necessary for fertilization.

The intimate contact involved in copulation creates a situation where disease organisms can readily be transmitted. Since the "sexual revolution," which began in the 1960s, many people have had multiple sexual partners, and the incidence of sexually transmitted diseases has greatly increased (see "Health Watch: Sexually Transmitted Diseases").

During Fertilization, the Sperm and Egg Nuclei Unite

Neither sperm nor egg lives very long. An egg may remain viable for a day, and sperm, under ideal conditions, may live for two. Therefore, fertilization can succeed only if copulation occurs within a couple of days before or after ovulation. You will recall that the egg leaves the ovary surrounded by follicle cells (Fig. 27-10a). These cells, now called the **corona radiata,** form a barrier between the sperm and the egg. A second barrier, the jellylike **zona pellucida** ("clear area"), lies between the corona radiata and the egg. Recent research suggests that the human egg releases a chemical attractant that lures the sperm toward it.

In the oviduct, hundreds of sperm reach the egg and encircle the corona radiata, each sperm releasing enzymes from its acrosome (Fig. 27-10b). These enzymes weaken both the corona radiata and the zona pellucida, allowing the sperm to wriggle through to the egg. If there aren't enough sperm, not enough enzymes are released, and none of the sperm will reach the egg. This may be the selective pressure for the ejaculation of so many sperm. Perhaps 1 in 100,000 reach the oviduct, and 1 in 20 of those find the egg, so only a few hundred of the 300 million sperm that were ejaculated join the attack on the barriers surrounding the egg.

When the first sperm finally contacts the surface of the egg, the cell membranes of egg and sperm fuse, and the sperm head is drawn into the cytoplasm of the egg. As the sperm enters, it triggers two vital changes in the egg. First, vesicles near the surface of the egg release chemicals into the zona pellucida, reinforcing it and preventing further sperm from entering the egg. Second, the egg undergoes its second meiotic division, producing a haploid gamete at last. Fertilization occurs as the haploid nuclei of sperm and egg fuse, forming a diploid nucleus that contains all the genes of a new human being.

Human males who have fewer than 20 million sperm per milliliter of semen (about one-fifth the normal amount) usually cannot fertilize a woman during intercourse because too few sperm reach the egg. If the sperm

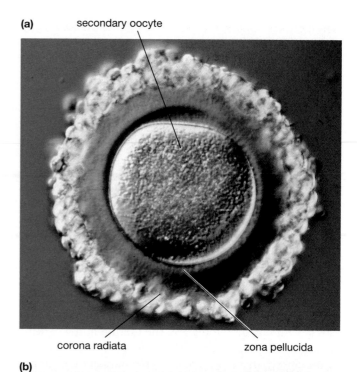

(a)   secondary oocyte

corona radiata   zona pellucida

(b)

Figure 27-10   The secondary oocyte and fertilization
(a) A human secondary oocyte shortly after ovulation. Sperm must digest their way through the small follicular cells of the corona radiata and the clear zona pellucida to reach the oocyte itself. **(b)** Sperm surround the oocyte, attacking its defensive barriers.

are otherwise normal, such men can father children by artificial insemination, in which a large quantity of their semen is injected directly into the oviduct. In other cases, a blocked oviduct may prevent sperm from reaching the egg. Today, some couples seek high-technology help in the form of *in vitro* fertilization (see "Scientific Inquiry: *In Vitro* Fertilization").

S C I E N T I F I C   I N Q U I R Y

In Vitro Fertilization

Louise Brown, the first "test-tube baby," was born in England in 1978. Since that well-publicized event, more than 200 centers for *in vitro* fertilization (IVF) have been established in the United States; these have produced over 26,000 healthy babies. The demand for IVF is fueled by an epidemic of infertility. Twenty years ago one out of every six American couples was unable to conceive after trying for 1 year or more; that rate has now tripled. One reason for the increase in infertility is that modern couples often delay childbearing, and fertility declines with age. A second reason is a higher incidence of sexually transmitted diseases such as chlamydia and gonorrhea that can scar and block the oviducts or sperm ducts. Both blocked ducts and low sperm counts can be overcome with IVF, because oocytes are removed directly from the ovaries, and their meeting with sperm is guaranteed within the confines of a small glass dish (*in vitro* means "in glass" in Latin). The popularity of IVF, which costs around $6000 per attempt, is a testimony to the strong biological drive to have children. With an average success rate of 14%, a typical conception via IVF costs over $40,000.

Although the technique is simple in concept, the procedure is complex and delicate. First, the woman is given daily injections of drugs or hormones or both to stimulate multiple ovulation. Using blood tests and ultrasound imaging of the ovaries, doctors determine when the time is ripe for ovulation. Then the woman is injected with human chorionic gonadotropin, which begins the process of expelling the oocytes from the follicles. Just before the oocytes are ejected, surgeons insert a thin fiber-optic viewing device through a small abdominal incision to locate the mature follicles. Next, they insert a long, hollow needle into each ripe follicle and suck out the follicular cells and fluid, which are examined under a microscope. With luck, an oocyte is present (Fig. E27-1). Usually, at least four oocytes are harvested and incubated in a glass dish to which freshly collected sperm are then added. In 48 hours, about two-thirds of the oocytes will have been fertilized and have reached the eight-cell stage. A few of these early embryos (usually two to four) are sucked into a tube and expelled very gently into the uterus. Transplanting multiple embryos increases the success rate for implantation, but at the same time it increases the probability for multiple births, which carry considerably higher risks than single births.

Recently, IVF has become a weapon in the fight to save endangered species. The National Zoo in Washington, D.C., has been working since 1984 to adapt IVF technology to help endangered species reproduce. They have developed a mobile IVF laboratory that can travel to zoos where endangered species are housed. A tremendous advantage of IVF is that it will allow sperm from a male of an endangered species to be transported between continents, if necessary, to fertilize an appropriate female. This method eliminates the danger and trauma of transporting the animals themselves. IVF also overcomes the very real probability that, once together, the animals will refuse to mate. In April 1990, the first "test-tube tiger" was born (Fig. E27-2). Only 200 individuals of this rare Siberian subspecies remain in the wild. Use of IVF increases the chances that we may save this and other precious forms of life on Earth.

Figure E27-1   Examining the harvested egg
The microscopic egg cell is found in the follicular fluid, and its image is projected onto a screen for examination.

Figure E27-2   The world's first test-tube tiger
Born in 1990, this test-tube tiger is the first successful use of IVF in the fight to save endangered species.

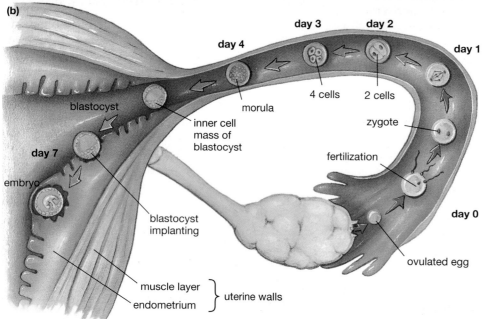

Figure 27-11   The journey of the egg

(a) The egg, surrounded by the corona radiata, travels down the oviduct toward the uterus. It emits chemicals that attract sperm, increasing its chances of being fertilized. (b) The egg is fertilized in the oviduct and slowly travels down to the uterus. Along the way, the zygote divides a few times, until a hollow blastocyst is formed. The inner cell mass will form the embryo, while the surrounding cells will adhere to the uterine endometrium, burrow in, and begin forming the placenta.

During Pregnancy, the Developing Embryo Grows within the Uterus

The zygote begins to divide while being carried by cilia down the oviduct to the uterus, a process that takes about 4 days (Fig. 27-11). By about 1 week after fertilization, the zygote has developed into a hollow ball of cells, the **blastocyst** (Fig. 27-11b). A thickened region of the blasto-cyst, called the **inner cell mass,** will become the embryo itself, while the sticky outer ball will adhere to the uterus and burrow into the endometrium, a process called **implantation.** Blood from ruptured uterine vessels plus glycogen secreted by glands in the endometrium nourish the growing embryo.

Obtaining nutrients directly from the nearby cells of the endometrium suffices only for the first week or two of embryonic growth. During this time, the placenta, composed of interlocking tissues of the embryo and the endometrium, begins to form. Through the placenta, the embryo will receive nutrients and oxygen and dispose of wastes into the maternal circulation. The details of embryonic development are presented later in this chapter.

On Limiting Fertility

Successful reproduction is essential if any species, including our own, is to endure. During most of human evolution, child mortality was high. Therefore, natural selection favored people who tended to produce many children. With a few tragic exceptions, people today enjoy low infant mortality and a life span triple that of ancient times. Although people today no longer need to have many children to ensure that a few survive to adulthood, we still have reproductive drives appropriate to prehistoric times. As a result, every 4 days a million new people are added to our increasingly overcrowded planet, and controlling human births has become a critical environmental issue. On the individual level, limiting reproduction allows people to plan their families and generally provide the best opportunities for themselves and their children.

Historically, limiting fertility has not been easy. In the past, some cultures have tried such inventive, if bizarre, techniques as swallowing froth from the mouth of a camel or placing crocodile dung in the vagina. Even 50 years ago there were no reliable methods of birth control. Since the 1970s, however, several effective techniques have been developed for **contraception**, or preventing pregnancy. Their mechanisms are described below, and their reliability and some possible side effects are summarized in Table 27-3. Of course, the choice of a contraceptive should be made only in consultation with a physician who can provide more complete information.

Permanent Contraception Can Be Achieved through Sterilization

The most effective, and, in the long run, most effortless method of contraception is **sterilization,** in which the pathways through which sperm or egg must travel are interrupted. In men, the vas deferens leading from each testis may be severed in an operation called a **vasectomy** (Fig. 27-12a). Sperm are still produced, but they cannot reach the penis during ejaculation. The surgery is performed under a local anesthetic, and vasectomy has no known physical side effects on health or sexual performance. The slightly more complex operation of **tubal ligation** renders a woman infertile by cutting her oviducts (Fig. 27-12b). About 37% of women of childbearing age in the United States have chosen this form of birth control. Ovulation still occurs, but sperm cannot travel to the egg, nor can the egg reach the uterus. Sterilization is generally permanent.

Sometimes, however, in a delicate and expensive operation, a surgeon can reconnect the vas deferens or oviducts.

There Are Three Major Approaches to Temporary Contraception

Temporary contraception techniques fall into three general categories: preventing ovulation, preventing sperm and egg from meeting when ovulation does occur, and preventing implantation of a fertilized egg in the uterus.

Synthetic Hormones Can Prevent Ovulation

As you learned earlier in this chapter, during a normal menstrual cycle, ovulation is triggered by a midcycle surge of LH. An obvious way to prevent ovulation is to suppress LH release by providing a continuing supply of estrogen and progesterone. Estrogen and progesterone (usually in synthetic form) are the components of **birth control pills.** "The Pill" is an extremely effective form of birth control, but it must be taken daily, usually for 21 days each menstrual period.

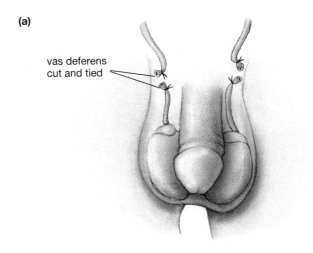

(a)

vas deferens cut and tied

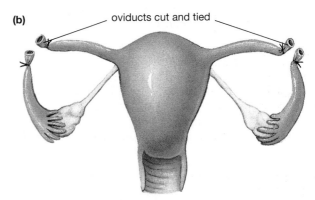

(b)

oviducts cut and tied

***Figure 27-12*   Sterilization**

Sterilization is the most effective form of birth control, but it is not dependably reversible. **(a)** In the male, the vas deferens is reached through a small slit in the scrotum. It is cut and each end is tied off. **(b)** In the female, the oviduct is reached through an abdominal incision. It is cut, and each end is tied off.

TABLE 27-3 :: *Birth Control Techniques*[a]

	What Is It?	*How Does It Work?*
NO! (negative response)	Saying NO! is deciding not to be sexually active until you're ready to take on that heavy a commitment.	Say NO! as often as necessary to get your point across. Wear a button that says "NO!" Guaranteed stopper: "If you really loved me, you wouldn't ask."
Natural Family Planning	Combination of cervical mucus (Billings) and basal body temperature (BBT) methods to determine fertile period when you can get pregnant. Billings: based on recognizing changes in mucus discharge that occur just BEFORE ovulation; BBT: based on body temperature changes just AFTER ovulation.	You must learn from a professional nurse practitioner or doctor how to predict your fertile period, using changes in cervical mucus and body temperature.
Condom (for men) Condom (for women)	For men: Thin rubber or latex (latex is strongest) disposable sheath (usually lubricated with non-oxynol-9) worn over penis during sex. Never use Vaseline as a lubricant. ("Natural" condoms don't protect against STDs.) For women: Soft, loose-fitting prelubricated polyurethane pouch with two rings: one inserted deep in vagina, the other, when in place, remains just outside the vagina.	For both: Placed correctly, catches sperm so they can't enter vagina. Also shields you from exposure to AIDS or other STDs. Protects your partner too.
Foam (spermicides)	Sperm-killing foam inserted into vagina before having sex. Choose foam that contains non-oxynol-9 for best protection against pregnancy and STDs.	Inserted deep into vagina with plastic applicator, forms chemical barrier over uterine entrance. Sperm die when they hit foam.
The Sponge[c]	Soft disposable polyurethane sponge filled with one gram of sperm-killing non-oxynol-9 that is activated by moistening sponge with water. Can be put in vagina hours before having sex, is effective immediately, and offers contraceptive protection for up to 24 hours. Some clinics say it's number 1 choice for young lovers.	Sponge moistened with water is placed far back in vagina up to 18 hours before having sex. Sponge fits around cervix, keeps sperm from entering uterus. Traps sperm, killing them on contact.
Diaphragm/ Cervical Cap	The diaphragm and cervical cap are birth control devices that fit over the cervix and prevent pregnancy by preventing sperm from entering uterus. Both must be individually fitted by a medical professional. Both are made of rubber-like materials. Cap is half the size of diaphragm, and can be left in place twice as long.	Place one teaspoonful of spermicide containing non-oxynol-9 inside the dome and another around the edge of diaphragm or cap. Spermicide seals barrier and kills sperm.
The Pill	A pill made of a combination of synthetic hormones almost like those produced by the ovaries. Take at the same time every day whether or not you have sex. It's the day-by-day action of the pills that protects you from pregnancy.	Prevents ovary from releasing an egg. With no egg present for a sperm to fertilize, a woman cannot become pregnant. Caution: Use a back-up method plus the pill in first couple months of use.
IUD (intrauterine device)	Small plastic device treated with copper or hormones. Nylon thread attached for easy checking. Fitted by medical professional.	Prevents egg from being implanted in uterine wall. Copper IUD is replaced every 6 years; hormonal IUD, every year.
Norplant	New! Long-lasting! Doctor inserts 6 tiny flexible progestogen-filled capsules under skin of upper arm to provide protection up to 5 years. Fertility returns when Norplant is removed.	Progestogen is released in steady low doses, preventing pregnancy by blocking ovulations.
Depo-Provera (injectable)	An injectable form of contraception (given as a shot) that protects you from pregnancy for a full 3 months. Contains a chemical similar to the hormone (progesterone) that ovaries produce during the second half of the menstrual cycle.	Depo-Provera works by preventing egg cells from ripening. If an egg is not released from your ovaries during a menstrual cycle, it can't be fertilized by sperm, so you can't get pregnant. But you have to be very sure to get a shot every 3 months—no more, no less.

[a]Modified with permission from *"No!" and Other Methods of Birth Control*, Private Line, P.O. Box 31, Kenilworth, Illinois 60043.

[b]All of these birth control methods (except Norplant and to a lesser extent, Depo-Provera) take effort on your part. Nothing works if you leave it in the dresser drawer!

[c]The company manufacturing the sponge stopped distributing it in 1995 for business reasons. The sponge may or may not be available when you read this. Ask at your drugstore.

Does It Have Any Side Effects?	Are There Any Dangers?	How Reliable Is It?[b]
Only positive ones. If you're not loaded down with a relationship you can't handle, you'll have time to make a life for *yourself* and you won't have to worry about getting pregnant or getting a sexually transmitted disease (STD) such as chlamydia, genital warts, herpes, or AIDS.	Only that people who put pressure on you might make fun of you or drop you cold. But if they do, ask yourself what they really want from you—love and friendship, or sex?	100% effective.
No—but it doesn't offer any protection against sexually transmitted diseases (STDs).	You could become pregnant unless you avoid sex during your fertile period. Some women have trouble identifying mucus changes.	10 to 15 women out of 100 become pregnant in a given year using the Billings or BTT methods. The calendar method alone is not a reliable form of birth control.
For both: Next to not having sex, condoms or pouch used along with spermicidal foam containing non-oxynol-9 are the best available way to protect you against AIDS and other STDs—or pregnancy. (One or both of you could be allergic to spermicidal foam or lubricant.)	NO. AIDS is no joke. Condoms could save your life. Use one every time you have sex even if your partner is on the pill, has an IUD, is wearing a diaphragm—or has had a vasectomy. For women: A pouch, properly used, provides even better protection than a condom against STDs.	2 to 10 women out of 100 become pregnant in a given year when the man uses a condom correctly every time. The time period for clinical testing of the pouch has been too short for accurate statistics. The manufacturer reports that 2 to 3 women out of 100 become pregnant in a given year when the woman uses a condom every time as directed.
Occasionally the foam you're using may cause a mild vaginal irritation, cause bladder or yeast infection, or irritate your partner. Usually a change to a different brand solves the problem, but be sure the foam you use contains non-oxynol-9. If necessary, change to a birth control method that doesn't require a spermicide.	No, except your partner should use a condom when you use foam to give you the best protection, next to not having sex, against STDs and pregnancy. Just remember that protection is not 100% guaranteed.	4 to 29 women out of 100 become pregnant in a given year when using foam alone.
You don't have to worry about hormonal side effects. In clinical trials small number of women tested discontinued using because of itching, irritation, rash or allergic reactions.	Manufacturer's instructions stress use for ONLY 24-hour period. Should NOT be used during menstrual period. Slight danger of toxic shock syndrome (TSS). Sponge does not stop flow of vaginal secretions.	8 to 10 women out of 100 become pregnant in a given year when using the sponge as directed.
None unless the spermicide you're using causes bladder or yeast infection. If it does, change to another brand.	Slight danger of toxic shock syndrome (TSS). Should be used for birth control only, not to control vaginal secretions. Be sure you have inserted it correctly. Don't forget the spermicidal foam or jelly.	5 to 10 women out of 100 in a given year become pregnant when using a diaphragm or cap consistently and correctly. (If you gain or lose weight, your diaphragm may no longer fit correctly and should be checked by a medical professional.)
Positive: regular periods, less anemia, less cramping, less benign breast disease. May inhibit some forms of cancer. Negative: (normally disappear within 3 months) may include nausea, spotting, missed periods, headaches, mood changes, dark skin areas. Major but rare: blood clots, high blood pressure, gallbladder disease, heart attacks, liver tumors.	Some women should not take the pill because of other health problems. A health professional will choose the best pill for you. Read carefully all material in pill package. Follow doctor's advice. Smoking increases the chance of blood clots or stroke, even in young people on the pill.	1 woman out of 300 becomes pregnant in a given year when using the pill correctly. Always check with your doctor about possible interactions between the pill and any other medications you are taking.
Possible cramps and heavy menstrual flow caused by the body's effort to push out IUD. Expect heavy menstrual periods the first few months.	Risk of pelvic inflammatory disease (PID) or tubal pregnancy. Tell doctor if you have fever or stomach pain.	1 to 5 women out of 100 in a given year become pregnant while using an IUD.
Irregular menstrual cycle for first 6 months with longer periods, spotting between periods or skipped periods. Possible headaches, weight gain.	The method is too new for accurate statistics. Counseling to understand method is extremely important. If pregnancy does occur, have Norplant removed at once.	1 woman out of 300 becomes pregnant in a given year when using Norplant for the first 2 years. After 2 years, less protection for overweight women.
Irregular menstrual bleeding at first, then little or no menstruation, weight gain, but not much. Possibly headache, nervousness, abdominal pain, dizziness, fatigue. Check with doctor if you experience any unusual symptoms.	Important that injection be given during first 5 days of menstrual period to avoid possibility of receiving the shot during a pregnancy—which could result in ectopic pregnancy. If you should get pregnant, see your doctor immediately.	1 woman out of 300 in a given year becomes pregnant while receiving Depo-Provera injections as prescribed.

Two new long-term contraceptives, Norplant and Depo-Provera, have been approved for use in the United States since 1990. Both contain synthetic hormones resembling progesterone that prevent ovulation. Norplant consists of six slim, 1.3-inch long silicone rubber rods inserted under the skin of the upper arm. The rods provide gradual, steady diffusion of hormone into the bloodstream for 5 years. In extensive tests, Norplant has proved slightly more effective than the birth control pill. Women using Norplant usually become fertile within months after the capsule is removed. Depo-Provera is injected once every 3 months.

Abstinence and Barrier Methods Prevent Sperm from Reaching the Egg

Abstaining from sexual intercourse is the only completely effective means (other than sterilization) of preventing sperm and egg from meeting. Abstinence also has the advantage of complete safety from possible contraceptive side effects and from sexually transmitted diseases.

There are several effective barrier methods that prevent the encounter of sperm and egg. One is the **diaphragm,** a rubber cap that fits snugly over the cervix, preventing sperm from entering the uterus. In conjunction with a **spermicide**, diaphragms are very effective and have no serious known side effects. Alternatively, a **condom** may be worn over the penis, preventing sperm from being deposited in the vagina. A female condom is now available that completely lines the vagina. Diaphragms and condoms must be applied shortly before intercourse, often at a time when the participants would rather be thinking about something else. Furthermore, if the diaphragm or condom happens to have even a small hole, sperm may still enter the uterus and fertilize the egg. (It seems a cruel irony that a man ejaculating 50 million sperm will often be infertile, but if just a few drops of semen escape past a diaphragm or condom, pregnancy might result.)

Other, less effective, procedures include using spermicides alone, **withdrawal** (removal of the penis from the vagina just before ejaculation), and **douching** (washing sperm out of the vagina before, it is hoped, they have had a chance to enter the uterus). Spermicides have some contraceptive effect, but withdrawal and douching are essentially useless. A final method of preventing fertilization is the **rhythm method:** abstinence from intercourse during the ovulatory period of the menstrual cycle. In practice, rhythm usually has a high failure rate, because of lack of discipline on the part of the users or inaccuracies in determining the menstrual cycle, which varies somewhat from month to month. Because a slight rise in body temperature and changes in the discharge of mucus from the cervix can be used to predict ovulation, the rhythm method can be made more effective if these indicators are monitored and sexual activity is regulated accordingly.

Intrauterine Devices Prevent Implantation

Even if an egg is fertilized, pregnancy will not occur unless the blastocyst implants in the uterus. The **intrauterine device (IUD)** is a small copper or plastic loop, squiggle, or shield that is inserted into the uterus and remains there until removed by a physician. Highly effective (if it stays in place), the IUD seems to work by irritating the uterine lining so that it cannot receive the embryo. A second method of preventing implantation is the "morning after" pill, which contains a massive dose of estrogen. For some women, preventing implantation has the major drawback that it is, in effect, an extremely early abortion.

Abortion Removes the Embryo from the Uterus

When contraception fails, pregnancy may be terminated by **abortion.** Abortion commonly involves dilating the cervix and removing the embryo and placenta by suction. The compound RU-486, taken as a pill, is now routinely used in France to terminate pregnancy up to about a month after conception and final approval is expected for use in the United States by late 1996. This substance binds to progesterone receptors and blocks the actions of progesterone, which are essential to the maintenance of pregnancy.

Because abortions are usually surgical procedures, they are potentially more dangerous to a woman's health than the contraceptive techniques described above. Although science can describe fetal development during pregnancy, it cannot provide judgments about when a fetus becomes a "person," or about the relative merits of fetal versus maternal rights. Therefore, abortion remains controversial.

Additional Contraceptive Methods Are under Development

Further advances in contraception are under development. For example, a vaginal ring that continuously releases synthetic progesterone has been tested in several countries but is not yet on the market. A pill that prevents ovulation by blocking receptors for gonadotropin-releasing hormone is in the testing stage. There are also once-a-month contraceptive pills under development, most of which prevent the uterus from becoming fully prepared for implantation. Another possibility is a contraceptive vaccine that would remain effective for one to several years. One such vaccine induces antibodies against human chorionic gonadotropin, which is essential for the implantation of the embryo in the uterus. Another induces the formation of antibodies to a protein called SP-10 that is unique to sperm.

You may have noticed that most of the contraceptive techniques are directed at the woman, not the man. It is much easier to interfere with ovulation than with sperm formation. A woman ovulates only once a month, and ovulation itself does not influence a woman's sexual drives. In contrast, testosterone is essential both for sperm for-

HEALTH WATCH

Sexually Transmitted Diseases

Sexually transmitted diseases (STDs), caused by viruses, bacteria, protists, or arthropods that infect the sexual organs and reproductive tract, are a serious and growing health problem worldwide. The World Health Organization estimates that there are 250 million new cases of STD each year. As the name implies, these diseases are transmitted either exclusively or primarily through sexual contact. Here we discuss some of the more common of these diseases.

Bacterial Infections

Gonorrhea, an infection of the genital and urinary tract, is one of the most common of *all* infectious diseases in the United States, estimated to infect at least 2 million people each year. The causative bacterium, which cannot survive outside the body, is transmitted almost exclusively by intimate contact. It penetrates the membranes lining the urethra, anus, cervix, uterus, oviducts, and throat. In males, inflammation of the urethra results in a discharge of pus from the penis and painful urination. About 10% of infected males and 50% of infected females have symptoms that are mild or absent. These people often do not seek treatment and become carriers who can readily spread the disease. Gonorrhea can lead to infertility by blocking the oviducts with scar tissue. Treatment by penicillin was formerly highly successful, but penicillin-resistant strains now require use of other antibiotics. Infants born to infected mothers may acquire the bacterium during delivery. The bacterium attacks the eyes of newborns and was once a major cause of blindness. Today, most newborns are immediately given antibiotic eyedrops to kill the bacterium.

Syphilis is a far more dangerous, though less prevalent, disease than gonorrhea. It is caused by a spiral-shaped bacterium that enters the mucous membranes of the genitals, lips, anus, or breasts. Like gonorrhea, it is readily killed by exposure to air and is spread only by intimate contact. Syphilis begins with a sore at the site of infection. Syphilis may be cured with antibiotics, but if untreated, syphilis bacteria spread through the body, multiplying and damaging many organs, including the skin, kidneys, heart, and brain, sometimes with fatal results. About 4 out of every 1000 newborns in the United States have been infected with syphilis before birth. The skin, teeth, bones, liver, and central nervous system of such infants may be damaged.

Chlamydia causes inflammation of the urethra in males and the urethra and cervix in females, but in many cases there are no obvious symptoms, so the infection goes un-

treated and is spread. Like the gonorrhea bacterium, *Chlamydia* can infect and sometimes block the oviducts, resulting in sterility. Chlamydial infection can cause eye inflammations in infants born to infected mothers.

Viral Infections

Acquired immunodeficiency syndrome, or AIDS, is caused by the HIV virus. Because the virus does not survive exposure to air, it is spread primarily by sexual activity and by contaminated blood and needles. The HIV virus attacks the immune system, leaving the victim vulnerable to a variety of infections, which almost invariably prove fatal. Children born to mothers with AIDS sometimes become infected before or during birth. There is no cure, although certain drugs, such as AZT, can prolong life. AIDS is discussed in detail in Chapter 24.

Genital herpes reached epidemic proportions during the 1970s and continues to spread to more than a million new victims yearly. It causes painful blisters on the genitals and surrounding skin and is transmitted primarily when blisters are present. The herpes virus never leaves the body but resides in certain nerve cells, emerging unpredictably, possibly in response to stress. The first outbreak is the most serious; subsequent outbreaks produce fewer blisters and may be quite infrequent. The drug acyclovir, which inhibits viral DNA replication, may reduce the severity of outbreaks. Pregnant women with an active case of genital herpes may transmit the virus to the developing child, causing severe mental or physical disability or stillbirth. Herpes may also be transmitted from mother to infant if the infant contacts blisters as it is born.

Protists and Arthropod Infections

Trichomoniasis is caused by *Trichomonas*, a flagellated protist that colonizes the mucous membranes lining the urinary tract and genitals of both males and females. The symptoms are a discharge caused by inflammation in response to the parasite. The protist is spread by intercourse but can also be acquired through contaminated clothing and toilet articles. Lengthy untreated infections can result in sterility.

Crab lice, also called pubic lice, are microscopic arachnids that live and lay their eggs in pubic hair. Their mouthparts are adapted for penetrating the skin and sucking blood and body fluids, a process that causes severe itching. "Crabs" are not only irritating; they may also spread infectious diseases. They can be controlled through careful hygiene and chemical treatments.

mation and sexual performance; early "male pills" caused not only infertility but also impotence. Nevertheless, there is a major research effort under way to develop male contraceptives equivalent to The Pill. A promising contraceptive is a daily dose of testosterone and a modified form of gonadotropin-releasing hormone, which together seem to block sperm production without affecting sexual performance. Clinical trials are now under way.

Human Development

Human development is controlled by the same mechanisms that control the development of other animals. In fact, our development strongly reflects our evolutionary heritage. Mammals (as well as birds) evolved from reptiles. The embryos of reptiles and birds (which develop within shelled eggs) and of mammals (which develop within the uterus) produce four membranes called **extraembryonic membranes**. In reptiles and birds, the **chorion** lines the shell and exchanges oxygen and carbon dioxide through the shell. In mammals the chorion is the outermost membrane of the placenta. The **amnion** encloses the embryo in a watery environment, the **allantois** surrounds wastes, and the **yolk sac** contains the stored food. Although mammalian eggs contain almost no yolk, much of the reptilian genetic program for development still persists, including the four extraembryonic membranes. Table 27-4 compares the structures and functions of these extraembryonic membranes in reptiles and mammals.

TABLE 27-4 ⁛ *Vertebrate Embryonic Membranes*

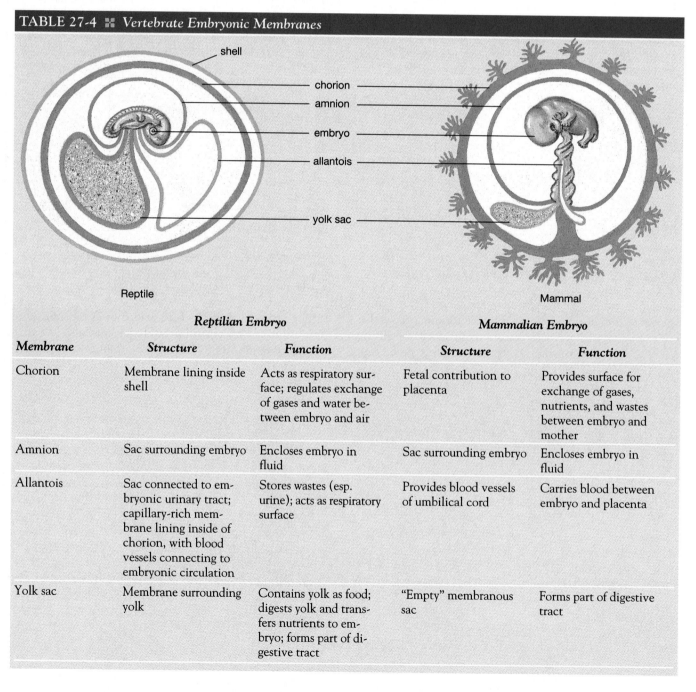

| Membrane | Reptilian Embryo | | Mammalian Embryo | |
	Structure	*Function*	*Structure*	*Function*
Chorion	Membrane lining inside shell	Acts as respiratory surface; regulates exchange of gases and water between embryo and air	Fetal contribution to placenta	Provides surface for exchange of gases, nutrients, and wastes between embryo and mother
Amnion	Sac surrounding embryo	Encloses embryo in fluid	Sac surrounding embryo	Encloses embryo in fluid
Allantois	Sac connected to embryonic urinary tract; capillary-rich membrane lining inside of chorion, with blood vessels connecting to embryonic circulation	Stores wastes (esp. urine); acts as respiratory surface	Provides blood vessels of umbilical cord	Carries blood between embryo and placenta
Yolk sac	Membrane surrounding yolk	Contains yolk as food; digests yolk and transfers nutrients to embryo; forms part of digestive tract	"Empty" membranous sac	Forms part of digestive tract

Age (weeks)

1	2	3	4	5	6
← zygote to formation of embryonic disc →		← embryo			
		2–3 mm	4 mm	8 mm	13 mm
Zygote cleaves; blastocyst implants	Two-layered embryo forms; amniotic cavity and yolk sac open	Gastrulation occurs; notochord and beginning of neural tube form	Neural tube closes; heart beats; arm buds, tail, and gill grooves form	Optic cup and lens pits form; leg buds form; brain enlarges	Finger rays and external ear form; pigment appears in retina; tail and gill grooves disappearing

Age (weeks)

7	8	9	10	11	12
embryo →		← fetus			
18 mm	30 mm	50 mm	61 mm	73 mm	87 mm
Toe rays form; bones begin to harden; back straightens; eyelids form	Upper limbs bend at elbows; genitalia begin to differentiate; fingers are distinct	Toes separate; eyelids develop; major parts of brain are present	Chin grows; nostrils separate; face appears human; genitals appear male or female	Well-defined neck appears; genitalia are complete; sucking reflex appears	

Age (months)

4	5	6	7	8	9
fetus					→
140 mm	190 mm	230 mm	270 mm	300 mm	350 mm
Blood cells form; all major organs form; head and body hair appear; movements are felt by mother		Fetus may be viable if born; eyelids open; lungs and lung circulation develop; may suck thumb; fat deposited under skin		Fetus usually viable if born; fat deposits increase; body hair is lost; head hair is well developed; most senses are well developed; fetus turns head down in uterus	

Figure 27-13   **Human embryonic development**

A calendar of human embryonic development, from blastula to birth.

Figure 27-13 summarizes the stages of human embryonic development. You may want to refer to this figure as we go along and, perhaps years from now, if you have a child of your own.

During the First Two Months, Rapid Differentiation and Growth Occur

For the first 2 months of life, the developing human is called an **embryo**. A human egg is normally fertilized in a woman's oviduct and undergoes a few cleavage divisions on its way to the uterus. Just before being implanted in the uterine wall, the embryo, now called a blastocyst, consists of a thin-walled, hollow ball with a thicker inner cell mass on one side (Fig. 27-14). The thin outer wall becomes the chorion and will form the embryonic contribution to the placenta. The inner cell mass develops into the embryo and the three other extraembryonic membranes.

After implantation, the inner cell mass grows and splits, forming two fluid-filled sacs that are separated by a double layer of cells called the **embryonic disc** (Figs. 27-14 and 27-15). One sac, bounded by the amnion, forms the amniotic cavity. The amnion eventually grows around the embryo, enclosing it in what one author has called its "private aquarium," providing the watery environment needed by all animal embryos. The yolk sac, corresponding to the yolk sac of reptiles and birds, forms the second cavity, although in humans it contains no yolk because the small amount in the egg is used up before the yolk sac forms. At this stage, the embryonic disc consists of two layers of cells, one adjacent to the yolk sac and the other adjacent to the amniotic cavity.

A process called **gastrulation** begins about the fifteenth day after fertilization (Fig. 27-15). During gastrulation three tissue layers are produced: **ectoderm** (Greek for "outer skin), **mesoderm** ("middle skin), and **endoderm** ("inner skin). Each layer will give rise to specific adult tissues, summarized in Table 27-5. During human gastrulation, the two layers of the embryonic disc split apart slightly, and a slit, the **primitive streak**, appears in the center of the layer adjacent to the amnion (the blue upper layer in Fig. 27-15). Cells of this layer migrate through the primitive streak into the interior of the embryo, forming mesoderm (pink). The upper layer (blue) in Figure 27-15 is now called ectoderm and the lower layer (yellow) is called endoderm. One of the earliest mesoderm structures to develop is the **notochord**, a supporting rod found at some stage in all vertebrates (Fig. 27-15b).

During the third week of development, the embryo and its amniotic sac begin to curl ventrally (toward the yolk sac; Fig. 27-15c). As the embryo grows, the endoderm pinches, forming a tube that will become the digestive tract (Fig. 27-15d). Simultaneously, the notochord induces formation of a groove in the overlying ectoderm, which folds inward and then closes over to become the **neural tube,** the forerunner of the brain and spinal cord. By the end of the fourth week, the amnion completely surrounds

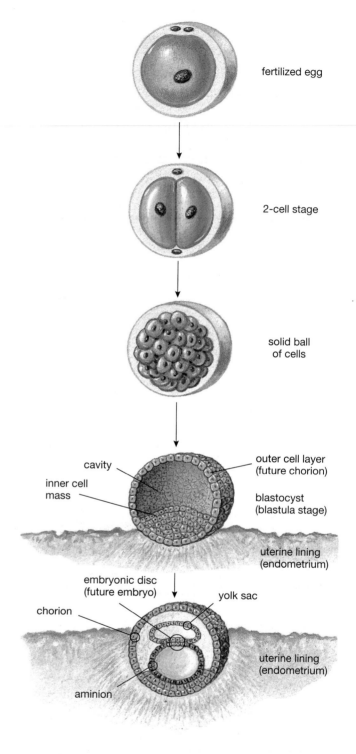

Figure 27-14   Human development during the first and second weeks

As it travels through the oviduct, the fertilized egg undergoes cleavage. In the uterus, it becomes a blastocyst and implants in the uterine lining. At this stage it consists of an outer layer of cells surrounding an inner cell mass. As it burrows into the uterine lining, the outer cell layer forms the chorion, the embryonic contribution to the placenta. The inner cell mass forms the amnion, yolk sac, and the embryonic disc, which will become the embryo.

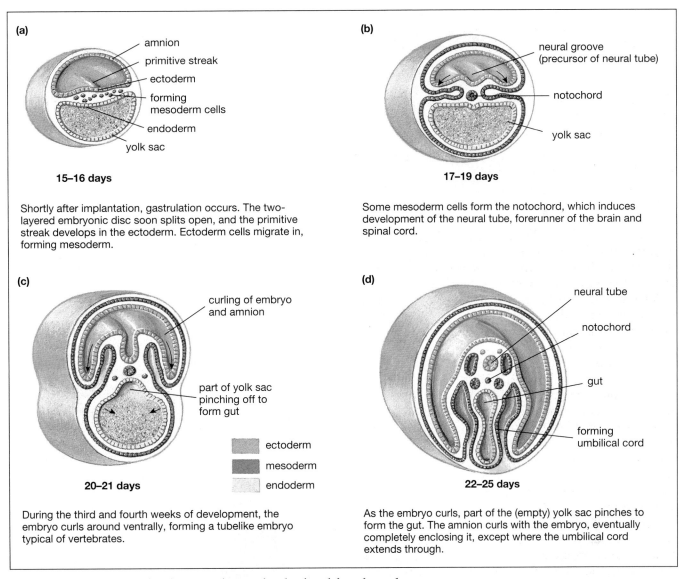

(a)

amnion
primitive streak
ectoderm
forming mesoderm cells
endoderm
yolk sac

15–16 days

Shortly after implantation, gastrulation occurs. The two-layered embryonic disc soon splits open, and the primitive streak develops in the ectoderm. Ectoderm cells migrate in, forming mesoderm.

(b)

neural groove (precursor of neural tube)
notochord
yolk sac

17–19 days

Some mesoderm cells form the notochord, which induces development of the neural tube, forerunner of the brain and spinal cord.

(c)

curling of embryo and amnion
part of yolk sac pinching off to form gut

ectoderm
mesoderm
endoderm

20–21 days

During the third and fourth weeks of development, the embryo curls around ventrally, forming a tubelike embryo typical of vertebrates.

(d)

neural tube
notochord
gut
forming umbilical cord

22–25 days

As the embryo curls, part of the (empty) yolk sac pinches to form the gut. The amnion curls with the embryo, eventually completely enclosing it, except where the umbilical cord extends through.

Figure 27-15 **Human development during the third and fourth weeks**

the embryo, punctured only by the umbilical cord connecting the embryo to the placenta.

By the end of the sixth week, the embryo has developed a notochord, primitive gill grooves, and a prominent tail (Fig. 27-16). These, of course, disappear as development continues. The embryo already has the rudimentary beginnings of the eyes, a beating heart, and separating fingers on its tiny hands. Especially notable at this stage is the rapid growth of the brain, which is nearly as large as the rest of the body. In fact, many of the structures of the adult brain are already recognizable.

As the second month draws to an end, nearly all the major organs have formed, and the embryo begins to look quite human (Fig. 27-17). The gonads appear and develop into testes or ovaries, depending on the presence or absence of the Y chromosome. Sex hormones are secreted, testosterone from the testes or estrogen from the ovaries, and these hormones affect the future development of the embryonic organs, including not only the reproductive or-

gans but also certain regions of the brain. After the second month of development, the embryo is called a **fetus,** denoting that it has taken on a generally human appearance.

TABLE 27-5 *Derivation of Adult Tissues from Embryonic Cell Layers*	
Embryonic Layer	*Adult Tissue*
Ectoderm	Epidermis of skin; lining of mouth and nose; hair; glands of skin (sweat, sebaceous, and mammary glands); nervous system; lens of eye; inner ear
Mesoderm	Dermis of skin; muscle, skeleton; circulatory system; gonads; kidneys; outer layers of digestive and respiratory tracts
Endoderm	Lining of digestive and respiratory tracts; liver; pancreas

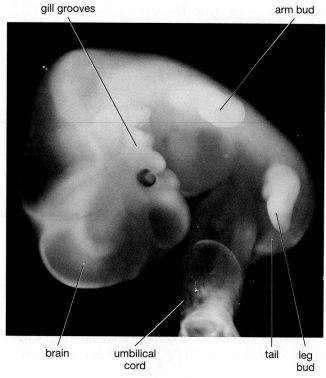

Figure 27-16   **The fifth week**

At the end of the fifth week, the human embryo is about half head. The feet and hands have begun to develop digits. A tail and gill grooves are clearly visible, evidence of our evolutionary relationship to other vertebrates.

These first 2 months of pregnancy are times of extremely rapid differentiation and growth for the embryo and also times of considerable danger. Although the fetus is vulnerable throughout development, rapidly developing organs are the most sensitive to environmental insults, such as drugs or certain medications taken by the mother.

The Placenta Secretes Hormones and Exchanges Materials between Mother and Embryo

During the first few weeks of pregnancy, embryonic cells burrow into the thickened lining of the uterus (the endometrium). The outer cells of the embryo form the chorion, which penetrates the endometrium with fingerlike projections called **chorionic villi.** From this complex interweaving of tissues arises the placenta, a marvelously intricate organ. The placenta has two major functions: It secretes hormones, and it allows the selective exchange of materials between the mother and the fetus.

As the placenta develops during the first 2 months of pregnancy, it begins secreting estrogen and progesterone. Estrogen stimulates growth of the uterus and mammary glands, while progesterone also stimulates the mammary glands and inhibits premature contractions of the uterus.

The placenta also regulates the exchange of materials between the blood of the mother and the blood of the fetus without allowing the two to mix. The chorionic villi contain a dense network of fetal capillaries and are bathed

Figure 27-17   **The eighth week**

At the end of the eighth week, the embryo is clearly human in appearance and is now called a fetus. Most of the major organs of the adult body have begun to develop.

in pools of maternal blood (Fig. 27-18). This arrangement permits diffusion of many small molecules between the fetal and maternal blood. Oxygen diffuses from maternal blood to fetal blood, and carbon dioxide from fetal blood to maternal blood. Nutrients travel from mother to fetus, some aided by active transport. Fetal urea diffuses into the mother's blood, to be filtered out by the mother's kidneys.

While allowing exchange by diffusion, the membranes of the capillaries and chorionic villi act as barriers to the passage of some large proteins and most cells. In spite of this barrier, some disease-causing organisms and many harmful chemicals can penetrate the placental barrier, as described in "Health Watch: The Placenta Provides Only Partial Protection."

Growth and Development Continue during the Last Seven Months

The fetus continues to grow and develop for another 7 months. Although the rest of the body is "catching up" with the head in size, the brain continues to develop rapidly, and the head remains disproportionately large. Nearly every nerve cell ever formed during the entire human life span develops during embryonic life. As the brain and spinal cord grow, they begin to generate noticeable behaviors. As early as the third month of pregnancy, the fetus begins to move about and respond to stimuli. Some instinctive behaviors appear, such as sucking, which will have obvious importance soon after birth. Structures that the fetus will need when it emerges from the womb, such as the lungs, stomach, intestine, and kidneys, enlarge and become functional, though they will not be used until after birth.

Fetuses 7 months or older can usually survive outside the womb, but larger and more mature fetuses have a much greater chance of survival. After about 9 months of development, the head is so large that it can barely fit through the mother's pelvis. The skull is compressed into a slightly conical shape as it passes through the birth canal.

Milk Secretion, or Lactation, Is Stimulated by Pregnancy

As the fetus grows, nourished by nutrients diffusing through the placenta, changes are occurring in the mother's breasts that prepare her to continue nourishing her child after it is born. The breast contains milk glands

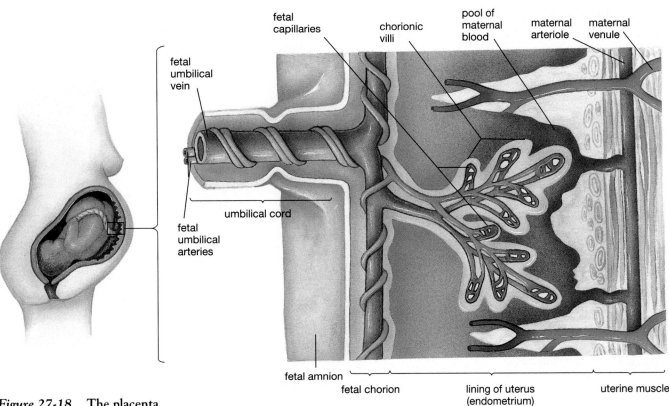

Figure 27-18   **The placenta**

The placenta is formed from both the chorion of the embryo and the endometrium of the mother. Capillaries of the endometrium break down, releasing blood to form pools within the placenta. Meanwhile, the chorion develops projections (the chorionic villi) extending into these pools of maternal blood. Blood vessels from the umbilical cord branch extensively within the villi. The resulting structure separates the maternal and fetal blood supplies, while generating a large surface area for diffusion of oxygen, carbon dioxide, nutrients, and wastes between the fetal capillaries and the maternal blood pools. Umbilical arteries carry deoxygenated blood from the fetus to the placenta, and umbilical veins carry oxygenated blood back to the fetus.

The Placenta Provides Only Partial Protection

When your grandparents were bearing children, doctors assumed that the placenta protected the developing fetus from most of the substances in maternal blood that could harm it. We know now that this is far from true. In fact, most medications and drugs and even some disease organisms readily penetrate the placental barrier and affect the fetus.

Infections May Cross the Placenta

The German measles virus can cross the placenta and attack the fetus, causing potentially severe retardation and other defects. The virus causing genital herpes (during active outbreaks) and the bacterium causing syphilis can cause mental or physical defects in the developing fetus. The virus causing AIDS may also cross the placenta, so infants may be born with this incurable disease.

Drugs Readily Cross the Placenta

A tragic example of drugs crossing the placenta is the tranquilizer thalidomide, commonly prescribed in Europe in the early 1960s (Fig. E27-3). Thalidomide's devastating effects on embryos were discovered only when many babies were born with missing or extremely abnormal limbs. More recently, in the late 1980s, the anti-acne drug Accutane was found to cause gross deformities in babies born to women using it. Although these are extreme examples, *any drug, including aspirin, has the potential to harm the fetus, and any woman who thinks she may be pregnant should seek medical advice about any drugs she takes.* The use of so-called recreational drugs such as heroin and cocaine has devastating effects on the developing fetus. Babies of heroin addicts are often born addicted. In the United States, hundreds of thousands of infants have been born in recent years to users of "crack" cocaine. Research suggests that these children may be impaired, both behaviorally and emotionally.

The Effects of Smoking

Probably the most common toxic substances to which fetuses are exposed are those in cigarette smoke. Because many women who smoke are so addicted that they don't stop smoking during pregnancy, nearly a million human embryos are exposed to the poisons (including nicotine, carbon monoxide, and a host of carcinogens) each year in the United States alone. Women who smoke have a higher incidence of miscarriages than do nonsmokers and they give birth to smaller infants who are more likely to

Figure E27-3   Drugs interfere with development
Children born to mothers taking the tranquilizer thalidomide cope heroically with missing and deformed limbs. Drugs with no obvious ill effects on the mother may have devastating impacts on her developing fetus.

die shortly after birth than infants of nonsmokers. There is evidence that some children born to heavy smokers also suffer behavioral and intellectual impairment.

Fetal Alcohol Syndrome

The effects of alcohol on the developing fetus can be devastating. When a pregnant woman drinks, alcohol in the blood of her unborn child reaches levels as high as in her own blood. Children born to alcoholic mothers often exhibit **fetal alcohol syndrome (FAS).** Such children are retarded and may be hyperactive and irritable. FAS children have small heads and abnormally small, improperly developed brains, (Fig. E27-4), characteristic facial features, and a higher-than-normal incidence of defects of the heart and other organs. Their growth is also inhibited. FAS is the third most common cause of mental retardation in the United States, with one out of every 500 to 750 children being affected. It is likely that far higher numbers of children suffer from *fetal alcohol effect*, a milder form of FAS caused by more moderate alcohol consumption. A woman who takes one or two alcoholic drinks a day during the first 3 months of pregnancy also significantly increases her chances of having a miscarriage. *Researchers have not established any safe level of alcohol consumption during any phase of pregnancy. The U.S. Surgeon General advises pregnant women and those who are likely to become pregnant to avoid all alcohol consumption.*

In summary, a pregnant woman should assume that whatever chemicals she ingests will find their way into the bloodstream of her developing infant. Women who are likely to become pregnant need to consider that crucial stages of development occur before most women even realize that they are pregnant. The mother's choices during this critical 9-month period can strongly influence her child's future well-being.

Figure E27-4   **Alcohol impairs brain development**

The devastating effects of alcohol on the developing brain are illustrated in these photos showing the brain of a child with fetal alcohol syndrome (left) and the brain of a normal child of the same age (right).

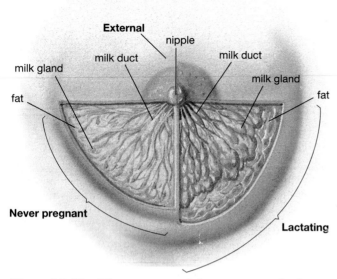

Figure 27-19   **The structure of the mammary glands**
During pregnancy, both fatty tissue and the milk-secreting glands and ducts increase in size.

arranged in a circle around the nipple, each with a duct leading to the nipple. Although the breasts begin enlarging at puberty under the influence of estrogen, their glandular structure does not fully develop until pregnancy occurs (Fig. 27-19). Then, large quantities of estrogen and progesterone (acting together with several other hormones) stimulate the milk glands to grow, branch, and develop the capacity to secrete milk. The actual secretion of milk is promoted by the pituitary hormone prolactin (see Chapter 25). The level of prolactin rises steadily from about the fifth week of pregnancy until birth.

Immediately after birth, estrogen and progesterone levels plummet, and prolactin, which stimulates milk secretion, takes over. Suckling stimulates nerve endings in the nipples. These signal the hypothalamus, which in turn triggers an extra surge of prolactin and oxytocin from the pituitary. Oxytocin causes muscles surrounding the milk glands to contract, ejecting the milk into the ducts leading to the nipples (see Chapter 25).

During the first few days after birth, the milk glands secrete a thin, yellowish fluid called **colostrum.** Colostrum is high in protein and contains antibodies that are absorbed directly through the infant's intestine and help protect the newborn against some diseases. Colostrum is gradually replaced by mature milk, which is higher in fat and milk sugar (lactose) and lower in protein.

Reproduction Culminates in Labor and Delivery

Near the end of the ninth month, give or take a few weeks, the process of birth normally begins (Fig. 27-20). Birth is the result of a complex interplay between uterine stretching caused by the growing fetus and fetal and maternal hormones that finally trigger **labor** (contractions of the uterus that result in the birth).

Unlike skeletal muscles, uterine muscles can contract spontaneously, and these contractions are enhanced by stretching. As the baby grows, it stretches the uterine muscles, which contract occasionally weeks before delivery. Recent research suggests that the final trigger for labor is provided by the fetus. The near-term fetus produces steroid hormones that cause increased estrogen and prostaglandin production by the placenta and uterus. These hormones make the uterus even more likely to contract. When the combination of hormones and stretching activate the uterus beyond some critical point, strong contractions begin, signaling the onset of labor. As the contractions proceed, the baby's head pushes against the cervix, making it expand in diameter (dilate). Stretch receptors in the walls of the cervix send signals to the hypothalamus, triggering oxytocin release. Under the dual stimulation of prostaglandin and oxytocin, the uterus contracts even more strongly. This positive feedback cycle is finally halted when

(a) The baby is oriented head downward, facing the mother's side. The cervix is beginning to thin (efface) and expand in diameter (dilate).

(b) The cervix is completely dilated to 10 centimeters (almost 4 inches) and the baby's head has entered the vagina, or birth canal. The baby has rotated to face the mother's back.

(c) The baby's head is emerging (crowning).

(d) The baby has rotated to the side once again as the shoulders emerge.

Figure 27-20   **Delivery**

the baby emerges. After a brief rest, uterine contractions resume, causing the uterus to shrink remarkably. During these contractions, the placenta is sheared off from the uterus and is expelled through the vagina as the "afterbirth."

Further release of prostaglandins in the umbilical cord causes the muscles surrounding fetal blood vessels in the umbilical cord to contract, shutting off blood flow. (Tying off the cord is standard practice but is not usually necessary; if it were, other mammals would not survive birth!) Although he or she is still intimately dependent on the parents for survival, a new human being has been born.

Usually delivered head first, the infant is in for a rude awakening. Formerly, all the world was soft, fluid-cushioned, and warm. Suddenly, the baby must obtain oxygen and eliminate carbon dioxide by breathing. It must regulate its own body temperature, and it must suckle to obtain food. Considering that they have recently experienced what by any standard is a change for the worse, babies are often remarkably cheerful (Fig. 27-21)!

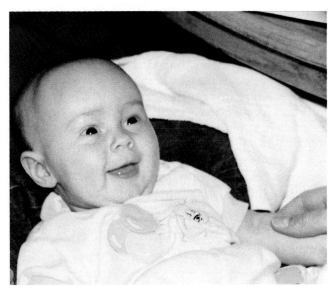

Figure 27-21   On her own

✖ SUMMARY OF KEY CONCEPTS

Mammalian Reproduction

The human male reproductive tract consists of paired testes that produce sperm and testosterone and accessory structures that conduct the sperm to the female's reproductive tract and secrete fluids that activate swimming by the sperm and provide energy. In human males, spermatogenesis and testosterone production are stimulated by FSH and LH, secreted by the anterior pituitary. Spermatogenesis and testosterone production are nearly continuous, beginning at puberty and lasting until death.

The human female reproductive tract consists of paired ovaries that produce eggs as well as the hormones estrogen and progesterone, and accessory structures that conduct sperm to the egg and receive and nourish the embryo during prenatal development. In human females, oogenesis, hormone production, and development of the lining of the uterus vary in a monthly menstrual cycle. The cycle is controlled by hormones from the hypothalamus (gonadotropin-releasing), anterior pituitary (FSH and LH), and ovaries (estrogen and progesterone).

During copulation, the male inserts his penis into the female's vagina and ejaculates semen. The sperm move through the vagina and uterus into the oviduct, where fertilization usually takes place. The unfertilized egg is surrounded by two barriers, the corona radiata and the zona pellucida. Enzymes released from the acrosomes at the tips of sperm digest these layers, permitting sperm to reach the egg. Only one sperm enters the egg and fertilizes it.

The fertilized egg undergoes a few cell divisions in the oviduct and then implants in the uterine lining. Implantation and subsequent release of chorionic gonadotropin by the embryo maintain the integrity of the corpus luteum and the endometrium during early pregnancy, preventing further menstrual cycles.

On Limiting Fertility

Permanent contraception can be achieved by sterilization: severing the vas deferens in males (vasectomy) or the oviducts in females (tubal ligation). Temporary contraception techniques include those that prevent ovulation: birth control pills, Norplant, and Depo-Provera. Barrier methods prevent sperm and egg from meeting. These include the diaphragm, the contraceptive sponge, and the condom, accompanied by spermicide. Spermicide alone is less effective. Withdrawal and douching are poor techniques. The rhythm method involves sexual abstinence around the time of ovulation. Intrauterine devices prevent implantation of the blastocyst. Abortion causes the expulsion of the developing embryo.

Human Development

After undergoing several divisions as it travels along the oviduct, the developing fertilized human egg becomes a blastocyst and implants in the thickened uterine wall. There it develops an amnion, a chorion, and a two-layered embryonic disc. During gastrulation, the layers of the disc give rise to three types of embryonic tissue—ectoderm, mesoderm, and endoderm—each of which will produce specific adult tissues (see Table 27-5).

The human developmental program clearly reflects our reptilian ancestry. Reptilian embryos develop four extraembryonic membranes—the chorion, amnion, allantois, and yolk sac—which function in gas exchange, provision of the watery environment needed for development, waste storage, and yolk storage, respectively. These membranes are retained by the human embryo but are modified to suit development inside the mother's womb (see Table 27-4).

The chorion invades the uterine lining (endometrium), and together these tissues form the placenta. Here the exchange of materials occurs by diffusion between maternal and fetal blood, while the blood supplies remain separate. By the end of the second month, the major organs have formed, the embryo appears human and is called a fetus. In the next 7 months, until birth, the fetus continues to grow, and the lungs, stomach, intestine, and kidneys enlarge and become functional. The stages of human development are summarized in Figure 27-13.

During pregnancy, milk glands in the mother's breasts enlarge under the influence of estrogen, progesterone, and other hormones. After birth, milk secretion is triggered by prolactin and oxytocin, whose release is triggered by suckling.

After about 9 months, uterine contractions are triggered by a complex interplay of uterine stretch and prostaglandin and oxytocin release. As a result, the uterus expels the baby and then the placenta.

▓ K E Y   T E R M S

THINKING THROUGH THE CONCEPTS

Multiple Choice

1. All of the following are barrier contraceptive devices EXCEPT
   a. IUD
   b. contraceptive sponge
   c. caps that fit over the cervix
   d. diaphragm
   e. male condom

2. In humans, spermatogenesis yields _____ sperm for each diploid germ cell, and oogenesis yields _____ secondary oocyte(s) for each germ cell.
   a. one, four
   b. two, one
   c. one, two
   d. four, one
   e. four, two

3. Which structure adds the *final* secretions to semen as it moves out of the human reproductive tract?
   a. epididymis
   b. bulbourethral gland
   c. seminal vesicle
   d. prostate gland
   e. interstitial cells

4. During fetal development, oogonia in human females halt meiosis at
   a. the secondary oocyte stage
   b. the polar body stage
   c. metaphase of meiosis II
   d. telophase of meiosis II
   e. prophase of meiosis I

5. The primary hormone that inhibits GnRH is
   a. FSH
   b. LH
   c. progesterone
   d. estrogen
   e. a hypothalamic releasing factor

6. In humans, the allantois
   a. forms blood vessels of the umbilical cord
   b. is the fetal contribution to the placenta
   c. stores wastes
   d. encloses the embryo in a watery environment
   e. contains stored food

7. The endoderm gives rise to the
   a. lining of the digestive tract
   b. epidermis of skin
   c. skeletal system
   d. muscles
   e. nervous system

8. In human development, ectoderm cells migrate through the primitive streak to form
   a. endoderm
   b. mesoderm
   c. the chorion
   d. the yolk sac
   e. the amnion

Review Questions

1. Compare the structures of the egg and sperm. What structural modifications are found in sperm to facilitate movement, energy use, and digestion?

2. What is the role of the corpus luteum in a menstrual cycle? In early pregnancy? What determines its survival after ovulation?

3. Construct a chart of common sexually transmitted diseases. List the disease's name, the cause (organism), symptoms, and treatment.

4. List the structures, in order, through which a sperm passes on its way from the seminiferous tubules of the testis to the oviduct of the female.

5. Name the three accessory glands of the male reproductive tract. What are the functions of the secretions they produce?

6. Diagram the menstrual cycle and describe the interactions among hormones produced by the pituitary gland and ovaries that produce the cycle.

7. Name two structures derived from each of the three embryonic germ layers—endoderm, ectoderm, and mesoderm.

8. What is gastrulation? Describe gastrulation in humans.

9. Describe the structure and function of four extraembryonic membranes found in reptiles and birds. Are these four present in placental mammals? Explain.

10. Explain how the structure of the placenta prevents mixing of fetal and maternal blood while allowing the exchange of substances between the mother and the fetus.

11. List and explain two very different functions of the placenta.

12. Is the placenta an effective barrier against substances that can harm the fetus? Describe two different types of harmful agents that can cross the placenta and their effects on the fetus.

13. Describe the changes in the breast that prepare a mother to nurse her child. How do hormones influence these changes and stimulate milk production?

14. Describe the events that lead to the expulsion of the baby and placenta from the uterus. Explain why this is an example of positive feedback.

APPLYING THE CONCEPTS

1. Identify and discuss some of the ethical issues involved in *in vitro* fertilization.

2. Discuss the most effective or appropriate method of birth control for each of the following couples: Couple A has intercourse three times a week but does not ever want to have children; Couple B has intercourse only once a month, and may want to have children someday; and couple C has intercourse three times a week and wants to have children someday.

3. Female condoms were recently introduced in the United States. What advantages and disadvantages can you think of for this form of birth control?

4. Pelvic endometriosis is a relatively common disease of women in which bits of the endometrial lining find their way onto abdominal organs and respond in typical ways to

hormones during a menstrual month. When the uterine lining bleeds during menstruation, so do these implants. Common treatments are oral contraceptives, Danazol (a compound that inhibits gonadotropins), and synthetic GnRH analogues that are paradoxically powerful inhibitors of FSH and LH. How does each of these compounds provide relief?

5. Would contraceptive drugs that block cell receptors for FSH be useful in males and/or females? Explain. What side effects would such drugs have?

6. Think of all the choices a couple have to obtain a child, including: *in vitro* fertilization using the couple's eggs and sperm, *in vitro* fertilization using a donor's sperm or egg, and insemination of a surrogate mother with sperm from the couple's husband. Think of some more. Now describe what ethical issues these various options present?

7. On the basis of your knowledge of genetics (Unit II) and evolution (Unit III) explain why the human embryo passes through a developmental stage in which it has gill grooves and a tail.

8. Estimates of the number of U.S. women who self-medicate with over-the-counter (OTC) drugs during the first trimester or throughout pregnancy range from 65% to 95%. Evidently the public doesn't perceive nonprescription substances as drugs. Many subtle disorders, especially of the nervous system, will probably be connected to over-the-counter drug use in coming years. If you were engaged in pharmaceutical research, which OTC drugs would you focus on as possible culprits? What trimester of pregnancy would you study closely? Who would you chose as your study population?

✖ FOR MORE INFORMATION

Beaconsfield, P., Birdwood, G., and Beaconsfield, R. "The Placenta." *Scientific American*, August 1980 (Offprint No. 1478). The placenta is one of the most remarkable of mammalian structures, allowing internal development of offspring within the mother.

Caldwell, M. "How Does a Single Cell Become a Whole Body?" *Discover*, November 1992. Explores the miracles of development in layperson's terms.

Cooke, J. "The Early Embryo and the Formation of Body Pattern." *American Scientist*, January–February 1988. Although the developmental patterns of worms, fruit flies, and humans are superficially very different, they are probably based on common mechanisms.

Crews, D. "Animal Sexuality." *Scientific American*, January 1994. Explores the wide range of mechanisms that control male and female sexual development among different types of animals.

DeRobertis, E. M., Oliver, G., and Wright, C. V. E. "Homeobox Genes and the Vertebrate Body Plan." *Scientific American*, July 1990. Genes can be injected into amphibian eggs, allowing direct observation of the effects of specific gene products on development.

Dewitt, P. E. "Cloning: Where Do We Draw the Line?" *Time*, November 8, 1993. Does cloning of animals suggest that this technique will be applied to humans?

Eberhard, W. G. "Runaway Sexual Selection." *Natural History*, December 1987. Interesting article on the evolution of male genitalia.

Riddle, J. M., and Estes, J. W. "Oral Contraceptives in Ancient and Medieval Times." *American Scientist*, May–June 1992. How did women control their fertility before modern medicine stepped in?

Rusting, R. "Why Do We Age?" *Scientific American*, December 1992. The underlying causes of human aging are being explored using experimental animals as diverse as fruit flies and roundworms.

Ulmann, A., Teutsch, G., and Philibert, D. "RU-486." *Scientific American*, June 1990. Describes the abortion pill and other applications for this drug.

Wassarman, P. W. "Fertilization in Mammals." *Scientific American*, December 1988. Describes the events leading to the penetration of the egg by the sperm, with emphasis on the role and composition of the zona pellucida.

Wright, K. "The Sniff of Legend." *Discover*, April 1994. This article explores human pheromones, sex attractants, and a sixth sense organ in the nose.

NET WATCH

On-line resources for this chapter are on the World Wide Web at:
http://www.prenhall.com/~audesirk (click on the table of contents link and then select Chapter 27).

Unit V
Ecology

The beauty and interdependence of Earth's biosphere are illustrated in Suzanne Duranceau's painting "Paradise."

28 Population Growth and Regulation

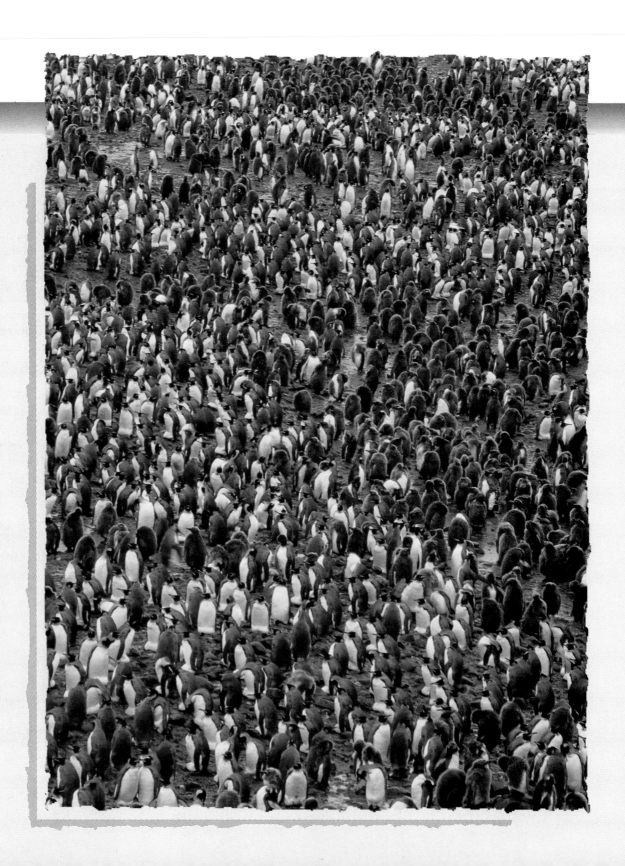

> *"There is no exception to the rule that every organic being naturally increases at so high a rate that, if not destroyed, the earth would soon be covered by the progeny of a single pair."*
>
> CHARLES DARWIN in *On the Origin of Species* (1859)

A densely aggregated breeding population of king penguins on South Georgia Island in the Atlantic Ocean off Antarctica. Brown chicks are guarded and fed by their white parents.

Introduction to Ecology

In eastern Colorado lies a remnant of shortgrass prairie, dominated by buffalo grass and blue grama grass. In spring and summer it is ablaze with wildflowers: paintbrush, vetch, sunflower, and bladderpod (Fig. 28-1). Prairie dogs stand upright by their burrows, alerting the "town" with vigorous high-pitched cries as a single hawk soars lazily overhead. The prairie is an **ecosystem,** a complex, interrelated network of living organisms and their nonliving surroundings. An ecosystem can be as small as a puddle or as large as an ocean. Within our prairie ecosystem, wildflowers, prairie dogs, the grass on which they feed, the hawk that preys on them, and the myriad microscopic organisms that keep the soil fertile, constitute a **community** of interacting populations. A **population** consists of all the members of a particular species, be they hawks, grasshoppers, buffalo grass, or bacteria, living in a given area.

Just to the north, a farm was abandoned nearly 2 years ago. The community here is quite different, with Russian thistle, pigweed, amaranth, and cheatgrass invading the new habitat. A "For Sale" sign in the field foreshadows the high-density housing developments that will soon displace both farmland and prairie. The first human residents will be delighted to see a hawk, a rare and magnificent bird of prey, practically in their backyards. But as the prairie dog town is bulldozed, this large predator will also disappear.

How has the prairie (in contrast to the failed farm) sustained itself for centuries without artificial fertilizer or irrigation? Why are predators such as the hawk rare relative to their prey? What keeps prairie dogs from overpopulating their habitat and starving? What happens when two organisms compete for the same resources? Why is the abandoned farm community different from that of the untouched prairie? What will the abandoned farm look like in 40 years, if left to itself? Will it ever again resemble the prairie community? Why has the human population continued to increase, while other populations remain stable or decline in the face of human expansion? These are a few of the questions of **ecology,** the science that deals with the interrelationships among living things and their environment. The environment consists of an **abiotic** (nonliving) component, including soil, water, and weather, and a **biotic** (living) component, including all forms of life

Figure 28-1   **The prairie ecosystem**

This montage shows representatives of interacting populations, which together constitute the prairie community. Insets show **(a)** a fly on a lichen-covered rock, **(b)** an Indian Paintbrush wildflower, **(c)** a prairie dog, and **(d)** a hawk.

within the ecosystem. Ecology (a term derived from the Greek word *oikos*, meaning "a place to live") is a tremendously diverse, complex, and relatively young scientific discipline.

In preceding chapters we have studied the anatomy, physiology, and behavior of individual organisms. The science of ecology begins at the next level—the population, which we examine in this chapter. In the following three chapters, we will proceed from this starting point to levels of increasing complexity: first to communities and the interactions within them, then to the organization of ecosystems, and finally to an exploration of the diversity of ecosystems that make up the biosphere.

How Populations Grow

Studies of undisturbed ecosystems show that many populations tend to remain relatively stable over time. Yet we are vividly aware from the human example that populations can readily increase. Let's first examine how and why populations grow, then look at the factors that normally control this growth.

Three factors determine whether and how much the size of a population will change: births, deaths, and migration. Organisms join a population through birth or **immigration** (migration in) and leave it through death or **emigration** (migration out). A population remains stable if, on the average, as many individuals leave as join. Population growth occurs when the number of births plus immigrants exceeds the number of deaths plus emigrants. Populations decline when the reverse occurs. A simple equation for the change in population size is:

$$\text{population change} = (\text{births} - \text{deaths}) + (\text{immigrants} - \text{emigrants})$$

In many natural populations, organisms moving in and out contribute relatively little to population change, leaving birth and death rates as the primary factors influencing population growth.

The ultimate size of any population (leaving out migration) is the result of a balance between two major opposing factors. The first is **biotic potential,** or the maximum rate at which the population could increase, assuming ideal conditions allowing a maximum birth rate and minimum death rate. Opposing this potential for growth are limits set by the living and non-living environment. These limits include the availability of food and space, competition with other organisms, and certain interactions among species, such as predation and parasitism. Collectively, these limits are called **environmental resistance.** Environmental resistance can both decrease the birth rate and increase the death rate. The interaction between biotic potential and environmental resistance usually results in a balance between population size and available resources. To understand how populations grow and how their size is regulated, we must examine each of these forces in more detail.

The Biotic Potential of a Population Produces Exponential Growth If Not Restrained

Changes in population size (ignoring migration) are functions of the birth rate, the death rate, and the number of individuals in the original population. Rates of change in population size can be measured as the changes in each of these variables for a given size of population during a given unit of time. For example, birth and death rates may be expressed as the number of births or deaths per 1000 individuals per year.

The **growth rate** (r) of a population is determined by subtracting the death rate (d) from the birth rate (b):

$$\begin{array}{ccc} r & = & b & - & d \\ \text{(growth rate)} & & \text{(birth rate)} & & \text{(death rate)} \end{array}$$

For example, the annual growth rate of a population of 10,000 in which 1500 births and 500 deaths occur each year (the given unit of time) can be calculated as follows:

$$r = \text{birth rate} - \text{death rate}$$

$$r = 1500/10{,}000 - 500/10{,}000 = 0.10, \text{ or } 10\%$$

or

$$r = 0.15 - 0.05 = 0.10, \text{ or } 10\%$$

To determine the number of individuals added to a population in a given time period, multiply the growth rate (r), which is a rate per individual, by the original population size (N):

$$\text{population growth} = rN$$

In this case, population growth (rN) equals 0.10 x 10,000 = 1000. If this growth rate is constant, then the following year, r must be multiplied by an even larger population size ($N + rN = 11{,}000$). This means that in the following year an additional 1100 individuals are added to the population size (N), which in turn increases the number of individuals added in the third year, and so on.

This is **exponential growth.** During exponential growth, the population grows (during a given time period) by a fixed percentage of its size at the beginning of that time period. Thus an increasing number of individuals is added to the population during each succeeding time period, causing population size to grow at an ever-accelerating pace. The curve formed by exponential population growth is often called a J-shaped growth curve, or a **J-curve,** after its shape, clearly illustrated in Figures 28-2 and 28-3.

Such growth occurs whenever births consistently exceed deaths, as they do if, on average, each individual produces more than one surviving offspring during its lifetime. Although the number of offspring produced by an individual each year varies from millions for an oyster to one or fewer for a human, each organism, whether working alone or as part of a sexually reproducing pair, has the potential to replace itself many times over during its lifetime. This capacity, called biotic potential, has evolved because natural selection has favored those individuals that leave the most offspring. Several factors influence the biotic potential of a particular species, including:

1. The age at which the average individual first reproduces
2. The average frequency with which reproduction occurs
3. The average number of offspring produced each time
4. The reproductive life span of the average organism
5. The average death rate under ideal conditions

We will use examples in which these factors differ to illustrate the concept of exponential growth (see Fig. 28-2).

The bacterium *Staphylococcus* is normally a harmless resident in and on the human body, where its population growth is restricted by environmental resistance. But in an ideal culture medium, such as warm custard, where it may occasionally be introduced accidentally, each bacterial cell can divide every 20 minutes, doubling the population three times each hour (the by-products of bacterial metabolism can result in food poisoning under these conditions). The larger the population grows, the more cells there are to divide. The biotic potential of bacteria is so great that, were nutrients unlimited, within 48 hours the offspring of a single bacterium would swamp Earth in a layer 7 feet deep!

In contrast, the bald eagle is a relatively long-lived, rather slowly reproducing species. Let's assume that the bald eagle can live 30 years, that it reaches sexual maturity at 4 years, and that each pair of eagles produces two off-

(a)

time (minutes)	number of bacteria
0	1
20	2
40	4
60	8
80	16
100	32
120	64
140	128
160	256
180	512
200	1024
220	2048

(b)

time (years)	number of eagles (a)	number of eagles (b)
0	2	2
2	2	2
4	4	2
6	8	4
8	14	8
10	28	12
12	52	18
14	100	32
16	190	54
18	362	86
20	630	142
22	1314	238
24	2504	392
26	4770	644
28	9088	1066
30	17314	1764

Figure 28-2 **Exponential growth curves**

All such curves share a similar J shape; the major difference is the time scale. **(a)** The growth of a population of bacteria, assuming one individual to start with and a doubling time of 20 minutes. **(b)** Growth of an eagle population, starting with a single pair of hatchlings, and assuming that the age at first reproduction is 4 years (red line) and 6 years (green line). Notice that after 26 years, the population of eagles that began reproducing at 4 years is seven times that of the eagles that began reproducing at 6 years.

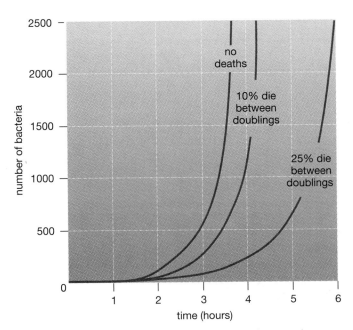

Figure 28-3 **The effect of death rates on bacterial population growth**

The graphs assume that the population doubles every 20 minutes. Notice that the population in which a quarter of the bacteria die every 20 minutes reaches 2500 only 2 hours and 20 minutes later than that in which no deaths occur.

spring per year for the remaining 26 years. Figure 28-2 compares the potential population growth of eagles with that of bacteria, assuming no deaths occur in either population during the time graphed. Notice that the shapes of the curves are virtually identical—both populations exhibit unrestricted exponential growth. Although the time scale differs, both population sizes eventually become astronomically large. Figure 28-2 also shows what happens if eagle reproduction begins at 6 years instead of 4. Under this condition, exponential growth still occurs, but the time required to reach a particular size increases considerably. This time factor has important implications for the human population: Delayed childbearing significantly slows population growth. If each woman bears three children in her early teens, the population will grow much faster than if each woman bears five children beginning at age 30.

So far we have looked only at birth rates. Even under ideal conditions, however, some mortality occurs. To illustrate the effect of differing death rates, we compare three bacterial populations in Figure 28-3. Again, the shapes of the curves are the same. In each case the population eventually approaches infinite size; but the time required to reach any given population size is increased by increasing mortality.

Exponential Growth Sometimes Leads to "Boom-and-Bust" Cycles

In nature, exponential growth occurs only under special circumstances and only for a limited time. For example, exponential growth is seen in populations that undergo

regular cycles, where rapid population growth is followed by a massive die-off. These **boom-and-bust cycles** occur in a variety of organisms for complex and varied reasons. Many short-lived, rapidly reproducing species, from algae to insects, have seasonal population cycles that are linked to predictable changes in rainfall, temperature, or nutrient availability (Fig. 28-4). In temperate climates, insect populations grow rapidly during the spring and summer, then crash with the killing hard frosts of winter. More-complex factors produce roughly 4-year cycles for small mammals, such as voles and lemmings, and much longer cycles in hare, muskrat, and grouse populations.

Lemming populations, for instance, may grow until they overgraze their fragile arctic tundra ecosystem. Lack of food, increasing populations of predators, and social stress caused by overpopulation may all contribute to a sudden high mortality. Many deaths occur as waves of lemmings emigrate from regions of high population density. During these dramatic mass movements, they are easy targets for predators, and many accidentally drown (some begin swimming when they encounter bodies of water, including the ocean). The reduced lemming population eventually contributes to a decline in predator numbers (see "Scientific Inquiry: Cycles in Predator and Prey Populations") and recovery of the plant community. This recovery, in turn, sets the stage for another round of exponential growth in the lemming population (Fig. 28-5).

In populations that do not show boom-and-bust cycles, exponential growth may occur temporarily under special circumstances, for example, if population-controlling factors, such as predators or parasites, are eliminated or if the food supply is increased. Exponential growth can also occur when individuals invade a new habitat where conditions are favorable and competition or predation is scarce. Many such invasions have happened when people have introduced foreign, or **exotic,** species into ecosystems, often with tragic results (see "Earth Watch: Introduced Species Disrupt Community Interactions," Chapter

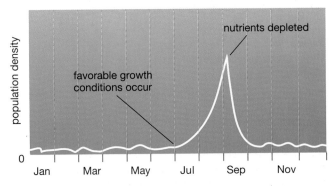

Figure 28-4 **A boom-and-bust population cycle**

Blue-green algae in a boom-and-bust cycle in Lake Erie. Algae survive at a low level through the fall, winter, and spring. Early in July, conditions become favorable for growth, and exponential growth occurs through August, after which nutrients become depleted and the population "goes bust."

SCIENTIFIC INQUIRY

Cycles in Predator and Prey Populations

If we assume that prey, such as rabbits, are eaten exclusively by a particular predator, such as the lynx, it seems logical that both populations might show cyclic changes, with increases in the predator population lagging behind increases in prey population size. For example, a large rabbit population will provide abundant food for lynx and their offspring, which will survive in large numbers. The lynx population will in turn increase and thus so will predation on the rabbits. Increased predation will cause a decline in the rabbit population, which will allow fewer lynx to survive and reproduce, forcing a decline in the lynx population shortly thereafter.

Does this out-of-phase population cycle of predators and prey actually occur in nature? A classic example of such a cycle was observed using the ingenious method of counting all the pelts from northern Canada lynx and snowshoe hares purchased by the Hudson Bay Company between 1845 and 1935. The availability of pelts (which presumably reflects population size) showed dramatic, closely linked cycles between these predators and their prey (Fig. E28-1). Unfortunately, as with any field investigation, many variables could influence the relationship between hare and lynx. One problem is that hare populations have been shown to fluctuate even without lynx present, possibly owing to overshooting carrying capacity and reducing their food supply (in this case, the important predator-prey interaction is between the hares and the plants they prey upon). Further, lynx do not feed exclusively on hares but can eat a variety of small mammals. Environmental variables such as exceptionally severe winters could adversely affect both populations and produce similar cycles.

To test this hypothesis more scientifically, investigators turned to controlled laboratory studies on populations of small predators and their prey. The study illustrated here involved only braconid wasp predators and their bean weevil prey. As predicted, the two populations showed regular cycles, with the predator population rising and falling slightly later than the prey population (Fig. E28-2). The wasps lay their eggs on weevil larvae, which provide food for the newly hatched wasps. A large weevil population assures a high survival rate for wasp offspring, increasing the predator population. Then, under intense predation pressure, the weevil population plummets, reducing the population size of the next generation of wasps. Reduced predation then allows the weevil population to increase rapidly, and so on.

It is highly unlikely that such a straightforward example will ever be found in nature, but this type of predator-prey interaction clearly can contribute to the fluctuations observed in many natural populations.

Figure E28-2   Experimental predator-prey cycles

Out-of-phase fluctuations in laboratory populations of the azuli bean beetle and its braconid wasp predator.

**Figure E28-1
Population
cycles**

Snowshoe hares and their lynx predators are graphed on the basis of the number of pelts received by the Hudson Bay Company.

Figure 28-5 **Boom-and-bust lemming population cycles**

Lemming population density follows roughly a 4-year cycle, as illustrated by these data from Point Barrow, Alaska.

29). As you will learn in the next section, all populations showing exponential growth must eventually either stabilize or crash.

Environmental Resistance Limits Population Growth

Exponential growth carries with it the seeds of its own destruction. As individuals join the population, competition for resources intensifies. Predators may increase in number or they may make this newly abundant prey a larger part of their diet. Parasites and diseases spread more readily because of crowding and weakness resulting from lack of food or from stress caused by adverse social interactions. Consequently, after a period of exponential growth, populations tend to stabilize at or below the maximum size that the environment can sustain, called carrying capacity (see next section). The rate of growth gradually declines and reaches a long-term state of equilibrium, fluctuating around a growth rate of zero. In this equilibrium, the birth rate is balanced by the death rate and population size is stabilized. This type of population growth, which is typical of long-lived organisms colonizing a new area, is represented graphically by an S-shaped growth curve, or **S-curve** (Fig. 28-6).

Populations Usually Stabilize at or below Carrying Capacity

Populations may stabilize at a level called the **carrying capacity** of the ecosystem. The carrying capacity is the maximum size of a given population of organisms that a particular area can support indefinitely. It is determined primarily by the sustained availability of two types of resources: renewable resources, such as nutrients, water, and light, that are replenished by natural processes and a nonrenewable resource, space, such as suitable nesting or denning areas.

Organisms will die if demands on renewable resources, such as food, water, and light (the energy source for plants), are too high. If space requirements are exceeded, animals may emigrate, often to less-suitable areas where their death rate will be higher. Reproduction will decline, because animals may not find adequate breeding sites or the seeds of plants may not reach a suitable place to germinate. Excess demands may damage ecosystems, reducing their carrying capacity. The result is either a population decline until the ecosystem recovers or a permanently reduced population. For example, overgrazing by cattle on dry western grasslands has reduced the grass cover and allowed sagebrush (which cattle will not eat) to thrive. Once established, sagebrush replaces edible grasses and reduces the carrying capacity of the land for cattle. Other dramatic cases of overgrazing have occurred when herbivores, such as reindeer, have been introduced onto islands without large predators, as shown in Figure 28-7.

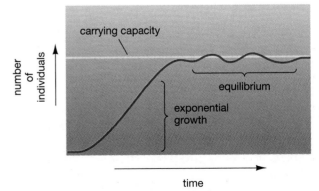

Figure 28-6   **The S-curve of population growth**

The population first grows exponentially, then fluctuates around carrying capacity.

Figure 28-7   **The effects of exceeding carrying capacity**

Exceeding carrying capacity can damage an ecosystem, reducing its ability to support the population. In 1911, 25 reindeer were introduced onto one of the Pribilof Islands (St. Paul) in the Bering Sea off Alaska. Food was plentiful, and there were no predators of reindeer on the island. The herd grew exponentially (note the initial J shape) until it reached 2000 reindeer in 1938. At that point, the small island was seriously overgrazed, food was scarce, and the population declined dramatically. By 1950, only eight reindeer survived.

In nature, environmental resistance maintains populations at or *below* the carrying capacity of the environment. Factors of environmental resistance may be classified into two broad categories. **Density-independent** factors limit population size regardless of the density (number of individuals per given area) of the population. **Density-dependent** factors increase in effectiveness as the population density increases. In the following sections, we will look more closely at these factors and how they control population growth.

Density-Independent Factors Limit Populations Regardless of Their Density

Density-independent factors kill individuals or reduce reproductive rates regardless of population density. Perhaps the most important density-independent factor is weather. For example, many insects and annual plant populations are limited in size by the number of individuals that can be produced before the first hard freeze. Such populations typically do not reach carrying capacity, because density-independent factors intervene first. Weather is largely responsible for the boom-and-bust population cycles described earlier. Human activities can also limit the growth of natural populations in ways that are independent of population density. Pesticides and pollutants may cause drastic declines in natural populations, as does habitat destruction caused by construction of farms, roads, and housing developments.

Density-Dependent Factors Become More Effective As Population Density Increases

Organisms that live several years have evolved various mechanisms to compensate for seasonal weather changes, thus circumventing this form of density-independent population control. Many mammals, for example, develop thick coats and store fat for the winter; some also hibernate. Other animals, including many birds, migrate long distances to find food and a hospitable climate. Plants may survive the rigors of winter by entering a period of dormancy, dropping their leaves and drastically slowing their metabolic activities.

For these long-lived species, by far the most important elements of environmental resistance are density-dependent factors. Density-dependent factors are those that become increasingly effective as population density increases, thus exerting a negative feedback effect on population size. The larger the population grows, the more changes are triggered that counteract this growth. Density-dependent factors include community interactions, such as predation and parasitism, as well as competition within the species or with members of other species. These factors are discussed below and in Chapter 29.

Predators Help Control Their Prey Populations

Predation is the act of killing and eating another living organism. This broad definition includes the swift and violent capture of elk by wolves (Fig. 28-8) as well as the less

Figure 28-8 **Predators help control prey populations**

Gray wolves have brought down an elk, who may have been weakened by old age or parasites. Predators will often switch to the most abundant prey, limiting prey populations.

dramatic munching on a cactus by a moth (Fig. 28-9) or a bird eating a seed. Organisms that kill and eat other organisms are **predators**. Predation controls prey populations in a density-dependent manner. It can become an increasingly important factor in population control as prey populations increase, simply because the denser the prey population, the more predator-prey encounters there will be. Many predators will eat a variety of prey depending on what is most abundant and easiest to find. Coyotes may eat more mice when the mouse population is high but switch to eating more ground squirrels as the mouse population declines, incidentally allowing the mouse population to recover.

Predators may also exert density-dependent effects by increasing in number as their prey population grows. For instance, predators that feed heavily on lemmings, such as the arctic fox and snowy owl, regulate the number of offspring they produce according to the abundance of lemmings. The snowy owl may produce up to 13 chicks when lemmings are abundant but not reproduce at all in years when lemmings are scarce. In other cases, an increase in predators may cause a crash of the prey population, which is then followed by a decline in the predator population. This series of events can result in **out-of-phase population cycles** of both predators and prey (see "Scientific Inquiry: Cycles in Predator and Prey Populations").

The influence of predators on prey populations varies considerably. Some predators feed primarily on prey made vulnerable because their populations have exceeded car-

(a)

(b)

Figure 28-9   **An introduced predator controls an introduced pest**

(a) A pasture in Queensland, Australia, is blanketed by the imported prickly pear cactus, whose spread was unchecked by predators. **(b)** The same site 3 years after the introduction of a predatory cactus moth appropriately named *Cactoblastis cactorum*. (Courtesy of the Department of Lands, Queensland, Australia.)

rying capacity. Such prey may be weakened by lack of food or may be exposed owing to lack of appropriate shelter. In such cases, predation may maintain prey populations near their sustainable density.

In other cases, predators may maintain the prey population at well below carrying capacity. A dramatic example is the prickly pear cactus, which was introduced into Australia. Lacking natural predators, it spread uncontrollably, destroying millions of acres of valuable pasture and range land. Finally, in the 1920s, a cactus moth (a predator of the prickly pear) was imported from Argentina and released to feed on the cacti. Within a few years, the cacti were virtually eliminated. Today the moth continues to maintain its prey at very low population densities (see Fig. 28-9).

Parasites Can Reduce Their Host Populations

In contrast to predation, **parasitism** is the process of feeding on another, usually larger, organism without killing it immediately or directly. **Parasites** often live on or in their prey, also called the **host**. Parasites include the familiar disease-causing microorganisms, mosquitoes, and intestinal worms, but also animals that graze on plants without killing them. Parasites may dramatically increase the death rate in a population in which individuals are already weakened by the effects of overcrowding. Further, infestation by parasites makes organisms more vulnerable to predators.

Parasitism is density dependent. Most parasites have limited motility and spread more readily between hosts at high population densities. For example, plant diseases spread readily through acres of densely planted crops, and childhood diseases spread rapidly through schools and daycare centers.

Both predators and parasites can have beneficial effects on the prey population as a whole. As you will learn in the next chapter, predators, parasites, and their prey coevolve. Parasites and predators often keep host populations within carrying capacity, They also frequently destroy the least fit of the prey, leaving the better-adapted prey to reproduce. The result is usually a balance in which the prey population is regulated but not eliminated.

When a predator or parasite is introduced into an area where it did not evolve, however, the local prey species often are not adapted for it, and the consequences can be disastrous. The successful conquest of much of Earth by Europeans can be traced partly to the disease organisms they carried with them as they invaded continents whose populations had no previous exposure or resistance to these parasites. Smallpox, for example, imported by Europeans, ravaged the native population of Hawaii, the Amerindians of Argentina, and the aborigines of Australia. The population balance in ecosystems can be destroyed when species are introduced into new areas where they have no natural predators or parasites (see "Earth Watch: Introduced Species Disrupt Community Interactions," Chapter 29).

Competition for Resources Helps Control Populations

The resources that determine carrying capacity (space, nutrients, light) are limited relative to the demand for them. Further, use by one individual limits their availability to another. For this reason, **competition** limits population size in a density-dependent manner. There are two major forms of competition: **interspecific competition** (between members of different species; see Chapter 29) and **intraspecific competition** (between members of the same species). Because the needs of members of the same species for water and nutrients, shelter, breeding sites, light, and other resources are almost identical, intraspecific competition is more intense. Organisms have evolved several ways to deal with intraspecific competition. Some organisms, including most plants and many insects, engage in **scramble competition**. This is a kind of free-for-all with resources as the prize. For example, when a plant disperses its seeds in a small area, hundreds may germinate. As they grow, however, the larger ones begin to shade the smaller, those with the most extensive roots absorb most of the water, and the weaker individuals eventually wither and die.

Many animals (and even a few plants) have evolved **contest competition,** which helps regulate population size and reduce direct competition. Contest competition consists of social or chemical interactions used to limit access to important resources. Territorial species—such as wolves, many fish, rabbits, and songbirds—defend an area containing important resources such as food or nesting sites. When the population begins to exceed the available resources, only the best-adapted individuals are able to defend adequate territories. Those without territories often do not reproduce and are also easy prey. The creosote bush, which secretes a chemical into the ground that prevents germination of seeds nearby, could be considered a territorial plant (see Fig. 28-11b).

Dominance hierarchies, or "pecking orders," are another form of contest competition observed in many species of social animals. High-ranking individuals have first access to food, breeding sites, and mates. When resources are limited, only dominant individuals obtain what they need to reproduce successfully. Thus, dominance hierarchies can limit population size.

As population densities increase and competition intensifies, some animals react by emigrating. Large numbers leave their homes to colonize new areas, and many, sometimes most, die in the quest. The massive movements of lemmings, which sometimes end in marches into the sea, are apparently in response to overcrowding. Migrating swarms of locusts plague the African continent, stripping all vegetation in their path (Fig. 28-10).

In laboratory studies, overcrowding of small mammals—such as mice, rats, and voles—causes **social stress.** Social stress results in a decrease in the size of reproductive organs, reduced reproductive rate, slower growth, reduced resistance to disease, and cannibalism of young, all

Figure 28-10   Emigration

In response to overcrowding, emigration can reduce local populations. Locust swarms provide a dramatic example.

of which would reduce population size. In the wild, however, population densities rarely reach the levels achieved in the laboratory, so these findings may have limited applicability to natural populations.

The size of any population is a result of complex interactions between density-independent and density-dependent forms of environmental resistance. For example, a stand of pines weakened by drought (a density-independent factor) may more readily fall victim to the pine bark beetle (density dependent). Likewise, a caribou weakened by parasites (density dependent) is likely to be killed by an exceptionally cold winter (density independent).

Population Patterns in Space and Time

Populations Distribute Themselves in Different Ways

Organisms may live in flocks, herds, pairs, or as solitary individuals, or they may cluster around resources such as water holes. The general pattern in which members of a population are dispersed within a given area is a population's distribution. Distribution may vary with time, changing with the breeding season, for example. Ecologists recognize three major types of spatial distribution: **clumped, uniform,** and **random** (Fig. 28-11).

Individuals in Many Populations Form Groups

There are many populations whose members live in groups and whose distribution can be described as clumped. Examples include family or social groupings, such as elephant herds, wolf packs, prides of lions, flocks

Figure 28-11   Population patterns in space

(a) Clumped, as illustrated by this gathering of caterpillars. (b) Uniform, seen in the spacing of these creosote bushes in the desert. (c) The random distribution of trees and plants in a rain forest.

of birds, or schools of fish (Fig. 28-11a). What are the advantages of clumping? Flocks provide many eyes that search for localized food, such as a tree full of fruit. Schooling fish avoid predation by confusing the predator with countless flashing bodies darting in all directions. Some species form temporary groups for mating and caring for their young. Other plant or animal populations cluster, not for social reasons, but because resources, such as nutrients, shelter, or water, are localized. Cottonwood trees, for example, aggregate along streams and rivers in grasslands. Animals also aggregate around water holes in the dry savanna of Africa.

Individuals of Some Populations Disperse Almost Uniformly

Uniformly distributed organisms maintain a relatively constant distance between individuals. This distribution occurs most frequently among animals that defend territories and exhibit territorial behaviors designed to protect scarce resources, such as breeding sites, nutrients, or water. Male Galapagos iguanas establish evenly spaced breeding territories. Shorebirds are also often found in evenly spaced nests, just out of reach of one another. Other territorial species, such as the Tawny Owl, mate for life and continuously occupy well-defined, relatively uniformly spaced territories (for breeding animals, the spacing refers to pairs, not individuals). Some plants growing in poor soil with limited water, such as the creosote bush (Fig. 28-11b), release chemicals that prevent germination of plants nearby. This mechanism helps assure adequate resources for each individual.

In a Few Populations, Organisms Are Randomly Distributed

In randomly distributed populations, which are relatively rare, individuals do not form social groups, the resources they need are more or less equally available throughout the area they inhabit, and resources are not scarce enough to require territorial spacing. Trees and other plants in rain forests come close to being randomly distributed (Fig. 28-11c).

There are probably no vertebrate species that maintain random distribution throughout the year because they must breed, a behavior that makes social interaction inevitable.

Populations Show Characteristic Patterns of Survivorship

Population patterns can be considered from the perspective of time as well as space. Over time, populations show characteristic patterns of deaths or (more optimistically) survivorship. These patterns, called **survivorship curves,** are revealed when the number of individuals of each age is graphed against time. Three different types of survivorship curve, which can be described as "early loss," "constant loss," and "late loss," based on the part of the life cycle during which most deaths occur, are shown in Figure 28-12.

Populations with convex survivorship curves (late loss) have relatively low infant death rates, and most individuals survive to old age. This curve is characteristic of humans and many other large animals, such as bighorn sheep. These species produce relatively few offspring, which are protected by the parents. Species with constant-loss survivorship curves have a fairly constant death rate; they have an equal chance of dying at any time during their life span. This phenomenon is seen in the American robin, the gull, and laboratory populations of organisms that reproduce asexually, such as *Hydra* and bacteria. The concave curve (early loss) is characteristic of organisms producing large numbers of offspring that have little parental care, being largely left to compete on their own. The death rate is very high among the offspring, but those that reach adulthood have a good chance of surviving to old age. Most invertebrates, most plants, and many fish exhibit concave survivorship curves. In addition, in some populations of black-tailed deer, 75% of the population dies within the first 10% of its life span, giving this mammalian population a concave curve as well.

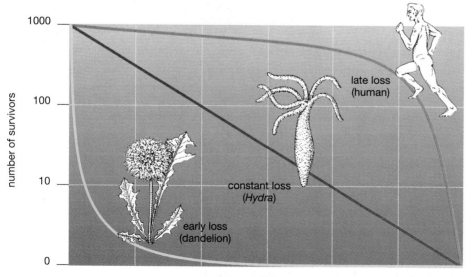

Figure 28-12   Population patterns in time

Three types of survivorship curve are shown. Because the life spans of these organisms differ so much, they can be compared on the same graph only if the percentages of survivors rather than the ages are used.

The Human Population

Your first child received free schooling and free medical care, and you were offered longer maternity leave and a larger pension if you signed a promise to have no more children. Pregnant with your second child, you face fines equivalent to all the benefits provided for the first child, up to a full year's pay. Each night, delegates come to your home to tell you of the harm you are doing to society. Your neighbors shun you, and the collective pressure finally drives you to the hospital and a state-sponsored abortion. A distant future scenario from a pessimistic science fiction novel? Hardly. Similar policies were instituted in China. With a population of over 1 billion and an increase of over 15 million people in 1983 alone, the government took drastic measures. Now, China's coercive policies have been eased somewhat, and birth rates are rising again. What is China's demographic future? What of the rest of the world? Let's examine the human population in light of what we know of exponential growth and carrying capacity.

The Human Population Is Growing Exponentially

Look at the graph of human population growth in Figure 28-13 and compare it with the exponential growth curves in Figure 28-2. The time span is different, but the growth curves are the same—exponential. It took over 1 million years for the human population to reach 1 billion, but the second billion was added in 100 years, the third in 30, the fourth in 15, and the fifth in 12 years. The population is projected to reach 6 billion in 1998. World population currently grows by about 88 million people yearly—a million people added every 4 days! Why hasn't environmental resistance put an end to our exponential growth? What is the carrying capacity of the world for humans?

Like all populations, humans have encountered environmental resistance, but unlike other populations, we have responded to environmental resistance by overcoming it. As a result, the human population has grown exponentially for an unprecedented time span; to accommodate our growing numbers, we have altered the face of the globe. Human population growth has been spurred by a series of "revolutions" that conquered various aspects of environmental resistance and increased Earth's carrying capacity for people.

Primitive people produced a **cultural revolution** when they discovered fire, invented tools and weapons, built shelters, and designed protective clothing. Weapons allowed more effective hunting and an increased food supply; clothing and shelter increased the habitable areas of the globe. Domesticated crops and animals supplanted hunting and gathering by about 8000 B.C. This **agricultural revolution** provided people with a larger, more dependable food supply and further increased Earth's carry-

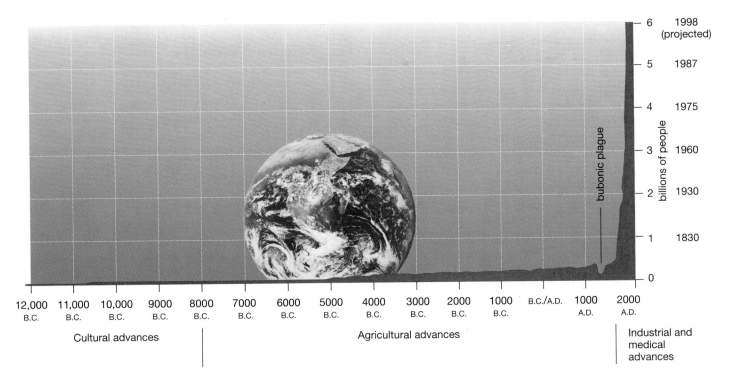

Figure 28-13   **Human population growth**

The human population from the Stone Age to the present has shown continued exponential growth. Note the dip in the fourteenth century caused by the bubonic plague. Note also the steadily decreasing intervals at which additional billions are added. (Inset) Earth is an island of life in a sea of emptiness; its space and resources are limited.

ing capacity for humans. Increased food resulted in longer lives and more childbearing years, but a high death rate from disease still restricted the population. Human population growth continued slowly for thousands of years until the **industrial-medical revolution** began in England in the mid-eighteenth century, spreading through Europe and North America in the nineteenth century. Medical advances dramatically decreased the death rate by reducing environmental resistance from disease. These advances included the discovery of bacteria and their role in infection, leading to control of bacterial disease through improved sanitation and the use of antibiotics, and the discovery of viruses, leading to the development of vaccines for diseases such as smallpox. The revolution continues today as research proceeds on vaccines against major killers, such as malaria and AIDS, and sophisticated medical advances, such as fertility drugs and *in vitro* fertilization, are developed.

In developed countries, such as those of western Europe, the industrial-medical revolution resulted in an initial rise in population due to decreased deaths, followed by a decline in birth rates. This decline can be attributed to many factors, including better education, increased availability of contraceptives, a shift to a primarily urban life-style, and more career options for women. In developed countries, such as Switzerland, Sweden, Austria, Germany, and England, populations have more or less stabilized. However, these countries are home to only about 4% of the world's population.

In less-developed countries, such as most of those in Central and South America, Africa, and Asia, medical advances have decreased death rates and increased the life span, but birth rates remain high. These countries have not experienced the increase in wealth that was partly responsible for the decline in birth rate of developed countries. Children serve as a form of social security in Third World nations because they may be the only support for parents in their old age. In agricultural societies, children are an important source of labor. Social traditions offer prestige to the man who fathers and the woman who bears many children. In Nigeria, the most crowded country in Africa, many men refuse to allow any form of birth control, and the average woman has more than six children. Nigeria is already suffering from loss of forests, the spread of deserts, soil erosion, and water pollution. With nearly half of its population of 93 million under the age of 15, continued growth is a certainty. Although a majority of African nations have concluded that their growth rates are too high and are working toward reducing them, lack of education and lack of access to contraceptives impede progress in curbing population growth. Of the projected 6.3 billion people on Earth in the year 2000, 5 billion will reside in less-developed countries. The prospects for population stabilization in the near future are nil, barring major catastrophes that might dramatically increase deaths. The reason can be seen clearly by looking at the age structures of the less-developed countries and comparing them with those of countries with stable populations, as described in the following section.

The Age Structure of a Population Predicts Its Future Growth

The **age structure** of a population is the distribution of males and females of various age groups, usually represented in a diagram. All age-structure diagrams come to a peak at the top, because relatively few people live into their nineties. The shape of the rest of the diagram, however, shows whether the population is expanding, stable, or shrinking (Fig. 28-14). If the number of children (age 0 to 14) exceeds the number of reproducing individuals (age 15 to 45), the population is expanding and the diagram resembles a pyramid. In the case of stable populations, the number of children is about equal to the number of reproducing adults, forming a straight-sided figure. In shrinking populations, there are fewer children than reproducing adults and the age-structure diagram is narrower at the base.

Figure 28-15 shows the average age structures of the populations of less-developed and developed countries. The outermost boundaries represent the projected population structure for the year 2025; the inner ones are for 1990. Each graph has been divided horizontally into three parts to show individuals who are prereproductive (0 to 14 years old), reproductive (15 to 45 years), and postreproductive (46 and older). In 1990, the less-developed countries (in Asia, Africa, India, and South and Central America) had an average annual growth rate of 1.9%, and the developed countries (in North America, Europe, Australia, Japan, and New Zealand) showed an average annual growth rate of 0.3%. In the developed countries, projections for the year 2025 are only slightly higher than present levels (see Fig. 28-15).

In contrast, each year in the less-developed countries, increasing numbers of people enter their reproductive years and give birth to an ever-increasing base of infants. Even if these countries were to reach **replacement-level fertility (RLF)** immediately (that is, if people of reproductive age had only sufficient children to replace them-

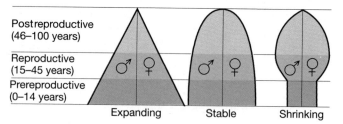

Figure 28-14   Age-structure diagrams

Here we show generalized figures for expanding, stable, and shrinking populations. The width of each figure is proportional to the number of individuals, with the male population on the left and the female population on the right. The vertical axis shows increasing age. The number of children (prereproductive, age 0 to 14 years) produced by the reproductive-age population determines whether the population is expanding (more children than reproductive-age adults), stable (number of children equal to number of reproductive-age adults), or shrinking (fewer children than reproductive-age adults).

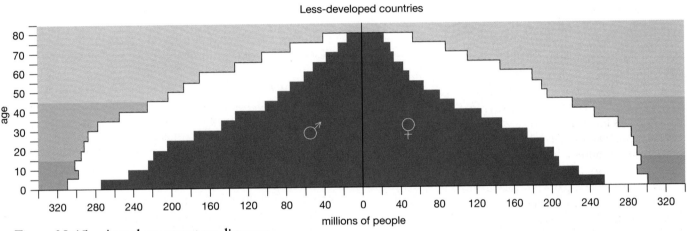

Figure 28-15   **Actual age-structure diagrams**

Here we show developed countries (top) and less-developed countries (bottom). Compare these with the idealized diagrams shown in Figure 28-14. The outer lines are projections to the year 2025. Notice how similar the current and projected age structures are. This similarity suggests that rapid population growth will continue in the less-developed countries for the foreseeable future.

selves), population growth would continue for decades because of built-in momentum produced by increasing numbers of people reaching reproductive age.

Notice, however, that predictions for these countries for the year 2025 show a structure very similar to the present one, the only difference being far more people in all age classes. Although several of these countries have recently initiated family-planning programs, parents in less-developed countries will probably continue to have larger than replacement-level families. The percent growth per year caused by natural increase (births – deaths) for various parts of the world is shown in Figure 28-16, p. 572.

Ironically, population growth in these countries is helping to perpetuate the poverty and lack of education that in turn tend to sustain high birth rates. The relationship of income and education to birth rate has been documented in many countries, including the United States. Here, on the average, women who do not complete high school have twice as many children as those with more than 4 years of college.

The Population of the United States Is Growing Rapidly

As shown in Figure 28-17, p. 572, the U.S. population is growing exponentially. In fact, at over 1% annually, our growth rate is over three times the average in the developed world. During the year ending in June 1995, for example, the U.S. population grew by over 2.5 million (nearly 300 people per hour), bringing the total over to 262 million. Recall the equation population change = (births – deaths) + (immigrants – emigrants). Let's examine each component of the equation to determine why the United States is growing so rapidly.

If each American woman had 2.1 children, we would have replacement-level fertility. (RLF is slightly higher than 2 because parents must replace both themselves and children who die before reaching maturity.) The U.S. birth rate in 1994 was exactly at replacement-level fertility, so why does our population continue to grow? Two factors

EARTH WATCH

Has the Human Population Exceeded Earth's Carrying Capacity?

A glance at the age structure of less-developed countries, where most of the world's population resides, shows a tremendous momentum for continued growth. World population in the year 2010 is predicted to be over 7 billion and growing. A modest United Nations (UN) projection is that the human population may stabilize in the year 2150 at 11.5 billion (Fig. E28-3). Can Earth support over twice its current population?

Earlier we defined carrying capacity as the maximum population that could be indefinitely sustained. Sustaining a population requires that the ecosystem not be damaged in ways that lower its ability to provide necessary resources. By this definition we have already exceeded Earth's carrying capacity for people. The upper limit of the planet's carrying capacity is determined by the ability of its plant life to harvest the energy from sunlight and produce high-energy molecules that organisms can use as food. Stanford University biologist Peter Vitousek estimates that human activities have already reduced the productivity of Earth's forests and grasslands by 12%. Each year, millions of acres of once-productive land are being turned into desert through overgrazing and deforestation, especially in less-developed countries. Earth's desert area is projected to increase by 20% by the year 2000 as a result of human activities. In a world where between 500 million and 1 billion people are chronically undernourished, 80% of the world's agricultural land is suffering moderate to severe erosion (Fig. E28-4).

In the years between 1990 and 2010, the world population is projected to increase by 33%. During this same period, the available fish, cropland, pastureland, and forests will all decline by 10% to 30% per person. The UN Food and Agricultural Organization reports that Africa's food production for each person has dropped by 20% since 1960; another 30% drop in the next 25 years is predicted. Each year the United States is losing 1 million acres of farmland to spreading urban development. The demand for wood in less-developed countries is far outstripping production, and large areas are being deforested each year. Deforestation in turn causes erosion of precious topsoil, runoff of much-needed fresh water, and the spread of deserts. The total world fish harvest peaked in 1989 and has been declining since, despite increased investments in fishing equipment and fish farming. Off the coast of New England, fish populations have plummeted since the early 1960s because of overfishing, and the fishing industry is collapsing. The destruction of tropical rain forests may exterminate 1 million species of animals and plants by the year 2000, as their habitats are destroyed. These are clear indications that our present population, at its present level of technology, is already "overgrazing" the world ecosystem and decreasing its ability to support all forms of life, including people.

Mass starvation, such as the famines in Somalia and Ethiopia, has occurred not because Earth lacks sufficient

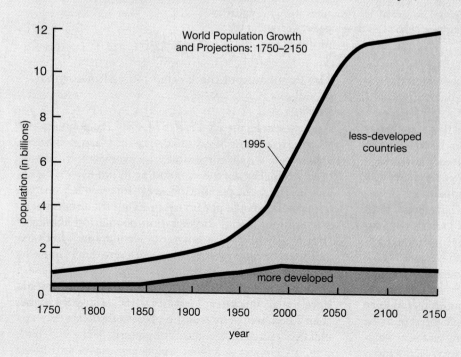

Figure E28-3   World population projected to the year 2150

The disparity in numbers between developed and less-developed countries is projected to increase even further than at present. (Modified from Population Bulletin, Population Ref. Bureau, Vol. 50, March 1995.)

(a)

(b)

Figure E28-4
Desertification

(a) Human activities, including overgrazing livestock, deforestation, and poor agricultural practices, convert once-productive land into barren desert. (b) The loss of productive land, when combined with an expanding human population, can lead to tragedy.

food for its current population, but because resources are unequally distributed among, and even within, countries. Each natural population must live within the carrying capacity of the ecosystem it occupies. In contrast, human populations often expand well beyond the ability of their local ecosystem to support them, relying on wealth derived from other sources (such as minerals) to allow them to import the food they need. When human populations exceed the carrying capacity of the soil on which they live, and also lack the wealth to import and distribute adequate food, disaster strikes. An exploding population, coerced by poverty and ignorance into using destructive farming techniques that damage the land, is trapped in a downward economic spiral driven and perpetuated by the demands of population growth.

In estimating how many people Earth can support, we must keep in mind that humans desire more out of existence than a minimum caloric intake each day. It has been estimated that for everyone on Earth to live as Americans do, world population would have to be *reduced* to 500 million, *less than one-tenth the present population*. For all to achieve a standard of living similar to that in Europe, where the life-style is less extravagant, *the present population would have to be reduced by half*. Merely "redistributing the wealth" is obviously not an adequate answer.

There are some who predict that technological advances will continue to increase Earth's carrying capacity for humans indefinitely. In evaluating this forecast, keep in mind that technology takes time, wealth, and education

to develop and implement. Rapid population growth in less-developed countries increases poverty, overloads the educational system, and hampers technological development, while simultaneously making its need more urgent. Ironically, the countries that so desperately need these predicted technological advances are often unable even to take advantage of today's level of technology. The high-yield crops developed in the 1960s (the so-called Green Revolution) have helped considerably, but their impact has been lessened because they require expensive equipment, fertilizer, pesticides, and irrigation water to cultivate. Less-developed countries often cannot afford these inputs, and donated farm equipment has been left to rust in the fields for lack of fuel as well as the parts and know-how to maintain it.

Hope for the future lies in using the intelligence that has allowed us to overcome environmental resistance to see signs of "overgrazing" and act before we have irrevocably damaged our world ecosystem. Human population *will* stop its exponential growth. Either we will voluntarily reduce our birth rate or various forces of environmental resistance will increase our death rate; the choice is ours. Facing the problem of how to limit births is politically and emotionally difficult, but continued failure to do so will be disastrous. Our dignity, our intelligence, and our role as self-appointed stewards of life on Earth all demand that we make the decision to halt population growth ourselves before we have permanently reduced Earth's ability to support all life, including our own.

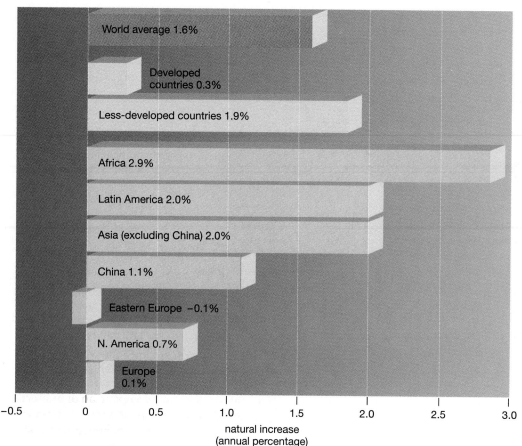

Figure 28-16   **Population growth by world regions**

Growth rates shown are due to natural increase (births – deaths) expressed as percent increase per year for various regions of the world. Europe now includes the former Soviet Union. These figures do not include immigration or emigration.

are contributing to the rapid growth of the U.S. population: the "baby boom" and immigration.

Part of our current growth rate is a legacy of our recent past. Parents of the late 1940s through the 1960s had families that were larger than replacement-level, resulting in a "baby boom" and a momentum in population growth that has not yet subsided. For example, over a million more women entered their reproductive years in 1980 than did so in 1970. Even though these women are averaging only 2.1 children each, because there are more women having children, our population swells.

A second crucial component of the U.S. population equation is immigration, which contributes more to population growth here than in any other nation in the world. Legal immigration now averages 880,000 per year, causing about 30% of our total growth. Illegal immigration is estimated to add an additional 300,000 people each year.

According to Census Bureau projections, the U.S. population will not stabilize in the foreseeable future. In the past few years, there have been jumps in fertility as well as immigration that have caused dramatic increases in future

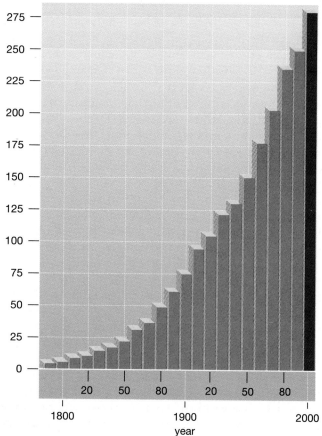

Figure 28-17   **Population growth in the United States**

Since 1790, our growth has shown the J-shaped curve characteristic of an exponentially growing population. The figure for the year 2000 is a conservative projection based on a 1% annual growth rate.

population projections. The Census Bureau projects that by the year 2050 there will be 392 million Americans (about 130 million more than today), and the population will still be growing.

The rapid growth of the U.S. population has major environmental implications both for us and for the world as a whole. The average American uses six times as much energy as the world average. With 4.7% of the world's population, we are responsible for 14% of the total methane output and 22% of carbon dioxide and chlorofluorocarbon emissions. These gases are major contributors to the greenhouse effect, described in Chapter 30. Because people in less-developed countries use considerably less energy than the world average, the 2.6 million Americans added each year contribute more to climate change than the 34 million people added each year by China and India combined.

When and how will human numbers ever stabilize? How many people can Earth support? There are no certain answers to these questions, but in "Earth Watch: Has the Human Population Exceeded Earth's Carrying Capacity?" we explore them in more detail.

✳ S U M M A R Y   O F   K E Y   C O N C E P T S

Introduction to Ecology

Ecology is the study of the interrelationships between organisms and their environment. Ecology may be approached at the level of the population, the members of a particular species in a given area; the community with its complex interactions of populations; or an entire ecosystem, including communities and their abiotic environment.

How Populations Grow

Individuals join populations through births or immigration and leave through death or emigration. Thus

$$\text{population change} = (\text{births} - \text{deaths}) + (\text{immigrants} - \text{emigrants})$$

The ultimate size of a stable population is the result of interactions between biotic potential, the maximum possible growth rate, and environmental resistance, which limits population growth.

All organisms have the biotic potential to more than replace themselves over their lifetime, resulting in population growth. Populations tend to grow exponentially, with increasing numbers of individuals added during each successive time period. The maximum rate of growth of a given population is influenced by the age at first reproduction, frequency of reproduction, number of offspring produced each time, length of reproductive life span, and the death rate under ideal conditions. Populations cannot continue to grow exponentially indefinitely; they either stabilize or undergo periodic boom-and-bust cycles as a result of environmental resistance.

Environmental resistance restrains population growth by increasing the death rate or decreasing the birth rate. The maximum size at which a population may be sustained indefinitely by an ecosystem is called the carrying capacity, determined by limited resources such as space, nutrients, and light.

Populations are maintained at or below carrying capacity by density-independent forms of environmental resistance, such as weather, and density-dependent forms of resistance, which include predation, parasitism, and competition. Predators often eat the most abundant prey and may switch prey on the basis of the size of prey populations. When predators concentrate on a single prey, both predator and prey populations may cycle. Parasites spread most rapidly in dense populations and can increase death rates when animals are stressed because of overcrowding. Interspecific competition limits both the size and distribution of populations. Intraspecific competition may be resolved directly through scramble competition or indirectly through chemical interactions or social behaviors, collectively called contest competition. Emigration is another possible response to overcrowding.

Population Patterns in Space and Time

The distribution of populations can be classified as clumped or uniformly or randomly distributed. Clumping may occur for social reasons or around limited resources. Uniform distribution is usually the result of territorial spacing. Random distribution is rare, occurring only when individuals do not interact socially and when resources are abundant and evenly distributed.

Survivorship in populations shows specific survivorship curves that express the likelihood of survival at a given age. Convex (late-loss) curves are characteristic of long-lived species with few offspring and whose offspring must receive parental care. Species with constant curves have an equal chance of dying at any age. Concave (early-loss) curves are typical of organisms that produce numerous offspring, most of which die.

The Human Population

The human population has exhibited exponential growth for an unprecedented time by overcoming certain aspects of environmental resistance and increasing Earth's carrying capacity for people. These feats have been accomplished by the use of tools, agriculture, industry, and medical advances. Age-structure diagrams depict numbers of males and females in various age groups constituting a population. Expanding populations have pyramidal age structures, stable populations show rather straight-sided figures, and decreasing populations are illustrated by figures that are constricted at the base.

Today, most of the world's people live in less-developed countries with rapidly expanding populations, where a variety of social and cultural conditions encourage large families. The United States is the fastest growing of the developed countries, because of high immigration rates and the "baby boom" of the 1940s through the 1960s. The carrying capacity of Earth for humans is unknown, but with a population of over 5.7 billion, resources are already too limited for all to be supported at a high living standard. A steady decline in productive land, wood, and fish harvests suggests that we are already damaging our world ecosystem and decreasing its ability to sustain us.

KEY TERMS

abiotic   p. 555
age structure   p. 568
agricultural revolution   p. 567
biotic   p. 555
biotic potential   p. 557
boom-and-bust cycle   p. 559
carrying capacity   p. 561
clumped distribution   p. 565
community   p. 555
competition   p. 564
contest competition   p. 564
cultural revolution   p. 567
density-dependent   p. 562
density-independent   p. 562

dominance hierarchy   p. 564
ecology   p. 555
ecosystem   p. 555
emigration   p. 556
environmental resistance   p. 557
exotic species   p. 559
exponential growth   p. 557
growth rate   p. 557
host   p. 564
immigration   p. 556
industrial-medical revolution   p. 568
interspecific competition   p. 564
intraspecific competition   p. 564
J-curve   p. 557

out-of-phase population cycle   p. 563
parasite   p. 564
parasitism   p. 564
population   p. 555
predation   p. 562
predator   p. 563
random distribution   p. 565
replacement-level fertility (RLF)   p. 568
scramble competition   p. 564
S-curve   p. 561
social stress   p. 564
survivorship curve   p. 566
uniform distribution   p. 565

THINKING THROUGH THE CONCEPTS

Multiple Choice

1. Why do many animals defend territories?
   a. to monopolize the resources within the territory
   b. to avoid predation
   c. to secure overwintering sites
   d. to advance within a dominance hierarchy
   e. to avoid each other
2. For exponential growth to occur, it is necessary that
   a. there is no mortality
   b. there are no density-independent limits
   c. the birth rate consistently exceed the death rate
   d. a species is very fast-reproducing
   e. the species is an exotic invader in an ecosystem
3. Which of the following currently contributes most to human population growth within the United States?
   a. the consequences of the "baby boom"
   b. immigration
   c. a birth rate above RLF
   d. average birth rate
   e. emigration
4. Which is the most common type of spatial distribution?
   a. logistic   b. uniform
   c. random   d. exponential
   e. clumped
5. Which continent has the highest rate of human population growth?
   a. North America   b. Africa
   c. Asia   d. Australia
   e. South America

6. If a population is above carrying capacity, what must happen?
   a. It must immediately crash.
   b. It can remain stable indefinitely.
   c. If the species is territorial, it can continue to increase.
   d. It must eventually decline.
   e. It must decline if it is not territorial.

Review Questions

1. Define ecosystem, community, and population, making clear the differences between the terms.
2. Draw the growth curve of a population before it encounters significant environmental resistance. What is the name of this type of growth, and what is its distinguishing characteristic?
3. Define biotic potential. Explain natural selection in terms of biotic potential and environmental resistance.
4. What distinguishes density-dependent from density-independent limits to growth? Provide a specific example of each kind of limit. Also, explain why density-independent limits are less likely to affect long-lived species.
5. What are the two types of intraspecific competition? What are the behaviors that lead to each type? Which type is most common for plants?
6. Describe (or draw a graph illustrating) what is likely to happen to a population that far exceeds the carrying capacity of its ecosystem. Explain your answer.
7. Outline the technological and social "revolutions" that have successively removed limits to human population growth.

8. There are two aspects of the reproduction of individual females that determine how fast a population will increase. What are these two aspects of individual reproduction? Why does each affect population growth?

✖ A P P L Y I N G   T H E   C O N C E P T S

1. Red tree squirrels live in many of the same areas as do snowshoe hares. But squirrel populations hardly fluctuate from year to year, and they do not show cycles like those of the hare population at all. A crucial difference between the two species is that hares are not territorial, whereas squirrels fiercely defend territories throughout the year. Explain how such territoriality could help prevent a population from ever exceeding its environment's carrying capacity and how prevention of excess population could in turn prevent marked population fluctuations.

2. In conservation, the concern is often with how fast and what patterns of population decline may be occurring, since we wish to stop the decline of rare species. Saying that a population is declining is equivalent to saying that r is negative. What does this expression tell us about birth rate relative to death rate? Assume that $r = -0.20$. If we have an initial population size of 1000 animals, calculate the population sizes over the next 10 years, and plot the results against time (as in Figure 28-3). This curve shows exponential decline, an all too common pattern for endangered species.

3. The United States has a long history of accepting large numbers of immigrants. Discuss the implications of immigration for population stabilization.

4. What factors encourage rapid population growth in less-developed countries? What will it take to change this?

5. Contrast age structure in rapidly growing versus stable human populations. Why is there a momentum in population growth built into a rapidly growing population?

6. There is clear evidence not only that the world's human population is growing beyond sustainable limits, but also that some resources (e.g., topsoil and drinkable water) are actually decreasing because of overuse. Despite the evidence, some people still argue that it is possible to support many more people by producing foods differently, living with fewer luxuries, and so on. List the resources that are currently overused or appear to be close to overuse and then suggest ways to overcome those limitations. Do you believe that all of the limitations can be avoided by efficiency and technological innovation?

✖ F O R   M O R E   I N F O R M A T I O N

Bongaarts, J. "Can the Growing Human Population Feed Itself?" *Scientific American*, March 1994. A balanced account of the arguments between the optimists and pessimists concerning the future world food supply.

Brown, L. R., and others. *State of the World*. New York: W. W. Norton & Company, Inc., 1994. Annually updated collection of articles concerning global resources, pollution, and population.

Ehrlich, P. R., and Erlich, A. *Extinction*. New York: Random House, Inc. 1981. The causes and consequences of the disappearance of species.

Kunzig, R. "Twilight of the Cod." *Discover*, April 1995. Overfishing has led to a precipitous decline of this previously abundant fish population and a collapse of this once-lucrative fishery.

Ludwig, D., Hilborn, R., and Walters, C. "Uncertainty, Resource Exploitation, and Conservation: Lessons from History." *Science* 260:17–36, 1993. An influential and highly controversial article in which three ecologists argue that there is no evidence that humans can sustainably use natural resources.

Meadows, D. H., Meadows, D. L., and Randers, J. *Beyond the Limits: Confronting Global Collapse, Envisioning a Sustainable Future*. Post Mills, VT: Chelsea Green Publishing, 1992. An up-to-date effort to explain the causes of population growth as well as an attempt to project the future of population and world resources.

Myers, J. H., and Krebs, C. J. "Population Cycles in Rodents." *Scientific American*, June 1974. Natural populations of small rodents show periodic 3- to 4-year cycles in population size.

Sisk, T. D., Launerm, A. E., Switky, K. R., and Ehrlich, P. R. "Identifying Extinction Threats." *BioScience* 44:592–604, 1994. Using information on human population densities and growth to identify particularly threatened communities and ecosystems.

N E T   W A T C H

On-line resources for this chapter are on the World Wide Web at:
http://www.prenhall.com/~audesirk (click on the table of contents link and then select Chapter 28).

29 Community Interactions

"When we try to pick out anything by itself, we find it hitched to everything else in the universe."

JOHN MUIR

This swallowtail butterfly larva shows an uncanny resemblance to a snake, discouraging would-be predators. The caterpillar's entire head is the "snake's nose," while the area behind the head is enlarged and bears markings that closely resemble eyes.

As winter wanes, spring rains create a small pond in the woods. Though it may last only a few months, the pond becomes a complete ecosystem. Its nonliving components define it physically: the depression in the forest floor holding the water, the fine grains of silt, dead leaves and twigs making up the complex bottom, and most of all the molecules of water. At first the ecosystem's living component may consist only of bacteria and fungi. But soon the bottom will teem with single-celled organisms; water-loving plants will sprout; and frogs, newts, and insects will lay eggs, populating the water with larvae. Snails will invade, and ducks may visit, leaving behind parasites that infect the snails. Soon dozens of species will inhabit the pond. The individuals of each species constitute a population. Although they each have their own way of life, the populations interact in three important ways. Populations may compete with other populations for limited resources. One population may prey upon another. And, in some cases, two species may live together in close association over an extended time. (This association may benefit one or both species and may work to the detriment of one or the other.) These three processes, competition, predation, and symbiosis, form the basis of a **community**.

Typically, populations within communities have evolved together, a process called **coevolution**. During coevolution, different species act as agents of natural selection on one another. Prey animals have evolved elaborate defenses that help them survive. Herbivores have digestive specializations that allow them to eat local plants. The plants in turn grow rapidly or defend themselves by chemical or physical means, keeping one step ahead of their predators. The whole community survives by a delicate balance between populations. This delicate balance may be toppled by the invasion of a single new species, as described in "Earth Watch: Introduced Species Disrupt Community Interactions."

The temporary pond quickly passes through a series of stages before reaching its mature late-summer form. The forest surrounding the pond, however, has remained stable for hundreds of years. Long-established communities are generally complex and self-sustaining and have interacting populations that remain relatively constant. Even these communities originate, however, over long periods of gradual change. In the case of the forest, this gradual change may have followed a fire, earthquake, or glacial scouring. During this development, one type of community gives way to another in a process called succession, which finally results in a stable, self-sustaining community.

In this chapter we focus on the complex web of community interactions that determine what species are found in stable ecosystems, and the community changes that occur during succession.

The Ecological Niche

Although the word *niche* may call to mind a small cubbyhole, in ecology it means much more. Each species occupies a unique **ecological niche** that encompasses all aspects of its way of life. The concept of the ecological niche includes the organism's physical home, or habitat. The habitat of a white-tailed deer, for example, is the eastern deciduous forest.

But the ecological niche goes far beyond habitat. It also specifies how the organism gets its supply of energy and materials—what might be called the organism's role or "occupation" within its ecosystem. An organism's predators, prey, and competitors as well as its behaviors and interactions are considered elements of its niche. In addition, the niche includes all the physical environmental factors necessary for survival, such as the range of temperatures under which the organism can survive, the amount of moisture it requires, the pH of the water or soil it may inhabit, the type of soil nutrients required, and the degree of shade it can tolerate. Although different types of organisms share many aspects of their niche with others, no two species ever occupy exactly the same ecological niche.

Competition between Species

Community interactions may be harmful or beneficial to the participating organisms (Table 29-1). During competition between different species, called **interspecific**

competition, two or more species attempt to use the same limited resources, particularly food or space or both. In the case of interspecific competition, both species are harmed in the sense that access to resources is limited. The intensity of interspecific competition depends on how similar the requirements of the two species are. In other words, the degree of competition is proportional to the amount of niche overlap between the competing species.

Just as no two organisms can occupy exactly the same physical space at the same time, no two species can inhabit the same ecological niche. This important concept, often called the **competitive exclusion principle**, was formulated in 1934 by the Russian microbiologist G. F. Gause. If two species with the same niches are placed together and forced to compete for limited resources, inevitably one will outcompete the other, and the less well-adapted of the two will die out. Gause demonstrated this principle using two species of the protist *Paramecium*: *P. aurelia* and *P. caudatum*. In separate laboratory flasks, the two species thrived on bacteria and fed in the same parts of the flask. Both populations flourished. But when Gause placed the two species together, one always eliminated, or "competitively excluded," the other (Fig. 29-1). Gause then repeated the experiment, replacing *P. caudatum* with another species, *P. bursaria*, which tended to feed

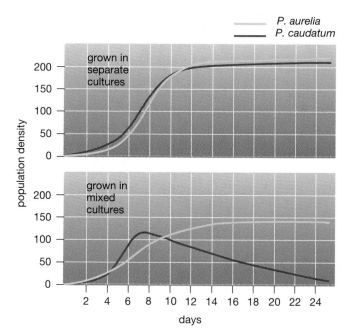

Figure 29-1   Competitive exclusion

(Top) Raised separately with a constant food supply, both *Paramecium aurelia* and *P. caudatum* show the S-curve typical of a population that initially grows rapidly, then stabilizes. (Bottom) Raised together and forced to occupy the same niche, *P. aurelia* consistently outcompetes *P. caudatum* and causes that population to die off. (Modified from G. F. Gause, *The Struggle for Existence*. Baltimore: Williams & Wilkins, 1934.)

TABLE 29-1 ⚏ *Interactions among Organisms*		
Type of Interaction	*Effect on Organism A*	*Effect on Organism B*
Competition between A and B	Harms	Harms
Predation by A on B	Benefits	Harms
Symbiosis		
Parasitism by A on B	Benefits	Harms
Commensalism of A with B	Benefits	No Effect
Mutualism between A and B	Benefits	Benefits

at the bottom of the flask. In this case, the two species of *Paramecium* were able to coexist indefinitely, because they occupied slightly different niches.

The ecologist R. MacArthur tested Gause's laboratory findings under natural conditions by investigating five species of North American warbler. These birds all hunt for insects and nest in the same type of spruce tree. Although the niches of these birds appear to overlap considerably, MacArthur found that the five species concentrated their search in specific areas of the tree, employed different hunting tactics, and nested at slightly different times. By partitioning the resource, the warblers minimize the overlap of their niches and reduce competition among the different species (Fig. 29-2).

As MacArthur discovered, when species with similar requirements coexist, they often occupy a smaller niche than either would if it were by itself. This phenomenon, called **resource partitioning**, is an evolutionary adaptation that reduces the harmful effects of interspecific competition. Resource partitioning is the outcome of the coevolution of species with extensive (but not total) niche overlap. Coevolution (introduced in Chapter 16) occurs when two or more species act as agents of natural selection on one another. Because natural selection favors individuals with fewer competitors, over evolutionary time the competing species develop physical and behavioral adaptations that minimize their competitive interactions. A dramatic example of this type of adaptation was discovered by Darwin among

finches of the Galapagos Islands. The finches had evolved different bill sizes and shapes and different feeding behaviors that reduced the competition among them (see Chapter 16).

Interspecific Competition Helps Control Population Size

Although coevolution tends to minimize niche overlap, closely related species still compete directly for limited resources. This competition may restrict the size and distribution of the competing populations. For example, barnacles of the genus *Chthamalus* share the rocky shores of Scotland with another genus of barnacle, *Balanus*, and their niches overlap considerably. The ecologist J. Connell found that *Chthamalus* dominates the drier upper shore and *Balanus* dominates the lower, which is more often submerged. When he scraped off *Balanus*, the *Chthamalus* population increased, spreading downward into the area its competitor had once inhabited. Because *Balanus* is intolerant of drying, removal of *Chthamalus* has little impact on the distribution of *Balanus*. Where the habitat is appropriate for both species, *Balanus* conquers, because it is larger and grows faster. But *Chthalamus* tolerates drier conditions, so on the upper shore, where only high tides submerge the barnacles, it has a competitive advantage. As this example illustrates, interspecific competition limits both the size and the distribution of the competing populations.

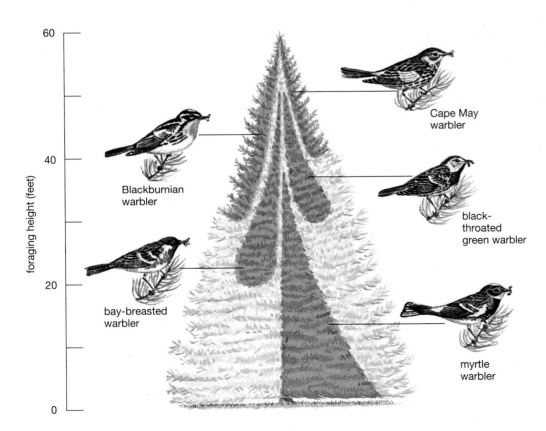

***Figure 29-2*   Resource partitioning**

These five insect-eating species of North American warbler search for food in different regions of spruce trees.

Blackburnian warbler

Cape May warbler

bay-breasted warbler

black-throated green warbler

myrtle warbler

foraging height (feet)

60

40

20

0

(a)

(b)

Figure 29-3   Forms of predation

(a) A predatory plant, the sundew, attracts and binds insects with its glistening sticky knobs. The insects provide supplemental nitrogen for the plant. (b) A few tropical bats feed on frogs or fish. Predatory bats home in on the mating calls of this tree frog.

Predators and Their Prey

Although predators are most commonly considered to be organisms that kill and eat other organisms, herbivorous animals are often included in this category even though they may kill only part of the plants they consume. For our purposes, then, predators include the antelope munching on sagebrush and the exotic sundew plant entangling and digesting its insect prey, as well as the bat swooping down on the frog and the bear catching a salmon (Fig. 29-3). Predators are usually either larger than their prey or hunt collectively, as wolves do when bringing down a moose. Predators are also less abundant than their prey, for reasons discussed in the next chapter.

Predator-Prey Interactions Shape Evolutionary Adaptations

To survive, predators must feed and prey must avoid becoming food. Therefore, these populations exert intense selective pressure on one another. As prey become more difficult to catch, predators must become more adept at hunting. Natural selection has endowed the cheetah with speed and camouflage spots and its zebra prey with speed and camouflage stripes. It has produced the keen eyesight of the hawk and the warning call of the Belding ground squirrel, the stealth of the jumping spider and the remarkable spider mimicry of the fly it stalks. In the following sections, we examine some of the results of predator-prey coevolution.

Bats and Moths Have Evolved Counteracting Strategies

Most bats are nighttime hunters that navigate and locate prey by echolocation. They emit extremely high-frequency and high-intensity pulses of sound and, by analyzing the returning echoes, create an "image" of their surroundings and nearby objects. In response to this unique prey-location system, certain moths (a favored prey of bats) have evolved simple ears that are particularly sensitive to the frequencies used by echolocating bats. When they hear a bat, these moths take evasive action, flying erratically or dropping to the ground. The bats may counter this defense by switching the frequency of their sound pulses away from the moth's sensitive range. Some moths have evolved a way to jam the bats' echolocation by producing their own high-frequency clicks. In response, when hunting a clicking moth, the bat may counter by turning off its own sound pulses temporarily and zeroing in on the moth by following the moth's clicks. These interactions dramatically illustrate the complexity of coevolutionary adaptations.

Camouflage Conceals Both Predators and Their Prey

An old saying of detective novels is that the best hiding place may be right out in plain sight! Both predators and prey have evolved colors and patterns that resemble their surroundings. Such coloration, called **camouflage**, renders animals inconspicuous even when they are in plain sight (Fig. 29-4). Some animals closely resemble specific (but uninteresting) objects such as leaves, twigs, or even bird droppings (Fig. 29-5). Camouflaged animals tend to remain motionless rather than fleeing their predators. (A fleeing bird dropping would be quite conspicuous!) While camouflaged animals often resemble plants, a few types of plants have evolved to resemble rocks, which are ignored by their herbivorous predators (Fig. 29-6).

Predators who ambush their prey rather than chasing it are also aided by camouflage. For example, the angler fish closely resembles the algae-covered rocks and algae on which it sits motionless, dangling a small lure from its upper lip (Fig. 29-7).

Figure 29-4   Camouflage by blending in

(a) The flounder's flat shape and its behavior (flipping sand over its back, then lying perfectly still) help the fish resemble a sandy bottom. The flounder can also alter its color and pattern somewhat to resemble its surroundings. (b) The tree bark mantis from Malaysia blends perfectly with the bark of this tree.

Figure 29-5   Camouflage by resembling specific objects

Resembling uninteresting parts of the environment allows some animals to avoid predation. (a) A leaf katydid from Costa Rica perfectly resembles a leaf, including an irregularity where the "leaf" has apparently been gnawed. (b) The leafy sea dragon (an Australian "seahorse" fish) has evolved extension of its body that precisely duplicates the algae in which it hides. (c) These Florida treehopper insects avoid detection by resembling thorns on a branch.

Figure 29-6   A plant that resembles a rock

This cactus of the American Southwest is appropriately called the living rock cactus.

Bright Colors Often Warn of Danger

Some animals have evolved very differently, exhibiting bright **warning coloration** (Fig. 29-8; see also Figs. 29-9, 29-13, and 29-14). These animals are usually distasteful and often extremely poisonous. Since poisoning your predator is small consolation if you have already been eaten, the bright colors declare: "Eat me at your own risk." A single unpleasant experience is enough to teach predators to avoid these conspicuous prey.

Some Organisms Gain Protection through Mimicry

Mimicry refers to a situation in which a species evolves to resemble something else, often another type of organism. For example, once warning coloration evolved, there arose a selective advantage for tasty, harmless animals to resemble poisonous ones. The deadly coral snake has brilliant warning coloration, and the harmless mountain king

Figure 29-7   Combining camouflage and aggressive mimicry

The angler fish waits in ambush, its body camouflaged to match the algae-encrusted rock on which it normally rests. Above its mouth dangles a lure, closely resembling a small fish. This will attract other would-be predators, who will find themselves prey.

snake avoids predation by resembling it (Fig. 29-9a, b). By resembling each other, two distasteful species may each benefit from predators' painful experience with the other. For example, predators rapidly learn to avoid the conspicuous stripes on bees, hornets, and yellowjackets. The monarch butterfly is poisonous and distasteful, and viceroy butterflies, which researchers have recently discovered are equally distasteful, have wing patterns strikingly similar to those of monarchs (Fig. 29-9c, d). A common color pattern results in faster learning by predators—and less predation on all similarly colored species.

Some predators have evolved **aggressive mimicry**, a "wolf-in-sheep's-clothing" approach to predation, in which they entice their prey to come close by resembling a harmless animal. For example, although the body of the angler fish is hidden by camouflage, it dangles a wriggling lure that

Figure 29-8   Warning coloration

The South American poison arrow frog, with its poisonous skin, advertises its unpleasant taste with bright and contrasting color patterns.

resembles a small fish just above its mouth. A curious fish attracted to the lure is quickly swallowed (see Fig. 29-7). The saber-toothed blenny fish imitates a harmless fish, called a cleaner wrasse, that picks parasites from the skin of larger fish. Parasite-ridden fish welcome the attention of the cleaner wrasse and are fooled by the saber-toothed blenny, who bites them instead of their parasites (Fig. 29-10).

Certain prey species use still another form of mimicry called **startle coloration**. Several insects, including certain moths and caterpillars, have evolved patterns of color that closely resemble the eyes of a much larger, and possibly dangerous, animal (Fig. 29-11). If a predator gets too close, the prey suddenly flashes its eye-spots, startling the predator enough to make an escape.

A sophisticated variation on the theme of prey who mimic predators is seen in snowberry flies, who are hunted by territorial jumping spiders. When a fly spots an approaching spider, it spreads its wings, moving them back and forth in a jerky dance. Seeing this display, the spider is likely to flee from the harmless fly. Why? Researchers recently observed that the markings on the fly's wings closely resemble the legs of another jumping spider. The jerky movements of the fly mimic those made by a jumping spider when it drives another spider from its territory (Fig. 29-12). Natural selection has finely tuned both the behavior and the appearance of the fly to avoid predation by jumping spiders.

Some Predators and Prey Engage in Chemical Warfare

Both predators and prey have evolved a variety of toxic chemicals for attack and defense. Many plants produce defensive toxins. For example, lupines produce chemicals called alkaloids, which deter attack by the blue butterfly, whose larvae feed on the lupine's buds. In fact, different individuals of the same species of lupine produce different forms of this toxin, thus making the evolution of resistance to it more difficult. The venom of spiders and poisonous snakes such as the coral snake (see Fig. 29-9a) serves both to paralyze prey and to deter predators.

Certain mollusks, including squid, octopuses, and some sea slugs, emit clouds of ink when attacked. These colorful chemical "smoke screens" confuse their predators and mask their escape. A dramatic example of chemical defense is seen in the bombardier beetle. In response to the bite of an ant, the beetle releases secretions from special glands into an abdominal chamber. Here, enzymes catalyze an explosive chemical reaction that shoots a toxic, boiling-hot spray onto the attacking ant (Fig. 29-13).

Plants and Herbivores Have Many Coevolutionary Adaptations

Plants too have evolved chemical adaptations that deter predators. Many, such as the milkweed, synthesize toxic and distasteful chemicals. Animals rapidly learn not to eat foods that make them sick, so milkweeds and other toxic

(a)

(b)

(c)

(d)

Figure 29-9   **Warning mimicry**

The warning coloration of the poisonous coral snake **(a)** is mimicked by the harmless king snake **(b)**. Mimicry of the warning coloration of the distasteful monarch butterfly **(c)** by the equally distasteful viceroy **(d)**. After attempting to eat either of these species, a bird will avoid both of them.

plants suffer little nibbling. Consequently, such plants are often very abundant; any animal immune to the plant poisons enjoys a bountiful food supply. As plants evolved toxic chemicals for defense, certain insects evolved increasingly efficient ways to detoxify or even make use of the chemicals. The result is that nearly every toxic plant is eaten by at least one type of insect. For example, monarch butterfly caterpillars consume the toxic milkweed. The caterpillars not only tolerate the milkweed poison but also store it in their tissues as a defense against their own predators. The stored toxin is even retained in the metamorphosed monarch butterfly (see Fig. 29-9c). Although the desert camphor weed oozes a sticky, noxious resin that prevents most animals from eating it, the predatory assassin bug collects the substance and smears it on her abdomen (Fig. 29-14). The camphor "glue" then coats her eggs as they emerge, making them distasteful to predators. Then, soon after hatching, young assassin bugs scrape the sticky substance from their discarded eggshells and transfer it to their forelegs, where it helps them capture their own insect prey. In this complex association, the assassin bug has evolved behaviors that allow it to utilize the defenses evolved by a plant against both its own predators and its prey.

Grasses have evolved tough silicon (glassy) substances in their blades, discouraging all predators except those with strong, grinding teeth and powerful jaws. Thus, grazing animals have come under selective pressure for longer, harder teeth. An example is the evolution of horses and the grass they eat. As grasses evolved tougher blades that reduce predation, horses evolved longer teeth with thicker enamel coatings that resist wear.

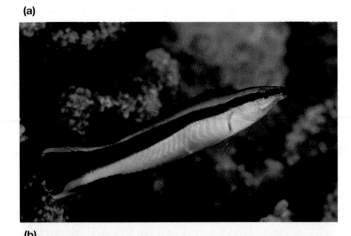

(a)

(b)

Figure 29-10 **Aggressive mimicry**

(a) Some fish, such as this marine wrasse, obtain food by eating parasites off the bodies of larger fish, a mutualistic relationship called a *cleaning symbiosis*. Fish welcome the approach of this predator of parasites. (b) The saber-toothed blenny has evolved to closely resemble the cleaner wrasse. Fish that allow the blenny to approach usually get a chunk bitten out of them—hence the term *aggressive mimicry*.

Keystone Predators Influence Community Structure

In some communities, a specific predator, described as a **keystone predator**, plays a major role in determining community structure. Removal of a keystone predator dramatically alters the community. For example, when the predatory sea star *Pisaster* was removed from sections of the rocky intertidal coast of the western United States, mussels (bivalve mollusks which are favored prey of *Pisaster*) became so abundant that they outcompeted algae and other invertebrates that normally coexist in intertidal communities. The African elephant is a keystone predator in the African savanna. By grazing on small trees and bushes, elephants prevent the encroachment of forests and help maintain the grassland community. It is difficult to identify keystone predators, since the predator must be selectively removed and the community studied for several years both before and after its removal. Therefore, ecologists have not yet determined whether communities dominated by keystone predators are rare or relatively common.

Symbiosis

Symbiosis, which literally means "living together," is defined as a close interaction between organisms of different species for an extended time. Considered in its broadest sense, symbiosis includes parasitism, commensalism, and mutualism. Although one species always benefits in symbiotic relationships, the second species may be harmed, not affected, or benefited (see Table 29-1).

Parasitism Harms, but Does Not Immediately Kill, the Host

Parasitism occurs when one species lives in or on its prey (also called host), often harming or weakening it but not

(a)   **(b)**   **(c)**

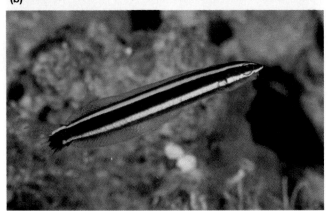

Figure 29-11 **Startle coloration**

(a) The peacock moth from Trinidad is well camouflaged, but should a predator approach too closely, it suddenly opens its wings to reveal spots resembling large eyes **(b).** The predator is startled, and the moth gets a chance to flee. (c) Would-be predators of this caterpillar larva of the swallowtail butterfly are deterred by its close resemblance to a snake. Note that the caterpillar's head is the snake's "nose."

Figure 29-12   Behavioral mimicry

In response to the approach of a jumping spider (top), the snowberry fly spreads its wings, revealing a pattern resembling spider legs (bottom). The fly enhances the effect by performing a jerky, side-to-side dance that resembles the leg-waving display of another jumping spider defending its territory.

immediately killing it. Although it is sometimes difficult to distinguish clearly between a predator and a parasite, parasites are generally much smaller and more numerous than their hosts. Familiar parasites include tapeworms, fleas, and numerous disease-causing protozoa, bacteria, and viruses. Many parasites, particularly worms and protozoa, have complex life cycles involving two or more hosts. There are few parasitic vertebrates; the lamprey eel, which attaches itself to a host fish and sucks its blood, is a rare example.

The variety of infectious bacteria and viruses and the precision of the immune system that counters their attacks are evidence of the powerful forces of coevolution between parasitic microorganisms and their hosts. Another example is the malaria parasite, which has provided strong selection pressure for a gene in humans that causes sickle-cell anemia. The parasites can't survive in the affected red blood cells. In certain areas of Africa where malaria is common, one-quarter of the human population carries the sickle-cell gene. *Trypanosoma*, a parasitic protozoan, causes both human sleeping sickness and a disease in cattle called nagana. African antelope, which coevolved with this parasite, are relatively unaffected by it. Cattle, a more recently introduced species, usually suffer but survive infection if they have been bred in an infested area for many generations. Newly imported cattle, however, generally die if not treated.

Commensalism Benefits One Species without Affecting the Other

Commensalism occurs when the relationship between two species benefits one without affecting the other. For example, as herds of bison graze, they disturb insects dwelling in the grass. Birds follow the bison, eating the insects. The

Figure 29-13   Chemical warfare

The bombardier beetle sprays a hot toxic brew in response to a leg pinch.

Figure 29-14   A complex coevolutionary relationship

A female assassin bug scrapes noxious resin from a camphor weed. Her warning coloration alerts potential predators of her painful sting.

(a)

(b)

Figure 29-15   **Commensal relationships**

(a) This tropical orchid clings to a rainforest tree. Its perch allows it access to sunlight without harming the tree. **(b)** Clusters of small barnacles hitchhike harmlessly on the hide of this manatee.

birds, which benefit, are considered commensal with the bison, who are unaffected. Birds and the trees in which they nest are involved in a commensal relationship. The birds obtain shelter and protection without affecting the trees. Many orchids attach to trees without harming them. The orchid obtains support and access to sunlight high above the dim tropical forest floor. Barnacles are harmless hitchhikers on marine animals, including whales and manatees, but benefit from a free ride through food-rich waters (Fig. 29-15).

In Mutualistic Interactions, Both Species Benefit

When two organisms interact in a way that benefits both, the relationship is called **mutualism**. Flowering plants and their pollinators have a mutualistic association. Mutualistic associations occur in the digestive tracts of cows and termites, where protists and bacteria find food and shelter while helping their hosts extract nutrients, and in our own intestines, where bacteria synthesize certain vitamins (see Chapter 23). The nitrogen-fixing bacteria inhabiting special chambers on the roots of legume plants are another important example. The bacteria obtain food and shelter from the plant and in return trap nitrogen in a form the plant can utilize. Some mutualistic partners have coevolved to the extent that neither can survive alone. An interesting example is the ant-acacia mutualism described in "Scientific Inquiry: Deciphering the Relationship between Ants and Acacias."

Mutualistic relationships involving vertebrates are rare and typically are less intimate and extended, as in the relationship of the cleaner wrasse (see Fig. 29-10) and the fish it cleans. Another is seen in the clown fish, which takes shelter among the venomous tentacles of the anemone. The fish derive shelter and protection and, at least occasionally, bring bits of food to their anemone host (Fig. 29-16).

Succession: Community Changes over Time

In a mature terrestrial ecosystem, the populations that make up the community interact with one another and with their nonliving environment in intricate ways. But this tangled web of life did not spring fully formed from bare rock or naked soil; rather, it emerged in stages over a long period, a process called succession. **Succession** is a change in a community and its nonliving environment over time. It is a kind of "community relay" in which assemblages of plants and animals replace one another in a sequence that is at least somewhat predictable.

Succession occurs under a variety of circumstances but is most easily observed in terrestrial and freshwater ecosys-

Figure 29-16   **Mutualism**

The clownfish snuggles unharmed among the stinging tentacles of the anemone. The fish obtains protection and sometimes brings food to the anemone in this mutualistic relationship.

SCIENTIFIC INQUIRY

Deciphering the Relationship between Ants and Acacias

Daniel Janzen, a doctoral student at the University of Pennsylvania, was walking down a road in Veracruz, Mexico, when he saw a flying beetle alight on a thorny tree, only to be driven off by an ant. Looking more closely, he saw that the tree, a bull's-horn acacia, was covered with ants. A large ant colony of the genus *Pseudomyrmex* made its home inside the enlarged thorns of the plant, whose soft pulpy interiors are easily excavated to provide shelter (Fig. E29-1).

To determine how important the ants are to the tree, Janzen began stripping the thorns by hand until he found and removed the thorn housing the ant queen, thus destroying the colony. He later turned to more efficient but dangerous methods, eliminating all the ants on a large stand of acacias with an insecticide. The acacias were unharmed by the poison, Janzen became ill from it, and the ants were all killed. Within a year of the spraying, Janzen found nearly all the acacia trees dead, consumed by insects and other herbivores and shaded out by competing plants. The ground surrounding the trees, which the ants normally kept trimmed, was overgrown. The trees were apparently dependent on their resident ants for survival.

Wondering if the ants could survive off the tree, Janzen painstakingly peeled the ant-inhabited thorns off 100 acacia trees, suffering multiple stings in the process. He housed each ant colony in a jar provided with local nonacacia vegetation and insects for food. The ant colonies all starved. Carefully examining the acacia, he found swollen structures filled with sweet syrup at the base of the leaves and protein-rich capsules on the leaf tips (Fig. E29-1, inset). Together, these provide a balanced diet for the ants.

Janzen's experiments strongly suggest that these species of ant and acacia have an obligatory mutualistic relationship: Neither can survive without the other. Of course, further observations were required to confirm this relationship. The fact that the ants starved in Janzen's jars did not rule out that they might survive successfully elsewhere, but, in fact, this species of ant is never found living independently of the acacia. Similarly, the bull's-horn acacia is never found without its resident ant colony. Thus, a chance observation followed by careful research led to the discovery of an important mutualistic association.

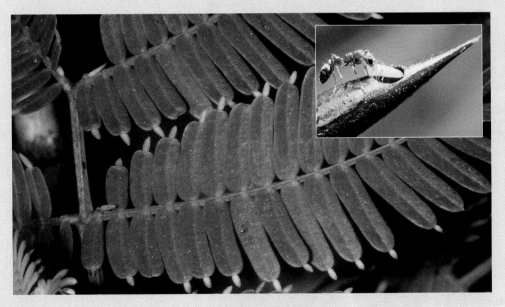

Figure E29-1   A mutualistic relationship

Yellow, protein-rich capsules are produced at the tips of certain acacia leaves. These provide food for the resident ants. (Inset) A hole in the enlarged thorn of the bull's-horn acacia provides shelter for members of the ant colony. The ant entering the thorn is carrying a food capsule produced by the acacia. As the ant colony grows, more thorns are invaded.

tems. Freshwater ponds and lakes tend to undergo a series of changes that transform them first into marshes and eventually to dry land. Shifting sand dunes are stabilized by creeping plants and may eventually support a forest. Volcanic eruptions may, as in the case of Mount St. Helens, wipe out previously existing ecosystems, or they may create new islands which are soon colonized. Forest fires create a nutrient-rich environment that encourages rapid invasion of new life (Fig. 29-17).

The precise changes occurring during succession are as diverse as the environments in which succession occurs, but certain general stages can be recognized. In each case, succession is begun by a few hardy invaders called **pioneers** and ends with a diverse and relatively stable **climax com-**

(a)

(b)

(c)

Figure 29-17   **Succession in progress**

(a) (Left) On May 18, 1980, the explosion of Mount St. Helens in Washington devastated the pine forest ecosystem on its sides. (Right) Nine years later, pioneering evergreens were about 4 to 6 inches tall, and shrubs carpeted the nutrient-rich ash. **(b)** (Left) In the summer of 1988, extensive fires swept through the forests of Yellowstone National Park in Wyoming. (Right) Two years later, flowering plants had sprung up in the sunlight, and wildlife populations were rebounding. **(c)** (Left) The Hawaiian volcano Mount Kilauea has erupted repeatedly during the past 8 years, sending rivers of lava over the surrounding countryside. (Right) An early colonizer of a lava flow.

munity. As the community progresses from the pioneers to the climax community, the organisms gradually alter the nonliving environment. Ironically, these changes often favor competitors, which displace the existing populations. The climax community differs from earlier successional stages because it no longer alters the environment significantly. The climax community will persist unless external forces, such as human activities, natural events such as fires, hurricanes, or floods, or a gradual change in climate alter it.

During terrestrial succession certain general trends occur in ecosystem structure:

1. The soil increases in depth, and its content of organic material (derived from the bodies of organisms) also increases.

2. The overall productivity (the amount of organic material produced in a given area over a given time) increases.

3. The number of species increases, as does the number of interactions within the community.

4. Longer-lived species come to dominate the ecosystem, and the rate at which populations replace one another slows. The presence of longer-lived species combined with more complex community interactions frequently results in a community that is more stable and resistant to change.

5. In the climax community, the total mass of living organisms reaches a maximum, and the species present no longer alter the ecosystem in ways that encourage the growth of their competitors.

There Are Two Major Forms of Succession: Primary and Secondary

Succession on dry land takes two major forms: primary and secondary. During **primary succession,** an ecosystem is forged from bare rock, sand, or a clear glacial pool where there is no trace of a previous community. The for- mation of an ecosystem "from scratch" is a process often requiring thousands or even tens of thousands of years. During **secondary succession,** a new ecosystem develops after an existing ecosystem is disturbed, as in the case of a forest fire or an abandoned farm field. It happens much more rapidly than primary succession because the previous community has left its mark in the form of soil and seeds. Succession in an abandoned farm field in the southeastern United States can reach its climax after two centuries. In the examples below, we examine these processes in more detail.

Primary Succession Can Begin on Bare Rock

Figure 29-18 illustrates primary succession as studied on Isle Royale, Michigan, an island in Lake Superior. Bare rock, such as that exposed by a retreating glacier, begins to liberate nutrients such as minerals by weathering. Cracks form as the rock alternately freezes and thaws, contracting and expanding. For lichens (symbiotic associations of fungi and algae), the weathered rock provides a place to attach where there are no competitors and plenty of sunlight. Lichens can photosynthesize, and they obtain minerals by dissolving some of the rock with an acid they secrete. As the pioneering lichens spread over the rock, drought-resistant, sun-loving mosses begin growing in the cracks. Fortified by nutrients liberated by the lichens, the moss forms a dense mat that traps dust, tiny rock particles, and bits of organic debris. The death of some of the moss adds to a growing nutrient base, while the moss mat itself acts as a sponge, trapping moisture. Within the moss, seeds of larger plants germinate. Eventually, their bodies contribute to a growing layer of soil. As woody shrubs such as blueberry and juniper take advantage of the newly formed soil, the moss and lichens may be shaded out and buried by decaying leaves and vegetation. Eventually, trees such as

Figure 29-18 **Primary succession**

Here, primary succession is shown occurring on bare rock in upper Michigan.

lichen on bare rock   bluebell, yarrow   blueberry, juniper   jack pine, black spruce, aspen   balsam fir, paper birch, white spruce, climax forest

time (years)

0   1000

E A R T H   W A T C H

Introduced Species Disrupt Community Interactions

Exotic species, or species introduced into ecosystems where they did not evolve, sometimes find no predators or parasites in their new environment, and they find new prey that have little defense against them. The unchecked population growth of these invaders may seriously damage the ecosystem as they displace, outcompete, and prey on native species. Both starlings and English sparrows have spread dramatically since their deliberate introduction to the eastern United States in the 1890s. Their success threatens our native bluebird. Aggressive African bees, dubbed "killer bees," were introduced into Brazil in 1959 and began spreading northward. They have now made their way into Texas and California and are continuing to spread, oucompeting local bee species and threatening the U.S. beekeeping industry. A Japanese vine called kudzu was introduced to the South as a decorative plant in 1876 and then planted extensively in the 1940s to control erosion. Today kudzu is a major pest, overgrowing and killing trees and underbrush, and occasionally engulfing small houses while their owners are vacationing (Fig. E29-2b). The water hyacinth, introduced from South America as an ornamental plant, now clogs about 2 million acres of southern lakes and waterways, slowing boat traffic and displacing natural vegetation (Fig. E29-2c).

In 1986, a trading vessel bringing cargo from Europe up the St. Lawrence Seaway discharged fresh water containing millions of larvae of the zebra mussel into the St. Clair River near Detroit. By 1989, the invading mussels had spread to Lake Erie. Now they inhabit all the Great

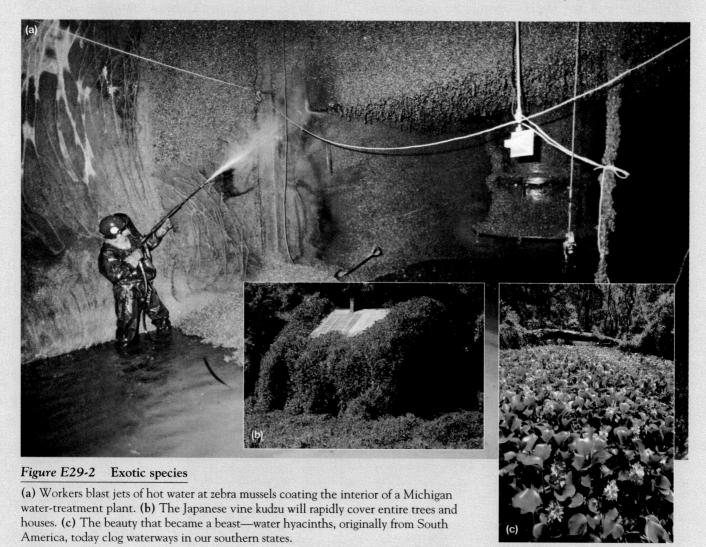

Figure E29-2   **Exotic species**

(a) Workers blast jets of hot water at zebra mussels coating the interior of a Michigan water-treatment plant. (b) The Japanese vine kudzu will rapidly cover entire trees and houses. (c) The beauty that became a beast—water hyacinths, originally from South America, today clog waterways in our southern states.

Lakes and the Upper Mississippi, and they are spreading throughout eastern waterways. They cover piers, boats, machinery, and beaches (Fig. E29-2a). At water-treatment plants on the Great Lakes, zebra mussel populations reach 600,000 per square yard, clogging pipes and reducing water flow to a trickle (Fig. E29-2a). The mussels cover and suffocate other forms of shellfish, threatening many endangered varieties with extinction. The U.S. Fish and Wildlife Service estimates that zebra mussel damage to fisheries, utilities, and industries dependent on the lakes they infest may cost $5 billion over the next 10 years.

By evading the checks and balances imposed by millennia of coevolution, exotic species are wreaking havoc on natural ecosystems throughout the world.

jack pine and blue spruce take root in the deeper crevices, and the sun-loving shrubs are shaded out. Within the depths of the forest, shade-tolerant seedlings of taller or faster-growing trees, such as balsam fir, paper birch, and white spruce, thrive. Eventually these overtower and replace the original trees, which are intolerant of shade. After a thousand years or more, a tall climax forest thrives on what was once bare rock.

An Abandoned Farm Field Will Undergo Secondary Succession

Figure 29-19 illustrates succession on an abandoned Southeastern farm. The pioneers are fast-growing annual weeds such as crabgrass, ragweed, and sorrel, which root in the rich soil already present and thrive in direct sunlight. A few years later, perennial plants such as asters, goldenrod, broom sedge grass, and woody shrubs, such as blackberry, invade. These plants multiply rapidly and dominate for the next few decades. Eventually, they are replaced by pines and fast-growing deciduous trees, such as tulip poplar and sweet gum, which sprout from windblown seeds. These trees become prominent after about 25 years, and a pine forest dominates the field for the rest of the first century. Meanwhile, shade-resistant, slow-growing hardwoods such as oak and hickory take root beneath the pines. After the first century, these begin to overtower and shade the pines, which eventually die from lack of sun. A relatively stable climax forest dominated by oak and hickory is present by the end of the second century.

Succession Also Occurs in Ponds and Lakes

In freshwater ponds or lakes, succession occurs both from changes within the pond or lake and as a result of an influx of nutrients from outside the ecosystem. Sediments and nutrients carried in by runoff from the surrounding land have a particularly large impact on small freshwater lakes, ponds, and bogs, which gradually undergo succession to dry land (Fig. 29-20).

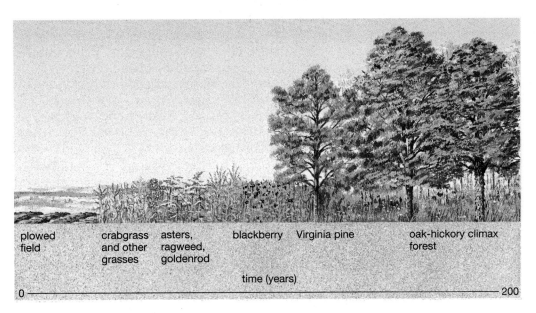

plowed field | crabgrass and other grasses | asters, ragweed, goldenrod | blackberry | Virginia pine | oak-hickory climax forest

time (years)

0   200

Figure 29-19
Secondary succession

Here, secondary succession is illustrated on a plowed, abandoned southeastern farm field.

(a)   **(b)**   **(c)**

Figure 29-20   **Succession in a freshwater pond**

In small ponds, succession is speeded by an influx of materials from the surroundings. **(a)** In this small pond, dissolved minerals carried by runoff from the surroundings support aquatic plants, whose seeds or spores were carried here by the winds or by birds and other animals. **(b)** Over time, the decaying bodies of aquatic plants build up soil that provides anchorage for more terrestrial plants. **(c)** Finally, the pond is entirely converted to dry land.

Succession Culminates in the Climax Community

Succession ends with a relatively stable climax community. In your travels, you have undoubtedly noticed that the type of climax community varies dramatically from one area to the next. For example, if you drive through Colorado, you will see a shortgrass prairie climax community on its eastern plains, spruce-pine forests in its mountains, tundra on their uppermost reaches, and sagebrush-dominated climax communities in its western valleys.

The exact nature of the climax community is determined by numerous geological and climatic variables, including temperature, rainfall, elevation, latitude, type of rock (which influences the type of nutrients available), exposure to sun and wind, and many more. Human activities may also dramatically alter the climax vegetation. Large stretches of grasslands in the West, for example, are now dominated by sagebrush because of overgrazing. The grass that normally outcompetes sagebrush is selectively eaten by cattle, allowing the sagebrush to prosper.

Some ecosystems are not allowed to reach the climax stage but are maintained in an earlier stage called a **subclimax.** The tallgrass prairie that once covered northern Missouri and Illinois is actually a subclimax of an ecosystem whose climax community is deciduous forest. The prairie was maintained by periodic fires, some set by lightning and others deliberately set by Native Americans to increase grazing land for buffalo. Forest now encroaches, and limited prairie preserves are maintained by carefully managed burning.

Agriculture also depends on the artificial maintenance of carefully selected subclimax communities. Grains are specialized grasses characteristic of the early stages of succession, and much energy goes into preventing competitors (weeds and shrubs) from taking over. The suburban lawn is also a painstakingly maintained subclimax ecosystem. Mowing destroys woody invaders, and selective herbicides kill pioneers such as crabgrass and dandelions.

To study succession is to study the variations in communities over *time*. The climax communities that form during succession are strongly influenced by climate and geology—the distribution of ecosystems in *space*. Deserts, grasslands, and deciduous forests are climax communities formed over broad geographical regions with similar environmental conditions; these regions of characteristic plant communities are called **biomes.** In Chapter 31 we explore some of the great biomes of the world. Although the communities that make up the various biomes differ radically in the types of populations they support, communities worldwide are structured according to general rules. These principles of ecosystem structure are described in Chapter 30.

✦ SUMMARY OF KEY CONCEPTS

The Ecological Niche

The ecological niche defines all aspects of a species' habitat and interactions with its living and nonliving environment. Each species occupies a unique ecological niche.

Competition between Species

Interspecific competition occurs when the niches of two populations within a community overlap and is proportional to the amount of niche overlap. When two species with the same niche are forced under laboratory conditions to occupy the same ecological niche, one species always outcompetes the other. Species within natural communities have evolved in ways that avoid excessive niche overlap, with behavioral and physical adaptations that allow resource partitioning. Interspecific competition limits both the size and the distribution of competing populations.

Predators and Their Prey

Predators eat other organisms and are usually both larger and less abundant than their prey. Predators and prey act as strong agents of selection on one another. Prey animals have evolved a variety of protective colorations that render them either inconspicuous (camouflage) or startling to their predators (startle coloration). Some prey have become poisonous and often exhibit warning coloration by which they are readily recognized and avoided. The situation in which one species has evolved to resemble another is called mimicry. Both predators and prey have evolved a variety of toxic chemicals for attack and defense. Plants that are preyed upon have evolved elaborate defenses ranging from poisons to thorns to overall toughness. These defenses, in turn, have selected for predators that can detoxify poisons, ignore thorns, and grind down tough tissues.

Symbiosis

Symbiotic relationships involve two different species that interact closely over an extended time. Symbiosis includes parasitism, in which the parasite feeds on a larger, less abundant host, usually harming it but not killing it immediately. In commensal symbiotic relationships, one species benefits, often by finding food more easily in the presence of another species, which is not affected by the association. Mutualism benefits both symbiotic species.

Succession: Community Changes over Time

A change in the types of populations inhabiting an ecosystem over time is called succession. Primary succession, which may take thousands of years, occurs where no remnant of a previous community exists. Primary succession could occur on rock scraped bare by a glacier or cooled from molten lava, a sand dune, or in a newly formed glacial lake. Secondary succession occurs much more rapidly, because it builds on the remains of a disrupted community, such as an abandoned field or the aftermath of a fire. During succession, pioneer organisms invade and alter the environment by their presence. Community changes occur as one group of organisms, particularly plants, alters the environment in ways that encourage growth of their competitors. These competitors replace the first colonizers and are subsequently replaced by others. Some ecosystems, including tallgrass prairie and farm fields, are maintained in relatively early stages of succession by periodic disruptions. Uninterrupted succession ends with a relatively stable group of organisms called the climax community.

✦ KEY TERMS

aggressive mimicry   p. 582
biome   p. 592
camouflage   p. 580
climax community   p. 587
coevolution   p. 577
commensalism   p. 585
community   p. 577
competitive exclusion principle   p. 578

ecological niche   p. 578
exotic species   p. 590
interspecific competition   p. 578
keystone predator   p. 584
mimicry   p. 581
mutualism   p. 586
parasitism   p. 584
pioneer   p. 587

primary succession   p. 589
resource partitioning   p. 579
secondary succession   p. 589
startle coloration   p. 582
subclimax   p. 592
succession   p. 586
symbiosis   p. 584
warning coloration   p. 581

✖ T H I N K I N G   T H R O U G H   T H E   C O N C E P T S

Multiple Choice

1. Which of the following is NOT usually true of climax communities? They have
   a. more species than early successional communities
   b. more organic matter in the soil
   c. more trophic levels
   d. more short-lived species
   e. more productivity

2. What were the differences in the niches of the *Paramecium* in Gause's second experiment that allowed them to coexist?
   a. food eaten
   b. body size
   c. feeding area
   d. preferred water temperature
   e. preferred pH

3. Which of the following is an example of a commensal relationship?
   a. orchids growing on tree limbs
   b. flowering plants and their pollinators
   c. lupines and blue butterflies
   d. bats and moths
   e. monarch and viceroy butterflies

4. What is the function of aggressive mimicry?
   a. to hide a prey from a predator
   b. to warn a predator that a prey is dangerous
   c. to warn a predator that a prey is distasteful
   d. to keep prey from recognizing a predator
   e. to startle a prey when it sees a predator

5. What is coevolution?
   a. two species selecting for different traits in each other
   b. individuals of two species living together
   c. the presence of two species in the same community
   d. two species evolving together through time

   e. individuals of two species learning how to coexist with or to hunt with each other

6. The competitive exclusion principle implies that
   a. coexisting species can use the same resources
   b. coexisting species cannot eat exactly the same things
   c. coexisting species cannot have identical ecological interactions
   d. coexisting species cannot be exactly the same size
   e. coexisting species cannot be closely related to one another

Review Questions

1. Define an ecological community, and list three important types of community interactions.

2. An ecologist visiting an island finds two very closely related species of birds, one of which has a slightly larger bill than the other. Interpret this finding with respect to the competitive exclusion principle and the ecological niche, defining both terms.

3. List four general ways in which plants and animals protect themselves against being eaten, and provide a specific example of each.

4. List three different types of symbiosis; define and provide an example of each.

5. Introduced species are most often invasive when their new home lacks effective predators. What does this situation imply about the factors controlling populations of these species in their native ecosystems?

6. List two subclimax and two climax communities. How do they differ?

7. Describe succession, and explain why it occurs. Detail one of the successional sequences in this chapter.

8. What type of succession would occur on a clear-cut section of a national forest, and why?

✖ A P P L Y I N G   T H E   C O N C E P T S

1. Herbivorous animals that eat seeds are considered by some ecologists to be predators of plants, while herbivorous animals that eat leaves are considered to be parasites of plants. Discuss the validity of this classification scheme.

2. There are many more cases in which mutualisms seem to have evolved from parasite-host interactions than from predator-prey interactions. Ecologists think the reason is that it is often in the best interests of a parasite not to hurt its host. Try to think of why this would be true—when is what is good for a host good for its parasite? When not? Contrast this situation to predation.

3. Restoration ecologists try to rebuild more natural communities in areas damaged by humans, such as mine tailings, overgrazed pastures, and dredged wetlands. Essentially, these scientists try to help or direct natural succession to reclaim damaged lands. Think about a situation that restoration ecologists might face: restoring an abandoned rock quarry (which is almost pure gravel) to forest. What kinds of plant species

would you first seed into the site? What could you do to alter the physical environment to hasten the speed of succession?

4. The successional sequence of small ponds depends on the inflow and outflow of the pond. In particular, ponds with no outflowing stream quite rapidly undergo succession to a marsh and then to dry land. This sequence is much slower, and may not happen at all, in ponds with outflows. Explain why, using your understanding of the changes that occur during succession.

5. Would you expect in primary succession or secondary succession that late-successional species would most rely on the growth of the pioneer species to make the site suitable for them? Explain why, and describe how this explanation accounts for the differences in the speed of the two kinds of succession.

6. Many natural communities rarely or never reach a climax community because of natural disturbances. Indeed, human prevention of natural disturbances, especially for-

est fires, is often as destructive to communities as are more direct human impacts. Fires typically create a patchwork of areas in many different successional stages. Explain why this kind of patchwork can often support many more species than a uniform landscape of either climax community or early successional areas. Argue for or against programs that suppress fires—should Smokey the Bear retire? Why or why not?

7. Few regions of the United States are unaffected by introduced exotic species. Find out about an introduced species in your area (e.g., through news articles, by consulting an ecologist in your Biology Department, or calling a local Fish and Wildlife Service office). Learn where it came from, how it got here, and what impact it has had or might have on local native species. Could its introduction have been prevented? Suggest several ways to limit the introduction of exotic species.

✼ FOR MORE INFORMATION

Amos, W. H. "Hawaii's Volcanic Cradle of Life." *National Geographic*, July 1990. A naturalist explores succession on lava flows.

Daniels, P. "How Flowers Seduce the Bugs and the Bees." *International Wildlife*, November/December 1984. Beautifully illustrated description of plants and their mutualistic pollinators.

Horn, H. S. "Forest Succession." *Scientific American*, May 1975 (Offprint No. 1321). Describes succession in eastern deciduous forests.

Mills, E. L., Leach, J. H., Carlton, J. T., and Secor, C. L. "Exotic Species and the Integrity of the Great Lakes: Lessons from the Past." *BioScience* 44:666–676, 1994. The history of invasions by various species and their consequences.

Moore, J. "The Behavior of Parasitized Animals." *BioScience* 45:89–96, 1995. The ways in which parasites have evolved to manipulate the behavior of their hosts and hosts resist these effects.

Rinderer, T. E., Oldroyd, B. P., and Sheppard, W. S. "Africanized Bees in the U.S." *Scientific American*, December 1993. What are the likely impacts of these aggressive invaders?

Williams, T. "Invasion of the Aliens." *Audubon*, September/October 1994. Exotic plants, imported deliberately, are changing the American landscape.

NET WATCH

On-line resources for this chapter are on the World Wide Web at:
http://www.prenhall.com/~audesirk (click on the table of contents link and then select Chapter 29).

30 How Ecosystems Work

> "A thing is right when it tends to preserve the integrity, stability, and beauty of the biotic community. It is wrong when it tends otherwise."
>
> ALDO LEOPOLD in *A Sand County Almanac* (1949)

A microcosm is reflected in a dewdrop on the spore case of a moss. The sparkling energy of sunlight, captured by photosynthetic organisms, powers nearly all life on Earth.

Nearly all life on Earth is powered by the energy of sunlight, from the leaping of the jackrabbit to the active transport of molecules through a cell membrane. Each time this energy is used, some of it is lost as heat. But while solar energy continuously bombards Earth and is continuously lost as heat, nutrients remain. They may change in form and distribution, but nutrients do not leave the world ecosystem and are continuously recycled. Two basic laws thus underlie ecosystem function. First, energy moves *through* ecosystems in a continuous one-way flow. Energy needs constant replenishment from an outside source, the sun. Second, in contrast to energy, nutrients constantly cycle and recycle in a circular flow *within* ecosystems (Fig. 30-1). These laws shape the complex interactions within living communities.

In Chapters 7 and 8, you learned how energy is trapped by photosynthesis, released by cellular respiration, and used to construct the complex molecules of life. In Chapter 29, you learned some of the complex ways in which organisms interact within ecological communities. In this chapter, we relate some of these basic principles to the way ecosystems work.

The Flow of Energy

The Energy Captured by Photosynthetic Organisms Is Called Primary Productivity

Ninety-three million miles away, the sun fuses hydrogen into helium, releasing tremendous quantities of energy. A tiny fraction of this energy reaches Earth in the form of electromagnetic waves, including heat, light, and ultraviolet energy. Of the energy that reaches Earth, much is reflected by the atmosphere, clouds, and Earth's surface. Still more is absorbed as heat by Earth and its atmosphere, leaving only about 1% to power all life on Earth. Of this 1%, green plants capture 3% or less. The teeming life on Earth is thus supported by less than 0.03% of the energy reaching it from the sun. But how does this energy enter the biological community? The energy that powers ecosystems enters through the process of photosynthesis.

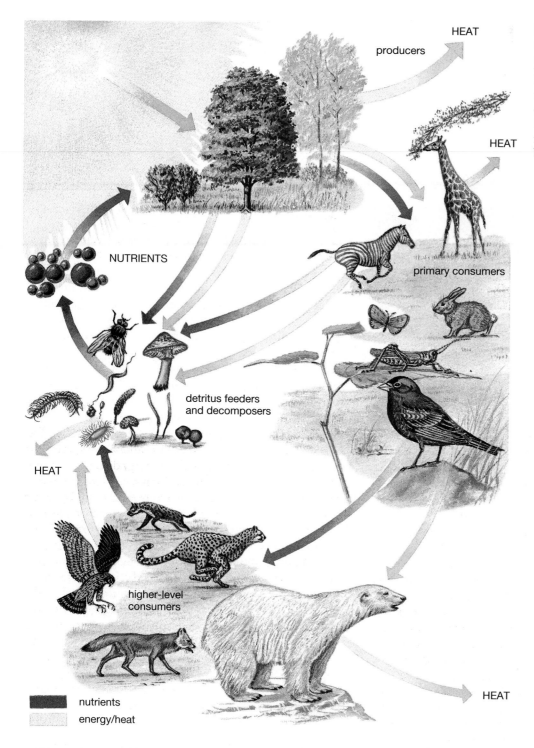

HEAT

producers

HEAT

primary consumers

NUTRIENTS

detritus feeders
and decomposers

HEAT

higher-level
consumers

HEAT

nutrients

energy/heat

Figure 30-1   **Energy flow,
nutrient cycling, and feeding
relationships in ecosystems**

Note that nutrients (purple)
neither enter nor leave the
cycle, whereas energy (yellow)
is lost at each level as heat and
must be constantly replenished
by sunlight.

In photosynthesis (performed by plants, plantlike protists, and cyanobacteria), pigments such as chlorophyll absorb certain wavelengths of sunlight. This solar energy is used to combine carbon dioxide and water into sugar, a compound that stores energy in chemical bonds. Respiration then uses the oxygen produced as a by-product of photosynthesis to break down glucose and capture the energy in its bonds in the form of ATP. Some of this energy

is used to power other chemical reactions, converting sugars into starches, cellulose, fats, vitamins, pigments, and proteins (Fig. 30-2). Photosynthetic organisms are called **autotrophs** (Greek, "self-feeders"), or **producers,** because they produce food for themselves. In addition, they directly or indirectly produce food for nearly all other life as well. Organisms that cannot photosynthesize cannot produce food for themselves; instead they must acquire ener-

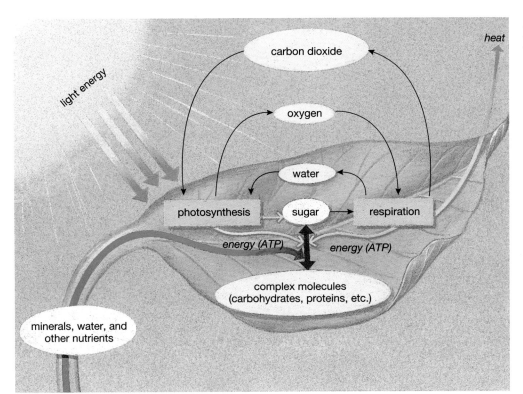

***Figure 30-2* Primary productivity: photosynthesis**

During photosynthesis, plants capture the energy of sunlight and store it in ATP, sugar, and other high-energy carbohydrates synthesized from carbon dioxide and water. Oxygen is released as a by-product. Some of the energy stored in sugar is liberated during respiration and used in the construction of more complex molecules. In the process, oxygen is used, water is released, and some energy is lost as heat.

gy prepackaged in the molecules of the bodies of other organisms. These organisms are called **heterotrophs** (Greek, "other-feeders"), or **consumers.**

The amount of life an ecosystem can support is determined by the energy captured by the producers. The energy that photosynthetic organisms store and make available to other members of the community over a given time period is called **net primary productivity.** Net primary productivity can be measured either in units of energy (calories) stored per unit area in a given time period or as the dry weight, also called **biomass,** of producers added to the ecosystem per unit area in a given time span. The productivity of an ecosystem is influenced by many environmental variables, including the amount of nutrients available to the producers, the amount of sunlight reaching them, the availability of water, and the temperature. In the desert, for example, lack of water limits productivity, while in the open ocean, light is limited in deep water and nutrients are limited in surface water. When resources are abundant, as in estuaries (regions where rivers meet the ocean) or in tropical rain forests, productivity is high. Some average productivities for a variety of ecosystems are shown in Figure 30-3.

Organisms Occupy Different Trophic Levels Based on How They Acquire Energy

Living things may be categorized according to their role in the flow of energy through communities. Energy flows through communities from photosynthetic producers through several levels of consumers. Each category of or-

ganism is called a **trophic** (literally "feeding") **level.** The producers, from redwood trees to cyanobacteria, form the first trophic level, obtaining their energy directly from sunlight (see Fig. 30-1). Consumers form several trophic levels, and certain consumers even switch trophic levels by eating organisms from different levels. Sparrows, for example, may eat either seeds or insects. Some consumers feed directly and exclusively on producers, the most abundant living energy source in the ecosystem. These **herbivores,** ranging from grasshoppers to giraffes, are called **primary consumers** and form the second trophic level. **Carnivores,** such as the spider, hawk, and wolf, are flesh-eaters, feeding primarily on herbivores; they are called **secondary consumers** and form the third trophic level. Some carnivores occasionally eat other carnivores, and when doing so they form the fourth trophic level: **tertiary consumers.** Depending on the nature of their prey, carnivores may also be **quaternary consumers** that occupy the fifth trophic level.

Food Chains and Food Webs Describe the Feeding Relationships within Ecosystems

To illustrate who feeds on whom in an ecosystem, ecologists commonly identify a representative of each trophic level that eats the representative of the level below it. This linear feeding relationship is called a **food chain.** As illustrated in Figure 30-4, different ecosystems have radically different food chains.

Natural communities rarely contain well-defined groups of primary, secondary, and tertiary consumers. A **food web** shows the many interconnecting food chains in an ecosystem, de-

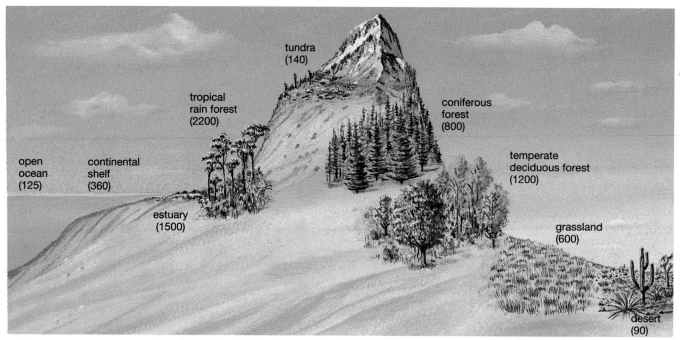

***Figure 30-3*   Ecosystem productivity compared**

Average net primary productivity in grams of organic material per square meter per year of some terrestrial and aquatic ecosystems.

scribing the actual feeding relationships within a given community much more accurately than does a food chain (Fig. 30-5). Some animals, such as raccoons, bears, rats, and humans, are omnivores (Latin, "eating all"), acting at different times as primary, secondary, and occasionally tertiary consumers. Many carnivores will eat either herbivores or other carnivores, thus acting either as secondary or tertiary consumers. An owl, for instance, is a secondary consumer when it eats a mouse, but a tertiary (third-level) consumer when it eats a shrew, which feeds on insects. If the shrew ate a carnivorous insect, it would be a tertiary consumer, and the owl that fed on it would then be a quaternary (fourth-level) consumer. When digesting a spider, a carnivorous plant, such as the sundew, can tangle the food web hopelessly by serving simultaneously as a producer and a secondary consumer!

Detritus Feeders and Decomposers Release Nutrients for Reuse

Among the most important strands in the food web are the detritus feeders and decomposers. The **detritus feeders** are an army of small and often unnoticed animals and protists that live on the refuse of life: molted exoskeletons, fallen leaves, wastes, and dead bodies. The network of detritus feeders is extremely complex, including earthworms, mites, protists, centipedes, some insects and crustaceans, nematode worms, and even a few large vertebrates such as vultures. These organisms consume dead organic matter, extract some of the energy stored within it, and excrete it in a still further decomposed state. Their excretory products serve as food for other detritus feeders

and decomposers, until most of the stored energy has been utilized. The **decomposers** are primarily fungi and bacteria that digest food outside their bodies, absorb the nutrients they need, and release the remaining nutrients.

Through the activities of detritus feeders and decomposers, the bodies and wastes of living organisms are reduced to simple molecules, such as carbon dioxide, water, minerals, and organic molecules, that return to the atmosphere, soil, and water. By liberating nutrients for reuse, detritus feeders and decomposers form a vital link in the nutrient cycles of ecosystems. In some ecosystems, such as deciduous forests, more energy passes through the detritus feeders and decomposers than through the primary, secondary, and tertiary consumers. Though inconspicuous, this portion of the food web is absolutely essential to life on Earth. If the detritus feeders and decomposers were to disappear suddenly, communities would gradually be smothered by accumulated wastes and dead bodies. The nutrients stored in these bodies would be unavailable to enrich the soil. Without these nutrients, the quality of the soil would become poorer and poorer until plant life could no longer be sustained. With plants eliminated, energy would cease to enter the community, and the higher trophic levels would disappear as well.

Energy Flows through Trophic Levels

As discussed in Chapter 4, a basic law of thermodynamics is that energy use is never completely efficient. For example, as your car converts the energy stored in gasoline to the energy of movement, about 75% is immediately lost

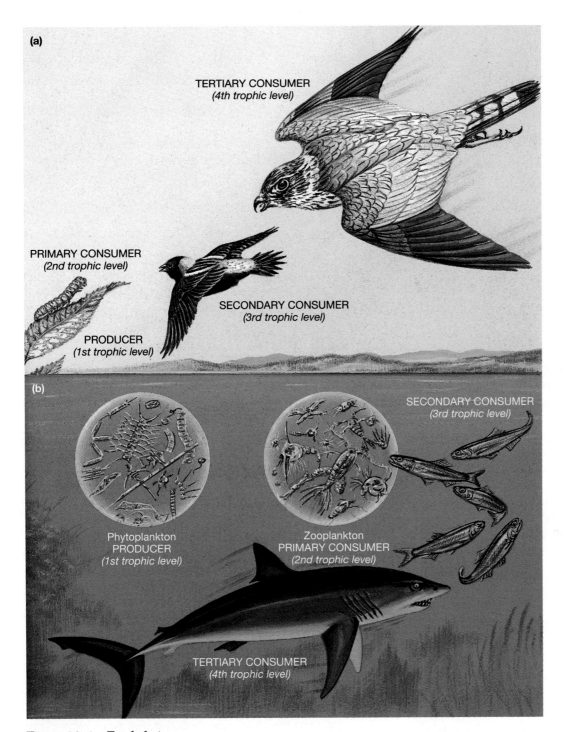

Figure 30-4 **Food chains**

(a) A simple terrestrial food chain. **(b)** A simple marine food chain.

as heat. So too in living systems; the splitting of the chemical bonds of ATP to power a muscular contraction also produces heat as a by-product, as anyone who has exercised vigorously is well aware. Small amounts of waste heat are also produced by the germination of a seed and the thrashing of the tail of a sperm.

The transfer of energy from one trophic level to the next is also quite inefficient. When a caterpillar (a primary consumer) eats the leaves of a tomato plant (a producer), only some of the solar energy originally trapped by the plant is available to the insect. Some was used by the plant to grow and maintain life. Some was converted into the chemical bonds of molecules such as cellulose, which the caterpillar cannot break down. Some energy remains in the plant. Therefore, only a fraction of the energy captured by the first trophic level is available to or-

Figure 30-5 A simple terrestrial food web

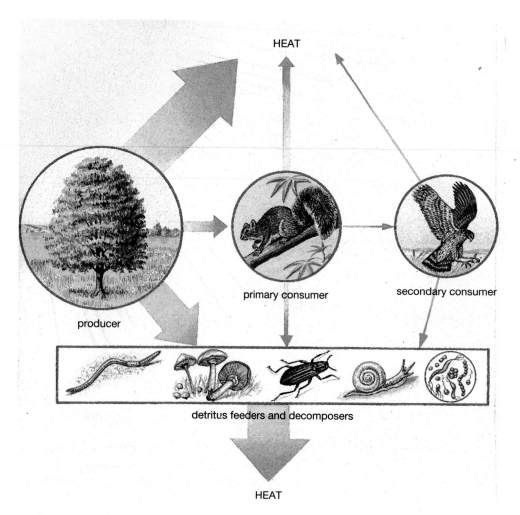

HEAT

producer

primary consumer

secondary consumer

detritus feeders and decomposers

HEAT

Figure 30-6   Energy loss
A diagram showing the loss of energy during its transfer between trophic levels in a forest community. The width of the arrows is roughly proportional to the quantity of energy transferred or lost.

ganisms in the second trophic level. The energy consumed by the caterpillar in turn is partially used to power crawling and the gnashing of mouthparts. Some is used to construct the indigestible exoskeleton, and much is given off as heat. All this energy is unavailable to the songbird in the third trophic level when it eats the caterpillar. The bird loses energy as body heat, uses more in flight, and converts a considerable amount into indigestible feathers, beak, and bone. All this energy will be unavailable to the hawk that catches it. A simplified model of energy flow through the trophic levels in a deciduous forest ecosystem is illustrated in Figure 30-6.

Energy Pyramids Illustrate Energy Transfer between Trophic Levels

Studies of a variety of ecosystems indicate that the net transfer of energy between trophic levels is roughly 10% efficient, although different ecosystems vary significantly. This means that, in general, the energy stored in primary consumers (herbivores) is only about 10% of the energy stored in the bodies of producers. The bodies of secondary consumers, in turn, possess roughly 10% of the energy stored in primary consumers. In other words, for every 100 calories of solar energy captured by grass, only about 10

calories are converted into herbivores, and only 1 calorie into carnivores. This inefficient energy transfer between trophic levels is called the "10% law." An **energy pyramid**, which shows maximum energy at the base and steadily diminishing amounts at higher levels, illustrates the energy relationships between trophic levels graphically (Fig. 30-7). Ecologists sometimes use biomass as a measure of the energy stored at each trophic level. Because the dry mass of the bodies of organisms at each trophic level is roughly proportional to the amount of energy stored in them, a **biomass pyramid** for a given ecosystem usually has the same general shape as its energy pyramid.

What does this mean for ecosystems? If you wander through an undisturbed ecosystem, you will notice that the predominant organisms are plants; these have the most energy available to them because they trap it directly from sunlight. The most abundant animals will be those feeding on plants, and carnivores will always be relatively rare. The inefficiency of energy transfer also has important implications for human food production. The lower the trophic level we utilize, the more food energy will be available to us; far more people can be fed on grain than on meat.

An unfortunate side effect of the inefficiency of energy transfer, coupled with human production and release of toxic chemicals, is the phenomenon of biological mag-

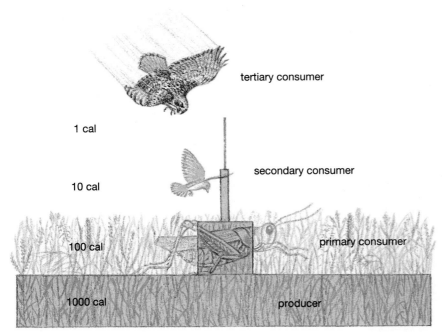

Figure 30-7 **An energy pyramid for a prairie ecosystem**

Each trophic level from producer to tertiary consumer has less energy stored in it. The width of each rectangle is proportional to the energy stored at that trophic level. A pyramid of biomass for this ecosystem would look quite similar.

nification. This is the concentration of certain persistent toxic chemicals in the bodies of carnivores, including ourselves, as described below.

Biological Magnification Occurs as Toxic Substances Are Passed through Trophic Levels

In the 1940s, the properties of the new insecticide DDT seemed close to miraculous. In the tropics, DDT saved millions of lives by killing the mosquitoes that spread malaria. Increased crop yields resulting from DDT's destruction of insect pests saved millions more from starvation. DDT is long lasting, so a single application keeps on killing. The inventor was awarded the Nobel Prize, and people looked forward to a new age of freedom from insect pests. Little did they realize that the indiscriminate use of this pesticide was unraveling the complex web of life.

For example, in the mid-1950s, the World Health Organization sprayed DDT on the island of Borneo to control malaria. A caterpillar that fed on the thatched roofs of houses was relatively unaffected, while a predatory wasp that fed on it was destroyed. Thatched roofs collapsed, eaten by the uncontrolled caterpillars. Gecko lizards that ate the poisoned insects accumulated high concentrations of DDT in their bodies. Both they, and the village cats that ate the geckos, died of DDT poisoning. With the cats eliminated, the rat population exploded and the village was threatened with an outbreak of plague, carried by the uncontrolled rats. The outbreak was avoided by airlifting new cats to the villages.

Meanwhile, in the United States, wildlife biologists during the 1950s and 1960s witnessed an alarming decline in populations of several predatory birds, especially fish-eaters, such as bald eagles, cormorants, and brown pelicans. These top predators are never abundant, and the decline pushed some, such as the brown pelican, close to extinction. The aquatic ecosystems supporting these birds had been sprayed with relatively low amounts of DDT to control insects. Scientists were amazed to find concentrations of DDT in the bodies of predatory birds up to *one million* times the concentration present in the water. This finding led to the discovery of **biological magnification,** or bioaccumulation, the process by which toxic substances accumulate in increasingly high concentrations in progressively higher trophic levels. In 1973, DDT was banned in the United States, although it is still used in some less-developed countries. In 1994, the bald eagle was removed from the endangered species list.

DDT and other substances that undergo biological magnification have two properties that make them dangerous: (1) They do not readily break down into harmless substances, and (2) they are fat soluble but not water soluble. Thus, they accumulate in the bodies of animals, particularly in the fat, rather than being broken down and excreted in the watery urine. Because the transfer of energy from lower to higher trophic levels is extremely inefficient, herbivores must eat large quantities of plant material (which may have been sprayed with DDT), carnivores must eat many herbivores, and so on. Because DDT is not excreted, the predator accumulates the substance from all its prey over many years. Thus, DDT reaches highest levels in top predators, such as predatory birds, where it interferes with the deposition of calcium into eggshells by mimicking and disrupting the functions of the natural sex hormone estrogen.

Today there is growing concern about the effects of a number of chlorinated compounds chemically related to DDT that also share its tendencies to persist, accumulate, and interfere with normal sex hormone functions. These chemi-

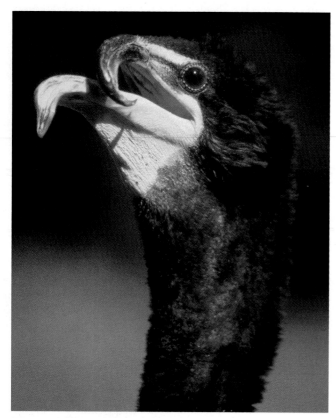

Figure 30-8   **The price of pollution**

Deformities such as the twisted beak of this double-crested cormorant from Lake Michigan have been linked to dioxin and related chemicals. Abnormalities of the reproductive and immune systems are also common in many different organisms exposed to these pollutants. Predatory animals are especially vulnerable because of bioaccumulation.

cals, which include dioxin and polychlorinated biphenyls (PCBs), are produced by a variety of manufacturing processes. Studies have focused on the Great Lakes, which until recently had relatively high levels of these chlorinated compounds. Here, populations of fish-eating river otters have declined sharply, and a variety of fish-eating birds, including the newly returned bald eagles, produce deformed offspring or eggs that never hatch (Fig. 30-8). Minute amounts of dioxin given to pregnant rats cause lowered sperm counts and genital deformities in their male offspring. In Florida's Lake Apopka, a spill of chlorinated chemicals including DDT in 1980 has recently been linked to a 90% decline in the birth rate of the lake's alligators and to abnormally small penis size in alligators reaching maturity in the lake. Some scientists suggest that human sperm counts have declined since the 1930s as a result of long-term exposure to pollution, although this effect is by no means proved.

A number of toxic substances are subject to biological magnification, including mercury and some radioactive compounds. In the 1960s, mercury from ocean pollution reached such high levels in swordfish (top predators) that the sale of swordfish meat was banned.

Understanding the properties of pollutants and the workings of food webs is crucial if we are to prevent both human health hazards and widespread loss of wildlife. Humans have considerable cause for concern, because we often feed high on the food chain. (When you eat a tuna sandwich, for example, you are a tertiary or even quaternary consumer.) In addition, our long life span provides more time for substances stored in the body to accumulate to toxic levels.

The Cycling of Nutrients

In contrast to the energy of sunlight, nutrients do not flow down onto Earth in a steady stream from above. For practical purposes, the same pool of nutrients has been supporting life for over 3 billion years. Nutrients are elements and small molecules that form all the chemical building blocks of life. Some, called **macronutrients,** are required by organisms in large quantities. These include water, carbon, hydrogen, oxygen, nitrogen, phosphorus, sulfur, and calcium. **Micronutrients,** including zinc, molybdenum, iron, selenium, and iodine, are required only in trace quantities. **Nutrient cycles,** also called **biogeochemical cycles,** describe the pathways these substances follow as they move from the living to the nonliving portions of ecosystems and back again to living tissues.

The major source, or **reservoir,** of important nutrients is generally in the nonliving environment. For example, the major reservoir of carbon and nitrogen is the atmosphere, so the cyclings of these elements are called **atmospheric cycles.** The reservoir of a **sedimentary cycle** is sediment—that is, soil or, as in the phosphorus cycle, rock. In the cycling of water, the major reservoir is the ocean. In the following section we briefly describe the cycles of carbon, nitrogen, phosphorus, and water.

The Carbon Cycle Is an Atmospheric Cycle

Chains of carbon atoms form the framework of all organic molecules. Carbon enters the living community through capture of carbon dioxide (CO_2) during photosynthesis. The major reservoir of this gaseous compound is the atmosphere, where it makes up 0.033% of the total gases.

Carbon enters food webs through producers, which trap CO_2 during photosynthesis. Some CO_2 is returned to the atmosphere through cellular respiration, and some that is incorporated into the plant body is later passed to herbivores. They, in turn, respire some of it and incorporate some into their tissues. All living things are eventually consumed by predators, detritus feeders, and decomposers, and ultimately most carbon is returned to the atmosphere as CO_2 (Fig. 30-9).

Some carbon is cycled much more slowly. For example, mollusks extract carbon dioxide dissolved in water and combine it with calcium to form calcium carbonate ($CaCO_3$), from which they construct their shells. Shells of dead mollusks collect in undersea deposits and may even-

carbon dioxide
in the atmosphere

carbon dioxide
dissolved in the
ocean

fire

producers

power plants
and factories

burning of
fossil fuels

consumers

detritus
feeders

fossil fuels

Figure 30-9   A simplified carbon cycle

Carbon is captured from the atmosphere during photosynthesis and passed up through the
trophic levels. It is released during respiration from all trophic levels and by the burning of
forests and fossil fuels.

tually be converted to limestone. Limestone may dissolve gradually as it is exposed to water, making the carbon available to living organisms once more. Another long-term part of the carbon cycle is the production of fossil fuels. **Fossil fuels** are formed from the remains of ancient plants and animals. The carbon found in the organic molecules of these ancient organisms remains in these deposits, transformed by high temperature and pressure over millions of years into coal, oil, or natural gas. The energy of prehistoric sunlight is also trapped in fossil fuels. Burning fossil fuel releases its energy. Human activities, including the extensive use of fossil fuels and the cutting and burning of Earth's great forests (where much carbon is stored), are increasing the amount of carbon dioxide in the atmosphere, as described later in this chapter.

The Nitrogen Cycle Is an Atmospheric Cycle

The atmosphere is about 79% nitrogen gas (N_2), but neither plants nor animals can use this gas directly. Instead, plants must be supplied with nitrates (NO_3^-) or ammonia (NH_3). But how is atmospheric nitrogen converted to these molecules? Ammonia is synthesized by certain bacteria and cyanobacteria that engage in **nitrogen fixation,** a process that combines nitrogen with hydrogen. Some of these bacteria are found living in water and soil; others have entered a symbiotic association with plants called **legumes** (a group including soybeans, clover, and peas), where they live in special swellings on the roots. Decomposer bacteria can also produce ammonia from amino acids and urea found in dead bodies and wastes. Still other bacteria convert ammonia to nitrates. Nitrates are also produced by electrical storms and by other forms of combustion that cause nitrogen to react with atmospheric oxygen. In human-dominated ecosystems, such as farm fields, gardens, and suburban lawns, ammonia and nitrates are supplied by chemical fertilizers. These fertilizers are produced by using the energy in fossil fuels to artificially "fix" atmospheric nitrogen.

Plants incorporate the nitrogen from ammonia and nitrates into amino acids, proteins, nucleic acids, and vitamins. These nitrogen-containing molecules from the plant are eventually consumed by primary consumers, detritus feeders, or decomposers. As it is passed through the food web, some of the nitrogen is liberated in wastes and dead bodies, which decomposer bacteria convert back to nitrates and ammonia. The cycle is balanced by a continuous return of nitrogen to the atmosphere by **denitrifying bacteria.** These residents of mud, bogs, and estuaries break down nitrates, releasing nitrogen gas back to the atmosphere (Fig. 30-10).

The Phosphorus Cycle Is a Sedimentary Cycle

Phosphorus is a crucial component of biological molecules, including the energy-transfer molecules ATP and NADP, nucleic acids, and the phospholipids of cell membranes. It is a major component of vertebrate teeth

and bones. The reservoir of phosphorus in ecosystems is rock, where it is found bound to oxygen in the form of phosphate. The phosphorus cycle is a sedimentary cycle, because unlike carbon and nitrogen, phosphorus does not enter the atmosphere. As phosphate-rich rocks are exposed and eroded, rainwater dissolves the phosphate. Dissolved phosphate is readily absorbed through the roots of plants and by other autotrophs such as photosynthetic protists and cyanobacteria, where it is incorporated into biological molecules such as ATP. From these producers, phosphorus is passed through food webs (Fig. 30-11). At each level, excess phosphate is excreted. Ultimately, decomposers return the phosphorus remaining in dead bodies back to the soil and water in the form of phosphate. Here it may be reabsorbed by autotrophs, or it may become bound to sediment and eventually reincorporated into rock. Some of the phosphate dissolved in fresh water is carried to the oceans. Although much of this phosphate ends up in marine sediments, some is absorbed by marine producers and eventually incorporated into the bodies of invertebrates and fish. Some of these, in turn, are consumed by seabirds, who excrete large quantities of phosphorus back onto the land. At one time, the guano (droppings) deposited by seabirds along the western coast of South America was mined and provided a major source of the world's phosphorus. Phosphate-rich rock is also mined, and the phosphate is incorporated into fertilizer. Soil that erodes from fertilized fields carries large quantities of phosphates into lakes, streams, and the ocean, where it stimulates the growth of producers.

Most Water Remains Chemically Unchanged during the Water Cycle

The water cycle, or **hydrologic cycle** (Fig. 30-12), differs from most other nutrient cycles because most water remains in the form of water throughout the cycle, and it is not used in the synthesis of new molecules. The major reservoir of water is the ocean, which covers about three-quarters of Earth's surface and contains over 97% of the available water. The hydrologic cycle is driven by solar energy, which evaporates water, and by gravity, which draws the water back to Earth in the form of precipitation (rain, snow, sleet, dew). Most evaporation occurs from the oceans, and much water returns directly to them by precipitation. Water falling on land takes more varied paths. Some is evaporated from the soil, lakes, and streams. A portion runs off the land back to the oceans, and a small amount enters underground reservoirs. Because the bodies of living things are roughly 70% water, some of the water in the hydrologic cycle enters the living communities of ecosystems. It is absorbed by the roots of plants, and much of this is evaporated back to the atmosphere from their leaves. A small amount is combined with carbon dioxide during photosynthesis to produce high-energy molecules. Eventually these are broken down

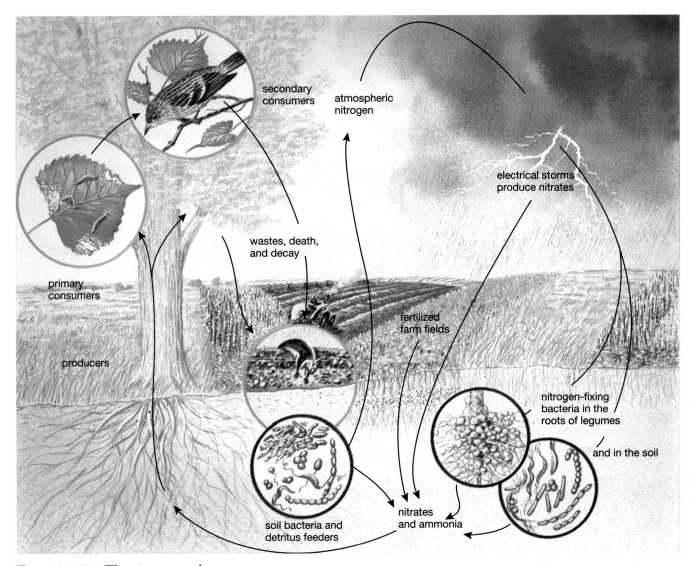

Figure 30-10   **The nitrogen cycle**

From its reservoir in the atmosphere, nitrogen gas is combined with oxygen to form nitrates
by lightning or by the burning of forests or fossil fuels and carried to Earth dissolved in rain.
Nitrogen-fixing bacteria produce ammonia. Nitrates and ammonia are also synthesized by
humans for use in fertilizers. These are absorbed by plants and other producers and
incorporated into biological molecules that are passed up through the trophic levels. Nitrates
and ammonia are released by excretion or by decomposer bacteria. Other bacteria convert
these molecules back to atmospheric nitrogen, completing the cycle.

during cellular respiration, releasing water back to the
environment. Heterotrophs get water from their food or
by drinking.

Although the hydrologic cycle would continue in the
absence of living things, the distribution of life and the
composition of biological communities depend on and,
to a great extent are determined by, patterns of precipitation and evaporation. For example, the operation of
the hydrologic cycle is considerably different over a
desert and a tropical rain forest, and this difference is reflected in the composition of the communities in these
ecosystems.

Human Intervention in Energy Flow and Nutrient Cycling

Many of the environmental problems that beset modern
society are the result of human interference in the way
ecosystems function. Primitive people were sustained solely by the energy flowing from the sun and produced wastes
that were readily taken back into the nutrient cycles. But
as the population grew and technology increased, humans
began to act more and more independently of these natural processes. We have mined from Earth substances, such

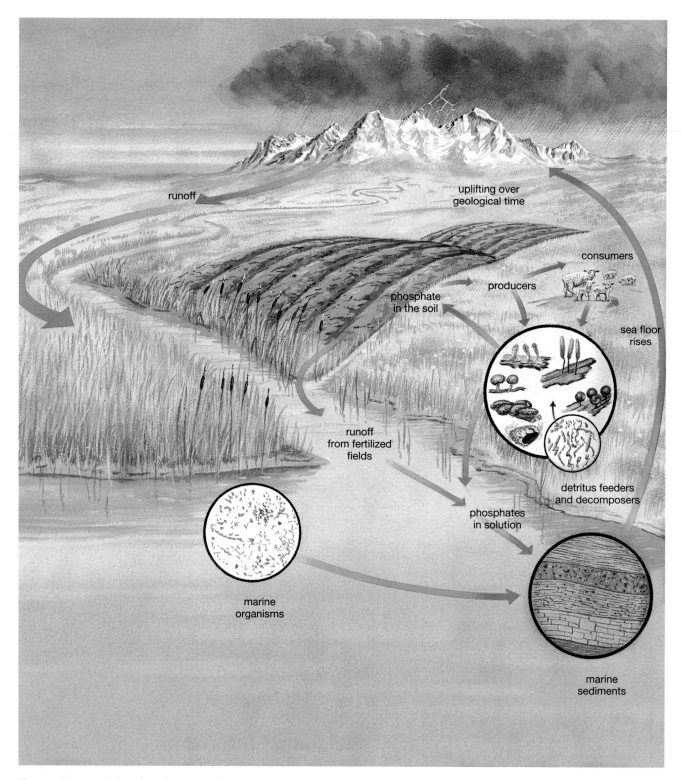

Figure 30-11   The phosphorus cycle

Phosphate dissolves from phosphate-rich rocks or from fertilizers and enters plants and other producers, where it is incorporated into biological molecules. These molecules are passed through the trophic levels. Phosphate is excreted or returned to the soil and water by decomposer bacteria. It may then be reused by producers or eventually incorporated into rock.

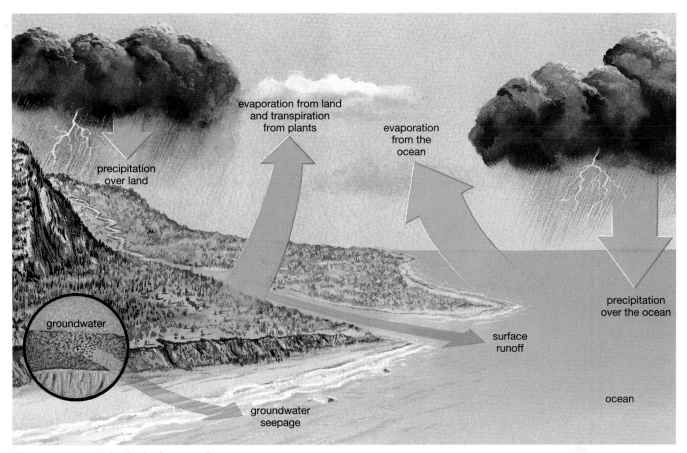

Figure 30-12   The hydrologic cycle

The hydrologic cycle is the simplest of the nutrient cycles.

as lead, arsenic, cadmium, mercury, oil, and uranium, that are foreign to natural ecosystems and toxic to many of the organisms in them (Fig. 30-13). In our factories, we synthesize substances never before found on Earth: pesticides, solvents, and a wide array of other industrial chemicals harmful to many forms of life. The Industrial Revolution, which began in the mid-nineteenth century, resulted in a tremendous increase in our reliance on energy from fossil fuels (rather than sunlight) for heat, light, transportation, industry, and even agriculture.

In this section we describe two environmental problems of global proportions that are a direct result of human reliance on fossil fuels: acid rain and the greenhouse effect.

Acid Rain Is Caused by Overloading the Nitrogen and Sulfur Cycles

Each year, the United States alone discharges about 30 million tons of sulfur dioxide into the atmosphere, two-thirds of it from power plants burning coal or oil (Fig. 30-14). The rest is largely a by-product of industrial boilers, smelters, and refineries. Although volcanoes and hot springs also release sulfur dioxide, human industrial activities account for 90% of the sulfur dioxide in the atmosphere. Twenty-five million tons of nitrogen oxides are also released by

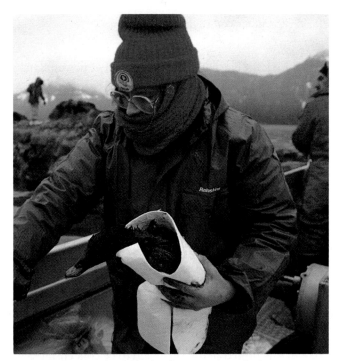

Figure 30-13   A natural substance out of place

An oil-soaked seabird is cleaned by rescuers after an oil spill off the coast of Alaska.

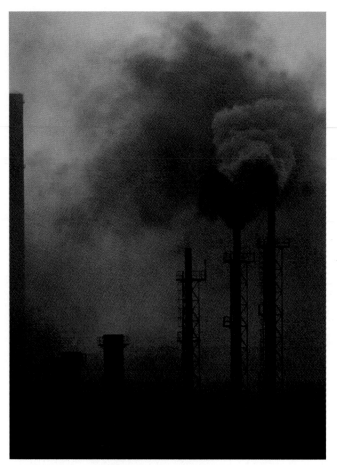

Figure 30-14   A source of acid rain

Power plants burning high-sulfur coal with inadequate emission controls are a major source of atmospheric sulfur dioxide, the prime contributor to acid rain.

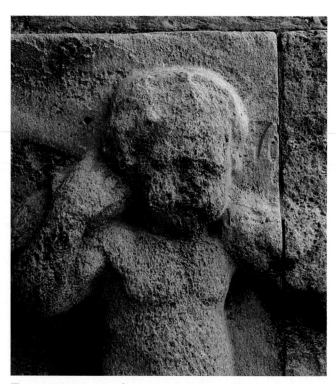

Figure 30-15   Acid rain is corrosive

This marble statue shows the corrosive effects of acid rain.

the United States each year, from vehicles, power plants, and industry.

Excess production of these substances was identified in the late 1960s as the cause of a growing environmental threat: **acid rain**. Combined with water vapor in the atmosphere, nitrogen oxides are converted to nitric acid, and sulfur dioxide is converted to sulfuric acid. Days later, and often hundreds or thousands of miles from the source, the acids fall, either dissolved in rain or as microscopic dry particles. The acids are corrosive, eating away statues and buildings (Fig. 30-15), damaging trees and crops, and turning lakes lifeless. Although both sulfuric and nitric acids form solutions in water vapor, sulfuric acid may also form particles under dry conditions. These particles, which may visibly cloud the air, are deposited even in the absence of rain, snow, or fog. **Acid deposition** is the term used to describe both the wet and dry acid assaults on the environment.

Although sources of sulfur dioxide are widely distributed, in the United States they tend to be concentrated in the upper Ohio Valley, Indiana, and Illinois, where high-sulfur coal is burned in old power plants with few emission controls. As complex chemical reactions in the atmosphere produce sulfuric and nitric acids, winds carry them toward New England, where lakes and forests are particularly vulnerable. Some areas of the country have rock rich in calcium carbonate, a natural buffer for acidity. In New England, however, hard rock, such as granite and basalt, predominates and cannot neutralize acid rain. Although the Northeast has been the most seriously damaged, it is not the only sensitive area; about 75% of the United States is at least moderately vulnerable to acid rain (Fig. 30-16).

Acid Rain Damages Life in Lakes, Farms, and Forests

In the Adirondack Mountains of New York, acid rain has made about 25% of all the lakes and ponds too acidic to support fish. But before the fish die, much of the food web that sustains them is destroyed. Clams, snails, crayfish, and insect larvae die first, then amphibians, and finally fish. The result is a crystal clear lake, beautiful—but dead. The impact is not limited to aquatic organisms. The loss of insect larvae and crustaceans has probably contributed to a dramatic decline in the population of black ducks, which feed on them.

Acid rain is not only a threat to lakes. Recent studies under laboratory conditions have shown that it also interferes with the growth and yield of many farm crops. Acid water draining through the soil washes out essential nutrients such as calcium and potassium. Acid rain may also kill decomposer microorganisms, thus preventing the return of nutrients to the soil. Plants, poisoned and deprived of nutrients, become weak and more vulnerable to infection and insect attack.

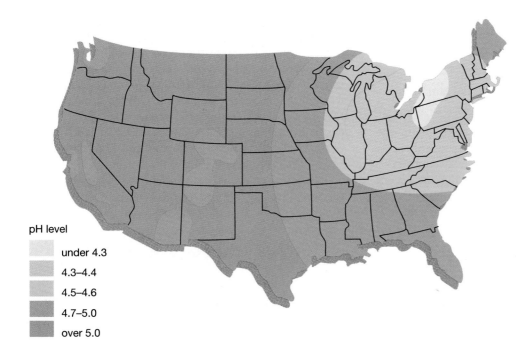

Figure 30-16 **Acid rain distribution**

The most acidic rain is concentrated in the Northeast, but thousands of lakes in the West are also vulnerable. (Source: National Acid Precipitation Assessment Program.)

pH level

- under 4.3
- 4.3–4.4
- 4.5–4.6
- 4.7–5.0
- over 5.0

High in the Green Mountains of Vermont, scientists have witnessed the death of about half of the red spruce and beech trees and one-third of the sugar maples since 1965. The snow, rain, and heavy fog that commonly cloak these eastern mountaintops is highly acidic. At a monitoring station atop Mount Mitchell in North Carolina, the pH of fog even on good days is 2.9, more acidic than vinegar. In this fragile and stressful environment, acid precipitation may well be tipping the balance against these forests (Fig. 30-17).

In the Black Forest of Germany, the devastation is even greater, with over half the trees affected. Many of the firs have turned yellow or brown, trunks have become gnarled, roots have shrunk, and plagues of insects and fungal diseases are attacking the weakened trees. German officials are convinced that acid rain from the burning of high-sulfur coal, in conjunction with other pollutants, is the cause. Vigorous pollution-control programs have begun, but many fear that the forests have been permanently damaged. In 1984, scientists confirmed that the growth rate of eastern U.S. forests had slowed dramatically over the past several decades. Germany's problems may foreshadow similar disasters for the forests of the United States.

Songbirds in the Netherlands are reproducing poorly, laying thin-shelled eggs in which develop embryos with bone deformities, evidence of calcium deficiency. Researchers have found that these birds get most of their calcium from eating snails, whose shells are a rich source of this nutrient. During the last 20 years, however, the snail populations in Dutch forests have nearly disappeared, a result, researchers believe, of acid rain. Songbird populations there may be the next to go.

Acid deposition increases the exposure of organisms to toxic metals, including aluminum, lead, mercury, and cadmium, which are extremely soluble in acidified water. Alu-minum may inhibit plant growth. In acid lakes and streams, aluminum dissolved from the surrounding rock and soil causes mucus to accumulate in the gills of fish, suffocating them. Drinking water in some households has been found dangerously contaminated with lead dissolved by acid water from lead pipes. Boston's water supply must be chemically treated to counteract the acidity. Fish in

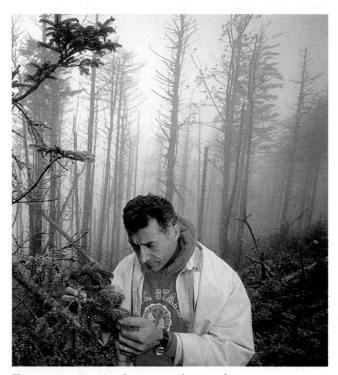

Figure 30-17 **Acid rain can destroy forests**

This forest atop Mount Mitchell in North Carolina was destroyed by acid deposition.

acidified water have been found to have dangerous levels of mercury in their bodies, which is subject to biological magnification as it is passed through trophic levels.

Interference with the Carbon Cycle Is Causing the Greenhouse Effect

Between 345 and 280 million years ago, under the unique conditions of the Carboniferous period, huge quantities of carbon were diverted from the carbon cycle when the bodies of plants and animals were buried in sediments, escaping decomposition. There, heat and pressure over time converted their bodies into fossil fuels such as coal, oil,

and natural gas. Without human intervention, this carbon would have remained essentially untouched for millions of years. Since the Industrial Revolution, however, modern cultures have increasingly relied on the energy stored in these fuels. As we burn them in our power plants, factories, and cars, we release CO_2 into the atmosphere. Since the start of the Industrial Revolution in 1850, the CO_2 content of the atmosphere has increased 25%. The burning of fossil fuels is the single greatest cause of this increase.

A second important source of additional CO_2 is global deforestation, the cutting of tens of millions of acres of forests each year. Deforestation is occurring principally in the tropics, where rain forests are rapidly being eliminat-

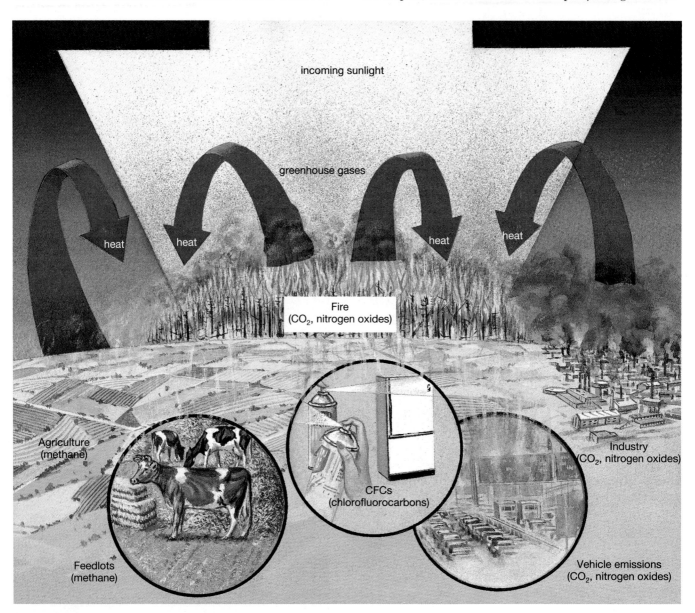

Figure 30-18   **The greenhouse effect**

Incoming sunlight warms the surface of Earth and is radiated back to the atmosphere. Greenhouse gases (such as CO_2 and nitrous oxides from combustion, chlorofluorocarbons from industrial activities, and methane from farms and feedlots) absorb some of this heat, trapping it in the atmosphere. The result is a gradual rise in average global temperatures.

"Living creatures press up against all barriers, they fill every possible niche all the world over. . . . We see life persistent and intrusive—spreading everywhere, insinuating itself, adapting itself, resisting everything, defying everything, surviving everything!"

Sir John Arthur Thompson, a Scottish biologist (1920)

In preceding chapters, we discussed the dynamics of population growth, community interactions, the flow of energy, and the cycling of nutrients, as well as the basic structure of ecosystems within which those processes occur. Ecosystems are extraordinarily diverse, yet clear patterns exist within this diversity. If you travel around the world, you may observe that, although the plant *species* may differ, very similar groups of plants are found wherever a particular climate exists. Deciduous forests, deserts, grasslands, tropical rain forests, coniferous forests, coral reefs, and estuaries all have distinguishing features that identify them no matter where they are found or what particular species inhabit them.

The reason we are able to place these communities into distinct categories is that each is dominated by organisms that are specifically adapted to the conditions of a particular environment. Variations in the availability of light, water, and in the range of temperatures shape the characteristic groupings of organisms that define the different types of ecosystems. The desert community, for example, is dominated by plants adapted to heat and drought. The cacti of the American Mojave Desert are strikingly similar to the euphorbia of South Africa, although these plants are in separate families and are only distantly genetically related. Their spinelike leaves and thick, green, water-storing stems are adaptations for water conservation (Fig. 31-1). Likewise, the plants of the arctic tundra and those of the alpine tundra of the Rocky Mountains show growth patterns clearly recognizable as adaptations to a cold, dry, windy climate. Thus, in regions of Earth where the environmental conditions are similar, we find similar types of organisms organized into similar types of communities.

Environments vary in the relative abundance of four basic resources that provide the requirements for life: nutrients, energy, water, and temperatures compatible with metabolic reactions. The availability of these basic resources shapes the different ecosystems that are the subject of this chapter.

Climate

Life on Earth, particularly life on land, is dramatically affected by both weather and climate. **Weather** refers to short-term fluctuations in temperature, humidity, cloud cover, wind, and precipitation over periods of hours or days.

Rain forest surrounds the Segama River in Borneo.

Figure 31-1
Environmental demands mold physical characteristics
Evolution in response to similar environments has molded the bodies of **(a)** American cacti and **(b)** South African euphorbia into nearly identical shapes, although they are in completely different families. Both of these desert plants have thick, water-conserving stems and leaves reduced to defensive spines that minimize evaporation.

Climate, in contrast, refers to overall patterns of weather that prevail from year to year and even century to century in a particular region. The amount of sunlight, water, and the range of temperatures determine the climate of a given region. Whereas weather affects individual organisms, climate influences and limits the overall distribution of entire species.

Both Climate and Weather Are Driven by the Sun

Both climate and weather are driven by a great thermonuclear engine: the sun. Solar energy reaches Earth in a range of wavelengths, from short, high-energy ultraviolet rays, through visible light, to the longer infrared wavelengths that produce heat. The solar energy that reaches Earth drives the wind, the ocean currents, and the global water cycle. Before it reaches Earth's surface, however, sunlight is modified by the atmosphere. A layer relatively rich in ozone (O_3) is located high in the upper atmosphere. This ozone layer absorbs much of the high-energy ultraviolet radiation from the sun that can damage biological molecules (see "Earth Watch: The Ozone Layer—We Have Punctured Our Protective Shield"). Dust, water vapor, and clouds scatter light, reflecting some of the energy back into space. Carbon dioxide, water vapor, ozone, methane, and other gases selectively absorb infrared wavelengths (heat), trapping warmth in the atmosphere. Human activities have intensified this natural greenhouse effect, as described in Chapter 30.

Only about half the solar energy reaching the atmosphere actually strikes Earth's surface. Of this, small fractions are immediately reflected directly back into space or used in photosynthesis, while the rest is absorbed as heat. Eventually, nearly all the incoming solar energy is sent back into space,

either as light or infrared radiation (heat). The solar energy temporarily absorbed and stored as heat by the atmosphere and Earth's surface maintains Earth's relative warmth.

Many Physical Factors Also Influence Climate

Many factors influence climate. Among the most important are latitude, air currents, ocean currents, and the presence of irregularly shaped continents.

Latitude Influences the Angle at Which Sunlight Strikes Earth

Latitude, measured in degrees, is the distance north or south from the equator, which is located at 0° latitude. The amount of sunlight that strikes a given area of Earth's surface has a major effect on average yearly temperatures. At the equator, sunlight hits Earth's surface nearly at a right angle, making the weather there consistently warm. Farther north or south, the sun's rays strike Earth's surface at a greater slant. This angle spreads the same amount of sunlight over a larger area, producing lower overall temperatures (Fig. 31-2).

Because Earth is tilted on its axis as it makes its yearly trip around the sun, the higher latitudes experience considerable variation in the directness of sunlight throughout the year, resulting in pronounced seasons. For example, in summer, the Northern Hemisphere is tilted toward the sun and thus receives sunlight more directly and experiences warmer weather; in our winter, the Southern Hemisphere is closer, as illustrated in Figure 31-2. The tilting hardly affects the angle of the equator to the sun's rays, so there is little seasonal variation at this latitude.

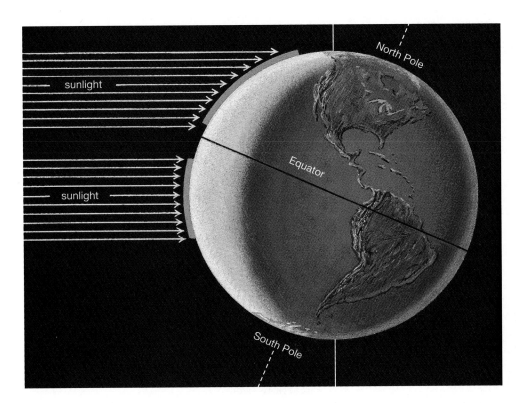

Figure 31-2   Earth's curvature and tilt produce seasons and climate

Seasons and climate are the result of an uneven distribution of solar energy. Near the equator, sunlight falls perpendicularly on Earth's surface, so its warmth is concentrated onto a relatively small area. Toward the poles, the same amount of sunlight falls over a much larger surface area. Temperatures are thus highest at the equator and lowest at the poles. The tilt of Earth on its axis causes seasonal variations in the directness of sunlight. In this figure, the Southern Hemisphere, which is tilted toward the sun, is receiving more direct sunlight than is the Northern Hemisphere, which is tilted away from the sun. Thus it is winter in the Northern Hemisphere and summer in the Southern Hemisphere.

Air Currents

Air currents are generated by Earth's rotation and differences in temperature between different masses of air. Because warm air is less dense than cold air, as the sun's direct rays fall on the equator, heated air rises. The warm air of the tropics is also laden with water evaporated by solar heat (Fig. 31-3a). As the water-saturated air rises, it cools somewhat. Because cool air cannot retain as much moisture as warm air, water condenses from the rising air and falls as rain. The direct rays of the sun and the rainfall produced when warm, moist air rises and is cooled create a band around the equator, called the tropics, which is both the warmest and the wettest region on Earth. The cooler dry air then flows north and south from the equator. At around 30° north and south latitudes, the cooled air begins to sink. As it sinks, the air is warmed by heat radiated from Earth. By the time it reaches the surface, it is both warm and very dry. Not surprisingly, the major deserts of the world are found at these latitudes (Fig. 31-3a, b). This air then flows back toward the equator. Farther north and south, this general circulation pattern is repeated, dropping moisture at around 60° north and south latitudes and creating extremely dry conditions at the North and South poles.

Ocean Currents Moderate Near-Shore Climates

Ocean currents are driven by Earth's rotation, the wind, and direct heating of the water by the sun. Continents interrupt the currents, breaking them into roughly circular patterns called **gyres,** which rotate clockwise in the Northern Hemisphere and counterclockwise in the Southern Hemisphere

(Fig. 31-4). Because water both heats and cools more slowly than land or air, ocean currents tend to moderate temperature extremes. Coastal areas, then, generally have less-variable climates than areas near the center of continents. For example, a gyre in the Atlantic Ocean brings warm water from equatorial regions north along the eastern coast of North America, creating a warmer, moister climate than is found farther inland. It then carries the still-warm water farther north and east, warming the western coast of Europe before returning south.

Continents and Mountains Complicate Weather and Climate

If Earth's surface were uniform, we would find climate zones occurring in bands corresponding to latitude. These zones would result from the interaction of temperature and rainfall, which are determined by the rise and fall of air masses (see Fig. 31-3a). The presence of irregularly shaped continents (which heat and cool relatively quickly) and oceans (which heat and cool more slowly) alters the flow of wind and water and contributes to the irregular distribution of ecosystems (see Fig. 31-7).

Variations in elevation within continents further complicate the picture. With increasing elevation, the atmosphere becomes thinner and retains less heat. The temperature drops approximately 2°C (3.5°F) for every 305 meters (1000 feet) elevation gain, so that snow-capped mountains may be found even in the tropics (Fig. 31-5).

Mountains also modify rainfall patterns. When water-laden air is forced to rise as it meets a mountain, it cools. Cooling reduces the air's ability to retain water, and the water

(a)

(b)

Figure 31-3   The distribution of rainfall

(a) Rainfall is determined primarily by the distribution of temperatures combined with Earth's rotation. The interaction of these two factors creates air currents that rise and fall predictably with latitude, producing broad general climatic regions. **(b)** Some of these regions are visible in this photograph of the African continent taken by *Apollo 11* astronauts. Along the equator, there are heavy clouds dropping moisture on the central African rain forests. Note the lack of clouds over the Sahara and Arabian deserts near 30° north latitude, and the South African desert near 30° south.

condenses as rain or snow on the windward (near) side of the mountain. The cool, dry air is warmed again as it travels down the far side of the mountain and absorbs water from the land, creating a local dry area called a **rain shadow.** For example, mountain ranges such as the Sierra Nevada of the western United States wring the moisture from the westerly winds coming off the Pacific Ocean, leaving deserts in the rain shadow on their eastern sides (Fig. 31-6).

The Requirements of Life

From the lichens on bare rock to the thermophilic (Greek for "heat-loving") algae in the hot springs of Yellowstone National Park to the bacteria thriving under the pressure-cooker conditions of a deep-sea vent, Earth teems with life. Underlying the diversity of habitats is the common ability to provide to varying degrees the four fundamental resources required for life on Earth. These resources are:

1. Nutrients from which to construct living tissue
2. Energy to power that construction
3. Liquid water to serve as a medium in which metabolic reactions occur
4. Appropriate temperatures in which to carry out these processes

As we shall see in the following sections, these resources are very unevenly distributed. Their availability limits the types of organisms that can exist within the various terrestrial and aquatic ecosystems on Earth.

Life on Land

Terrestrial organisms are restricted in their distribution largely by the availability of water and appropriate temperatures. Terrestrial ecosystems receive plenty of light, even on an overcast day, and the soil provides abundant nutrients. Water, however, is limited and very unevenly distributed, both in place and in time. Terrestrial organisms must be adapted to obtain water when it is available and conserve it when it is scarce.

Like water, favorable temperatures are very unevenly distributed in place and time. At the South Pole, even in summer, the average temperature is well below freezing, and (not surprisingly) life is scarce. Other places, such as central Alaska, have favorable temperatures only during certain seasons of the year, while the tropics have a uniformly warm, moist climate, and life abounds.

Terrestrial Biomes Support Characteristic Plant Communities

Terrestrial communities are dominated and defined by their plant life. Because plants can't escape from drought, sun, or winter weather, they tend to be precisely adapted

Figure 31-4 **Ocean circulation patterns are called gyres**

Gyres travel clockwise in the Northern Hemisphere and counterclockwise in the Southern Hemisphere. These currents tend to distribute warmth from the equator to northern and southern coastal areas.

to the climate of a particular region. Large land areas with similar environmental conditions and characteristic plant communities are called **biomes** (Fig. 31-7, p. 628). Biomes are usually named after the major type of vegetation found there. The predominant vegetation of each biome is determined by the complex interplay of rainfall and temperature (Fig. 31-8, p. 629). These factors determine the available soil moisture needed for plant growth as well as for compensation for water losses through evaporation. In addition to the total amount of rainfall and the overall yearly average temperature, the *variability* of rain and temperature over the year also determines which plants can grow in an area. Arctic tundra plants, for example, must be adapted to survive marshy conditions in the early summer but cold desertlike conditions for much of the rest of the year, when water is solidly frozen and unavailable. In the following sections we discuss the major biomes, beginning at the equator and working our way northward. We also discuss some of the effects human activities have on these biomes.

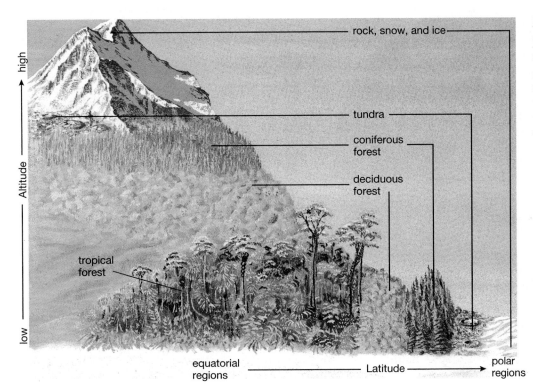

Figure 31-5 **Effects of elevation on temperature**

Climbing a mountain in some ways is like going north; in both cases, increasingly cool temperatures produce a similar series of biomes.

rock, snow, and ice

tundra

coniferous forest

deciduous forest

tropical forest

high

Altitude

low

equatorial regions ——————— Latitude ——————→ polar regions

E A R T H   W A T C H

The Ozone Layer—We Have Punctured Our Protective Shield

A small fraction of the radiant energy produced by the sun, called ultraviolet, or UV, radiation, is so highly energetic that it can damage biological molecules. In small quantities, UV radiation helps human skin to produce vitamin D and causes tanning in fair-skinned people. But in larger doses, UV causes sunburn and premature aging of skin, skin cancer, and cataracts, a condition in which the lens of the eye becomes cloudy.

Fortunately, most of the UV radiation is filtered out by ozone in the stratosphere, a layer of atmosphere extending from 10 to 50 kilometers (6 to 30 miles) above Earth. In pure form, ozone (O_3) is a bluish, explosive, and highly poisonous gas. In the stratosphere, the normal concentration of ozone is about 0.1 part per million, compared with 0.02 part per million in the lower atmosphere. This ozone-enriched layer is called the **ozone layer.** Ultraviolet light striking ozone and oxygen causes reactions that break down as well as regenerate ozone. In the process, the UV radiation is converted to heat, and the overall level of ozone remains constant—normally.

In 1985, British atmospheric scientists published a startling discovery: The springtime levels of stratospheric ozone over Antarctica had declined by over 40% since 1977. A hole had been punctured in Earth's protective shield. In the localized "ozone hole" over Antarctica, ozone now dips to about one-third of its predepletion levels, a decline of 66% (Fig. E31-1). Although depletion of the ozone layer is severest over Antarctica, the ozone layer is somewhat reduced over most of the world, including nearly all of the continental United States. Parts of the Northern Hemisphere experience up to 6% reductions at some times of the year. Evidence of ozone thinning over the Arctic has led researchers to predict that we will soon have an Arctic, as well as an Antarctic, ozone hole.

On the basis of increased UV exposure due to ozone loss, the Environmental Protection Agency increased its estimates of skin cancer deaths over the next 50 years from 9300 up to 200,000. Studies using mice show that exposure to UV light in dosages similar to a day at the beach without sunscreen can depress their immune systems, making them more vulnerable to a variety of diseases. Human health effects are only one cause for concern. Photosynthesis by phytoplankton, the producers for marine ecosystems, is reduced under the ozone hole above Antarctica. Some types of trees and farm crops are also harmed by increased UV radiation.

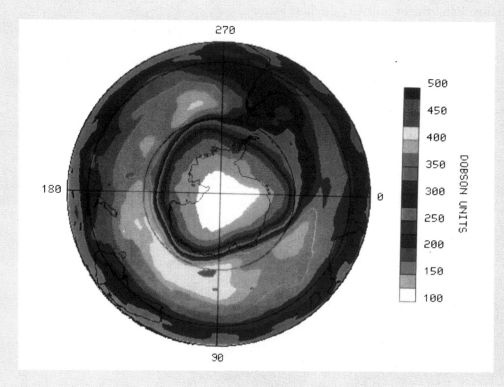

Figure E31-1   Satellite image of the Antarctic ozone hole

The ozone hole is seen in white and pink on this image recorded by a satellite in October 1993, when ozone levels dipped to less than one-third of normal. Dobson units measure the thickness of the ozone (in hundredths of millimeters) in a given layer of atmosphere if it were purified and held at standard temperature and pressure.

The decline in the thickness of the ozone layer is caused by increasing levels of chlorofluorocarbons (CFCs), which contain chlorine, fluorine, and carbon. Developed about 60 years ago, these gases were widely used as coolants in refrigerators and air conditioners, aerosol spray propellants, agents for producing polystyrene (Styrofoam), and cleansers for electronic parts. These chemicals are very stable and were considered safe. Their stability, however, proved to be a major problem, because they remain chemically unchanged as they gradually rise into the stratosphere. There, under intense bombardment by UV light, the CFCs break down, releasing chlorine atoms. Chlorine catalyzes the breakdown of ozone to oxygen gas (O_2) while remaining unchanged itself. Over the Arctic and Antarctic regions, the clouds are composed of ice particles that provide a surface on which the reaction can occur.

Although CFC production has not been halted, in a series of treaties beginning in 1987, industrialized nations throughout the world agreed to gradually phase out ozone-depleting chemicals, with the goal of eliminating their use entirely by 2030. Atmospheric CFCs are still increasing, but as a result of these treaties their *rate* of increase has dropped considerably. Because these highly stable compounds can persist 50 to 100 years and take decades to rise into the stratosphere, the millions of tons of CFCs already released will continue to erode our protective ozone shield, with possibly severe consequences.

Tropical Rain Forests

Large areas of South America and Africa lie along the equator. Here the temperature averages between 25° and 30° C (77° and 86° F) with little variation, and rainfall ranges from 250 to 400 centimeters (100 to 160 inches) each year. These evenly warm, evenly moist conditions combine to create the most diverse biome on Earth, the **tropical rain forest,** dominated by huge broadleaf evergreen trees (Fig. 31-9, p. 630).

Although rain forests cover only 6% of Earth's total land area, they are believed to be home to between 5 and 8 million species, representing half to two-thirds of the world's total. A recent survey of a 2.5-acre site in the upper Amazon basin revealed 283 species of trees, most of which were represented by a single individual.

Tropical rain forests typically have several layers of vegetation. The tallest trees reach 50 meters (150 feet) and tower above the rest of the forest. Below is a fairly continuous canopy of treetops at about 30 to 40 meters (90 to 120 feet). Below these is often another layer of shorter trees. Huge woody vines, often 100 meters (328 feet) or more in length, grow up the trees, reaching the sunlight far above. These layers of vegetation block out most of the sunlight. Plants that live in the dim green light that filters through to the forest floor often have enormous leaves to trap the little available energy.

Because edible plant material close to the ground is relatively scarce, much of the animal life in tropical rain forests is arboreal (living in the trees), including numerous birds, monkeys, and insects. Competition for the nutrients that do reach the ground is intense, among both

Figure 31-6   The rain shadow of the Sierra Nevada

Here, rainfall is graphed against altitude. Westerly (eastward-moving) winds deposit their moisture on the western slope, leaving a desert to the east.

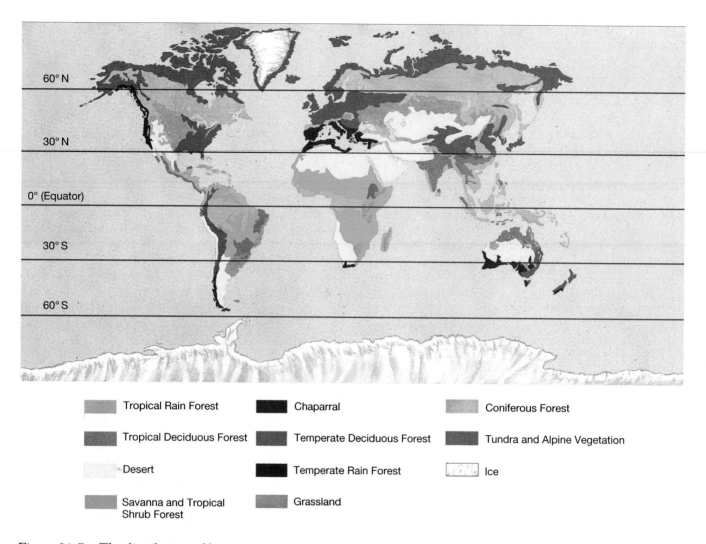

60° N

30° N

0° (Equator)

30° S

60° S

▨ Tropical Rain Forest	■ Chaparral	▨ Coniferous Forest
▨ Tropical Deciduous Forest	▨ Temperate Deciduous Forest	▨ Tundra and Alpine Vegetation
▨ Desert	■ Temperate Rain Forest	▨ Ice
▨ Savanna and Tropical Shrub Forest	▨ Grassland	

Figure 31-7   **The distribution of biomes**

Although mountain ranges and the sheer size of the continents complicate their pattern, note the overall consistencies. Tundra and coniferous forest always occur in the northernmost parts of the Northern Hemisphere, while the deserts of Mexico, the Sahara, Saudi Arabia, South Africa, and Australia are located around 20° to 30° north and south latitude.

animals and plants. Even such unlikely sources of food as the droppings of monkeys are in great demand. For example, dung beetles feed and lay their eggs on monkey droppings. When ecologists attempted to collect droppings of the South American howler monkeys to find out what the monkeys had been eating, they found themselves in a race with the beetles, hundreds of which would arrive within minutes after a dropping hit the ground!

Almost as soon as bacteria or fungi release any nutrients from dead plants or animals into the soil, rainforest trees and vines absorb the nutrients. This is one of the reasons why, despite the teeming vegetation, agriculture is very risky and usually destructive in rain forests. Virtually all the nutrients in a rain forest are tied up in the vegetation, so the soil is very infertile. If extensive land is cut down and the trees are carried away for lumber, few nutrients remain to support crops. Further, even if the nutrients are released by burning the veg-

etation, the heavy year-round rainfall quickly dissolves and erodes them away, leaving the soil infertile after a few seasons of cultivation. The exposed soil, which is rich in iron and aluminum, then takes on an impenetrable, bricklike quality as it bakes in the tropical sun. As a result, secondary succession on cleared rainforest land is slowed significantly. Even small forest cuttings take about 70 years to regenerate.

Human Impact   In spite of their unsuitability for agriculture, rain forests are being felled for lumber and agriculture at a rate of roughly 38 million acres per year, or about *75 acres per minute* (Fig. 31-10). About 40% of the original rain forests are now gone. Harvard biologist Edward O. Wilson estimates that the destruction of tropical rain forests may be driving 27,000 species to extinction *each year*. With the loss of biodiversity, humanity loses access to a wealth of potential drugs and raw materials. For example, in

Figure 31-8   **Temperature and rainfall influence biome distribution**

Together, these factors determine the available soil moisture needed for plant growth.

Malaysia, the sap of a rare species of tree was discovered to yield a compound called calanolide A, which, in test tubes, completely blocks replication of the HIV virus. The removal of these forests is also contributing significantly to the buildup of carbon dioxide in the atmosphere, making the greenhouse effect more severe. It is estimated that one-quarter of the carbon released into the atmosphere during the past decade came from the loss of tropical rain forests. In West Africa, burning the forests is also causing acid rain, which is damaging the remaining trees.

Although disastrous losses are continuing, recognition of the problem is growing. Some areas have been set aside as protected preserves, and some reforestation efforts are under way. Local residents are becoming more involved in conservation efforts. In Brazil, a union of people who collect natural rubber from tree sap is fighting to preserve large tracts of land open only to rubber production and to the harvesting of fruits and nuts. This approach suggests the ultimate solution, which is tragically slow in coming—sustainable use,

the harvesting of products without permanently damaging the trees or the ecosystem they support.

Tropical Deciduous Forests

Slightly farther away from the equator, the rainfall is not nearly as constant, and there are pronounced wet and dry seasons. In these areas, which include India, much of Southeast Asia, and parts of South and Central America, **tropical deciduous forests** grow. During the dry season, the trees cannot get enough water from the soil to compensate for evaporation from their leaves. As a result, the plants have adapted to the dry season by shedding their leaves, thereby minimizing water loss. If the rains fail to return on schedule, the trees delay forming new leaves until the drought passes.

Savanna

Along the edges of the tropical deciduous forest, the trees gradually become more widely spaced, with grass-

(a)

(b)

(c)

(d)

Figure 31-9   **The tropical rain forest**

Towering trees draped with vines reach for the light in the dense tropical rain forest. Because food and light energy are found high in trees, amid their branches dwells the most diverse assortment of animals on Earth, including **(a)** a fruit-eating toucan, **(b)** a shimmering butterfly, **(c)** a tree-climbing orchid, and **(d)** a howler monkey.

es growing between them. Eventually, grasses become the dominant vegetation, with only scattered trees and thorny scrub forests here and there; this is the **savanna** (Fig. 31-11). Savanna grasslands typically have a rainy season during which virtually all of the year's precipitation falls—30 centimeters (12 inches) or less. When the dry season arrives, it comes with a vengeance. No rain may fall for months, and the soil becomes hard, dry, and dusty. Grasses are well adapted to this type of climate, growing very rapidly during the rainy season and dying back to drought-resistant roots during the dry times.

Only a few specialized trees, such as the thorny acacia or the water-storing baobab, can survive the devastating savanna dry seasons. In areas in which the dry season becomes even more pronounced, virtually no trees at all can grow, and the savanna imperceptibly grades into tropical grassland.

The African savanna has probably the most diverse and impressive array of large mammals on Earth, including numerous herbivores, such as antelope, wildebeest, buffalo, elephants, and giraffes, and carnivores, such as the lion, leopard, hyena, and wild dog.

Figure 31-10   **Fire engulfs a tropical rain forest in Brazil**

The cleared area will be converted to ranching or agriculture, doomed to failure because of the poor quality of the soil.

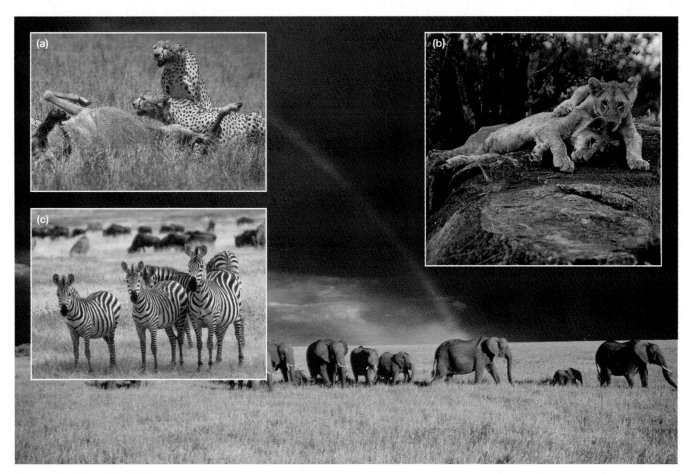

Figure 31-11   **The African savanna**

Elephants roam while a cheetah feasts on its prey **(a)**. Large herds of grazing animals, such as zebras **(c)** can still be seen on African preserves. The herds of herbivores provide food for the greatest assortment of large carnivores on the planet, including lions **(b)**, leopards, cheetahs, hyenas, and wild dogs.

Figure 31-12   Poaching threatens African wildlife
Rhinoceros horns, believed by some to have aphrodisiac properties, fetch staggering prices and encourage poaching. The black rhino is now nearly extinct.

Human Impact Africa's rapidly expanding human population threatens the wildlife of the savanna. Poaching has driven the black rhinoceros to the brink of extinction (Fig. 31-12) and also endangers the African elephant. The abundant grasses that make the savanna a suitable habitat for so much wildlife also make it suitable for grazing domestic cattle. As the human population of East Africa increases, so does the pressure of cattle grazing upon the savanna. Fences increasingly disrupt the migration of the great herds of herbivores in search of food and water. Ecologists have discovered that the native herbivores are much more efficient at converting grass into meat than are cattle. Perhaps the future African savanna may support herds of domesticated antelope and other large native grazers in place of cattle.

Deserts

Even drought-resistant grasses need at least 25 to 50 centimeters (10 to 20 inches) of rain a year, depending on its seasonal distribution and the average temperature. When less rain than this falls, many grasses fail and **desert** biomes occur. These biomes are found on every continent, often around 20° to 30° north and south latitude, and also in the rain shadows of major mountain ranges. As with all biomes, the designation *desert* includes a variety of environments. At one extreme are certain areas of the Sahara or Chile, where it virtually never rains and there is no vegetation at all (Fig. 31-13a). The more common deserts, however, are characterized by widely spaced vegetation and large areas of bare ground. The plants are often spaced evenly, as if planted by hand (Fig. 31-13b). Frequently, the perennial plants are bushes or cacti with large, shallow root systems. The shallow roots quickly soak up

the soil moisture after the infrequent desert storms, while the rest of the plant is typically covered with a waterproof, waxy coating to prevent evaporation of precious water. Water is stored in the thick stems of cacti and other succulents (Fig. 31-13c). The spines of cacti are leaves modified for protection and water conservation, presenting almost no surface area for evaporation. In many deserts, all the rain falls in just a few storms, and specialized annual wildflowers take advantage of the brief period of moisture to race through germination, growth, flowering, and seed production in a month or less (Fig. 31-14).

The animals of the deserts, like the plants, are specially adapted to survive on little water. Most deserts appear to be almost completely devoid of animal life during the day, because the animals seek relief from the sun and heat in cool underground burrows. After dark, when the deserts cool down considerably, horned lizards, snakes, and other reptiles emerge to feed, as do mammals, such as the kangaroo rat (see Fig. 31-13d), and birds, such as the burrowing owl. Most of the smaller animals survive without ever drinking at all, getting all the water they need from their food and that produced during cellular respiration in their tissues. Larger animals, such as desert bighorn sheep, are dependent on permanent water holes during the driest times of the year.

Human Impact While human activities are reducing the extent of many biomes, they are causing the spread of deserts, a process call **desertification.** A dramatic example is occurring in the Sahel, which borders the southern edge of the Sahara desert in Africa. Twenty-five years of below-average rainfall, coupled with rapid growth of the human population, have caused a steady southward spread of the desert. Western technology allowed the Sahelians to drill deep wells, abandon their nomadic way of life, and increase their livestock herds. The natural vegetation was destroyed as livestock competed for forage during the drought (Fig. 31-15). Native shrubs were uprooted by goats, preventing regeneration. The growing human population began cultivating increasingly fragile land without allowing time for the soil to recover between crops. As the desert spread southward, so did the people, with their livestock and crops, causing further deforestation, soil destruction, and desertification of once-productive land. The Sahel is an example of a human population's exceeding the carrying capacity of the land. The loss of productivity of the ecosystem is nearly irreversible, and massive famines, such as occurred in Ethiopia in the mid-1980s, are a tragic result.

Chaparral

In many coastal regions that border on deserts, such as southern California and much of the Mediterranean, a unique type of vegetation grows, called **chaparral.** The annual rainfall in these regions is similar to that of a desert, but the proximity of the sea provides a slightly longer rainy season in the winter and frequent fogs during the spring and fall, which re-

(a)

(b)

(c)

(d)

Figure 31-13   **Desert biomes**

(a) Under the most extreme conditions of heat and drought, deserts can be almost devoid of life, such as these sand dunes of the Sahara desert in Africa. **(b)** Throughout much of Utah and Nevada, the Great Basin Desert presents a monotonous landscape of widely spaced shrubs, such as sagebrush and greasewood. These shrubs often secrete a growth inhibitor from their roots, preventing germination of nearby plants and thus reducing competition for water. **(c)** The Sonoran Desert in southern Arizona and northern Mexico is typified by the saguaro cactus, prickly pear, and the spindly ocotillo, whose red flowers are pollinated by hummingbirds. The body of the saguaro is pleated and can expand after a rain to store water. **(d)** The kangaroo rat is an elusive inhabitant of the deserts of North America.

duce evaporation. Chaparral consists of small trees or large bushes with thick waxy or fuzzy evergreen leaves that conserve water. These shrubs are also able to withstand the frequent summer fires started by lightning (Fig. 31-16).

Human Impact As is often the case with deserts, grasslands, and forests, the extent of chaparral vegetation is influenced by human activities. About 400 B.C., the Greek philosopher Plato wrote that "there are mountains in Attica [southern Greece] which can now keep nothing but bees, but which were clothed, not so very long ago, with timber suitable for roofing very large buildings." The original Greek forests were cut down not only for ceiling beams but also for the great Athenian naval fleets. Subsequently, heavy grazing by goats prevented regrowth of the forests. Only the chaparral plants, which could survive in the hot sun and were distasteful or poisonous to goats, regrew. The forests have never returned.

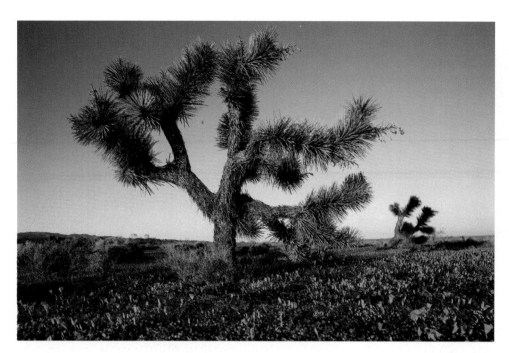

Figure 31-14   **The Mojave Desert**

In spring, this southern California desert is carpeted with wildflowers. Through much of the year, annual wildflower seeds lie dormant waiting for the spring rains to fall.

Grasslands

In the temperate regions of North America, deserts occur in the rain shadows east of the mountain ranges, such as the Sierra Nevada and Rocky Mountains. Farther east, as the rainfall gradually increases, the land begins to support more and more grasses, giving rise to the prairies of the Midwest. These **grassland,** or prairie, biomes are usually concentrated in the centers of continents. You can see a large grassland in the center of the Eurasian continent in Figure 31-7. In general, they have a continuous cover of grass and virtually no trees at all except along the rivers. From the tallgrass prairies of Iowa and Illinois (Fig. 31-17)

to the shortgrass prairies of eastern Colorado, Wyoming, and Montana (Fig. 31-18), the North American grassland once stretched across almost half the continent.

Water and fire are the crucial factors in the competition between grasses and trees. The hot dry summers and frequent droughts of the shortgrass prairies can be tolerated by grass but are fatal to trees. Trees can grow in the more eastern prairies, but historically the trees were destroyed by frequent fires, often set by Native Americans to maintain grazing land for the bison. Although the tops of the grasses are destroyed by fire, their root systems usually survive; trees, on the other hand, are killed outright. The resulting grasslands of North America once supported huge

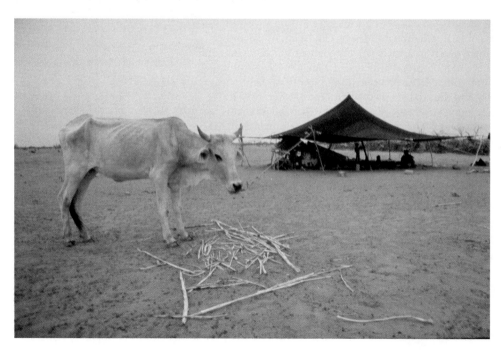

Figure 31-15   **Overgrazing**

In the Sahel region of Africa, overgrazing is degrading the already limited productivity of the land and contributing to the southward spread of the Sahara desert.

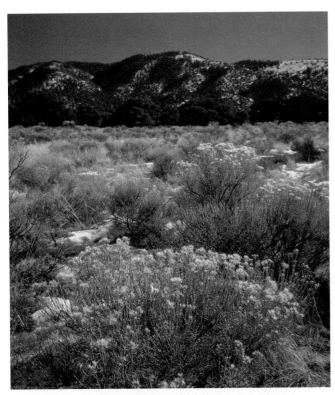

Figure 31-16   The chaparral biome

This biome is limited primarily to coastal mountains in dry regions, such as the San Gabriel Mountains in southern California. Chaparral is maintained by frequent fires set by summer lightning. Although the tops of the plants may be burned off, the roots send up new sprouts the next spring.

Figure 31-17   Tallgrass prairie in Missouri

In the central United States, moisture-bearing winds out of the Gulf of Mexico produce summer rains, allowing a lush growth of tall grasses and wildflowers such as these coneflowers.

herds of bison; as many as 60 million in the early nineteenth century. Grasses growing and decomposing for thousands of years produced what may be the most fertile soil in the world.

Human Impact With the elimination of the bison and the development of plows that could break the dense turf, the former prairie has become the "breadbasket" of North America, so named because enormous quantities of grain are cultivated in its fertile soil. The tallgrass prairie has been converted to agricultural land, except for tiny protected remnants.

On the western shortgrass prairie, cattle and sheep have replaced the bison and pronghorn antelope. As a result of their overgrazing the grasses, the boundary between the cool deserts and the grassland has often been altered in favor of desert plants. Much of the sagebrush desert of the American West is actually the result of overgrazing shortgrass prairie (Fig. 31-19). Cattle prefer grass to sagebrush, so heavy grazing destroys the grass. Consequently moisture that the grass would have absorbed is left in the soil, encouraging the growth of the woody sagebrush. Because the cattle will not eat sagebrush, it soon becomes the dominant vegetation; thus the prairie grasses are replaced by plants characteristic of the cool desert.

Temperate Deciduous Forests

At their eastern edge the North American grasslands merge into the **temperate deciduous forest** biome (Fig. 31-20). Higher precipitation occurs there than in the grasslands (75 to 150 centimeters; 30 to 60 inches) and, in particular, more summer rainfall occurs. The soil therefore retains enough moisture for trees to grow, and the resulting forest shades out grasses. In contrast to the tropical forests, the temperate deciduous forest biome has cold winters, usually with at least several hard frosts and often long periods of below-freezing weather. Winter in this biome has an effect on the trees similar to that of the dry season in the tropical deciduous forests: During periods of subfreezing temperatures, liquid water is not available to the trees. To reduce evaporation when water is in short supply, the trees drop their leaves in the fall. They produce leaves again in the spring, when liquid water becomes available. During the brief time in spring when the ground has thawed but the trees have not yet blocked off all the sunlight, numerous wildflowers grace the forest floor.

Insects and other arthropods are numerous and conspicuous in deciduous forests. The decaying leaf litter on the forest floor also provides food and habitat for bacteria, earthworms, fungi, and small plants. Many arthropods feed on these or on each other. A variety of vertebrates, including mice, shrews, squirrels, raccoons, and many species of birds, dwell in the deciduous forests.

Human Impact Large mammals such as black bear, deer, wolves, bobcats, and mountain lions were formerly abundant, but the predators have been largely wiped out

Figure 31-18   Shortgrass prairie

The lands east of the Rocky Mountains receive relatively little rainfall, and shortgrass prairie results, characterized by low-growing bunch grasses such as buffalo grass and grama grass, and a wealth of wildflowers such as this coneflower **(d)**. Pronghorn antelope **(a)**, prairie dogs **(b)**, and protected bison herds **(c)** occupy this biome.

Figure 31-19   Sagebrush desert or shortgrass prairie?

Biomes are influenced not only by temperature, rainfall, and soil but also by human activities. The shortgrass prairie field on the right has been overgrazed by cattle, causing the grasses to be replaced by sagebrush.

Figure 31-20   **Temperate deciduous forest**

In the eastern United States, the white-tailed deer **(a)** is the largest herbivore, and birds such as the Blue Jay **(c)** are abundant. In spring, a profusion of woodland wildflowers, such as these hepaticas **(b)**, blooms briefly before the trees produce leaves.

by humans. Clearing for lumber and agriculture has dramatically reduced America's deciduous forests, and virgin forests are now almost nonexistent.

Temperate Rain Forests

On the Pacific coast, from the lowlands of the Olympic peninsula in Washington to southeast Alaska, is the **temperate rain forest** biome (Fig. 31-21). As in the tropical rain forest, there is no shortage of liquid water year-round. This abundance of water is due to two factors. First, there is a tremendous amount of rain. The Hoh River rain forest in Olympic National Park receives over 400 centimeters (160 inches) of rain each year, over 60 centimeters (24 inches) in the month of December alone. Second, the moderating influence of the Pacific Ocean prevents severe frost from occurring along the coast, so the ground seldom freezes.

The abundance of water means that the trees have no need to shed their leaves in the fall, and almost all the trees

are evergreens. In contrast to the broadleaf evergreen trees of the tropics, the temperate rain forests are dominated by conifers. The ground and often the trunks of the trees are covered with mosses and ferns. As in the tropical rain forests, so little light reaches the forest floor that tree seedlings usually cannot become established. Whenever one of the forest giants falls, however, it opens up a patch of light, and new seedlings quickly sprout, often right atop the fallen log. This event produces a "nurse log," as shown in Figure 31-21b.

Taiga

North of the grasslands and temperate forests, the **northern coniferous forest**, also called the **taiga** (Fig. 31-22), stretches horizontally across the entire continent of North America, mainly in southern Canada. Conditions in the taiga are harsher than those in the temperate deciduous forest. In the taiga, the winters are longer and colder, and the growing season is shorter. The few months of warm

Figure 31-21   **Temperate rain forest**

The Hoh River temperate rain forest in Olympic National Park. The coniferous trees do not block the light as effectively as do broadleaf trees, so ferns, mosses, and wildflowers grow in the pale green light of the forest floor. The dead feeds the living, as new trees grow from the decay of this fallen giant, called a "nurse log" **(b)**, and flowering foxglove **(a)** and fungi **(c)** find ideal conditions amid the moist, decaying vegetation.

weather are too short to allow trees the luxury of regrowing leaves in the spring. As a result, the taiga is populated almost entirely by evergreen coniferous trees with small, waxy needles. The waxy coating and small surface area of the needles reduce water loss by evaporation during the cold months, and the leaves remain on the trees year-round. Thus the tree is instantly ready to take advantage of good growing conditions when spring arrives, and can continue slow growth late into the fall.

Because of its harsh climate, the diversity of life in the taiga is much lower than in many other biomes. Vast stretches of central Alaska, for example, are covered by a somber forest composed almost exclusively of black spruce and an occasional birch. Large mammals, such as the wood bison, grizzly bear, moose, and wolf, which have mostly been wiped out in the southern regions of their original range, still roam the taiga, as do smaller animals, such as the wolverine, fox, snowshoe hare, and deer.

Human Impact Owing to the remoteness of the northernmost taiga and the severity of its climate, a greater per-

centage of the taiga remains in undisturbed condition than any other North American biome except the tundra. Nonetheless, the taiga is a major source of lumber for construction, and huge expanses of forest have been clearcut (Fig. 31-23).

Tundra

The last biome encountered before reaching the polar ice caps is the arctic **tundra**, a vast treeless region bordering the Arctic Ocean (Fig. 31-24). Conditions in the tundra are severe. Winter temperatures in the arctic tundra often reach −55° C (−40° F) or below, winds howl at 50 to 100 kilometers (30 to 60 miles) per hour, and precipitation averages 25 centimeters (10 inches) or less per year, making this a "freezing desert." Even during the summer, the temperatures can drop to freezing, and the growing season may last only a few weeks before a hard frost occurs. Somewhat less cold but similar conditions produce alpine tundra on mountaintops above the altitude where trees can grow.

The cold climate of the arctic tundra results in **perma-frost,** a permanently frozen layer of soil often no more than half a meter (about 1.5 feet) below the surface. As a result, when summer thaws come, the water from melted snow and ice cannot soak into the ground and the tundra becomes a huge marsh. Trees cannot survive in the tundra; the permafrost limits root growth to the topmost meter or so of soil.

Nevertheless, the tundra supports a surprising abundance and variety of life. The ground is carpeted with small perennial flowers and dwarf willows no more than a few centimeters tall and often with a large lichen called reindeer moss, a favorite food of caribou. The standing pools provide superb mosquito habitat. The mosquitoes and other insects provide food for numerous birds, most of which migrate long distances to nest and raise their young during the brief summer feast. The tundra vegetation supports lemmings, which are eaten by wolves, snowy owls, arctic foxes, and even grizzly bears.

Human Impact The tundra is perhaps the most fragile of all the biomes because of its short growing season. A

willow 10 centimeters (4 inches) high may have a trunk 7 centimeters (3 inches) in diameter and be 50 years old. Human activities in the tundra leave scars that persist for centuries. Fortunately for the tundra inhabitants, the impact of civilization is localized around oil drilling sites, pipelines, mines, and military bases.

Rainfall, Temperature, and Vegetation

Terrestrial biomes are greatly influenced by both temperature and rainfall, whose effects interact. Temperature strongly influences the effectiveness of rainfall in providing soil moisture for plants and standing water for animals to drink. The hotter it is, the more rapidly water evaporates, both from the ground and from plants. As a result of this interaction of temperature with rainfall (and to a lesser extent, the distribution of rain throughout the year), areas that receive almost exactly the same rainfall can have startlingly different vegetation, all the way from desert to taiga. Take a trip with us from southern Arizona to northern Alaska as we visit ecosystems

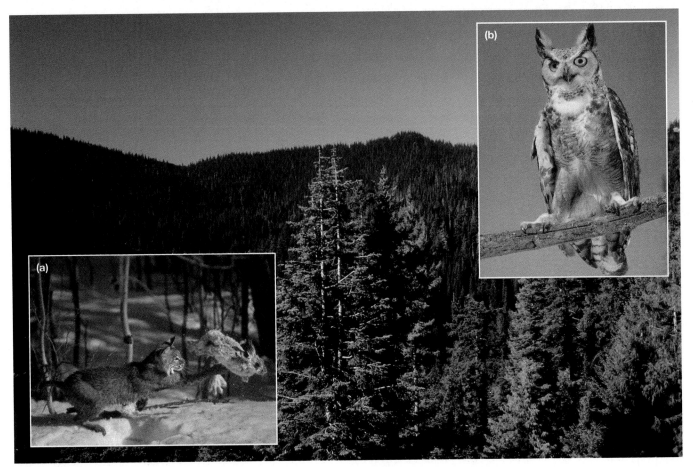

Figure 31-22   **The taiga**
The small needles and pyramidal shape of conifers allow them to shed heavy snows. Winter is a challenge not only for the trees but also for animals such as this snowshoe hare and the bobcat that preys on it **(a)**. The hare is also prey for the great horned owl **(b)**. Taiga animals face diminished food supply but increased energy requirements during subfreezing weather.

Figure 31-23   Clearcutting

Clearcutting (the removal of all the trees in a given area) as seen in this Oregon forest, is relatively simple and cheap. But its environmental costs are high. Erosion will diminish the fertility of the soil, slowing new growth. Further, the dense stands of even-age trees that often regrow are more vulnerable to attack by parasites than a natural stand of trees of various ages would be.

that each receive around 28 centimeters (12 inches) of rain annually.

The Sonoran Desert near Tucson, Arizona (see Fig. 31-13c), has an average annual temperature of 20°C (68°F) and gets about 28 centimeters (a foot) of rain each

year. The landscape is dominated by giant saguaro cactus and low-growing, drought-resistant bushes. Fifteen hundred kilometers (931 miles) north, in eastern Montana, rainfall is about the same, but we have passed into the shortgrass prairie biome (see Fig. 31-18), largely because the average temperature is much lower, about 7°C (45°F). Central Alaska receives about the same annual rainfall (28 centimeters), yet is covered with taiga forest (see Fig. 31-22). As a result of the low average annual temperature (about −4°C, or 25°F), permafrost underlies much of the ground. During the summer thaw, the taiga earns its Russian name "swamp forest," although its rainfall is about the same as that of the Sonoran Desert.

Life in Water

Although this chapter has emphasized terrestrial biomes, the saltwater oceans and seas are the largest ecosystems on Earth, covering about 71% of its surface. Freshwater ecosystems, in contrast, cover less than 1%.

The unique properties of water lend some common features to aquatic ecosystems. First, because water is slower to heat and cool than air, temperatures in aquatic ecosystems are more moderate than in terrestrial ecosystems. Second, although water may appear quite transparent, it absorbs a considerable amount of the light

Figure 31-24   Tundra

Life on the tundra is adapted to cold. Plants such as dwarf willows and perennial wildflowers (such as this dwarf clover [c]) grow low to the ground, escaping the chilling tundra wind. Tundra animals, such as caribou (a) and arctic foxes (b), can regulate blood flow in their legs, keeping them just warm enough to prevent frostbite, while preserving precious body heat for the brain and vital organs.

energy that sustains life. Even in the clearest water, the intensity of light decreases rapidly with depth, so that at depths of 200 meters (600 feet) or more, little light is left to power photosynthesis. If the water is at all cloudy—for example, because of suspended sediment or micro-organisms—the depth to which light can penetrate is greatly reduced. Third, nutrients in aquatic ecosystems tend to be concentrated near the bottom sediments where light levels often are too low to support photosynthesis. This separation of energy and nutrients limits aquatic life. Of the four requirements for life, aquatic ecosystems provide abundant water and appropriate temperatures. Thus, the major factors that determine the quantity and type of life in aquatic ecosystems are energy and nutrients.

Although they share some common features, aquatic ecosystems are extremely diverse. Freshwater ecosystems include rivers, streams, ponds, lakes, and marshes; marine (saltwater) ecosystems include estuaries, tide pools, the open ocean, and coral reefs. In the following sections, we look more closely at a few of these important aquatic ecosystems.

Freshwater Lakes Have Distinct Regions of Life and Temperature

Freshwater lakes vary tremendously in size, depth, and nutrient content. Although each lake is unique, moderate to large lakes in temperate climates share some common features, including distinct life zones and temperature stratification.

Life Zones Are Based on Access to Light and Nutrients

The distribution of life in lakes depends largely on access to light, to nutrients, and to a place for attachment (the bottom). The life zones of lakes, then, correspond to specific locations within the lake.

Near the shore is the **littoral zone.** In this zone, the water is shallow, and plants find abundant light, anchorage, and adequate nutrients from the bottom sediments. Not surprisingly, littoral-zone communities are the most diverse. Cattails and bulrushes abound nearest the shore, and water lilies and entirely submerged vascular plants and algae flourish at the deepest reaches of the littoral zone (Fig. 31-25). The plants of the littoral zone trap sediments carried in by streams and by runoff from the surrounding land, increasing the nutrient content in this region. Living among the anchored plants are microscopic organisms called **plankton**. There are two forms of plankton, **phytoplankton** (Greek, "drifting plants"), which includes photosynthetic protists, bacteria, and algae, and **zooplankton** (Greek, "drifting animals"), such as protozoa and tiny crustaceans. The greatest diversity of animals in the lake is also found in this zone. Littoral invertebrate animals include small crustaceans, insect larvae, snails, flatworms, and *Hydra*; vertebrates include frogs, minnows, and aquatic snakes and turtles.

As the water increases in depth farther from shore, plants are unable to anchor to the bottom and still collect enough light for photosynthesis. This open-water area is divided into two regions: the upper limnetic zone and the lower profundal zone (see Fig. 31-25).

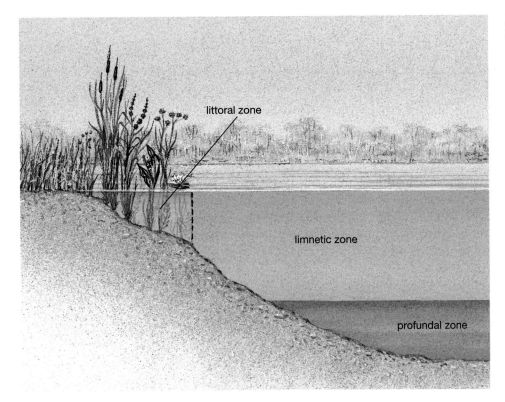

Figure 31-25   Lake life zones
There are three life zones in a "typical" lake: a near-shore littoral zone with rooted plants, an open-water limnetic zone, and a deep, dark profundal zone.

littoral zone

limnetic zone

profundal zone

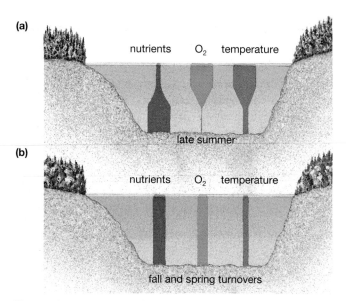

Figure 31-26   **Temperature stratification in lakes**
(a) Occurring during the summer, stratification results in high surface temperatures and oxygen levels, while surface nutrients become depleted because of a lack of mixing.
(b) During the spring and fall turnovers, temperature stratification breaks down, and water from all levels mixes, resulting in uniform temperature, oxygen, and nutrient levels.

In the **limnetic zone,** enough light penetrates to support photosynthesis. Here phytoplankton including cyanobacteria (also called blue-green algae) serve as producers. These are eaten by protozoa and small crustaceans, which in turn are consumed by fish.

Below the limnetic zone lies the **profundal zone,** where light is insufficient to support photosynthesis. This area is nourished mainly by detritus that falls from the littoral and limnetic zones and by incoming sediment. It is inhabited primarily by decomposers and detritus feeders, such as snails and certain insect larvae and fish.

Temperature Stratification Occurs in Summer in Temperate Lakes

Lakes in regions that experience seasonal temperature changes become stratified during the summer, with a layer of relatively warm, less-dense water "floating" atop a layer of denser cold water (around 4° C, or 40° F; water reaches its highest density at 4° C; Fig. 31-26). The layers persist through the summer and restrict the mixing of surface and bottom waters. Because nutrients are concentrated in the bottom sediments, the upper region, where photosynthetic organisms flourish, becomes depleted of nutrients as the summer progresses. Simultaneously, the heterotrophic organisms near the bottom use up available oxygen, which is not replenished from the surface. So by late summer, the surface populations are limited by lack of nutrients, while those in the profundal zone are limited by lack of oxygen. (The boom-and-bust popula-

tion cycle of blue-green algae on the surface of Lake Erie, rapid growth followed by a die-off due to nutrient depletion, is shown in Figure 28-4.)

In the fall, surface temperatures drop and the stratification breaks down, allowing mixing of the nutrient-rich lower layer and the oxygen-rich upper layer. A similar situation occurs in the spring as the floating ice (at 0° C, or 32° F) melts, and surface waters warm to 4° C and mix with the deep water. Fall and spring turnovers maintain the lake's capacity to support its natural communities and cause spurts of growth among phytoplankton and zooplankton, particularly in the spring, when temperature and hours of sunlight are increasing.

Freshwater Lakes Are Classified According to Their Nutrient Content

Although each lake is unique, freshwater lakes may be classified on the basis of their nutrient content as either oligotrophic (Greek, "poorly fed") or eutrophic (Greek, "well fed").

Oligotrophic lakes are very low in nutrients. They often are formed by glaciers that scrape depressions in bare rock, and they are fed by mountain streams carrying little sediment. Because there is little sediment or microscopic life to cloud the water, oligotrophic lakes are clear, and light penetrates deeply. Therefore, photosynthesis (and by extension oxygenation, as oxygen is produced as a by-product of photosynthesis) in deeper water is possible, and the limnetic zone may extend all the way to the bottom. Oxygen-loving fish, such as trout, thrive in oligotrophic lakes.

Eutrophic lakes receive larger inputs of sediments, organic material, and inorganic nutrients (such as phosphorus) from their surroundings, allowing them to support dense communities. They are murkier, both from suspended sediment and dense phytoplankton populations, so the lighted limnetic zone is shallower. Dense "blooms" of algae occur seasonally in the limnetic zone. Their dead bodies fall into the profundal zone, where they are used as food by decomposer organisms. The metabolic activities of many of these decomposers use oxygen, depleting the oxygen content of the profundal zone.

Lakes are transient ecosystems, although very large lakes may persist for millions of years. Over time, they gradually fill with sediment, undergoing succession to dry land (see Chapter 29). As nutrient-rich sediment accumulates, oligotrophic lakes tend to become eutrophic, a process called eutrophication.

Human Impact   Human activities may greatly accelerate the process of eutrophication, because nutrients are carried into lakes from farms, feedlots, and sewage. Even if solid wastes are removed, water discharged from sewage-treatment plants is often rich in phosphates and nitrates dissolved from wastes and detergents. Excessively enriched water also washes off fertilized farm fields and feedlots, where the manure of thousands of cattle accumulates. These added nutrients support excessive growth of phytoplankton. These producers, particularly blue-green algae, form a scum on the lake surface, depriving the submerged plants of sunlight. The plant bodies are decomposed by bacteria, depleting the dissolved oxygen. Deprived of oxygen, fish, snails, and insect larvae die, and their bodies fuel more bacterial growth, further depleting oxygen. Even without oxygen, certain bacteria that produce foul-smelling gases thrive. Although it is full of life, the nutrient-polluted lake smells dead. Most of the trophic levels, including the fish, have been eliminated, and the community is dominated by bacteria and microscopic algae.

Acid rain, caused by the burning of fossil fuels, poses a very different threat to freshwater ecosystems. Few organisms can withstand the low pH of acidified lakes, which often appear clear and oligotrophic because of the absence of life. Acid rain was discussed in Chapter 30.

Marine Ecosystems Cover Much of Earth

In the oceans, the upper layer of water, where the light is strong enough to support photosynthesis, is called the **photic zone.** Below the photic zone lies the **benthic zone,** where the only energy comes from the excrement and bodies of organisms that sink or swim down.

As in lakes, most of the nutrients in the oceans are at or near the bottom where there is not enough light for photosynthesis. Nutrients dissolved in the water of the photic zone are constantly being incorporated into the bodies of living organisms. When these organisms die, some sink into the benthic zone, providing its organisms with nutrients. If no new nutrients entered the photic zone, life (in both the photic and the benthic zones) would eventually cease.

Fortunately, there are two sources of new nutrients: the land, from which rivers constantly remove nutrients and carry them to the oceans, and **upwelling,** which brings cold, nutrient-laden water from the ocean to the surface. Upwelling often occurs along coastlines, as in California, Peru, and West Africa, where prevailing winds displace surface water, causing it to be replaced by water from below. Upwelling also occurs around Antarctica. Not surprisingly, the major concentrations of life in the oceans are found where abundant light is combined with a source of nutrients, which happens most often in regions of upwelling and in shallow coastal waters, including coral reefs.

Coastal Waters Support the Most Abundant Marine Life

The most abundant life in the oceans is found in a narrow strip surrounding Earth's land masses, where the water is shallow and a steady flow of nutrients is washed off the land. Coastal waters include the **intertidal zone,** the area that is alternately covered and uncovered by water with the rising and falling of the tides, and the **near-shore zone,** relatively shallow but constantly submerged areas, including bays and coastal wetlands such as salt marshes and estuaries (Fig. 31-27). The near-shore zone is the only part of the ocean where large plants can grow, anchored to the bottom. In addition, the abundance of nutrients and sunlight promotes the growth of a veritable soup of photosynthetic phytoplankton. Associated with these plants and protists are animals from nearly every phylum: annelid worms, sea anemones, jellyfish, sea urchins, sea stars, mussels, snails, fish, and sea otters, to name just a few. A large number and variety of organisms live permanently in coastal waters, but many that spend most of their lives in the open ocean come into the coastal waters to reproduce. Bays, salt marshes, and estuaries in particular are the breeding grounds for a wide variety of organisms, such as crabs, shrimp, and an array of fish, including most of our commercially important species.

Human Impact   Coastal regions are of great importance not only to the organisms that live or breed there but also to humans who wish to use them for food sources, recreation, mineral and oil extraction, or living places. Like rain forests, wetlands of all types once covered 6% of Earth's surface, but nearly half of them have been dredged or filled in. In the United States, 100 million of our 215 million acres of wetlands have been lost. As populations increase in our coastal states and resources such as oil become increasingly scarce, the conflict between preservation of our coastal wetlands as wildlife and animal habitat and development of these areas for housing, marinas, and energy extraction will become increasingly intense. Because much of the life of the entire ocean is dependent upon the well-being of the coastal waters, it is essential that we protect these fragile, vital areas.

Coral Reefs

Coral reefs are actually created by animals and plants. In warm tropical waters, with just the right combination of bottom depth, wave action, and nutrients, specialized algae and corals (types of cnidarians) build reefs from their own calcium carbonate skeletons. Coral reefs are most abundant in tropical waters of the Pacific and Indian oceans, the Caribbean, and the Gulf of Mexico as far north as southern Florida, where the maximum water temperatures range between 22° C and 28° C (72° F and 82° F).

Reef-building corals are involved in a mutualistic relationship with unicellular algae called dinoflagellates, who live embedded in the coral tissue. These corals grow best at depths of less than 40 meters (130 feet), where light can penetrate and allow their algal partners to photosynthesize. The algae benefit from the high nitrogen, phosphorus, and carbon dioxide levels in the coral tissues. In return, algae provide food for the coral and help produce calcium carbonate, which forms the coral skeleton. Coral reefs provide an anchoring place for many other algae, a

Figure 31-27   **Near-shore ecosystems**

(a) A salt marsh in the eastern United States. Expanses of shallow water fringed by marsh grass (*Spartina*) provide excellent habitat and breeding grounds for many marine organisms and shorebirds. (b) Although the shifting sands present a challenge to life, grasses stabilize them, and animals such as this *Emerita* crab (inset) burrow in the sandy intertidal zone. (c) A rocky intertidal shore in Oregon, where animals and algae grip the rock against the pounding waves and resist drying during low tide. (Inset) Colorful sea stars cling to the rocks surrounded by a seaweed, fucus. (d) Towering kelp sway through the clear water off southern California, providing the basis for a diverse community of invertebrates, fishes, and an occasional sea otter (inset).

home for bottom-dwelling animals, and shelter and food for the most diverse collection of invertebrates and fish to be found in the oceans (Fig. 31-28). The Great Barrier Reef in Australia is home to more than 200 species of coral alone, and a single reef may harbor 3000 species of fish, invertebrates, and algae.

Human Impact   Coral reefs are extremely sensitive to certain types of disturbance, especially silt caused by soil eroding from nearby land. As silt clouds the water, light is diminished and photosynthesis reduced, hampering growth of the corals. The reef may eventually become buried in mud, the corals smothered, and the entire marvelous community of diverse organisms destroyed. Ero-

sion from construction, roadways, and poor land management has produced enough silt to ruin several reefs near Honolulu, Hawaii. The reef inhabitants have been replaced by large numbers of sediment-feeding invertebrates such as sea cucumbers. In the Philippines, logging has dramatically increased erosion, so coral reefs (as well as rain forests) are being destroyed. Another hazard is sewage and runoff from agriculture, which fertilize near-shore ocean water, causing eutrophication and a dense growth of algae. This blocks sunlight from the coral's dinoflagellates, and decaying algal bodies deplete the water of oxygen, killing the coral.

Still another threat to the reef communities is over-fishing. In at least 80 countries, a variety of species, in-

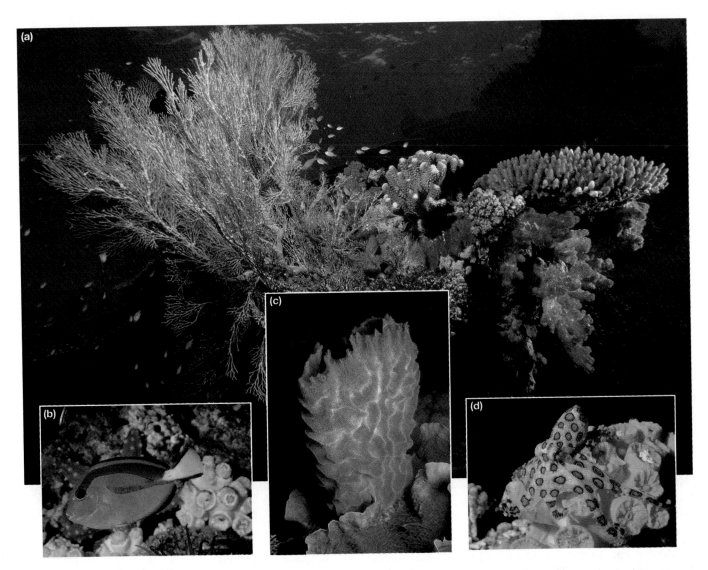

Figure 31-28   **Coral reefs**
(a) Coral reefs, composed of the bodies of corals and algae, provide habitat for an extremely diverse community of extravagantly colored animals. Many fish, including this blue tang (b), feed on coral (note the bright yellow corals in background). A vast array of invertebrates such as this sponge (c) and blue-ringed octopus (d) live among the corals of Australia's Great Barrier Reef. This tiny (15 centimeters, or 6 inches, fully extended) octopus is one of the world's most venomous creatures, possessing enough venom to kill 10 adult humans.

cluding mollusks, turtles, fish, crustaceans, and even corals, are being harvested from reefs faster than they can replace themselves. Many of these are sold to shell collectors and aquarium owners in developed countries. In some tropical countries, dynamite is used to kill coral reef fish, destroying entire sections of the coral reef community in the process. Tropical fish collectors often use poison to stun the fish before collecting them, leaving most dead. Removal of predators from reefs may disrupt the ecological balance of the community, allowing an explosion in populations of coral-eating sea urchins, which are destroying reefs, for example, along the Kenyan coast in Africa.

As with rain forests, both protection and sustainable use are crucial to the survival of these fragile and diverse underwater ecosystems. Carefully regulated harvesting and tourism produce far more economic benefits than activities that destroy the reefs. Some countries have recognized this, and there are now many areas where reefs are protected.

The Open Ocean

Beyond the coastal regions lie vast areas of the ocean in which the bottom is too deep to allow plants to anchor and still receive enough light to grow. Most life in the open ocean is limited to the upper photic zone, in which

Figure 31-29   The open ocean

Porpoises skim the surface **(a)**, rare humpback whales leap clear of the water **(b)**, and fish such as this blue jack swim **(c)**. The photosynthetic phytoplankton **(d)** are the producers on which most other life ultimately depends. Phytoplankton are eaten by zooplankton **(e)**, represented by this microscopic crustacean, a copepod. The spiny projections on these planktonic creatures help keep them from sinking below the photic zone.

the life forms are **pelagic**—that is, free-swimming or floating—for their entire lives. The food web of the open ocean is dependent on phytoplankton consisting of microscopic photosynthetic protists, mainly diatoms and dinoflagellates (Fig. 31-29d). These organisms are consumed by zooplankton, such as tiny crustaceans that are relatives of crabs and lobsters (Fig. 31-29e). Zooplankton in turn serve as food for larger invertebrates, small fish, and even marine mammals such as the humpback whale (Fig. 31-29b).

One challenge faced by inhabitants of the open ocean is to remain afloat in the photic zone where sunlight and food are abundant. Many members of the planktonic community have elaborate flotation devices, such as oil droplets in their cells or long projections, to slow down their rate of sinking (see Fig. 31-29d,e). Most fish have swim bladders that can be filled with gas to regulate their buoyancy. Some animals, and even some of the phytoplankton, actively swim to stay in the photic zone. Many small crustaceans migrate to the surface at night to feed,

then sink into the dark depths during daylight, thus avoiding visual predators such as fish. The amount of pelagic life varies tremendously from place to place. The blue clarity of tropical waters is a result of a lack of nutrients, which limits the concentration of plankton in the water. Nutrient-rich waters that support a large plankton community are greenish and relatively murky.

Below the photic zone, the only available energy in most cases comes from the excrement and dead bodies drifting down from above. Nevertheless, a surprising quantity and variety of life exists in the benthic zone, including fishes of bizarre shapes, worms, sea cucumbers, sea stars, and mollusks.

Human Impact   Two major threats to the open ocean are pollution and overfishing. Open ocean pollution takes several forms. Ocean-going vessels dump millions of plastic containers overboard daily. Plastic six-pack holders, foam cups, and packing material wash and blow off the

Figure 31-30   Vent communities
Located in the ocean depths, vent communities include giant worms nearly 4 meters (12 feet) long. These worms are red with oxygen-trapping hemoglobin. They have no digestive tract but are host to sulfur bacteria, which provide the worms with energy by oxidizing hydrogen sulfide.

land, collecting on parts of the ocean surface. The plastic looks like food to unsuspecting sea turtles, gulls, porpoises, seals, and whales, who often die after trying to consume it. Until 1992, New York city barged its refuse and sewage sludge out to sea, creating a heavily contaminated area covering 40 square miles of open floor. The open ocean has served as a dumping ground for radioactive wastes. Oil contaminates the open ocean from many sources, including tanker spills, runoff from improper disposal on land, leakage from offshore oil wells, and natural seepage.

The increasing demand for fish to feed a growing human population, coupled with increasingly efficient fishing technologies, has resulted in depletion of many important fisheries. The northeastern U.S. cod fishery has collapsed due to overfishing. Populations of lobsters, salmon, haddock, king crab, and many other types of seafood have also declined dramatically because of overfishing.

Vent Communities

In 1977, a new and unusual source of nutrients, forming the basis of a spectacular undersea community, was discovered in the deep ocean. Geologists exploring the Galapagos Rift (an area of the Pacific floor where plates forming Earth's crust are separating) found vents spewing superheated water, black with sulfur and minerals. Surrounding these vents was a rich community of pink fish, blind white crabs, enormous mussels, white clams, sea anemones, and tube worms (Fig. 31-30). Twenty-two previously undescribed families and 284 new species of animals have been found in these novel communities.

In this unique ecosystem, sulfur bacteria serve as the primary producers, harvesting the energy stored in hydrogen sulfide spewed from beneath Earth's crust. This process, called chemosynthesis, replaces photosynthesis at these depths, over 2500 meters (7500 feet) below the surface. The bacteria proliferate in the warm water surrounding the vents, covering nearby rocks with thick, mat-like colonies. These colonies provide the food on which the animals of the vent community thrive. Many vent animals consume the bacteria directly. Others, such as the giant tube worm (which lacks a digestive tract), harbor the bacteria within cells in their bodies (see Fig. 31-30). In this mutualistic association, the bacteria provide high-energy carbon compounds, while the tube worm provides hydrogen sulfide, which it transports to the bacteria in its bloodstream using a unique form of hemoglobin.

The world still holds wonders and mysteries for those who seek them, and we have only begun to explore the versatility and diversity of life on Earth.

E A R T H   W A T C H

Humans and Ecosystems

The expanding human population has left relatively few ecosystems undisturbed. Our impacts on natural ecosystems are so diverse and wide ranging that they far exceed the scope of this book. However, we can identify some general characteristics of ecosystems dominated by humans and contrast them to the characteristics of undisturbed ecosystems. Below are listed six characteristic differences between these two types of ecosystems and a few ideas for minimizing them.

First, ecosystems dominated by people tend to be simpler—that is, to have fewer species and fewer community interactions—than undisturbed ecosystems. Although a city street bustles with life and apparent complexity, count the number of species you encounter in an average block and compare it with the number you encounter while hiking in a wilderness for a similar distance. As humans enter an ecosystem, animals in the highest trophic levels are the first to go. Carnivores are always relatively rare, and their specialized needs are most easily disrupted. Many big carnivores, such as the wolf and mountain lion, require large undisturbed hunting territories. Human beings often selectively destroy large predators, believing them a threat to people or their livestock. As a result, even in relatively undisturbed areas they have often been eliminated. Grizzly bears and wolves no longer roam the Rocky Mountains. On the western shortgrass prairie, prairie dog towns fall as human towns and ranches rise (Fig. E31-2). The black-footed ferret, a predator of prairie dogs, faces possible extinction. Farm fields have been deliberately simplified from the original prairie biome to eliminate competition and predation and allow the maximum productivity of a single crop species. Nowhere is the contrast between human and natural ecosystems greater than in the tropical rain forests, whose unparalleled diversity is replaced by failed attempts at farming.

There is probably no practical way to restore to human ecosystems the great diversity found in undisturbed areas, nor is doing so always desirable from a human standpoint. Agriculture, for example, demands a simplified ecosystem (although not all biomes are suited to it). Cities concentrate human activities and culture and may lessen the human impact on the surrounding countryside. However, as we recognize the benefits of artificially simplified ecosystems, we must be aware of the need to preserve intact as many natural communities as possible. The undisturbed forest traps and purifies water and can reduce air pollution. Swamps and estuaries, when not dredged or filled, contain a wealth of detritus feeders and decomposers that purify water. Coastal wetlands are breeding places for millions of birds and spawning sites for the majority of our commercially important fish and crustacean species. In

Figure E31-2   Habitat destruction

Loss of habitat due to human activities is a major threat to most of Earth's wildlife.

undisturbed, diverse ecosystems we find aesthetic pleasure, as well as a storehouse of species whose commercial or medicinal values are not yet recognized. For example, nearly half the medicines in use today were originally discovered in plants, and human beings have examined only a small fraction of existing plants for possible medical uses.

Second, whereas natural ecosystems run on sunlight, human ecosystems have become dependent on nonrenewable energy from fossil fuels. From the suburbanite pouring gas into a lawn mower to the farmer driving a tractor, managing a simplified ecosystem is an energy-intensive proposition. Energy must be expended to oppose the tendency of the natural system to restore complexity. Fertilizers also require large amounts of energy to produce, and farms must be heavily fertilized because nutrient cycles have been disrupted.

To counteract this trend, some farmers are returning to organic farming. Plant and animal waste and natural nutrient cycles are used to maintain soil fertility. Alternating legume crops, such as soybeans and alfalfa, with other crops helps maintain nitrogen in the soil. Mulching can help retain fertility, reduce water loss, and control weeds. The use of natural insect predators and limited pesticide spraying at critical times during a pest's life cycle can dramatically reduce the need for poisons. Organic farms can be as productive as more conventional farms while using 15% to 50% less energy from fossil fuels per quantity of food produced.

In our homes and commercial structures, better insulation and increased use of solar heat can result in dramatic energy savings. By increasing our use of renewable energy sources, such as wind and especially sunlight, we can conserve fossil fuels, dramatically reduce pollution, and move human ecosystems a step closer to those that occur naturally.

Third, natural ecosystems recycle nutrients, whereas human ecosystems tend to lose nutrients. Walk around a suburban neighborhood on trash-pickup day. Grass clippings and leaves are packed in plastic bags to be hauled away. To compensate for this loss of natural nutrients, suburbanites often heavily fertilize their lawns and gardens. A similar trend has occurred in modern farming. The exposed soil is eroded away by wind and water, removing crucial nutrients and requiring large inputs of fertilizer to replace them. Runoff from the field, carrying fertile topsoil and artificial fertilizers, may pollute nearby rivers, streams, and lakes. Pesticides may kill detritus feeders and decomposers, further disrupting natural nutrient cycles. Thus, whereas fertile soil accumulates in many natural ecosystems, it tends to be lost in those dominated by human beings. Some 3 billion tons of topsoil are eroded from farms each year in the United States. The Mississippi River alone carries off 40 tons each hour.

Again, organic farming can help reverse this trend, and we can apply the same principles in our own lawns and gardens. Farmers can counteract erosion by using contour planting, in which row crops are oriented so that they slow the flow of water instead of funneling it down a slope. Row crops, which tend to allow erosion, can be alternated in strips with dense, soil-catching crops such as wheat. Planting rows of trees as windbreaks helps prevent soil loss from blowing wind, and it creates a more diverse ecosystem and a nesting place for insect-eating birds. Farm fields need not be plowed in the fall and left unplanted to erode during the winter; in fact, an increasing number of farmers are planting crops in the stubble from the previous season, reducing erosion.

Fourth, natural ecosystems tend to store water and purify it through biological processes, whereas human ecosystems tend to pollute water and shed it rapidly. A thundershower strikes a forest, an adjacent city, and a farm. The rich soil of the forest sponges up the water, which gradually drains into the ground, filtered by the soil and purified by decomposers that break down organic contaminants. In the nearby city, water pours from sidewalks, rooftops, and streets, picking up soot, silt, oil, lead, and garbage. It races down gutters into storm sewers, and a weakly toxic soup gushes into the nearest stream or river. Farm runoff carries priceless topsoil, expensive fertilizer, and animal manure into rivers and lakes, where these resources become pollutants.

Preventing erosion will simultaneously conserve water and reduce water pollution from farm runoff. Manure from cattle feedlots, which is a significant source of both groundwater and surface-water pollution, could be placed on fields, where it would restore needed nutrients.

Although water will continue to run off our cities, the pollutants it carries can be reduced by minimizing our reliance on fossil fuels. We must eliminate leaded gasoline and tighten standards for emissions from diesel and gasoline engines and smokestacks. Efficient public transportation systems will reduce pollution and the massive frustration caused by congested traffic. Insulation will reduce power consumption in our homes and offices, as will increased reliance on solar heat.

Fifth, simple human ecosystems such as farms tend to be extremely unstable. Natural ecosystems have many species and tend to remain stable over time. Herbivorous insects are controlled by natural predators, including other insects, birds, and shrews. Insect populations are also limited because their preferred plants are scattered among many others rather than growing in a pure stand, as on a farm. On farms, both pest insects and their natural predators are exposed to pesticides. Unfortunately, the pests often develop resistance to the poison, while their predators are killed. In these simplified communities, unfavorable weather conditions or the introduction of an exotic species (see Chapter 29) can be disastrous.

Farmers can counteract this trend by planting smaller fields with a wider variety of crops. Alternating crops not only helps maintain soil fertility but also helps prevent the proliferation of disease and insect pests that are specialized for a particular crop. Populations of corn borers, for example, will die off during years when the field is planted in alfalfa. Use of biological controls such as natural insect predators and insect diseases can reduce reliance on pesticides.

Finally, human ecosystems are characterized by continuously growing populations, whereas nonhuman populations in natural ecosystems are relatively stable. As our population expands, the spread of human-dominated ecosystems presents a growing threat to the diversity of species and to the delicate balance that has evolved over the 3-billion-year history of life on Earth.

In summary, natural ecosystems tend to be complex, stable, and self-sustaining, powered by solar energy and nourished by recycled nutrients. They provide diverse habitats for wildlife, purify contaminants through the action of decomposers, and build up nutrient-rich soil. Modern human ecosystems are relatively simple and are sustained by large inputs of energy from fossil fuels. They tend to minimize wildlife habitat, contaminate soil and water, and lose nutrients and fertile soil. These problems are compounded by continued population growth, which is causing expansion of human-dominated ecosystems at the expense of undisturbed ones.

As we have pointed out, human ecosystems do not have to be as disruptive and alien to the operation of natural ecosystems as we have allowed them to become. Through understanding, education, and commitment, appropriate use of technology, and stabilization of our population, we can reverse many of these destructive trends.

✷ S U M M A R Y   O F   K E Y   C O N C E P T S

Throughout the world, areas with similar climates have similar groups of organisms. Although species differ, adaptations to climate create recognizable distinguishing features, particularly among plants.

Climate

The availability of sunlight, water, and appropriate temperatures determines the climate of a given region. Sunlight that reaches Earth may be either reflected or absorbed by the atmosphere, land, and water, which radiates it back as heat, with a small amount captured by photosynthetic organisms. This heat maintains the temperature of Earth. Equal amounts of solar energy are spread over a smaller surface at the equator than farther north and south, making the equator relatively warm, while higher latitudes have lower overall temperatures. Earth's tilt on its axis causes dramatic seasonal variation at northern and southern latitudes.

Warm air currents at the equator and at 60° north and south latitudes rise and release moisture as they cool, increasing precipitation at these latitudes. Cooler, dry air sinks at around 30° north and south latitudes and at the poles, creating areas of very low moisture. These patterns are modified by the continents. Ocean gyres carry warm equatorial waters into northern and southern latitudes, moderating the climate of coastal areas.

Climate is modified by the rapid heating and cooling of continents as compared with the oceans, and by elevation. Air cools as it rises, so high elevations are cooler than low ones. Cool air holds less moisture, so rain falls on the windward side of mountain ranges, and rain shadows occur on the far sides. Deserts are often found in the rain shadows of mountain ranges. Because temperature decreases both with increasing latitude and increasing elevation, the vegetation changes observed going up a mountain are similar to those seen while traveling northward in the Northern Hemisphere.

The Requirements of Life

The requirements for life on Earth include nutrients, energy, liquid water, and a reasonable temperature. The differences in the form and abundance of living things in various locations on Earth are largely attributable to differences in the interplay of these four factors.

Life on Land

On land, the crucial limiting factors are temperature and liquid water. Large regions of the continents that have similar climates will have similar vegetation, determined by the interaction of temperature and rainfall (availability of water). These regions are called biomes.

Tropical forests are warm and wet, dominated by huge broadleaf evergreen trees. Most nutrients are tied up in vegetation, and most animal life is arboreal. Rain forests, home to at least 50% of all species, are rapidly being cut for agriculture, although the soil is extremely poor.

The African savanna is an extensive grassland with pronounced wet and dry seasons. It is home to the world's most diverse and extensive herds of large mammals.

Deserts, hot and dry, are found primarily between 20° and 30° of latitude and in the rain shadows of mountain ranges. Plants are widely spaced and have adaptations to conserve water. Animals tend to be small and nocturnal, also adapted to drought.

Chaparral exists in desertlike conditions, which are moderated by their proximity to a coastline, allowing small trees and bushes to thrive.

Grasslands, concentrated in the centers of continents, have a continuous grass cover and few trees. They produce the world's richest soils and have largely been converted to agriculture.

Temperate deciduous forests, whose broadleaf trees drop their leaves in winter to conserve moisture, dominate the eastern half of the United States. The wet temperate rain forests, dominated by evergreens, are found on the northern Pacific coast of the United States.

The taiga, or northern coniferous forest, covering much of the northern United States and southern Canada, is dominated by conifers whose small waxy needles are adapted for water conservation and year-round photosynthesis.

The tundra is a frozen desert where permafrost prevents the growth of trees and bushes remain stunted. Nonetheless, a diverse array of animal life and perennial plants flourishes in this fragile biome.

Life in Water

Energy and nutrients are the major limiting factors in the distribution and abundance of life in aquatic ecosystems. Nutrients are found in bottom sediments and are washed in from surrounding land, concentrating them near shore and in deep water.

Freshwater lakes have three life zones. The littoral zone, near shore, is rich in energy and nutrients and supports the most diverse community. The limnetic zone is the lighted region of open water where photosynthesis can occur. The profundal zone is the deep water, where light is inadequate for photosynthesis and the community is dominated by heterotrophic organisms.

Lakes in temperate climates become stratified during the summer and winter, with a layer of cold dense water at the bottom and less dense, warmer water (in summer) or colder water (in winter) in an upper layer. During the spring and fall turnovers when temperatures equalize, the layers mix, distributing oxygen and nutrients.

Oligotrophic lakes are clear, low in nutrients, and support sparse communities. Eutrophic lakes are rich in nutrients and support dense communities. During succession to dry land, lakes tend to go from an oligotrophic to a eutrophic condition.

Most life in the oceans is found in shallow water, where sunlight can penetrate, and is concentrated near the continents and in areas of upwelling, where nutrients are most plentiful.

Coastal waters, consisting of the intertidal zone and the near-shore zone, contain the most abundant life. Producers consist of aquatic plants anchored to the bottom and photosynthetic protists called phytoplankton.

Coral reefs are confined to warm, shallow seas. The calcium carbonate reefs form a complex habitat supporting the most diverse undersea ecosystem.

In the open ocean, most life is found in the photic zone, where light supports phytoplankton. In the lower benthic zone, life is supported by nutrients that drift down from the photic zone. Specialized vent communities, supported by bacteria, thrive at great depths in the superheated waters where Earth's crustal plates are separating.

KEY TERMS

benthic zone p. 643
biome p. 625
chaparral p. 632
climate p. 622
coral reef p. 643
desert p. 632
desertification p. 632
eutrophic lake p. 642
grassland p. 634
gyre p. 623
intertidal zone p. 643
latitude p. 622

limnetic zone p. 642
littoral zone p. 641
near-shore zone p. 643
northern coniferous forest p. 637
oligotrophic lake p. 642
ozone layer p. 626
pelagic p. 646
permafrost p. 639
photic zone p. 643
phytoplankton p. 641
plankton p. 641
profundal zone p. 642

rain shadow p. 624
savanna p. 630
taiga p. 637
temperate deciduous forest p. 635
temperate rain forest p. 637
tropical deciduous forest p. 629
tropical rain forest p. 627
tundra p. 638
upwelling p. 643
weather p. 621
zooplankton p. 641

THINKING THROUGH THE CONCEPTS

Multiple Choice

1. Most plant species fit into only a few morphological types. Which type seems most restricted by temperature and rainfall?
   a. trees
   b. shrubs
   c. grasses
   d. perennial herbs
   e. annual weeds
2. In which biome is the smallest fraction of carbon and nutrients present in the soil?
   a. tropical rain forest
   b. savanna
   c. tundra
   d. grassland
   e. coniferous forest
3. How do mountain ranges create deserts?
   a. by lifting land up into colder, drier air
   b. by completely blocking the flow of air into desert areas, thus preventing clouds from getting there
   c. by forcing air to first rise and then fall, thus causing rain on one side of the mountains and desert on the other
   d. by causing the world wind patterns that make certain latitudes very dry
   e. by causing very steep slopes that are subject to erosion
4. What is the primary reason that plants from distant, but climatically similar, places often look the same?
   a. common ancestry

   b. adaptation to the same physical conditions
   c. adaptation to similar herbivores
   d. continental drift
   e. effects of past climate change
5. Which of these biomes has been increased in area by human activities?
   a. savanna
   b. temperate rain forest
   c. grassland
   d. coniferous forest
   e. desert
6. What biome has the richest soil and has largely been converted to agriculture?
   a. tundra
   b. coniferous forest
   c. grassland
   d. tropical rain forest
   e. deciduous forest

Review Questions

1. Explain how air currents contribute to the formation of the tropics and the large deserts.
2. What are large, roughly circular ocean currents called? What effect do they have on climate, and where is that effect strongest?
3. What are the four major requirements for life? Which two are most often limiting in terrestrial ecosystems? In ocean ecosystems?

4. Explain why traveling up a mountain takes you through biomes similar to those you would encounter traveling north for a long distance.

5. Where are the nutrients of the tropical forest biome concentrated? Why is life in the tropical rain forest concentrated high above the ground?

6. Explain two different undesirable effects of agriculture in the tropical rain forest biome.

7. List some adaptations of (a) desert plants and (b) desert animals to heat and drought.

8. How and where does desertification occur?

9. How are trees of the taiga adapted to a lack of water and a short growing season?

10. How do deciduous and coniferous biomes differ?

11. What single environmental factor best explains why there is shortgrass prairie in Colorado, tallgrass prairie in Illinois, and deciduous forest in Ohio?

12. Where are the world's largest populations of large herbivores and carnivores found?

13. Where is life in the oceans most abundant, and why?

14. Why is the diversity of life so high in coral reefs? What human impacts threaten them?

15. Distinguish between the limnetic, littoral, and profundal zones of lakes in terms of their location and the communities they support.

16. Distinguish between oligotrophic and eutrophic lakes. Describe (a) a natural scenario and (b) a human-created scenario under which an oligotrophic lake might be converted to a eutrophic lake.

17. Explain the reason and importance of the spring and fall overturn in temperate lakes.

18. Distinguish between the photic and benthic zones. How do organisms in the photic zone obtain nutrients? How are nutrients obtained in the benthic zone?

19. What unusual primary producer forms the basis for rift-vent communities?

20. On the basis of the location of the worst atmospheric ozone depletion, which biomes are likely to be most affected by increased UV penetration?

✖ A P P L Y I N G   T H E   C O N C E P T S

1. List at least six differences between human-dominated and undisturbed ecosystems, and discuss in some detail how these differences can be minimized.

2. In which terrestrial biome is your college or university located? Discuss similarities and differences between your location and the general description of that biome in the text. If you are living in a city, how has the urban environment modified your interaction with the biome?

3. During the 1960s and 1970s, many parts of the United States and Canada banned the use of detergents containing phosphates. Until that time, almost all laundry detergents and many soaps and shampoos had high concentrations of phosphates. What environmental concern do you think prompted these bans, and what ecosystem has benefited most from the bans?

4. Because ozone depletion is expected to get worse, not better, for decades to come, biologists have tried to assess which types of species will be most susceptible to increased UV penetration. Two of the groups that may be most vulnerable are quite different from each other: ocean plankton and long-lived birds and mammals. Try to think of what makes each of these groups so susceptible to increased UV radiation.

5. Although temperature and moisture are the primary determinants of terrestrial ecosystems, both natural and human-induced differences in soil types can strongly affect plant communities. Both naturally rocky soils (e.g., serpentine barrens and granitic outcrops) and the inorganic, gravelly soils produced by mining operations often support only grasses and shrubs, even in biomes that are largely dominated by trees. How could these different soils affect the ability of trees to grow? In particular, how could rocky soils affect the temperature or moisture experienced by plants growing on them, as opposed to the temperature and moisture experienced by plants growing nearby in soils with more organic matter?

6. Understanding the ways in which the four basic requirements for life determine where different biomes occur can help us predict the consequences of global warming. Global warming is expected to make most areas warmer, but it is also expected to change rainfall in ways that are hard to predict—some areas will get wetter and others drier. Our ignorance of how rainfall will change is not very important in understanding shifts in more northerly biomes, but it is very important for our understanding of changes in tropical areas. Look at Figure 31-8 and explain why this is true.

7. More northerly forests are far better able to regenerate after logging than are tropical rain forests. Try to explain why this would be true. HINT: The cold soils of northern climates greatly slow down decomposition rates.

✖ F O R   M O R E   I N F O R M A T I O N

Allan, J. D., and Flecker, A. S. "Biodiversity Conservation in Running Waters: Identifying the Major Factors That Threaten Destruction of Riverine Species and Ecosystems." *BioScience* 43:32–42, 1993. The major human impacts on river ecosystems and how to restore these communities.

Bell, R. H. V. "A Grazing Ecosystem in the Serengeti." *Scientific American*, July 1971. Describes the great migrations of the herbivores of the African savanna.

Dodd, J. L. "Desertification and Degradation in Sub-Saharan Africa: The Role of Livestock." *BioScience* 44:28–34, 1994. An argument that the process of desert formation is not as clearly linked to human-caused degradation as is usually portrayed.

Goreau, T. F., Goreau, N. I., and Goreau, T. J. "Corals and Coral Reefs." *Scientific American*, August 1979. Discusses the ecology of the great reefs built by tiny polyps harboring symbiotic protists.

Holloway, M. "Sustaining the Amazon." *Scientific American*, July 1993. Describes the threats to the Amazon rain forest and innovative ways to preserve tropical forests and their biodiversity.

Kusler, J. A., Mitsch, W. J., and Larson, J. S. "Wetlands." *Scientific American*, January 1994. Describes the characteristics and values of these threatened, but highly important ecosystems.

Mares, M. A. "Desert Rodents, Seed Consumption, and Convergence: Evolutionary Shuffling of Adaptations." *BioScience* 43:372–379, 1993. An example of how animal as well as plant species and communities living in similar climates can show convergence.

Milton, S. J., Dean, R. J., du Plessis, M. A., and Siegfried, W. R. "A Conceptual Model of Arid Rangeland Degradation." *BioScience* 44:70–76, 1994. Rangeland degradation proceeds in steps, each of them increasingly difficult and costly to reverse.

Toon, B., and Turco, R. P. "Polar Stratospheric Clouds and Ozone Depletion." *Scientific American*, June 1991. Describes the unique chemistry of the Antarctic clouds that causes ozone breakdown.

Tunnicliffe, V. "Hydrothermal-Vent Communities of the Deep Sea." *American Scientist*, July–August 1992. A comprehensive look at the structure and function of undersea vent communities, with excellent illustrations.

Wilson, E. O. "Threats to Biodiversity." *Scientific American*, September 1989. The elimination of habitats, particularly in the rain forest, is driving plants and animals to extinction in record numbers.

N E T   W A T C H

On-line resources for this chapter are on the World Wide Web at:
http://www.prenhall.com/~audesirk (click on the table of contents link and then select Chapter 31).

Answers to Multiple-Choice Questions

Chapter 1
1:a 2:c 3:d 4:c 5:e 6:e

Chapter 2
1:d 2:b 3:c 4:b 5:a 6:a

Chapter 3
1:a 2:e 3:a 4:b 5:c 6:c

Chapter 4:
1:d 2:c 3:b 4:a 5:c 6:e

Chapter 5
1:a 2:c 3:d 4:a 5:a 6:c 7:b

Chapter 6
1:b 2:e 3:e 4:a 5:d 6:c

Chapter 7
1:d 2:c 3:b 4:e 5:d 6:b

Chapter 8
1:a 2:c 3:d 4:c 5:a 6:c

Chapter 9
1:c 2:c 3:a 4:c 5:b 6:d

Chapter 10
1:b 2:c 3:b 4:d 5:a 6:d

Chapter 11
1:c 2:a 3:a 4:b 5:a 6:c

Chapter 12
1:c 2:e 3:b 4:a 5:d 6:c

Chapter 13
1:c 2:c 3:b 4:d 5:a 6:e

Chapter 14
1:b 2:c 3:c 4:c 5:b

Chapter 15
1:c 2:d 3:b 4:a 5:e 6:a

Chapter 16
1:e 2:d 3:e 4:e 5:a 6:d

Chapter 17
1:a 2:d 3:b 4:c 5:c 6:e

Chapter 18
1:e 2:d 3:d 4:a 5:b 6:c

Chapter 19
1:e 2:a 3:d 4:b,d,e 5:e 6:d

Chapter 20
1:d 2:b 3:e 4:c 5:a 6:d

Chapter 21
1:e 2:d 3:e 4:d 5:c 6:a

Chapter 22
1:d 2:a 3:c 4:d 5:e 6:d 7:c
8:e 9:b 10:e

Chapter 23
1:e 2:b 3:c 4:a 5:c 6:a 7:d
8:b 9:d 10:e 11:b

Chapter 24
1:e 2:c 3:b 4:a 5:c 6:e

Chapter 25
1:e 2:e 3:b 4:c 5:d

Chapter 26
1:c 2:c 3:e 4:d 5:d 6:e 7:c
8:e 9:d 10:e

Chapter 27
1:a 2:d 3:b 4:e 5:c 6:a 7:a
8:b

Chapter 28
1:a 2:c 3:a 4:e 5:b 6:d

Chapter 29
1:d 2:c 3:a 4:d 5:a 6:c

Chapter 30
1:a 2:c 3:b 4:d 5:e 6:d

Chapter 31
1:a 2:a 3:c 4:b 5:e 6:c

Glossary

abiotic (ā-bī-ah´-1tik): nonliving.

abortion: the procedure for terminating pregnancy in which the cervix is dilated and the embryo and placenta removed by suction.

abscisic acid (ab-sis´-ik): a plant hormone that generally inhibits the action of other hormones, enforcing dormancy in seeds and buds and causing closing of stomata.

abscission layer: a layer of thin-walled cells at the base of the petiole of a leaf that produces an enzyme that digests the cell wall holding leaf to stem, allowing the leaf to fall off.

absorption: the process by which nutrients are taken into cells.

accessory pigments: colored molecules other than chlorophyll a that absorb light energy and pass it to chlorophyll.

acetylcholine (ah-sēt´-il-kō´-lēn): a neurotransmitter found in the brain and in synapses of motor neurons onto skeletal muscles.

acid: a substance that releases hydrogen ions (H+) into solution; a solution with a pH of less than 7.

acid deposition: the deposition of nitric or sulfuric acid, either in dissolved in rain (acid rain) or in the form of dry particles, as a result of the production of nitrogen oxides or sulfur dioxide through burning, primarily of fossil fuels.

acrosome (ak´-rō-sōm): a vesicle located at the tip of an animal sperm, which contains enzymes needed to dissolve protective layers around the egg.

actin (ak'-tin): a major muscle protein, whose interactions with myosin produce contraction; found in the thin filaments of the muscle fiber.

action potential: a rapid change from a negative to a positive electrical potential in a nerve cell. This signal travels along an axon without change in size.

activation energy: in a chemical reaction, the energy needed to force the electron clouds of reactants together, prior to the formation of products.

active site: the region of an enzyme molecule that binds substrates and performs the catalytic function of the enzyme.

active transport: the movement of materials across a membrane through the use of cellular energy, usually against a concentration gradient.

active visual signal: a specific movement or posture that communicates information.

adaptation: a characteristic of an organism that helps it to survive and reproduce in a particular environment; also, the process of acquiring such characteristics.

adaptive radiation: The rise of many new species in a relatively short time as a result of a single species invading different habitats and evolving under different selective pressures in those habitats.

adenosine triphosphate (a-den´-ō-sen trīfos´-fāt; ATP): a molecule composed of ribose sugar, adenine, and three phosphate groups. The last two phosphate groups are attached by energy-carrier "high-energy bonds." ATP is the major energy carrier in cells.

adipose tissue (a´-di-pōse): tissue composed of fat cells.

adrenal cortex: the outer part of the adrenal gland, which secretes steroid hormones that regulate metabolism and salt balance.

adrenal medulla: the inner part of the adrenal gland, which secretes epinephrine (adrenalin) and norepinephrine (noradrenalin).

aerobic: using oxygen.

age structure: the distribution of males and females in a population according to age groups.

agglutination (a-glu-tin-ā´-shun): clumping of foreign substances or microbes, caused by binding with antibodies.

aggression: antagonistic behavior, usually between members of the same species, often resulting from competition for resources.

aggressive mimicry (mim´ik-ré): the evolution of a predatory organism to resemble a harmless one, thus gaining access to its prey.

algae (al´-gē; sing. alga): a general term for simple aquatic plants lacking vascular tissue.

allantois (al-an-tō´-is): one of the embryonic membranes of reptiles, birds, and mammals. In reptiles and birds, the allantois serves as a waste-storage organ. In mammals, the allantois forms most of the umbilical cord.

allele (al-ēl´): one of several alternative forms of a particular gene.

allele frequency: for any given gene, the relative proportion of each allele of that gene found in a population.

allergy: an inflammatory response produced by the body in response to invasion by foreign materials, such as pollen, which are themselves harmless.

allopatric speciation (al-ō-pat´-rik): speciation that occurs when two populations are separated by a physical barrier that prevents gene flow between them (geographical isolation).

allosteric regulation (al-ō-ster´-ik): enzyme regulation in which an inhibitor molecule binds to an enzyme at a site away from the active site, changing the shape or charge of the active site so that it can no longer bind substrate molecules.

alternation of generations: a life cycle typical of plants in which a diploid sporophyte (spore-producing) generation alternates with a haploid gametophyte (gamete-producing) generation.

altruism: a type of behavior that may decrease the reproductive success of the individual performing it but benefits that of other individuals.

alveolus (al-vē´-ō-lus; pl. alveoli): a tiny air sac within the lungs surrounded by capillaries where gas exchange with the blood occurs.

amino acid: the individual subunit of which proteins are made, composed of a central carbon atom to which is bonded an amino group (—NH$_2$), a carboxyl group (—COOH), a hydrogen atom, and a variable group of atoms denoted by the letter "R."

ammonia: NH$_3$; a highly toxic nitrogen-containing waste product of amino acid breakdown. In the mammalian liver, it is converted to urea.

amniocentesis (am-nē-ō-sen-tē´-sis): a procedure for sampling the amniotic fluid surrounding a fetus. A sterile needle is inserted through the abdominal wall, uterus, and amniotic sac of a pregnant woman, into the amniotic fluid. Ten to 20 milliliters of amniotic fluid is withdrawn. Various tests may be performed on the fluid and the fetal cells suspended in it to provide information on the developmental and genetic state of the fetus.

amnion (am´-nē-on): one of the embryonic membranes of reptiles, birds, and mammals, enclosing a fluid-filled cavity that envelops the embryo.

amniotic egg (am-nē-ot´-ik): the egg of reptiles and birds. It contains an amnion that encloses the embryo in a watery environment; this allows the egg to be laid on dry land.

amplexus (am-pleck´-sus): a form of external fertilization found in amphibians, in which the male holds the female during spawning and releases his sperm directly onto her eggs.

amygdala (am-ig´-da-la): part of the forebrain of vertebrates, involved in production of appropriate behavioral responses to environmental stimuli.

amylase (am´-ē-lās): an enzyme found in saliva and pancreatic secretions that catalyzes the breakdown of starch.

anaerobic: not using oxygen.

analogous structures: structures that have similar functions and superficially similar appearance but very different anatomy, such as the wings of insects and birds. The similarities are due to similar selective pressures rather than common ancestry.

anaphase (an´-a-fāz): the stage of mitosis in which the sister chromatids of each chromosome separate from one another and are moved to opposite poles of the cell. In meiosis I, the stage in which homologous chromosomes, consisting of two sister chromatids, are separated. In meiosis II, the stage in which the sister chromatids of each chromosome separate from one another and are moved to opposite poles of the cell.

androgens: the collective term for male sex hormones.

angina pectoris (an-jī´-na pek-tōr´-is): chest pain associated with reduced blood flow to the heart muscle caused by obstruction of coronary arteries.

angiosperm (an´-gē-ō-sperm): a flowering vascular plant.

annual ring: pattern of alternating light (early) and dark (late) xylem of woody stems and roots, formed because of unequal availability of water in different seasons of the year, usually spring and summer.

antagonistic muscles: a pair of muscles, one of which contracts and in so doing extends the other, an arrangement that makes possible movement of the skeleton at joints.

anterior (an-tēr´-ē-ur): the front, forward, or head end of an animal.

anterior pituitary: a lobe of the pituitary gland that produces prolactin and growth hormone as well as hormones that regulate hormone production in other glands.

anther (an´-ther): the uppermost part of the stamen in which pollen develops.

antheridium (an-ther-id´-e-um): a structure in which male sex cells are produced, found in the bryophytes and certain seedless vascular plants.

antibiotic resistance: the ability of a mutated pathogen to resist the effects of an antibiotic that normally kills it.

antibody: a protein produced by cells of the immune system that combines with a specific antigen and usually facilitates its destruction.

anticodon: a sequence of three bases in transfer RNA that is complementary to the three bases of a codon of messenger RNA.

antidiuretic hormone (an-tē-dī-ūr-et´-ik): also called ADH; a hormone produced by the hypothalamus and released into the bloodstream by the posterior pituitary gland when blood volume is low. It increases the permeability of the distal tubule and the collecting duct to water, allowing more water to be reabsorbed into the bloodstream.

antigen: a complex molecule, usually protein or polysaccharide, that stimulates the production of a specific antibody.

apical dominance: the phenomenon whereby a growing shoot tip inhibits the sprouting of lateral buds.

apical meristem (āp´-i-kul mer´-i-stem): the cluster of meristematic cells found at the tip of a shoot or root (or one of their branches).

appendicular skeleton (ap-pen-dik´-ū-lur): that portion of the skeleton consisting of the bones of the extremities and their attachments to the axial skeleton; the pectoral and pelvic girdles, the arms, legs, hands, and feet.

aqueous humor (ā´-kwē-us): clear, watery fluid between the cornea and lens of the eye.

archegonium (ar-ke-gō´-nē-um): a structure in which female sex cells are produced, found in the bryophytes and certain seedless vascular plants.

arteriole (ar-tēr´-ē-ōl): a small artery that empties into capillaries. Contraction of the arteriole regulates blood flow to various parts of the body.

artery (ar´-tur-ē): a vessel with muscular, elastic walls that conducts blood away from the heart.

ascus (as´-kus): a saclike case in which sexual spores are formed by members of the fungal division Ascomycota.

asexual reproduction: reproduction that does not involve the fusion of haploid sex cells. The parent body may divide and new parts regenerate, or a new, smaller individual may be formed attached to the parent, to drop off when complete.

association neuron: In neural networks, a nerve cell that is postsynaptic to a sensory neuron and presynaptic to a motor neuron. In actual circuits, there maybe many association neurons between individual sensory and motor neurons.

aster: in cell division of animal cells, a star-shaped array of microtubules extending in all directions outward from the centriole.

atherosclerosis (ath´-er-ō-skler-ō´-sis): a disease characterized by obstruction of arteries by cholesterol deposits and thickening of the arterial walls.

atmospheric cycle: a nutrient cycle for a substance whose primary reservoir is the atmosphere.

atom: the smallest particle of an element that retains the properties of the element.

atomic number: the number of protons in the nuclei of all atoms of a particular element.

atrioventricular node (ā´-trē-ō-ven-trik´-ū-lar nōd): a specialized mass of muscle at the base of the right atrium through which the electrical activity initiated in the sinoatrial node is transmitted to the ventricles.

atrioventricular valve: heart valves between the atria and the ventricles that prevent backflow of blood into the atria during ventricular contraction.

atrium (ā´-trē-um): a chamber of the heart that receives venous blood and passes it to a ventricle.

auditory canal (aw´-dih-tōry): a canal within the outer ear that conducts sound from the external ear to the eardrum.

auditory nerve: the nerve leading from the mammalian cochlea to the brain, carrying information about sound.

autoimmune disease: a disorder in which the immune system produces antibodies against the body's own cells.

autonomic nervous system: part of the peripheral nervous system of vertebrates that synapses on glands, internal organs, and smooth muscle and produces largely involuntary responses.

autosome (aw´-tō-sōm): a chromosome found in homologous pairs in both males and females, and which does not bear the genes determining sex.

autotroph (aut´-ō-trof): a "self-feeder," usually meaning a photosynthetic organism.

auxin (awk´-sin): a plant hormone that influences many plant functions, including phototropism, apical dominance, and root branching. Auxin generally stimulates cell elongation and, in some cases, cell division and differentiation.

axial skeleton: the skeleton forming the body axis, including the skull, vertebral column, and rib cage.

axon: a long process of a nerve cell, extending from the cell body to synaptic endings on other nerve cells or on muscles.

B cell: a type of lymphocyte that participates in humoral immunity; gives rise to plasma cells that secrete antibodies into the circulatory system and to memory cells.

bacillus (buh-sil´-us; pl. bacilli): a rod-shaped bacterium.

bacterial conjugation: the exchange of genetic material between bacteria.

bacteriophage (bak-tir´-ē-ō-fāj): a virus specialized to attack bacteria.

bacterium (bak-tir´-ē-um; pl. bacteria): an organism consisting of a single prokaryotic cell surrounded by a complex polysaccharide coat.

balanced polymorphism: the prolonged maintenance of two or more alleles in a population, usually because each allele is favored by a separate selective force.

ball-and-socket joint: a joint in which the rounded end of one bone fits into a hollow depression in another, as in the hip; allows movement in several directions.

bark: the outer layer of a woody stem, consisting of phloem, cork cambium, and cork cells.

Barr body: an inactivated X chromosome found in cells of female mammals, which have two X chromosomes. The Barr body usually appears as a dark spot in the nucleus.

basal body: the organelle, structurally identical to a centriole, that gives rise to the microtubules of cilia and flagella.

base: a substance that is capable of combining with and neutralizing H+ ions. In molecular genetics, one of the nitrogen-containing, single- or double-ringed structures that distinguish one nucleotide from another. In DNA, the bases are adenine, guanine, cytosine, and thymine.

basidiospore (ba-sid´-ē-ō-spōr): a sexual spore formed by members of the fungal division Basidiomycota.

basidium (bas-id´-ē-um): a diploid cell, often club-shaped, formed on the basidiocarp of members of the fungal division Basidiomycota, which produces basidiospores by meiosis.

basilar membrane (bas´-eh-lar): a membrane in the cochlea that bears hair cells that respond to the vibrations produced by sound.

behavior: as any observable response to external or internal stimuli.

behavioral isolation: lack of mating between species of animals that differ substantially in courtship and mating rituals.

bilateral symmetry: body plan in which only a single plane drawn through the central axis will divide the body into mirror-image halves.

bile (bīl): a liquid secretion of the liver stored in the gallbladder and released into the small intestine during digestion. Bile is a complex mixture of bile salts, water, other salts, and cholesterol.

bile salts: substances synthesized in the liver from cholesterol and amino acids, that assist in the breakdown of lipids by dispersing them into small particles on which enzymes may act.

binary fission: The process by which a single bacterium divides in half, producing two identical offspring.

binocular vision: The ability to see objects simultaneously through both eyes, providing greater depth perception and more accurate judgment of the size and distance of an object from the eyes.

biodiversity: all living things within a given geographical area and the interrelationships among them.

biological clock: a metabolic timekeeping mechanism found in most organisms, whereby the organism measures the approximate length of a day (24 hours) even without external environmental cues such as light and dark.

biological magnification: the increasing accumulation of a toxic substance in progressively higher trophic levels.

biome (bī´-ōm): a general type of ecosystem occupying extensive geographical areas, characterized by similar plant communities: for example, deserts.

biosphere (bī´-ō-sfēr): that part of the Earth inhabited by living organisms; includes both the living and nonliving components.

biotic (bī-ah´-tik): living.

biotic potential: maximum rate at which a population could increase, assuming a maximum birth rate and minimum death rate.

birth control pill: a temporary contraceptive method that prevents ovulation by providing a continuing supply of estrogen and progesterone, which in turn suppresses LH release; must be taken daily, usually for 21 days each menstrual cycle.

bladder: a hollow muscular storage organ for urine.

blade: the flat part of a leaf.

blastocyst (blas´-tō-sist): an early stage of human embryonic development, consisting of a hollow ball of cells, enclosing a mass of cells attached to its inner surface, which becomes the embryo.

blastopore: the site at which a blastula indents to form a gastrula.

blastula: in animals, the embryonic stage attained at the end of cleavage, in which the embryo usually consists of a hollow ball with a wall one or several cell layers thick.

blood: A fluid consisting of plasma in which blood cells are suspended, carried within the circulatory system.

Bohr effect (bor): the tendency of hemoglobin to release oxygen more readily when the blood is slightly more acidic than normal, as is caused by an increase in dissolved carbon dioxide.

book lungs: thin layers of tissue resembling pages in a book, enclosed in a chamber and used as a respiratory organ by certain types of arachnids.

boom-and-bust cycle: a population cycle characterized by rapid exponential growth followed by a sudden major decline in population size, seen in seasonal species and some populations of small rodents, such as lemmings.

Bowman's capsule: the cup-shaped portion of the nephron in which blood filtrate is collected from the glomerulus.

bradykinin (brā-dē-kī´-nin): a chemical formed during tissue damage that binds to receptor molecules on pain nerve endings, giving rise to the sensation of pain.

brain: the part of the central nervous system of vertebrates enclosed within the skull.

branch root: a root that arises as a branch of a preexisting root, through divisions of pericycle cells and subsequent differentiation of the daughter cells.

bronchiole (bron´-kē-ōl): a narrow tube formed by repeated branching of the bronchi, which conducts air into the alveoli.

bronchus (bron´-kus): a tube that conducts air from the trachea to each lung.

bryophyte (brī´-ō-fīt): a division of simple nonvascular plants, including mosses and liverworts.

bud: in animals, a small copy of an adult that develops on the body of the parent. It eventually breaks off and becomes independent. In plants, an embryonic shoot, usually very short and consisting of an apical meristem with several leaf primordia.

budding: asexual reproduction by growth of a miniature copy, or bud, of the adult animal on the body of the parent. The bud breaks off to begin independent existence.

buffer: a compound that minimizes changes in pH by reversibly taking up or releasing H+ ions.

bulbourethral gland (bul-bō-ū-rē´-thrul): in male mammals, a gland that secretes a basic, mucus-containing fluid that forms part of the semen.

bulk flow: the movement of many molecules of a gas or fluid in unison from an area of higher pressure to an area of lower pressure.

bundle-sheath cells: cells that surround the veins of plants. In C_4 plants, but not C_3 plants, these cells contain chloroplasts.

C_3 cycle: the cyclic series of reactions whereby carbon dioxide is fixed into carbohydrates during the light-independent reactions of photosynthesis. Also called Calvin-Benson cycle.

C_4 pathway: the series of reactions in certain plants that fixes carbon dioxide into oxaloacetic acid, which is later broken down for use in the C_3 cycle of photosynthesis.

callus: a porous mass of bone and cartilage secreted by osteoblasts in response to a break in bone; it holds together the ends of the bones while repair takes place.

calmodulin (kal-mod´-ū-lin): a cytoplasmic protein that acts as a second messenger by binding calcium ions and changing configuration. The configuration change may activate enzymes, initiating a series of biochemical reactions.

calorie (kal´-ōr-ē): a measure of the energy derived from food. When capitalized (that is, Calorie), this unit is the amount of energy required to raise the temperature of 1 liter of water 1 degree Celsius. It represents 1000 calories (with a lowercase "c"). The energy content of foods is measured in Calories.

Calvin-Benson cycle: see C_3 cycle.

cambium (kam´-bē-um; pl. cambia): a lateral meristem that runs parallel to the long axis of roots and stems that causes secondary growth of woody plant stems and roots. See cork cambium; vascular cambium.

camera eye: the type of eye found in vertebrates and cephalopod mollusks, consisting of a light-sensitive retina, a lens that focuses images on the retina, and muscles that adjust the shape of the lens.

camouflage (cam'-a-flage): coloration and\or shape that renders an organism inconspicuous in its environment.

cancer: a disease in which some of the body's cells escape from normal regulatory processes and divide without control.

capillary: the smallest type of blood vessel, capillaries connect arterioles with venules. Capillary walls, through which exchange of nutrients and wastes occurs, are only one cell thick.

capsule: a polysaccharide or protein coating that surrounds the cell walls of some bacteria.

carbohydrate: a compound composed of carbon, hydrogen, and oxygen, with the approximate chemical formula (CH2O)n; includes sugars and starches.

carbon fixation: the initial steps in the C^3 cycle, in which carbon dioxide reacts with ribulose bisphosphate to form a stable organic molecule.

carbonic anhydrase (car-bon´-ik an-hī´-drās): an enzyme found in red blood cells that catalyzes the formation of bicarbonate ions from dissolved carbon dioxide and water.

cardiac cycle: the alternation of contraction (systole) and relaxation (diastole) of the heart chambers; systole and diastole.

cardiac muscle (kar´-dē-ak): specialized muscle of the heart, able to initiate its own contraction independent of the nervous system.

carnivore (kar´-neh-vōr): literally "meat eater," a predatory organism feeding on other heterotrophs; a secondary (or higher) consumer.

carnivorous (kar-niv´-e-rus): feeding on the bodies of other animals.

carotenoid (ka-rot´-en-oyd): red, orange, or yellow pigments found in chloroplasts that serve as accessory light-gathering molecules in thylakoid photosystems.

carpel (kar´pel): the female reproductive structure of a flower, composed of stigma, style, and ovary.

carrier: an individual who is heterozygous for a recessive condition. Carriers display the dominant phenotype but can pass on their recessive allele to their offspring.

carrier protein: a membrane protein that facilitates diffusion of specific substances across the membrane. The molecule to be transported binds to the outer surface of the carrier protein; the protein then changes shape, allowing the molecule to move across the membrane through the protein.

carrying capacity: the maximum population size that an ecosystem can maintain on a sustained basis. Determined primarily by the availability of space, nutrients, water, and light.

cartilage (kar´-teh-lij): a form of connective tissue forming portions of the skeleton, consisting of chondrocytes and their extracellular secretion of collagen. Cartilage resembles flexible bone.

Casparian strip (kas-par´-ē-an): a waxy, waterproof band in the cell walls between endodermal cells in a root, which prevents the movement of water and minerals in and out of the vascular cylinder through the extracellular space.

catalyst (cat´-a-list): a substance that speeds up a chemical reaction without itself being permanently changed in the process. Catalysts lower the activation energy of a reaction.

catastrophism: the hypothesis that Earth has experienced a series of geological catastrophes, probably imposed by a supernatural being, which accounts for the multitude of species, both extinct and modern, while preserving creationism.

cell: the smallest unit of life, consisting, at a minimum, of an outer membrane enclosing a watery medium containing organic molecules, including genetic material composed of DNA.

cell body: part of a nerve cell in which most of the common cellular organelles are located. Also often a site of integration of inputs to the nerve cell.

cell cycle: the sequence of events in the life of a cell, from one division to the next.

cell division: in eukaryotes, the process of reproduction of single cells, usually into two identical daughter cells, by mitosis accompanied by cytokinesis.

cell-mediated immunity: an immune response in which foreign cells or substances are destroyed by contact with T cells.

cell plate: in plant cell division, a series of vesicles that fuse to form the new plasma membranes and cell wall separating the daughter cells.

cell wall: a layer of material, usually made up of cellulose or celluloselike materials, found outside the plasma membranes of plants, fungi, bacteria, and some protists.

cellular respiration: the oxygen-requiring reactions occurring in mitochondria that break down the end products of glycolysis into carbon dioxide and water, while capturing large amounts of energy as ATP.

cellular slime mold: a funguslike protist consisting of individual amoeboid cells that can aggregate to form a sluglike mass, which in turn forms a fruiting body.

cellulase: an enzyme that catalyzes the breakdown of the carbohydrate cellulose into its component glucose molecules. Cellulase is almost entirely restricted to microorganisms.

cellulose: an insoluble carbohydrate composed of glucose subunits; forms the cell wall of plants.

central nervous system: in vertebrates, the brain and spinal cord.

central vacuole: a large, fluid-filled vacuole occupying most of the volume of many plant cells; performs several functions, including maintaining turgor pressure.

centriole (sen´-trē-ōl): in animal cells, a microtubule-containing structure found at the microtubule organizing center and the base of each cilium and flagellum. Gives rise to the microtubules of cilia and flagella and may be involved in spindle formation during cell division.

centromere (sen´-trō-mēr): the region of a replicated chromosome at which the sister chromatids are held together.

cephalization (sef-al-ī-zā´-shun): the increasing concentration over evolutionary time of sensory structures and nerve ganglia at the anterior end of animals.

cerebellum (ser-uh-bel´-um): part of the hindbrain of vertebrates, concerned with coordinating movements of the body.

cerebral cortex (ser-ē´-brel kōr´-tex): a thin layer of neurons on the surface of the vertebrate cerebrum, in which most neural processing and coordination of activity occurs.

cerebral hemisphere: one of two nearly symmetrical halves of the cerebrum, connected by a broad band of axons, the corpus callosum.

cerebrospinal fluid: a clear fluid produced within the ventricles of the brain that fills the ventricles and cushions the brain and spinal cord.

cerebrum (ser-ē´-brum): part of the forebrain of vertebrates concerned with sensory processing, direction of motor output, and coordination of most bodily activities. The cerebrum consists of two nearly symmetrical halves (the hemispheres) connected by a broad band of axons, the corpus callosum.

cervix (ser´-vicks): a ring of connective tissue at the outer end of the uterus, leading into the vagina.

channel protein: a membrane protein that forms a channel or pore completely through the membrane and that is usually permeable to one or a few water-soluble molecules, especially ions.

chemical bond: the force of attraction between neighboring atoms that holds them together in a molecule.

chemical equilibrium: the condition in which the "forward" reaction of reactants to products proceeds at the same rate as the "backward" reaction from products to reactants, so that no net change in chemical composition occurs.

chemiosmosis (kē-mē-os-mō´-sis): a process of ATP generation in chloroplasts and mitochondria. The movement of electrons down an electron transport system is used to pump hydrogen ions across a membrane, thereby building up a concentration gradient of hydrogen ions across the membrane. The hydrogen ions diffuse back across the membrane through the pores of ATP-synthesizing enzymes. The energy of their movement down their concentration gradient drives ATP synthesis.

chemoreceptor: a sensory receptor that responds to chemicals from the environment; used in the chemical senses of taste and smell.

chemosynthetic (kēm´-ō-sin-the-tic): capable of oxidizing inorganic molecules to obtain energy.

chiasma (kē-as´-ma; pl. chiasmata): during prophase I of meiosis, a point at which a chromatid of one chromosome crosses with a chromatid of the homologous chromosome. Exchange of chromosomal material between chromosomes takes place at a chiasma.

chitin (kī´-tin): a compound found in the cell walls of fungi and the exoskeletons of insects and some other arthropods, composed of chains of nitrogen-containing, modified glucose molecules.

chlorophyll (klōr´-ō-fil): a pigment found in chloroplasts that captures light energy during photosynthesis; absorbs violet, blue, and red light but reflects green.

chloroplast (klōr´-ō-plast): the organelle of plants and plantlike protists that is the site of photosynthesis; surrounded by a double membrane and containing an extensive internal membrane system bearing chlorophyll.

cholecystokinin (kō´-lē-sis-tō-ki´-nin): a digestive hormone produced by the small intestine that stimulates release of pancreatic enzymes.

chondrocyte (kon´-drō-sīt): the living cells of cartilage, which with its extracellular secretions of collagen, form cartilage.

chorion (kor´-ē-on): the outermost embryonic membrane in reptiles, birds, and mammals. In birds and reptiles, the chorion functions mostly in gas exchange. In mammals, the chorion forms most of the embryonic part of the placenta.

chorionic gonadotropin (CG): a hormone secreted by the chorion (one of the fetal membranes), which maintains the integrity of the corpus luteum during early pregnancy.

chorionic villi (kor-e-on-ik): in mammals, fingerlike projections of the chorion of the embryo that penetrate the uterine lining and form the embryonic portion of the placenta.

chorionic villus sampling: a procedure for sampling cells from the chorionic villi produced by a fetus. A tube is inserted into the uterus of a pregnant woman, and a small sample of villi are suctioned off for genetic and biochemical analysis.

choroid (kōr´-oyd): a darkly pigmented layer of tissue behind the retina that contains blood vessels and pigment that absorbs stray light.

chromatid (krō´-ma-tid): one of the two identical strands of DNA and protein forming a replicated chromosome. The two sister chromatids are joined at the centromere.

chromatin (krō´-ma-tin): the complex of DNA and proteins that makes up eukaryotic chromosomes.

chromosome (krō´-mō-sōme): in eukaryotes, a linear strand composed of DNA and protein, found in the nucleus of a cell, that contains the genes; in prokaryotes, a circular strand composed solely of DNA.

chylomicron (kī-lō-mī´-kron): a droplet consisting of triglycerides, cholesterol, and phospholipids, and coated with a thin layer of protein. Chylomicrons are formed in the cells lining the small intestine, enter the lymphatic system, and eventually reach the circulatory system.

chyme (kīme): an acidic, souplike mixture of partially digested food, water, and digestive secretions that is released from the stomach into the small intestine.

ciliate (sil´-ē-et): a category of protozoan characterized by cilia and a complex unicellular structure, including harpoonlike organelles called trichocysts. Members of the genus Paramecium are well-known ciliates.

cilium (sil´-ē-um; pl. cilia): a short, hairlike projection from the surface of certain eukaryotic cells, containing microtubules in a 9+2 arrangement. Movement of cilia may propel cells through a fluid medium or move fluids over a stationary surface layer of cells.

circadian rhythm (sir-kā´-dē-un): an event that recurs with a period of about 24 hours, even in the absence of environmental cues.

citric acid cycle: a cyclic series of reactions in which the acetyl groups from the pyruvic acids produced by glycolysis are broken down to CO_2, accompanied by the formation of ATP and electron carriers. Occurs in the matrix of mitochondria.

class: the taxonomic category composed of related genera. Closely related classes form a division or phylum.

classical conditioning: a training procedure in which an animal learns to perform a response (such as salivation) to a new stimulus that did not elicit that response originally (such as a sound). This is accomplished by pairing a stimulus that elicits the response automatically (in this case, food) with the new stimulus.

cleavage: the early cell divisions of embryos, in which little or no growth occurs between divisions. Cleavage reduces the cell size and distributes gene-regulating substances to the newly formed cell.

climax community: a relatively stable community that forms the endpoint of succession.

cloning: the process of producing many identical copies of a gene; also used to denote producing many genetically identical copies of an organism.

closed circulatory system: the type of circulatory system found in certain worms and vertebrates in which the blood is always confined within the heart and vessels.

clumped distribution: characteristic distribution of populations in which individuals are clustered into groups. These may be social or based on the need for a localized resource.

coccus (ka´-kus; pl. cocci): a spherical bacterium.

cochlea (kōk´-lē-uh): a coiled, bony, fluid-filled tube found in the mammalian inner ear, which contains receptors (hair cells) that respond to the vibration of sound.

codominance: the relation between two alleles of a gene, such that both alleles are phenotypically expressed in heterozygous individuals.

codon: a sequence of three bases of messenger RNA that specifies a particular amino acid to be incorporated into a protein. Certain codons also signal the beginning and end of protein synthesis.

coelom (sē´-lōm): a space or cavity within the body separating the body wall from the inner organs.

coenzyme: an organic molecule, often derived from a water-soluble vitamin, that is bound to certain enzymes and is required for their proper functioning.

coevolution: the evolution of adaptations in two species due to their extensive interactions with one another, so that each species acts as a major force of natural selection upon the other.

cohesion: the tendency of the molecules of a substance to hold together.

cohesion-tension theory: a model for transport of water in xylem, which states that water is pulled up the xylem tubes, powered by the force of evaporation of water from the leaves (producing tension) and held together by hydrogen bonds between nearby water molecules (cohesion).

coleoptile (kō-lē-op´til): a protective sheath surrounding the shoot in monocot seeds, allowing the shoot to push aside soil particles as it grows.

collagen (kol´-uh-gen): a fibrous protein found in connective tissue such as bone and cartilage.

collar cells: specialized cells lining the inside channels of sponges. Flagella extend from a sievelike collar, creating a water current that draws microscopic organisms through the collar to be trapped.

collecting ducts: conducting tubes within the kidney that collect urine from many nephrons and conduct it through the medulla into the renal pelvis. Urine may become concentrated in the collecting ducts if ADH is present.

collenchyma (kol-en´-ki-ma): an elongated, polygonal plant cell type with irregularly thickened primary cell walls that is alive at maturity and that provides support to the plant body.

colostrum (ko-los´-trum): a yellowish fluid high in protein and containing antibodies that is produced by the female breasts before milk secretion begins.

commensalism (kum-en'-sal-ism): a symbiotic relationship between two species in which one benefits while the other is neither harmed nor benefited.

communication: the act of producing a signal that causes another animal, usually of the same species, to change its behavior in a way beneficial to one or both participants.

community: all the interacting populations within an ecosystem.

compact bone: the hard and strong outer bone; composed of Haversian systems.

companion cell: a cell adjacent to a sieve-tube element in phloem, involved in control and nutrition of the sieve-tube element.

competition: interaction that occurs between individuals when both attempt to utilize a resource (for example, food or space) that is limited relative to the demand for it.

competitive exclusion principle: the concept that no two species can simultaneously and continuously occupy the same ecological niche.

competitive inhibition: in enzyme-catalyzed reactions, a condition in which two molecules (at least one a substrate for the enzyme) compete for entry into the active site of the enzyme, thus slowing the rate of reaction.

complement: a group of blood-borne proteins that participate in the destruction of foreign cells to which antibodies have bound.

complement reactions: interactions among foreign cells, antibodies, and complement proteins, resulting in the destruction of the foreign cells.

complementary base pair: in nucleic acids, bases that pair by hydrogen bonding; in DNA, adenine is complementary to thymine, and guanine is complementary to cytosine.

complete flower: a flower that has all four floral parts (sepals, petals, stamens, and carpels).

compound: a substance composed of two or more elements that can be broken into its constituent elements by chemical means.

compound eye: a type of eye found in arthropods, composed of numerous independent subunits, called ommatidia. Each ommatidium apparently contributes a single piece of a mosaic-like image perceived by the animal.

concentration: the number of particles of a dissolved substance in a given unit of volume.

concentration gradient: the difference in concentration of a substance between two parts of a fluid or across a barrier such as a membrane.

condom: a contraceptive sheath that may be worn over the penis during intercourse, preventing sperm from being deposited in the vagina.

cone: a cone-shaped photoreceptor cell in the vertebrate retina, not as sensitive to light as the rods. The three types of cones are most sensitive to different colors of light and provide color vision. See also rod.

conifer (kon´-eh-fer): a class of tracheophyte that reproduces using seeds formed inside cones and retains its leaves throughout the year.

connective tissue: a tissue type consisting of diverse tissues, including bone, fat, and blood, which generally contain large amounts of extracellular material.

constant region: part of an antibody molecule that is similar in all antibodies.

consumer: an organisms that eats other organisms; a heterotroph.

contest competition: a mechanism for resolving intraspecific competition using social or chemical interactions.

contractile vacuole: a fluid-filled vacuole found in certain protists that takes up water from the cell cytoplasm, contracts, and expels the water outside the cell through a pore in the plasma membrane.

control: that portion of an experiment in which all possible variables are held constant; in contrast to the "experimental" portion, in which a particular variable is altered.

convergence: a condition in which a large number of nerve cells provide input to a smaller number of cells.

convergent evolution: the independent evolution of similar structures among unrelated organisms, due to similar selective pressures. See analogous structures.

convolutions: foldings of the cerebral cortex of the vertebrate brain.

copulation: reproductive behavior in which the penis of the male is inserted into the body of the female, where it releases sperm.

cork cambium: a lateral meristem in woody roots and stems that gives rise to cork cells.

cork cell: a protective cell of the bark of woody stems and roots; at maturity, cork cells are dead, with thick, waterproofed cell walls.

cornea (kor´-nē-uh): the clear outer covering of the eye in front of the pupil and iris.

corona radiata (ka-rō´-na rā-dē-a´-ta): the layer of cells surrounding an egg after ovulation.

corpus callosum (kor´pus kal-ō´-sum): the band of axons that connect the two cerebral hemispheres of vertebrates.

corpus luteum (kor´-pus lū´-tē-um): in the mammalian ovary, a structure derived from the follicle after ovulation, which secretes the hormones estrogen and progesterone.

cortex: the part of a primary root or stem located between the epidermis and the vascular cylinder.

cotyledon (kot-ul-ē´don): also called a seed leaf; a leaflike structure within a seed that absorbs food molecules from the endosperm and transfers them to the growing embryo.

countercurrent flow: a structural arrangement that enhances diffusion between two fluids that differ in their concentration of dissolved substances by moving the fluids past one another in opposite directions, separated by semipermeable membranes.

coupled reaction: a pair of reactions, one exergonic and one endergonic, that are linked together so that the energy produced by the exergonic reaction provides the energy needed to drive the endergonic reaction.

covalent bond (ko-vā´-lent): a chemical bond between atoms in which electrons are shared.

creationism: the hypothesis that all species of organisms on Earth were created in essentially their present form by a supernatural being, and that significant modification of those species, specifically their transformation into new species, cannot occur through natural processes.

crista (kris´-ta; pl. cristae): a fold in the inner membrane of a mitochondrion.

crop: an organ found in both earthworms and birds in which ingested food is temporarily stored before being passed to the gizzard, where it is pulverized.

cross-bridge: in muscles, an extension of myosin that binds to and pulls upon actin to produce contraction of the muscles.

cross-fertilization: union of sperm and egg from two different individuals of the same species.

crossing over: the exchange of corresponding segments of the chromatids of two homologous chromosomes during meiosis.

cultural evolution: changes in the behavior of a population of animals, especially humans, by learning behaviors acquired by members of previous generations.

cupula (kūp´-ū-luh): a gelatinous structure found in the lateral line system of fish and the semicircular canals of the vestibular apparatus of other vertebrates. Hairs are embedded in the cupula, which is deflected by movement of the fluid filling the canals.

cuticle (kū´-ti-kul): a waxy or fatty coating on the exposed surfaces of epidermal cells of many land plants, which aids in the retention of water.

cyanobacteria: photosynthetic prokaryotic cells, utilizing chlorophyll and releasing oxygen as a photosynthetic by-product, sometimes called "blue-green algae."

cyclic AMP: a cyclic nucleotide formed within many target cells as a result of the reception of modified amino acid or peptide hormones, which causes metabolic changes in the cell; often called a second messenger.

cyclic nucleotide (sik´-lik nū-klē-ō-tīd): a nucleotide in which the phosphate group is bonded to the sugar at two points, forming a ring. Cyclic nucleotides serve as intracellular messengers.

cyst (sist): an encapsulated resting stage in the life cycle of certain invertebrates, such as parasitic flatworms and roundworms.

cysteine amino acids: by attraction and repulsion among amino acid side groups, and by interaction between the cellular environment (water or lipids) and the amino acid side groups.

cytokinesis (sī-tō-ki-nē´-sis): division of the cytoplasm and organelles into two daughter cells during cell division. Usually cytokinesis occurs during telophase of mitosis.

cytokinin (sī-tō-kī´-nin): a plant hormone that promotes cell division, fruit growth, sprouting of lateral buds, and prevents aging of plant parts, especially leaves.

cytoplasm (sī´-tō-plazm): the material contained within the plasma membrane of a cell, exclusive of the nucleus.

cytoskeleton: a network of protein fibers in the cytoplasm that gives shape to a cell, holds and moves organelles, and is often involved in cell movement.

cytosol (sī´-tō-sol): the fluid part of the cytoplasm.

cytotoxic T cell: a type of T cell that directly destroys foreign cells upon contacting them.

day-neutral plant: a plant in which flowering occurs as soon as it has grown and developed, regardless of length of day.

decomposers: a group of organisms, mainly fungi and bacteria, that digest organic material by secreting digestive enzymes into the environment. In the process they liberate nutrients into the environment.

degeneracy: the property of the genetic code whereby several codons may specify the same amino acid.

dehydration synthesis: a chemical reaction in which two molecules are joined by a covalent bond with the simultaneous removal of a hydrogen from one molecule and a hydroxyl group from the other, forming water.

deletion: a mutation in which one or more pairs of nucleotides are removed from a gene.

dendrite (den´-drīt): branched tendrils that extend outward from the cell body, specialized to respond to signals from the external environment or from other neurons.

denitrifying bacteria (dē-nī´-treh-fī-ing): bacteria that break down nitrates, releasing nitrogen gas to the atmosphere.

density-dependent: description of any factor, such as predation, that limits population size more effectively as the population density increases.

density-independent: description of any factor, such as freezing weather, that limits a sensitive population's size and growth regardless of its density.

deoxyribonucleic acid (dē-ox-ē-rī-bō-nū- klā´-ik; DNA): a molecule composed of deoxyribose nucleotides; the genetic information of all living cells.

dermal tissue system: a plant tissue system that makes up the outer covering of the plant body.

dermis (dur´-mis): the layer of skin lying beneath the epidermis, composed of connective tissue and containing blood vessels, muscles, nerve endings, and glands.

desmosome (dez´-mō-sōm): a strong cell-to-cell junction that functions in attaching adjacent cells to one another.

detritus feeders (de-trī´-tus): a diverse group of organisms ranging from worms to vultures that live off the wastes and dead remains of other organisms.

development: the process by which an organism proceeds from fertilized egg through adulthood to eventual death.

diabetes mellitus (di-a-bē´-tes mel-ī´-tus): a disease characterized by defects in the production, release, or reception of insulin, characterized by high blood glucose levels that fluctuate with sugar intake.

diaphragm (dī´uh-fram): a dome-shaped muscle forming the floor of the chest cavity. Contraction of this muscle pulls it downward, enlarging the cavity and causing air to be drawn into the lungs. Also, a contraceptive rubber cap that fits snugly over the cervix, preventing the sperm from entering the uterus and thereby preventing pregnancy.

diastole (di-as´-tō-lē): the portion of the cardiac cycle during which the ventricles are relaxed; the lower of the two blood pressure readings.

diatom (dī´-e-tom): a category of protist that includes photosynthetic forms with two-part glassy outer coverings. Diatoms are important photosynthetic organisms in fresh and salt water.

dicot: short for dicotyledon; a type of flowering plant characterized by embryos with two cotyledons, or seed leaves, modified for food storage.

differential permeability: the property of a membrane by which some substances can permeate more readily than other substances.

differential reproduction: differences in reproductive output among individuals of a population, usually as a result of genetic differences.

differentiated cell: a mature cell specialized for a specific function; in plants, differentiated cells usually do not divide.

differentiation: the process whereby relatively unspecialized cells, especially of embryos, become specialized into particular tissue types.

diffusion: the net movement of particles from a region of high concentration of that particle to a region of low concentration, driven by the concentration gradient. May occur entirely within a fluid or across a barrier such as a membrane.

digestion: the process by which food is physically and chemically broken down into molecules that can be absorbed by cells.

dihybrid cross: a breeding experiment involving parents that differ in two distinct, genetically determined traits.

dinoflagellate (dī-nō-fla´-gel-et): a category of protist that includes photosynthetic forms in which two flagella project through armorlike plates. Abundant in oceans, these sometimes reproduce rapidly, causing "red tides."

dioecious (dī-ē´-shus): pertaining to organisms in which male and female gametes are produced by separate individuals.

diploid (dip´-loyd): referring to a cell with pairs of homologous chromosomes.

direct development: a developmental pathway in which the offspring is born as a miniature version of the adult and does not radically change its body form as it grows and matures.

directional selection: a type of natural selection in which one extreme phenotype is favored over all others.

disaccharide (dī-sak´-a-rīd): a carbohydrate formed by the covalent bonding of two monosaccharides.

disruptive selection: a type of natural selection in which both extreme phenotypes are favored over the average phenotype.

disulfide bridge: the covalent bond formed between the sulfur atoms of two cysteines in a protein; often causes the protein to fold, by bringing otherwise distant parts of the protein close together.

divergence: a condition in which a small number of nerve cells provide input to a larger number of cells.

divergent speciation: speciation in which a single parental species splits to form two or more new species that exist simultaneously. See phyletic speciation.

division: the taxonomic category contained within a kingdom and consisting of related classes of plants, fungi, bacteria, or plantlike protists.

DNA hybridization: a technique by which DNA from two species is separated into single strands and then allowed to reform. Hybrid double-stranded DNA from the two species can occur where the sequence of nucleotides is complementary. The greater the degree of hybridization, the closer the evolutionary relatedness of the two species.

DNA library: a readily accessible, easily duplicable complete set of all the DNA of a particular organism, usually cloned into bacterial plasmids.

DNA polymerase: an enzyme that bonds DNA nucleotides together into a continuous strand, using a preexisting DNA strand as a template.

dominance hierarchy: a social arrangement in which animals, usually through aggressive interactions, establish a rank for some or all of the members that determines access to resources.

dominant: an allele that can determine the phenotype of heterozygotes completely, so that they are indistinguishable from individuals homozygous for the allele. In the heterozygotes, the expression of the other (recessive) allele is completely masked.

dopamine (dōp´-uh-mēn): a transmitter in the brain whose actions are largely inhibitory. Loss of dopamine-containing neurons causes Parkinson's disease.

dormancy: a state in which an organism does not grow or develop; usually marked by lowered metabolic activity and resistance to adverse environmental conditions.

dorsal (dōr´-sul): the top, back, or uppermost surface of an animal oriented with its head forward.

dorsal root ganglion: a ganglion located on the dorsal (sensory) branch of each spinal nerve, containing the cell bodies of sensory neurons.

double covalent bond: a covalent bond that occurs when two atoms share two pairs of electrons.

double fertilization: in flowering plants, a phenomenon in which two sperm nuclei fuse with the nuclei of two cells of the female gametophyte. One sperm fuses with the egg to form the zygote, while the second sperm nucleus fuses with the two haploid nuclei of the primary endosperm cell to form a triploid endosperm cell.

douching: washing the vagina after intercourse in an attempt to remove sperm before they enter the uterus; an ineffective contraceptive method.

Down syndrome: a genetic disorder caused by the presence of three copies of chromosome 21. Common characteristics include mental retardation, distinctively shaped eyelids, a small mouth with protruding tongue, heart defects, and low resistance to infectious diseases.

duct: a tube or opening through which exocrine secretions are released.

echolocation: use of ultrasonic sounds, which bounce back from nearby objects, to produce an "auditory image" of nearby surroundings; used by bats and porpoises.

ecological isolation: lack of mating between organisms belonging to different populations that occupy distinct habitats within the same general area.

ecological niche (nitch): the role of a particular species within an ecosystem, including all aspects of its interaction with the living and nonliving environment.

ecology (ē-kol´-uh-gē): the study of the interrelationships of organisms with each other and with their nonliving environment.

eecosystem (ē´kō-sis-tem): all the organisms and their nonliving environment within a defined area.

ectoderm (ek´-tō-derm): the outermost embryonic tissue layer, which gives rise to structures such as hair, the epidermis of the skin, and the nervous system.

effector (ē-fek´-tōr): a part of the body (usually a muscle or gland) that carries out responses as directed by the nervous system.

egg: the haploid female gamete, usually large and nonmotile, containing food reserves for the developing embryo.

electron: a subatomic particle, found in the electron shells outside the nucleus of an atom, bearing a unit of negative charge and very little mass.

electron carrier: a molecule that can reversibly gain and lose electrons. Electron carriers generally accept high-energy electrons produced during an exergonic reaction and donate the electrons to acceptor molecules that use the energy to drive endergonic reactions.

electron shell: a region within which electrons orbit corresponding to a fixed energy level within which at a given distance from the nucleus of an atom.

electron transport system: a series of electron carrier molecules found in the thylakoid membranes of chloroplasts and the inner membrane of mitochondria, which extract energy from electrons and generate ATP or other energetic molecules.

electrophoresis: a biochemical technique that separates molecules according to their electrical charge and molecular weight.

element: a substance that cannot be broken down to a simpler substance by ordinary chemical means.

embryo: the stages of development of animals that begin with fertilization of the egg cell and end with hatching or birth. In mammals, embryo usually refers to the early stages in which the developing animal does not yet resemble the adult of the species.

embryo sac: the haploid female gametophyte of flowering plants.

embryonic disk: in human embryonic development, the flat, two-layered group of cells that separates the amniotic cavity from the yolk sac.

emergent property: an intangible attribute that arises as the result of complex ordered interactions among individual parts.

emigration (em-uh-grā´shun): movement of individuals out of an area.

emphysema (em-fuh-sē´-muh): a condition in which the alveoli become brittle and rupture, causing decreased area for gas exchange.

endergonic (en-der-gon´-ik): pertaining to a chemical reaction that requires an input of energy to proceed; an "uphill" reaction.

endocrine gland: a ductless, hormone-producing gland consisting of cells that release their secretions into the extracellular fluid from which the secretions diffuse into nearby capillaries.

endocytosis (en-dō-sī-tō´-sis): the movement of material into a cell by a process in which the plasma membrane engulfs extracellular material, forming membrane-bound sacs that enter the cytoplasm.

endoderm (en´-dō-derm): the innermost embryonic tissue layer, which gives rise to structures such as the lining of the digestive and respiratory tracts.

endodermis (en-dō-der´-mis): the innermost layer of small, close-fitting cells of the cortex of a root that form a ring around the vascular cylinder.

endogenous pyrogen: a chemical produced by the body that stimulates the production of a fever (elevated body temperature).

endometrium (en-dō-mē´-trē-um): the nutritive inner lining of the uterus.

endoplasmic reticulum (ER) (en-dō-plaz´-mik re-tik´-ū-lum): a system of membranous tubes and channels within eukaryotic cells; the site of most protein and lipid synthesis.

endorphin (en-dōr´-fin): one of a group of peptide neuromodulators in the vertebrate brain that, by reducing the sensation of pain, mimics some of the actions of opiates.

endoskeleton (en´-dō-skel´-uh-tun): a rigid internal skeleton with flexible joints to allow for movement.

endosperm: a triploid food storage tissue found in the seeds of flowering plants used to nourish the developing plant embryo.

endospore (spōr): a resistant or resting structure that disperses readily and withstands unfavorable environmental conditions.

endosymbiotic hypothesis: the hypothesis that certain organelles, especially chloroplasts and mitochondria, arose as mutually beneficial associations between the ancestors of eukaryotic cells and captured bacteria living within the cytoplasm of the pre-eukaryotic cell.

energy: the capacity to do work.

energy carrier: a molecule that stores energy in "high-energy" chemical bonds and releases the energy again to drive coupled endothermic reactions. ATP is the most common energy carrier in cells.

energy level: the specific amount of energy characteristic of a given electron shell in an atom.

energy pyramid: a graphical representation of the energy contained in succeeding trophic levels, with maximum energy at the base (primary producers) and steadily diminishing amounts at higher levels.

enhancer: a part of chromosome, usually located close to a gene, that binds proteins that alter the rate of gene transcription.

entropy (en´-trō-pē): a measure of the amount of randomness and disorder in a system.

environmental resistance: any factor that tends to counteract biotic potential, limiting population size.

enzyme (en´zīm): a protein catalyst that speeds up the rate of specific biological reactions.

epicotyl (ep´-ē-kot-ul): the part of the embryonic shoot located above the cotyledons but below the tip of the shoot.

epidermis (ep-uh-der´-mis): specialized epithelial tissue that forms the outer layer of skin. In plants, the outermost layer of cells of a leaf, young root, or young stem.

epididymis (e-pi-di´-dē-mus): tubes that connect with and receive sperm from the seminiferous tubules of the testis.

epiglottis (ep-eh-gla´-tis): a flap of cartilage in the lower pharynx that covers the opening to the larynx during swallowing. In doing so, the epiglottis directs the food down the esophagus.

epistasis (ep-i-stā´-sis): a pattern of inheritance in which the expression of one gene depends on, or is modified by, the expression of another gene.

epithelial tissue (eh-puh-thē´-lē-ul): a tissue type that forms membranes that cover the body surface and line body cavities, and that also gives rise to glands.

equilibrium population: a population in which allele frequencies and the distribution of genotypes do not change from generation to generation.

erythroblastosis fetalis (ē-rith´-rō-blas-tō´-sis fē-tal´-is): a condition in which the red blood cells of a newborn Rh-positive baby are attacked by antibodies produced by its Rh-negative mother, causing jaundice and anemia. Retardation and death are possible consequences if treatment is inadequate.

erythrocytes (ē-rith´-rō-sītes): red blood cells active in oxygen transport, which contain the red pigment hemoglobin.

erythropoietin (ē-rith´-rō-pō-ē´-tin): a hormone produced by the kidneys in response to oxygen deficiency that stimulates production of red blood cells by the bone marrow.

esophagus (eh-sof´-eh-gus): a muscular passageway connecting the pharynx to the next chamber of the digestive tract, the stomach in humans and other mammals.

essential amino acids: amino acids that are required nutrients. The body is unable to manufacture them, and so they must be supplied in the diet.

essential fatty acids: fatty acids that are required nutrients. The body is unable to manufacture them, and so they must be supplied in the diet.

estrogen: in vertebrates, a female sex hormone produced by follicle cells of the ovary, which stimulates follicle development, oogenesis, development of secondary sex characteristics, and growth of the uterine lining.

ethology (ē-thol´-ō-gē): the study of animal behavior under natural or near-natural conditions.

ethylene: a plant hormone that promotes ripening of fruits and leaf and fruit drop.

euglenoid (yū´-gle-nōyd): a category of protist characterized by one or more whiplike flagella used for locomotion and a photoreceptor for detecting light. Euglenoids are photosynthetic, but some are capable of heterotrophic nutrition if deprived of chlorophyll.

eukaryotic (ū-kar-ē-ot´-ik): referring to cells of organisms of the kingdoms Protista, Fungi, Plantae, and Animalia. Eukaryotic cells have their genetic material enclosed within a membrane-bound nucleus and contain other membrane-bound organelles.

eustachian tube (ū-stā´-shin): a tube connecting the middle ear with the pharynx that allow pressure between the middle ear and the atmosphere to equilibrate.

evolution: the descent of modern organisms from preexisting life forms; strictly speaking, any change in the proportions of different genotypes in a population from one generation to the next.

excitatory synapse: a synapse between two nerve cells in which the resting potential of the postsynaptic cell becomes less negative and more likely to fire an action potential due to the activity of the presynaptic cell.

excretion: Elimination of waste substances from the body. Excretion can occur from the digestive system, skin glands, urinary system, or lungs.

excretory pore: Opening in the body wall of certain invertebrates, such as the earthworm, through which urine is excreted.

exergonic (ex-er-gon´-ik): pertaining to a chemical reaction that liberates energy (either heat energy or in the form of increased entropy); a "downhill" reaction.

exhalation (ex-per-ā´-shun): the act of releasing air from the lungs, which results from relaxation of the respiratory muscles.

exocrine gland: a gland that releases its secretions into ducts that lead to the outside of the body or into the digestive tract.

exocytosis (ex-ō-sī-tō´-sis): the movement of material out of a cell by a process in which intracellular material is enclosed within a membrane-bound sac that moves to the plasma membrane and fuses with it, releasing the material outside the cell.

exon: a segment of DNA in a eukaryotic gene that codes for amino acids in a protein (see also intron).

exoskeleton (ex´-ō-skel´-uh-tun): a rigid external skeleton that supports the body, protects the internal organs and has flexible joints to allow for movement.

exotic species: A species introduced into an ecosystem where it did not evolve. Occasionally such species may flourish and outcompete native species.

exponential growth: a continuously accelerating increase in population size.

extensor muscle: a muscle that straightens a joint.

external ear: the fleshy portion of the ear that extends outside the skull.

external fertilization: union of sperm and egg outside the body of either parental organism.

extinction: the death of all members of a species.

extracellular digestion: the physical and chemical breakdown of food that occurs outside of a cell, usually in a digestive cavity.

extraembryonic membranes: in reptile, bird, and mammal embryonic development, the chorion, amnion, allantois, and yolk sac that function in gas exchange, provision of the watery environment needed for development, waste storage, and storage of the yolk, respectively.

eyespot: a simple, lensless eye found in various invertebrates, including flatworms and jellyfish. Eyespots can distinguish light from dark and sometimes the direction of light, but they cannot form an image.

facilitated diffusion: diffusion of molecules across a membrane, assisted by protein pores or carriers embedded in the membrane.

family: the taxonomic category contained within an order and consisting of related genera.

fat: a lipid composed of three saturated fatty acids covalently bonded to glycerol; fats are solid at room temperature. Also, fat-storing connective tissue whose cells are packed with triglycerides (fats); also called adipose tissue.

fatty acid: an organic molecule composed of a long chain of carbon atoms, with a carboxylic acid (—COOH) group at one end. Fatty acids may be saturated (all single bonds between the carbon atoms) or unsaturated (one or more double bonds between the carbon atoms).

feedback inhibition: in enzyme-mediated chemical reactions, the condition in which the product of a reaction inhibits one or more of the enzymes involved in synthesizing the product.

fermentation: anaerobic reactions that convert the pyruvic acid produced by glycolysis into lactic acid or alcohol and CO_2.

fertilization: the fusion of male and female haploid gametes to form a zygote.

fetal alcohol syndrome (FAS): a cluster of symptoms including retardation and physical abnormalities that occur in infants born to mothers who consumed large amounts of alcoholic beverages during pregnancy.

fetus: the later stages of mammalian embryonic development (after the second month for humans), when the developing animal has come to resemble the adult of the species.

fever: an elevation in body temperature caused by chemicals (pyrogens) released by white blood cells in response to infection.

fibrillation: rapid, uncoordinated, and ineffective contractions of heart muscle cells.

fibrin (fī´-brin): a clotting protein formed in the blood in response to a wound. Fibrin binds with other fibrin molecules and provides a matrix around which a blood clot forms.

fibrous root system: a root system, commonly found in monocots, characterized by many roots of approximately the same size arising from the base of the stem.

filament: in flowers, the stalk of a stamen, which bears an anther at its tip.

filtrate: Fluid produced by filtration; in the kidneys, fluid produced by filtration of the blood through the glomerular capillaries.

filtration: within Bowman's capsule in each nephron of a kidney, the process by which blood is pumped under pressure through permeable capillaries of the glomerulus, forcing out water, dissolved wastes, and nutrients.

fimbria (fim´-brē-a; pl. fimbriae): in female mammals, the ciliated, fingerlike projections of the oviduct that sweep the ovulated egg from the ovary into the oviduct.

first law of thermodynamics: a principle of physics that within any isolated system, energy can be neither created nor destroyed, but can be converted from one form to another.

fission: asexual reproduction by dividing the body into two smaller, complete organisms.

fitness: the reproductive success of an organism, usually expressed in relation to the average reproductive success of all individuals in the same population.

fixed action pattern: stereotyped, rather complex behavior that is genetically programmed (innate); often triggered by a stimulus called a releaser.

flagellum (fla-jel´-um; pl. flagella): a long, hairlike extension of the plasma membrane. In eukaryotic cells it contains microtubules arranged in a 9+2 pattern. Movement of flagella propel some cells through fluid media.

flame cell: in flatworms, a specialized cell containing beating cilia that conducts water and wastes through the branching tubes that serve as an excretory system.

flexor muscle: a muscle that flexes (decreases the angle of) a joint.

flower: the reproductive structure of an angiosperm plant.

fluid: a liquid or gas.

fluid mosaic model: a model of membrane structure. According to this model, membranes are composed of a double layer of phospholipids in which is embedded a variety of proteins. The phospholipid bilayer is a somewhat fluid matrix that allows movement of proteins within it.

fluid-phase endocytosis: nonselective movement of extracellular fluid into a cell, enclosed within a vesicle formed from the plasma membrane.

follicle: in the ovary of female mammals, the oocyte and its surrounding accessory cells.

follicle-stimulating hormone (FSH): a hormone produced by the anterior pituitary gland that stimulates spermatogenesis in males and development of the follicle in females.

food chain: an illustration of feeding relationships in an ecosystem using a single representative from each of the trophic levels.

food vacuole: a membrane-bound space within a single cell in which food is enclosed. Digestive enzymes are released into the vacuole, where intracellular digestion occurs.

food web: a relatively accurate representation of the complex feeding relationships within an ecosystem, including many organisms at various trophic levels, with many of the consumers occupying more than one level simultaneously.

forebrain: during development, the anterior portion of the brain. In mammals, the forebrain differentiates into the thalamus, the limbic system, and the cerebrum. In humans, the cerebrum contains about half of all the neurons in the brain.

fossil: the remains of an organism, usually preserved in rock. Fossils include petrified bones or wood; shells; impressions of body forms such as feathers, skin, or leaves; and markings made by organisms such as footprints.

fossil fuels: fuels such as coal, oil, and natural gas, derived from the bodies of prehistoric organisms.

founder effect: a type of genetic drift in which an isolated population founded by a small number of individuals may develop allele frequencies that are very different from those of the parent population, because of chance inclusion of disproportionate numbers of certain alleles in the founders.

fovea (fō´-vē-uh): the central region of the vertebrate retina, upon which images are focused. The fovea contains closely packed cones.

free-living: not parasitic.

free nerve endings: finely-branched endings of some receptor neurons, some of which respond to touch and pressure, others to heat and cold, and others to pain; also produce sensations of itching and tickling.

fruit: in flowering plants, the ripened ovary (plus, in some cases, other parts of the flower) that contains the seeds.

fruiting body: a spore-forming reproductive structure found in certain protists, bacteria, and fungi.

functional group: one of several groups of atoms commonly found in organic molecules, including hydrogen, hydroxyl, amino, carboxyl, and phosphate groups.

gallbladder: a small sac next to the liver in which the bile secreted by the liver is stored and concentrated. Bile is released from the gallbladder to the small intestine through the bile duct.

gamete (gam´-ēt): a haploid sex cell formed in sexually reproducing organisms.

gametic incompatibility: the inability of sperm from one species to fertilize eggs of another species.

gametophyte (ga-mēt´-ō-fīt): the multicellular haploid stage in the life cycle of plants.

ganglion (gang´-lē-un): a cluster of neurons.

ganglion cells: a type of cell comprising the innermost layer of the vertebrate retina whose axons form the optic nerve.

gap junction: a type of cell-to-cell junction in animals in which channels connect the cytoplasm of adjacent cells.

gastric inhibitory peptide: A hormone produced by the small intestine that inhibits the activity of the stomach.

gastrin: A hormone produced by the stomach that stimulates acid secretion in response to food.

gastrovascular cavity: a sac like chamber with digestive functions, found in simple invertebrates. A single opening serves as both mouth and anus, while the chamber provides direct access of nutrients to the cells.

gastrula (gas´-trū-la): in animal development, a three-layered embryo with ectoderm, mesoderm, and endoderm cell layers. The endoderm layer usually encloses the primitive gut.

gastrulation (gas-trū-lā´-shun): the process whereby a blastula develops into a gastrula, including the formation of endoderm, ectoderm, and mesoderm;

gene: a unit of heredity encoding the information needed to specify the amino acid sequence of proteins and hence particular traits. A gene is a segment of DNA located at a particular place on a chromosome.

gene flow: the movement of alleles from one population to another owing to migration of individual organisms.

gene pool: the total of all alleles of all genes in the population; for a single gene, the total of all the alleles of that gene that occur in a population.

generative cell: in flowering plants, one of the haploid cells of a pollen grain. The generative cell undergoes mitosis to form two sperm cells.

genetic code: the collection of codons of mRNA, each of which directs the incorporation of a particular amino acid into a protein during protein synthesis.

genetic drift: a change in the allele frequencies of a small population purely by chance.

genetic recombination: the generation of new combinations of alleles on homologous chromosomes, due to exchange of DNA during crossing over.

genotype (jēn´-ō-tīp): the genetic composition of an organism; the actual alleles of each gene carried by the organism.

genus (jē-nis): the taxonomic category consisting of very closely related species.

geographical isolation: the separation of two populations by a physical barrier.

germ layer: A tissue layer formed during early embryonic development.

germination: the growth and development of a seed, spore, or pollen grain.

gibberellin (jib-er-el´-in): a plant hormone that stimulates seed germination, fruit development, and cell division and elongation.

gills: in aquatic animals, a branched tissue richly supplied with capillaries around which water is circulated for gas exchange.

gizzard: a muscular organ found in earthworms and birds in which food is mechanically broken down prior to chemical digestion.

glomerulus (glō-mer´-ū-lus): a dense network of thin-walled capillaries located within the Bowman's capsule of each nephron. Here blood pressure forces water and dissolved nutrients through capillary walls for filtration by the nephron.

glucose: the most common monosaccharide, with the molecular formula $C_6H_{12}O_6$. Most polysaccharides, including cellulose, starch, and glycogen, are made of glucose subunits covalently bonded together.

glycerol (glis´-er-ol): a three-carbon alcohol to which fatty acids are covalently bonded to make fats and oils.

glycogen (glī´-kō-jen): a long, branched polymer of glucose that is stored by animals in the muscles and liver and metabolized as a source of energy.

glycolysis (glī-kol´-i-sis): reactions carried out in the cytosol that break down glucose into two molecules of pyruvic acid, producing two ATP molecules. Glycolysis does not require oxygen but can proceed when oxygen is present.

glycoprotein: a protein to which a carbohydrate is attached.

goiter: a swelling of the neck caused by iodine deficiency, which affects the functioning of the thyroid gland and its hormones.

Golgi complex (gōl´-jē): a stack of membranous sacs found in most eukaryotic cells, which is the site of processing and separation of membrane components and secretory materials.

gonadotropin-releasing hormone (GnRH): a hormone produced by the neurosecretory cells of the hypothalamus, which stimulates cells in the anterior pituitary to release FSH and LH. GnRH is involved in the menstrual cycle and spermatogenesis.

gonads: organs where reproductive cells are formed; in males, the testes, and in females, the ovaries.

gonorrhea (gon-a-rē´-uh): a sexually transmitted bacterial infection of the reproductive organs. Untreated gonorrhea may result in sterility.

gradient: a difference in concentration, pressure, or electrical charge between two regions of space.

gradualism: a model of evolution, stating that morphological change and speciation are slow, gradual processes that are not necessarily simultaneous or linked.

Gram stain: A stain which is selectively taken up by the walls of ceratin types of bacteria (gram positive bacteria) and rejected by the cell walls of others called (gram negative bacteria). The stain is used to distinguish bacteria based on their cell wall construction.

granum (gra´-num; pl. grana): in chloroplasts, a stack of thylakoids.

gravitropism: growth with respect to the direction of gravity.

gray crescent: in frog embryonic development, an area of intermediate pigmentation in the fertilized egg containing gene-regulating substances required for the normal development of the tadpole.

gray matter: Outer portions of the brain and inner region of the spinal cord composed largely of neuron cell bodies which give these ares a gray color.

greenhouse effect: the ability of certain gases such as carbon dioxide and methane to trap sunlight energy in the atmosphere as heat. The glass in a greenhouse does the same, hence the name. This warming of the atmosphere is being enhanced by human production of these gases.

ground tissue system: a plant tissue system consisting of parenchyma, collenchyma, and sclerenchyma cells that makes up the bulk of a leaf or young stem, excluding vascular or dermal tissues. Most ground tissue cells function in photosynthesis, support, or carbohydrate storage.

guard cell: one of a pair of specialized epidermal cells surrounding the central opening of a stoma of a leaf, which regulates the size of the opening.

gymnosperms (jim´-nō-sperms): non-flowering seed plants such as conifers, cycads, and gingkos.

habituation (heh-bich-ū-ā´-shun): simple learning characterized by a decline in response to a harmless, repeated stimulus.

hair cell: the type of receptor cell found in the inner ear. Hair cells bear hairlike projections, the bending of which between two membranes causes the receptor potential.

hair follicle: a gland in the dermis of mammalian skin, formed from epithelial tissue, that produces a hair.

halophile (hā´-lō-fīl): leterally "salt-loving," a type of archaebacterium that thrives in concentrated salt solutions.

haploid (hap´-loyd): referring to a cell that has only one member of each pair of homologous chromosomes.

Haversian system (ha-ver´-sē-un): unit of hard bone consisting of concentric layers of bone matrix with embedded osteocytes surrounding a small central canal containing a capillary.

heart: A muscular organ within the circulatory system responsible for pumping blood throughout the body.

heart attack: a situation in which blood flow through a coronary artery is severaly reduced or blocked, depriving some of the heart muscle of its blood supply.

heat of fusion: the energy that must be removed from a compound to transform it from a liquid into a solid at its freezing temperature.

heat of vaporization: the energy that must be supplied to a compound to transform it from a liquid into a gas at its boiling temperature.

helix (hē´-licks): a spiral structure similar to a corkscrew or a spiral staircase; a type of secondary structure of a protein.

helper T cell: a type of T cell that aids other immune cells to recognize and act against antigens.

hemocoel (hē´-mō-sēl): a blood cavity within the bodies of certain invertebrates in which blood bathes tissues directly. A hemocoel is part of an open circulatory system.

hemoglobin (hē´mō-glō-bin): an iron-containing protein that gives red blood cells their color. Hemoglobin binds to oxygen in the lungs and releases it to the tissues.

hemophilia: a recessive, sex-linked disease in which the blood fails to clot normally.

herbivore (erb´-i-vōr): literally "plant-eater," an organism that eats feed directly and exclusively on producers; a primary consumer.

hermaphroditic (her-maf´-ruh-dit´-ik): possessing both male and female sexual organs. Some hermaphroditic animals can fertilize themselves; others must exchange sex cells with a mate.

heterotroph (het´-er-ō-trōf´): literally "other-feeder," meaning an organism that eats other organisms.

heterozygous (het-er-ō-zī´-gōt): carrying two different alleles of a given gene; sometimes called a hybrid.

hindbrain: the posterior portion of the brain, containing the medulla, pons, and cerebellum

hinge joint: a joint at which one bone is moved by muscle and the other bone remains fixed, such as in the knee, elbow, or fingers; allows movement in only two dimensions.

hippocampus (hip-ō-cam´-pus): part of the forebrain of vertebrates, important in emotion and especially learning.

histamine: a substance released by certain cells in response to tissue damage and invasion of the body by foreign substances. Histamine promotes dilation of arterioles, leakiness of capillaries and triggers some of the events of the inflammatory response.

homeobox (hō´-mē-ō-box): a sequence of DNA coding for special, 60 amino acid proteins. These proteins activate or inactivate genes that control development. Homeoboxes thus specify embryonic cell differentiation.

homeostasis (hōm-ē-ō-stā´sis): the maintenance of relatively constant environment required for optimal functioning of cells, maintained by the coordinated activity of numerous regulatory mechanisms, including the respiratory, endocrine, circulatory, and excretory systems. The precise regulation of the composition of fluid bathing the body cells is a homeostatic function performed largely by the urinary system.

hominid: A general term applied to all humans and their prehistoric relatives, beginning with the Australopithecines, whose fossils date back at least 4.4 million years.

homologous structures: structures that may differ in function but that have similar anatomy, presumably because of descent from common ancestors.

homologue (hō-´mō-log): a chromosome that is similar in appearance and genetic information to another chromosome with which it pairs during meiosis. Also called homologous chromosome.

homozygous (hō-mō-zī´-gōt): carrying two copies of the same allele of a given gene; also called truebreeding.

hormone: a chemical synthesized by one group of cells, secreted, and carried in the bloodstream to other cells, whose activity is influenced by reception of the hormone.

humoral immunity: an immune response in which foreign substances are inactivated or destroyed by antibodies circulating in the blood.

hybrid: an organism that is the offspring of parents differing in at least one genetically determined characteristic; also used to refer to the offspring of parents of different species.

hybrid infertility: reduced fertility (often complete sterility) in hybrid offspring of two different species.

hybrid inviability: the failure of a hybrid offspring of two different species to survive to maturity.

hydrogen bond: the weak attraction between a hydrogen atom bearing a partial positive charge (due to polar covalent bonding with another atom) and another atom, usually oxygen or nitrogen, bearing a partial negative charge. Hydrogen bonds may form between atoms of a single molecule or of different molecules.

hydrolysis (hī-drol´-i-sis): the chemical reaction that breaks a covalent bond through the addition of hydrogen to the atom forming one side of the original bond, and a hydroxyl group to the atom on the other side.

hydrophilic (hī-drō-fil´-ik): pertaining to a substance that dissolves readily in water, or to parts of a large molecule that form hydrogen bonds with water.

hydrophobic (hī-drō-fō´-bik): pertaining to a substance that does not dissolve in water.

hydrophobic interaction: the tendency for hydrophobic molecules to cluster together when immersed in water.

hydrostatic skeleton (hī-drō-stat´-ik): the use of fluid contained in body compartments to provide support for the body and mass against which muscles can contract.

hypertension: arterial blood pressure that is chronically elevated above the normal level.

hypertonic (hī-per-ton´-ik): referring to a solution that has a higher concentration of dissolved particles (and therefore a lower free water concentration) than the cytoplasm of a cell.

hypha (hī´-pha; pl. hyphae): a threadlike structure, many of which make up the fungal body, that consists of elongated cells, often with many haploid nuclei.

hypocotyl (hī´-pō-kot-ul): the part of the embryonic shoot located below the cotyledons but above the root.

hypothalamus (hī-pō-thal´-a-mus): a region of the brain that controls the secretory activity of the pituitary gland, and also synthesizes, stores, and releases certain peptide hormones, and directs autonomic nervous system responses.

hypothesis (hī-poth´-e-sis): in science, a supposition based on previous observations, which is offered as an explanation for an event, and used as the basis for further observations or experiments.

hypotonic (hī-pō-ton´-ik): referring to a solution that has a lower concentration of dissolved particles (and therefore a higher free water concentration) than the cytoplasm of a cell.

immigration (im-uh-grā´-shun): movement of individuals into an area.

immune response: a specific response by the immune system to invasion of the body by a particular foreign substance or microorganism, characterized by recognition of the foreign material by immune cells and its subsequent destruction by antibodies or cellular attack.

implantation: the process whereby the early embryo embeds itself within the lining of the uterus.

imprinting: the process by which an animal forms an association with another animal or object in the environment during a sensitive period.

inactivation: a process whereby certain chromosomes or parts of chromosomes are converted into a dense mass, preventing transcription.

inclusive fitness: the reproductive success of all organisms bearing a given allele, usually expressed in relation to the average reproductive success of all individuals in the same population. Compare with fitness.

incomplete dominance: a pattern of inheritance in which the heterozygous phenotype is intermediate between the two homozygous phenotypes.

incomplete flower: a flower that is missing one of the four floral parts (sepals, petals, stamens, or carpels).

independent assortment: a pattern of inheritance of multiple traits, in which the distribution of alleles for one trait into the gametes does not affect the distribution of alleles for other traits. Occurs with genes that are located on different chromosomes.

indirect development: a developmental pathway in which an offspring goes through radical changes in body form as it matures.

induction: the process by which a group of cells causes other cells to differentiate into a specific tissue type.

inflammatory response: a nonspecific, local response to injury to the body, characterized by phagocytosis of foreign substances and tissue debris by white blood cells, and "walling off" of the injury site by clotting of fluids escaping from nearby blood vessels.

inhalation: the act of drawing air into the lungs by enlarging the chest cavity.

inheritance of acquired characteristics: the hypothesis that organisms' bodies change during their lifetimes by use and disuse, and that these changes are inherited by their offspring.

inhibiting hormone: a hormone secreted by the neurosecretory cells of the hypothalamus that inhibits the release of specific hormones from the anterior pituitary gland.

inhibitory synapse: a synapse between two nerve cells in which the resting potential of the postsynaptic cell becomes more negative and less likely to fire an action potential due to the activity of the presynaptic cell.

innate (in-nāt´): inborn; instinctive; determined by the genetic makeup of the individual.

inner cell mass: in human embryonic development, the cluster of cells on one side of the blastocyst, which will develop into the embryo.

inner ear: the innermost part of the mammalian ear, composed of the bony, fluid-filled tubes of the cochlea and the vestibular apparatus.

inorganic molecule: any molecule that does not contain both carbon and hydrogen.

insertion: the site of attachment of a muscle to the relatively moveable bone on one side of a joint. In genetics, mutation in which one or more pairs of nucleotides are inserted within a gene.

insight: a complex form of learning that requires manipulation of mental concepts to arrive at adaptive behavior.

instinctive: innate; inborn; determined by the genetic makeup of the individual.

integration: in nerve cells, the process of adding up electrical signals from sensory inputs or other nerve cells, to determine the appropriate outputs.

integument (in-teg´-ū-ment): the outer layers of cells of the ovule surrounding the embryo sac; develops into the seed coat.

intensity: the strength of stimulation or response.

intercalated discs (in-tur´-cal-ā-ted): specialized interlocking membranes containing large numbers of gap junctions that connect adjacent cardiac muscle fibers.

interferon: a protein released by certain virus-infected cells that increases the resistance of other, uninfected, cells to viral attack.

intermediate filament: part of the cytoskeleton of eukaryotic cells, probably functioning mainly for support, composed of several types of proteins.

intermembrane compartment: the fluid-filled space between the inner and outer membranes of a mitochondrion.

internal fertilization: union of sperm and egg inside the body of the female.

internode: the part of a stem between two nodes.

interphase: the stage of the cell cycle between cell divisions. During interphase, chromosomes are replicated, and other cell functions occur, such as growth, movement, and acquisition of nutrients.

interspecific competition: competition between individuals of different species.

interstitial cells (in-ter-sti´-shul): in the vertebrate testis, testosterone-producing cells located between the seminiferous tubules.

interstitial fluid (in-tur-sti´-shul): fluid similar in composition to plasma (except lacking large proteins) which leaks from the capillaries and acts as a medium of exchange between the body cells and the capillaries.

intervertebral discs (in-tur-ver-te´-brul): pads of cartilage found between the vertebrae that act as shock absorbers.

intracellular digestion: the chemical breakdown of food that occurs within single cells.

intraspecific competition: competition between individuals of the same species.

intrauterine device (IUD): a small copper or plastic loop, squiggle, or shield that is inserted in the uterus; a contraceptive method that works by irritating the uterine lining so that it cannot receive the embryo.

intron: a segment of DNA in a eukaryotic gene that does not code for amino acids in a protein.

invertebrate (in-vert´-uh-bret): a category of animals that never possess a vertebral column.

ion (ī´-on): an atom or molecule that has either an excess of electrons (and hence is negatively charged) or has lost electrons (and is positively charged).

ionic bond: a chemical bond formed by the electric attraction between positively and negatively charged ions.

iris: the pigmented muscular tissue of the vertebrate eye that surrounds and controls the size of a central opening, the pupil, through which light enters.

islet cells: clusters of cells in the endocrine portion of the pancreas that produce insulin and glucagon.

isotonic (ī-sō-ton´-ik): referring to a solution that has the same concentration of dissolved particles (and therefore the same free water concentration) as the cytoplasm of a cell.

isotope: one of several forms of a single element, the nuclei of which contain the same number of protons but different numbers of neutrons.

J-curve: the shape of the growth curve of an exponentially growing population, in which increasing numbers of individuals join the population with each succeeding time period.

joint: a flexible region between two rigid units of an exoskeleton or endoskeleton, to allow for movement between the units.

keratin (ker´-uh-tin): a fibrous protein found in hair, nails, and the epidermis of skin.

kidney: one of a pair of organs of the excretory system located on either side of the spinal column. Kidneys filter blood, removing wastes and regulating the composition and water content of the blood.

kin selection: selection favoring a certain allele because it increases the survival or reproductive success of relatives bearing the same allele.

kinesis (kin-ne´-sis): an innate behavior in which an organism changes the speed of its random movement in response to an environmental stimulus.

kinetic energy: the energy of movement; includes light, heat, mechanical movement, and electricity.

kinetochore (kī-net´-ō-kōr): a protein structure that forms at the centromere regions of chromosomes; attaches the chromosomes to the kinetochore microtubules of the spindle.

kinetochore microtubule: a microtubule of the spindle apparatus that connects the kinetochore of a chromosome with a spindle pole; involved in movement of the chromosome to the spindle pole during anaphase.

kingdom: the broadest taxonomic category, consisting of phyla or divisions. We recognize five kingdoms in this text: Monera, Protista, Fungi, Plantae, and Animalia.

Klinefelter syndrome: a set of characteristics typically found in individuals who have two X chromosomes and one Y chromosome. These individuals are phenotypically males, but sterile, and have several femalelike traits, including broad hips and partial breast development.

Krebs cycle: the citric acid cycle (in honor of Hans Krebs, who discovered many of its biochemical details).

lacteal (lak-t´el): a single lymph capillary that penetrates each villus of the small intestine.

lactose (lak´-tōs): a disaccharide composed of glucose and galactose; found in mammalian milk.

large intestine: The final section of the digestive tract consisting of the colon and the rectum, where feces are formed and stored.

larva (lar´-vuh): an immature form of an organism with indirect development prior to metamorphosis into its adult form. The caterpillars of moths and butterflies, and the maggots of flies, are larvae.

larynx (lār´-inx): that portion of the air passage between the pharynx and the trachea. The larynx contains the vocal cords.

lateral bud: a cluster of meristemic cells at the node of a stem; under appropriate conditions, it grows into a branch.

lateral line organ: a receptor organ in fish and aquatic amphibians that detects water movement. It utilizes clusters of hair cells whose hairs are bent by the deflection of a gelatinous cupula.

lateral meristem: also called cambium; a meristematic tissue that forms cylinders that run parallel to the long axis of roots and stems; usually found between the primary xylem and primary phloem (vascular cambium) and just outside the phloem (cork cambium).

leaf: an outgrowth of a stem, usually flattened and photosynthetic.

leaf primordium (pri-mōr´-de-um; pl. primordia): a cluster of meristem cells located at the node of a stem that develops into a leaf.

learning: an adaptive change in behavior as a result of experience.

legume (leg´-yoom): a family of plants characterized by root swellings which nitrogen-fixing bacteria are housed. Includes soybeans. lupines, alfalfa, and clover.

lens: a clear object that bends light rays; in eyes, a flexible or movable structure used to focus light upon a layer of photoreceptor cells.

leukocyte (loo´-kō-sīt): any of the white blood cells circulating in the blood.

lichen (lī´-ken): a symbiotic association between an alga or cyanobacterium and a fungus, resulting in a composite organism.

life cycle: the events in the life of an organism from one generation to the next.

ligament: a tough connective tissue band connecting two bones.

light-dependent reactions: the first stage of photosynthesis, in which the energy of light is captured as ATP and NADPH; occurs in thylakoids of chloroplasts.

light-harvesting complex: in photosystems, the assembly of pigment molecules (chlorophyll and accessory pigments) that absorb light energy and transfer the energy to electrons.

light-independent reactions: the second stage of photosynthesis, in which the energy obtained by the light-dependent reactions is used to fix carbon dioxide into carbohydrates; occurs in the stroma of chloroplasts.

limbic system: a diverse group of brain structures, mostly in the lower forebrain, including the thalamus, hypothalamus, amygdala, hippocampus, and parts of the cerebrum, involved in basic emotion, drives, and behaviors, and in learning.

linkage: the inheritance of certain genes as a group because they are parts of the same chromosome. Linked genes do not show independent assortment.

lipase (lī-pāse): an enzyme that catalyzes the breakdown of lipids such as fats.

lipid (li´-pid): one of a number of water-insoluble organic molecules, containing large regions composed solely of carbon and hydrogen. Lipids include oils, fats, waxes, phospholipids, and steroids.

liver: An organ with varied functions including bile production, glycogen storage, and detoxification of poisons.

locus: the physical location of a gene on a chromosome.

long-day plant: a plant that will flower only if the length of daylight is greater than some species-specific duration.

loop of Henle (hen´-lē): a specialized portion of the tubule of the nephron in birds and mammals that creates an osmotic concentration gradient in the fluid immediately surrounding it. This gradient in turn makes possible the production of urine more osmotically concentrated than blood plasma.

lungs: respiratory organs consisting of inflatable chambers within the chest cavity in which gas exchange occurs.

luteinizing hormone (LH): a hormone produced by the anterior pituitary gland that stimulates testosterone production in males and development of the follicle, ovulation, and production of the corpus luteum in females.

lymph (limf): pale fluid within the lymphatic system composed primarily of interstitial fluid and lymphocytes.

lymph nodes: small structures that act as filters for lymph. They contain lymphocytes and macrophages, which inactivate foreign particles such as bacteria.

lymphatic system: a system consisting of lymph vessels, lymph capillaries, lymph nodes, and the thymus and spleen. The system helps protect the body against infection, absorbs fats, and returns excess fluid and small proteins to the blood circulatory system.

lymphocyte (lim´-fō-sīt): a white blood cell type important in the immune response.

lysosome (lī´-sō-sōm): a membrane-bound organelle containing intracellular digestive enzymes.

macronutrient: a nutrient needed in relatively large quantities (often defined as composing more than 0.1% of an organism's body).

macrophage (mak´-rō-faj): a type of white blood cell that engulfs microbes. Macrophages destroy microbes by phagocytosis and also present microbial antigens to T cells, helping to stimulate the immune response.

major histocompatibility complex (MHC): proteins, usually located on the surfaces of body cells, that identify the cell as "self"; MHC proteins are also important in stimulating and regulating the immune response.

maltose (mal´-tōs): a disaccharide composed of two glucose molecules.

mammary glands (mam´-uh-rē): milk-producing organs used by female mammals to nourish their young.

mantle (man´-tul): an extension of the body wall in certain invertebrates, such as molluscs. It may secrete a shell, protect the gills, and, as in cephalopods, aid in locomotion.

marsupial (mar-sū´-pē-ul): a type of mammal whose young are born at an extremely immature stage and undergo further development in a pouch while they remain attached to a mammary gland. Includes kangaroos, opossums, and koalas.

matrix: the fluid contained within the inner membrane of a mitochondrion.

matter: the material of which the universe is made.

mechanical incompatibility: the inability of male and female organisms to exchange gametes, usually because of incompatibility of the reproductive structures.

mechanoreceptor: a receptor that responds to mechanical deformation, such as is caused by pressure, touch, or vibration.

medulla (med-oo´-la): part of the hindbrain of vertebrates that controls automatic activities such as breathing, swallowing, heart rate, and blood pressure.

medusa (meh-dū´-suh): a bell-shaped, often free-swimming stage in the life cycle of many cnidarians. Jellyfish are one example.

megakaryocyte (meg-a-kār´-ē-ō-sīt): a large cell type that remains in the bone marrow, pinching off pieces of itself. These fragments enter the circulation as platelets.

megaspore: a haploid cell formed by meiosis from a diploid megaspore mother cell. Through mitosis and differentiation, the megaspore develops into the female gametophyte.

megaspore mother cell: a diploid cell contained within the ovule of a flowering plant, which undergoes meiosis to produce four haploid megaspores.

meiosis (mī-ō´-sis): a type of cell division found in eukaryotic organisms, in which a diploid cell divides twice to produce four haploid cells.

membrane: continuous sheets of epithelial cells that cover the body and line body cavities. In cells, a thin sheet of lipids and proteins that surrounds the cell or its organelles, separating them from their surroundings.

memory cell: a long-lived descendant of a B or T cell that has been activated by contact with antigen. Memory cells are a reservoir of cells that rapidly respond to reexposure to the antigen.

meninges (men-in´-ges): three layers of connective tissue that surround the brain and spinal cord.

menstrual cycle: in human females, a complex 28-day cycle during which hormonal interactions among the hypothalamus, pituitary gland, and ovary coordinate ovulation and the preparation of the uterus to receive and nourish the fertilized egg. If pregnancy does not occur, the uterine lining is shed during menstruation.

menstruation: in human females, the monthly discharge of uterine tissue and blood from the uterus.

meristem cell (mer´-i-stem): an undifferentiated cell that remains capable of cell division throughout the life of a plant.

mesoderm (mes´-ō-derm): the middle embryonic tissue layer, lying between the endoderm and ectoderm, and usually the last to develop. Mesoderm gives rise to structures such as muscle and skeleton.

mesoglea (mez-ō-glē´-uh): a middle, jellylike layer within the body wall of cnidarians.

mesophyll (mez´-ō-fil): loosely packed parenchyma cells located beneath the epidermis of a leaf.

messenger RNA (mRNA): a strand of RNA, complementary to the DNA of a gene, that conveys the genetic information in DNA to the ribosomes to be used during protein synthesis. Sequences of three bases (codons) in mRNA specify particular amino acids to be incorporated into a protein.

metabolic pathway: a sequence of chemical reactions within a cell, in which the products of one reaction are the reactants for the next.

metabolism: the sum of all chemical reactions occurring within a single cell or within all the cells of a multicellular organism.

metamorphosis (met-a-mōr´-fō-sis): in animals with indirect development, a radical change in body form from larva to sexually mature adult, as seen in amphibians (tadpole to frog) and insects (caterpillar to butterfly).

metaphase (met´-a-fāz): the stage of mitosis in which the chromosomes, attached to kinetochore microtubules, are lined up along the equator of the cell.

methanogen (me-than´-ō-gen): a type of anaerobic archaebacterium capable of converting carbon dioxide to methane.

micelle (mī´-cēl): an aggregation of bile salts with its lipid-soluble ends facing inward. Fatty acids and monoglycerides dissolve in the hydrophobic portions and are carried to the cells of the lining of the small intestine.

microfilament: part of the cytoskeleton of eukaryotic cells, composed of the proteins actin and (sometimes) myosin; functions in the movement of cell organelles and in locomotion by extension of the plasma membrane.

micronutrient: a nutrient needed in relatively small quantities (often defined as composing less than 0.01% of an organism's body).

microsphere: a small, hollow sphere formed from proteins or proteins complexed with other compounds.

microspore: a haploid cell formed by meiosis from a microspore mother cell. Through mitosis and differentiation, the microspore develops into the male gametophyte.

microspore mother cell: a diploid cell contained within an anther of a flowering plant, which undergoes meiosis to produce four haploid microspores.

microtubule: a hollow, cylindrical strand found in eukaryotic cells, composed of the protein tubulin; part of the cytoskeleton used in movement of cell organelles, cell growth, and construction of cilia and flagella.

microtubule organizing center: the region(s) of a eukaryotic cell at which tubulin subunits are assembled into microtubules.

microvilli (mī-krō-vi´-lī): a fringe of microscopic projections of the cell membrane of each villus that serve to increase its surface area.

midbrain: during development, the central portion of the brain, which contains an important relay center, the reticular formation.

middle ear: part of the mammalian ear composed of the tympanic membrane, the eustachian tube, and three bones that transmit vibrations from the auditory canal to the oval window.

middle lamella: a thin layer of sticky polysaccharides, such as pectin, and other carbohydrates that separates and sticks together the primary cell walls of adjacent plant cells.

migration: in population genetics, the flow of genes between populations.

mimicry (mim´-ik-re): the situation in which a species has evolved to resemble another type of organism.

mineral: an inorganic substance, especially one found in rocks or soil.

mitochondrion (mī-tō-kon´-drē-un): an organelle, bounded by two membranes, that is the site of the reactions of aerobic metabolism.

mitosis (mī-tō´-sis): a type of nuclear division found in eukaryotic cells. Chromosomes are duplicated during interphase before mitosis. During mitosis, one copy of each chromosome moves into each of two daughter nuclei. The daughter nuclei are therefore genetically identical to each other.

mixture: a substance composed of two or more elements in variable proportions.

molecule (mol´-e-kūl): a particle composed of one or more atoms held together by chemical bonds. A molecule is the smallest particle of a compound that displays all the properties of that compound.

molt: to shed an external body covering, such as an exoskeleton, skin, feathers, or fur.

Monera (mō´-ne-ra): a taxonomic kingdom consisting of unicellular prokaryotic organisms, including bacteria, archaebacteria, and cyanobacteria.

monocot: short for monocotyledon; a type of flowering plant characterized by embryos with one seed leaf, or cotyledon.

monoecious (mon-ē´-shus): pertaining to organisms in which male and female gametes are produced in the same individual.

monohybrid cross: a breeding experiment in which the parents differ in only one genetically determined trait.

monomer (mo´-nō-mer): a small organic molecule, several of which may be bonded together to form a chain called a polymer.

monosaccharide (mo-nō-sak´-a-rīd): the basic molecular unit of all carbohydrates, usually composed of a chain of carbon atoms to which are bonded hydrogen and hydroxyl groups.

monotreme: a type of mammal that lays eggs; for example, the platypus.

morula (mor´-ū-la): in animals, an embryonic stage during cleavage, when the embryo consists of a solid ball of cells.

motor neuron: a neuron that receives instructions from the association neurons and activates effector organs such as muscles or glands.

motor unit: a single motor neuron and all the muscle fibers on which it synapses.

mouth: the opening of a tubular digestive system into which food is first introduced.

mucous membrane: the lining of the inside of the respiratory and digestive tracts.

muscle fiber: an individual muscle cell.

mutation: a change in the base sequence of DNA in a gene; usually refers to a genetic change that is significant enough to change the appearance or function of the organism.

mutualism (mū´-chū-al-ism): a symbiotic relationship in which both participating species benefit.

mycelium (mī-sēl´-ē-um): the body of a fungus, consisting of a mass of hyphae.

mycorrhiza (mi-ko-ri´za; pl. mycorrhizae): a symbiotic relationship between a fungus and the roots of a land plant that facilitates mineral extraction and absorption.

myelin (mī´-eh-lin): a wrapping of insulating membranes of specialized nonneural cells around the axon of a vertebrate nerve cell. Myelin increases the speed of conduction of action potentials.

myofibril (mī-ō-fī´-bril): a cylindrical subunit of each muscle cell, consisting of a series of sarcomeres. Myofibrils are surrounded by sarcoplasmic reticulum.

myometrium (mī-ō-mē´-trē-um): the muscular outer layer of the uterus.

myosin (mī´-ō-sin): one of the major proteins of muscle, whose interaction with actin produces contraction; found in the thick filaments of the muscle fiber. See also actin.

natural causality: the scientific principle that natural events occur as a result of preceding natural causes.

natural killer cell: a type of white blood cell that destroys some virus-infected cells and cancerous cells on contact. Part of the nonspecific internal defense against disease.

natural selection: the unequal survival and reproduction of organisms due to environmental forces, resulting in the preservation of favorable adaptations. Usually, natural selection refers specifically to differential survival and reproduction based on genetic differences among individuals.

negative feedback: a situation in which a change initiates a series of events that tend to counteract the change and restore the original state. Negative feedback in physiological systems maintains homeostasis.

nematocyst (nēm-āt´-ō-sist): a specialized cell found in cnidarians which, when disturbed, ejects a sticky or poisoned thread. Used by cnidarians to trap and sting prey.

nephridium (nef-rid´-ē-um): a type of excretory organ found in earthworms, mollusks, and certain other invertebrates. A nephridium somewhat resembles a single vertebrate nephron.

nephron (nef´-ron): the functional unit of the kidney, where blood is filtered and urine formed.

nephrostome: the funnel shaped opening of the nephridium of some invertebrates such as earthworms, into which coelomic fluid is drawn for filtration.

nerve: a bundle of axons of nerve cells, bound together in a sheath.

nerve cord: also called the spinal cord of vertebrates, a nervous structure lying along the dorsal side of the body of chordates.

nerve net: a simple form of nervous systems consisting of a network of neurons that extend throughout the tissues of an organism, such as a cnidarian.

net primary productivity: the energy stored in the primary producers of an ecosystem over a given time period.

neural tube: a structure derived from ectoderm during early embryonic development that later becomes the brain and spinal cord.

neuromodulator: a chemical, usually a peptide, released by neurons that alter the activity of groups of neurons over longer periods of time.

neuromuscular junction: the synapse formed between a motor neuron and muscle fiber.

neuron (nur´-on): a single nerve cell.

neurosecretory cell: a specialized nerve cell that synthesizes and releases hormones.

neurotransmitter: a chemical released by a nerve cell close to a second nerve cell, a muscle, or a gland cell, that influences the activity of the second cell.

neutral mutation: a mutation (change in DNA sequence) that has little or no effect on the function of the encoded protein.

neutralization: the process of covering up or inactivating a toxic substance with antibody.

neutron: a subatomic particle found in the nuclei of atoms, bearing no charge and having mass approximately equal to that of a proton.

nitrogen fixation: the process of combining atmospheric nitrogen with hydrogen to form ammonium (NH_4^+).

nitrogen-fixing: possessing the ability to remove nitrogen from the atmosphere and combine it with hydrogen to produce ammonium (NH_4^+).

nitrogen-fixing bacteria: bacteria that possess the ability to remove nitrogen (N_2) from the atmosphere and combine it with hydrogen to produce ammonium (NH_4^+).

node: a region of a stem at which leaves and lateral buds are located. Also, an interruption of the myelin on a vertebrate myelinated axon, exposing naked membrane at which action potentials are generated.

nodule: a swelling on the root of a legume or other plant that consists of cortex cells inhabited by nitrogen-fixing bacteria.

nondisjunction: an error in meiosis in which chromosomes fail to segregate properly into the daughter cells.

nonpolar covalent bond: a covalent bond with equal sharing of electrons.

norepinephrine(nor-epi-nef-rin´): also called noradrenaline, a neurotransmitter which is released by neurons of the parasympathetic nervous system and prepares the body to respond to stressful situations.

notochord (nōt-ō-kōrd): a stiff but somewhat flexible, supportive rod found in all members of the phylum Chordata at some stage of development.

nuclear envelope: the double membrane system surrounding the nucleus of eukaryotic cells. The outer membrane is often continuous with the endoplasmic reticulum.

nucleic acid (nū-klā´-ik): an organic molecule composed of nucleotide subunits. The two common types of nucleic acids are ribonucleic acids (RNA) and deoxyribonucleic acids (DNA).

nucleoid (nū-klē-oyd): the location of the genetic material in prokaryotic cells; not membrane-enclosed.

nucleolus (nū-klē´-ō-lus): the region of the eukaryotic nucleus engaged in ribosome synthesis, consisting of the genes encoding ribosomal RNA, newly synthesized ribosomal RNA, and ribosomal proteins.

nucleotide: an individual subunit of which nucleic acids are composed. A single nucleotide consists of a phosphate group bonded to a sugar (deoxyribose in DNA), which is in turn bonded to a nitrogen-containing base (adenine, guanine, cytosine, or thymine). Nucleotides are linked together by bonds between the phosphate of one and the sugar of the next, to form a strand of nucleic acid.

nucleus: the membrane-bound organelle of eukaryotic cells that contains the cell's genetic material. Also, the central region of an atom, composed of protons and neutrons.

nutrient: a substance acquired from the environment and needed for survival, growth, and development of an organism.

nutrient cycle: a description of the movement of a specific nutrient (carbon, nitrogen, phosphorus, water, etc.) through the living and nonliving portions of an ecosystem.

nutrition: the process of acquiring nutrients from the environment and, if necessary, processing them into a form that can be used by the body.

oil: a lipid composed of three fatty acids, some of which are unsaturated, covalently bonded to a molecule of glycerol. Oils are liquid at room temperature.

olfaction (ōl-fak´-shun): a chemical sense, the sense of smell; in terrestrial vertebrates the result of detection of airborne molecules.

ommatidium (ōm-ma-tid´-ē-um): individual light-sensitive subunit of a compound eye. Each ommatidium consists of a lens and several receptor cells.

oncogene: a gene that, when transcribed, causes a cell to become cancerous.

one-gene, one-protein hypothesis: the proposition that each gene encodes the information for the synthesis of a single protein.

oogenesis: the process by which egg cells are formed.

oogonium (ō-ō-gō´-nē-um; plural, oogonia): a diploid cell in female animals that gives rise to a primary oocyte.

open circulatory system: a type of circulatory system found in some invertebrates, such as arthropods and molluscs, that includes an open space (the hemocoel) in which blood directly bathes body tissues.

operant conditioning: a laboratory training procedure in which an animal learns to perform a response (such as pressing a lever) through reward or punishment.

operator: in prokaryotes, a segment of DNA that governs access of RNA polymerase to the promoter and thus regulates transcription.

operculum: an external flap, supported by bone, that covers and protects the gills of most fish.

operon: a unit of organization of prokaryotic DNA, in which several genes that specify related functions (for example, enzymes in the same biosynthetic pathway) are grouped together on the chromosome, are transcribed at the same time, and are regulated together.

opioids (ōp´-ē-ōyd): a group of peptide neuromodulators found in the vertebrate brain that mimic some of the actions of opiates (such as opium) and also seem to influence many other processes, including emotion and appetite.

optic disk (op´-tik): the area of the retina at which the axons of the ganglion cell merge to form the optic nerve; the blind spot of the retina.

optic nerve (op´-tik): the nerve leading from the eye to the brain, carrying visual information.

orbital: the region of an atom, outside the nucleus, in which an electron is likely to be found.

order: the taxonomic category contained within a class and consisting of related families.

organ: a structure (such as the liver, kidney, or skin) composed of two or more distinct tissue types that function together.

organ system: two or more organs that work together to perform a specific function, i.e., the digestive system.

organelle (or-ga-nel´): a structure found in the cytoplasm of eukaryotic cells that performs a specific function; sometimes used to refer specifically to membrane-bound structures such as the nucleus or endoplasmic reticulum.

organic molecule: a molecule that contains both carbon and hydrogen.

organism (ōr´-ga-niz-em): an individual living thing.

organogenesis (or-gan-ō-jen´-i-sis): the process by which the layers of the gastrula (endoderm, ectoderm, mesoderm) rearrange themselves into organs.

origin: the site of attachment of a muscle to the relatively stationary bone on one side of a joint.

osculum (os´-kū-lum): a relatively large opening in the sponge body through which water is expelled.

osmosis (oz-mō´-sis): the diffusion of water across a differentially permeable membrane, usually down a concentration gradient of free water molecules. Water moves into the solution that has a lower free water concentra-tion from the solution with the higher free water concentration.

osmotic pressure: The pressure required to counterbalance the tendency of water to move from a solution with a higher concentra tion of free water molecules into one with a lower concentration of free water molecules.

osteoblast (os´-tē-ō-blast): a cell type that produces bone.

osteoclast: a cell type that dissolves bone.

osteocyte: a mature bone cell.

osteoporosis (os´-tē-ō-pōr-ō´-sis): a condition in which bones become porous, weak, and easily fractured; most common in elderly women.

otolith membrane (ō´-tō-lith): a gelatinous membrane in which are embedded the hairs of hair cells of the utricle and saccule of the inner ear. The otolith membrane is heavier than the surrounding fluid due to calcium carbonate crystals embedded in it and therefore responds to the pull of gravity.

outer ear: the outermost part of the mammalian ear, including the external ear and auditory canal leading to the tympanic membrane.

oval window: the membrane-covered entrance to the inner ear.

ovary: the gonad of female animals. Also, in flowering plants, a structure at the base of the carpel containing one or more ovules; develops into the fruit.

oviduct: in mammals, the tube leading from the ovary to the uterus.

ovulation: The release of a secondary oocyte, ready to be fertilized, from the ovary.

ovule: a structure within the ovary of a flower, inside which the female gametophyte develops. After fertilization, the ovule develops into the seed.

pacemaker: a cluster of specialized muscle cells in the upper right atrium of the heart that produce spontaneous electrical signals at a regular rate; the sino-atrial node.

Pacinian corpuscle (pas-in´-ē-an): a receptor of the skin and joints whose ending is wrapped in a layered capsule of connective tissue, and that responds to rapid movement.

palisade cells: columnar mesophyll cells containing chloroplasts just beneath the upper epidermis of a leaf.

pancreas (pan´-krē-as): a combined exocrine and endocrine gland located in the abdominal cavity next to the stomach. Its endocrine portion secretes the hormones insulin and glucagon, which regulate glucose concentrations in the blood. The exocrine portion secretes enzymes for fat, carbohydrate, and protein digestion into the small intestine and neutralizes the acidic chyme.

pancreatic juice: A mixtureof water, sodium bicarbonate, and enzymes released by the pancreas into the small intestine.

parasite (par´-uh-sīt): an organism that lives in or on the body of another organism, called a host, causing it harm as a result.

parasitism: a symbiotic relationship in which one organism (usually smaller and more numerous than its host) benefits by feeding on the other, which is usually harmed but not immediately killed.

parasympathetic nervous system: the division of the autonomic nervous system that produces largely involuntary responses related to maintenance of normal body functions, such as digestion.

parathyroids: a set of four small endocrine glands embedded in the surface of the thyroid gland that produce parathormone, which (with calcitonin from the thyroid) regulates calcium ion concentration in the blood.

parenchyma (par-en´-ki-ma): a plant cell type that is alive at maturity, usually with thin primary cell walls, that carries out most of the metabolism of a plant. Most dividing meristem cells in a plant are parenchyma.

parthenogenesis (par-the-nō-gen´-i-sis): a specialization of sexual reproduction, in which a haploid egg undergoes development without fertilization.

passive transport: movement of materials across a membrane down a gradient of concentration, pressure, or electrical charge without using cellular energy.

passive visual signal: a body shape, color, or size that communicates information, even in the absence of specific behaviors.

pathogen (path´-ō-gen): an organism capable of producing disease.

pedigree: a diagram showing genetic relationships among a set of individuals, usually with respect to a specific genetic trait.

peptide (pep´-tīd): a chain composed of two or more amino acids linked together by peptide bonds.

peptide bond: the covalent bond between the amino group nitrogen of one amino acid and the carboxyl group carbon of a second amino acid, which joins the two amino acids together in a peptide or protein.

peptide hormone: A hormone consisting of a chain of amino acids. Small proteins that function as hormones are included in this category.

peptidoglycan (pep-tid-ō-glī´-can): material found in prokaryotic cell walls consisting of chains of sugars cross-linked by short chains of amino acids called peptides.

pericycle (per´-i-sī-kul): the outermost layer of cells of the vascular cylinder of a root.

periderm: the outer cell layers of roots and a stem that have undergone secondary growth, consisting primarily of cork cambium and cork cells.

periosteum: a thin layer of connective tissue, rich in capillaries, osteoblasts, and osteoblast-forming cells that covers healthy bone. Rupture of the periosteum cause by a fracture stimulates production and release of osteoblasts.

peripheral nervous system: in vertebrates, that part of the nervous system connecting the central nervous system to the rest of the body.

peristalsis: rhythmic coordinated contractions of the smooth muscles of the digestive tract that move substances throughthe digestive tract.

permeate (per´-mē-āt): to pass through, as through pores in a membrane.

petal: part of a flower, often brightly colored and fragrant, serving to attract potential animal pollinators.

petiole (pet´-ē-ōl): the stalk that connects the blade of a leaf to the stem.

pH scale: a scale with values from 0 to 14, used for measuring the relative acidity of a solution. At pH 7 a solution is neutral, pH 0 to 7 is acidic, and pH 7 to 14 is basic. Each unit on the pH scale represents a tenfold change in the concentration of hydrogen ions.

phagocytosis (fa-gō-sī-tō´-sis): a type of endocytosis in which extensions of a plasma membrane engulf extracellular particles and transport them into the interior of the cell.

pharynx (fār´-inx): in vertebrates, a chamber at the back of the mouth shared by the digestive and respiratory systems.

phenotype (fēn´-ō-tīp): the physical characteristics of an organism. Phenotype can be defined as outward appearance (such as flower color), as behavior, or in molecular terms (such as glycoproteins on red blood cells).

phenylketonuria (fen-ul-kē-tō-nū´-rē-a): a recessive disease in which the enzyme that catalyzes the conversion of the amino acid phenylalanine to tyrosine is defective.

pheromone (fār´-uh-mōn): a chemical produced by an organism that alters the behavior or physiological state of another of the same species.

phloem (flō´-um): a conducting tissue of vascular plants that transports a concentrated sugar solution up and down the plant.

phospholipid (fos-fō-li´-pid): a lipid consisting of glycerol to which two fatty acids and one phosphate group are bonded. The phosphate group bears another group of atoms, often containing nitrogen, and often bearing an electric charge. A double layer of phospholipids is found in all cell membranes.

phospholipid bilayer: a double layer of phospholipids that forms the basis of all cellular membranes; the phospholipid heads, which are hydrophilic, face the water of extracellular fluid or the cytoplasm, and the tails, which are hydrophobic, are buried in the middle of the bilayer.

photon (fō´-ton): The smallest unit of light energy.

photophosphorylation (fō-tō-fos-for-i-lā´-shun): ATP synthesis in the light-dependent reactions of photosynthesis; light energy ("photo") is used to add a phosphate group to ("phosphorylate") ADP to yield ATP.

photopigment (fō´-tō-pig-ment): a chemical substance in photoreceptor cells that changes molecular conformation when struck by light.

photoreceptors: receptor cells that respond to light; in vertebrates, rods and cones.

photorespiration: A series of reactions in plants in which O2 replaces CO2 during the C3 cycle, preventing carbon fixation. This wasteful process dominates when C3 plants are forced to close their stomata to prevent water loss.

photosynthesis: the complete series of chemical reactions in which the energy of light is used to synthesize high-energy organic molecules, usually carbohydrates, from low-energy inorganic molecules, usually carbon dioxide and water.

photosystem: in thylakoid membranes, a light-harvesting complex and its associated electron transport system.

phototropism: growth with respect to the direction of light.

phycocyanin (fī-kō-sī´-a-nin): a blue or purple pigment found in the membranes of chloroplasts and used as an accessory light-gathering molecule in thylakoid photosystems.

phyletic speciation: a form of evolution in which the gene pool of a species changes so greatly over time that a new species is formed. Old and new species do not exist simultaneously. See divergent speciation.

phylum (fī-lum): the taxonomic category of animals and animallike protists contained within a kingdom and consisting of related classes.

phytochrome (fī´-tō-krōm): a light-sensitive plant pigment that mediates many plant responses to light, including flowering, stem elongation, and seed germination.

phytoplankton (fī´-tō-plank-ten): a general term describing photosynthetic protists that are abundant in marine and freshwater environments.

pili (pil´-ī; sing. pilus): hair-like projections made of protein and found on the surface of certain bacteria, often used to attach the bacterium to another cell.

pineal gland (pī-nē´-al): a small gland within the brain that secretes melatonin. The pineal controls the seasonal reproductive cycles of some mammals.

pinocytosis (pī-nō-sī-tō´-sis): nonselective movement of extracellular fluid into a cell, enclosed within a vesicle formed from the plasma membrane.

pioneer: an organism that is among the first to colonize an unoccupied habitat in the first stages of succession.

pit: an area in the cell walls between two plant cells in which secondary walls did not form, so that the two cells are separated only by a relatively thin and porous primary cell wall.

pit organs: thermoreceptive organs found in certain snakes (vipers) which consist of sensory cells protected within a pair of pits located between each eye and the nose; used to detect warm-blooded prey.

pith: cells forming the center of a root or stem.

pituitary: an endocrine gland located at the base of the brain that produces several hormones, many of which influence the activity of other glands.

placenta (pluh-sen´-ta): In mammals, a structure formed by a complex interweaving of the uterine lining and the embryonic membranes, especially the chorion; functions in gas, nutrient, and waste exchange between embryonic and maternal circulatory systems and secretes hormones.

plant hormone: the plant-regulating chemicals auxin, gibberellins, cytokinins, ethylene, and abscisic acid; plant hormones somewhat resemble animal hormones in that they are chemical produced by cells in one location that influence the growth or metabolic activity of other cells, often some distance away in the plant body.

plaque (plak): a deposit of cholesterol and other fatty substances within the wall of an artery.

plasma: the fluid, noncellular portion of the blood.

plasma cell: an antibody-secreting descendant of a B cell.

plasma membrane: the outer membrane of a cell, composed of a bilayer of phospholipids in which proteins are embedded.

plasmid (plaz´-mid): a small, circular piece of DNA found in the cytoplasm of many bacteria; usually does not carry genes required for the normal functioning of the bacterium but may carry genes that assist bacterial survival in certain environments, such as antibiotic resistance.

plasmodesma (plaz-mō-dez´-ma; pl. plasmodesmata): a cell-to-cell junction in plants that connects the cytoplasm of adjacent cells.

plasmodial slime mold: a funguslike protist consisting of a multinucleate mass or plasmodi-um that crawls and ingests decaying organic matter in amoeboid fashion and forms a fruiting body under adverse conditions. Also called acellular slime mold.

plasmodium (plas-mō´-dē-um): a sluglike mass of cytoplasm containing thousands of nuclei that are not confined within individual cells.

plastid (plas´-tid): in plant cells, an organelle bounded by two membranes that may be involved in photosynthesis (chloroplasts), pigment storage, or food storage.

platelets (plāt´-lets): cell fragments formed from megakaryocytes in bone marrow. Platelets, which lack nuclei, circulate in the blood and play a role in blood clotting.

pleated sheet: a type of secondary structure of a protein. In a pleated sheet, protein chains lie side-by-side, bonded to one another by hydrogen bonds.

pleiotropy (plē´-ō-trō-pē): a situation in which a single gene influences more than one phenotypic characteristic.

pleural membranes: Membranes that line the chest and surround the lungs.

point mutation: a mutation in which a single base pair in DNA has been changed.

polar body: in oogenesis, a small cell containing a nucleus but virtually no cytoplasm, produced by the first meiotic division of the primary oocyte.

polar covalent bond: a covalent bond with unequal sharing of electrons, so that one atom is relatively negative while the other is relatively positive.

polar microtubule: a microtubule of the spindle apparatus that extends from one spindle pole to the equator of the cell; involved in moving the spindle poles farther apart during anaphase.

pollen: the male gametophyte of a seed plant.

pollination: in flowering plants, when pollen lands on the stigma of a flower of the same species; in conifers, when pollen lands within the pollen chamber of a female cone of the same species.

polygenic inheritance: a pattern of inheritance in which the interactions of two or more functionally similar genes determine phenotype.

polymer (pa´-li-mer): a molecule composed of three or more (may be thousands) smaller subunits called monomers. The monomers that make up a single polymer may be identical (for example, the glucose monomers of starch) or different (for example, the amino acids of a protein).

polymerase chain reaction (PCR): a method of producing virtually unlimited numbers of copies of a specific piece of DNA, starting with as little as one copy of the desired DNA.

polyp (pol´-ip): the sedentary, vase-shaped stage in the life cycle of many cnidarians. Hydra and sea anemones are examples.

polyploid (pol´-ē-ployd): having more than two homologous chromosomes of each type.

polysaccharide (pol-ē-sak´-a-rīd): a large carbohydrate molecule composed of branched or unbranched chains of repeating monosaccharide subunits, usually glucose or modified glucose molecules. Polysaccharides include starches, cellulose, and glycogen.

pons: a portion of the hindbrain just above the medulla containing neurons that influence sleep and the rate and pattern of breathing.

population: a group of individuals of the same species, found in the same time and place, and actually or potentially interbreeding.

population bottleneck: a form of genetic drift in which a population becomes extremely small, which may lead to differences in allele frequencies as compared with other populations of the species, and to a loss in genetic variability.

population genetics: the study of the frequency, distribution, and inheritance of alleles in a population of organisms.

positive feedback: a situation in which a change initiates events that tend to amplify the original change.

posterior (post-tēr´-ē-ur): the tail, hindmost, or rear end of an animal.

posterior pituitary: a lobe of the pituitary gland that is an outgrowth of the hypothalamus and releases antidiuretic hormone and oxytocin.

postmating isolating mechanism: any structure, physiological function, or developmental abnormality that prevents organisms of two different populations, once mating has occurred, from producing vigorous, fertile offspring.

postsynaptic neuron: the nerve cell at a synapse that changes its electrical potential in response to a chemical (the neurotransmitter) released by another (presynaptic) cell.

postsynaptic potential: an electrical signal produced in a postsynaptic cell by transmission across the synapse. It may be excitatory (EPSP), making the cell more likely to produce an action potential, or inhibitory (IPSP), tending to inhibit an action potential.

potassium channel: a protein that spans a nerve cell membrane, forming a pore, and which allows only potassium ions to flow through the pore.

potential energy: "stored" energy, usually chemical energy or energy of position within a gravitational field.

preadaptation: a feature evolved under one set of environmental conditions that, purely by chance, helps an organism adapt to new environmental conditions.

prebiotic evolution: evolution before life existed; especially abiotic synthesis of organic molecules.

precapillary sphincter (sfink´-ter): a ring of smooth muscle between an arteriole and a capillary that regulates the flow of blood into the capillary bed.

predation (pre-dā´-shun): the act of killing and eating another living organism.

predator: an organism that kills and eats other organisms.

premating isolating mechanism: any structure, physiological function, or behavior that prevents organisms of two different populations from exchanging gametes.

pressure flow theory: a model for transport of sugars in phloem, which states that movement of sugars into a phloem sieve tube causes water to enter the tube by osmosis, while movement of sugars out of another part of the same sieve tube causes water to leave by osmosis; the resulting pressure gradient causes bulk movement of water and dissolved sugars from the end of the tube into which sugar is transported toward the end of the tube from which sugar is removed.

presynaptic neuron: a nerve cell that releases a chemical (the neurotransmitter) at a synapse, which causes changes in the electrical activity of another (postsynaptic) cell.

primary cell wall: cellulose and other carbohydrates secreted by a young plant cell between the middle lamella and the plasma membrane.

primary consumer: an organism that feeds on producers; an herbivore.

primary endosperm cell: the central cell of the female gametophyte of a flowering plant, containing the polar nuclei (usually two). After fertilization, the primary endosperm cell undergoes repeated mitotic divisions to produce the endosperm of the seed.

primary growth: growth in length and development of initial structures of plant roots and shoots, due to cell division of apical meristems and differentiation of the daughter cells.

primary oocyte (ō´-ō-sīt): a diploid cell, derived from the oogonium by growth and differentiation, which undergoes meiosis to produce the egg.

primary phloem: phloem in young stems produced from an apical meristem.

primary root: the first root that develops from a seed.

primary spermatocyte (sper-ma´-tō-sīt): a diploid cell, derived from the spermatogonium by growth and differentiation, which undergoes meiosis to produce four sperm.

primary structure: the amino acid sequence of a protein.

primary succession: succession that occurs in an environment, such as bare rock, in which no trace of a previous community is present.

primary xylem: xylem in young stems produced from an apical meristem.

primer pheromone: a chemical produced by an organism that alters the physiological state of another of the same species.

primitive streak: in reptiles, birds, and mammals, the region of the ectoderm of the two-layered embryonic disk through which cells migrate to form mesoderm.

prion (prē´-on): a protein which, in mutated form acts as an infectious agent causing certain neurodegenerative diseases, including kuru and scrapie.

producer: a photosynthetic organism; an autotroph.

product: an atom or molecule resulting from a chemical reaction.

progesterone (prō-ge´-ster-ōn): a hormone produced by the corpus luteum that promotes development of the uterine lining.

prokaryotic (prō-kar-ē-ot´-ik): referring to cells of the kingdom Monera. Prokaryotic cells do not have their genetic material enclosed within a membrane-bound nucleus and also lack other membrane-bound organelles.

promoter: a specific sequence of DNA to which RNA polymerase binds, initiating gene transcription.

prophase (prō´-fāz): the first stage of mitosis, in which the chromosomes first become visible in the light microscope as thickened, condensed threads and the spindle begins to form. As the spindle is completed, the nuclear envelope breaks apart. The spindle invades the nuclear region and attaches to the kinetochores of the chromosomes. Also, the first stage of meiosis. In meiosis I, the homologous chromosomes pair up and exchange parts at chiasmata. In meiosis II, the spindle reforms and chromosomes attach to the microtubules.

proprioceptor (prō´-prē-ō-cep-tōr): a receptor that monitors the position of the parts of the body and their direction of movement by responding to stretch and pressure.

prostaglandin (pro-sta-glan´-din): a family of modified fatty acid hormones manufactured by many cells of the body.

prostate gland (prō´-stāt): a gland that produces part of the fluid component of semen. The prostate fluid is basic and contains a chemical that activates sperm movement.

protease (prō´-tē-ās): an enzyme that digests proteins.

protein: an organic molecule composed of one or more chains of amino acids.

protista (prō-tis´-tuh): a taxonomic kingdom including unicellular, eukaryotic organisms.

proton: a subatomic particle found in the nuclei of atoms, bearing a unit of positive charge and a relatively large mass roughly equal to the mass of the neutron.

protozoan (prō-te-zō´-an; pl. protozoa): a nonphotosynthetic or animallike protist.

pseudocoel (sū´-dō-sēl): "false coelom"; a body cavity with a different embryological origin than a coelom, but serving a similar function; found in roundworms.

pseudoplasmodium (sū´-dō-plas-mō´-dē-um): an aggregation of individual amoeboid cells that form a sluglike mass.

pseudopod (sūd´-ō-pod): extension of the cell membrane by which certain cells, such as amoebae, locomote and engulf prey.

punctuated equilibrium: a model of evolution, stating that morphological change and speciation are rapid, simultaneous events (the "punctuation"), separated by long periods during which a species remains unchanged (the "equilibrium").

pupil: the adjustable opening in the center of the iris through which light enters the eye.

purine: a type of nitrogen-containing base found in nucleic acids that consists of two fused rings; includes adenine and guanine in both DNA and RNA.

pyloric sphincter (pī-lor´-ik sfink´-ter): a circular muscle at the base of the stomach that regulates the passage of chyme into the small intestine.

pyrimidine: a type of nitrogen-containing base found in nucleic acids that consists of a single ring; includes cytosine (both DNA and RNA), thymine (DNA only) and uracil (RNA only).

quaternary structure (kwat´-er-na-rē): the complex three-dimensional structure of a protein that is composed of more than one peptide chain.

queen substance: a chemical produced by a queen bee that can act as both a primer and a releaser pheromone.

radial symmetry: a body plan in which any plane drawn along a central axis will divide the body into approximately mirror-image halves. Cnidarians and many adult echinoderms show radial symmetry.

radioactive: pertaining to an atom with an unstable nucleus that spontaneously disintegrates with the emission of radiation.

radula (ra´-dū-luh): a ribbon of tissue in the mouth of gastropod mollusks that bears numerous teeth on its outer surface and is used to scrape and drag food into the mouth.

random distribution: spacing in which the probability of finding an individual is equal in all parts of an area.

reactant: an atom or molecule that is used up in a chemical reaction to form a product.

reaction center: in the light-harvesting complex of a photosystem, the chlorophyll molecule to which light energy is transferred by the antenna pigments. The captured energy ejects an electron from the reaction center chlorophyll, and the electron is transferred to the electron transport system.

receptor: (1) a cell that responds to an environmental stimulus (chemicals, sound, light, pH, etc.) by changing its electrical potential; (2) a protein molecule in a cell membrane that binds to another molecule (hormone, neurotransmitter) triggering metabolic or electrical changes in a cell.

receptor-mediated endocytosis: selective uptake of molecules from the extracellu-lar fluid, by binding to a receptor located at a coated pit on the plasma membrane, and pinching off the coated pit into a vesicle that moves into the cytosol.

receptor potential: an electrical potential change in a receptor cell produced in response to reception of an environmental stimulus (chemicals, sound, light, heat, etc.). The size of the receptor potential is proportional to the intensity of the stimulus.

receptor protein: a protein located on a membrane (or in the cytoplasm), that recognizes and binds to specific molecules. Binding by receptors often triggers a response by a cell, such as endocytosis, increased metabolic rate, or cell division.

recessive: an allele expressed only in homozygotes, and which is completely masked in heterozygotes.

recognition protein: a protein or glycoprotein protruding from the outside surface of a plasma membrane that identifies a cell as belonging to a particular species, to a specific individual of that species, and often to a specific organ within the individual.

reflex: a simple, stereotyped movement of part of the body that occurs automatically in response to a stimulus.

regeneration: (1) regrowth of a body part after loss or damage; (2) asexual reproduction by regrowth of an entire body from a fragment.

regulatory gene: a gene that controls the timing or rate of transcription of other genes.

releaser: a stimulus that triggers a fixed action pattern.

releaser pheromone: a chemical produced by one organism that alters the behavior of another of the same species.

releasing hormone: a hormone secreted by the hypothalamus that causes the release of specific hormones by the anterior pituitary gland.

renal artery: The artery carrying blood to each kidney.

renal cortex: the outer layer of the kidney where nephrons are located.

renal medulla: the layer of the kidney just inside the renal cortex where loops of Henle produce a highly concentrated interstitial fluid that is important for producing concentrated urine.

renal pelvis: The inner chamber of where urine from the collecting ducts accumulates before entering the ureters.

renal vein: The vein carrying cleansed blood away from each kidney

replacement-level fertility (RLF): the average birth rate at which a reproducing population exactly replaces itself during its lifetime.

replication fork: in DNA replication, a Y-shaped opening in the double helix of DNA, at which new DNA strands are synthesized.

repression: in the lactose operon, the condition that the genes of the operon are not transcribed unless specifically activated by the presence of lactose.

repressor protein: in the lactose operon, a protein that binds to the operator site, thereby preventing access of RNA polymerase to the promoter.

reproductive isolation: the failure of organisms of one population to breed successfully with members of another population; may be due to premating or postmating isolating mechanisms.

reservoir: the major source of any particular nutrient in an ecosystem, usually in the abiotic portion.

respiratory center: a cluster of neurons in the brainstem that sends rhythmic bursts of nerve impulses to the respiratory muscles, resulting in breathing.

resting potential: a negative electrical potential found in unstimulated nerve cells.

restriction enzyme: an enzyme, usually isolated from bacteria, that cutss double-stranded DNA at a specific nucleotide sequence; the cleavage sequence differs for different restriction enzymes.

restriction fragment: a piece of DNA that has been isolated by cleaving a larger piece of DNA with restriction enzymes.

restriction fragment length polymorphisms (RFLPs): differences in the lengths of restriction fragments, produced by cutting samples of DNA from different individuals of the same species with the same set of restriction enzymes; occurs because of differences in nucleotide sequences among different individuals of the same species.

reticular formation (reh-tik´-ū-lar): a diffuse network of neurons extending from the hindbrain, through the midbrain, and into the lower reaches of the forebrain, involved in filtering sensory input and regulating what information is relayed to higher brain centers for further attention.

retina (ret´-in-a): a multilayered sheet of nerve tissue at the rear of camera-type eyes, composed of photoreceptor cells plus associated nerve cells that refine the photoreceptor information and transmit it to the optic nerve.

retrovirus: a virus that uses RNA as its genetic material. When a retrovirus invades a eukaryotic cell, it "reverse transcribes" its RNA into DNA, which then directs the synthesis of more viruses, using the transcription and translation machinery of the cell.

reverse transcriptase: an enzyme found in retroviruses that catalyzes the synthesis of DNA from an RNA template.

Rh factor: a protein found on the red blood cells of some people (Rh-positive) but not others (Rh-negative). Exposure of Rh-negative individuals to Rh-positive blood triggers antibody production to Rh-positive blood cells.

rhizoid (rī´-zōyd): a rootlike structure found in bryophytes that anchors the plant and absorbs water and nutrients from the soil.

rhizome (rī´-zōm): an undergound stem, usually horizontal, that stores food.

rhythm method: a contraceptive method involving abstinence from intercourse during ovulation.

ribonucleic acid (rī-bō-nū-klā´-ik; RNA): a molecule composed of ribose nucleotides, each of which consists of a phosphate group, the sugar ribose, and one of the bases adenine, cytosine, guanine, or uracil; transfers hereditary instructions from the nucleus to the cytoplasm; also the genetic material of some viruses.

ribosomal RNA (rRNA): a type of RNA that combines with proteins to form ribosomes.

ribosome: an organelle consisting of two subunits, each composed of ribosomal RNA and protein. Ribosomes are the site of protein synthesis, during which the sequence of bases of messenger RNA is translated into the sequence of amino acids in a protein.

ribozyme: an RNA molecule that can catalyze certain chemical reactions, especially those involved in synthesis and processing of RNA itself.

RNA polymerase: in RNA synthesis, an enzyme that catalyzes the bonding of free RNA nucleotides into a continuous strand, using RNA nucleotides that are complementary to those of a strand of DNA.

rod: a rod-shaped photoreceptor cell in the vertebrate retina, sensitive to dim light but not involved in color vision. See also cone.

root cap: a cluster of cells at the tip of a growing root, derived from the apical meristem. The root cap protects the growing tip from damage as it burrows through the soil.

root hair: a fine projection from an epidermal cell of a young root that increases the absorptive surface area of the root.

root system: the part of a plant, usually below ground, that anchors the plant in the soil, absorbs water and minerals, stores food, transports water, minerals, sugars, and hormones, and produces certain hormones.

rough endoplasmic reticulum: endoplasmic reticulum lined on the outside with ribosomes.

runner: a horizontally growing stem that may develop new plants at nodes that touch the soil.

S-curve: the growth curve that describes a population of long-lived organisms introduced into a new area. It consists of an initial period of exponential growth, followed by decreasing growth rate, and finally, relative stability around a growth rate of zero.

saccule (sak´-ūle): a portion of the vestibular apparatus responsible for detecting gravity.

saltatory conduction (sal´-ta-tōr-ē): literally "jumping" conduction; the transmission of an action potential along a myelinated axon, during which the action potential seems to jump from one node to the next.

saprobe (sap´-rōbe): an organism that derives its nutrients from the bodies of dead organisms.

sarcodine (sar-kō´-dīn): a category of nonphotosynthetic protist (protozoa) characterized by the ability to form pseudopodia. Some, such as amoebae, are naked, while others have elaborate shells.

sarcomere (sark´-ō-mēr): the unit of contraction of a muscle fiber; a subunit of the myofibril, consisting of actin and myosin filaments and bounded by Z-lines.

sarcoplasmic reticulum (sark´-ō-plas´-mik re-tik´-ū-lum): specialized endoplasmic reticulum found in muscle cells and forming interconnected hollow tubes. The sarcoplasmic reticulum stores calcium ions and releases them into the interior of the muscle cell to initiate contraction.

saturated: referring to a fatty acid with as many hydrogen atoms as possible bonded to the carbon backbone; a fatty acid with no double bonds in its carbon backbone.

sclera: a tough white connective tissue layer that covers the outside of the eyeball and forms the white of the eye.

sclerenchyma (skler-en´-ki-ma): a plant cell type with thick, hardened secondary cell walls, that usually dies as the last stage of differentiation and provides both support and protection for the plant body.

scramble competition: direct interactions between individuals attempting to acquire the same limited resources.

scrotum (skrō´-tum): the pouch of skin containing the testes of male mammals.

sebaceous gland (se-bā´-shus): a gland in the dermis of skin, formed from epithelial tissue, that produces an oily substance called sebum that lubricates the epidermis.

second law of thermodynamics: a principle of physics that any change in an isolated system causes the quantity of concentrated, useful energy to decrease, and the amount of randomness and disorder (entropy) to increase.

second messenger: a term applied to intracellular chemicals, such as cyclic AMP, that are synthesized or released within a cell in response to the binding of a hormone or neurotransmitter (the first messenger) to receptors on the cell surface. Second messengers bring about specific changes in the metabolism of the cell.

secondary cell wall: a thick layer of cellulose and other polysaccharides secreted by certain plant cells between the primary cell wall and the plasma membrane.

secondary consumer: an organism that feeds on primary consumers; a carnivore.

secondary growth: growth in diameter of a stem or root due to cell division in lateral meristems and differentiation of their daughter cells.

secondary oocyte: a large haploid cell derived from the first meiotic division of the diploid primary oocyte.

secondary phloem: phloem produced from the cells arising toward the outside of the vascular cambium.

secondary spermatocyte (ō´-ō-sīt): a large haploid cell derived by meiosis I from the diploid primary spermatocyte.

secondary structure: a repeated, regular structure assumed by protein chains, held together by hydrogen bonds; usually either a helix or a pleated sheet.

secondary succession: succession that occurs after an existing community is disturbed, in an environment modified by that community, for example, after a forest fire.

secondary xylem: xylem produced from cells arising at the inside of the vascular cambium.

secretin: hormone produced by the small intestine that stimulates production and release of digestive secretions by the pancreas and liver.

sedimentary cycle: a nutrient cycle for a substance whose primary reservoir is sediment, such as soil or rock.

seed: the reproductive structure of a seed plant. The seed is protected by a seed coat and contains an embryonic plant and a supply of food for it.

seed coat: the thin, tough, and waterproof outermost covering of a seed, formed from the integuments of the ovule.

segmentation (seg-men-tā´-shun): division of the body into repeated, often similar units.

segmentation movements: unsynchronized contractions of the small intestine that result in mixing of partially digested food and digestive enzymes. The movements also bring nutrients into contact with the absorptive intestinal wall.

segregation: a principle of inheritance, that the two alleles of a given gene separate from each other during gamete formation, with the result that each gamete receives one allele; occurs because of the separation of homologous chromosomes during meiosis.

self-fertilization: union of sperm and egg from the same individual.

selfish gene: the concept that genes promote the survival of themselves in individuals through innate self-sacrificing behavior that enhances the survival of others carrying the same genes; helps explain the evolution of altruism.

semen: the sperm-containing fluid produced by the male reproductive tract.

semicircular canals: three fluid-filled, semicircular tubes of the inner ear that function in the detection of movements of the head in all directions.

semiconservative replication: the process of replication of the DNA double helix; the two DNA strands separate, and each is used as a template for the synthesis of a complementary DNA strand. Each daughter double helix therefore consists of one parental strand and one new strand.

semilunar valve: the type of valve between the ventricles of the heart and the pulmonary artery and aorta. Semilunar valves prevent backflow of blood into the ventricles when they relax.

seminal vesicle: in male mammals, a gland that produces a basic, fructose-containing fluid that forms part of the semen.

seminiferous tubules (semi-i-ni´-fer-us): a series of tubes in the vertebrate testis in which sperm are produced.

senescence: in plants, a specific aging process, often including deterioration and dropping of leaves and flowers.

sensitive period: the particular stage in an animal's life during which it imprints.

sensory neuron: a nerve cell that responds to a stimulus, either from the internal or external environment.

sensory receptor: a cell specialized to respond to particular internal or external environmental stimuli by producing an electrical potential.

sepal (sē´-pul): set of modified leaves that surround and protect a flower bud, often opening into green, leaflike structures when the flower blooms.

septum (pl. septa): a partition that separates the fungal hypha into individual cells. Pores in the septa allow transfer of materials between cells.

serotonin (ser-uh-tō´-nin): a neurotransmitter in the central nervous system, which is involved in mood, sleep, and the inhibition of pain.

Sertoli cell: a large cell in the seminiferous tubule that regulates spermatogenesis and nourishes the developing sperm.

sessile (ses´-ul): not free to move about, usually permanently attached to a surface.

severe combined immune deficiency (SCID): a disorder in which no immune cells, or very few, are formed; the immune system is incapable of responding properly to invading disease organisms, and the victim is very vulnerable common infections.

sex chromosome: one of the pair of chromosomes that differ between the sexes and usually determine the sex of an individual; for example, human females have similar sex chromosomes (XX) while males have dissimilar ones (XY).

sex-influenced traits: a set of traits that occur more commonly or strongly in a single sex, often due to differing levels of sex hormones, but are not coded by genes on the sex chromosomes.

sex linkage: a pattern of inheritance characteristic of genes located on one type of sex chromosome (for example, X) and not found on the other type (for example, Y).

sex-linked inheritance: inheritance of traits controlled by genes carried on the X chromosome. Females show the dominant trait unless they are homozygous recessive, whereas males will express whatever allele is found on their single X chromosome.

sexual recombination: during sexual reproduction, the formation of new combinations of alleles in offspring, due to the inheritance of one homologous chromosome from each of two, genetically different parents.

sexual reproduction: a form of reproduction in which genetic material from two parental organisms is combined in the offspring. Usually, two haploid gametes fuse to form a diploid zygote.

sexual selection: a type of natural selection in which the choice of mates by one sex is the selective agent.

shoot system: all the parts of a vascular plant exclusive of the root. Usually aboveground, it consists of stem, leaves, buds, and (in season) flowers and fruits. Its functions include photosynthesis, transport of materials, reproduction, and hormone synthesis.

short-day plant: a plant that will flower only if the length of daylight is shorter than some species-specific duration.

sickle-cell anemia: a recessive disease caused by a single amino acid substitution in the hemoglobin molecule. Sickle-cell hemoglobin molecules tend to cluster together, distorting the red blood cell shape and causing them to break and clog the capillaries.

sieve plate: structues between two adjacent sieve-tube elements in phloem, where holes formed in the primary cell walls interconnect the cytoplasm of the elements.

sieve tube: in phloem, a single strand of sieve-tube elements, which transports sugar solutions.

sieve-tube element: one of the cells of a sieve tube, which form the phloem.

simple diffusion: diffusion of water, dissolved gases, or lipid- soluble molecules through the phospholipid bilayer of a membrane.

sinoatrial node (sī´-nō-āt´-rē-ul nōd): also called the SA node, a small mass of specialized muscle in the wall of the right atrium. It generates electrical signals rhythmically and spontaneously and serves as the heart's pacemaker.

skeletal (striated) muscle: the type of muscle that is attached to and moves the skeleton and is under the direct, usually voluntary, control of the nervous system.

skeleton: a supporting structure for the body, upon which muscles act to change the body configuration.

small intestine: portion of the digestive tract between the stomach and large intestine in which most digestion and absorption of nutrients occurs.

smooth endoplasmic reticulum: endoplasmic reticulum without ribosomes.

smooth muscle: type of muscle found around hollow organs, such as the digestive tract, bladder, and blood vessels, normally not under voluntary control.

sodium channel: a protein that spans a nerve cell membrane, forming a pore, and which allows only sodium ions to flow through the pore.

sodium-potassium pump: a set of active transport molecules in nerve cell membranes that use the energy of ATP to pump sodium ions out of the cell and potassium ions in, maintaining the concentration gradients of these ions across the cell membrane.

solvent: a liquid that is capable of dissolving (uniformly dispersing) other substances in itself.

somatic nervous system: that portion of the peripheral nervous system that controls voluntary movement by activating skeletal muscles.

spawning: a method of external fertilization in which male and female parents shed gametes into the water, and sperm must swim through the water to reach the eggs.

speciation: the process whereby two populations achieve reproductive isolation.

species (spē´-sēs): all of the organisms that are potentially capable of interbreeding under natural conditions or, if asexually reproducing, are more closely related to one another than to other organisms within the genus.

specific heat: the amount of energy required to raise the temperature of 1 gram of a substance 1° C.

sperm: the haploid male gamete, usually small, motile, and containing little cytoplasm.

spermatid: a haploid cell derived from the secondary spermatocyte by meiosis II. The mature sperm is derived from the spermatid by differentiation.

spermatogenesis: the process by which sperm cells are formed.

spermatogonium (pl. spermatogonia): a diploid cell lining the walls of the seminiferous tubules that gives rise to a primary spermatocyte.

spermatophore: in some animals, in a variation on internal fertilization, the males package their sperm in a container that can be inserted in the female reproductive tract.

spermicide: a sperm-killing chemical; used for contraceptive purposes.

spicule (spik´-ūle): subunits of the endoskeleton of sponges, made of protein, silica, or calcium carbonate.

spinal cord: part of the central nervous system of vertebrates, extending from the base of the brain to the hips, protected by the bones of the vertebral column; contains the cell bodies of motor neurons synapsing on skeletal muscles, the circuitry for some simple reflex behaviors, and axons communicating with the brain.

spindle apparatus: a football-shaped array of microtubules that moves the chromosomes to opposite poles of a cell during anaphase of meiosis and mitosis.

spiracles (spī´-re-kuls): openings in the abdominal segments of insects through which air enters the tracheae.

spirilla (spī´-rilla): a spiral shaped bacterium.

spleen: An organ of the lymphatic system in which lymphocytes are produced, and blood is filtered past lymphocytes and macrophages which remove foreign particles and aged red blood cells.

spongy cells: irregularly shaped mesophyll cells containing chloroplasts located just above the lower epidermis of a leaf.

spontaneous generation: the proposal that living organisms can arise from nonliving matter.

sporangium (spōr-an´-gē-um; pl. sporangia): a structure in which spores are produced.

spore: in the alternation of generation life cycle of plants, a haploid cell produced through meiosis, that then undergoes repeated mitotic divisions and differentiation of daughter cells to produce a multicellular, haploid organism called the gametophyte.

sporophyte (spōr´-ō-fīt): the diploid form of a plant that produces haploid, asexual spores through meiosis.

sporozoan (spōr-ō-zō´-en): a category of parasitic protist. Sporozoans have complex life cycles often involving more than one host, and are named for their ability to form infectious spores. A well-known member (genus Plasmodium) causes malaria.

stabilizing selection: a type of natural selection in which those organisms displaying extreme phenotypes are selected against.

stamen (stā´-men): the male reproductive structure of a flower, consisting of a filament and an anther in which pollen grains develop.

starch: a polysaccharide composed of branched or unbranched chains or glucose molecules, used by plants as a carbohydrate- storage molecule.

start codon: a codon in messenger RNA that signals the beginning of protein synthesis on a ribosome.

startle coloration: a form of mimicry in which a color pattern (often resembling large eyes) can be displayed suddenly by a prey organism when approached by a predator.

statoliths: In plants, starch-filled organelles in the stem and root cap that sink to the downward side of cells and act as a sensor for gravity.

stem: the normally vertical, aboveground part of a plant body that bears leaves.

sterilization: the method of contraception in which the pathways through which the sperm (vas deferens) or egg (oviducts) must travel are interrupted. Generally permanent, it is the the most form of contraception.

steroid: a class of hormone whose chemical structure resembles cholesterol that is secreted by the ovaries and placenta, the testes, and the adrenal cortex.

stigma (stig´-ma): the pollen-capturing tip of a carpel.

stoma (stō´-ma; pl. stomata): an adjustable opening in the epidermis of a leaf, surrounded by a pair of guard cells, that regulates the diffusion of carbon dioxide and water into and out of the leaf.

stomach: muscular sac between the esophagus and small intestine where food is stored and mechanically broken down, and in which protein digestion begins.

stop codon: a codon in messenger RNA that stops protein synthesis and causes the completed protein chain to be released from the ribosome.

stroke: an interruption of blood flow to part of the brain, caused by the rupture of an artery or the blocking of an artery by a blood clot. Loss of blood supply leads to rapid death of the area of the brain affected.

stroma (strō´-ma): the semi-fluid material inside chloroplasts in which the grana are embedded.

structural gene: a gene that codes for a protein used by the cell for purposes other than gene regulation. Structural genes code for enzymes or for proteins that are structural parts of a cell.

style: a stalk connecting the stigma of a carpel with the ovary at its base.

subatomic particle: the particles of which atoms are made: electrons, protons, and neutrons.

subclimax: a community where succession is stopped before the climax community is reached and is usually maintained by regular disturbance, for example, tallgrass prairie maintained by regular fires.

substrate: the atoms or molecules that are the reactants for an enzyme-catalyzed chemical reaction.

subunit: a small organic molecule, several of which may be bonded together to form a larger molecule. See also monomer.

succession (suk-se´-shun): the change in the community structure of an ecosystem over time. Community changes alter the ecosystem in ways that favor competitors, and species replace one another in a somewhat predictable manner until a stable, self-sustaining climax community is reached.

sucrose: a disaccharide composed of glucose and fructose.

sugar: a simple carbohydrate molecule, either a monosaccharide or a disaccharide.

supernormal stimulus: a stimulus that exaggerates crucial elements of the releaser, making it more effective than the normal releaser.

suppressor T cell: a type of T cell that depresses the response of other immune cells to foreign antigens.

surface tension: the property of a liquid to resist penetration by objects at its interface with the air, due to cohesion between molecules of the liquid.

survivorship curve: a curve resulting when the number of individuals of each age in a population is graphed against their age, usually expressed as a percentage of their maximum life span.

symbiosis (sim'-bī-ō'sis): a close association between different species over an extended period. Either or both may benefit from the association, or, in the case of parasitism, one of the participants is harmed. Symbiosis includes mutualism, commensalism, and parasitism.

sympathetic nervous system: the division of the autonomic nervous system that produces largely involuntary responses that prepare the body for stressful or highly energetic situations.

sympatric speciation (sim-pat´-rik): speciation of populations that are not physically divided; usually due to ecological isolation or chromosomal aberrations (such as polyploidy).

synapse (sin´-apz): the site of communication between nerve cells. One cell (presynaptic) usually releases a chemical (the neurotransmitter) that changes the electrical potential of the second (postsynaptic) cell.

synaptic cleft: a tiny space that separates the pre- and postsynaptic cells.

synaptic terminal: swellings at the branched endings of axons, where the axon forms a synapse.

syphilis (si´-ful-is): a sexually transmitted bacterial infection of the reproductive organs which, if untreated, can damage the nervous and circulatory systems.

systole (sis´-tō-lē): the portion of the heart cycle during which the ventricles contract; the higher of the two blood pressure readings.

T cell: a type of lymphocyte that recognizes and destroys specific foreign cells or substances, or that regulates other cells of the immune system.

T-cell receptor: a protein receptor located on the surface of a T cell, which binds a specific antigen and triggers the immune response of the T cell.

taproot system: a root system commonly found in dicots, consisting of a long, thick main root that develops from the primary root, and many smaller lateral roots that grow from the primary root.

target cell: a cell upon which a particular hormone exerts its effect.

taste: a chemical sense for substances dissolved in water or saliva; in mammals, perceptions of sweet, sour, bitter, or salt produced by stimulation of receptors on the tongue.

taste bud: a cluster of taste receptor cells and supporting cells located in a small pit beneath the surface of the tongue and communicating with the mouth through a small pore. The human tongue has about 10,000 taste buds.

taxis (tax´-is; pl. taxes): innate behavior that is a directed movement of an organism toward or away from a stimulus such as heat, light, or gravity.

taxonomy (tax-on-uh-mē): the science by which organisms are classified into hierarchically arranged categories that reflect their evolutionary relationships.

Tay-Sachs disease: a recessive disease caused by a deficiency in enzymes regulating lipid breakdown in the brain.

tectorial membrane (tek-tōr´-ē-ul): one of the membranes of the cochlea, in which the hairs of the hair cells are embedded. During sound reception, movement of the basilar membrane relative to the tectorial membrane bends the cilia.

telophase (tēl´-ō-fāz): the last stage of mitosis, in which a nuclear envelope reforms around each new daughter nucleus, the spindle disappears, and the chromosomes relax from their condensed form. In meiosis I, the stage during which the spindle disappears and the chromosomes usually relax from their condensed form. In meiosis II, the stage during which chromosomes relax into their extended state, nuclear envelopes reform, and cytokinesis occurs.

template strand: the strand of the DNA double helix from which RNA is transcribed.

temporal isolation: the inability of organisms to mate if they have significantly different breeding seasons.

temporal lobe: part of a cerebral hemisphere of the human brain, involved in recall of learned events.

tendon: a tough connective tissue band connecting a muscle to a bone.

tendril: a slender outgrowth of a stem that coils about external objects and supports the stem; usually a modified leaf or branch.

tentacle (ten´-te-kul): an elongate, extensible projection of the body of cnidarians and cephalopod mollusks that may be used for grasping, stinging, and immobilizing prey, and locomotion.

terminal bud: the bud at the extreme end of a stem or branch.

territoriality: the defense of an area in which important resources are located.

tertiary consumer (ter´-she-ār-ē): a carnivore that feeds on other carnivores.

tertiary structure (ter´-she-ār-ē): the complex three-dimensional structure of a single peptide chain. The tertiary structure is held in place by disulfide bonds between

test cross: a breeding experiment in which an individual showing the dominant phenotype is mated with an individual that is homozygous recessive for the same gene. The ratio of offspring with dominant versus recessive phenotypes can be used to determine the genotype of the phenotypically dominant individual.

testis (pl. testes): the gonad of male mammals.

testosterone: in vertebrates, a hormone produced by the interstitial cells of the testis; stimulates spermatogenesis and the development of male secondary sex characteristics.

tetany: smooth, sustained maximal contraction of a muscle in response to rapid firing by its motor neuron.

thalamus: part of the forebrain, the thalamus relays sensory information to many parts of the brain.

theory: in science, an explanation for natural events that is based on a large number of observations and is in accord with scientific principles, especially causality.

thermoacidophile (ther-mō-a-sid´-eh-fīl): a form of archaebacterium that thrives in hot, acidic environments.

thermoreceptor: a sensory receptor that responds to changes in temperature.

thick filaments: bundles of myosin protein within the sarcomere that interact with thin filaments to produce muscle contraction.

thin filaments: protein strands within the sarcomere that interact with thick filaments to produce muscle contraction. Composed primarily of actin, with accessory proteins.

thorn: a hard, pointed outgrowth of a stem; usually a modified branch.

threshold: the electrical potential (less negative than the resting potential) at which an action potential is triggered.

thrombin: an enzyme produced in the blood as a result of injury to a blood vessel. Thrombin catalyzes the production of fibrin, a protein that assists in blood clot formation.

thylakoid (thī´-la-koyd): a disk-shaped, membranous sac found in chloroplasts, the membranes of which contain the photosystems and ATP-synthesizing enzymes used in the light-dependent reactions of photosynthesis.

thymus (thī´-mus): a gland, considered part of the lymphatic system, located in the upper chest in front of the heart. The thymus secretes thymosin, which stimulates lymphocyte maturation. It begins to degenerate at puberty and has little function in the adult.

thyroid: an endocrine gland, located in front of the larynx in the neck, that secretes the hormones thyroxine (affecting metabolic rate) and calcitonin (regulating calcium ion concentration in the blood).

tight junction: a type of cell-to-cell junction in animals that prevents the movement of materials through the spaces between cells.

tissue: a group of (usually similar) cells that together carry out a specific function, i.e., muscle. The tissue may include extracellular material produced by its cells.

tonsils: patches of lymphatic tissue consisting of connective tissue containing many lymphocytes, located in the pharynx and throat.

trachea (trā-kē-uh): a rigid but flexible tube supported by rings of cartilage, which conducts air between the larynx and the bronchi.

tracheae (trā-kē): elaborately branching tubes that ramify through the bodies of insects and carry air close to each body cell. Air enters the tracheae through openings called spiracles.

tracheid (tra-kē-id): an elongated xylem cell with tapering ends containing pits in the cell walls; forms tubes that transport water.

tracheophyte (trā´-kē-ō-fīt): a category of plant that has conducting vessels; a vascular plant.

transcription: the synthesis of an RNA molecule from a DNA template.

transducer: a device that converts signals from one form to another. Sensory receptors are transducers that convert environmental stimuli, such as heat, light, or vibration, into electrical signals (such as action potentials) recognized by the nervous system.

transfer RNA (tRNA): a type of RNA that (1) binds to a specific amino acid and (2) bears a set of three bases (the anticodon) complementary to the mRNA codon for that amino acid. Transfer RNA carries its amino acid to a ribosome during protein synthesis, recognizes a codon of mRNA, and positions its amino acid for incorporation into the growing protein chain.

transformation: a method of acquiring new genes, whereby DNA from one bacterium (usually released after the death of the bacterium) becomes incorporated into the DNA of another, living, bacterium.

translation: the process whereby the sequence of bases of messenger RNA is converted into the sequence of amino acids of a protein.

transpiration (trans´-per-ā-shun): evaporation of water through the stomata of a leaf.

transport protein: a protein that regulates the movement of water-soluble molecules through the plasma membrane.

transverse (T) tubules: deep infoldings of the muscle cell membrane that conduct the action potential inside the cell.

trial-and-error learning: a process by which adaptive responses are learned through rewards or punishments provided by the environment.

trichocyst (trik´-eh-sist): a stinging organelle of protists.

triglyceride (trī-gliss'-er-īd): a lipid composed of three fatty acid molecules bonded to a single glycerol molecule.

triple covalent bond: a covalent bond that occurs when two atoms share three pairs of electrons.

trisomy 21: see Down syndrome.

trisomy X: a condition of females who have three X chromosomes instead of the normal two. Most of these women are phenotypically normal and are fertile.

trophic level: literally "feeding level"; the position of an organism in a food chain, defined by the organisms's source of energy. trophic levels include producers, primary consumers, secondary consumers, and so on.

true-breeding: pertaining to an individual all of whose offspring produced through self-fertilization are identical to the parental type. True-breeding individuals are homozygous for a given trait.

tubal ligation: surgical procedure in which a woman's oviducts are cut so that the egg cannot reach the uterus, which makes her infertile.

tube cell: the outermost cell of a pollen grain; the tube cell digests a tube through the tissues of the carpel, ultimately penetrating into the female gametophyte.

tube feet: cylindrical extensions of the water-vascular system of echinoderms, used for locomotion, grasping food, and respiration.

tubular reabsorption: the process by which cells of the tubule of the nephron remove water and nutrients from the filtrate within the tubule and return them to the blood.

tubular secretion: the process by which cells of the tubule of the nephron remove additional wastes from the blood, actively secreting them into the tubule.

tubule (tūb´-ūle): the tubular portion of the nephron. It includes a proximal portion, the loop of Henle, and a distal portion. Urine is formed from the blood filtrate as it passes through the tubule.

tumor-suppressor gene: a gene that encodes information for a protein that inhibits cancer formation, probably by regulating cell division in some way.

turgor pressure: pressure developed within a cell (especially the central vacuole of plant cells) as a result of osmotic water entry.

Turner syndrome: a set of characteristics typical of a woman with only one X chromosome. These women are sterile, with a tendency to be very short and to lack normal female secondary sexual characteristics.

tympanic membrane (tim-pan´-ik): the eardrum; a membrane stretched across the opening of the ear, which transmits vibration of sound waves to bones of the middle ear.

uniform distribution: a relatively regular spacing of individuals within a population, often as a result of territorial behavior.

uniformitarianism: the hypothesis that Earth developed gradually through natural processes similar to those at work today occurring over long periods of time.

unsaturated: referring to a fatty acid with fewer than the maximum number of hydrogen atoms bonded to its carbon backbone; a fatty acid with one or more double bonds in its carbon backbone.

urea (ū-rē´-uh): a water-soluble, nitrogen-containing waste product of amino acid breakdown that is one of the principal components of mammalian urine.

ureter (ū´-re-tur): a tube that conducts urine from each kidney to the bladder.

urethra (ū-rē´-thra): the tube leading from the urinary bladder to the outside of the body; in males, the urethra also receives sperm from the vas deferens and conducts both sperm and urine (at different times) to the tip of the penis.

uric acid (ūr´-ik acid): a nitrogen-containing waste product of amino acid breakdown, which is a relatively insoluble white crystal. Uric acid is excreted by birds, reptiles, and insects.

urine: fluid produced and excreted by the urinary system containing water and dissolved wastes, such as urea.

uterus: in female mammals, the part of the reproductive tract that houses the embryo during pregnancy.

utricle (ū´-tri-cul): a portion of the vestibular apparatus responsible for detecting gravity.

vaccination: an injection into the body that contains antigens characteristic of a particular disease organism and that stimulates an immune response.

vacuole (vak´-ū-ōl): a vesicle, often large, consisting of a single membrane enclosing a fluid-filled space.

vagina: the passageway leading from the outside of the body to the cervix of the uterus.

variable: a condition, particularly in a scientific experiment, that is subject to change.

variable region: part of an antibody molecule that differs among antibodies; the ends of the variable regions of the light and heavy chains form the specific binding site for antigen.

vas deferens (vas de´-fer-ens): the tube connecting the epididymis of the testis with the urethra.

vascular bundle (vas´-kū-lar): a strand of xylem and phloem found in leaves and stems; in leaves, commonly called a vein.

vascular cambium: a lateral meristem located between the xylem and phloem of a woody root or stem and gives rise to secondary xylem and phloem

vascular cylinder: the centrally located conducting tissue of a young root, consisting of primary xylem and phloem.

vascular tissue system: a plant tissue system consisting of xylem (which transports water and minerals from root to shoot) and phloem (which transports water and sugars throughout the plant).

vasectomy: a surgical procedure in which a man's vas deferens are cut, preventing sperm from reaching the penis during ejaculation, which makes him infertile.

vein: a large-diameter, thin-walled vessel that carries blood from venules back to the heart.

ventral (ven´-trul): the lower, or underside of an animal whose head is oriented forward.

ventricle (ven´-trē-kul): the lower muscular chamber on each side of the heart, which pumps blood out through the arteries. The right ventricle sends blood to the lungs, and the left to the rest of the body.

venule (ven´-yul): a narrow vessel with thin walls that carries blood from capillaries to veins.

vertebrate: an animal that possesses a vertebral column.

vesicle (ves´-i-kul): a small membrane-bound sac within the cytoplasm.

vessel: a tube of xylem composed of vertically stacked vessel elements, with perforated or missing end walls, leaving a continuous, uninterrupted hollow cylinder.

vessel element: one of the cells of a xylem vessel; elongated, dead at maturity, with thick lignified lateral cell walls for support but with end walls either lacking entirely or heavily perforated.

vestibular apparatus (ves-tib´-ū-lar): the portion of the inner ear of mammals that responds to gravity and changes in the direction and speed of movement.

vestigial structures (ves-tij´-ē-ul): structures that serve no apparent purpose, but which are homologous to functional structures in related organisms and evidence of evolution.

villus (vi´-lus): fingerlike projections of the wall of the small intestine that increase its absorptive surface area.

viroid (vī-rōyd): a particle of RNA capable of infecting a cell and directing the production of more viroids; responsible for certain plant diseases.

virus (vī´-rus): a noncellular parasitic particle consisting of a protein coat surrounding a strand of genetic material. Viruses multiply only within cells of living organisms.

vitamin: any one of a group of diverse chemicals that must be present in trace amounts in the diet to maintain health. Vitamins are used by the body in conjunction with enzymes in a variety of metabolic reactions.

vitreous humor (vit´-rē-us): a clear jellylike substance that fills the large chamber of the eye between the lens and retina.

vocal cords: folds of tissue that extend across the opening of the larynx and produce sound when air is forced between them. Muscles alter the tension on the vocal cords and control the size and shape of the opening, which in turn determines whether sound is produced and what its pitch will be.

waggle dance: symbolic communication used by honeybee foragers to communicate the location of a food source to their hivemates.

water-vascular system: a system in echinoderms consisting of a series of canals through which seawater is conducted and used to inflate tube feet for locomotion, grasping food, and respiration.

wax: a lipid composed of fatty acids covalently bonded to long-chain alcohols.

white matter: Portions of the brain and spinal cord consisting largely of myelin-covered axons that give these areas a white appearance

withdrawal: removal of the penis from the vagina just before ejaculation in an attempt to avoid pregnancy; an ineffective contraceptive method.

X-linked inheritance: see sex-linked inheritance.

xylem (zī´-lum): a conducting tissue of vascular plants that transports water and minerals from root to shoot.

yolk: protein- or lipid-rich substances contained in eggs as food for the developing embryo.

yolk sac: one of the embryonic membranes of reptile, bird, and mammalian embryos. In birds and reptiles, the yolk sac is a membrane surrounding the yolk in the egg. In mammals, the yolk sac is empty but forms part of the umbilical cord and the digestive tract.

Z lines: fibrous protein structures to which the thin filaments of skeletal muscle are attached, forming the boundaries of sarcomeres.

zona pellucida (pel-ū´-si-da): a clear, noncellular layer between the corona radiata and the egg.

zooflagellate (zō-ō-fla´-gel-et): a category of nonphotosynthetic protist that move using flagella.

zoospore (zō´-ō-spōr): a nonsexual reproductive cell that swims using flagella, such as is formed by members of the fungal division Oomycota.

zygospore (zī´-gō-spōr): produced by the division Zygomycota, a fungal spore surrounded by a thick, resistant wall, which forms from a diploid zygote.

zygote (zī´-gōt): in sexual reproduction, a diploid cell (the fertilized egg) formed by the fusion of two haploid gametes.

Photo Credits

Frontmatter
p. ii Dwight Kuhn Photography **p. v** Audesirk

Chapter 1
Opener NASA Headquarters **1-1a** Dr. Jeremy Burgess/Science Photo Library/Photo Researchers, Inc. **1-1b** Craig Tuttle/The Stock Market **1-1c** Kim Taylor/Bruce Coleman, Inc. **1-3** E. H. Newcomb/Biological Photo Service **1-4** Anua and Mahoj Shah/Animals Animals/Earth Scenes **1-5** Tom McHugh/Photo Researchers, Inc. **1-6** Will & Deni McIntyre/Photo Researchers, Inc. **1-7a** CNRI/Science Photo Library/Photo Researchers, Inc. **1-7b** Eric V. Grave/Photo Researchers, Inc. **1-7c** Richard L. Carlton/Photo Researchers, Inc. **1-7d** Patti Murray/Animals Animals/Earth Scenes **1-7e** William Townsend/Photo Researchers, Inc. **1-8** John Durham/Science Photo Library/Photo Researchers, Inc. **1-9** Doug Perrine/DRK Photo **1-10** Tom Algire/Tom Stack & Associates **1-11** Chip Clark/Smithsonian Institution **E1-2** Luiz C. Marigo/Peter Arnold, Inc. **E1-2** (inset) Gunter Ziesler/Peter Arnold, Inc. **1-13** Audesirk

Unit I
Opener (p. xx) Secchi, Iecaque, Roussel, Uclaf, CNRI/Photo Researchers, Inc.

Chapter 2
Opener Blaine Harrington III/The Stock Market **2-1** Jim Anderson/Woodfin Camp & Associates **E2-1c** Courtesy of Drs. M. J. Fulham & Giovanni Di Chiro, The Neuroimaging Section, NINDS, National Institutes of Health, Bethesda, MD. **E2-2** NIH/Science Source/Photo Researchers, Inc. **2-11a** David Dennis/Tom Stack & Associates **2-11b** Audesirk

Chapter 3
Opener Ken Edward/Biografix/Science Source/Photo Researchers, Inc. **3-4a** Dr. Jeremy Burgess/Science Photo Library/Photo Researchers, Inc. **3-5** (left) Larry Ulrich/DRK Photo **3-5** (middle) Jeremy Burgess, M.D./Photo Researchers, Inc. **3-5** (right) Biophoto Associates/Photo Researchers, Inc. **3-6** Richard Kolar/Animals Animals/Earth Science **3-11a** Robert Pearcy/Animals Animals/Earth Science **3-11b** Jeff Foott/DRK Photo **3-11c** Nuridsany et Perennou/Photo Researchers, Inc.

Chapter 4
Opener Bruce Davidson/Animals Animals/Earth Science **4-1** Focus on Sports **4-7a,b** T. A. Steitz/Yale University

Chapter 5
Opener Moredun Animal Health Ltd/Science Photo Library/Photo Researchers, Inc. **5-1** (upper) Biophoto Associates/Photo Researchers, Inc. **5-1** (lower) Cecil Fox/Science Source/Photo Researchers, Inc. **E5-1a,b,c** M. I. Walker/Photo Researchers, Inc. **E5-1d** David M. Phillips/Visuals Unlimited **E5-2** Dr. Don Fawcett/Photo Researchers, Inc. **E5-3a** Biophoto Associates/Photo Researchers, Inc. **E5-3b** Karl Aufderheide/Visuals Unlimited **E5-3c** Dr. K. Tanaka **5-3** Dr. Stanley L. Erlandsen/PH College Archives **5-5** John Cardomona/Biological Photo Service **5-8a** Richard Rodewald/Biological Photo Service **5-8b** Don W. Fawcett/Photo Researchers, Inc. **5-9** Phototake/Carolina Biological Supply Co. **5-10** Omikron/Science Source/Photo Researchers, Inc. **5-12** Dr. Barry F. King/Biological Photo Service **5-13** D. W. Fawcett/Photo Researchers, Inc. **5-14** Zdenek Hruban, M.D. **E5-4** Dr. Don Fawcett/T. Kanaseki/Photo Researchers, Inc. **5-16** W.P. Wergin/Biological Photo Service **5-17** Keith R. Porter, Laboratory of Cell Biology University of Maryland **5-18** Biophoto Associates/Photo Researchers, Inc. **5-19a** (left) Dr. Gopal Murti/Science Photo Library/

Science Source/Photo Researchers, Inc. **5-19a** (middle) Dr. Peter Dawson/Science Photo Library/Photo Researchers, Inc. **5-19a** (right) Francis LeRoy, Biocosmos/Science Photo Library/Photo Researchers, Inc. **5-20** Keith R. Porter/Laboratory of Cell Biology University of Maryland **5-21** (upper) William L. Dentler/Biological Photo Service **5-21** (lower) Dr. de Harven/Photo Researchers, Inc.

Chapter 6
Opener Jan Hinsch/Science Photo Library/Photo Researchers, Inc. **6-1** Farrell Grehan/PhotoResearchers, Inc. **6-2** Biophoto Associates/Science Source/Photo Researchers, Inc. **6-8a,b,c** Dr. Joseph Kuranstin-Mills **6-11a-d** M. M. Perry & A. B. Gilbert, *The Journal of Cell Science*, 39:257-272 (1979) **6-12** Linda Hufnagel **6-13a,b** Thomas Eisner **6-15** Dr. Don Fawcett/G. Raviola; P. Elias; D. Friend/Photo Researchers, Inc. **6-16** L. Andrew Staehelin **6-17** Photo Researchers, Inc. **6-18** W.P. Wergin/Biological Photo Service **6-19** Steven J. Krasemann/DRK PHoto

Chapter 7
Opener Alfred Pasieka/Peter Arnold, Inc. **7-2a** Ken W. Davis/Tom Stack & Associates **7-2b,c** E. H. Newcome/Biological Photo Service

Chapter 8
Opener Francois Gohier/Photo Researchers, Inc. **8-3a** Focus on Sports **8-3b** Jonathan Watts/Science Photo Library/Photo Researchers, Inc.

Unit II
Opener (p. xxi) CNRI/Science Photo Library/Photo Researchers, Inc.

Chapter 9
Opener Dr. Gopal Murti/Science Photo Library/Photo Researchers, Inc. **9-2** Dr. Gopal Murti/Science Photo Library/Photo Researchers, Inc. **9-3** CNRI/Science Photo Library/Photo Researchers, Inc. **9-4** Biophoto Associates/Photo Researchers, Inc. **9-5** CNRI/Science Photo Library/Photo Researchers, Inc. **9-7a-h** Dr. Andrew S. Bayer, University of Oregon **9-9a** Biological Photo Service **9-9b** John D. Cunningham/Visuals Unlimited **9-12a** T. E. Schroeder/Biological Photo Service/ **9-14a** M. Abbey/Photo Researchers, Inc. **9-14b** Carolina Biological Supply Co./Phototake NYC **9-15** Audesirk

Chapter 10
Opener Dr. Richard J. Feldman/Molecular Graphics/National Institutes of Health **10-3** Thomas Broker/CNRI/Phototake NYC **10-6a** Rosalind Franklin/Photo Researchers, Inc. **10-6b** CSHL Archives/Peter Arnold, Inc. **E10-1** A. Barrington Brown/Photo Researchers, Inc.

Chapter 11
Opener E. V. Kiseleva, 1989, *FEBS LETTS*, 25:251-253 **11-5** Dr. Oscar L. Miller, Jr. **11-16** Science Source/Photo Researchers, Inc. **11-17** Dr. Murray L. Barr

Chapter 12
Opener Minden Pictures **12-1** C. C. Brinton, Jr. & J. Carnahan from *Principles of Population Genetics* 2/e, 1989, edited by Daniel L. Hartl and Andrew C. C.ark, Sunderland, MA:Sinaver. **12-5** James Kezer

Chapter 13
Opener Ron Kimball Studios **13-1a,c** Joe McDonald/Animals Animals/Earth Scenes **13-1b** Tom Stack & Associates **13-1c** Beth Tobias **13-2** Photo Researchers, Inc. **13-3** Audesirk **13-14** Carolina Bio-

logical Supply Co./Phototake NYC **13-18** (p. xxiii) Jane Burton/Bruce Coleman, Inc.

Chapter 14
Opener Keith V. Wood/Science VU/Visuals Unlimited **14-1** Professor Stanley N. Cohen/Science Photo Library/Photo Researchers, Inc. **14-6** Hank Morgan/Photo Researchers, Inc. **14-8** Applied Biosystems/Peter Arnold, Inc. **E14-1** Kenneth W. Fink/Photo Researchers, Inc. **E14-1** (inset) Gilbert Grant/Photo Researchers, Inc. **14-9** Steve Grove/UPI/Bettmann **E14-2a** Gary Braasch Photography **E14-2b** Dr. Joseph Newhouse, Ph.D.

Chapter 15
Opener David Stoecklein/The Stock Market **15-2a** (left) Linda Rose/Globe Pictures, Inc. **15-2a** (right) Paramount Pictures/Photofest **15-2b** Audesirk **15-4a** Marlene and Bob Lippman **15-5a,b** Photo Researchers, Inc. **15-6a** Bettmann **15-6b** Susan E. Folstein, *Huntington's Disease: A Disorder of Families*.Fig. 5.6, p. 78. Johns Hopkins University Press, 1989. **15-7a** Hart-Davis/Science Photo Library/Photo Researchers, Inc. **E15-1** Nick Kelsh/Peter Arnold, Inc. **15-8** Gunn & Stewart/Mary Evans/Photo Researchers, Inc. **15-10a** CNRI/Science Photo Library/Photo Researchers, Inc. **15-10b** Lawrence Midgale **E15-2a** CNRI/Science Photo Library/Photo Researchers, Inc.

Unit III
Opener (p. xxii) Martin Dohm/Photo Researchers, Inc.

Chapter 16
Opener Ed Reschke/Peter Arnold, Inc. **16-1a** Wayne Lankinen/Valan Photos **16-1b** Edgar T. Jones/Bruce Coleman, Inc. **16-2** Jim Steinberg/Photo Researchers, Inc. **16-4a** John Cancalosi/DRK Photo **16-4b** Chip Clark,National Museum of Natural History, Smithsonian Institution **16-4c** American Museum of Natural History **16-6** Bettmann **16-7** (left) Christian Grzimek/OKAPIA/Photo Researchers, Inc. **16-7** (right) Frans Lanting/Photo Researchers, Inc. **16-8b** E. C. Williams/Visuals Unlimited **16-8c** Norman D. Tomalin/Bruce Coleman, Inc. **16-11a,b** Stephen Dalton/Photo Researchers, Inc. **16-11c** Douglas T. Cheeseman, Jr./Peter Arnold, Inc. **16-11d** Frans Lanting/Minden Pictures, Inc. **16-14a** Keith Gillett/Animals Animals/Earth Scenes **16-14b** Dan McCoy/Rainbow **16-14c** C. Eldeman/Petit Format/Science Source/Photo Researchers, Inc. **16-14d** Carolina Biological Supply Company/Phototake NYC **16-15a** Stephen J. Krasemann/DRK Photo **16-15b** Timothy O'Keef/Tom Stack & Associates **16-16a,b** M. W. F. Tweedle/Bruce Coleman, Inc. **16-17** Audesirk

Chapter 17
Opener Phil Degginger/Bruce Coleman, Inc. **17-3b** (left) Luiz C. Marigo/Peter Arnold, Inc. **17-3b** (right) Y. R. Tymstra/Valan Photos **17-4** Gregory Dimijian/Photo Researchers, Inc. **17-5** Joe McDonald/Visuals Unlimited **17-6** Pat & Tom Leeson/Photo Researchers, Inc. **17-7** D. Cavagnaro/DRK Photo **17-11** Robert & Linda Mitchell Photography **17-12a** M. P. L. Fogden/Bruce Coleman, Inc. **17-12b** Tim Davis/Photo Researchers, Inc. **17-13** Jeff Lepore/Photo Researchers, Inc. **17-14** Tom McHugh/Steinhart Aquarium/Photo Researchers, Inc. **17-15** Jim Kern Expeditions **E17-1** Frans Lanting/Minden Pictures **17-18** NASA Headquarters **17-19** Thomas A. Wiewandt, Ph.D.

Chapter 18
Opener Paul A. Johnsgard/Photo Researchers, Inc. **18-1** Michael Fogden/DRK Photo **18-3a,b** Pat & Tom Leeson/Photo Researchers, Inc. **18-4** Guy L. Bush **18-5** Joy Spurr/Bruce Coleman, Inc.

Chapter 19

Opener Royal Observatory, Edinburgh/AATB/Science Photo Library/Photo Researchers, Inc. **19-4** Sidney Fox/Science VU/Visuals Unlimited **19-6** M. Abbey/Visuals Unlimited **19-8a** Field Museum of Natural History **19-8b** James L. Amos/Photo Researchers, Inc. **19-8c** Carolina Biological Supply Co./Phototake NYC **19-8d** Douglas Faulkner/Photo Researchers, Inc. **19-9** Rudolph F. Zallinger/Peabody Museum of Natural History **19-11** Bob Gossington/Bruce Coleman **19-12** Rudolph F. Zallinger/Peabody Museum of Natural History **19-13a** Tom McHugh/Chicago Zoological Park/Photo Researchers, Inc. **19-13b** Frans Lanting/Minden Pictures, Inc. **19-13c** Nancy Adams/Tom Stack & Associates **19-14** Tim D. White/Brill Atlanta **19-17** Chip Clark, National Museum of Natural History, Smithsonian Institution **19-18** Jerome Chatin/Gamma-Liaison, Inc.

Chapter 20

Opener (top) Shattil/Rozinski/Tom Stack & Associates (left) S. J. Krasemann/Peter Arnold, Inc. (right) E.R. Digginger/Animals Animals/Earth Scenes **20-2a** Francois Gohier/Photo Researchers, Inc. **20-2b** Kelvin Aitken/Peter Arnold, Inc. **20-3a** C. Stephen Murphree/Dept. of Entomology, Auburn University/Biological Photo Service **20-3b** Dr. Greg Rouse, Dept. of Invertebrate Zoology/National Museum of Natural History **20-3c** Dr. Jeremy Burgess/Science Photo Library/Photo Researchers, Inc. **20-5** Zig Koch/Kino Fotoarquivo/Discover Text & Photo Syndication

Unit IV

Opener (p. xxiii) Alfred Pasieka/Photo Researchers, Inc.

Chapter 21

Opener Michio Hoshiro/Minden Pictures **21-5 (right)** Manfred Kage/Peter Arnold, Inc. **E21-1a** Michael Klein/Peter Arnold, Inc. **E21-1b** Larry Mulvehill/Science Source/Photo Researchers, Inc.

Chapter 22

Opener Professor P. Motta, Dept. of Anatomy, University 'La Sapienza' Roma/Science Photo Library/Photo Researchers, Inc. **22-7** Bill Longcore/Science Source/Photo Researchers, Inc. **22-10** Lennart Nilsson/Bonnier Alba AB **22-11** Discover Text & Photo Syndication **22-12** CNRI/Science Photo Library/Photo Researchers, Inc. **22-15** Lennart Nilsson **E22-1a** Lou Lainey/Time Picture Syndicate **E22-1c** Biophoto Associates/Photo Researchers, Inc. **22-29** John D. Cuningham/Visuals Unlimited **E22-2** Bill Travis, M.D./National Cancer Institute **E22-3** O. Auerbach/Visuals Unlimited

Chapter 23

Opener Haroldo Palo Jr./Natural History Photographic Agency **23-1** Johnny Johnson/Tony Stone Images **23-3** Tobi Zausner **23-8e** Lennart Nilsson **E23-1** St. Bartholomew's Hospital/Science Photo Library/Photo Researchers, Inc. **23-15** Tony Stone Images

Chapter 24

Opener Lennart Nilsson/Bonnier Alba AB **24-1** CNRI/Science Photo Library/Photo Researchers, Inc. **24-2** Lennart Nilsson/Bonnier Alba AB **24-4** ©NIBSC/Science Photo Library/Photo Researchers, Inc. **24-6** Leonard Lessin/Peter Arnold, Inc. **24-11** (upper) Secchi/Lecaque/Roussel/CNRI/Science Source/Photo Researchers, Inc. (lower) Dr. Brian Eyden/Science Photo Library/Photo Researchers, Inc. **24-12** Dr. Andrejs Liepins/Science Photo Library/Photo Researchers, Inc. **24-15b** Dr. Matthew A. Gonda, Ph.D./National Cancer Institute **24-15c** Lennart Nilsson/Bonnier Alba AB **E24-2** CNRI/Science Photo Library/Photo Researchers, Inc.

Chapter 25

Opener Werner Bokelberg/The Image Bank **25-5** Bettmann **25-7b** Biophoto Associates/Photo Researchers, Inc.

Chapter 26

Opener Secci-Lecaque/Roussel-Oclaf/CNRI/Science Photo Library/Photo Researchers, Inc. **E26-1** Dr. Marcus Raichle/Washington University School of Medicine **E26-2** Dr. William J. Powers, Dept. of Neurology, Washington University School of Medicine, St. Louis **E36-3a,b** Monte S. Buchsbaum, M.D., Mount Sinai School of Medicine, New York, NY **26-17** Hanna Damasio, M.D./Science Magazines **26-19** Robeer S. Preston/Professor J.E. Hawkins **26-22** Eichorst/Susan Eichorst **26-23a** L. West/Photo Researchers, Inc. **26-23b** Frans Lanting/Minden Pictures **26-24a** Dwight Kuhn Photography **26-24b** Audesirk **26-28a** Charles E. Mohr/Photo Researchers, Inc. **26-28b** Brian Parker/Tom Stack & Associates

Chapter 27

Opener Petit Format/Nestle/Photo Researchers, Inc. **27-3** CNRI/Science Photo Library/Photo Researchers, Inc. **27-7a,b** Gary Martin/Visuals Unlimited **27-10a** Lennart Nilsson **27-10b** Lennart Nilsson, *A Child is Born*, Dell Publishing Company. **E27-1** Hank Morgan/Science Source/Photo Researchers, Inc. **E27-2** Juana Anderson/AP/Wide World Photos **27-11a** Lennart Nilsson **27-16** Lennart Nilsson **27-17** Lennart Nilsson **E27-3** Leonard McCombe/Life Magazine, Time Warner, Inc. **E27-4** Sterling K. Clarren, M.D., Children's Hospital & Medical Center, University of Washington School of Medicine **27-21** Audesirk

Unit V

Opener (p. xxiv) Suzanne Duranceau/Illustratice, Inc.

Chapter 28

Opener Kin Heacox/Peter Arnold, Inc. **28-1** (background) W. Perry Conway **28-1a** Jeff Foott/Tom Stack & Associates **28-1b** Gary Milburn/Tom Stack & Associates **28-1c** Rod Planck/Tom Stack & Associates **28-1d** Gary R. Zahm/DRK Photo **28-2a** G. Musil/Visuals Unlimited **28-2b** Thomas Kitchin/Tom Stack & Associates **28-7** Bryan & Cherry Alexander **28-8** Tom McHugh/Photo Researchers, Inc. **28-9a,b** Department of Lands, Queensland, Australia **28-10** Gianni Tortoli/Photo Researchers, Inc. **28-11a** Robert & Linda Mitchell Photography **28-11b** Tom Bean/DRK Photo **28-11c** Gary Braasch **28-13** Manned Spacecraft Center, Houston, TX/NASA/Johnson Space Center **E28-4a** Frans Lanting/Minden Pictures **E28-4b** Steve Holbrooke/The Stock Market

Chapter 29

Opener James L. Castner **29-3a** Carolina Biological Supply Co./Phototake NYC **29-3b** Merlin D. Tuttle/Bat Conservation International/Photo Researchers, Inc. **29-4a** Marty Snyderman/Planet Earth Pictures **29-4b** Robert & Linda Mitchell Photography **29-5a** Michael Fogden/DRK Photo **29-5b** Paul A. Zahl/Photo Researchers, Inc. **29-5c** Ray Coleman/Photo Researchers, Inc. **29-6** Robert & Linda Mitchell Photography **29-7** Andrew J. Martinez/Photo Researchers, Inc. **29-8** James L. Castner **29-9a** Zig Leszczynski/Animals Animals/Earth Scenes **29-9b** Breck P. Kent **29-9c,d** Dr. E. R. Degginger **29-10a,b** Zig Leszczynski/Animals Animals/Earth Scenes **29-11a,b** James L. Castner **29-11c** Jeff Lepore/Photo Researchers, Inc. **29-12** (upper) Robert P. Carr/Bruce Coleman, Inc. **29-12** (lower) Dr. Monica Mather/Simon Fraser University **29-13** Dan Aneshausley, Cornell University/Thomas Eisner **29-14** Thomas Eisner **29-15a** Frans Lanting/Minden Pictures **29-15b** Jeff Foott Productions **29-16** Charles Seaborn/Odyssey Productions **29-17a** Gary Braasch Photography **29-17b** (left) Wendy Shattil/Bob Rozinski/Tom Stack & Associates **29-17b** (right) Jeff Foott Productions **29-17c** (left) John I. Kjargaard/KAIO Productions **29-17c** (right) Krafft-Explorer/Photo Researchers, Inc. **E29-1** Gilbert Grant/Photo Researchers, Inc. **E29-1** (inset) Carol Hughes/Bruce Coleman, Inc. **E29-2a** Ron Peplowski/The Detroit Edison Company **E29-2b** Chuck Pratt/Bruce Coleman, Inc. **E29-2c** Robert & Linda Mitchell Photography **29-20a,b,c** Robert & Linda Mitchell Photography

Chapter 30

Opener Dwight Kuhn Photography **30-8** Thomas A. Schneider **30-13** Bob Hallinen/Gamma-Liaison, Inc. **30-14** V. Leloup/Figaro/Gamma-Liaison, Inc. **30-15** Doug Plummer/Photo Researchers, Inc. **30-17** Will McIntyre/Photo Researchers, Inc.

Chapter 31

Opener Frans Lanting/Minden Pictures **31-1a,b** Audesirk **31-3b** Manned Spacecraft Center, Houston, TX/NASA/Johnson Space Center **E31-1** NASA/Science Photo Library/Photo Researchers, Inc. **31-9** (background) Loren McIntyre **31-9a** Schafer & Hill/Peter Arnold, Inc. **31-9b** Bruce Lyon/Valan Photos **31-9c** Michael Fogden/DRK Photo **31-9d** Frans Lanting/Minden Pictures, Inc. **31-10** Peter May/Peter Arnold, Inc. **31-11** (background) W. Perry Conway **31-11a** Aubry Lang/Valan Photos **31-11b** Wendy Bass **31-11c** Stephen J. Krasemann/DRK Photo **31-12** James Hancock/Photo Researchers, Inc. **31-13a** Don & Pat Valantiss/DRK Photo **31-13b** Brian Parker/Tom Stack & Associates **31-13c** Audesirk **31-13d** Tom McHugh/Photo Researchers, Inc. **31-14** Jeff Foott Productions **31-15** Thierry Rannou/Gamma-Liaison, Inc. **31-16** Brian Parker/Tom Stack & Associates **31-17** Harvey Payne **31-18** (background) Tom Bean/DRK Photo **31-18a** W. Perry Conway/Tom Stack & Associates **31-18b** Nicholas Davore III/Photographers/Aspen **31-18c** Bob Gurr/Valan Photos **31-18d** Jim Brandenburg/Minden Pictures **31-19** Audesirk **31-20** (background) Gary Braasch Photography **31-20a** Thomas Kitchin/ Tom Stack & Associates **31-20b** Joe McDonald/Animals Animals/Earth Scenes **31-20c** Stephen J. Krasemann/DRK Photo **31-21** (background) Robert A. Jureit/The Stock Market **31-21a** Jeff Foott Productions **31-21b** Audesirk **31-21c** Richard Thom/Tom Stack & Associates **31-22** (background) Audesirk **31-22a** Marty Stouffer Productions/Animals Animals/Earth Scenes **31-22b** Wm. Bacon III/National Audubon Society/Photo Researchers, Inc. **31-23** Gary Braasch Photography **31-24** (background) S. J. Krasemann/Valan Photos **31-24a** Michael Giannechini/Photo Researchers, Inc. **31-24b** James Simon/Photo Researchers, Inc. **31-24c** Jeff Foott/Tom Stack & Associates **31-27a** T. P. Dickinson/Comstock **31-27b,c,d** Audesirk **31-27b** (inset) Walter Dawn/National Audubon Society/Photo Researchers, Inc. **31-27c** (inset) Thomas Kitchin/Tom Stack & Associates **31-27d** (inset) Stan Wayman/Photo Researchers, Inc. **31-28a,c** Charles Seaborn/Odyssey Productions **31-28b,d** (b Alex Kerstitch **31-29** (background) Stephen J. Krasemann/DRK Photo **31-29a** Paul Humann/Jeffrey L. Rotman **31-29b** James D. Watt/Planet Earth Pictures **31-29c** Fred McConnaughey/Photo Researchers, Inc. **31-29d** Biophoto Associates/Science Source/Photo Researchers, Inc. **31-29e** Robert Arnold/Planet Earth Pictures **31-30** Dr. Frederick Grassle **E31-2** Roy Morsch/The Stock Market

Index